Life Histories

The Natural History of the Crustacea Series

Functional Morphology and Diversity (Volume 1)
Edited by Les Watling and Martin Thiel

Lifestyles and Feeding Biology (Volume 2)
Edited by Martin Thiel and Les Watling

Nervous Systems and Control of Behavior (Volume 3)
Edited by Charles Derby and Martin Thiel

Physiology (Volume 4)
Edited by Ernest S. Chang and Martin Thiel

Life Histories (Volume 5)
Edited by Gary A. Wellborn and Martin Thiel

Life Histories

The Natural History of the Crustacea
Volume 5

EDITED BY GARY A. WELLBORN AND MARTIN THIEL

OXFORD
UNIVERSITY PRESS

OXFORD
UNIVERSITY PRESS

Oxford University Press is a department of the University of Oxford. It furthers the University's objective of excellence in research, scholarship, and education by publishing worldwide. Oxford is a registered trade mark of Oxford University Press in the UK and certain other countries.

Published in the United States of America by Oxford University Press
198 Madison Avenue, New York, NY 10016, United States of America.

Library of Congress Cataloging-in-Publication Data
Names: Wellborn, Gary A., editor. | Thiel, Martin, 1962– editor.
Title: Life histories / edited by Gary Wellborn and Martin Thiel.
Description: New York, NY : Oxford University Press, [2018] |
Series: The natural history of the crustacea series ; Volume 5 |
Includes bibliographical references and index.
Identifiers: LCCN 2017056192 | ISBN 9780190620271
Subjects: LCSH: Crustacea—Life cycles. | Crustacea—Adaptation.
Classification: LCC QL435 .L53 2018 | DDC 595.3—dc23
LC record available at https://lccn.loc.gov/2017056192

PREFACE

Life Histories is the fifth volume of a ten-volume series titled *The Natural History of the Crustacea*. It follows *Volume 1: Functional Morphology and Diversity, Volume 2: Lifestyles and Feeding Biology, Volume 3: Nervous Systems and Control of Behavior*, and *Volume 4: Physiology*. The remaining five volumes will explore additional aspects of crustacean natural history, including behavioral ecology, reproduction, development, evolution and biogeography, fisheries and aquaculture, and ecology and conservation biology.

Chapters in this volume synthesize our current understanding of diverse topics in crustacean life histories. Two central themes unite the chapters: (1) an underlying cost-benefit perspective to illuminate the evolution and ecology of life histories and behaviors and (2) exploration of this perspective across the breathtaking variety of crustacean ecologies, morphologies, life cycles, habitats, and taxonomic diversity.

In this volume, Olesen examines the diversity of crustacean life cycles from the perspective of environmental adaptation, and Baeza, Ocampo, and Luppi consider the specialized life cycles of symbiotic crustaceans. Strathmann describes constraints and adaptations shaping offspring developmental strategies. Several chapters review and synthesize current literature on fundamental life history traits. Glazier discovers interesting patterns in an analysis of body size scaling relationships of clutch mass, offspring mass, and clutch size across several crustacean groups, and Maszczyk and Brzeziński review current science on ecological determinants of body size, maturation size, and growth rate. Varpe and Ejsmond integrate theoretical and empirical research on semelparity and iteroparity in crustaceans. San Vicente organizes and explores patterns of voltinism, and Vogt brings together disparate sources of research on aging and life span in crustaceans. Predation, a powerful agent of life history evolution, is the central theme of three chapters. Weiss and Tollrian provide a current and comprehensive review of predator-induced defenses, Wellborn discusses the manifold antipredator adaptations of prey species, and Bleakley presents the evolutionary and ecological consequences of cannibalism. Several chapters provide life history perspectives on salient behavioral topics. Bauer examines the often-remarkable seasonal and life cycle migrations of crustaceans, and Dawidowicz and Pijanowska review and synthesize current science on diel vertical migrations of planktonic crustaceans. Laidre considers the evolutionary ecology of construction and defense of burrows, and Hughes and Heuring review territoriality in crustaceans. Finally, Walsh and coauthors present the topic of ecoevolutionary dynamics.

Collectively, these 16 chapters provide a thorough exposition of present knowledge across the major themes in crustacean life histories. We expect this volume will be valuable to scholars and students of both life histories and crustaceans, and we also hope its syntheses and thoughtful ideas spur new avenues of research on life histories within the Crustacea and beyond.

ACKNOWLEDGMENTS

We thank our contributors for graciously sharing their time, energy, knowledge, and insights to make this volume possible—it has been both a pleasure and honor to work with each of them. Our editorial assistants, Annie Mejaes, Mika Tan, Tim Kiessling, and Miles Abadilla were impeccably skilled and organized, and they always kept us moving forward. We thank our external reviewers for their valuable and generous feedback. Finally, we express our appreciation to our publisher, Oxford University Press, for its commitment to this project.

Editing of this book was generously supported by Universidad Católica del Norte, Chile.

CONTRIBUTORS

EDITORS

Gary A. Wellborn
Department of Biology
University of Oklahoma
730 Van Vleet Oval
Norman, OK 73019
USA

Martin Thiel
Facultad Ciencias del Mar
Universidad Católica del Norte
Larrondo 1281
Coquimbo 1781421
Chile

AUTHORS

J. Antonio Baeza
Department of Biological Sciences
Clemson University
132 Long Hall
Clemson, SC 29634
USA
Smithsonian Marine Station at Fort Pierce
701 Seaway Drive
Fort Pierce, FL 34949
USA
Departamento de Biología Marina
Facultad de Ciencias del Mar
Universidad Católica del Norte
Larrondo 1281
Coquimbo
Chile

Raymond T. Bauer
Department of Biology
University of Louisiana, Lafayette
300 E. St. Mary Boulevard
Lafayette, LA 70504-3602
USA

Shannon Beston
Department of Biology
University of Texas at Arlington
701 S. Nedderman Drive
Arlington, TX 76019

Bronwyn Bleakley
Department of Biology
Stonehill College
320 Washington Street
Easton, MA 02357
USA

Tomasz Brzeziński
Department of Hydrobiology
Faculty of Biology
University of Warsaw
Żwirki i Wigury 101
02-089 Warsaw
Poland

Piotr Dawidowicz
Department of Hydrobiology
Faculty of Biology
University of Warsaw
Żwirki i Wigury 101
02-089 Warsaw
Poland

Maciej J. Ejsmond
Institute of Environmental Sciences
Jagiellonian University
Ulica Gronostajowa 7
30-387 Krakow
Poland
Department of Arctic Biology
The University Centre in Svalbard
P.O. Box 156
9171 Longyearbyen
Norway

Collin Funkhouser
Department of Biology
University of Texas at Arlington
701 S. Nedderman Drive
Arlington, TX 76019
USA

Michael Gillis
Department of Biology
University of Texas at Arlington
701 S. Nedderman Drive
Arlington, TX 76019
USA

Douglas S. Glazier
Biology Department
Juniata College
1700 Moore Street
Huntingdon, PA 16652
USA

Jared Goos
Department of Biology
University of Texas at Arlington
701 S. Nedderman Drive
Arlington, TX 76019
USA

Whitney L. Heuring
Department of Biology
College of Charleston
66 George Street
Charleston, SC 29424
USA

Julian Holmes
Department of Biology
University of Texas at Arlington
701 S. Nedderman Drive
Arlington, TX 76019
USA

Melissa Hughes
Department of Biology
College of Charleston
66 George Street
Charleston, SC 29424
USA

Mark E. Laidre
Department of Biological Sciences
Dartmouth College
Hanover, NH 03755
USA

Tomás A. Luppi
Instituto de Investigaciones Marinas y Costeras (IIMyC)
Universidad Nacional de Mar del Plata—Consejo Nacional de Investigaciones Científicas y Técnicas (CONICET)
CC 1260, Rodriguez Peña 4046
Mar del Plata 7600
Argentina

Piotr Maszczyk
Department of Hydrobiology
Faculty of Biology
University of Warsaw
Żwirki i Wigury 101
02-089 Warsaw
Poland

Emiliano H. Ocampo
Instituto de Investigaciones Marinas y Costeras (IIMyC)
Universidad Nacional de Mar del Plata—Consejo Nacional de Investigaciones Científicas y Técnicas (CONICET)
CC 1260, Rodriguez Peña 4046
Mar del Plata 7600
Argentina

Jørgen Olesen
Natural History Museum of Denmark (Zoological Museum)
University of Copenhagen
Universitetsparken 15
DK-2100 Copenhagen
Denmark

Michelle Packer
Department of Biology
University of Texas at Arlington
701 S. Nedderman Drive
Arlington, TX 76019
USA

Joanna Pijanowska
Department of Hydrobiology
Faculty of Biology
University of Warsaw
Żwirki i Wigury 101
02-089 Warsaw
Poland

Richard R. Strathmann
Friday Harbor Laboratories and Department of Biology
University of Washington
Friday Harbor, WA 98250
USA

Ralph Tollrian
Department of Animal Ecology, Evolution and Biodiversity
Faculty of Biology and Biotechnology
Ruhr University Bochum
NDEF 05/754
Universitätsstraße 150
44780 Bochum
Germany

Øystein Varpe
Department of Arctic Biology
The University Centre in Svalbard
P.O. Box 156
9171 Longyearbyen
Norway
Akvaplan-niva
Fram Centre
9296 Tromsø
Norway

Carlos San Vicente
Calle Nou 8,
43839 Creixell, Tarragona
Spain

Günter Vogt
Faculty of Biosciences
University of Heidelberg
Im Neuenheimer Feld 234
69120 Heidelberg
Germany

Matthew R. Walsh
Department of Biology
University of Texas at Arlington
701 S. Nedderman Drive
Arlington, TX 76019
USA

Linda C. Weiss
Department of Animal Ecology, Evolution and Biodiversity
Faculty of Biology and Biotechnology
Ruhr University Bochum
NDEF 05/751
Universitätsstraße 150
44780 Bochum
Germany

Gary A. Wellborn
Department of Biology
University of Oklahoma
730 Van Vleet Oval
Norman, OK 73019
USA

CONTENTS

1

CRUSTACEAN LIFE CYCLES—DEVELOPMENTAL STRATEGIES AND ENVIRONMENTAL ADAPTATIONS

Jørgen Olesen

Abstract

Crustacea (or Pancrustacea) have explored virtually all possible milieus in different parts of their life cycle, including freshwater, marine, and terrestrial habitats, and even the air (pterygote insects). Many crustacean taxa display complex life cycles that involve prominent shifts in environment, lifestyle, or both. In this chapter, the overwhelming diversity of crustacean life cycles will be explored by focusing on *changes* in the life cycles, and on how different phases in a life cycle are adapted to their environment. Shifts in crustacean life cycles may be dramatic such as those seen in numerous decapods and barnacles where the development involves a change from a pelagic larval phase to an adult benthic phase. Also, taxa remaining in the same environment during development, such as holoplanktonic Copepoda, Euphausiacea, and Dendrobranchiata, undergo many profound changes in feeding and swimming strategies. Numerous taxa shift from an early larval naupliar (anterior limbs) feeding/swimming system using only cephalic appendages to a juvenile/adult system relying almost exclusively on more posterior appendages. The chapter focuses mainly on nondecapods and is structured around a number of developmental concepts such as anamorphosis, metamorphosis, and epimorphosis. It is argued that few crustacean taxa can be characterized as entirely anamorphic and none as entirely metamorphic. Many taxa show a combination of the two, even sometimes with two distinct metamorphoses (e.g., in barnacles), or being essentially anamorphic but with several distinct jumps in morphology during development (e.g., Euphausiacea and Dendrobranchiata). Within the Metazoa the Crustacea are practically unrivalled in diversity of lifestyles involving, in many taxa, significant changes in milieu (pelagic versus benthic, marine versus terrestrial) or in feeding mode. Probably such complex life cycles are among the key factors in the evolutionary success of Crustacea.

Life Histories. Edited by Gary A. Wellborn and Martin Thiel.

INTRODUCTION

Crustaceans only rarely employ the same lifestyle during their entire life span and stay in the same general habitat (e.g., benthic, planktonic, parasitic). In most crustacean taxa, there is a shift (or more shifts) in environment and/or lifestyle that play a prominent role in the life cycle. The unfolding of animal life as discrete developmental phases that exhibit contrasting morphological, physiological, behavioral, or ecological attributes has been termed *complex life cycles* (see Moran 1994). This chapter explores crustacean life cycles with particular focus on smaller or larger changes in lifestyle (e.g., in locomotion and feeding) during development and their link to shifts in environment (e.g., from pelagic to benthic or from free-living to parasitic).

Classical examples of dramatic life cycle shifts are seen in numerous decapods (e.g., crabs, lobsters), where development involves a change from a pelagic larval phase to a benthic phase with bottom-associated, typically crawling adults (Felder et al. 1985, Charmantier et al. 1991, Anger 2001, Rötzer and Haug 2015), or in barnacles where the adults settle irreversibly (Glenner and Høeg 1993, Anderson 1994, Høeg and Møller 2006, Maruzzo et al. 2012). Other examples of spectacular shifts in environment during development are seen in river-inhabiting amphidromous caridean shrimps, the life cycle of which often involves upstream hatching followed by river drift of larvae to the sea, again followed by larval/juvenile upstream migration (crawling/swimming along banks at night) to the adult habitat (Bauer 2011, 2013; see Chapter 8 in this volume). The life cycle of taxa such as terrestrial hermit crabs and land crabs even requires a shift from a marine larval phase to a terrestrial adult phase (Bliss 1968, Greenaway 2003, Wang et al. 2007). The most spectacular changes in morphology and lifestyle during the life span are unarguably seen in the parasitic barnacles (Rhizocephala) and tantulocarids, where development includes complicated shifts in lifestyle involving, for example, larval stages specialized for settling and adults modified for absorbing nutrients from mostly crustacean hosts (Huys et al. 1993, Høeg 1995, Glenner et al. 2000).

Crustacean life cycles may also involve changes that are less dramatic but still important. Taxa staying roughly in the same environment during their entire life cycle may undergo many changes in feeding/swimming strategy. Numerous taxa shift from an early larval naupliar feeding/swimming system using only cephalic appendages to a juvenile/adult system relying almost exclusively on more posterior appendages. Examples are large branchiopods such as fairy shrimps (Anostraca) and tadpole shrimps (Notostraca), where early larvae exclusively use naupliar appendages for both feeding and locomotion, while these functions are fully shifted to more posterior appendages during further development (Fryer 1983, 1988, Schrehardt 1987). Other taxa such as the interstitial/benthic cephalocarids, mystacocarids, and harpacticoid copepods start with a naupliar feeding and locomotory system that is basically maintained during development, also in adults, but supplemented by a posterior system resulting in the two systems operating concomitantly from late larvae into the adult (Sanders 1963, Walossek 1993, Olesen 2001). In yet other taxa, the early developmental phase is characterized by the complete absence of feeding (lecithotrophy). This is the case in the naupliar phase of both euphausids (krill) and dendrobranchiate shrimps, the only two taxa within Malacostraca where development involves a naupliar phase (Martin and Gómez-Gutiérrez 2014, Martin et al. 2014a, Akhter et al. 2015), but also in the entire naupliar phase of parasitic barnacles (Rhizocephala; Walossek et al. 1996), Remipedia (Koenemann et al. 2009), and in the earliest naupliar stages of some branchiopods and copepods (Anderson 1967, Dahms 1989, Olesen 2004, Ivanenko et al. 2007). In many cases, the nonfeeding phase in the early part of crustacean life cycles has been taken to an extreme in which free-living larvae have been skipped completely. Instead, larvae are brooded in various types of brooding chambers or are attached to specialized limbs of the female (e.g., peracarid malacostracans, cladoceromorphan branchiopods, ascothoracican thecostracans, or astacid decapods; Martin et al. 2014b).

This chapter takes a broad look at crustacean life cycles. The subject is large, because a life cycle in biology can be broadly defined as a series of changes in form that an organism undergoes, returning to the starting state in the next generation, implying that virtually any aspects of the life history of a given species can be relevant. Here, the overwhelming diversity of crustacean life cycles will be explored by focusing on *changes* in the life cycles, and on how different phases in a life cycle are adapted to their environment, and, to narrow the scope, most focus has been on nondecapods (for decapod larval biology, see Volume 7 of this series). All imaginable adaptions in the life cycle to dispersal, settling, infesting, and swimming/feeding are necessarily intimately linked to the distinct and sometimes peculiar morphology of many crustacean larval types. Hence, familiarity with some of the unique larval types, both their names and morphology, is unavoidable. Therefore, at various places in this chapter short overviews of general crustacean types of development are given, spanning from the more gradual (anamorphic) type, involving a long series of stages changing only slightly during development (e.g., fairy shrimps in temporary ponds, and the "living fossils," cephalocarids), to the more modified types of development involving abrupt, metamorphic changes with distinct larval types sometimes adapted for specific functions such as locating a host and settling in cirripedes and tantulocarids (cyprid and tantulus larvae).

For a full understanding of crustacean life cycle specializations, it is important to note that a given life cycle also is the result of its ancestry, meaning that it is best interpreted by implementing not only ecological but also evolutionary interpretations. Moran (1994) pointed out that the description of complex life cycles of animals is divided into two largely discrete bodies of literature with different foci: (1) ecologically oriented studies emphasizing ecological reasons behind the diversification of life cycles into discrete phases, and (2) developmentally oriented studies more concerned with evolutionary and phylogenetic explanations focused on the relative degree of conservativeness in aspects of various life cycle stages. In reality, these two different ways of understanding life cycle adaptions are not in conflict but rather are relevant at two different time scales. Whereas the ecological approach seeks shorter-term explanations, such as food availability, for particular adaptions (e.g., a certain feeding behavior), an evolutionary/developmental approach operates at much longer evolutionary scales using shared ancestry as a tool for understanding the sometimes profound conservatism of particular developmental phases in the life cycle. One key question when dealing with Crustacea, many groups of which have complex life cycles, is the extent to which different development phases are evolutionarily linked to each other. For example, to what degree are larvae and adults free to evolve independently of one another? Or, as many developmental biologists have been concerned with, are early stages of ontogeny more conserved than late stages? These and many other aspects will be integrated into the following treatment of crustacean life cycles.

Due to the long and successful evolution of Crustacea, a multitude of developmental types and life cycles exist, and these are accompanied by a likewise diverse and not always straightforward terminology. Much terminology was defined in the *Atlas of Crustacean Larvae* (Martin et al. 2014b), but due to the complicated nature of the subject, and because definition of terms is an ongoing process that goes hand in hand with an increased understanding of homologies and phylogeny, some definitions will be repeated or modified in this chapter. A number of developmental concepts (anamorphosis, metamorphosis, epimorphosis) and their interrelationships are of particular importance, and consequently the chapter is structured around these concepts, with each section beginning with a more precise definition and discussion followed by examples from various crustacean taxa.

GRADUAL (ANAMORPHIC) DEVELOPMENT—ANCESTRAL OR ADAPTATION TO STABLE ENVIRONMENT?

Anamorphosis, or anamorphic development, is defined by Martin et al. (2014b) as "a type of indirect development in which a hatchling or larva with few segments develops into the adult gradually, via a series of small stepwise changes in body morphology, with progressively more segments and appendages added during morphogenesis." This type of development, where a nauplius-type larva develops into the adult, is seen in some Branchiopoda (e.g., Anostraca) (Fig. 1.1) and Cephalocarida (Fig. 1.2H–L) and was already present early in crustacean evolution, as evidenced by the Cambrian *Rehbachiella kinnekullensis* (Walossek 1993). As many as 25 preadult stages have been reported for *Artemia salina* (Anostraca; Benesch 1969; also see Walossek 1993), and Fryer (1983) identified 20 instars of *Branchinecta ferox* before a full complement of functional trunk limbs was acquired. Common for these "classic" anamorphically developing taxa is that no profound changes in morphology and locomotion/feeding take place between 2 consecutive stages during the development (no metamorphosis). All major changes are stepwise and involve multiple instars. The development starts with a naupliar (cephalic) locomotory/feeding system that gradually develops into an adult thoracic one. The two systems operate concomitantly (e.g., in late anostracan larva in Fig. 1.1D) across many intermediate stages of all three taxa (Cephalocarida, Anostraca, *Rehbachiella*). In adult anostracans, the thoracic system has taken over both functions entirely (except for the coxal part of the mandible, which is obviously still involved in feeding; Barlow and Sleigh 1980, Fryer 1983, Olesen and Møller 2014), while in both Cephalocarida and *Rehbachiella kinnekullensis*, the "naupliar two" second antennae maintain a role in locomotion (Sanders 1963, Walossek 1993) alongside a fully functional adult thoracopodal system. In accordance with a number of previous authors (Sanders 1963, Fryer 1983, Walossek 1993, Haug and Haug 2015), I view this specific type of strict anamorphic development, starting with a nauplius-type larva to which somites and functional limbs are added gradually posteriorly, as the most likely candidate for an ancestral type of crustacean development. However, in addition to being ancestral, the anamorphic development may also be an adaption to stable environments, which, in the case of most anostracans are found in the water phase of small freshwater temporary pools, and in the case of cephalocarids as bottom dwellers both as larvae and adults (Sanders 1963). Some constancy in environment is probably a prerequisite for anamorphic development.

The anamorphosis concept, however, is considerably more complex than implied previously. Closer scrutiny provides several examples of types of gradual development that do not apply strictly to the definition. For example, in some taxa the degree of anamorphosis depends on which part of the body is considered (i.e., body somites, limbs, or internal anatomy). When comparing anostracan and cephalocaridan development, the latter has an essentially gradual development of body somites but shows a considerable "delay" in the appearance of limbs and their further differentiation (Olesen et al. 2011, Haug and Haug 2015) (Fig. 1.2H–L). Also, because crustaceans molt their external cuticle during development, the development is stepwise externally but gradual internally. Complicating things even further is the concept of complex life cycles where more than one phase of gradual development can be present during the entire course of development, separated by one or more smaller or larger, nongradual jumps in morphology, or metamorphoses, between 2 stages in the series. Well-known examples are copepods in which two anamorphic phases (naupliar and copepodite) are separated by a metamorphic shift in morphology, and dendrobranchiate decapods, where more anamorphic phases (naupliar, protozoeal, and mysis phases) are separated by profound jumps in morphology (both treated in detail below).

Used in the strict sense of the definition above, anamorphic development applies to only a few taxa such as certain branchiopods (anostracans), cephalocarids, perhaps remipedes, and certain Cambrian "Orsten" fossils such as *Rehbachiella kinnekullensis*, that all start with an orthonauplius

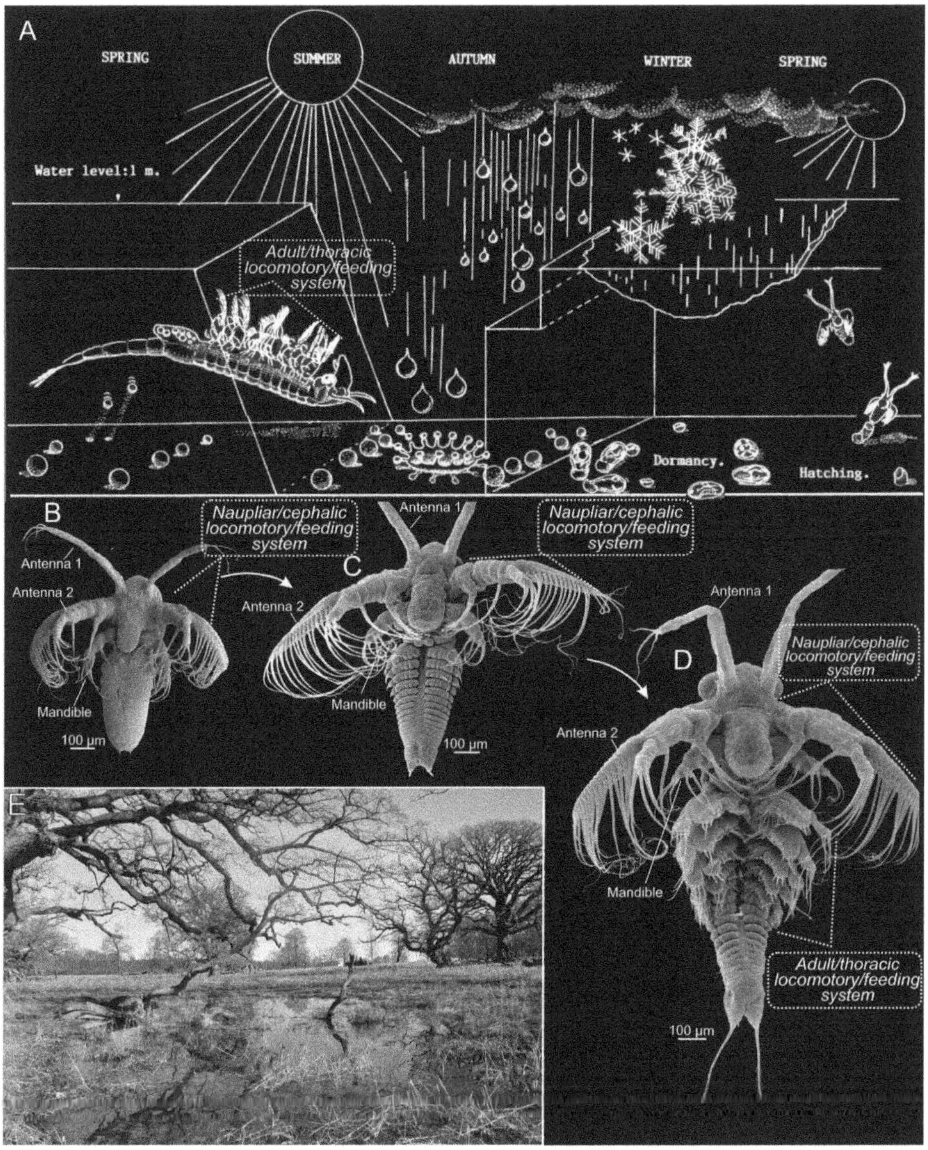

Fig. 1.1.
Life cycle and larvae of anamorphically developing fairy shrimps (Branchiopoda: Anostraca). (A) Schematized year cycle from spring to spring of *Eubranchipus grubii* in ephemeral pool followed from hatching of resting eggs to adult. Modified from Mossin (1986), with permission from Oxford University Press. (B–D) Three larval stages of *E. grubii*, and early stage, an intermediate stage, and a late stage. Modified from Møller et al. (2004), from Springer. (E) Ephemeral pool in Denmark inhabited by *E. grubii*. Photograph by Jørgen Olesen ©.

that during development adds one or few somites in each molt until adulthood. Such strict definition is unpractical for Crustacea, which display many other types of gradual development not involving clear addition of somites. Therefore, in this chapter, "anamorphic development" is applied in a broader sense, essentially as a synonym for gradual development. This implies that anamorphic development in Crustacea does not necessarily concern the entire development but often

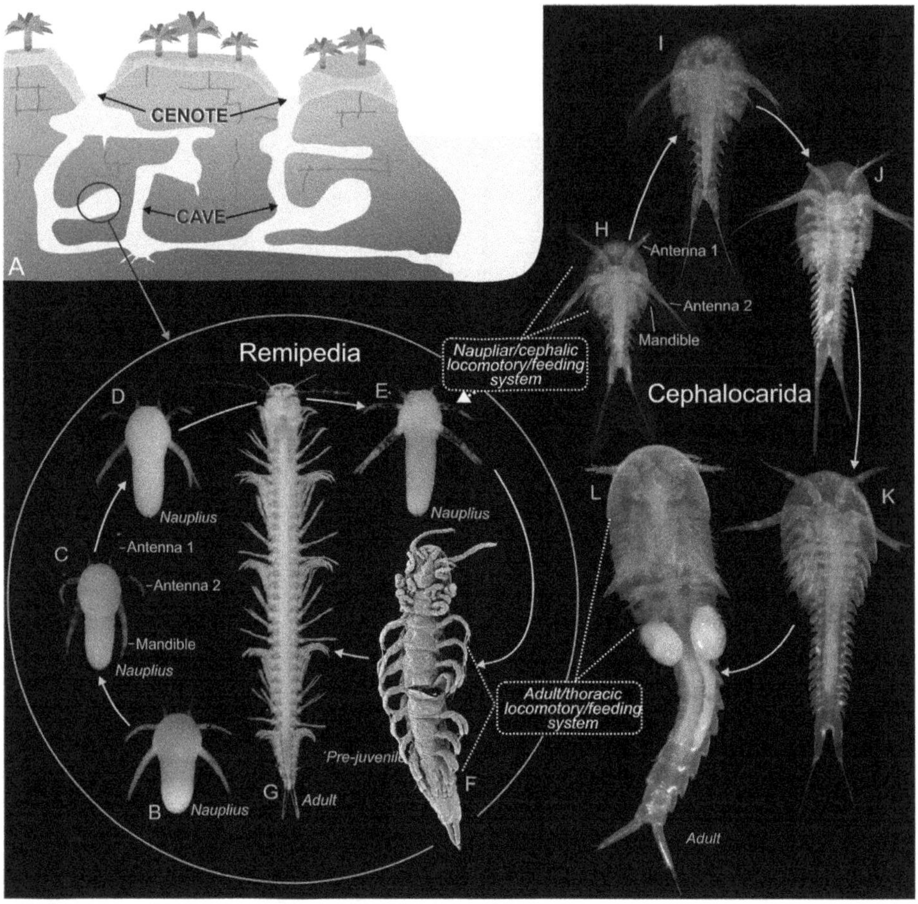

Fig. 1.2.
Life cycles of anamorphically developing Remipedia and Cephalocarida. (A) Schematized representation of typical remipede-inhabited anchialine cave system at the Yucatan Peninsula, Mexico. Illustrated by Tim Kiessling and Jørgen Olesen, under Creative Commons license (BY). (B–E) Four early lecithotrophic nauplius larvae of *Pleomothra apletocheles* (Remipedia). (F) Late larval/early juvenile of *P. apletocheles* (Remipedia). (G) Adult of *Xibalbanus tulumensis* (Remipedia) from "Cenote Crustacea," Yucatan Peninsula, Mexico. Photograph by Jørgen Olesen ©. (H–K) Four larval stages of *Hutchinsoniella macracantha* (Cephalocarida). (L) Adult of *H. macracantha* (Cephalocarida). (B–F) Modified from Koenemann et al. (2009), with permission from Springer. (H–L) Modified from Olesen et al. (2014c), with permission from Johns Hopkins University Press.

only parts (e.g., only the naupliar phase in cirripedes), and that it can concern a broad spectrum of body parts not all of which develop at the same speed. Hence, a development is anamorphic if one or more of the following applies: (1) somites are added gradually (one or few somites per molt); (2) appendages are added gradually; (3) appendages gradually become more developed; (4) internal structures such as anlagen to somite or appendages gradually appear despite not being clearly indicated externally; (5) gradual addition of setation, spines, or scales; and (6) general enlargement of body or its parts.

Within branchiopods, at least some anostracan branchiopods are strictly anamorphic during their entire development (Fig. 1.1; Fryer 1983, Schrehardt 1987). However, the larval development of branchiopods such as laevicaudatan and spinicaudatan clam shrimps can also be characterized as anamorphic despite their lack of a clear addition of externally identifiable somites (Fig. 1.3G–L; Olesen and Grygier 2004, Olesen 2005). The hatchling of *Lynceus brachyurus* (Laevicaudata) is a peculiar dorsoventrally flattened "UFO"-shaped larva, consisting of a large, rounded dorsal shield dorsally and an almost as large ventral labrum, between which the swimming antennae and mandibles are concealed (except when swimming; Fig. 1.3G). The ensuing larvae are very similar and essentially only enlarge and change shape slightly during development (Olesen 2005). The last larva in this series is followed by a juvenile very different from the larvae (metamorphosis). Development in the larval phase in the Spinicaudata is characterized by larger morphological modifications than in the Laevicaudata. The development of notostracans (tadpole shrimps) is essentially gradual during its entire development, although there are significant jumps in morphology during certain stages (but none metamorphic) (Møller et al. 2003, Olesen and Møller 2014). As a whole, the development of Branchiopoda can be characterized as anamorphic, but with important exceptions in clam shrimps and cladoceromorphans (Cyclestherida and Cladocera), which are described below. The predominantly anamorphic development in branchiopods as seen in recent anostracans may be inherited from ancestral crustaceans because much evidence suggests anamorphic development was already present in the Cambrian, around 500 Ma (Walossek 1993, Haug and Haug 2015). It was retained in branchiopods because of an early successful colonization of inland ephemeral pools, which, despite their fluctuating nature, seem to have been a very constant environment present since the Devonian, around 400 Ma (Scourfield 1926, Gueriau et al. 2016, Strullu-Derrien et al. 2016).

Also the cave-dwelling Remipedia, one of the classical candidates as "primitive" crustaceans because of their long, homonomous bodies (Schram 1983, Brusca and Brusca 1990), but now finding a phylogenetic position close to the Hexapoda (Regier et al. 2010, von Reumont et al. 2012, Lozano-Fernandez et al. 2016), seemingly undergo anamorphic development (Fig. 1.2A–G). The development is only known for one species (*Pleomothra apletocheles*), and all that is known is based on 14 specimens, which were individually collected by cave divers (by hand) in an anchialine cave in the Bahamas (Koenemann et al. 2007, 2009). The fact that the limited number of larvae found to date represents as many as 9 different developmental stages suggests that more stages remain to be found. It appears that a very long part of remipede development is nonfeeding/lecithotrophic, starting with yolk-containing nauplius-type larvae that gradually add limb buds and increase in size (Fig. 1.2B–E). Among the few specimens collected was a late specimen (termed *prejuvenile* by Olesen et al. 2014a), in which the antennae and mandibles had atrophied or become modified and no longer act in locomotion, a function transferred to the about 10 trunk limbs that are developed at this stage (Fig. 1.2F). Most likely, the transfer from a naupliar locomotory system to a "thoracic" system is gradual as in the Anostraca/Cephalocarida/*Rehbachiella* system, but the limited specimens so far collected do not allow for a firm conclusion. Certainly an important difference is that the anterior naupliar system in Remipedia is used for locomotion only (not feeding), which is in contrast to the combined locomotory/feeding function in Anostraca/Cephalocarida/*Rehbachiella*. Probably, the presence of anamorphic development in Remipedia is facilitated by the relatively stable environment of anchialine caves. The long phase of lecithotrophy is certainly derived and may be an adaptation to poor nutrition in many anchialine caves.

Yet another group that essentially undergoes anamorphic development (but with few species studied), are the Mystacocarida, a species-poor group of tiny (< 1 mm) crustaceans that undergo their entire life cycle interstitially, for example, among sand grains just a dozen centimeters below the surface of sandy Mediterranean beaches (e.g., southern France; Fig. 1.3A–F). As above, their anamorphic development (lack of metamorphoses) may be linked to the lack of shift in environment during their life cycle. As in Anostraca/Cephalocarida/*Rehbachiella*, the hatchling in *Derocheilocaris remanei* and *D. typica* (Olesen 2001, Haug et al. 2011) is a naupliar-type larva (metanauplius), with three pairs of naupliar appendages (and weakly developed maxillules; Fig. 1.3C). The naupliar

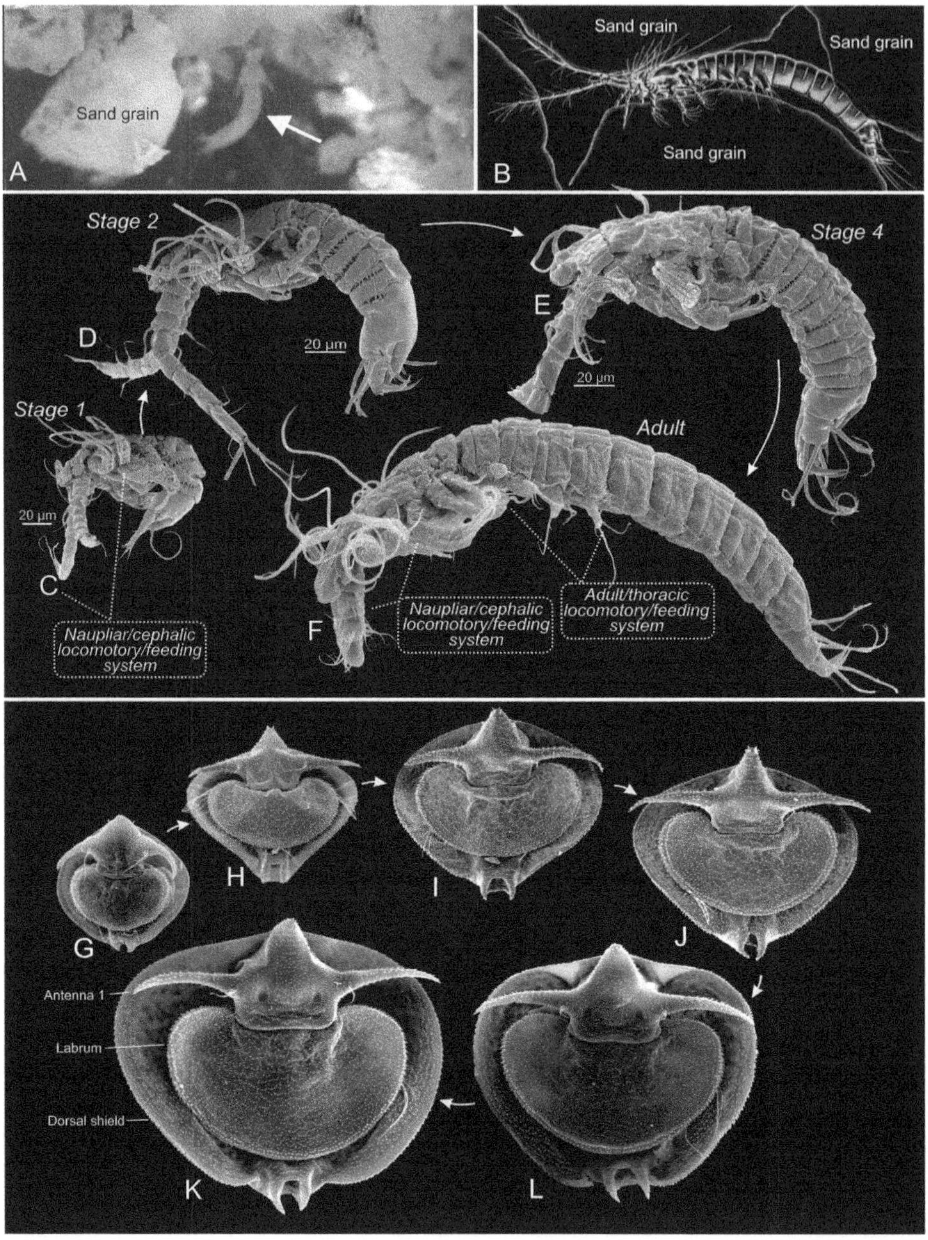

Fig. 1.3.

Life cycle and larval stages of Mystacocarida and Branchiopoda. (A) Adult *Derocheilocaris remanei* (Mystacocarida) among sand grains from beach at Canet Plage near Perpignan, France. Photograph by Jørgen Olesen ©. (B) *Derocheilocaris typica* (Mystacocarida) crawling in interstitium between sand grains. Modified from Lombardi and Ruppert (1982), with permission from Springer. (C–F) Developmental sequence of *D. typica* (Mystacocarida). Modified from Olesen (2001), with permission from The Royal Danish Academy of Sciences and Letters. (G–L) Anamorphic development of larval sequence of *Lynceus brachyurus* (Branchiopoda: Laevicaudata). Modified from Olesen (2005), with permission from John Wiley and Sons. See color version of figure part 1.3A in the centerfold.

locomotion/feeding apparatus (biramous mandibles and antennae) is retained into adulthood in mystacocarids, which has often been assumed to indicate a neotenic origin (e.g., Hessler and Newman 1975, Hessler 1992), but may be interpreted more simply as retained primitive features similar to that seen in the Cephalocarida. Many other aspects of the type of anamorphosis seen in *D. remanei* and *D. typica* (Olesen 2001, Haug et al. 2011) are rather different from the assumed ancestral type in Anostraca/Cephalocarida/*Rehbachiella*, such as the fact that the hatchling is rather "advanced" (with three somites), body development is "speeded up" with mostly two somites added per molt (Olesen and Haug 2014, Haug and Haug 2015), and trunk limbs are modified to four pairs of setose lobes in adults (Fig. 1.3B, F), the latter being a clear adaption to interstitial life among sand grains.

What most of these anamorphically developing taxa have in common is the lack of abrupt changes in environment during their development. Of the taxa mentioned above, large branchiopods such as most anostracans and laevicaudatans undertake their entire life cycle in the water phase of small temporary freshwater pools, remipedes develop in the marine water phase of anchialine caves, cephalocarid larvae and adults are associated to marine sediments, and mystacocarids are interstitial in sandy beaches for their entire life. Hence, the presence of a constant environment during development seems to be a prerequisite for anamorphic development. But there is no inverse relation. Pelagic copepods, for example, despite developing in a constant environment (the pelagic water phase) are known to have a distinct metamorphosis in their development (see the following section). It is striking that most of the anamorphically developing taxa are archaic-looking for a number of other reasons than their gradual development, most notably the anostracans and cephalocarids, which have stayed morphologically constant since the Cambrian. Therefore, the type of anamorphic development seen in these taxa seems to be evolutionarily very old, possibly without any particular adaptive value. In other cases, however, the extended anamorphic phases in a number of pelagic crustaceans such as copepods, krill, and dendrobranchiate shrimps may be adaptions to dispersal, while at the same time having old evolutionary roots.

CRUSTACEAN METAMORPHOSES, DEVELOPMENTAL "JUMPS," AND THEIR RELATION TO LIFE HISTORY

On "Metamorphosis" Within Crustacea

In only a few crustaceans, such as anostracans, is development strictly anamorphic, that is, gradual across their entire development. Instead, in many taxa, such as thecostracans, branchiurans, and malacostracans, development can be characterized as predominantly anamorphic but interrupted by molts between two instars that involve significant changes in morphology or lifestyle. Often the term *metamorphosis* is applied to these abrupt changes (e.g., Haug and Haug 2013). Well-known examples of crustaceans that undergo metamorphic changes during their development include, among many others, the shift from a naupliar phase to the cyprid (a settling stage) in most cirripedes, and the characteristic shift from a zoeal phase to the decapodid (e.g., megalopa in Brachyura), a transitional stage between the planktonic and benthic phase in the life cycle.

The balance between anamorphic and metamorphic development will be scrutinized in a number of crustacean examples. The possible adaptive value of metamorphosis in particular taxa will also be evaluated. But first it is worth considering what is actually meant by *metamorphosis*. Martin et al. (2014b) broadly referred to it as a "profound change in morphology (typically accompanied by behavioral and functional changes) during the life cycle of an organism." This definition wisely does not address the very pertinent question: Exactly *how* profound should the change in morphology be to be characterized as a metamorphosis? It also does not address a challenge specific for arthropods that relates to their molting nature. Specifically, the internal anatomical development may be more gradual than what is expressed in the external cuticle (e.g., when internal changes begin well ahead of an eventual metamorphic molt and continue afterward).

Here, for practical reasons, the term *metamorphosis* is suggested to be primarily applied on the basis of changes in the external cuticle. Cuticular information is simpler to obtain than information on internal anatomy and is available for a large assemblage of crustaceans in the literature. Furthermore, it is suggested that the type of "profound change in morphology" that qualifies to be termed "a metamorphosis" should involve an abrupt shift from one stage to another (or within a few stages) regarding which appendages specifically are involved in feeding or locomotion. Other types of smaller changes that only involve adding more somites, appendages, and so on, between two molts, or accelerating or delaying the development of some morphological aspects but not others (heterochrony), but which do not involve any major shifts in feeding or locomotion, can, if they are significant over a short period, be regarded as "developmental jumps."

The Branchiura Life Cycle: A Combination of Anamorphic and Metamorphic Development

The Branchiura, ectoparasites on mainly freshwater fish, use in most species spectacular suction disks (modified first maxillae) for temporary attachment to the fish surface (Fig. 1.4). Their development combines anamorphic and metamorphic development (Møller et al. 2007, Olesen and Møller 2014). In some species, for example *Argulus foliaceus*, the hatchling is very different from the following stages and the separating molt can be called a metamorphosis because it involves a significant shift in which appendages are used for locomotion (Fig. 1.4B,C). The *A. foliaceus* hatchling has a naupliar type of swimming using exopods of second antennae and mandibles (Møller et al. 2007; Fig. 1.4B), but in the following stage the mode of swimming changes completely. The naupliar exopods have atrophied significantly (Møller and Olesen 2014) and play no role in swimming, a function taken over by the more posteriorly placed trunk limbs, which are directed laterally as four pairs of large oars similar to the situation in the adults (Fig. 1.4C,D). Clearly, one result of this metamorphic transfer of locomotion from naupliar appendages to trunk limbs is enhanced swimming speeds and thereby more efficient host location. Feeding apparently does not change markedly during development and is accomplished by a mouth cone consisting of modified mouthparts (Martin 1932, Gresty et al. 1993), which is in contrast to many other crustaceans where structures of the more anteriorly positioned antennae 2 are involved in feeding during early development. Branchiurans deviate much from other crustaceans, but it is interesting to note that the naupliar-style of locomotion (using antennae and mandibles) in the hatchling of *A. foliaceus* and a few other species (Møller et al. 2007) is an ancestral (plesiomorphic) characteristic, providing a link to many other crustaceans starting their development with nauplii that use a similar form of locomotion. In the majority of branchiurans, however, the naupliar-style–swimming hatchling has been lost so that the development starts with a more advanced stage, basically like the second stage of *A. foliaceus*.

Copepoda—Anamorphic Crustaceans with Distinct Metamorphosis

A treatment of crustacean life cycles is not complete without the Copepoda. In almost every measurable way, including habitats, lifestyles, morphological variation, and abundance, the Copepoda are among the most diverse groups of multicellular animals on Earth (Huys and Boxshall 1991, Huys 2014). Most significantly, copepods are dominant members of both the marine and freshwater holoplankton, but they have also colonized numerous interstitial habitats, and many forms are parasitic on a variety of other organisms. Despite their large diversity, the development of many groups

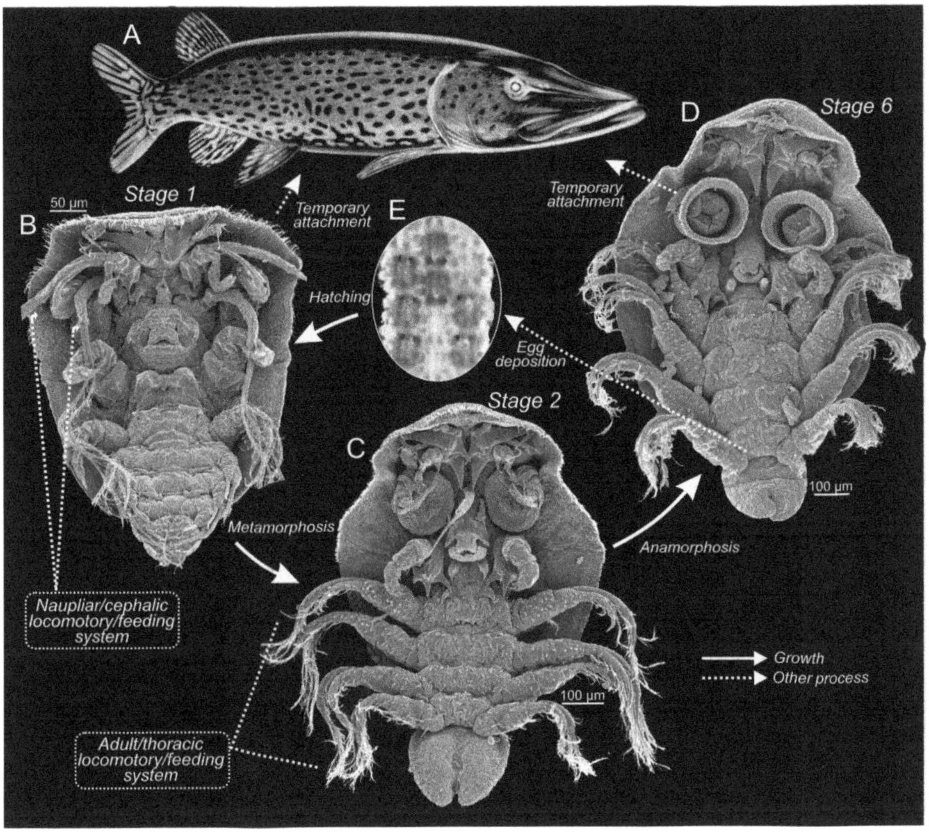

Fig. 1.4.
Life cycle and larval stages of Branchiura. (A) Pike, at typical fish host of *Argulus foliaceus* (Branchiura). Illustrated by Robin Woolnough ©. (B–D) Three developmental stages of *A. foliaceus* (Branchiura). (E) Close-up sediment-attached egg string of *A. foliaceus* (Branchiura). (B–E) Modified from Møller and Olesen (2014), with permission from Johns Hopkins University Press.

of copepods can be divided into an early naupliar phase and a later copepodite phase (leading to the adult), with, in many groups, a constant number of stages (six nauplii and five copepodites). However, there are numerous modifications of this general scheme, such as reductions in the number of nauplii, or brooding of the naupliar phase (see Ferrari and Dahms 2007, Huys 2014). The focus here is on a "standard" type of copepod development, with its two different developmental phases (naupliar and copepodite), and, with a few examples from the literature as background, I explore the extent to which development of copepods can be categorized as anamorphic, metamorphic, or a combination of these (example in Fig. 1.5). The focus will be on which significant changes in locomotion/feeding take place at the transition to naupliar and copepodite phases. Gurney (1942) stated that "the most complete example of metamorphosis is shown by the Copepoda." This seems to be an exaggeration, however, considering, for example, the profound shift in morphology and lifestyle between the free-living nauplii and the settling cyprid in cirripedes and the almost as significant change in morphology between the planktonic zoea phase and the transitional (before settling) decapodids of decapods (see subsequent discussion).

Considering first the typical naupliar development as seen in many species, this phase is easily characterized as anamorphic. In the case of, for example, *Canuella perplexa* (Harpacticoida), the

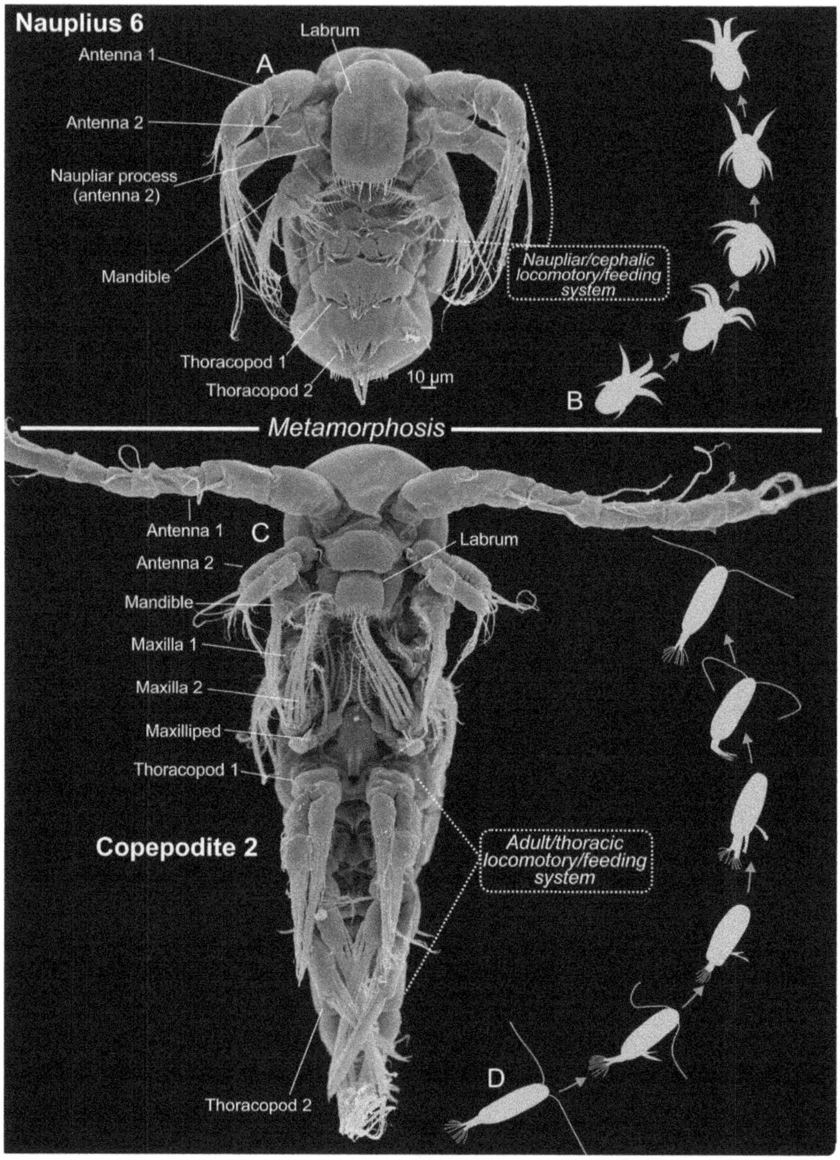

Fig. 1.5.
Developmental phases and metamorphosis in Copepoda. (A) Late nauplius of *Acartia tonsa* (Copepoda). (B) Snapshots of swimming sequence of nauplius of *A. tonsa.* (C) Copepodite 2 of *A. tonsa.* (D) Snapshots of swimming sequence of copepodite/adult of *A. tonsa.* (A, C) Photograph by Jørgen Olesen © of cultured material (originally from Øresund, Denmark; specimens generously provided by Benni W. Hansen). (B, D) Illustrations created on basis of a video from Kiørboe et al. 2014, with permission from The National Academy of Sciences (PNAS).

hatchling is a short, ovoid, typical copepod-type nauplius, with only three pairs of appendages and no additional limb rudiments (Vincx and Heip 1979). See also the nauplius of *Acartia tonsa* in Fig. 1.5A), which is typical of many groups (Salvacion et al. 2004, Chullasorn et al. 2009, Moon et al. 2015). The further development is very gradual and involves primarily only general enlargement, addition of setae on limbs, appearance of more setae/spines terminally at the hind body, and the addition of some postmandibular limb buds (starting in N2 with the anlage of maxilla 1 as two long setae, and ending in N6 with rather well-developed maxillae 1) and 3 additional limb buds. Because of the presence of postmandibular limb anlagen in N5–N6, these stages can, by definition, be called "metanauplii" as in Haug and Haug (2015). In some other taxa, there are no distinct postmandibular limb buds late in the naupliar phase (e.g., Carton 1968).

Naupliar swimming in calanoids has been categorized as comprising of at least 2 types (after Gauld 1959, van Duren and Videler 1995, Mauchline 1998, Andersen Borg et al. 2012): (1) slow gliding movement performed by antennae and mandibles (antennules in "resting" position), and (2) rapid, darting movements in which all three pairs of appendages sweep backward and forward rapidly in what is considered to be a metachronal rhythm (termed "fast swimming" or "relocation jumps" after Andersen Borg et al. 2012) (Fig. 1.5B). These 2 motility modes have since been confirmed in several studies, with calanoid nauplii exhibiting both types and cyclopoid nauplii only moving in the jerky mode (Bruno et al. 2012). In an early study, Storch (1928) observed two different modes of feeding: (1) nauplii of the calanoid *Diaptomus gracilis* creating a feeding current using the antennae and the mandibles, and (2) nauplii of the cyclopoid *Cyclops strenuus* grasping food particles. High-speed cinematography applied to the study of naupliar locomotion and feeding in 3 species of Calanoida (*Temora longicornis, Oithona davisae, Acartia tonsa*) clarified many details (Andersen Borg et al. 2012, Bruno et al. 2012). Fast swimming, involving fast metachronal movements of the naupliar appendages, was found in all three species, but in *A. tonsa* it was mainly caused by beats of antenna 1 (A1) and antenna 2 (A2) (mandible, Md, relatively immobile), in contrast to the other 2 species where all naupliar limbs are involved. Nauplii of two of the species (*T. longicornis* and *A. tonsa*) are ambush feeders that feed on motile prey during fast swimming, whereas one species (*O. davisae*) produces a feeding current while swimming slowly ("gliding") using the antennae and mandibles (Bruno et al. 2012).

Between the naupliar phase and the following copepodite phase, there is a profound modification in morphology, involving a shift in mode of locomotion, which clearly can be categorized as a metamorphosis (as in the calanoid copepod *A. tonsa* in Fig. 1.5). Dahms (1992), who explored this metamorphic shift in copepod development in detail for a number of species of harpacticoid copepods, identified a suite of significant changes both at the general and more detailed level. Not least significantly, three thoracopods appear in one step, changing the mode of locomotion from a naupliar-based system to a thoracopod-based one (Fig. 1.5), which, as in adults, are responsible for the fast, jumping mode of swimming so typical of copepods (Fig. 1.5C,D). The evolutionary origin of this locomotory system has been suggested as one of the explanations for the great evolutionary success of copepods (Kiørboe 2011). The mode of swimming is essentially the same in early copepodites and the adult, but because more thoracopods appear gradually during development, the swimming speed is constantly increased in successive stages (Mauchline 1998). In the raptorial cyclopoids, swimming is almost exclusively by use of the thoracopods, the swimming legs (Strickler 1975, Alcaraz and Strickler 1988). In calanoids, the dominant way of locomotion is swimming at a constant speed by setting up a feeding/swimming current by "vibration" of mouthparts, but this is regularly broken up by escape reactions that consist of a series of power strokes of the thoracopods (swimming legs) (Alcaraz and Strickler 1988, Kiørboe et al. 2010).

In summary, based on a combination of morphological and functional criteria, it is clear that the shift between the naupliar and copepodid phase can be characterized as a metamorphosis. It is striking, however, that the copepod metamorphosis does not involve a shift in environment (e.g., from a planktonic to a benthic/parasitic phase), which is often coupled to metamorphic changes in other crustacean taxa. Many copepods, such as the calanoids focused on previously, are holoplanktonic and therefore essentially spend their entire life cycle in the same environment. Hence, it is not straightforward to explain the presence of metamorphosis in copepods. It is likely, however, that its presence is linked to the profound differences in feeding and/or locomotion in the naupliar and copepodid phase, because these are the parts of the behavior that change significantly during metamorphosis. In a number of decapods with distinct metamorphoses such as scyllarids, fossil specimens that are intermediate between well-known developmental phases (larval and juvenile) have appeared in recent years (Haug and Haug 2013), indicating a gradual evolutionary appearance of a metamorphosis in these taxa. Despite absence of developmentally intermediate copepod fossil specimens (copepods fossilize poorly, Selden et al. 2010), it is likely that metamorphosis in copepods also has appeared gradually. This more anamorphic developmental series would have included stages that combined naupliar and adult locomotion/feeding, as is seen in intermediate stages of the Anostraca/Cephalocarida/*Rehbachiella* type of development. It may be that such mechanically more complicated transitional stages, involving 2 different locomotory systems operating concomitantly, has been selected against. Favored instead are 2 distinctly different and better adapted phases (naupliar and copepodid phase).

Malacostraca—Large Diversity in Life Cycles and Development

The diverse Malacostraca display numerous developmental types spanning from a mainly anamorphic type of development (but with metamorphosis-like jumps in morphology) with free larvae hatching at the nauplius stage in the mainly holoplanktonic taxa Euphausiacea (krill) and Dendrobranchiata (Decapoda), to strongly metamorphic development (but with anamorphic sequences) with distinct larval types in Stomatopoda and some pleocyematan decapods (Martin 2014; see the following section). Many taxa such as Leptostraca, Peracarida (e.g., Isopoda, Amphipoda, Thermosbaenacea), and some crayfish (Olesen and Walossek 2000, Scholtz and Kawai 2002, Vogt and Tolley 2004, Boyko and Wolff 2014, Wolff 2014, Olesen et al. 2014b, Goy 2014, Olesen et al. 2015) have skipped free larvae completely and development is epimorphic, or "direct," with the part of the developmental phase corresponding to the free larval phase in other crustaceans passed inside an "egg" in various types of brood chambers (peracarids) or associated with various appendages of the female (see the following section). It has been discussed whether the presence of free-living nauplii in euphausiids and dendrobranchiates is an ancestral feature for the Malacostraca (see Akhter et al. 2015), retained in these 2 taxa, or whether the nauplii have become secondarily free-living again from "egg-nauplii," as are those of many other malacostracans (see Scholtz 2000, Jirikowski et al. 2013). Here the former hypothesis (free nauplii as ancestral) has been favored (see arguments in Akther et al. 2015). The following will show that euphausiids and dendrobranchiates occupy unique positions as morphological links between malacostracans and nonmalacostracans. In particular, the naupliar phase of these taxa has much resemblance with the early development of many nonmalacostracans, whereas stages in the later phases (e.g., calyptopis, furcilia, protozoea) have much resemblance to larval types found in some other malacostracans (Decapoda, Pleocyemata) in which the naupliar phase is lacking.

Euphausiacea (Krill)—Life Cycle and Development

The Euphausiacea (krill) are a relatively small (86 species) but ecologically important group of holoplanktonic crustaceans found in all world oceans and are important links between trophic levels (Boden et al. 1955, Mauchline 1980, Martin and Gómez-Gutiérrez 2014). The euphausid life cycle (Fig. 1.6) is well known and comprises a short developmental phase of lecithotrophic nauplii (Fig. 1.6B), followed by a lecithotrophic metanauplius (Fig. 1.6C), again followed by a phase of more advanced feeding larvae, the calyptopis larvae (Fig. 1.6D,E), finally followed by a phase of stages called furcilia larvae during which the adult morphology is gradually reached (Sars 1898, Knight 1976, 1978, 1980, Marschall and Hirche 1984, Martin and Gómez-Gutiérrez 2014). However,

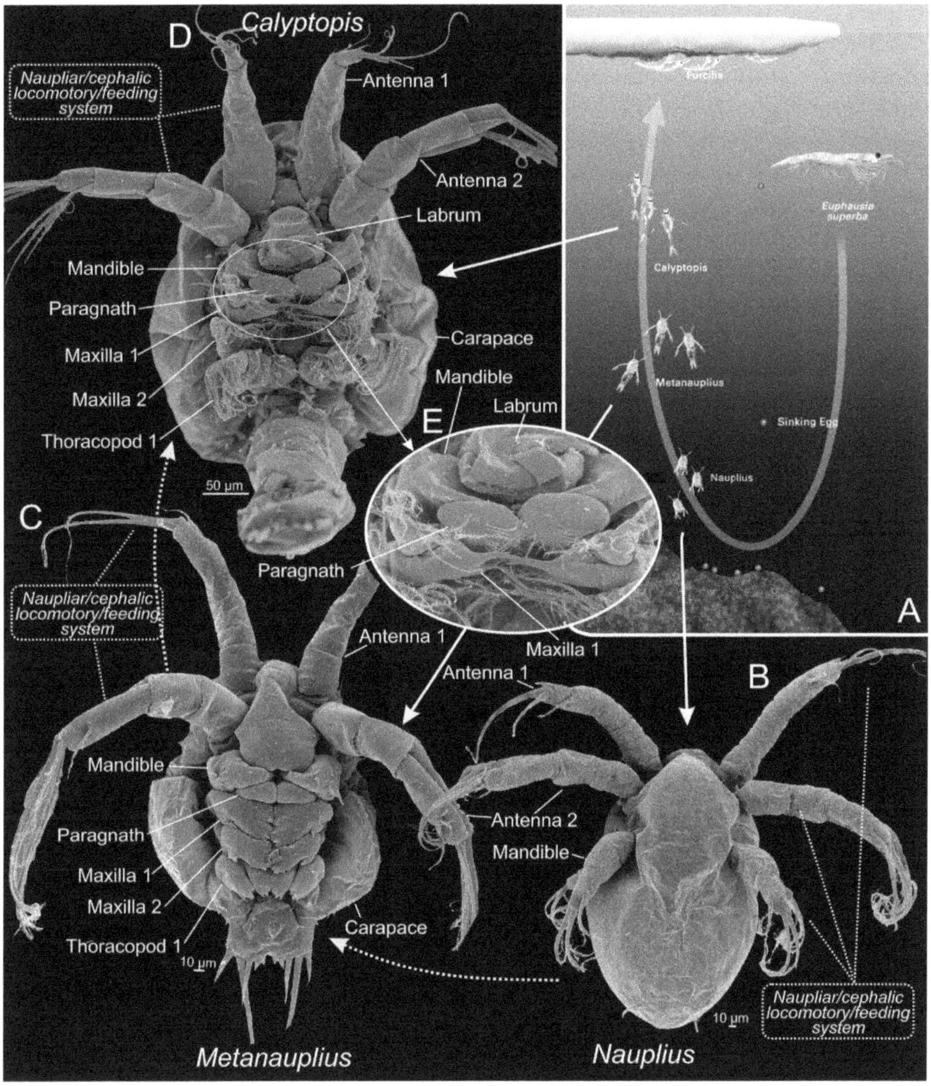

Fig. 1.6.
Life cycle and larval stages of Euphausiacea (Malacostraca). (A) Generalized life cycle of a euphausiacean. Modified from an illustration by Karen Carr, with permission from Sant Ocean Hall/Smithsonian Institution. (B) Nauplius of *Thysanoessa raschii* (Euphausiacea). (C) Metanauplius of *T. raschii.* (D) Calyptopis larva of *T. raschii.* (E) Closeup of mouth parts of calyptopis larva. (B–E) Modified from Akther et al. (2015).

variation to this pattern exists because some species are so-called sac spawners, meaning that they, instead of releasing their fertilized eggs into the free water, spawn into a sac attached to the posterior pairs of thoracic legs in which the early part of the development takes place (Gómez-Gutiérrez and Robinson 2005). These species are essentially brooders. Larvae are then released from the sac at a progressed stage (e.g., as a pseudometanauplius; Boden 1951, Martin and Gómez-Gutiérrez 2014). Most species spawn eggs directly into the water in which the embryos, during the sinking phase of the eggs, undergo considerable development before hatching. The hatched lecithotrophic nauplii are motile but clearly nonfeeding because they lack feeding structures (Fig. 1.6B). Mostly two naupliar stages are reported in this developmental phase (Lebour 1926a, Macdonald 1928, Fraser 1936, Knight 1975, Suh et al. 1993, Jia et al. 2014), but Akther et al. (2015) found evidence for three nauplii in *Thysanoessa raschii* that differed from each other in only minute details. The development in this phase is anamorphic, because only minute changes take place. Swimming is accomplished by three pairs of naupliar appendages (antennae 1 and 2 and mandibles; Marschall 1984).

The metanauplius differs in significant ways from the preceding nauplius (Fig. 1.6C), including (1) loss of mandibular palp; (2) appearance of anlagen to paragnaths and three pairs of limbs (maxillae 1 and 2 and thoracopod 1); and (3) appearance of a dorsal shield, which is the precursor of the carapace (e.g., in *T. raschii*; see Akther et al. 2015). Swimming now takes place by the use of A1 and A2 only (Md palp reduced). Following the metanauplius are three feeding calyptopis larvae (Fig. 1.6D), characterized by the presence of compound eyes (but imperfectly developed), a more developed carapace and well-developed mouthparts (mandible, maxillae 1 and 2, thoracopod 1) being active in feeding (Marschall 1984, Martin and Gómez-Gutiérrez 2014). Based on general morphology of the calyptopis (Fig. 1.6D), it would seem that antennae 1 and 2 are the main propulsive organs, but locomotory functions need clarification based on studies of live specimens. After this comes a phase of 3–6 furcilia larvae (number is species-dependent), which are characterized by fully developed compound eyes, and during which the thoracopods and pleopods develop (Martin and Gómez-Gutiérrez 2014). In the first furcilia stages the antennae are still natatory, but in the later larvae of the phase this function is lost (Martin and Gómez-Gutiérrez 2014), and apparently it is gradually taken over by the thoracopod exopods and the pleopods. The last couple of furcilia stages were originally named *cyrtopia larvae* based on a reduction of the natatory function of the second antennae (Sars 1885, Macdonald 1928), but later authors have considered these changes too gradual to justify separate naming of two phases (Fraser 1936, Boden 1950, Silas and Mathew 1977, Martin and Gómez-Gutiérrez 2014). Pleopodal swimming begins in the furcilia phase and its significance is increased during furcilia development as more pleopods are added (number varies much between species, see Lebour 1926b), but is probably insignificant in furcilia 1–2 as the pleopods are only buds. In adult *Euphausia superba,* the thoracopodal exopods move independently of the endopods (which form the filter basket) and together with the pleopods are part of a metachronically moving limb series that serve for locomotion (pleopods) and possibly produce a feeding current (thoracopodal; Boyd et al. 1984). It is uncertain how early in the furcilia phase the thoracopodal exopods function as in adults, but this function is probably attained alongside with the development of the filtering chamber starting in furcilia 2 in *E. superba* (Marschall 1984).

The overall impression of the development of Euphausiacea as outlined above is that it is anamorphic. None of the changes between various larval types can be characterized as clearly metamorphic because no abrupt change in the mode of locomotion takes place. Naupliar type of locomotion (but with mandibular exopod already reduced in metanauplius) is kept until late in the furcilia phase but is then gradually taken over by the thoracopodal exopods and the pleopods, which is also the adult mode of locomotion (together with "escape" swimming). It is known that the Antarctic krill (*E. superba*) go through a characteristic vertical migration during their development ("ontogenetic migration"), which can take place over a time span of 2 years (Fig. 1.6A). Generally, gravid *E. superba* are found offshore in deeper water than the rest of the adult population, which, after spawning, ensures that the eggs are in suitable waters for their further development (Nicol 2006). Eggs of *E. superba* sink, but eggs of other species may float (Nicol 2006).

The embryos of *E. superba* hatch into nonfeeding, free-swimming larvae at depths of 700 to 1,000 meters (Fig. 1.6A). The larvae swim upward, developing as they swim, and reach the surface some 30 days after the eggs are laid. The first feeding stage (the first calyptopis) is reached after 30 days, and it is critical for survival that the larvae find food within six days (Nicol 2006). Late larval stages and juveniles spend a part of their development associated with sea ice and feed on ice algae (Nicol 2006) (Fig. 1.6A). A practical advantage of having the eggs and developing larvae laid offshore from the main body of the krill population may be avoidance of cannibalism on larvae from the swarming adults, which are dominant consumers of suspended material (Nicol 2006).

Lecithotrophy is generally considered an adaptation to low or unpredictable food production as seen in, for example, the deep sea (Thorson 1961), and is congruent with lecithotrophy in the early phase of krill development (nauplii and metanauplius). Nicol (2006) remarked that there is an obvious need to better understand the early life history stages of krill. Locomotory and feeding mechanisms of a broad suite of developmental stages need more study, because this is important for a full understanding of the ecological roles of various krill species. Currently, only a rough picture can be put together based on scattered information from different sources. Larval locomotion is accomplished by the antennae 1 and 2 during the naupliar development, but after this phase locomotion is gradually transferred to the thoracopodal exopods and the pleopods. But how does this transition exactly take place? Feeding is known to start in the first calyptopis, and is accomplished by the mandibles, maxillae 1 and 2, and the thoracopods 1 apparently operating as a filtration system (Fig. 1.6D,E), for example, feeding on diatoms as shown by analyses of gut contents by Marschall (1984). According to Marschall (1984), the number of feeding appendages and their essential morphology remain relatively constant between the first calyptopis and first furcilia stages, and the feeding basket present in adults apparently becomes functional from second furcilia. But no detailed information on how the filtration system operates in calyptopis larvae seems available, and a complete understanding on how it transforms into the complex filtering basket in adults is lacking (see adult function in Boyd et al. 1984).

Dendrobranchiata—Life Cycle and Development

Dendrobranchiata (Fig. 1.7), containing the economically important penaeoid and sergestoid shrimps, are commonly classified as sister group to all decapods (Richter and Scholtz 2001, Tavares and Martin 2010). They are widely thought to be the least derived of the extant Decapoda, for example, because they, as euphausiaceans, shed their eggs directly into the water from which nauplii hatch (except Luciferidae, which have a short brooding phase), and because the larval development overall is anamorphic (Tavares and Martin 2010, Martin et al. 2014b). Actually, as will be shown in the following, the development is in many ways more anamorphic than that of Euphausiacea to which many similarities exist. There are approximately 500 extant species, nearly all of which are marine with species from shallow tropical waters to depths of about 1,000 m on the continental slopes (Pérez Farfante and Kensley 1997, Tavares and Martin 2010, Martin et al. 2014c). Almost half of the known species are members of the Penaeidae, which inhabit shallow and inshore tropical and subtropical waters, some species of which form the basis of massive seafood industries. Other dendrobranchiate families are predominantly in deep water, while others again are either deep benthic dwellers, members of the meso- and bathypelagic fauna, or entirely planktonic. Most information on life cycles is available for coastal penaeids, but from what can be inferred from the few available studies on deep benthic or pelagic species, many aspects of the life cycles seem the same, such as duration of larval phases and feeding strategies (Martin et al. 2014c).

Development of dendrobranchiates can be divided into a number of phases, during each of which the larval stages in the series changes only relatively little (Fig. 1.7). These phases are (following Martin et al. 2014c): a phase of nonfeeding nauplii (five to eight stages), a protozoeal phase (three stages), a mysis phase (variable number mentioned in literature), and a decapodid phase. Although practical, this division into distinctly named phases, reusing terms from nonmalacostracan (naupliar phase) and nondendrobranchiate decapod (postnaupliar phase) ontogenies, somewhat hides the gradual nature of dendrobranchiate development. The division into phases has traditionally been based on a combination of somite and limb development, and on limb functionality (as in other Crustacea), some details of which will be reviewed here. In the naupliar phase of species, such as *Metapenaeopsis*

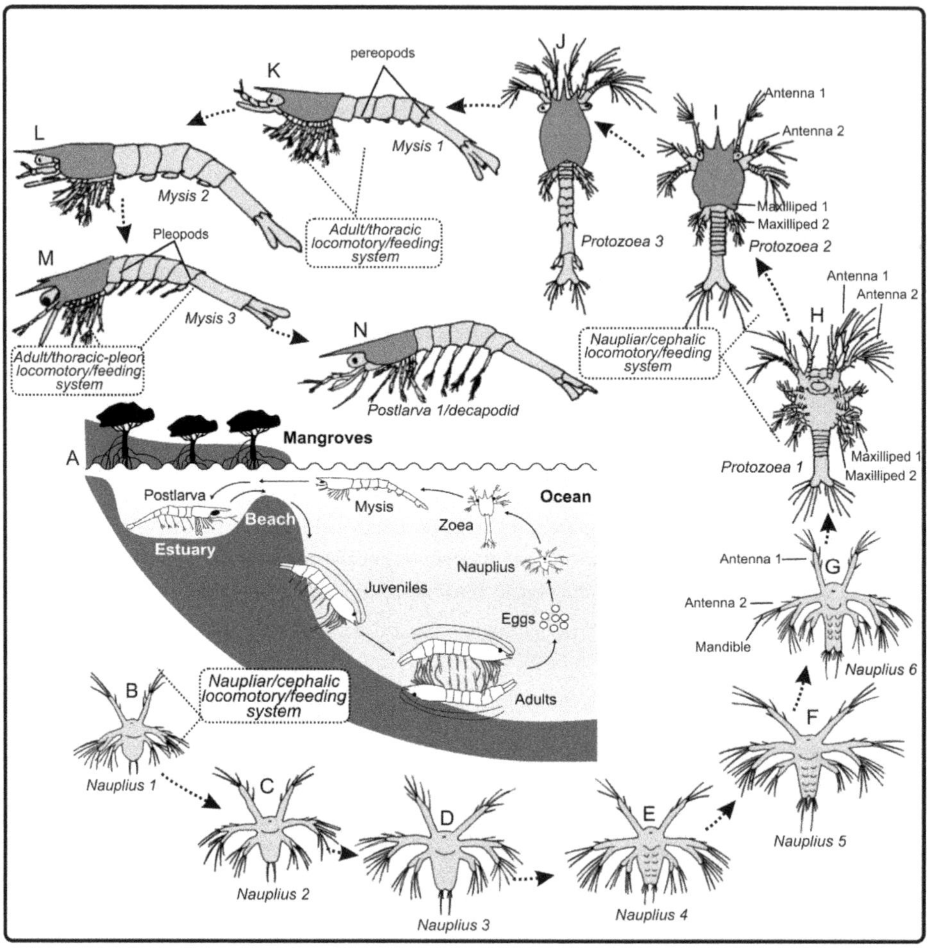

Fig. 1.7.
Life cycle and larval stages of Dendrobranchiata (Malacostraca: Decapoda). (A) Generalized life cycle of typical penaeid shrimp. (B–N) Complete developmental sequence of *Penaeus monodon* (Decapoda: Penaeidae). Modified from Kungvankij et al. 1985, with permission from FAO.

dalei and *Penaeus semisulcatus*, six lecithotrophic naupliar stages (N1–N6) have been recognized (Fig. 1.6A–F) (Choi and Hong 2001, Ronquillo et al. 2006).

The development in this phase is essentially anamorphic. During the entire naupliar phase, only the naupliar appendages (A1, A2, and Md) are functional. Among the gradual changes are that antennae 1 and 2 become more segmented and more setose, caudal spines are added, and, perhaps most significantly, anlagen to carapace and four postmandibular limbs (mx1, mx2, mxp1, and mxp2) are present in N5 and further developed in N6. Actually, by definition, because distinct anlagen to postmandibular limbs are present, N5 and N6 could be termed *metanauplii*. For nauplii of *Metapenaeus ensis*, Chu et al. (1996) reported that they alternate between swimming and resting, that the swimming movement is rather jerky, and that most of the thrust seems to be provided by movements of the first and second antennae. Between N6 and the first protozoea of *M. ensis* (and other dendrobranchiatans, see Martin et al. 2014c), a significant jump in morphology takes place and feeding starts (compare Fig. 1.7G with Fig. 1.7H). The changes in morphology include elongation and segmentation of the body, the appearance of compound eyes and a free carapace, the loss of the mandibular palps, and the activation of mx1, mx2, mxp1, and mxp2, which were only buds in N6. The 2 pairs of antennae are still important swimming appendages in *M. ensis* but now also thoracopods 1 and 2 (mxp 1 and 2) are functional, resulting in continuous swimming (Chu et al. 1996). In the third protozoea the remaining thoracopods are prominent and biramous but still not functional (Williamson 1982, Martin et al. 2014c) and probably therefore not the explanation for the higher swimming speed noted in this stage of *M. ensis* by Chu et al. (1996).

With the molt to the first mysis stage, the penaeid larvae undergo major changes in their appearance, and their bodies begin to look shrimp-like (Chu et al. 1996, Martin et al. 2014c) (compare Fig. 1.7J with Fig. 1.7K). The most significant change is the development of pereopods with large exopods, which, according to Chu et al. (1996), replace the first and second antennae as functional locomotory appendages. All three mysis stages swim with the dorsal side of their body upward. The third mysis has the ability of swimming both in forward and backward direction unlike previous stages, which can only swim in forward direction (Chu et al. 1996). From the last mysis stage to the decapodid (= first postlarva), all exopods are reduced or lost; the uniramous pleopods become large, setose, and functional, taking over as the sole locomotory appendages (Martin et al. 2014c); and the larva again swims forward with a rhythmic beating of the pleopods (Dall et al. 1990).

The overall development of Dendrobranchiata can best be characterized as anamorphic but with significant jumps in morphology in relation to the molt between various phases, that, depending on the criteria, may be characterized as metamorphic. The jump in morphology between the naupliar and the protozoeal phase results in significant changes in both feeding (from lecithotrophy to feeding), involving activation of mandibles and four postmandibular limbs (mx1–mxp2), and change in swimming mode, which now involves maxillipeds 1 and 2 in addition to the antennae 1 and 2 in the nauplii (compare Fig. 1.7H with Fig. 1.7I). However, there is also functional continuity between the two phases because antennae 1 and 2 in both phases play a significant role in locomotion. The shift in morphology and functionality between the protozoeal and the mysis phase is also profound, involving a complete posterior transfer of locomotion from an anterior antenna/maxillipedal system to a system based on thoracopodal exopods (compare Fig 1.7J with Fig. 1.7K). Also, the jump in morphology and functionality between the mysis and the decapodid is clear-cut, involving the reduction of thoracopodal exopods as organs for locomotion replaced by an even more posterior pleopodal system (compare Fig 1.7M with Fig. 1.7N).

Certainly the summarized changes in feeding and locomotion during the life cycle of peneaid shrimps (and other dendrobranchiatans) are intimately linked to their life history, but the details are difficult to elucidate. The life history of penaeid shrimps in general involves an offshore planktonic larval phase, and estuarine, benthic postlarval/juvenile phase, and an inshore adult and spawning phase (e.g., Dall et al. 1990). For the individual specimen, this involves (1) a transition from a marine planktonic larva to a benthic estuarine juvenile, representing a major morphological, physiological and behavioral transformation, and (2) the return of the juveniles or subadults from the estuarine nursery grounds to the offshore adult breeding grounds. After spawning, the eggs have been reported to sink. Hence, the first task of the newly hatched lecithotrophic nauplii are to work themselves upward in the pelagic phase, helped by the fact that among all the larvae in the developmental sequence studied by Chu et al. (1996), the nauplii were the fastest swimmers. The lecithotrophy in this phase obviously sets the larva free of an eventual variable food supply until it reaches shallow water with more food, after which metamorphosis to the feeding protozoea takes place.

Decapod Metamorphoses

One of the more striking metamorphoses in Crustacea is that between the zoeal phase and the decapodid phase (followed by benthic juveniles) in the predominantly benthic (as adults) pleocyematan decapods (e.g., crabs, lobsters), as seen in Fig. 1.8. Decapoda (nondendrobranchiatan) are immensely diverse and, when including larvae, its variation and complexity is practically unrivaled in the sea. Decapod larval patterns are much too diverse to treat in detail in this chapter (see e.g., Gurney 1942, Anger 2001). Martin (2014) says, "concerning their larval development, it is fair to say that more is known about decapods than for all other groups of crustaceans combined." The development of pleocyematan decapods can roughly be divided into two phases: (1) an early zoeal phase in which thoracopods (maxillipeds and pereopods) are used for locomotion and feeding, and (2) a later decapodid phase in which the pleopods have become the limbs responsible for locomotion. This second phase is often described as a transitional phase between the planktonic zoea and a benthic juvenile phase. In many taxa, the decapodid larvae have special names in various taxa such as "Glaucothoe" in Anomura, "Megalopa" in Brachyura (see summaries in Martin et al. 2014b). Much variation exists to this generalized pattern, including that the morphological transition between the two phases rarely is fully distinct. In most caridean shrimps for example, the transition between the zoeal and the decapodid phase is actually rather gradual. Anger (2001) reported that in *Palaemonetes argentinus* late zoea already have setose and weakly moving pleopods, foreshadowing the decapodid condition, and that the natatory exopods of the pereopods and maxillipeds remain

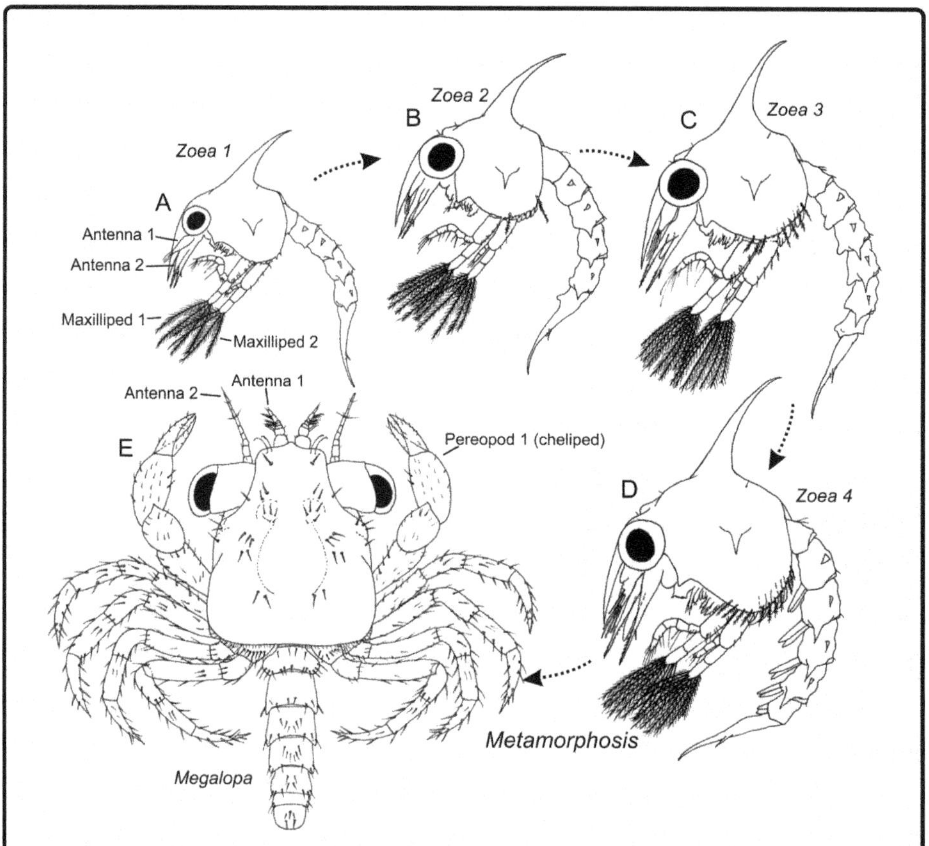

Fig. 1.8.
Larval stages of Brachyura (Decapoda). (A–D) Zoea 1 to 4 of *Pilumnus reticulatus* (Brachyura). (E) Megalopa of *P. reticulatus*. Modified from Spivak and Rodríguez (2002).

present and functional throughout the decapodid phase, representing a retained zoeal character, the functionality of which is not lost before the juvenile phase. Anger (2001) summarized the development of caridean shrimps as typically being gradual. The caridean larval development has sometimes been termed *hemimetabolic* (incomplete metamorphic), but it is at least as correct to term it *anamorphic*, thereby highlighting its essentially plesiomorphic gradual nature. Aspects of the development of many other pleocyematan decapods, such as lobsters (Homaridae), spiny lobsters and slipper lobsters (Achelata), and crabs (Brachyura), appear in general more metamorphic between the zoeal and decapodid phase, but also here many anamorphic aspects remain, making the categorization into distinct phases sometimes challenging; this has resulted in numerous and sometimes conflicting naming schemes in different papers (see also Volume 7 on Development and Larval Ecology).

Thecostraca and Tantulocarida—Distinct Metamorphoses and Permanent Attachments

Among the most intriguing life cycles in Crustacea are those of various groups of the Thecostraca, involving, in the majority of species, two metamorphic shifts in morphology and lifestyle during development (Anderson 1994, Grygier 1996, Høeg and Møller 2006, Høeg et al. 2009, 2015) (Fig. 1.9A–C). The early development of the majority of thecostracans goes through an anamorphic larval phase with 4–6 free-living nauplii that are either feeding (most thoracicans) (Fig. 1.9B) or lecithotrophic (most rhizocephalans), followed by a metamorphosis into a distinct type of larvae, the cyprid (or cypridoid larvae) (Fig. 1.9C), which is a fast swimming nonfeeding larval type adapted for attachment with a pair of specialized first antennae. Settling is followed by another, even more profound metamorphosis into an irreversibly attached juvenile/adult (Fig. 1.9A).

In the Thoracica (sessile and stalked barnacles) and Acrothoracica (boring barnacles), which includes most thecostracans, juveniles settle on substrata such as rocks, piers, and driftwood, and from this point initiate filtratory feeding with specialized trunk limbs (cirri) (Fig. 1.9A). In Rhizocephala (parasitic barnacles) and Ascothoracida, however, the settled juveniles/adults attain a parasitic lifestyle using other crustaceans (rhizocephalans) or cnidarians or echinoderms (ascothoracidans) as hosts (see Chapter 15 in this volume). Thecostracan nauplii generally swim as typical nauplii, using their naupliar appendages as oars; they are slow swimmers compared to the later cyprids as well as when compared to other nauplii, such as those of copepods (Gauld 1959, Walker 2004). On the contrary, the cyprid is a much-specialized nonfeeding fast-swimming type of larva well adapted to finding a proper place for settling, either on hard substratum in the case of free-living forms, or on other crustacean host organisms in the case of parasites. The cyprid adaptions involve a transfer of locomotory function from the anterior three pairs of naupliar appendages to the posterior 6 pairs of thoracopods (Fig. 1.9C), resulting in much increased swimming speed facilitated by the generally torpedolike shape of the cyprid carapace (Walker 2004). A significant adaption of the cyprids are the specialized first antennae adapted to search for proper substratum and for attachment; for example, this involves bipedal walking on the substratum while temporarily attaching left and right first antennae alternatingly, ending with permanent attachment facilitated by a cementing substance produced by first antennae "cement glands" (Lagersson and Høeg 2002, Høeg et al. 2015). From this point in development, various taxa differ significantly in their subsequent development. In many taxa, the cyprid settles irreversibly as small hermaphroditic calcareous mollusk-like filtering organisms, sometimes stalked, and sometimes, as in *Scalpellum scalpellum*, going through very complicated life cycles involving hermaphroditic organisms (as typical for barnacles) but with the occurrence of dwarf males. This type of reproductive system is termed *androdioecy* and characterized by the coexistence of males and hermaphrodites (Chan and Høeg 2015), a sexual system elsewhere among crustaceans only seen in branchiopods. Even more

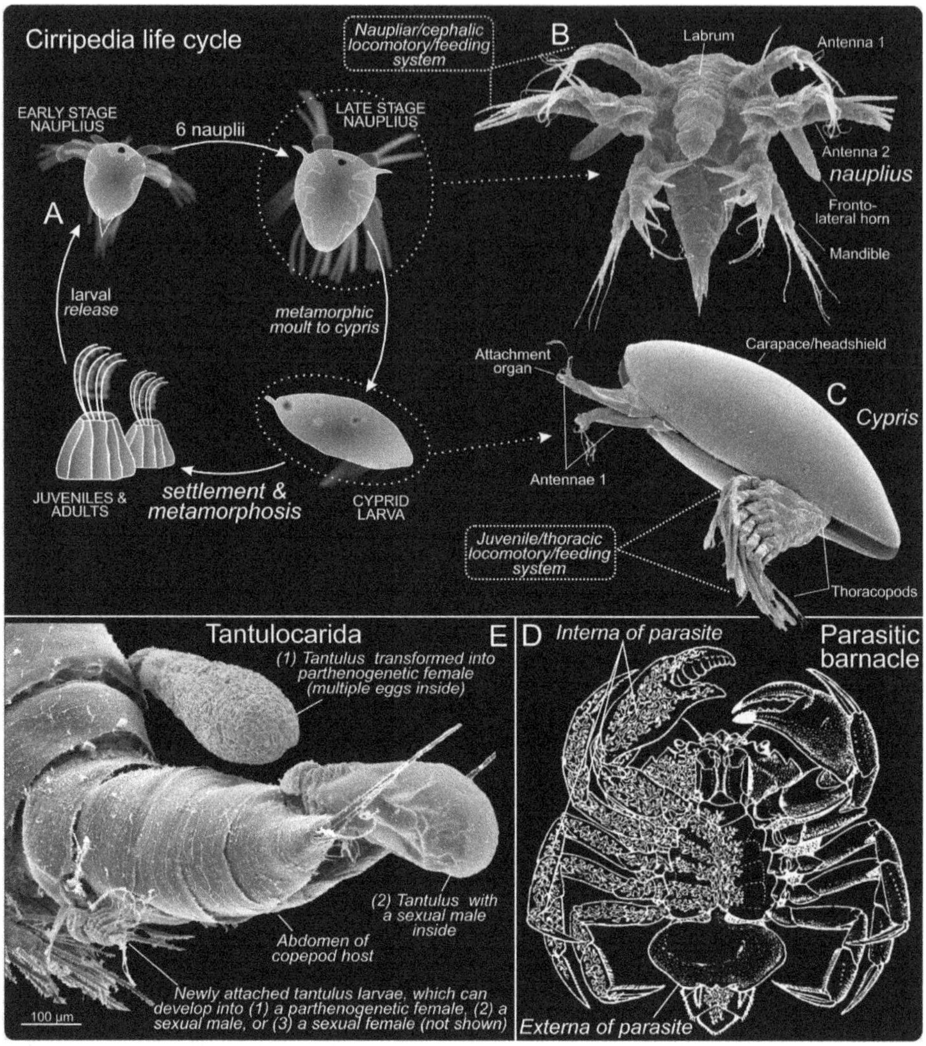

Fig. 1.9.
Life cycle and developmental stages of Cirripedia and Tantulocarida. (A) Generalized life cycle of barnacles from nauplius to permanently settled adult. Modified from Chen et al. 2011. (B) Nauplius of *Pollicipes pollicipes* (Cirripedia: Pendunculata). Photograph by Jørgen Olesen ©. (C) Cypris larvae of unknown cirripede. Photograph by Jens T. Høeg ©. (D) Schematic illustration of internal and external habitus of parasitic barnacle associated with host brachyuran crab. Modified from Delage (1884). (E) Three life stages of *Arcticotantulus kristensenii* (Tantulocarida) attached to hind body of copepod host. Modified from Knudsen et al. (2009), with permission from Zootaxa. See color version of figure part 1.9E in the centerfold.

complicated are the life cycles of parasitic barnacles, involving a number of larval types not seen in the nonparasitic barnacles. In some taxa, for example, the cyprid settles on the gills of its future host (e.g., a porcellanid crab; Ritchie and Høeg 1981) and metamorphoses into a small, round, limbless type of larva called a kentrogon, which develops a stylet, through which primordial cell material is injected into the hemocoel of the host. In at least some species, the cell material is worm-shaped and therefore has been termed a *vermigon* (Glenner et al. 2000, Pérez-Losada et al. 2009). The vermigon develops into a root system inside the host, called an interna, and emerges later with one or more external reproductive part(s), called an externa(e), for example, in the case of a porcellanid or brachyuran host under the reduced tail flap (see Fig. 1.9D).

Even more complex is the life cycle of the Tantulocarida. Tantulocarids are minute parasites on other small crustaceans such as copepods or tanaids (Boxshall and Lincoln 1983, Huys et al. 1993, 2014, on which much of the following is based). Molecular evidence suggests they are closely related to Thecostraca, perhaps as an ingroup, which is supported by a number of morphological similarities (Petrunina et al. 2013). Their life cycle, which has been pieced together from indirect evidence from a number of species, appears to form a complex dual life cycle, combining a sexual phase (with free-swimming adults) and a presumed parthenogenetic multiplicative phase that takes place on the host (the parthenogenetic female shown in Fig. 1.9E) (Huys et al. 1993, 2014). A central larval type in the tantulocarid life cycle is the tantulus larva, because it occurs in both life cycles. The tantulus larva is the infective stage and it permanently attaches to the host by a frontal oral disk through which its cephalic stylet penetrates into the host (e.g., a copepod; Fig. 1.9E). Prior to attachment by the tantulus, at least in the sexual part of the cycle, the larva goes through a benthic nonfeeding phase as part of the temporary meiofauna (Huys 1991). Attached tantulus larvae undergo development on the host without conventional molting. In the parthenogenetic pathway, the tantulus forms a large dorsal trunk sac immediately behind the cephalon (Huys et al. 2014). The contents of the sac differentiate into eggs that are later released as fully developed tantulus larvae. The sexual cycle involves a unique type of metamorphosis in which a free-swimming adult (male or female) is formed within the expanded trunk sac of the preceding tantulus larva (Fig. 1.9). Mating between free-swimming adults (female and male) has never been observed. A benthic nonfeeding tantulocarid nauplius-type larva has been reported recently but not formally described (Martínez Arbizu 2005, Mohrbeck et al. 2010), suggesting that eggs produced by the sexual female hatch as nauplii rather than as tantulus larvae (Huys et al. 2014), into which they presumably develop (see Chapter 15 in this volume).

CRUSTACEAN DIRECT (EPIMORPHIC) DEVELOPMENT AND ITS RELATION TO LIFE HISTORY

Among the more significant adaptions in crustacean life history is the complete lack of free-living larvae in some taxa. This type of development is commonly referred to as direct development, brooding, epimorphic development, or parental care (see terminology in Glossary of Martin et al. 2014b). Embryonized larvae are brooded in association with specialized structures/organs of the mother animal (e.g., under carapace or attached to limbs) (Fig. 1.10). Because such modifications of crustacean life cycles are seen in numerous taxa that are not closely related, it is clear that life cycles with direct development have evolved independently a number of times in crustaceans. This is also supported by the fact that very different parts of the mother animal are responsible for enclosing or holding the developing embryos. Sometimes the carapace is modified as a brooding chamber, which is seen in Thermosbaenacea (Fig. 1.10C,D), cladoceromorphan branchiopods (*Cyclestheria* and Cladocera) (Fig. 1.10E,F), in some ostracods, or in some ascothoracid thecostracans (Olesen 1999, Olesen 2013, Olesen et al. 2015). In Peracarida (Malacostraca) (except Thermosbaenacea), parts (presumably epipods) of the thoracopods form large, overlapping plates termed *oostegites* which, together with the ventral side of the female body proper, form a cavity ventral to the female in which brooding takes place (e.g., Fig. 1.10A,B). Another case of direct development is seen in freshwater crayfish in which the developing specimens cling to the pleopods of the female during development (Scholtz and Kawai 2002, Vogt and Tolley 2004). Yet another example is seen in Leptostraca where the full development takes place in a kind of chamber formed between the two rows of female thoracopods (Manton 1934, Olesen and Walossek 2000). In some taxa, brooding takes place only in the early part of development, such as in pleocyematan decapods in general, where the early part of

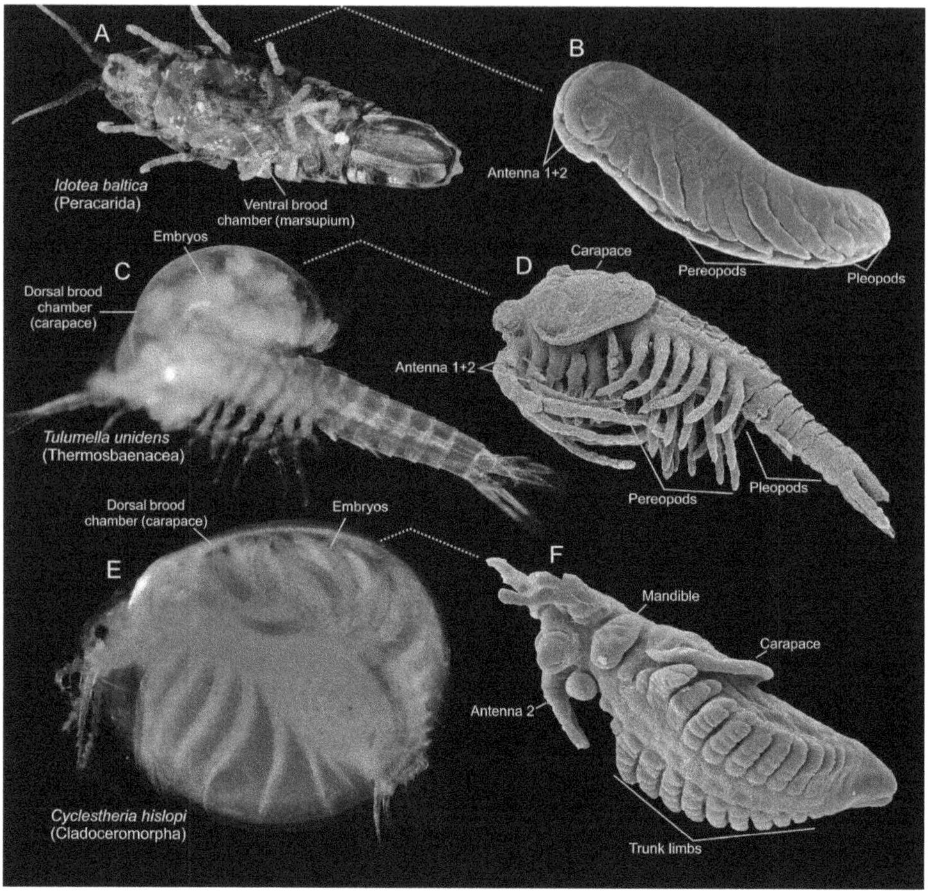

Fig. 1.10.
Brooding crustaceans. (A) *Idotea balthica* (Isopoda), ventral side showing brood pouch between thoracopods. Photograph by David Fenwick ©. (B) Embryo of *I. balthica*. Photograph by Jørgen Olesen ©, modified from Akther et al. 2015. (C) *Tulumella unidens* (Thermosbaenacea) with brood pouch under carapace. (D) Embryo of *T. unidens*. (E) *Cyclestheria hislopi* (Branchiopoda, Cyclestherida) with brood under carapace. Photograph by Jørgen Olesen © (Thailand). (F) Embryo of *Cyclestheria hislopi*. From Olesen (1999), with permission from John Wiley and Sons. (C, D) Photographs by Jørgen Olesen ©, modified from Olesen et al. (2015).

development is passed while the eggs are attached to pleopods, after which release occurs at an advanced larval stage, a zoeal type of larva. Although it is clear that direct development has appeared convergently in Crustacea, the precise adaptive value in different taxa is less clear. Direct development in crayfish has been interpreted as an adaptation to the freshwater environment in which larval dispersal may be useless, and attachment of juveniles to their mother may be advantageous by preventing them from being dislodged by water currents (Scholtz and Kawai 2002, Vogt 2013). Similar explanations may be applied to the absence of free larvae in freshwater branchiopods such as practically all cladocerans (water fleas) and their sister group, the clam shrimp *Cyclestheria hislopi* (Fig. 1.10E,F). However, even in the case of branchiopods, the picture is not as simple as the large branchiopods (e.g., Anostraca, Notostraca), which are primarily found in temporary ponds, but have all have retained free larvae (e.g., Martin et al. 2014b). Generally speaking, it would seem that consequences of brooding/parental care can be both positive (e.g., improved offspring survival, no dislodgements by water currents) or negative (reduced dispersal) (see Thiel 2000).

CONCLUSIONS AND FUTURE DIRECTIONS

The diversity of crustacean (mainly nondecapod) life cycles has been explored in this chapter with a focus on *changes* in the life cycles, and on how different phases in a life cycle are adapted to their environment. Among Metazoa, the Crustacea display a practically unrivalled diversity in lifestyles throughout their life cycle. For example, during development, many crustacean taxa greatly modify their feeding strategy and mode of locomotion, something often coupled to marked shifts in environment (e.g., from pelagic to benthic or from free-living to parasitic). However, even taxa that stay roughly in the same environment (e.g., living freely in the water or interstitially among sand grains) and develop gradually (anamorphic) often change significantly in feeding mode and type of locomotion.

Crustacean development very often involves strong elements of anamorphic development (Figs. 1.11 and 1.12). A general trend in crustacean development is that locomotion and feeding are gradually transferred posteriorly (e.g., from naupliar appendages to thoracopods) and, in the case of locomotion of some malacostracans, even to the pleopods in the tail region. It is striking that abandonment of the anterior naupliar locomotion/feeding system takes place alongside an increasingly developed and active postnaupliar system (e.g., thoracopodal), so that the 2 systems function simultaneously over much of the development. This is seen not only in classically anamorphic taxa such as Anostraca and Cephalocarida but also in the holoplanktonic Euphausiacea and Dendrobranchiata (Decapoda), even though the development here can be categorized into distinct larval phases (Fig. 1.12B,C). In these 2 malacostracan taxa, the naupliar type of locomotion persists very long during development and does not disappear before the end of the calyptopis (Euphausiacea) or prozoea (Dendrobranchiata) phases, before which, at least in Dendrobranchiata, a more posterior locomotory system (mxp1 and mxp2) has long been well developed and active. Also, the development of some decapod taxa, such as some carideans, is essentially anamorphic, but in this case not involving naupliar development or naupliar locomotion/feeding (Fig. 1.12D).

Full metamorphosis, previously defined as an abrupt shift from one stage to another (or within a few stages), regarding which appendages specifically are involved in feeding or locomotion, is only seen in a few places within the Crustacea. Striking examples of crustacean metamorphoses are copepods where profound shifts in locomotion and feeding take place (between the naupliar and copepodite phase), and in thecostracans, where the development even involves 2 metamorphoses (between nauplius phase and cyprid, and between cyprid and settled juvenile) (Fig. 1.12A,F).

This great evolutionary plasticity through crustacean life cycles in general, with different developmental phases being adapted for different lifestyles and different milieus, undoubtedly is one of the key factors behind the evolutionary success of the Crustacea. A holistic approach including a variety of developmental phases is necessary in the study of Crustacea, not least in ecological studies where feeding in different developmental phases of the life cycle can be completely different. From the literature survey conducted in relation to this chapter, it is clear that much is still to be learned about basic functional aspects of feeding in a variety of developmental stages in even well-known crustaceans such as krill and dendrobranchiate shrimps. And last but not least, it has long been assumed, and in later years confirmed, that comparative studies on ontogeny/development hold crucial information for the understanding of metazoan evolution, simply because the development of a given animal is an important toolbox on which evolution works. It is clear that the Crustacea, with their compartmentalized developmental sequences, have been affected by heterochrony (a change in the relative timing of events during development). Much crustacean evolution can be explained by either larval characters moved forward in development or adult characters moved backward, highlighting the need for better understanding of both morphological and molecular aspects of crustacean ontogeny and life cycles in general.

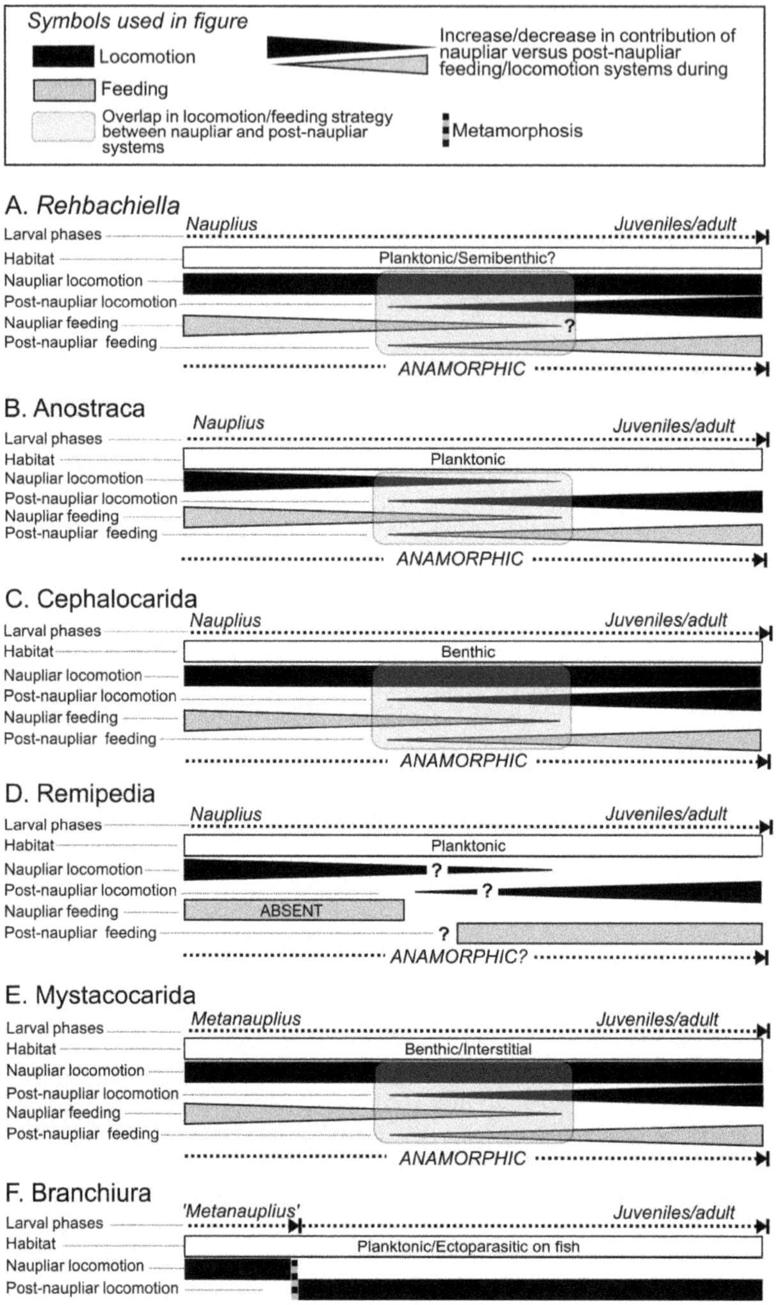

Fig. 1.11.
Summary of developmental strategies within Crustacea and their relation to environment. See text for details on taxa. The figure summarizes the following: (1) generalized habitat occurrence (e.g., planktonic/benthic); (2) style of locomotion (e.g., naupliar, postnaupliar) during development); (3) type of feeding during development (e.g., naupliar, postnaupliar); (4) overlap in locomotion of feeding style between, for example, naupliar and postnaupliar systems; and (5) cases of profound developmental changes that can be characterized as metamorphoses. The representations are schematic so no homology between development phases in various taxa can be inferred from their relative position in the scheme.

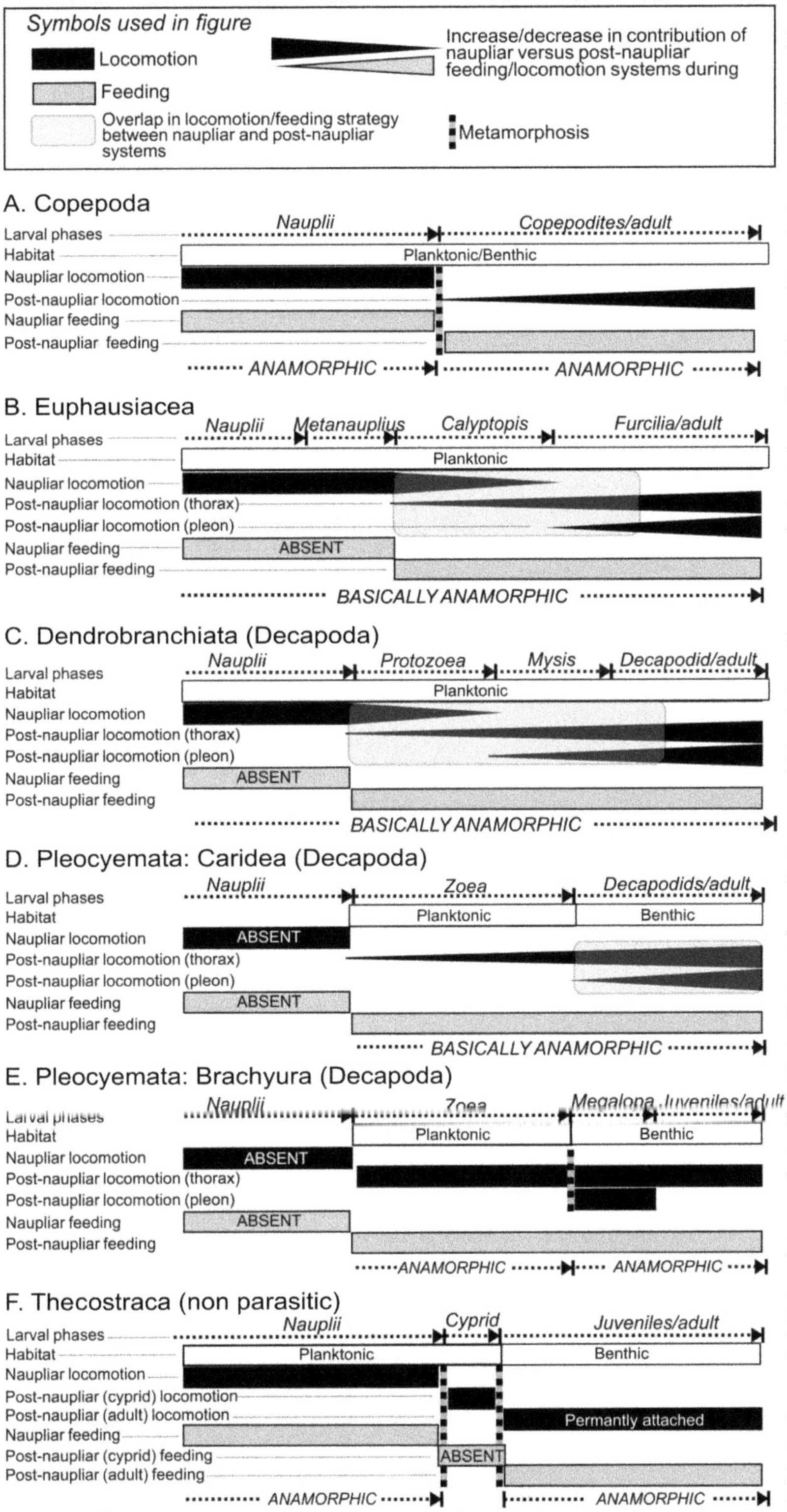

Fig. 1.12.
Summary of developmental strategies within Crustacea and their relation to environment. See text and caption of Fig. 1.11 for details.

REFERENCES

Akther, H., M.A. Agersted, and J. Olesen. 2015. Naupliar and metanaupliar development of *Thysanoessa raschii* (Malacostraca, Euphausiacea) from Godthåbsfjord, Greenland, with a reinstatement of the ancestral status of the free-living nauplius in malacostracan evolution. PLoS ONE 10:e0141955.

Alcaraz, M., and J.R. Strickler. 1988. Locomotion in copepods: pattern of movements and energetics of *Cyclops*. Hydrobiologia 167/168:409–414.

Andersen Borg, C.M., E. Bruno, and T. Kiørboe. 2012. The kinematics of swimming and relocation jumps in copepod nauplii. PLoS ONE 7:e47486.

Anderson, D.T. 1967. Larval development and segment formation in the branchiopod crustaceans *Limnadia stanleyana* King (Conchostraca) and *Artemia salina* (L.) (Anostraca). Australian Journal of Zoology 15:47–91.

Anderson, D.T. 1994. Barnacles: structure, function, development and evolution. Chapman and Hall, London, England.

Anger, K. 2001. The biology of decapod crustacean larvae. A.A. Balkema, Rotterdam, The Netherlands.

Barlow, D.I., and M.A. Sleigh. 1980. The propulsion and use of water currents for swimming and feeding in larval and adult *Artemia*. Pages 61–73 *in* G. Persoone, P. Sorgeloos, O. Roels, and E. Jaspers, editors. The brine shrimp *Artemia*, 1, morphology, genetics, radiobiology, toxicology. Universa Press, Wetteren, Belgium.

Bauer, R.T. 2011. Amphidromy and migrations of freshwater shrimps. I. Costs, benefits, evolutionary origins, and an unusual case of amphidromy. Pages 145–156 *in* C. Fransen, and A. Asakura, editors. New frontiers in crustacean biology. Proceedings of the TCS Summer Meeting, Tokyo, 20–24 September 2009. Brill, Leiden, The Netherlands.

Bauer, R.T. 2013. Amphidromy in shrimps: a life cycle between rivers and the sea. Latin American Journal of Aquatic Research 41:633–650.

Benesch, R. 1969. Zur Ontogenie und Morphologie von *Artemia salina* L. Zoologische Jahrbücher, Abteilung der Anatomie und Ontogenie der Tiere 86:307–458.

Bliss, D. 1968. Transition from water to land in decapod crustaceans. American Zoologist 8:355–392.

Boden, B.P. 1950. The post-naupliar stages of the Crustacean *Euphausia pacifica*. Transactions of the American Microscopical Society 69:373–386.

Boden, B.P. 1951. The egg larval stages of *Nyctiphanes simplex*, a euphausiid crustacean from California. Journal of Zoology 121:515–527.

Boden, B.P., Johnson, M. W., and E. Brinton. 1955. The Euphausiacea (Crustacea) of the North Pacific. Bulletin of the Scripps Institution of Oceanography of the University of California 6:287–398.

Boxshall, G.A., and R.J. Lincoln. 1983. Tantulocarida, a new class of Crustacea ectoparasitic on other crustaceans. Journal of Crustacean Biology 3:1–16.

Boyd, C.M., M. Heyraud, and C.N. Boyd. 1984. Feeding of the antarctic krill *Euphausia superba*. Journal of Crustacean Biology 4:123–141.

Boyko, C.B., and C. Wolff. 2014. Chapter 40. Isopoda and Tanaidacea. Pages 210–215 *in* J.W. Martin, J. Olesen, and J.T. Høeg, editors. Atlas of crustacean larvae. Johns Hopkins University Press, Baltimore, Maryland.

Bruno, E., C.M. Andersen Borg C.M., and T. Kiørboe. 2012. Prey detection and prey capture in copepod nauplii. PLoS ONE 7:e47906.

Brusca, R.C., and G.J. Brusca. 1990. Invertebrates. Sinauer Associates, Inc., Sunderland, Massachusetts.

Carton, Y. 1968. Développement de *Cancerilla tubulata* Dalyell, parasite de l'ophiure *Amphipholis squamata* Della Chiaje. Crustaceana 1 (Suppl.):11–12.

Chan, B.K.K., and J.T. Høeg. 2015. Diversity of lifestyles, sexual systems and larval development pattern in sessile crustaceans. Pages 14–34 *in* M. Thiel, and L. Watling, editors. The natural history of the Crustacea, volume 2: Lifestyles and feeding biology. Oxford University Press, New York.

Charmantier, G., M. Charmantier-Daures, and D.E. Aiken. 1991. Metamorphosis in the lobster *Homarus* (Decapoda): A review. Journal of Crustacean Biology 11:481–495.

Chen, Z.F., K. Matsumura, H. Wang, S.M. Arellano, X. Yan, I. Alam, J.A. Archer, V.B. Bajic, and P.Y. Qian. 2011. Toward an understanding of the molecular mechanisms of barnacle larval settlement: a comparative transcriptomic approach. PLoS One 6:e22913.

Choi, J.H., and S.Y. Hong. 2001. Larval development of the kishi velvet shrimp, *Metapenaeopsis dalei* (Rathbun) (Decapoda: Penaeidae), reared in the laboratory. Fishery Bulletin 99:275–291.
Chu, K.H., C.C. Sze, and C.K. Wong. 1996. Swimming behaviour during the larval development of the shrimp *Metapenaeus ensis* (de Haan, 1844) (Decapoda, Penaeidae). Crustaceana 69:368–378.
Chullasorn, S., W.-X. Yang, H.-U. Dahms, P. Kangtia, M. Holynska, W. Anansatitporn, L.O. Sanoamuang, and J.-S. Hwang. 2009. Naupliar development of *Eucyclops* cf. *serrulatus tropicalis*, *Euc.* cf. *spatulatus*, and *Ectocyclops medius* Kiefer, 1930 (Copepoda: Cyclopidae). Zoological Studies 48:12–32.
Dahms, H.-U. 1989. First record of a lecithotrophic nauplius among Harpacticoida (Crustacea, Copepoda) collected from the Weddell Sea (Antarctica). Polar Biology 10:221–224.
Dahms, H.-U. 1992. Metamorphosis between naupliar and copepodit phases in the Harpacticoida. Philosophical Transactions of the Royal Society B 335:221–236.
Dall, W., B.J. Hill, P.C. Rothlisberg, and D.J. Sharples. 1990. The biology of the Penaeidae. Vol. 27. Academic Press, London, England.
Delage Y. 1884. Évolution de la Sacculine (*Sacculina carcini* Thomps.), crustacé endoparasite de l'ordre nouveau des Kentrogonides. Archives de zoologie expérimentale et générale 2e série, t. 2:417–736.
Felder, D.L., J.W. Martin, and J.W. Goy. 1985. Patterns in early postlarval development of decapods. Pages 163–225 *in* A. Wenner, editor. Crustacean issues 2. Larval growth. A.A. Balkema, Rotterdam, The Netherlands.
Ferrari, F.D., and H.U. Dahms. 2007. Post-embryonic development of the Copepoda. Crustaceana Monographs 8. Brill, Leiden, The Netherlands.
Fraser, F.C. 1936. On the development and distribution of the young stages of krill (*Euphausia superba*). Discovery Reports 14:1–192.
Fryer, G. 1983. Functional ontogenetic changes in *Branchinecta ferox* (Milne-Edwards) (Crustacea, Anostraca). Philosophical Transactions of the Royal Society London B 303:229–343.
Fryer, G. 1988. Studies on the functional morphology and biology of the Notostraca (Crustacea: Branchiopoda). Philosophical Transactions of the Royal Society London B 321:27–124.
Gauld, D.T. 1959. Swimming and feeding in crustacean larvae: the nauplius larva. Proceedings of the Zoological Society of London 132:31–50.
Glenner, H., and J.T. Høeg. 1993. Scanning electron microscopy of metamorphosis in four species of barnacles (Cirripedia Thoracica Balanomorpha). Marine Biology 117:431–439.
Glenner, H., J.T. Høeg, J.J. O'Brien, and T.D. Sherman. 2000. Invasive vermigon stage in the parasitic barnacles *Loxothylacus texanus* and *L. panopaei* (Sacculinidae): closing of the rhizocephalan life-cycle. Marine Biology 136:249–257.
Gómez-Guitiérrez, J., and C.J. Robinson. 2005. Embryonic, early larval development time, hatching mechanism and interbrood period of the sac-spawning euphausiid *Nyctiphanes simplex* Hansen. Journal of Plankton Research 27:279–295.
Goy, J.W. 2014. Chapter 49. Astacidea. Pages 256–262 *in* J.W. Martin, J. Olesen, and J.T. Høeg, editors. Atlas of crustacean larvae. Johns Hopkins University Press, Baltimore, Maryland.
Greenaway, P. 2003. Terrestrial adaptations in the Anomura (Crustacea: Decapoda). Memoirs of Museum Victoria 60:13–26.
Gresty, K.A., G.A. Boxshall, and K. Nagasawa. 1993. The fine structure and function of the cephalic appendages of the branchiuran parasite, *Argulus japonicus* Thiele. Philosophical Transactions of the Royal Society of London B 339:119–135.
Grygier, M.J. 1996. Classe des Thécostracés (Thecostraca Gruvel, 1905). Sous casse des Facetotecta (Facetotecta Grygier, 1985). Pages 425–432 *in* J. Forest, editor. Traité de Zoologie, Vol. 7 (2), Crustacés: Géneralités (suite) et Systématique (1re partie). Masson, Paris, France.
Gueriau, P., N. Rabet, G. Clément, L. Lagebro, J. Vannier, D.E.G. Briggs, S. Charbonnier, S. Olive, and O. Béthoux. 2016. A 365-million-year-old freshwater community reveals morphological and ecological stasis in branchiopod crustaceans. Current Biology 26:383–390.
Gurney, R. 1942. The larvae of the decapod Crustacea. The Ray Society, London, England.
Haug, J.T., and C. Haug. 2013. An unusual fossil larva, the ontogeny of achelatan lobsters, and the evolution of metamorphosis. Bulletin of Geosciences 88:195–206.

Haug, J.T., and C. Haug. 2015. Chapter 1. "Crustacea": Comparative aspects of larval development. Pages 1–37 *in* A. Wanninger, editor. Evolutionary developmental biology of invertebrates 4: Ecdysozoa II: Crustacea. Springer-Verlag, Wien, Austria.

Haug, J.T., J. Olesen, A. Maas, and D. Waloszek. 2011. External morphology and post-embryonic development of *Derocheilocaris remanei* (Mystacocarida) revisited, with a comparison to the Cambrian taxon *Skara*. Journal of Crustacean Biology 32:668–692.

Hessler, R.R. 1992. Reflections on the phylogenetic position of the Cephalocarida. Acta Zoologica 73:315–316.

Hessler, R.R., and W.A. Newman. 1975. A trilobitomorph origin for the Crustacea. Fossils and Strata 4:437–59.

Høeg, J.T. 1995. The biology and life cycle of the Rhizocephala (Cirripedia). Journal of the Marine Biological Association of the United Kingdom 75:517–550.

Høeg, J.T., and O.S. Møller. 2006. When similar beginnings leads to different ends: Constraints and diversity in cirripede larval development. Invertebrate Reproduction and Development 49:125–142.

Høeg, J.T., M. Pérez-Losada, H. Glenner, G.A. Kolbasov, and K.A. Crandall. 2009. Evolution of morphology, ontogeny and life cycles within the Crustacea Thecostraca. Arthropod Systematics and Phylogeny 67:199–217.

Høeg, J.T., J. Deutsch, B.K.K. Chan, and H. Semmler Le. 2015. "Crustacea": Cirripedia. Pages 153–181 *in* A. Wanninger, editor. Evolutionary developmental biology of invertebrates 4: Ecdysozoa II: Crustacea. Springer-Verlag, Wien, Austria.

Huys, R. 1991. Tantulocarida (Crustacea: Maxillopoda): a new taxon from the temporary meiobenthos. Marine Ecology 12:1–34.

Huys, R. 2014. Chapter 27. Copepoda. Pages 144–163 *in* J.W. Martin, J. Olesen, and J.T. Høeg, editors. Atlas of crustacean larvae. Johns Hopkins University Press, Baltimore, Maryland.

Huys, R., and G.A. Boxshall. 1991. Copepod evolution. The Ray Society, London, England.

Huys, R., G.A. Boxshall, and R.J. Lincoln. 1993. The Tantulocaridan life cycle: the circle closed? Journal of Crustacean Biology 13:432–442.

Huys, R., J. Olesen, A.S. Petrunina, and J.W. Martin. 2014. Chapter 23. Tantulocarida. Pages 122–127 *in* J.W. Martin, J. Olesen, and J.T. Høeg, editors. Atlas of crustacean larvae. Johns Hopkins University Press, Baltimore, Maryland.

Ivanenko, V.N., P.M. Arbizu, and J. Stecher. 2007. Lecithotrophic nauplius of the family Dirivultidae (Copepoda; Siphonostomatoida) hatched on board over the Mid-Atlantic Ridge (5°S). Marine Ecology 28:49–53.

Jia, Z., P. Virtue, K.M. Swadling, and S. Kawaguchi. 2014. A photographic documentation of the development of Antarctic krill (*Euphausia superba*) from egg to early juvenile. Polar Biology 37:165–179.

Jirikowski, G.J., S. Richter, and C. Wolff. 2013. Myogenesis of Malacostraca—the "egg-nauplius" concept revisited. Frontiers in Zoology 10:76.

Kiørboe, T. 2011. What makes pelagic copepods so successful? Journal of Plankton Research 33:677–785.

Kiørboe, T., A. Andersen, V.J. Langlois, and H.H. Jakobsen. 2010. Unsteady motion: escape jumps in planktonic copepods, their kinematics and energetics. Journal of the Royal Society Interface 7: 1591–1602.

Kiørboe, T., H. Jiang, R.J. Gonçalves, L.T. Nielsen, and N. Wadhwa. 2014. Flow disturbances generated by feeding and swimming zooplankton. Proceedings of the National Academy of Sciences 111:11738–11743.

Knight, M.D. 1975. The larval development of Pacific *Euphausia gibboides* (Euphausiacea). Fishery Bulletin 73:145–168.

Knight, M.D. 1976. Larval development of *Euphausia sanzoi* Torelli (Crustacea: Euphausiacea). Bulletin of Marine Science 26:538–557.

Knight, M.D. 1978. Larval development of *Euphausia fallax* Hansen (Crustacea: Euphausiacea) with a comparison of larval morphology within the *E. gibboides* species group. Bulletin of Marine Science 28:255–281.

Knight, M.D. 1980. Larval development of *Euphausia eximia* (Crustacea: Euphausiacea) with notes on its vertical distribution and morphological divergence between populations. Fishery Bulletin 78:313–335.

Knudsen, S. W., Kirkegaard, M., and Olesen, J. 2009. The tantulocarid genus *Arcticotantalus* removed from Basipodellidae into Deoterthridae (Crustacea: Maxillopoda) after the description of a new species from Greenland, with first live photographs and an overview of the class. Zootaxa 2035: 41–68.

Koenemann, S., F.R. Schram, A. Bloechl, M. Hoenemann, T.M. Iliffe, and C. Held. 2007. Post-embryonic development of remipede crustaceans. Evolution and Development 9:117–121.

Koenemann, S., J. Olesen, F. Alwes, T.M. Iliffe, M. Hoenemann, P. Ungerer, C. Wolff, and G. Scholtz. 2009. The post-embryonic development of Remipedia (Crustacea)—additional results and new insights. Development Genes and Evolution 219:131–145.

Kungvankij, P., L.B. Tiro, Jr, B.J. Pudadera, Jr., I.O. Potestas, K.G. Corre, E. Borlongan, G.A. Talean, L.F. Bustilo, E.T. Tech, A. Unggui, and T.E. Chua. 1985. Shrimp hatchery design, operation and management—training manual. FAO Fisheries and Aquaculture Department. http://www.fao.org/docrep/field/003/AC232E/AC232E04.htm (accessed April 26 2017).

Lagersson, N.C., and J.T. Høeg. 2002. Settlement behavior and antennulary biomechanics and in cypris larvae of *Balanus amphitrite* (Crustacea: Thecostraca: Cirripedia). Marine Biology 141:513–526.

Lebour, M.V. 1926a. The Euphausiidæ in the neighbourhood of Plymouth III. *Thysanoessa inermis*. Journal of the Marine Biological Association of the UK 14:1–21.

Lebour, M.V. 1926b. A general survey of larval euphausiids, with a scheme for their identification. Journal of the Marine Biological Association of the United Kingdom 14:519–527.

Lombardi, J., and E. Ruppert. 1982. Functional morphology of locomotion in *Derocheilocaris typica*. Zoomorphology 100:1–10.

Lozano-Fernandez, J., R. Carton, A.R. Tanner, M.N. Puttick, M. Blaxter, J. Vinther, J. Olesen, G. Giribet, G. Edgecombe, and D. Pisani. 2016. A molecular palaeobiological exploration of arthropod terrestrialization. Philosophical Transactions of the Royal Society London B 371:20150133.

Macdonald, R. 1928. The life history of *Thysanoessa raschii*. Journal of the Marine Biological Association of the United Kingdom 15:57–79.

Manton, S.M. 1934. On the embryology of the crustacean *Nebalia bipes*. Philosophical Transactions of the Royal Society of London B223:163–238.

Marschall, H.P. 1984. Development, swimming, and feeding of early stages of krill. Antarctic Journal of the United States 195:137–138.

Marschall, H.P., and H.J. Hirsche. 1984. Development of eggs and nauplii of *Euphausia superba*. Polar Biology 2:245–250.

Martin, J.W. 2014. Chapter 45. Introduction to the Decapoda. Pages 230–231 *in* J.W. Martin, J. Olesen, and J.T. Høeg, editors. Atlas of crustacean larvae. Johns Hopkins University Press, Baltimore, Maryland.

Martin, J.W., and J. Gómez-Gutiérrez. 2014. Chapter 43. Euphausiacea. Pages 220–225 *in* J.W. Martin, J. Olesen, and J.T. Høeg, editors. Atlas of crustacean larvae. Johns Hopkins University Press, Baltimore, Maryland.

Martin, J.W., J. Olesen, and J.T. Høeg. 2014a. Chapter 2. The crustacean nauplius. Pages 8–16 *in* J.W. Martin, J. Olesen, and J.T. Høeg, editors. Atlas of crustacean larvae. Johns Hopkins University Press, Baltimore, Maryland.

Martin, J.W., J. Olesen, and J.T. Høeg. 2014b. Atlas of crustacean larvae. Johns Hopkins University Press, Baltimore, Maryland.

Martin, J.W., M.M. Criales, and A. dos Santos. 2014c. Chapter 46. Dendrobranchiata. Pages 235–242 *in* J.W. Martin, J. Olesen, and J.T. Høeg, editors. Atlas of crustacean larvae. Johns Hopkins University Press, Baltimore, Maryland.

Martin, M.F. 1932. On the morphology and classification of *Argulus* (Crustacea). Proceedings of the Zoological Society of London 102:771–806.

Martínez Arbizu, P. 2005. Additions to the life cycle of Tantulocarida. Sixth International Crustacean Congress, 18–22 July 2005, Glasgow, Scotland.

Maruzzo, D., N. Aldred, A.S. Clare, and J.T. Høeg. 2012. Metamorphosis in the cirripede crustacean *Balanus amphitrite*. PLoS ONE 7:e37408.

Mauchline, J. 1980. The biology of mysids and euphausiids. Advances in Marine Biology 18:1–681.

Mauchline, J. 1998. The biology of calanoid copepods. Advances in Marine Biology 33:1–710.

Mohrbeck, I., P.M. Martínez, and T. Glatzel. 2010. Tantulocarida (Crustacea) from the Southern Ocean deep sea, and the description of three new species of *Tantulacus* Huys, Andersen & Kristensen, 1992. Systematic Parasitology 77:131–151.

Mossin, J. 1986. Physiochemical factors inducing embryonic development and spring hatching of the European fairy shrimp *Siphonophanes grubei* (Dybowsky) (Crustacea: Anostraca). Journal of Crustacean Biology 6:693–704.

Møller, O.S., and J. Olesen. 2014. Chapter 24. Branchiura. Pages 128–134 *in* J.W. Martin, J. Olesen, and J.T. Høeg, editors. Atlas of crustacean larvae. Johns Hopkins University Press, Baltimore.
Møller, O.S., J. Olesen, and J.T. Høeg. 2003. SEM studies on the early larval development of *Triops cancriformis* (Bosc) (Crustacea: Branchiopoda: Notostraca). Acta Zoologica 84:267–284.
Møller, O.S, J. Olesen, and J.T. Høeg. 2004. On the larval development of *Eubranchipus grubii* (Crustacea, Branchiopoda, Anostraca), with notes on the basal phylogeny of the Branchiopoda. Zoomorphology 123:107–123.
Møller, O.S., J. Olesen, and D. Waloszek. 2007. Swimming and cleaning in the free-swimming phase of *Argulus* larvae (Crustacea, Branchiura)—appendage adaptation and functional morphology. Journal of Morphology 268:1–11.
Moon, S.Y., S.-H. Youn, H.-J. Oh, H.Y. Soh, S.D. Choi, and H.S. Yoon. 2015. Description of postembryonic developmental stages of *Pseudodiaptomus koreanus* Soh, Kwon, Lee & Yoon, 2012 (Copepoda, Calanoida, Pseudodiaptomidae). Crustaceana 88:1387–1419.
Moran, N.A. 1994. Adaptation and constraint in the complex life cycles of animals. Annual Review of Ecology, Evolution, and Systematics 25:573–600.
Nicol, S. 2006. Krill, currents, and sea ice: *Euphausia superba* and its changing environment. BioScience 56:112–120.
Olesen, J. 1999. Larval and post-larval development of the branchiopod clam shrimp *Cyclestheria hislopi* (Baird, 1859) (Crustacea, Branchiopoda, Conchostraca, Spinicaudata). Acta Zoologica 80:163–184.
Olesen, J. 2001. External morphology and larval development of *Derocheilocaris remanei* Delamare-Deboutteville & Chappuis, 1951 (Crustacea, Mystacocarida), with a comparison of crustacean segmentation and tagmosis patterns. Det Kongelige Danske Videnskabernes Selskab. Biologiske Skrifter 53:1–59.
Olesen, J. 2004. On the ontogeny of the Branchiopoda (Crustacea): contribution of development to phylogeny and classification. Pages 217–269 *in* G. Scholtz, editor. Evolutionary developmental biology of crustacea. A.A. Balkema Publishers, Lisse, The Netherlands.
Olesen, J. 2005. Larval development of *Lynceus brachyurus* (Crustacea, Branchiopoda, Laevicaudata): Redescription of unusual crustacean nauplii, with special attention to the molt between last nauplius and first juvenile. Journal of Morphology 264:131–148.
Olesen, J. 2013. The crustacean carapace—morphology, function, development, and phylogenetic history. Pages 103–139 *in* M. Thiel, and L. Watling, editors. The natural history of the Crustacea, volume 1: Functional morphology and diversity of crustaceans. Oxford University Press, New York.
Olesen, J., and M.J. Grygier. 2004. Larval development of Japanese 'conchostracans': part 2, larval development of *Caenestheriella gifuensis* (Crustacea, Branchiopoda, Spinicaudata, Cyzicidae), with notes on homologies and evolution of certain naupliar appendages within the Branchiopoda. Arthropod Structure and Development 33:453–469.
Olesen, J., and J.T. Haug. 2014. Chapter 26. Mystacocarida. Pages 138–143 *in* J.W. Martin, J. Olesen, and J.T. Høeg, editors. Atlas of crustacean larvae. Johns Hopkins University Press, Baltimore, Maryland.
Olesen, J., and O.S. Møller. 2014. Chapter 7. Notostraca. Pages 40–46 *in* J.W. Martin, J. Olesen, and J.T. Høeg, editors. Atlas of crustacean larvae. Johns Hopkins University Press, Baltimore, Maryland.
Olesen, J., and D. Walossek. 2000. Limb ontogeny and trunk segmentation in *Nebalia* species (Crustacea, Malacostraca, Leptostraca). Zoomorphology 120:47–64.
Olesen, J., J.T. Haug, A. Maas, and D. Waloszek. 2011. External morphology of *Lightiella monniotae* (Crustacea, Cephalocarida) in the light of Cambrian 'Orsten' crustaceans. Arthropod Structure and Development 40:449–478.
Olesen, J., S.V. Martinsen, T.M. Iliffe, and S. Koenemann. 2014a. Chapter 15. Remipedia. Pages 84–89 *in* J.W. Martin, J. Olesen, and J.T. Høeg, editors. Atlas of crustacean larvae. Johns Hopkins University Press, Baltimore, Maryland.
Olesen, J., T. Boesgaard, T.M. Iliffe, and L. Watling. 2014b. Thermosbaenacea, Spelaeogriphacea, and "Mictacea." Pages 195–198 *in* J.W. Martin, J. Olesen, and J.T. Høeg, editors. Atlas of crustacean larvae. Johns Hopkins University Press, Baltimore, Maryland.
Olesen, J., S.V. Martinsen, G. Hampson, A. Addis, M. Carcupino. 2014c. Cephalocarida. Pages 90–96 *in* J.W. Martin, J. Olesen, and J.T. Høeg, editors. Atlas of crustacean larvae. Johns Hopkins University Press, Baltimore, Maryland.

Olesen, J., T. Boesgaard, and T.M. Iliffe. 2015. The unique dorsal brood pouch of Thermosbaenacea (Crustacea, Malacostraca) and description of a late embryonic stage of *Tulumella unidens* from the Yucatan Peninsula (Mexico), with a discussion of mouth part homologies to other Malacostraca. PLoS ONE:e0122463.
Pérez Farfante, I, and B. Kensley. 1997. Penaeoid and sergestoid shrimps and prawns of the world. Keys and diagnoses for the families and genera. Mémoires du Muséum National d'Histoire Naturelle 175:1–233.
Pérez-Losada, M., J.T. Høeg, and K.A. Crandall. 2009. Remarkable convergent evolution in specialized parasitic Thecostraca (Crustacea). BMC Biology 7:15.
Petrunina, A.S., T.V. Neretina, N.S. Mugue, and G.A. Kolbasov. 2013. Tantulocarida versus Thecostraca: inside or outside? First attempts to resolve phylogenetic position of Tantulocarida using gene sequences. Journal of Zoological Systematics and Evolutionary Research 52:95–176.
Regier, J.C., J.W. Shultz, A. Zwick, A. Hussey, B. Ball, R. Wetzer, J.W. Martin, and C.W. Cunningham. 2010. Arthropod relationships revealed by phylogenomic analysis of nuclear protein-coding sequences. Nature 463:1079–1084.
Richter, S., and G. Scholtz. 2001. Phylogenetic analysis of the Malacostraca (Crustacea). Journal of Zoological Systematics and Evolutionary Research 39:113–36.
Ritchie, L.E., and J.T. Høeg. 1981. The life history of *Lernaeodiscus porcellanae* (Cirripedia: Rhizocephala) and co-evolution with its porcellanid host. Journal of Crustacean Biology 1:334–347.
Ronquillo, J.D., T. Saisho, and R.S. McKinley. 2006. Early developmental stages of the green tiger prawn, *Penaeus semisulcatus* de Haan (Crustacea, Decapoda, Penaeidae). Hydrobiologia 560:175–196.
Rötzer, M.A.I.N., and J.T. Haug. 2015. Larval development of the European lobster and how small heterochronic shifts lead to a more pronounced metamorphosis. International Journal of Zoology 2015:345172.
Salvacion, M, N. Golez, T. Takahashi, T. Ishimaru, and A. Ohno. 2004. Post-embryonic development and reproduction of *Pseudodiaptomus annandalei* (Copepoda: Calanoida). Plankton Biology and Ecology: 51:15–25.
Sanders, H.L. 1963. The Cephalocarida. Functional morphology, larval development, comparative external anatomy. Connecticut Academy of Arts and Sciences: 15: 1–80.
Sars, G.O. 1885. Report on the Schizopoda collected by H.M.S. Challenger, during the years 1873–76. Vol. 37. Pages 1–228 *in* C.W. Thomson, and J. Murray, editors. Reports on the scientific results of the voyage of H.M.S. Challenger during the years 1873–76, Zoology, London, England.
Sars, G.O. 1898. On the propagation and early development of Euphausiidæ. Archiv for Mathematik og Naturvidenskab 20:1–41.
Scholtz, G. 2000. Evolution of the nauplius stage in malacostracan crustaceans. Journal of Zoological Systematics and Evolutionary Research 38:175–187.
Scholtz, G., and T. Kawai. 2002. Aspects of embryonic and postembryonic development of the Japanese freshwater crayfish *Cambaroides japonicus* (Crustacea, Decapoda) including a hypothesis on the evolution of maternal care in the Astacida. Acta Zoologica (Stockholm) 83:203–212.
Schram, F.R. 1983. Remipedia and crustacean phylogeny. Pages 23–28 *in* F.R. Schram, editor. Crustacean issues 1. Crustacean phylogeny. A.A. Balkema, Rotterdam, The Netherlands.
Schrehardt, A. 1987. A scanning electron-microscope study of the post-embryonic development of *Artemia. Artemia* Research and its Applications 1:5–32.
Scourfield, D.J. 1926. On a new type of crustacean from the Old Red Sandstone (Rhynie Chert Bed, Aberdeenshire)—*Lepidocaris rhyniensis* gen. et sp. nov. Philosophical Transactions of the Royal Society London B 214:153–187.
Selden, P.A., R. Huys, M.H. Stephenson, H.A. P., and P.N. Taylor. 2010. Crustaceans from bitumen clast in Carboniferous glacial diamictite extend fossil record of copepods. Nature Communications 1:50.
Silas, E.G., and K.J. Mathew. 1977. A critique to the study of larval development in Euphausiacea. Proceedings of the Symposium on Warm Water Zooplankton Special Publication:571–582.
Spivak, E.D., and A. Rodríguez. 2002. *Pilumnus reticulatus* Stimpson, 1860 (Decapoda: Brachyura: Pilumnidae): a reappraisal of larval characters from laboratory reared material. Scientia Marina 66:5–19.
Storch, O. 1928. Der Nahrungserwerb zweier Copepoden-nauplien (*Diaptomus gracilis* und *Cyclops strenuus*). Zoologische Jahrbücher, Abteilung für Allgemeine Zoologie und Physiologie der Tiere 45:385–436.

Strickler, J.R. 1975. Swimming of planktonic *Cyclops* species (Copepoda, Crustacea): Pattern, movements and their control. Pages 599–613 *in* T.Y.-T. Wu, C.J. Brokaw, and C. Brennen, editors. Swimming and flying in nature, volume 2. Springer US, Baltimore, Maryland.

Strullu-Derrien, C., T. Goral, J.E. Longcore, J. Olesen, P. Kenrick, and G.E. Edgecombe. 2016. A new chytridiomycete fungus intermixed with crustacean resting eggs in a 407-million-year-old continental freshwater environment. PLoS ONE 11: e0167301.

Suh, H.-L., H.Y. Soh, and S.Y. Hong. 1993. Larval development of the euphausiid *Euphausia pacifica* in the Yellow Sea. Marine Biology 115:625–633.

Tavares, C., and J.W. Martin. 2010. Suborder Dendrobranchiata Bate, 1888. Pages 99–164 *in* S. Schram, J.C. von Vaupel Klein, J. Forest, and M. Charmantier-Daures, editors. Eucarida: Euphausiacea, Amphionidacea, and Decapoda (partim). Treatise on zoology—anatomy, taxonomy, biology—the Crustacea. Brill, Leiden, The Netherlands.

Thiel, M. 2000. Extended parental care behavior in crustaceans–a comparative overview. Crustacean Issues 12:211–226.

Thorson, G. 1961. Length of pelagic larval life in marine bottom invertebrates as related to larval transport by ocean currents. American Association for the Advancement of Science Publication 67:455–474.

van Duren, L.A., and J.J. Videler. 1995. Swimming behaviour of developmental stages of the calanoid copepod *Temora longicornis* at different food concentrations. Marine Ecology Progress Series 126:153–161.

Vincx, M., and C.H.R. Heip. 1979. Larval development and biology of *Canuella perplexa* T. and A. Scott, 1893 (Copepoda, Harpacticoida). Cahiers de Biologie Marine 20:281–299.

Vogt, G. 2013. Abbreviation of larval development and extension of brood care as key features of the evolution of freshwater Decapoda. Biological Reviews 88:81–116.

Vogt, G., and L. Tolley. 2004. Brood care in freshwater crayfish and relationship with the offspring's sensory deficiencies. Journal of Morphology 262:566–582.

von Reumont, B.M., R.A. Jenner, M.A. Wills, E. Dell'Ampio, G. Pass, I. Ebersberger, B. Meyer, S. Koenemann, T.M. Iliffe, A. Stamatakis, O. Niehuis, K. Meusemann, and B. Misof. 2012. Pancrustacean phylogeny in the light of new phylogenomic data: support for Remipedia as the possible sister group of Hexapoda. Molecular Biology and Evolution 29:1031–1045.

Walker, G. 2004. Swimming speeds of the larval stages of the parasitic barnacle, *Heterosaccus lunatus* (Crustacea: Cirripedia: Rhizocephala). Journal of the Marine Biological Association of the United Kingdom 84:737–742.

Walossek, D. 1993. The Upper Cambrian *Rehbachiella* and the phylogeny of Branchiopoda and Crustacea. Fossils and Strata 32:1–202.

Walossek, D., J.T. Høeg, and T.C. Shirley. 1996. Larval development of the rhizocephalan cirripede *Briarosaccus tenellus* (Maxillopoda: Thecostraca) reared in the laboratory: a scanning electron microscopy study. Hydrobiologia 328:9–47.

Wang, F.-L., H.-L. Hsieh, and C.-P. Chen. 2007. Larval growth of the coconut crab *Birgus latro* with a discussion on the development mode of terrestrial hermit crabs. Journal of Crustacean Biology 27:616–625.

Williamson, D.I. 1982. Larval morphology and diversity. Pages 43–110 *in* L.G. Abele, editor. The biology of Crustacea, volume 2: Embryology, morphology, and genetics. Academic Press, New York.

Wolff, C. 2014. Chapter 39. Amphipoda. Pages 206–207 *in* J.W. Martin, J. Olesen, and J.T. Høeg, editors. Atlas of crustacean larvae. Johns Hopkins University Press, Baltimore, Maryland.

2

BODY SIZE, MATURATION SIZE, AND GROWTH RATE OF CRUSTACEANS

Piotr Maszczyk and Tomasz Brzeziński

Abstract

Crustaceans present a remarkable variety of forms that differ greatly in body size and growth strategies (determinate or indeterminate). This diversity reflects the long evolutionary history of this group and the variety of environments a crustacean may inhabit. It is rooted in a wide array of internal (physiological, structural) growth constraints and different extrinsic ecological factors determining the extent to which the body size of an individual crustacean attains its upper limit. We briefly review the combined effects of these factors with a focus on the effects of food quality and quantity, predation, and temperature on life histories in the context of an individual, as well as at the population and community levels. We discuss the discrepancy between the possible and the attained body size in an attempt to resolve the extent to which the observed pattern (1) is genetically based, (2) reflects the adaptive plasticity of the phenotype, and (3) is driven by global environmental changes.

A review of recent publications shows that freshwater and marine crustaceans grow faster but attain smaller size at elevated temperature (the pattern follows the temperature size rule). It also reveals an apparent decrease in mean body size of crustacean populations and communities at an elevated temperature, both spatially or temporally (at least 79% of studies at each level). However, it is not clear whether these patterns are a result of direct effects of temperature or of the interaction of temperature with other extrinsic factors. Studies that have sought to disentangle the direct and indirect effects of temperature have suggested that increased oxygen demands, food quality constraints, and the effects of size-selective predators can be complementarily or exclusively responsible for the patterns. With the advent of global warming, elucidation of mechanisms responsible for the effects of temperature on body size of animals at different levels of organization seems to be an issue of major importance.

Life Histories. Edited by Gary A. Wellborn and Martin Thiel.

INTRODUCTION

Body size has long been considered the most important trait of an organism, influencing nearly every aspect of its biology and ecology (Peters 1983). Almost all characteristics of an individual, as well as of a population and community, predictably vary with body size. On the one hand, body size is positively correlated with longevity and fecundity at the individual level, with intraspecific competitive abilities at the population level, and with interspecific competitive abilities at the community level. On the other hand, it is negatively correlated with the mass-specific metabolic rate, population growth, and population size. Crustaceans display great variability of size at various stages of ontogeny, even within closely related species (Hartnoll 1982). Size ranges from a fraction of a millimeter for copepods and ostracods (e.g., a harpacticoid copepod *Stygotantulus stocki* has a body length of 0.1 mm), to decimeters in the American lobster (*Homarus americanus*), with a carapace length of up to 40 cm and weight of over 20 kg (making it the heaviest of all living arthropod species), or the Japanese spider crab (*Macrocheira kaempferi*), with a carapace length of up to 40 cm and a chelar span of nearly 400 cm (making it the longest of all extant arthropod species; Fig. 2.1). As with all other arthropods, tissue growth of crustaceans is a continuous process, but the increase in their external dimensions is discontinuous due to molting, when the old integument is cast off and a rapid increase in size occurs before the new integument hardens and becomes inextensible (Hartnoll 1982). Molts are separated by an intermolt, a period when the integument is hardening and size remains the same despite increase in body mass (Fig. 2.2).

There are 2 distinctly different growth strategies among crustaceans: indeterminate and determinate (Fig. 2.2). When indeterminate, the organism continues to molt indefinitely, with the percentage of increase in body size that occurs within a molt (i.e., molt increment) declining with time and the intermolt period increasing with size (Fig. 2.2A). The total number of molts in a lifetime is species specific, but the number of molts within each larval stage (and juvenile stage) can vary with ambient conditions. For instance, the number of juvenile stages of the water

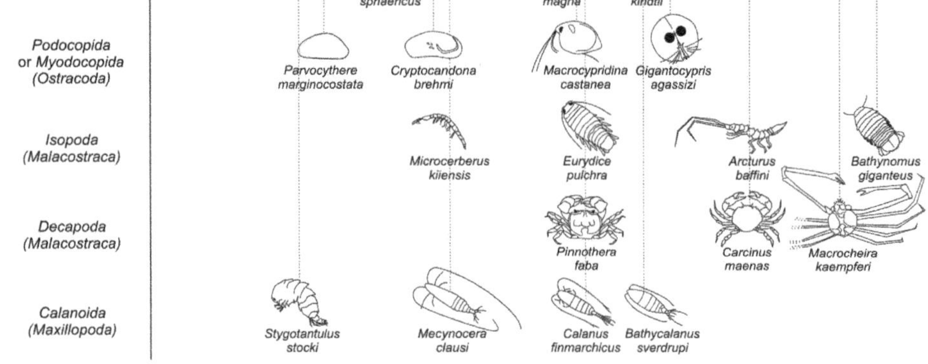

Fig. 2.1.
Variability of maximum body size (carapace length) among crustaceans (note that the *x* axis is logarithmic).

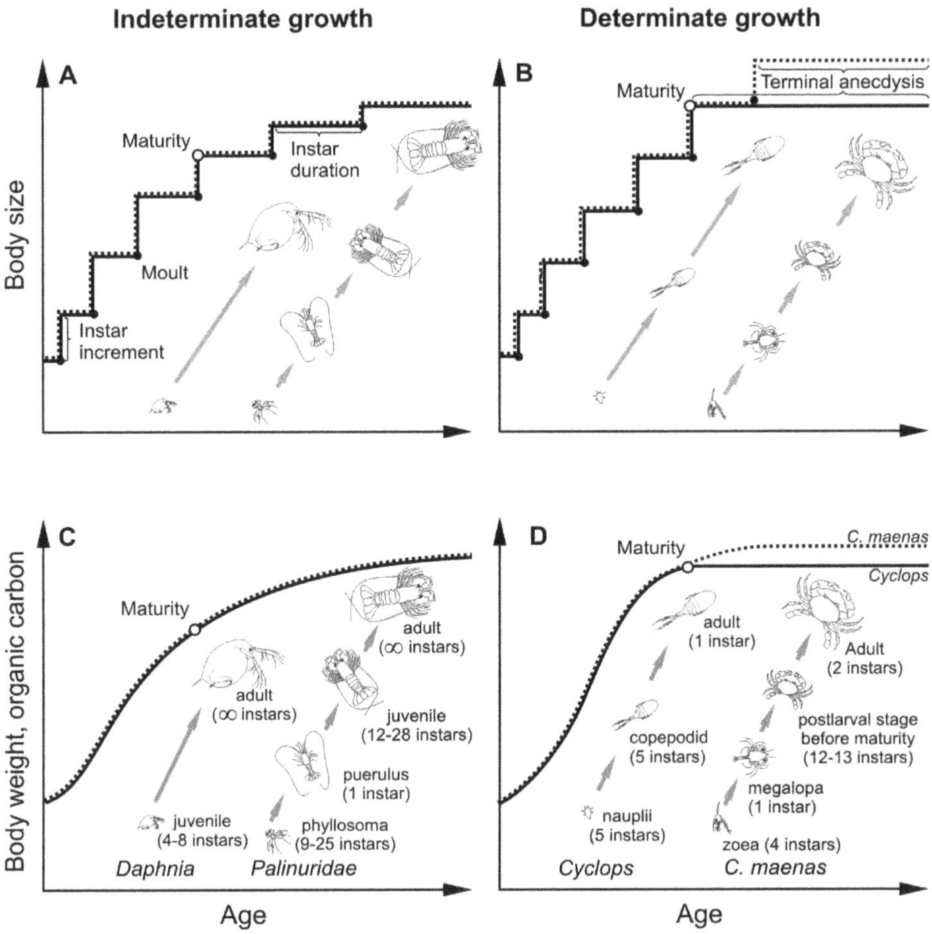

Fig. 2.2.
Schematic representation of individual growth curves indicating the sequence of instars and developmental stages for two example groups of taxa. (A–B) Body length-based growth curves of crustaceans with (A) indeterminate growth, as in *Daphnia* (continuous line) and *Palinuridae* (spiny lobster, dashed line), and (B) determinate growth, as in *Cyclops* (continuous line) and *Carcinus maenas* (dashed line). (C–D) Corresponding weight-based growth curves. Please note that for clarity, the scale on each axis is omitted and may be different for each species.

flea *Daphnia cucullata* varies between 4–8 at different food levels (Maszczyk and Bartosiewicz 2012) and the number of larval phyllosoma instars of the spotted spiny lobster (*Panulirus guttatus*) varies between 9 and 25, plus a single puerulus stage and 12–28 juvenile stages (Fig. 2.2C; Wahle and Fogarty 2006). When growth is determinate, the organism undergoes a terminal molt beyond which no further molting occurs (anecdysis), and stops growing usually at the time of maturity. Some shrimps (Rhynchocinetidae) are notable exceptions: males reach sexual maturity in a female-like stage, and then continue to grow until reaching the final molt stage (Correa et al. 2000). Among other exceptions are *Carcinus* and *Portunus*, which have the terminal anecdysis but still grow and molt after becoming sexually mature (Hartnoll 1982). The number of preadult instars in determinate growers would be either variable (e.g., in *Maja* and *Carcinus*) or invariant (e.g., in Ostracoda and Copepoda; Hartnoll 1982); moreover it varies significantly among species. For example, Cyclopoida have five naupliar and five copepodid stages, and the crab *Carcinus maenas* has

4 zoea stages, 1 megalopa stage, 12–13 postlarval instars before sexual maturity, and 2 adult stages (Fig. 2.2D, Mohamedeen and Hartnoll 1989, Hartnoll 2015). In each of the two growth strategies, the intermolt period changes with age. For instance, the determinate grower *C. maenas* requires 3 days to complete the first zoea and 170 days to complete the 13th postlarval (puberty) instar (Mohamedeen and Hartnoll 1989), and the large-bodied lobster *H. americanus*, an indeterminate grower, requires 3 days to complete the first larval stage and 3 years to complete subsequent stages in the adult female (Wahle and Fogarty 2006, Hartnoll 1982).

As well, no obvious correlation can be observed between growth format and phylogenetic position. Among the evolutionarily older Entomostracans, the Branchiopoda are known to have indeterminate growth, whereas Ostracoda and Copepoda have determinate growth. Both growth strategies are also present in the most advanced Decapoda (e.g., the lobster *H. americanus* has indeterminate growth, and the spider crab *Maja squinado* has determinate growth; Hartnoll 1982).

Both body size and growth rate of crustaceans are affected by a combination of intrinsic (physiological and structural) constraints and extrinsic (environmental) factors. The most important of the intrinsic factors affecting body size and growth rate is the synchronization between the processes of molting and reproduction, as well as size-related growth rate (Hartnoll 2001). Three extrinsic factors and their interactions have the greatest impact on both body size and individual growth rate: (1) competition for limited food, with winners able to grow and reproduce at lower food levels; (2) predation, with survivors being best in avoiding predation risk through morphological or behavioral adjustments; and (3) ambient temperature, which sets the rate of body growth (assimilation and respiration rates) and population increases (growth and birth rates). Where predation and food limitations act as the principal factors of natural selection (favoring individuals that are more evasive or more efficiently exploiting food, or both), temperature acts merely as a controller of the velocities for predator avoidance and resource exploitation. The effect of other factors on crustaceans' body size and growth rates, such as parasitism, salinity, toxicity, pH, and competition for space, have been widely reviewed in Anger (2001) and Hart and Bychek (2011). Here we discuss the different ecological and evolutionary issues related to the effect of extrinsic factors on the phenotypic plasticity of body size and growth rate in a variety of crustaceans, as well as the consequences at the individual, population, and community level.

BODY SIZE AND GROWTH RATE OF AN INDIVIDUAL

Food Quantity and Quality

Food Quantity

A reduced food supply can delay maturity and reduce somatic growth rate and body size at a given stage (by reducing the molt increment and increasing the number of molts; Fig. 2.3A) in a variety of crustacean species (Hartnoll 2001, Anger 2001), including decapods (Oh and Hartnoll 2000), cladocerans (Maszczyk and Bartosiewicz 2012), and copepods (Hart and Bychek 2011). Food supply is most important in the juvenile stages of a crustacean's life as a key factor controlling survival, intermolt period, and individual growth rate (Wahle and Fogarty 2006). In their early stages, many crustaceans are at least partially protected from starvation risk by energy reserves originally contained in the egg. Therefore, until these resources are depleted, their growth is largely independent of food supply in the environment. For instance, the larval (zoea) stage of *H. americanus* (Wahle and Fogarty 2006), as well as the first two intermolts of *Daphnia cucullata* and *D. longispina* (Maszczyk and Bartosiewicz 2012), were observed to be partially independent of food supply from the environment. The lithodid crab (*Paralomis granulosa*) and the southern king crab (*Lithodes*

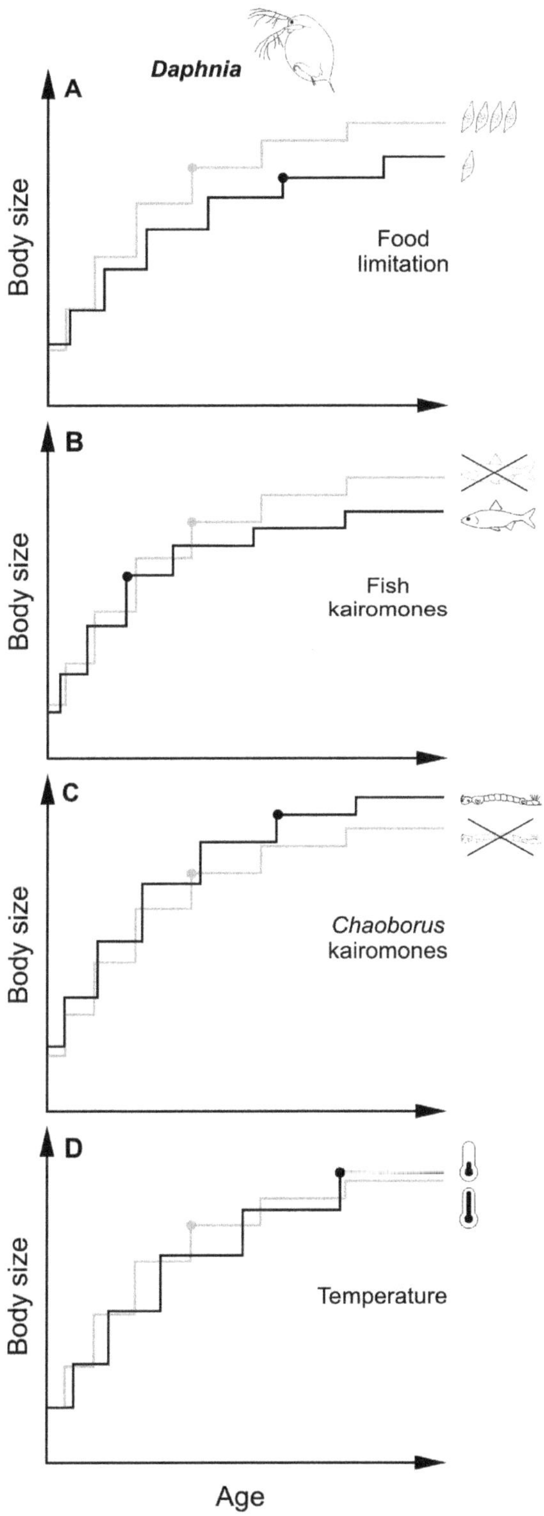

Fig. 2.3.
Schematic representation of the effect of food limitation on body length-based growth curve (number of preadult instars, molt increment, and instar duration) and age at maturity (circles) for *Daphnia* under different predation conditions. (A) Control versus under food-limited conditions. (B) Control versus in the presence of kairomones produced by fish. (C) Control versus in the presence of kairomones produced by an invertebrate predator. (D) Control versus under warmer temperatures.

santolla) exhibited full independence of food supply (the phenomenon termed *lecithotrophy*) from hatching through metamorphosis (Anger 2001). In contrast, the zoea larvae of spider crabs (*Hyas araneus*) and larvae of other brachyuran crabs must feed from hatching in order to survive (Anger 2001). Also, cladocerans produce fewer but larger neonates when food levels are low, which could be an adaptive phenotypic response that enhances neonate survival. Later maturation and production of fewer but larger neonates in response to low food supply has been also observed in the presence of crowding chemicals (Gliwicz et al. 2012). These chemicals provide information about high population density of conspecifics, and also carry information about oncoming starvation.

Food Quality

The effect of food quality on crustacean growth and other life history parameters is a complex issue. It includes the effect of dietary composition, stoichiometry (elemental ratios in food) and biochemical composition (e.g., dietary proteins, lipids, carbohydrates, and vitamins), as well as ingestibility, digestibility, and the toxicity of food particles (DeMott 1989, Anger 2001). The effects of low food quality on body size and life histories of crustaceans are usually similar to the effects of limitation by quantity of food. Crustaceans fed a diet of inadequate quality typically grow more slowly and mature later at a smaller body size than those consuming a more nutrient-rich diet (Hart and Bychek 2011).

Crustaceans, like all other heterotrophic organisms, regularly face an imbalanced diet composition in terms of stoichiometry and biochemical components in relation to their requirements. This may negatively affect their growth, development, and reproduction due to the inefficient utilization of all available dietary components, as well as the increased metabolic costs of the energy expended to acquire elements that are limited in the diet and to excrete waste products (DeMott et al. 1998, Plath and Boersma 2001, Acharya et al. 2004). The problem with low-quality food resulting from an imbalanced diet particularly relates to primary consumers, because their food (algae and detritus) is high in carbon and relatively low in phosphorus and nitrogen. Contrary to marine environments, in most freshwater systems, phosphorus rather than nitrogen is the limiting nutrient for phytoplankton growth, and hence most studies focus on the effect of the limited availability of this element, although nitrogen has also received attention (Plath and Boersma 2001). The carbon-to-phosphorus (C:P) ratio in freshwater phytoplankton appears to be lower compared to zooplankton and shows considerable variation (DeMott et al. 1998). Therefore, the elemental content of algae limits the growth of *Daphnia* and other herbivorous zooplankton in freshwater, at least when the C:P ratio exceeds a threshold of 300 (Sterner and Hessen 1994).

During ontogeny, growth and body stoichiometry change. Usually, growth decreases as the nitrogen-to-phosphorus (N:P) ratio in tissues increases (Elser et al. 1996), which is consistent with the predictions of the growth rate hypothesis that states higher growth rate is associated with higher phosphorus body content and lower C:P and N:P ratios, reflecting the allocation of phosphorus to the relatively phosphorus-rich RNA, the expression of which is required to support fast somatic growth (Acharya et al. 2004). This indicates that small-bodied, fast-growing (younger) individuals have greater phosphorus requirements and are more susceptible to its limitation.

As with all heterotrophic organisms, crustaceans require dietary amino acids, lipids, and carbohydrates for successful growth and development, and not all required compounds can be synthesized de novo. Therefore, limitation of these nutrients in the diet hampers growth and induces changes in other life history traits. A variety of lipids is essential for crustaceans, including sterols (e.g., cholesterol, as a structural component of cell membranes and an essential precursor for the biosynthesis of molting hormones), triacylglycerides (as a major energy source and the predominant form of storage) and long-chain polyunsaturated fatty acids (PUFAs; Müller-Navarra et al. 2000, Martin-Creuzburg et al. 2009). Dietary cholesterol, for example, has a strong positive influence on the growth and fecundity of daphniids (Martin-Creuzburg et al. 2009, Martin-Creuzburg et al. 2012), and food rich in PUFAs enhances growth in decapods (Anger 2001, Wahle and Fogarty 2006), daphniids (Müller-Navarra et al. 2000, von Elert 2002), and copepods (Søreide et al. 2010).

The nutritional value of food can be affected by an insufficient elemental and biochemical composition as well as by its ingestibility, digestibility, or toxicity. Algae differ in food quality for herbivorous zooplankton. Cyanobacteria have long been recognized as a low-quality food for herbivorous zooplankton, hampering their growth, size at maturity, neonate size, and instar-specific growth rate (DeMott 1989). Three major food quality constraints have been identified in cyanophytes: (1) grazing resistance, (2) a deficiency in essential nutrients, and (3) toxicity. Grazing resistance consists mainly of the disruption in the feeding process by filamentous and large colonial cyanophyte species, which mechanically interfere with food collection (Porter and Orcutt 1980). *Daphnia* can diminish the strength of this interference by changing the morphology of their feeding apparatus and cleaning it more frequently. Mechanical interference with the filtering process not only hampers ingestion but is also associated with higher rejection and respiration rates and, thus, increased energy costs. Apart from the negative effect of cyanophytes on feeding rate due to interference, they are widely considered to be low-quality food in terms of biochemical composition, deficient in PUFAs (Müller-Navarra et al. 2000) and sterols (Martin-Creuzburg et al. 2009, Martin-Creuzburg et al. 2012). It was also found that toxins occurring in particular strains of cyanobacteria may negatively influence *Daphnia* life histories (Lampert 1981).

It seems that there is no universal pattern between the body size of crustaceans and their susceptibility to food quality limitations. Rather, food quality limitation depends on the feeding mode, with selective feeders less affected in nature than nonselective filter feeders (e.g., no conclusive evidence for PUFA limitation in selectively feeding copepods—Jónasdóttir et al. 2005). However, according to DeMott et al. (2010), large-bodied individuals may be favored over small-bodied ones when feeding on poorly digestible seston, because the time it takes for food to pass through the gut in large-bodied individuals is longer, facilitating higher assimilation efficiency.

Predation

Predation also may affect body size either through direct selection caused by selective mortality or by phenotypic plasticity in response to reduced mortality risk. This could be so even in the case of large and long-lived decapods, which are vulnerable to predation (and cannibalism) in some circumstances, especially in the larval and in the settling stage, as well as during molting when their body is not protected by a hardened carapace. In the case of marine and freshwater planktonic animals living offshore in the absence of complex architecture of inshore and terrestrial areas providing antipredator refugia, predation undoubtedly has a greater effect on body size (Gliwicz 2003, Hart and Bychek 2011). Vertebrate predators (mainly fish) usually select large-bodied prey, as opposed to gape-limited predatory invertebrates (e.g., *Chaoborus*, cyclopoids, *Bythothrephes*, and mysid shrimps), which are restricted to consuming smaller-bodied prey (Lampert and Sommer 2007).

Theoretical predictions on phenotypic plasticity in response to mortality risk assume that under visually oriented predation, optimization of fitness in planktonic prey causes them to reproduce at a small size and early age (Fig. 2.3B), and under tactile and chemosensory predation, planktonic prey invest more in juvenile growth at the expense of delayed reproduction (Fig. 2.3C). These predictions have been verified by many studies in which the inducible reaction of phenotypic plasticity to the presence of a selective factor has been tested using kairomones, which are chemical cues that provide prey with reliable information about predation risk (Wyatt 2011, Hazlett 2011). The majority of these experiments have been performed with *Daphnia* as a model organism (Lass and Spaak 2003a, Maszczyk and Bartosiewicz 2012; see Chapter 12 in this volume), but effects of kairomones on behavior and development of copepods (Hart and Bychek 2011), decapods (Anger 2001, Hazlett

1999) have also been reported. For instance, copepods exposed to exudates from planktivorous fish reduced their feeding activity, and juvenile lobsters (*H. americanus*) and the megalopae of the mud crab (*Panopeus herbstii*) perceive chemical signals from benthic predators and respond with a delay in settlement and metamorphosis (Anger 2001, Lass and Spaak 2003a); these responses decrease mortality risk at the cost of reduced growth.

Temperature

Temperature has a direct primary influence on the biochemical and physiological processes of crustaceans, which are ectotherms, and, in turn, on their life history characteristics such as growth rate and development time (Bottrell et al. 1976). Although growth rate of crustaceans increases with temperature faster in their early stages compared to later ones, it generally increases with temperature at a rate slightly above the $Q_{10} = 2$ rule (Panov and McQueen 1998). Accordingly, the intensity of each metabolic process in poikilotherms should double with a temperature increase of 10°C. Although growth rate is slower at lower temperatures, body size at a given stage, as well as maximal body size, of the majority of crustaceans is larger in cooler environments (Hartnoll 2001). This pattern is the result of both an increased intermolt time, and an increase in size at each molt (Fig. 2.3D).

This observation is consistent with the temperature-size rule (TSR; Atkinson 1994, Kozlowski et al. 2004), which is observed for most other ectotherms. The TSR suggests that the rate of functional bioenergetic processes responsible for growth and development differs in reaction to temperature change—that is, lower temperatures prolong developmental time more than they hamper growth rates (at least later in ontogeny; Atkinson 1994, Kozlowski et al. 2004). The rule seems to be more apparent in aquatic than terrestrial species (Forster et al. 2012), and it is considered a universal phenotypic response to climate change and the warming of aquatic environments (Daufresne et al. 2009). The ubiquitous pattern consistent with TSR within crustaceans is summarized in Table 2.1. Recent studies on TSR revealed that crustaceans with either determinate (*Heterocypris bosniaca*; Aguilar-Alberola and Mesquita-Joanes 2014) or indeterminate (*Artemia franciscana*; Forster and Hirst 2012) growth respond to temperature increases according to TSR only in their later stages (Table 2.1). The cellular basis of body size changes in a wide range of ectotherms more often appears to be due to alterations in cell size rather than number (Czarnoleski et al. 2015). Jalal et al. (2013) investigated the cellular basis of greater body size at a lower temperature in crustaceans, and found that *Daphnia magna* and *D. pulex* have an elevated nucleus size at lower temperatures, which also provides indirect evidence of greater cell size because nucleus size positively correlates with cell size.

Studies on TSR have been mainly directed toward finding a single mechanism to explain the phenomenon (Angilletta et al. 2004, Forster et al. 2012). One of the fundamental explanations of greater stage-specific body size (e.g., size at maturity) at lower temperatures is based on biophysical factors; it is a consequence of a disproportionately rapid decrease in catabolism over anabolism (i.e., energy utilization over energy uptake; von Bertalanffy 1960). Because Angilletta and Dunham (2003) and Kozlowski et al. (2004) provided strong theoretical evidence against this explanation, other hypotheses have suggested that patterns of temperature-dependent adult body size may be driven by other factors correlated with ambient temperature, such as oxygen availability (Atkinson et al. 2006, Forster et al. 2012, Klok and Harrison 2013) or predation risk (Kozlowski et al. 2004, Weetman and Atkinson 2004).

Interaction of Temperature with Other Factors

Temperature can affect body size and the somatic growth rate of crustaceans (1) directly through impacts on physiological and ecological processes as well as (2) indirectly by affecting oxygen demands, predation risk, and food quantity and quality (or their demands).

Table 2.1. Summary of published studies on direct and indirect (interaction with other factors) effects of temperature on body size at stage of an individual in crustacean species (numbers in parentheses indicate notes at the bottom of the table).

Order or subclass	Species	Response of body size to increased temperature		Reference
		Temperature alone	**Temperature × other factor**	
Copepoda	6 species	5 reductions, 1 increase	–	Atkinson 1994 (1)
Copepoda	*Scottolana canadensis*	Reduction	–	Lonsdale and Levinton 1985
Copepoda	*Mesocyclops edax*	Reduction (2)	–	Wyngaard 1986
Copepoda	*Pseudocalanus* sp.	Reduction	–	Daufresne et al. 2009
Decapoda	*Callinectes sapidus*	Reduction	–	Atkinson 1994 (1)
Decapoda	*Cancer irroratus*	Reduction	–	Atkinson 1994 (1)
Mysidacea	*Neomysis mercedis*	Reduction	–	Atkinson 1994 (1)
Cladocera	*Daphnia pulex* and *Daphnia curvirostris*	Increase (3)	Temperature. × fish predation No interaction	Weetman and Atkinson 2004 (4)
Cladocera	*Daphnia mendotae* and *Daphnia pulicaria*	Reduction (5)	Temperature × fish predation Significant interaction	Bernot et al. 2006
Cladocera	*Daphnia hyalina* × *galeata*	Increase	Temperature × fish predation No interaction	Doksæter and Vijverberg 2001
Cladocera	*Daphnia galeata*	No effect	Temperature × fish predation No interaction	Lass and Spaak 2003b
Cladocera	*Simocephalus vetulus*	Reduction	–	Perrin 1988
Cladocera	*Daphnia pulex* and *Daphnia magna*	Reduction	–	Jalal et al. 2013

(continued)

Table 2.1. ***(Continued)***

Order or subclass	**Species**	**Response of body size to increased temperature**		**Reference**
		Temperature alone	**Temperature × other factor**	
Amphipoda	*Caprella mutica*	Reduction (6)	–	Hosono 2011
Isopoda	*Asellus aquaticus*	Increase Reduction only in anoxia	Temperature × oxygen	Hoefnagel and Verberk 2015
Isopoda	*Idotea balthica*	Increase	–	Manyak-Davis et al. 2013
Anostraca	*Artemia franciscana*	Reduction only in later instars	Temperature × ontogeny	Forster and Hirst 2012
Ostracoda	*Heterocypris bosniaca*	Reduction only in later instars	Temperature × ontogeny	Aguilar-Alberola and Mesquita-Joanes 2014

(1) See the Appendix in Atkinson 1994 for references.
(2) Increase to 25°C and a decrease from 25°C.
(3) *D. pulex*—increased maturation size, decreased asymptotic size; *D. curvirostris*—increased maturation size, increased asymptotic size.
(4) See text for other references for copepods and cladocerans.
(5) *Daphnia mendotae*—reduction only in the presence of kairomones.
(6) Reduction only in females—maturation in earlier instars.

Temperature × Oxygen

The discrepancy between oxygen supply and demand, the extent of which depends on thermal conditions, has been frequently invoked to explain differential growth patterns described by the TSR. Broadly speaking, increasing temperatures elevate an individual's oxygen demand more than its oxygen supply, and small-bodied individuals are less susceptible to this stress than large-bodied ones, because, theoretically, they should be able to deliver more oxygen to tissues (Atkinson et al. 2006, Czarnoleski et al. 2015). At least 3 mechanisms may contribute to the observed advantage of small over large body size at temperature-related oxygen deficiencies: (1) the respiratory surface area for oxygen uptake in a smaller body is expected to be challenged to a lesser degree in supplying oxygen due to its putative greater surface-to-body volume ratio (Atkinson et al. 2006, Aguilar-Alberola and Mesquita-Joanes 2014); (2) once inside a smaller body oxygen is transported over shorter distances (Chapelle and Peck 1999, Czarnoleski et al. 2015); and (3) a smaller body, being composed of smaller cells, contains relatively more membranes, which allows for faster diffusion because oxygen diffuses more readily through membranes than through the cytoplasm (Czarnoleski et al. 2015). If oxygen limitation is responsible for the TSR pattern, it should be more apparent in larger-bodied species (and more apparent in older than in younger conspecifics). Moreover, it should also be more apparent in aquatic species than in terrestrial ones because oxygen uptake is more challenging under water than in the air. Both of these predictions have been confirmed by Forster et al. (2012), who provided strong evidence in support of oxygen availability as a major driver of TSR, at least in aquatic organisms. Hoefnagel and Verberk (2015) as well revealed that both temperature and oxygen affected age at maturity and growth of the aquatic crustacean *Asellus aquaticus*, but the relationship between developmental and growth rate as expected by the TSR assumption was observed only under oxygen limitation.

Temperature × Predation Risk

Although the literature on the interactive effect of temperature and predation risk on crustaceans in marine environments is rather scarce, a variety of studies on the effect of temperature on fish communities in lakes, particularly in the context of global warming due to climate change, have revealed that warming might cause a greater mortality risk to freshwater zooplankton due to enhanced predation, particularly from planktivorous fish (Wagner and Benndorf 2007, Brucet et al. 2010, Jeppesen et al. 2010, Iglesias et al. 2011, Vadadi-Fülöp et al. 2012). Studies on fish foraging behavior indicated that an elevated temperature increases capture probability, instantaneous search rate, and swimming speed of planktivorous fish, causing a greater than linear increase in the capture rate, and therefore, a greater than linear increase in mortality risk of plankton (Persson 1986). Moreover, the effect of increased temperature on mortality could be even stronger due to the reduction of time needed by fish to find a patch of crustacean prey (Gliwicz and Maszczyk 2016).

Because mortality risk caused by planktivorous fish seems to be strongly positively correlated with temperature, it could be expected that temperature can be used as an indirect measure to indicate the increase of mortality risk for plankton. If so, elevated temperatures would cause earlier maturation at a smaller body size and the production of larger clutches in relation to body size. Moreover, it could also be expected that such life history adjustments would be more apparent in the presence of fish kairomones. Many studies document earlier maturation at a smaller body size of *Daphnia* and other planktonic crustaceans exposed to fish kairomones (Bernot et al. 2006, Lass and Spaak 2003b, Weetman and Atkinson 2004), which is consistent with TSR and the life history adjustments observed in a majority of other studies on crustaceans (Table 2.1). However, only Bernot et al. (2006) have observed the interactive effect of temperature and kairomones.

Temperature × Food

Although food quantity interacts with temperature to affect crustacean growth rate and body size (e.g., temperature affects the neonate size of *Daphnia* to a greater extent at a high algal food level), much more attention has been paid to the combined effects of temperature and food quality. Temperature may indirectly affect growth rate of crustaceans and other ectotherms either by modifying their requirements for elemental food quality or by altering the quality of the food itself.

Elevated temperatures may alter food quality requirements by affecting life history traits. For instance, it may force an increase in relative reproductive investment, which could modify phosphorus requirements, when different phosphorus investments for somatic growth and egg production are required (Frost et al. 2008). Temperature can also alter food quality requirements by constraining biochemical pathways and cellular processes. An elevated temperature enhances growth rate, which is linked to greater phosphorus requirements through greater demands for phosphorus-rich ribosomal RNA (Acharya et al. 2004). Moreover, changes of ambient temperature can negatively affect crustacean growth by increasing demand for essential lipids, such as cholesterol and PUFAs, and availability of these lipids frequently limits growth of crustaceans. At high temperatures crustaceans are more prone to limitation by cholesterol (Martin-Creuzburg et al. 2009), whereas under cold conditions, the requirements of crustaceans for PUFAs increase (Sperfeld and Wacker 2011). In both cases, the reason postulated for an increased demand for the particular lipid is to maintain membrane fluidity, so that membranes can continue to function unimpaired at different ambient temperatures (Hazel 1995).

Elevated environmental temperature can reduce the nutritional quality of algal food for crustaceans in different ways. For instance, temperature can modify the biochemical composition of algae, mainly due to the reduction of phosphorus (decreased phosphorus-to-carbon [P:C] ratios) and PUFA contents (Woods et al. 2003). Decreases of P:C ratios in algal food are accompanied by reduced phosphorus content in the bodies of many *Daphnia* species (Seidendorf et al. 2010). Warmer temperatures can also change the composition of an algal community in a lake from relatively edible species to less edible cyanobacteria (Kosten et al. 2012). The nutritional quality of phytoplankton can also be reduced by temperature through changes in the morphology of algal cells and colonies, which makes them less ingestible and digestible. For instance, at elevated temperatures, green algae form barely digestible colonies more frequently (Lürling and van Donk 1999), and the filaments of cyanobacteria become shorter, thicker, and less breakable (Soares et al. 2013).

The strength of other harmful effects of cyanobacteria on zooplankton is also dependent on temperature. Rising water temperatures may increase the interference of cyanobacterial filaments with the filtering apparatus of zooplankton by decreasing water viscosity (filaments are getting stacked in the mesh of the filter screens). This may cause an increase in the vulnerability of herbivorous zooplankton to toxins produced by cyanobacteria (Bednarska et al. 2011). In turn, somatic growth of animals limited by quality of food is hampered, and as shown by studies on biochemical limitation, this indirect effect of temperature may enhance the direct effects of temperature on growth of crustaceans.

Predation × Food Supply

The majority of studies on growth rate and other life history adjustments of crustaceans show stronger effects of kairomones at high food levels. Gliwicz and Maszczyk (2007) found that the difference in somatic growth and body size at maturity of hybrid *D. hyalina × galeata* in the presence and absence of fish kairomones was negligible at low levels of algal food and gradually increased as food concentration increased. In the case of these planktonic crustaceans, the stronger manifestation of the phenotypic plasticity response at a high food level could be because the animals are able to invest more energy in antipredation defenses when reproduction is less constrained by resource acquisition. It may also be attributed to a greater size-selective mortality risk, as well-fed animals at high food concentrations are more conspicuous (larger and less transparent). This explanation is consistent with observations of other groups of terrestrial and aquatic animals, showing that satiated animals are ready to pay greater costs for predation risk avoidance when food levels are high. On the contrary, other studies of *Daphnia* have found enhanced antipredator defenses and more pronounced effects on life histories at a low food level (e.g., Pauwels et al. 2010). These authors hypothesized that such life history adjustments reflect a relatively stronger investment in defense when resources are limited, preventing higher investment in reproduction. Similar reasoning was also proposed by Glazier (1999) for benthic crustaceans.

FACTORS AFFECTING SIZE DISTRIBUTION IN CRUSTACEAN POPULATIONS

Size distribution and mean body size of individuals in crustacean populations vary temporally (from year to year or throughout seasons) and spatially (with depth, among sites, and among distant populations). Such variability has been best documented for planktonic cladocerans and copepods as seasonal succession, cline change from littoral to pelagic zone, and latitudinal- or depth-related clines in several crustaceans (Pfrender and Lynch 2000, Lonsdale and Levinton 1985, Manyak-Davis et al. 2013). The variation in the mean body size at the population level could be due not only to greater or smaller sizes at the same physiological stage but also to changes in the population age structure that affect relative proportions of differently sized individuals at different stages of ontogenesis.

The 2 most intriguing questions are how such differences arise and how they are maintained. Variation in mean body size may result from changes in the genetic structure of a population or from phenotypic plasticity expressed by individuals in a heterogeneous environment, or by a combination of both, contributing to the process of local adaptation. Crustaceans express high phenotypic plasticity when exposed to varying environmental conditions (described in the previous section). This plasticity also affects the mean size of individuals in a natural population (Stibor and Lampert 2000, Hairston et al. 2001, Brzeziński et al. 2010), and causes microevolutionary changes in the genetic structure of a population, which are associated with changes of mean body size and growth rate. Disentangling the relative importance of the effects of phenotypic plasticity on mean body size in natural populations from the effects driven by changes in genetic structure, as well as distinguishing the direct effects of a particular factor from the indirect effects that are mediated by other, covarying environmental factors, would require careful examination, using a combination of transplant (common-garden) and life table experiments.

Internal Constraints

One may expect that, as with any other trait, variability in body size and somatic growth reflects the genetic diversity existing in a population. This was found not only for benthic crustaceans with a sexual reproductive mode living in heterogeneous habitats (Brian et al. 2006) but also for copepods or parthenogenetic branchiopods occupying more homogeneous pelagial habitats (Carvalho and Crisp 1987).

The genetic basis of observed variation in body size remains largely unknown. Some authors suggest that it may result from the modulation of ploidy level. Nearly half of *Daphnia* DNA is in the endopolyploid state, varying among tissues (from 2 to 2048 C, where C represents a ploidy level), which may affect the size of cells and body structures (Gregory and Hebert 1999, Neiman et al. 2017). It is hypothesized that differences in level of ploidy within an individual may be responsible for the differential regulation of gene expression and modulation of cell division, and it seems that they may be critical for development of inducible defense in *Daphnia* (Neiman et al. 2017). The chromatin diminution (deletion of large amounts of DNA from early somatic cell lines) of copepods seems to be related to body size; the smaller genome allows faster development, but at a smaller body size (Gregory and Hebert 1999). Polyploid populations of cladocerans are more frequently found at higher rather than at lower latitudes, and polyploid clones from northern populations seem to attain a larger body size (Beaton and Hebert 1988). The prevalence of polyploids at higher latitudes is frequently attributed to their (hypothesized) greater tolerance to extreme environmental conditions. It seems that indeed, in some cases, polyploids are better adapted to low temperatures or to low salinities, but it has not yet been proven that they always perform better in extreme habitats (Jose and Dufresne 2010). Other studies found associations between growth rate, length of ribosomal DNA, and food stoichiometry for cladocerans, suggesting that high growth may result from the more efficient expression of genetic information, which depends on the amount of rDNA and, in turn, on the availability of phosphorus, which is an essential structural element of nucleic acids (Weider et al. 2004). Identification of functional polymorphism in the genomes of crustaceans requires further study, as does the unraveling of links between genes and somatic growth and development in crustaceans. This process of discovery may be facilitated by progress in ecological genomics. Genomes of the two model species of *Daphnia* (*D. pulex* and *D. magna*) have been sequenced recently, but a large fraction of the sequenced genes lacks homology to the genes of other arthropods, and the functions of these genes remain unknown (Orsini et al. 2011).

Extrinsic Factors

Extrinsic factors determine the extent to which body size of individuals in a population attains its maximum under the constraints of intrinsic limitations. Mean body size in natural populations is nearly always lower than that of the animals raised in close-to-optimal laboratory conditions (Stibor and Lampert 2000), suggesting that extrinsic factors have widespread impacts on crustacean body size in natural populations.

Food Quality and Quantity

The discrepancy between the possible and attained body size may result from the fact that animals in the field most often are hungry. However, the direct effects of food limitation are not easy to discriminate from other factors, particularly the impact of size-selective predation by fish (e.g., threatened crustacean prey reduce foraging), which maintains population density at a level far from the carrying capacity provided by available resources (Gliwicz 2003). Unique examples of food limitation are provided by studies of zooplankton populations in predator-free high mountain lakes or saline lakes (for benthic crustaceans, see also Glazier 1999). In these environments, intraspecific competition leads to a permanent state of low food levels (Gliwicz 2003). The interesting issue is the extent to which the effects of food quantity on body size influence the genetic composition of a population. It is assumed that a larger body size provides an individual with increased competitive abilities, which contribute to the seasonal succession of genotypes in pelagic *Daphnia*, and to the domination of large-bodied clones during the clear-water phase, when food is limited (Sommer et al. 1986). There is debate about the extent to which the observed pattern of intraspecific changes in mean body size reflects phenotypic plasticity or, instead, the selection of genotypes that differ based on body size. Both generalist genotypes with high phenotypic plasticity and specialist genotypes superior in conditions specific for a particular time of the season usually contribute to the observed changes in mean body size and the genetic structure of a population (Carvalho and Crisp 1987). Another aspect of the effect of body size limitations of crustaceans related to food quantity is the effect on juveniles as compared to adults. This is particularly true for cladocerans, with small-bodied neonates being less efficient at food collection and less resistant to starvation than adults. Increased mortality of neonates suffering from food shortages has been suggested as the reason for particular demographic structures and the phenology of daphnid populations in predator-free habitats (Gliwicz 2003). Similar effects are observed in benthic decapods, where large-bodied adults become cannibalistic when overcrowded, causing mortality of newly molted, smaller individuals (Wickins and Lee 2002).

Food quality also affects size structure of crustacean populations in nature (Gliwicz 2003). This is either due to the fact that malnourished individuals are not able to sustain body growth compared to well-fed ones (as in case of biochemical limitation; Müller-Navarra et al. 2000), or due to size-selective mortality, acting against either small- (DeMott et al. 2010) or large-bodied individuals (DeMott et al. 2001). Microevolutionary changes in the genetic structure of a population and corresponding changes of mean body size (somatic growth) driven by changes in food quality have been observed. For instance, Hairston et al. (2001) showed that a cladoceran population coexisting with a high abundance of cyanobacteria shifted toward dominance of clones resistant to the negative effects of lower quality cyanobacteria: they achieved a higher growth rate and their growth rate was hampered to a lesser extent than clones from the period when cyanobacteria were absent. Some authors believe that there is also some sort of specialization of genotypes with regard to their ability to exploit habitats of different food qualities. Filter feeders from resource-deprived environments are superior to their conspecifics from environments with high quality food, but they do relatively poorer in resource-rich environments (Tessier et al. 2000, Sarnelle and Wilson 2005). However, Tessier et al. (2000) did not find evidence for a relationship between body size and susceptibility to hunger, and this result is also supported by a meta-analysis suggesting that susceptibility to the effects of cyanobacteria is not related to body size (Tillmanns et al. 2008).

This conclusion, however, may not be correct because it is difficult to separate the effects of food quality from those of other environmental factors, particularly predation, which may overwhelm the effect of food quality (Brzeziński et al. 2010). Moreover, the effects of low food quality may be a trigger that initiates a cascade of indirect effects. For example, lack of essential lipids (PUFAs, sterols) may not only affect the somatic growth of crustaceans but also their ability to deal with thermal stress (Hazel 1995) or to avoid size-selective predation (Brzeziński and von Elert 2015).

Predation

Predation often has a major impact on populations. Because predators are usually size selective, they can affect the demographic and body size structure of a population either directly, by removing individuals of one particular size spectrum of crustacean prey, or indirectly, through the induction of defense mechanisms in crustacean prey mediated by chemical cues (Table 2.2; Gliwicz 2003, Hart and Bychek 2011). Crustaceans are frequently exposed simultaneously to both vertebrate (mainly fish) and invertebrate predators. More often, the relative strength of different types of predators varies either temporally or spatially, and generates shifts in mean body size and the demographic structure of crustacean prey (e.g., Boersma et al. 1999, Stibor and Lampert 2000, Benndorf et al. 2000). In summer, strong fish predation on freshwater pelagic *Daphnia* populations may result in a decreased proportion of adults and a reduction of adult body size, whereas, in autumn, predation by invertebrate predators leads to a decrease in proportion of young individuals and an increase in body size.

Chemical cues released by predators may also induce antipredation defenses in crustacean prey, including adaptive modifications of life histories affecting body size. These indirect effects of predators (via the plasticity of morphological traits) on attained mean body size and the demographic structure of prey populations may be significant (Stibor and Lampert 2000). Nevertheless, some studies show that changes of the mean or the reaction norm, or both, of body-size related traits could be interpreted as the result of genetic differentiations between populations, reflecting adaptation to local (seasonal) predation regimes, and not phenotypic plasticity (Boersma et al. 1999, Brzeziński et al. 2010). Moreover, antipredation defenses exhibited by threatened prey involve not only modifications of life history or morphology but also behavioral changes (e.g., reducing foraging behavior and migration; Gliwicz 2003, Lampert and Sommer 2007), which may trigger a cascade of effects on body size driven by other extrinsic factors. Reduced foraging time or migration to predator-free but suboptimal, food-depleted or colder habitats may affect body size either negatively (due to food limitation) or positively (due to residence in lower temperatures providing a longer life).

Direct and Indirect Effects of Temperature

It is widely recognized that mean body size of a population generally declines with temperature either spatially (along a latitudinal gradient or depth), or temporally (across seasons, or in paleoecological data; Table 2.2). However, most of these studies rely on a correlational approach, simply noting the fact that body size covaries with temperature change without identifying mechanisms, such as phenotypic plasticity or microevolutionary changes in the population. Results of common-garden and transplant experiments designed to explore these relationships and mechanisms are inconsistent. Some studies showed thermal tolerance in cladocerans is not related to body size and found no evidence for local thermal adaptation along a latitudinal gradient (Mitchell and Lampert 2000, Chopelet et al. 2008). Others found intraspecific variation with regard to thermal tolerance and microevolutionary changes in response to warming, with genotypes especially fit at a narrow temperature range ("thermal specialists") dominating the population at particular times of the season (Carvalho 1987, Wiggins and Frappell 2000). Changes in the thermal tolerance of natural populations as a response to recent temperature shifts increase on the scale of decades (Geerts et al. 2015). In copepods spatial adaptation of growth rate to environmental temperature was found in *Scottolana canadensis* along the east coast of North America (Lonsdale and Levinton 1985). Both northern and southern populations of this species had higher performance (i.e., higher growth rate) in laboratory experiments at temperatures prevailing in their specific environments.

These inconsistencies may stem from the fact that observed patterns of body size variation may result from a complex interplay between physiological, demographic, and ecological factors, all of which are also dependent on temperature. For example, it is assumed that higher water temperatures induce a greater risk of mortality due to size-selective predation. This assumption is corroborated by finding that changes related to climate warming increase the risk of predation by prolonging the period in which predators forage, increasing the abundance of predators and their efficiency in capturing prey, and, consequently, increasing the size-dependent mortality risk of *Daphnia* (Wagner and Benndorf 2007). Moreover, despite findings that body size in populations is inversely correlated with temperature, Manyak-Davis et al. (2013) discovered that the latitudinal gradient of predation pressure, not temperature itself, is the main factor determining body size trends along a latitudinal cline in benthic isopods. Seasonal shifts in the genetic composition of *Daphnia* populations are also attributed to seasonally varying temperature-dependent predation. *Daphnia* clones with different body size, life histories, and phenotypic plasticity of these traits are adapted to specific predation regimes, and dominate in particular seasons (Stibor and Lampert

Table 2.2. Summary of published studies on effects of spatial and temporal variation of temperature and other factors correlated with temperature on intra- or interpopulation variation in mean body size of selected crustacean taxa (numbers in parentheses indicate notes at the bottom of the table).

Order or subclass	Species	Source of variation	Response of body size to increased temperature		Variability in age structure (AS) or in life history adjustments (LH)	Reference
			Temperature	**Other factor**		
Cladocera	*Daphnia galeata*	Seasonality	Reduction	Reduction due to increased predation risk	AS–not tested; LH–genetic	Wagner and Benndorf 2007
Cladocera	*Daphnia hyalina*	Seasonality	Reduction	Reduction due to increased predation risk	AS–not tested; LH–genetic	Stibor and Lampert 2000
Isopoda	*Idotea balthica*	Latitude	Increase	Reduction due to increased predation risk	AS–decrease in relative number of the largest individuals; LH–genetic	Manyak-Davis et al. 2013
Isopoda	*Asellus aquaticus*	Seasonality	Reduction		AS–not tested; LH–genetic	Thompson 1986
Copepoda	*Arctodiaptomus salinus*	Latitude and altitude	Reduction		AS–not tested; LH–not tested	Anufriieva and Shadrin 2014
Copepoda	*Scottolana canadensis*	Latitude	Reduction		AS–not tested; LH–genetic	Lonsdale and Levinton 1985
Cladocera	*Daphnia magna*	Latitude	Reduction		AS–not tested; LH–plastic	Mitchell and Lampert 2000
Cladocera	*Daphnia magna*	Seasonality	Reduction		AS–not tested; LH–genetic	Carvalho 1987
Amphipoda	*Hyalella azteca*	Seasonality	Reduction		AS–decrease in relative number of the largest individuals; LH–not tested	Panov and McQueen 1998
Decapoda	*Jasus edwardsii*	Latitude	Reduction		AS–not tested; LH–not tested	Wahle and Fogarty 2006
Decapoda	*Homarus americanus*	Latitude	Reduction		AS–not tested; LH–not tested	Wahle and Fogarty 2006
Decapoda	*Procambarus clarkii*	Seasonality	Increase		AS–not tested; LH–not tested	Bonvillain et al. 2015

(*continued*)

Table 2.2. ***(Continued)***

Order or subclass	Species	Source of variation	Response of body size to increased temperature		Variability in age structure (AS) or in life history adjustments (LH)	Reference
			Temperature	Other factor		
Ostracoda	20 species	Bathymetry	Reduction in 14; no in 5		AS–not tested; LH–not tested	Hunt and Roy 2006 (2)
Ostracoda	3 species	Latitude	Reduction in 2; no in 1		AS–not tested; LH–not tested	Hunt and Roy 2006 (2)
Ostracoda	2 species	Seasonality	Reduction		AS–not tested; LH–not tested	Hunt and Roy 2006 (2)
Ostracoda	*Alteratrachyleberis scrobiculata*	Paleobathymetry	Reduction		AS– not tested; LH – not tested	Hunt and Roy 2006 (2)
Ostracoda	19 species of *Poseidonamicus*	Paleobathymetry	Reduction		AS–not tested; LH–not tested	Hunt and Roy 2006 (2)

(1) See Table 1 in Watt et al. 2010 for references.
(2) See Hunt and Roy's (2006) Table 3 in the Supporting Text published on the PNAS web site for references.

2000). One may expect that combined effects of temperature with other factors (e.g., oxygen availability, food quality) may also influence mean body size in crustacean populations, although studies sought to disentangle effects of these factors are scarce. Finally, variation of mean body size within a population may result from the effects of temperature on its age structure. The population structure could be modified, for example, either by elongation of life span or decelerated developmental rate of individuals in low temperature, both of which may alter relative proportions of individuals at different stages of ontogenesis. Nevertheless, this effect is rarely assessed in field studies and common-garden experiments (Table 2.2).

BODY SIZE COMPOSITION OF A COMMUNITY

A useful metric for size structure at the community level is the combination of mean body size and proportional abundance of each population within the community. For herbivorous zooplankton, proportional abundance of species is mainly the outcome of 2 biotic factors: species-specific competitive abilities and species-specific vulnerabilities to predation, particularly from planktivorous fish (Gliwicz 2003, Lampert and Sommer 2007). Temperature is identified as the most important abiotic factor directly (through the rate of physiological processes) and indirectly (through interaction with other factors, such as oxygen concentration, predation, and interspecific competition) affecting seasonal as well as spatial changes in the mean body size of zooplankton communities.

The Size-Efficiency Hypothesis

Generally, the zooplankton community in lakes is dominated by small-bodied species when there is a high abundance of planktivorous fish, as compared to the prominence of large-bodied species in predator-free environments (Fig. 2.4). To account for this pattern, Brooks and Dodson (1965) formulated the size-efficiency hypothesis (SEH). This hypothesis predicts that the composition of zooplankton in a habitat is the result of the combined effects of competition and predation. For

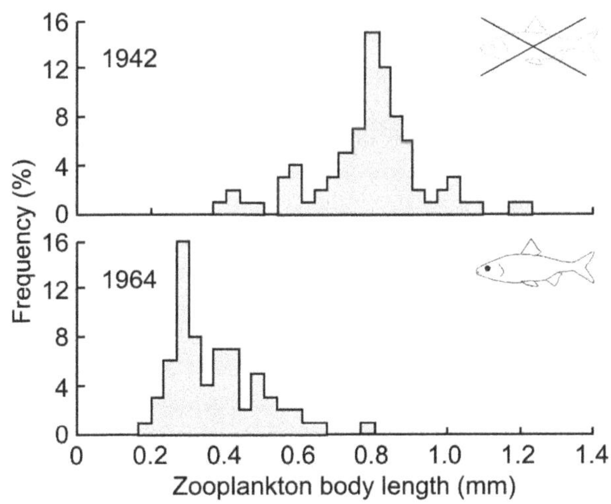

Fig. 2.4.
Size distribution of the zooplankton community in a lake in New England before (1942) and after (1964) the immigration of a planktivorous fish, *Alosa pseudoharengus*. Redrawn from Brooks and Dodson (1965).

example, when planktonic animals compete for small particles, large zooplankton are able to filter more efficiently; hence, large zooplankton dominate at low fish predation, whereas under high fish predation, small zooplankton dominate because large zooplankton are more readily consumed by fish. A moderate risk of fish predation permits the coexistence of differently sized species (see Gliwicz 2003 and Lampert and Sommer 2007 for reviews). Since Brooks and Dodson (1965) proposed the SEH, studies of the ecology and evolution of freshwater zooplankton have focused intensely on the role of body size in aquatic communities. Although it is clearly established that small-bodied zooplankton species are favored when fish abundance is high, the mechanism(s) responsible for the dominance of large-bodied species in the absence of fish predation seems to be more complex.

Interspecific Competition of Zooplankton in the Absence of Fish Predation

Zooplankton communities from fish-free habitats usually are monopolized by a single large-bodied planktonic herbivore, such as a cladoceran (*Daphnia pulicaria* in Lake Czarny, Tatra Mountains, Poland; Gliwicz 2003) or a branchiopod (*Artemia franciscana* in Great Salt Lake, Utah, United States; Gliwicz 2003). Although the SEH predicts that dominance of large-bodied species in the absence of fish predation is the result only of large-bodied species' more efficient filtration of food particles and their ability to eat larger particles, reasons for this dominance could be expanded to include the costs of food acquisition. Specifically, the rate of energy gained by assimilating larger particles likely increases more rapidly with body size than the rate of energy loss due to respiration. In other words, energy gained by larger-bodied species is higher at the same level of food supply.

Monopolization of the community by a large-bodied species is also consistent with Tilman's (1982) resource-based competition theory, which states that the outcome of exploitative competition is determined by the minimal resource level at which only a superior competitor is able to maintain positive population growth. The competitive superiority of larger planktonic animals has been expressed by using the threshold food concentration (TFC), which is the concentration of food at which the metabolic expenditure is balanced with net food intake (or more simply, at which the net growth rate is zero). For instance, Gliwicz (1990) found that the TFC was consistently lower for larger than for smaller cladoceran species of the genus *Daphnia*, although some recent studies did not confirm this relationship (Iwabuchi and Urabe 2012). Additionally, the food threshold for egg production (i.e., the food concentration above which eggs are produced) is lower in larger-bodied *Daphnia* species than in smaller ones, and larger-bodied species have a higher individual growth rate at a given food concentration (Hart and Bychek 2011).

Although all these lines of evidence indicate large zooplankton species can maintain their population density when food supplies are low while smaller species cannot (or their growth is more hampered), they consider only the effect of food quantity, but not food quality, on competitive abilities. Thus, this evidence seems to be insufficient in explaining the competitive superiority of larger-bodied species when the magnitude of the change in production (and therefore superiority) related to food quality differs among species (Hart and Bychek 2011).

According to ecological stoichiometry concepts (as discussed in previous sections), it could be expected that food quality, notably the C:P ratio, influences interspecific competitive abilities of planktonic animals (Iwabuchi and Urabe 2012). The effect of low-quality food on interspecific competitive abilities is also important in the context of the modification of zooplankton communities due to the presence of cyanobacterial blooms. The mechanical interference of blooming filamentous and colonial cyanobacteria decreases competitive abilities of large-bodied species of cladocerans (Porter and Orcutt 1980, Gliwicz and Lampert 1990). Filtration rate is hampered in the presence of cyanobacterial filaments, and there is an increase in respiration of large-bodied *Daphnia* species (Sikora and Dawidowicz 2014). In turn, the presence of cyanobacterial filaments caused a reversal of the rank order of TFC such that the thresholds were lower for larger *D. pulicaria* than for smaller *D. cucullata* in the absence of cyanobacteria, and they became lower for *D. cucullata* than *D. pulicaria* in their presence (Gliwicz and Lampert 1990).

Other studies argue that the competitive abilities of small-bodied species could be underestimated when competitive abilities are examined only with one type of algal food. Competitive abilities of small-bodied species could be further enhanced by their ability to adjust their filter apparatus to collect smaller particles of bacterial and detrital food (Maszczyk and Bartosiewicz 2012), and indeed smaller-bodied *Ceriodaphnia quadrangula* outcompeted larger *D. pulex* in the presence of mixed algal and bacterial food, but not in the presence of algal food alone (Iwabuchi and Urabe 2010).

Effect of Temperature on Predation and Interspecific Competition of Zooplankton

Generally, across species from a variety of crustacean groups mean body size is positively correlated with latitude, and so, inversely correlated with temperature (Table 2.3). This pattern is well established in a variety of other ectotherms, and is usually assigned to Bergmann's rule that body mass increases with higher latitude, originally only applied to the interspecific variability in body size of endotherms (Watt et al. 2010). The high-latitude endpoint of body sizes is called polar gigantism (Chapelle and Peck 1999) and is observed in many distinct taxa, especially in marine crustaceans, including amphipods, copepods, isopods, and decapods. According to a review by Moran and Woods (2012) and Sainte-Marie (1991), possibly the clearest example of a cline exhibiting a latitudinal pattern and ending with giant species are amphipods of the suborder Gammaridae, which are diverse and have numerous representatives in both polar and nonpolar regions. Moreover, San Martin et al. (2006) revealed that larger copepods dominate in cooler, higher latitudinal regions (see Table 2.3 for other examples of latitudinal clines).

This pattern is also well documented in freshwater communities, especially in zooplankton (Fig. 2.5). In tropical and subtropical lakes (with subsurface water temperatures in the range of 24°C – 35°C), small-bodied zooplankton species dominate, whereas temperate and subpolar lakes (with subsurface water temperatures in the range of 8°C –24°C) are inhabited by relatively large-bodied species (Gillooly and Dodson 2000). The trend toward a large body size across species in a variety of crustacean groups at cooler ambient temperatures is also apparent in other geographical clines (i.e., along altitude and depth), either in marine or freshwater environments (Table 2.3). Moreover, the negative correlation between temperature and proportion of large-bodied species has long been recognized in the seasonally variable size structure of the zooplankton community (Table 2.3), where large-bodied species tend to decline during summer (Sommer et al. 1986). Therefore, it seems very likely that temperature rather than any other environmental factor (e.g., primary production, nutrient regeneration, ultraviolet radiation, or photoperiod) is responsible for these temporal and spatial patterns in crustacean body size, either directly or indirectly through effects on other environmental factors.

Elevated temperatures could cause a decrease in mean body size at the community level, not only due to the decreased mean body size of individuals in each population, but also because of the increased proportion of small-bodied species (Daufresne et al. 2009), although most studies have not disentangled these two sources of variability. It can be assumed that, in the case of zooplankton (and likely also in all other animal communities), two general and nonexclusive hypotheses may explain the greater proportion of small-bodied species at elevated temperature. The first hypothesis is based on the assumption that higher water temperature may increase predation mortality that selectively eliminates larger species. The second hypothesis maintains that warming increases the competitive abilities of smaller species compared to larger ones (causing elimination or decrease in abundance of the latter) due to their physiological adaptations to higher temperatures. Increased

Table 2.3. Summary of published correlational studies on the effects of spatial and temporal variation of temperature on body size (body length—BL, body weight—BW or valve length—VL) at the community level across different crustacean taxa (numbers in parentheses indicate notes at the bottom of the table).

Order or subclass	No. of species investigated	Source of variation	Environment	Response of body size to increased temperature	Reference
Amphipoda	214	Latitude and bathymetry	Freshwater, brackish, marine	Reduction (BL)	Sainte-Marie 1991
Amphipoda	993	Latitude and bathymetry	Freshwater, marine, terrestrial	Reduction (BL)	Poulin and Hamilton 1995
Copepoda	1,038	Latitude	Freshwater, marine	Reduction (BL)	Poulin 1995a
Calanoida	18	Latitude	Lakes, reservoirs	No difference (BW)	Havens et al. 2015 (1)
Cyclopoida	27	Latitude	Lakes, reservoirs	Reduction (BW)	Havens et al. 2015
Cladocera	85	Latitude	Lakes, reservoirs	Reduction (BW)	Havens et al. 2015
Cladocera	49	Latitude, Seasonality	Lakes, ponds, reservoirs	Reduction (BL)	Gillooly and Dodson 2000
Cladocera	18	Altitude	Lakes	Reduction (BL)	Green 1995
Isopoda	746	Latitude and bathymetry	Freshwater, terrestrial, marine	Reduction (BL)	Poulin 1995b
Isopoda	41	Latitude	Marine	Reduction (BL)	Luxmoore 1982
Mysidacea	683	Bathymetry	Marine	Reduction (BL)	Mauchline 1972 (2)
Euphausiacea	85	Bathymetry	Marine	Reduction (BL)	Mauchline 1972 (2)
Ostracoda	12	Bathymetry	Marine	Reduction (VL)	Hunt and Roy 2006 (3)
Ostracoda	9	Long-term	Marine	Reduction (VL)	Hunt and Roy 2006
Copepoda	52	Long-term	North Pacific	Variable (BL)	Chiba et al. 2015
Cladocera	2	Long-term	Lake Washington	Reduction (BL)	Winder and Schindler 2004

(1) Havens et al. 2015 also contains references for marine copepods.
(2) Mauchline (1972) presents many other examples of the relationship with depth and latitude.
(3) See Hunt and Roy's (2006) Table 3 in the supporting text published on the PNAS website for references (www.PNAS.org).

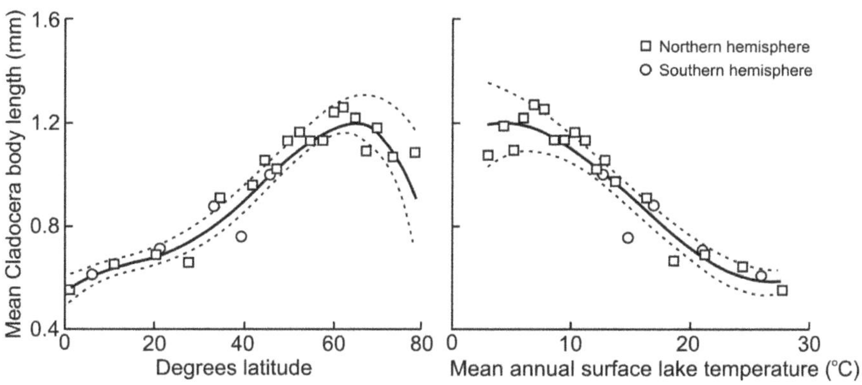

Fig. 2.5.
Mean body size of cladocerans along latitude (left panel) and corresponding mean annual surface lake temperature (right panel) in the northern (North and Central America), and southern (South, Central America and the Caribbean) hemisphere. Each point represents the mean total body length of all limnetic herbivorous cladocerans reported for water bodies occurring with a given latitudinal interval. Redrawn from Gillooly and Dodson (2000).

competitive abilities at elevated temperatures of a species could be the result of any combination of various mechanisms, including direct (affecting physiological processes) or indirect (affecting oxygen availability, food quantity, or food quality) effects of temperature that favor small-bodied species.

Determining the role of elevated temperatures in shaping the species composition of different communities is a complex issue because different combinations of direct and indirect effects of temperature might affect communities along different temporal (seasonal and long-term) and spatial (local, bathymetric, altitudinal, and latitudinal) scales. A correlational approach frequently used to explore this complex issue may have ignored the fact that combinations of extrinsic factors other than the chosen combination could possibly provide a better explanation of the observed trends (Gillooly and Dodson 2000). However, it is generally accepted that seasonal changes in the composition of zooplankton species in freshwater communities are attributed to the combined effects of temperature and biotic factors, mainly food availability or the impact of predation, or both. For instance, midsummer declines in daphnid abundances and shifts in the zooplankton size spectra from large *Daphnia* species to smaller genera, including other crustaceans, such as bosminids, are attributed to the appearance of newly hatched planktivorous fish, and changes in abundance and taxonomic composition of phytoplankton, dominated during the summer by cyanobacteria and algae that are poor-quality food (DeMott 1989, Müller-Navarra et al. 2000). A few studies suggested that oxygen deficiencies during warm seasons also play an important role in the dominance of smaller-bodied species (Hu and Tessier 1995).

Although studies about the effect of climate change on seasonal and long-term changes in mean body size of marine zooplankton communities are rather scarce and inconclusive (Chiba et al. 2015), long-term studies on freshwater ecosystems indicated that the combined effects of temperature and biotic factors on seasonal changes in zooplankton communities can intensify during warm years through increased seasonal growth patterns and activity at different trophic levels. For instance, earlier ice breakup and warmer water temperatures caused a prolonged duration of midsummer decline and a decreased height of spring peak densities of large-bodied zooplankton. This decline was due either to rapid overexploitation of food resources (Benndorf et al. 2001) or increased fish foraging activity (a relative increase of newly hatched fish during early summer) and a demographic shift in zooplankton communities (Benndorf et al. 2001, Iglesias et al. 2011, Jeppesen et al. 2010, Vadadi-Fülöp et al. 2012). Moreover, warmer winters have increased the competitive dominance of small-bodied species due to increased harmful cyanobacterial blooms in the summer, which hamper food particle filtration rates of large-bodied species (Adrian and Deneke 1996). Also, elevated temperatures during winter and early spring have generated shifts in zooplankton composition due to a variation in the magnitude of phenological responses between phyto- and zooplankton. For instance, Winder and Schindler (2004) have found that during warmer springs in Lake Washington, the phytoplankton bloom occurred sooner but was not followed by increased densities of large-bodied *D. pulicaria,* whose populations were declining (in relation to smaller-bodied *D. thorata*) because their emergence no longer corresponded with high algal abundance.

Although the negative correlation between temperature and mean body size of zooplankton communities has been widely recognized within spatial clines (Table 2.3), a full explanation for this phenomenon is still lacking. Long-term studies on the effects of global warming revealed another yet unexplained phenomenon: the shifts from cold-adapted to warm-adapted calanoid species in the North Sea (Beaugrand et al. 2002) and from larger to smaller barnacles on the rocky shores of the Atlantic along latitudinal gradients with progressive warming (Hawkins et al. 2008). These shifts may be due to the indirect effect of temperature in shaping the body size of communities (by affecting competitive abilities), in which thermal tolerance was observed to decrease in both upper and lower thermal limits with latitude, at least in marine ectotherms (Sunday et al. 2011).

Tolerance of low oxygen levels may also play an important role in favoring smaller species at lower latitudes. For instance, Chapelle and Peck (2004) analyzed 2,092 species of benthic gammaridean amphipods and found a positive correlation between oxygen concentration and maximum body size along a latitudinal cline in freshwater and marine habitats. Moreover, the dominance of small-bodied zooplankton species at low latitudes is usually attributed to the more frequent and more abundant cyanobacterial blooms in tropical and subtropical waters. In a recent study, however, Sikora and Dawidowicz (2014) found that neither the direct effects of elevated temperature nor the combined effect of temperature and the presence of filamentous cyanobacteria caused the reversal in competitive superiority between larger- and smaller-bodied *Daphnia* species. This finding suggests that other factors (e.g., increased fish predation; Iglesias et al. 2011) complementarily or exclusively must be responsible for the elimination of large zooplankton taxa at elevated temperatures in freshwater environments at spatial (also in temporal) scales.

FUTURE DIRECTIONS

Although there is a large body of literature investigating the effects of global climate change on the phenotype of individuals, we still lack a thorough understanding of the complex interplay between the factors and constraints affecting body size of an individual crustacean. Surprisingly little is known about the genetic and cellular background of the observed changes in body size and life histories, either at the individual or population level. Recent progress in ecological genomics with *Daphnia* offers a promising insight into the genetic and physiological processes contributing to the evolution of body size. Still, unraveling the links between genes, cell development, and the somatic growth of crustaceans is in its early stages.

It remains largely unexplored if and how cell size may contribute to the variation of individual body size in heterogeneous environments. One question that requires further study is the extent to which changes in body size observed in natural populations reflect the outcome of selection versus phenotypic plasticity. This issue may have important consequences, particularly for proper ecosystem management to maintain genetic diversity in a changing world. Another important challenges lies in disentangling the effects of particular environmental factors on body size, as debate on the relative importance of temperature, food availability, and predation seems to be far from reaching a widely accepted consensus. Future research may involve the evolution of crustacean body size in the context of the effects of food limitation and selective predation using paleolimnological and paleoecological data reexamined with barcoding-based taxonomy. Such a taxonomic approach could soon be developed and integrated with the prospect of new discoveries on the genomic response of crustaceans, such as *Daphnia*, to different ecological landscapes and water temperature increases resulting from global warming.

CONCLUSIONS

With the advent of global warming, current literature focuses on temperature as a factor moderating the impact of other factors affecting the body size of crustaceans. The review of recent publications shows an apparent decrease of individual body size at stage (the pattern according to TSR) of freshwater and marine crustaceans (in 81% of studies; Table 2.1), as well as the mean body size of their populations (in 79% of studies; Table 2.2) and communities (89% of studies; Table 2.3) at an elevated temperature. It is less clear whether these patterns are a result of the direct or indirect effects of temperature, and which mechanisms are responsible for them. Most studies have been designed to test the effect of temperature alone and have ignored the fact that the interaction of temperature with other extrinsic factors (either complementary or exclusive) could possibly provide a better explanation of the observed trends.

Despite this, the indirect effects responsible for the body size patterns at each of the ecological levels have been identified as prevailing in several cases. At the individual level, a few studies identified the interactive effect of temperature either with oxygen demands or food requirements, but not with predation risk by visual predators. Although correlational studies at the population level have so far identified an interactive effect of temperature with predation risk, studies at the community level identified an interactive effect of temperature with either oxygen demands or predation risk. In nature, several environmental factors act simultaneously on the body size of crustaceans; thus there are many possible interactions between particular factors, and more efforts should be made to disentangle these interactive effects. The complex interplay between internal constraints and external factors affecting the body size of crustaceans is still far from being fully understood.

ACKNOWLEDGMENTS

We thank E. Babkiewicz, M. Czarnocka-Cieciura, Z.M. Gliwicz, and J. Kozlowski for their insightful comments that helped to improve this manuscript. PM was supported by the Polish National Science Centre (grants no. 2014/13/N/NZ8/02462, 2016/23/D/NZ8/03532, 2014/15/B/NZ8/00245) and the Ministry of Science and Higher Education (grant no. IP2015 048974). TB was supported by the Polish National Science Centre (grant no. 2016/21/D/NZ8/01298).

REFERENCES

Acharya K., M. Kyle, and J.J. Elser. 2004. Biological stoichiometry of *Daphnia* growth: an ecophysiological test of the growth rate hypothesis. Limnology and Oceanography 49:656–665.

Adrian, R., and R. Deneke. 1996. Possible impact of mild winters on zooplankton succession in eutrophic lakes of the Atlantic European area. Freshwater Biology 36:757–770.

Aguilar-Alberola, J.A., and F. Mesquita-Joanes. 2014. Breaking the temperature-size rule: thermal effects on growth, development and fecundity of a crustacean from temporary waters. Journal of Thermal Biology 42:15–24.

Anger, K. 2001. The biology of decapod crustacean larvae. Crustacean issues. Vol. 14. A.A. Balkema, Lisse, The Netherlands.

Angilletta, M.J., Jr., and A.E. Dunham. 2003. The temperature-size rule in ectotherms: simple evolutionary explanations may not be general. The American Naturalist 162:332–342.

Angilletta, M.J., Jr., T.D. Steury, and M.W. Sears. 2004. Temperature, growth rate, and body size in ectotherms: fitting pieces of a life-history puzzle. Integrative and Comparative Biology 44:498–509.

Anufriieva, E.V., and N.V. Shadrin. 2014. Factors determining the average body size of geographically separated *Arctodiaptomus salinus* (Daday, 1885) populations. Zoological Research 35:132–141.

Atkinson, D. 1994. Temperature and organism size—a biological law for ectotherms? Advances in Ecological Research 3:1–58.

Atkinson, D., S.A. Morley, and R.N. Hughes. 2006. From cells to colonies: at what levels of body organization does the 'temperature-size rule' apply? Evolution and Development 8:202–214.

Beaton, M.J., and P.D.N. Hebert. 1988. Geographical parthenogenesis and polyploidy in *Daphnia pulex*. The American Naturalist 132:157–163.

Beaugrand, G., P.C. Reid, F. Ibanez, J.A. Lindley, and M. Edwards. 2002. Reorganization of North Atlantic marine copepod biodiversity and climate. Science 296:1692–1694.

Bednarska, A., J. Łoś, and P. Dawidowicz. 2011. Temperature-dependent effect of filamentous cyanobacteria on *Daphnia magna* life history traits. Journal of Limnology 70:353–358.

Benndorf, J., B. Wissel, A.F. Sell, U. Hornig, P. Ritter, and W. Böing. 2000. Food web manipulation by extreme enhancement of piscivory: an invertebrate predator compensates for the effects of planktivorous fish on a plankton community. Limnologica-Ecology and Management of Inland Waters 30: 235–245.

Benndorf, J., J. Kranich, T. Mehner, and A. Wagner. 2001. Temperature impact on the summer decline of *Daphnia galeata*: an analysis of long-term data from the biomanipulated Bautzen Reservoir (Germany). Freshwater Biology 46:199–211.

Bernot, R.J., W.K. Dodds, M.C. Quist, and C.S. Guy. 2006. Temperature and kairomone induced life history plasticity in coexisting *Daphnia*. Aquatic Ecology 40:361–372.

Boersma, M., L. de Meester, and P. Spaak. 1999. Environmental stress and local adaptation in *Daphnia magna*. Limnology and Oceanography 44:393–402.

Bonvillain, C.P., D.A. Rutherford, and W.E. Kelso. 2015. Effects of environmental hypoxia on population characteristics of red swamp crayfish *Procambarus clarkii* in the Atchafalaya River Basin, Louisiana. Hydrobiologia 743:309–319.

Bottrell, H.H., A. Duncan, Z.M. Gliwicz, E. Grygierek, A. Herzig, A. Hillbricht-Ilkowska, H. Kurasawa, P. Larsson, and T. Weglenska. 1976. A review of some problems in zooplankton production studies. Norwegian Journal of Zoology 24:419–456.

Brooks, J.L., and S.I. Dodson. 1965. Predation, body size, and composition of plankton. Science 150:28–35.

Brzeziński, T., and E. von Elert. 2015. Predator evasion in zooplankton is suppressed by polyunsaturated fatty acid limitation. Oecologia 179:687–697.

Brzeziński, T., P. Dawidowicz, and E. von Elert. 2010. The role of food quality in clonal succession in *Daphnia*: an experimental test. Oecologia 164:379–388.

Brian, J.V., T. Teresa Fernandes, R.J. Ladle, and P.A. Todd. 2006. Patterns of morphological and genetic variability in UK populations of the shore crab, *Carcinus maenas* Linnaeus, 1758 (Crustacea: Decapoda: Brachyura). Ecology 329:47–54.

Brucet, S., D. Boix, X.D. Quintana, E. Jensen, L.W. Nathansen, C. Trochine, M. Meerhoff, S. Gascón, and E. Jeppesen. 2010. Factors influencing zooplankton size structure at contrasting temperatures in coastal shallow lakes: implications for effects of climate change. Limnology and Oceanography 55:1697–1711.

Campbell, R.G., M.M. Wagner, G.J. Teegarden, C.A. Boudreau, and E.G. Durbin. 2001. Growth and development rates of the copepod *Calanus finmarchicus* reared in the laboratory. Marine Ecology Progress Series 221:161–183.

Carvalho, G.R. 1987. The clonal ecology of *Daphnia magna* (Crustacea: Cladocera) II. Thermal differentiation among seasonal clones. Journal of Animal Ecology 56:469–478.

Carvalho, G.R., and D.J. Crisp. 1987. The clonal ecology of *Daphnia magna* (Crustacea: Cladocera) I. Temporal changes in clonal structure of a natural population. Journal of Animal Ecology 56:453–468.

Chapelle, G., and L.S. Peck. 1999. Polar gigantism dictated by oxygen availability. Nature 399:114–115.

Chapelle, G., and L.S. Peck. 2004. Amphipod crustacean size spectra: new insights in the relationship between size and oxygen. Oikos 106:167–175.

Chiba, S., S.D. Batten, T. Yoshiki, Y. Sasaki, K. Sasaoka, H. Sugisaki, and T. Ichikawa. 2015. Temperature and zooplankton size structure: climate control and basin-scale comparison in the North Pacific. Ecology and Evolution 5:968–978.

Chopelet, J., P.U. Blier, and F. Dufresne. 2008. Plasticity of growth rate and metabolism in *Daphnia magna* populations from different thermal habitats. Journal of Experimental Zoology 309:553–562.

Correa, C., J.A. Baeza, E. Dupré, I.A. Hinojosa, and M. Thiel. 2000. Mating behavior and fertilization success of three ontogenetic stages of male rock shrimp *Rhynchocinetes typus* (Decapoda: Caridea). Journal of Crustacean Biology 20:628–640.

Czarnoleski, M., J. Ejsmont-Karabin, M.J. Angilletta, Jr., and J. Kozlowski. 2015. Colder rotifers grow larger but only in oxygenated waters. Ecosphere 6:1–5.

Daufresne, M., K. Lengfellner, and U. Sommer. 2009. Global warming benefits the small in aquatic ecosystems. Proceedings of the National Academy of Sciences of the USA 106:12788–12793.

DeMott, W.R. 1989. The role of competition in zooplankton succession—responses to food quality. Pages 195–252 *in* U. Sommer, editor. Plankton ecology: succession in plankton communities. Springer-Verlag, Berlin, Germany.

DeMott, W.T., R.D. Gulati, and K. Siewertsen. 1998. Effects of phosphorus-deficient diets on the carbon and phosphorus balance of *Daphnia magna*. Limnology and Oceanography 43:1147–1161.

DeMott, W.R., R.D. Gulati, and E. van Donk. 2001. *Daphnia* food limitation in three hypereutrophic Dutch lakes: evidence for exclusion of large-bodied species by interfering filaments of cyanobacteria. Limnology and Oceanography 46:2054–2060.

DeMott, W.R., E.N. McKinney, and A.J. Tessier. 2010. Ontogeny of digestion in *Daphnia*: implications for the effectiveness of algal defenses. Ecology 91:540–548.

Doksæter, A., and J. Vijverberg. 2001. The effects of food and temperature regimes on life-history responses to fish kairomones in *Daphnia hyalina* × *galeata*. Hydrobiologia 442:207–214.

Elser, J.J., D.R. Dobberfuhl, N.A. Mackay, and J.H. Schampel. 1996. Organism size, life history, and N:P stoichiometry. Toward a unified view of cellular and ecosystem processes. Bioscience 46:674–684.

Forster, J., and A.G. Hirst. 2012. The temperature-size rule emerges from ontogenetic differences between growth and development rates. Functional Ecology 26:483–492.

Forster, J., A.G. Hirst, and D. Atkinson. 2012. Warming-induced reductions in body size are greater in aquatic than terrestrial species. Proceedings of the National Academy of Sciences of the USA 109:19310–19314.

Frost, P.C., D Ebert, and V.H. Smith. 2008. Bacterial infection changes the elemental composition of *Daphnia magna*. Journal of Animal Ecology 77:1265–1272.

Geerts, A.N., J. Vanoverbeke, B. Vanschoenwinkel, W. van Doorslaer, H. Feuchtmayr, D. Atkinson, B. Moss, T.A. Davidson, C.D. Sayer, and L. de Meester. 2015. Rapid evolution of thermal tolerance in the water flea *Daphnia*. Nature Climate Change 5:665–668.

Gillooly, J.F., and S.I. Dodson. 2000. Latitudinal patterns in the size distribution and seasonal dynamics of Cladocera. Limnology and Oceanography 45:22–30.

Glazier, D.S. 1999. Variation in offspring investment within and among populations of *Gammarus minus* Say (*Crustacea: Amphipoda*) in ten mid-Appalachian springs (USA). Archiv fuer Hydrobiologie 146:257–283.

Gliwicz, Z.M. 1990. Food thresholds and body size in cladocerans. Nature 343:638–640.

Gliwicz, Z.M. 2003. Between hazards of starvation and risk of predation: the ecology of offshore animals. International Ecology Institute, Oldendorf/Luhe, Germany.

Gliwicz, Z.M., and W. Lampert. 1990. Food thresholds in *Daphnia* species in the absence and presence of blue-green filaments. Ecology 71:691–702.

Gliwicz, Z.M., and P. Maszczyk. 2007. *Daphnia* growth is hindered by chemical information on predation risk at high but not at low food levels. Oecologia 150:706–715.

Gliwicz, Z.M., and P. Maszczyk. 2016. Heterogeneity in prey distribution allows for higher food intake in planktivorous fish, particularly when hot. Oecologia 180:383–399.

Gliwicz, Z.M., P. Maszczyk, and W. Uszko. 2012. Enhanced growth at low population density in *Daphnia*: the absence of crowding effects or relief from visual predation? Freshwater Biology 57:1166–1179.
Green, J. 1995. Altitudinal distribution of tropical planktonic Cladocera. Hydrobiologia 307:75–84.
Gregory, T.R., and P.D.N. Hebert. 1999. The modulation of DNA content: proximate causes and ultimate consequences. Genome Research 9:317–324.
Hairston, N.G., Jr., C.L. Holtmeier, W. Lampert, L.J. Weider, D.M. Post, J.M. Fischer, C.E. Cáceres, J.A. Fox, and U. Gaedke. 2001. Natural selection for grazer resistance to toxic cyanobacteria: evolution of phenotypic plasticity? Evolution 55:2203–2214.
Hart, R.C., and E.A. Bychek. 2011. Body size in freshwater planktonic crustaceans: an overview of extrinsic determinants and modifying influences of biotic interactions. Hydrobiologia 668:61–108.
Hartnoll, R.G. 1982. Growth. Pages 111–196 *in* D.E. Bliss, and L.G. Abele, editors. The biology of Crustacea, 2. Embryology, morphology and genetics. Academic Press, New York.
Hartnoll, R.G. 2001. Growth in Crustacea—twenty years on. Hydrobiologia 449:111–122.
Hartnoll, R.G. 2015. Postlarval life histories of Brachyura. Pages 375–416 *in* P. Castro, P. Davie, D. Guinot, F.R. Schram, and J.C. von Vaupel Klein, editors. Decapoda: Brachyura, Treatise on zoology—anatomy, taxonomy, biology. Vol. 9C-I. Brill, Leiden, The Netherlands.
Hawkins, S.J., P.J. Moore, M.T. Burrows, E. Poloczanska, N. Mieszkowska, R.J.H. Herbert, S.R. Jenkins, R.C. Thompson, M.J. Genner, and A.J. Southward. 2008. Complex interactions in a rapidly changing world: responses of rocky shore communities to recent climate change. Climate Research 37:123–133.
Havens, K.E., R.M. Pinto-Coelho, M. Beklioğlu, K.S. Christoffersen, E. Jeppesen, T.L. Lauridsen, A. Mazumder, G. Méthot, B.P. Alloul, U.N. Tavşanoğlu, Ş. Erdoğan, and J. Vijverberg. 2015. Temperature effects on body size of freshwater crustacean zooplankton from Greenland to the tropics. Hydrobiologia 743:27–35.
Hazel, J.R. 1995. Thermal adaptation in biological membranes: is homeoviscous adaptation the explanation? Annual Review of Physiology 57:19–42.
Hazlett, B.A. 1999. Responses to multiple chemical cues by the crayfish *Orconectes virilis*. Behaviour 136:161–177.
Hazlett, B.A. 2011. Chemical cues and reducing the risk of predation. Pages 355–370 *in* T. Breithaupt, and M. Thiel. Chemical Communication in Crustaceans. Springer, New York.
Hoefnagel, K.N., and W.C.E.P. Verberk. 2015. Is the temperature-size rule mediated by oxygen in aquatic ectotherms? Journal of Thermal Biology 54:56–65.
Hosono, T. 2011. Effect of temperature on growth and maturation pattern of *Caprella mutica* (*Crustacea, Amphipoda*): Does the temperature-size rule function in caprellids? Marine Biology 158:363–370.
Hu, S.S., and A.J. Tessier. 1995. Seasonal succession and the strength of intra- and interspecific competition in a *Daphnia* assemblage. Ecology 76:2278–2294.
Hunt, G., and K. Roy. 2006. Climate change, body size evolution, and Cope's Rule in deep-sea ostracodes. Proceedings of the National Academy of Sciences of the USA 103:1347–1352.
Iglesias, C., N. Mazzeo, M. Meerhoff, G. Lacerot, J.M. Clemente, F. Scasso, C. Kruk, G. Goyenola, J. García-Alonso, S.L. Amsinck, J.C. Paggi, S.J. de Paggi, and E. Jeppesen. 2011. High predation is of key importance for dominance of small-bodied zooplankton in warm shallow lakes: evidence from lakes, fish exclosures and surface sediments. Hydrobiologia 667:133–147.
Iwabuchi, T., and J. Urabe. 2010. Phosphorus acquisition and competitive abilities of two herbivorous zooplankton, *Daphnia pulex* and *Ceriodaphnia quadrangula*. Ecological Research 25:619–627.
Iwabuchi, T., and J. Urabe. 2012. Food quality and food threshold: implications of food stoichiometry to competitive ability of herbivore plankton. Ecosphere 3:51.
Jalal, M., M.W. Wojewodzic, C.M.M. Laane, and D.O. Hessen. 2013. Larger *Daphnia* at lower temperature: a role for cell size and genome configuration? Genome 56:511–519.
Jeppesen, E., M. Meerhoff, K. Holmgren, I. González-Bergonzoni, F. Teixeira-de Mello, S.A.J. Declerck, L. de Meester, M. Søndergaard, T.L. Lauridsen, R. Bjerring, J.M. Conde-Porcuna, N. Mazzeo, C. Inglesias, M. Reizenstein, H.J. Malmquist, Z. Liu, D. Balayla, and X. Lazzaro. 2010. Impacts of climate warming on lake fish community structure and potential effects on ecosystem function. Hydrobiologia 646:73–90.
Jónasdóttir, S.H., N.H Trung, F. Hansen, and S. Gärtner. 2005. Egg production and hatching success in the calanoid copepods *Calanus helgolandicus* and *Calanus finmarchicus* in the North Sea from March to September 2001. Journal of Plankton Research 27:1239–1259.

Jose, C., and F. Dufresne. 2010. Differential survival among genotypes of *Daphnia pulex* differing in reproductive mode, ploidy level, and geographic origin. Evolutionary Ecology 24:413–421.

Klok, C.J., and J.F. Harrison. 2013. The temperature size rule in arthropods: independent of macro-environmental variables but size dependent. Integrative and Comparative Biology 53:557–570.

Kosten, S., V.L.M. Huszar, E. Bécares, L.S. Costa, E. van Donk, L.-A. Hansson, E. Jeppesen, C. Kruk, G. Lacerot, N. Mazzeo, L. de Meester, B. Moss, M. Lürling, T. Nõges, S. Romo, and M. Scheffer. 2012. Warmer climates boost cyanobacterial dominance in shallow lakes. Global Change Biology 18:118–126.

Kozlowski, J., M. Czarnoleski, and M. Dańko. 2004. Can optimal resource allocation models explain why ectotherms grow larger in cold? Integrative and Comparative Biology 44:480–493.

Lampert, W. 1981. Inhibitory and toxic effects of blue-green algae on *Daphnia*. Internationale Revue der gesamten Hydrobiologie 66:285–298.

Lampert, W., and U. Sommer. 2007. Limnoecology: the ecology of lakes and streams. Oxford University Press, New York.

Lass, S., and P. Spaak. 2003a. Chemically induced antipredator defenses in plankton: a review. Hydrobiologia 491:221–236.

Lass, S., and P. Spaak. 2003b. Temperature effects on chemical signaling in a predator-prey system. Freshwater Biology 48:669–677.

Lonsdale, D.J., and J.S. Levinton. 1985. Latitudinal differentiation in copepod growth: an adaptation to temperature. Ecology 66:1397–1407.

Lürling, M., and E. van Donk. 1999. Grazer-induced colony formation in *Scenedesmus acutus* (Chlorophyceae): ecomorph expression at different temperatures. Journal of Phycology 35:1120–1126.

Luxmoore, R.A. 1982. Molting and growth in serolid isopods. Journal of Experimental Marine Biology and Ecology 56:63–85.

Manyak-Davis, A., T.M. Bell, and E.E. Sotka. 2013. The relative importance of predation risk and water temperature in maintaining Bergmann's rule in a marine ectotherm. The American Naturalist 182:347–358.

Martin-Creuzburg, D., E. Sperfeld, and A. Wacker. 2009. Colimitation of a freshwater herbivore by sterols and polyunsaturated fatty acids. Proceedings of the Royal Society B: Biological Sciences 276:1805–1814.

Martin-Creuzburg, D., A. Wacker, C. Ziese, and M.J. Kainz. 2012. Dietary lipid quality affects temperature-mediated reaction norms of a freshwater key herbivore. Oecologia 168:901–912.

Maszczyk, P., and M. Bartosiewicz. 2012. Threat of treat: the role of fish exudates in the growth and life history of *Daphnia*. Ecosphere 3:1–19.

Mauchline, J. 1972. The biology of bathypelagic organisms, especially Crustacea. Deep-Sea Research 19:753–780.

Mitchell, S.E., and W. Lampert. 2000. Temperature adaptation in a geographically widespread zooplankter, *Daphnia* magna. Evolutionary Biology 13:371–382.

Mohamedeen, H., and R.G. Hartnoll. 1989. Larval and postlarval growth of individually reared specimens of the common shore crab *Carcinus maenas* (L.). Journal of Experimental Marine Biology and Ecology 134:1–24.

Moran, A.L., and H.A. Woods. 2012. Why might they be giants? Towards an understanding of polar gigantism. The Journal of Experimental Biology 215:1995–2002.

Müller-Navarra, D.C., M.T. Brett, A.M. Liston, and C.R. Goldman. 2000. A highly unsaturated fatty acid predicts carbon transfer between primary producers and consumers. Nature 403:74–77.

Neiman, M., M.J. Beaton, D.O. Hessen, P.D. Jeyasingh, and L.J. Weider. 2017. Endopolyploidy as a potential driver of animal ecology and evolution. Biological Reviews 92:234–247.

Oh, C.W., and R.G. Hartnoll. 2004. Reproductive biology of the common shrimp *Crangon crangon* (Decapoda: *Crangonidae*) in the central Irish sea. Marine Biology 144:303–316.

Orsini, L., E. Decaestecker, L. de Meester, M.E. Pfrender, and J.K. Colbourne. 2011. Genomics in the ecological arena. Biology Letters 7:2–3.

Panov, V.E., and D.J. McQueen. 1998. Effects of temperature on individual growth rate and body size of a freshwater amphipod. Canadian Journal of Zoology 76:1107–1116.

Pauwels, K., R. Stoks, and L. de Meester. 2010. Enhanced antipredator defence in the presence of food stress in the water flea *Daphnia magna*. Functional Ecology 24:322–329.

Perrin, N. 1988. Why are offspring born larger when it is colder? Phenotypic plasticity for offspring size in the cladoceran *Simocephalus vetulus* (Muller). Functional Ecology 2:283–288.
Persson, L. 1986. Temperature-induced shift in foraging ability in two fish species, roach (*Rutilus rutilus*) and perch (*Perca fluviatilis*): implications for coexistence between poikilotherms. The Journal of Animal Ecology 55:829–839.
Peters, R.H. 1983. The ecological implications of body size. Cambridge University Press, Cambridge, England.
Pfrender, M.E., and M. Lynch. 2000. Quantitative genetic variation in *Daphnia*: temporal changes in genetic architecture. Evolution 54:1502–1509.
Plath, K., and M. Boersma. 2001. Mineral limitation of zooplankton: stoichiometric constraints and optimal foraging. Ecology 82:1260–1269.
Porter, K.G., and J.D. Orcutt. 1980. Nutritional adequacy, manageability, and toxicity as factors that determine the food quality of green and blue-green algae for *Daphnia*. Pages 268–281 *in* W.C. Kerfoot, editor. Evolution and ecology of zooplankton communities. University Press of New England, Hanover, New Hampshire.
Poulin, R. 1995a. Clutch size and egg size in free-living and parasitic copepods: a comparative analysis. Evolution 49:325–336.
Poulin, R. 1995b. Evolutionary influences on body size in free-living and parasitic isopods. Biological Journal of the Linnean Society 54:231–244.
Poulin, R., and W.J. Hamilton. 1995. Ecological determinants of body size and clutch size in amphipods: a comparative approach. Functional Ecology 9:364–370.
Sainte-Marie, B. 1991. A review of the reproductive bionomics of aquatic gammaridean amphipods: variation of life history traits with latitude, depth, salinity and superfamily. Hydrobiologia 223:189–227.
San Martin, E., R.P. Harris, X. Irigoien. 2006. Latitudinal variation in plankton size spectra in the Atlantic. Deep-Sea Research Part II: Topical Studies in Oceanography 53:1560–1572.
Sarnelle, O., and A.E. Wilson. 2005. Local adaptation of *Daphnia pulicaria* to toxic cyanobacteria. Limnology and Oceanography 50:1565–1570.
Seidendorf, B., N.M. Meier, A. Petrusek, M. Boersma, B. Streit, and K. Schwenk. 2010. Sensitivity of *Daphnia* species to phosphorus-deficient diets. Oecologia 162:349–357.
Sikora, A., and P. Dawidowicz. 2014. Do the presence of filamentous cyanobacteria and an elevated temperature favor small-bodied *Daphnia* in interspecific competitive interactions? Fundamental and Applied Limnology 185:307–314.
Soares, M.C.S., M. Lürling, and V. Huszar. 2013. Growth and temperature-related phenotypic plasticity in the cyanobacterium *Cylindrospermopsis raciborskii*. Phycological Research 61:61–67.
Sommer, U., Z.M. Gliwicz, W. Lampert, and A. Duncan. 1986. The PEG-model of seasonal succession of planktonic events in fresh water. Archiv für Hydrobiologie 106:433–471.
Søreide, J.E., E. Leu, J. Berge, M. Graeve, and S. Falk-Petersen. 2010. Timing of blooms, algal food quality and *Calanus glacialis* reproduction and growth in a changing Arctic. Global Change Biology 16:3154–3163.
Sperfeld, E., and A. Wacker. 2011. Temperature- and cholesterol-induced changes in eicosapentaenoic acid limitation of *Daphnia magna* determined by a promising method to estimate growth saturation thresholds. Limnology and Oceanography 56:1273–1284.
Sterner, R.W., and D.O. Hessen. 1994. Algal nutrient limitation and the nutrition of aquatic herbivores. Annual Review of Ecology and Systematics 25:1–29.
Stibor, H., and W. Lampert. 2000. Components of additive variance in life-history traits of *Daphnia hyalina*: seasonal differences in the response to predator signals. Oikos 88:129–138.
Sunday, J.M., A.E. Bates, and N.K. Dulvy. 2011. Global analysis of thermal tolerance and latitude in ectotherms. Proceedings of the Royal Society B: Biological Sciences 278:1823–1830.
Tessier, A.J., M.A. Leibold, and J. Tsao. 2000. A fundamental trade-off in resource exploitation by *Daphnia* and consequences to plankton communities. Ecology 81:826–841.
Thompson, P.J. 1986. Heritability for body size in the isopod *Asellus aquaticus* (L.). Crustaceana 51:241–244.
Tilman, D. 1982. Resource competition and community structure. Princeton University Press, Princeton, New Jersey.
Tillmanns, A.R., A.E. Wilson, F.R. Pick, and O. Sarnelle. 2008. Meta-analysis of cyanobacterial effects on zooplankton population growth rate: species-specific responses. Fundamental and Applied Limnology 171:285–295.

Vadadi-Fülöp, C., C. Sipkay, G. Meszaros, and L. Hufnagel. 2012. Climate change and freshwater zooplankton: what does it boil down to? Aquatic Ecology 46:501–519.
von Bertalanffy, L. 1960. Principles and theory of growth. Pages 137–259 *in* W.W. Nowinski, editor. Fundamental aspects of normal and malignant growth. Elsevier, Amsterdam, The Netherlands.
von Elert, E. 2002. Determination of limiting polyunsaturated fatty acids in *Daphnia galeata* using a new method to enrich food algae with single fatty acids. Limnology and Oceanography 47:1764–1773.
Wagner, A., and J. Benndorf. 2007. Climate-driven warming during spring destabilizes a *Daphnia* population: a mechanistic food web approach. Oecologia 151:351–364.
Wahle, R.A., and M.J. Fogarty. 2006. Growth and development: understanding and modeling growth variability in lobsters. Pages 1–44 *in* B. Phillips, editor. Lobsters: biology, management, aquaculture and fisheries. Blackwell Publishing, Oxford, England.
Watt, C., S. Mitchell, and V. Salewski. 2010. Bergmann's rule; a concept cluster? Oikos 119:89–100.
Weetman, D., and D. Atkinson. 2004. Evaluation of alternative hypotheses to explain temperature-induced life history shifts in *Daphnia*. Journal of Plankton Research 26:107–116.
Weider, L. J., K.L. Glenn, M. Kyle, and J.J. Elser. 2004. Associations among ribosomal (r)DNA intergenic spacer length variation, growth rate and C:N:P stoichiometry in the genus *Daphnia*. Limnology and Oceanography 49:1417–1423.
Wickins, J.F., and D.O.C. Lee. 2002. Crustacean farming: ranching and culture. Blackwell Science, Oxford, England.
Wiggins, P.R., and P.B. Frappell. 2000. The influence of hemoglobin on behavioural thermoregulation and oxygen consumption in *Daphnia carinata*. Physiological and Biochemical Zoology 73:153–160.
Winder, M., and D.E. Schindler. 2004. Climate change uncouples trophic interactions in an aquatic ecosystem. Ecology 85:2100–2106.
Woods, H.A., W. Makino, J.B. Cotner, S.E. Hobbie, J.F. Harrison, K. Acharya, and J.J. Elser. 2003. Temperature and the chemical composition of poikilothermic organisms. Functional Ecology 17:237–245.
Wyatt, T.D. 2011. Pheromones and behavior. Pages 23–38 *in* T. Breithaupt, and M. Thiel, editors. Chemical communication in crustaceans. Springer, New York.
Wyngaard, G.A. 1986. Genetic differentiation of life history traits in populations of *Mesocyclops edax* (Crustacea: Copepoda). The Biological Bulletin 170:279–295.

3

CLUTCH MASS, OFFSPRING MASS, AND CLUTCH SIZE: BODY MASS SCALING AND TAXONOMIC AND ENVIRONMENTAL VARIATION

Douglas S. Glazier

Abstract

In this chapter, I show how clutch mass, offspring (egg) mass, and clutch size relate to body mass among species of branchiopod, maxillipod, and malacostracan crustaceans, as well as how these important life history traits vary among major taxa and environments independently of body size. Clutch mass relates strongly and nearly isometrically to body mass, probably because of physical volumetric constraints. By contrast, egg mass and clutch size relate more weakly and curvilinearly to body mass and vary in inverse proportion to one another, thus indicating a fundamental trade-off, which occurs within many crustacean taxa as well. In general, offspring (egg) size and number and their relationships to body mass appear to be more ecologically sensitive and evolutionarily malleable than clutch mass. The body mass scaling relationships of egg mass and clutch size show much more taxonomic and ecological variation (log-log scaling slopes varying from near 0 to almost 1 among major taxa) than do those for clutch mass, a pattern also observed in other animal taxa. The curvilinear body mass scaling relationships of egg mass and number also suggest a significant, size-related shift in how natural selection affects offspring versus maternal fitness. As body size increases, selection apparently predominantly favors increases in offspring size and fitness up to an asymptote, beyond which increases in offspring number and thus maternal fitness are preferentially favored. Crustaceans not only offer excellent opportunities for furthering our general understanding of life history evolution, but also their ecological and economic importance warrants further study of the various factors affecting their reproductive success.

Life Histories. Edited by Gary A. Wellborn and Martin Thiel.

INTRODUCTION

The Crustacea are a useful taxonomic group for studying life history evolution because these animals exhibit extraordinary diversity in body sizes, lifestyles, behaviors, and habitats. Life histories involve the timing and relative amount of allocation of resources to growth, development, reproduction, and survival during an organism's lifetime. Life history theory is devoted to explaining the enormous diversity of life histories (Sibly and Calow 1986, Stearns 1992, Roff 2002). Understanding life history variation is fundamental because life history traits relate intimately to evolutionary fitness with important consequences for population dynamics and species survival. This chapter focuses on three important, data-rich life history traits: clutch mass, offspring (egg) mass, and clutch size (offspring number per breeding event) in various branchiopod, maxillipod, and malacostracan crustaceans. Clutch mass and clutch size represent the amount of resources allocated to total offspring production, and thus are critical for maternal fitness. Offspring mass represents the amount of resources allocated to individual offspring, which is critical for both offspring and maternal fitness. As will be seen, key factors related to variation in these life history traits are body mass, taxonomic affiliation, and environment.

BODY MASS SCALING

General Patterns

Life history traits typically relate strongly to body size in animals and plants (Peters 1983, Reiss 1989, Visman et al. 1996, Ramirez Llodra 2002, Hendriks and Mulder 2008). Across species of various crustacean taxa, both egg size and number usually correlate positively with female size (Lynch 1980, Kiørboe and Sabatini 1995, Johnson et al. 2001, Achouri et al. 2008, Neuheimer et al. 2015), but there are many exceptions (Corey 1981, Fukui 1988, Ross and Quetin 2000, Johnson et al. 2001). Therefore, body size effects are usually accounted for first, before examining the effects of other factors.

I compared clutch mass, egg mass, and clutch size with maternal body mass by using an original survey of 313 species (Table 3.1). Maternal body mass ranged more than 9 orders of magnitude (~2 μg in the copepod *Oithona davisae* to ~8 kg in the red king crab *Paralithodes camtschaticus*, a decapod), which is greater than the body mass range of mammals from tiny bats and shrews to huge whales (~8 orders of magnitude). Egg mass ranged more than 5 orders of magnitude (~0.1 μg to > 10 mg), whereas egg number per clutch ranged more than 6 orders of magnitude (~2 to > 8,000,000).

All three life history traits correlate significantly with maternal body mass (Fig. 3.1). Clutch mass increases strongly, linearly, and nearly isometrically with increasing body mass (scaling slope ≈ 1) (Fig. 3.1A), whereas egg mass and clutch size increase relatively weakly, allometrically (scaling slopes < 1) and curvilinearly in log-log space (Fig. 3.1B,C). The relationship between egg mass and body mass is concave downward, whereas the relationship between egg number and body mass is concave upward. These curvilinear relationships are almost mirror images of each other, suggesting that egg mass and number do not vary independently of one another (also see below). At the small end of the body mass range, increases in body mass tend to be associated with relatively large increases in egg mass, but little or no change in egg number, whereas the opposite is observed at the large end of the body mass range. The instantaneous scaling slope for egg mass approaches 1 at 10–100 μg body mass, but is near or less than 0 at 0.1–1 kg body mass. By contrast, the instantaneous slope for clutch size is near or less than 0 at 10–100 μg body mass, but near or greater than 1 at 0.1–1 kg body

Table 3.1. Sources of data on wet clutch mass, wet egg mass and clutch size given in Fig. 3.1, Fig. 3.3, Fig. 3.4, and Fig. 3.5. Dry mass was converted to wet mass by dividing by 0.2 (Hendriks and Mulder 2008). For copepods, carbon mass was converted to dry mass by dividing by 0.4 (Kiørboe and Sabatini 1995), and to wet mass by dividing by 0.1 (also see Andersen et al. 2016). Mass data allow comparisons among crustaceans with diverse body shapes.

Taxon	Number of species	Sources
Branchiopoda:		
Anostraca	1	Hildrew 1985
Cladocera	18	Green 1956, Ivanova and Vassilenko 1987, Tessier and Consolatti 1989, Glazier 1992, Tollrian 1995, Dudycha and Lynch 2005
Maxillipoda:		
Cirripedia	7	Barnes and Barnes 1968
Copepoda	76	Hutchinson 1967, Sazhina 1985, Kiørboe and Sabatini 1995, Lee et al. 2003, Kosobokova et al. 2007, Forster et al. 2011, Zamora-Terol and Saiz 2013
Malacostraca:		
Decapoda	122	Shakuntala 1977, Hines 1982, 1988, 1991, Ivanova and Vassilenko 1987, Paul and Fuji 1989, Thessalou-Legaki 1992, Clarke 1993, Thessalou-Legaki and Kiortsis 1997, Anger and Moreira 1998, Austin 1998, Diesel et al. 2000, Pinheiro and Terceiro 2000, Ramirez Llodra et al. 2000, Kotb and Hartnoll 2002, Henmi 2003, Arcos et al. 2004, Brante et al. 2004, Harlıoğlu and Barim 2004, Litulo 2004, Müller et al. 2004, Brillon et al. 2005, Maguire et al. 2005, Leme 2006, Hartnoll et al. 2007, 2010, Hernaez et al. 2008, Amin et al. 2009, Béguer et al. 2010, Briones-Fourzán et al. 2010, Janas and Mańkucka 2010, Johnson et al. 2010, Powell and Watts 2010, Echeverría-Sáenz and Wehrtmann 2011, Veríssimo et al. 2011, Pérez-González et al. 2012, Tropea et al. 2012, Meireles et al. 2013, Przemysław and Marcello 2013, Rashid et al. 2013, López-Sánchez and Quintero-Torres 2015, Swetha et al. 2015, Zairion et al. 2015
Euphausiacea	1	Iguchi and Ikeda 1994
Peracarida:		
Amphipoda	35	Dagg 1976, Ivanova and Vassilenko 1987, Glazier 1999, Poltermann et al. 2000, Węsławski and Legeżyńska 2002, Žganec et al. 2011
Cumacea	18	Ivanova and Vassilenko 1987
Isopoda	34	Phillipson and Watson 1965, Saito 1969, Watanabe 1980, Ivanova and Vassilenko 1987, Willows 1987, Briones-Fourzán and Lozano-Alvarez 1991, Ma et al. 1991, Hornung and Warburg 1993, 1994, Warburg et al. 1993, Tsai and Chen 1997, Johnson et al. 2001, Tsai and Dai 2001, Glazier et al. 2003, Lardies et al. 2004a, b, Quadros et al. 2008
Mysida	1	Clutter and Theilacker 1971
TOTAL	313	

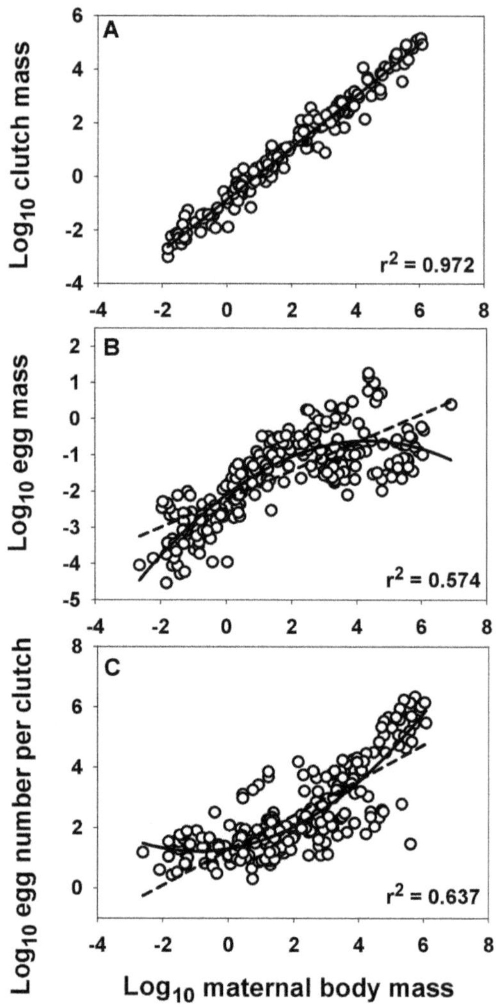

Fig. 3.1.
Scaling of 3 reproductive traits in relation to $\log_{10}$ maternal wet body mass (mg) (and associated r^2 values) of diverse species of crustaceans in the classes Branchipoda, Maxillipoda, and Malacostraca. (A) Log_{10} total wet mass of eggs (or embryos) in a clutch (mg): least squares regression equation is Y = 0.968 ± 0.012(X) – 0.864 [± 95% confidence intervals (CI) for the slope are given; $r = 0.986$; $P < 0.00001$; n = 196]. (B) Log_{10} egg (or embryo) wet mass (mg): the least squares regression (dashed line) is Y = 0.390 ± 0.021(X) – 2.214 [$r = 0.758$; $P < 0.00001$; n = 265], and the polynomial (quadratic) regression equation (solid line) is $Y = 0.686(X) - 0.079(X^2) - 2.101$ [$r = 0.830$; $P < 0.00001$ for both the X and X^2 terms]. (C) Log_{10} egg number in a clutch: the least squares regression equation (dashed line) is Y = 0.581 ± 0.028(X) + 1.252 [$r = 0.798$; $P <$ 0 .00001; n = 250], and the polynomial (quadratic) regression equation (solid line) is $Y = 0.179(X) + 0.095(X^2)$ [$r = 0.850$; $P = 0.00068$ for X term and < 0.00001 for X^2 term]. Data sources are in Table 3.1.

mass (instantaneous slopes were calculated as the first derivatives of the quadratic equations given in the Fig. 3.1 legend). Both small and large crustaceans increase their clutch mass with increasing body mass in a similar, nearly isometric way, but to do so, small crustaceans rely more on increases in egg mass, whereas large crustaceans rely more on increases in egg number, a newly discovered pattern that I attempt to explain later in this chapter.

For animals and plants generally, Visman et al. (1996) showed that variation in offspring number per clutch is well explained (92%) by both maternal and offspring dry masses in a single multiple regression model. For crustaceans specifically (n = 91), they showed that 72% of the variation in offspring number per clutch was explained by maternal mass, but this increased to 84% if offspring mass was also included as an independent variable. My analysis (using a larger database based on wet masses) revealed a similar pattern. In other words, 63.7% of the variation in egg number per clutch is explained by variation in maternal body mass (n = 250; Fig. 3.1C), whereas 94.8% of the variation is explained when egg mass is also included in the multiple regression model [$\log_{10}$ egg number per clutch = −0.724 + 0.947 ($\log_{10}$ maternal body mass)—0.935 ($\log_{10}$ egg mass); $r = 0.974$; $P < 0.00001$; n = 196]. This result is important for three reasons. First, it strongly supports the view of Visman et al. (1996) that both maternal and offspring masses should be included in predictive models of offspring production. Second, it indicates that clutch size varies nearly isometrically with body mass (slope = 0.947) after accounting for the effects of egg mass. Third, it suggests that egg size and number are strongly related: increases in one tend to be associated with nearly proportional decreases in the other (as indicated by a slope = −0.935). Negative interspecific correlations between offspring size and number have also been observed within various crustacean taxa, including cladocerans (Lynch 1980), copepods (Poulin 1995), barnacles ($r = -0.467$, $P = 0.0093$, n = 30: using data on egg diameter vs. number of eggs per brood from Barnes 1989), amphipods (Sainte-Marie 1991, Sutcliffe 1993, Poulin and Hamilton 1995), isopods (Glazier et al. 2003), and brachyuran crabs (Hines 1982, 1986, Rabalais and Gore 1985, Diesel et al. 2000).

Structural and Resource Limits

The body mass scaling of total clutch mass and egg size and number per clutch may be the result of structural and (or) resource (energetic) constraints. Here I argue that structural constraints primarily affect the scaling of clutch mass, whereas both structural and resource constraints may influence the scaling of egg size and number and the trade-offs between them. In short, clutch mass appears to be strongly related to available body space, whereas the size and number of eggs appear to be affected by not only available body space, but also the amount of resources available to produce and maintain eggs and embryos.

Spatial Constraints on Clutch Mass (Making Room for Eggs)

Total clutch mass (volume) tends to scale isometrically (proportionately) with body mass or nearly so, not only in crustaceans as a whole (Fig. 3.1A), but also within most major taxa, as will be seen. Metabolic constraints cannot explain the scaling of clutch mass because its log-log slope (0.968 ± 0.012, 95% confidence interval [CI]; Fig. 3.1A) differs significantly from that shown by resting metabolic rate in crustaceans (0.74 ± 0.06, n = 225; Glazier 2010; for further explanation, see Reiss 1989, Glazier 1999). If metabolic constraints were involved, the scaling of clutch mass would be negatively allometric (slope ~ 0.75), like that of metabolic rate. The near-isometric scaling of clutch mass (slope ~ 1) is more likely due to body space constraints within a confining exoskeleton. Because the volume of the body cavity, ovaries, or brood pouch tends to scale isometrically with body mass, so should the mass or volume of eggs that these spaces or structures contain (Hines 1982, Steele and Steele 1991a, Ramirez-Llodra 2002). Energetic constraints may be indirectly involved as well, because the amount of body space available for energy-storage compounds (e.g., lipids) used for egg production likely also scales isometrically with body mass (see Lease and Wolf 2011). However, body space constraints appear to be preeminent, as shown by a recent experimental study using two *Daphnia* species with different adult body sizes. The volume of eggs produced in early clutches was not only directly proportional to brood space volume, but also this relationship was unaffected by food level, and it occurred despite the presence of unused somatic energy (lipid) reserves after producing a clutch of eggs (Bartosiewicz et al. 2015).

Another physical factor that may limit total clutch mass, at least in some cases, is surface area-to-volume constraints on oxygen uptake (Fernández et al. 2006). The number of eggs produced by a female, and how this number scales with body mass, may be limited by a female's ability to ventilate (and thus oxygenate) eggs deep inside relatively large, thick clutch masses (Strathmann and Strathmann 1995, Baeza et al. 2015). This problem may be especially severe in large crustaceans that produce large clutches and live in warm tropical waters with relatively low oxygen levels. For example, the very large Caribbean king crab *Maguimithrax spinosissimus* shows unusually shallow scaling of clutch mass with female mass (slope = 0.54; Baeza et al. 2015), compared to crustaceans generally (slope ≈ 1; Fig. 3.1A). However, this effect of oxygen limitation may not occur in all large crustaceans, especially those that live in cool, high latitude seas with relatively high oxygen levels. Indeed, my analysis of data in Hines (1982, 1991) revealed that larger crab species inhabiting cool temperate waters do not exhibit shallower scaling of clutch mass in relation to maternal body mass than do smaller species. The body mass scaling slope for clutch mass is not significantly correlated with adult body mass among 28 brachyuran crab species ($r = 0.314$, $P = 0.104$) or nine *Cancer* species ($r = -0.146$, $P = 0.708$).

Spatial and Resource Constraints on Egg Mass and Number

Both physical and energetic constraints may directly influence the allometric scaling of egg size and number per clutch (Fig. 3.2 shows photographs of egg-bearing females representing the crustacean classes analyzed in this chapter). Because of body space limits on the mass (volume) of a clutch of eggs and the storage materials used to produce it, any increase in individual egg mass (volume) should entail a decrease in egg number and vice versa. Consistent with this view is the nearly proportional (1 to 1) negative relationship between crustacean egg size and number already described. A structural constraint on the body mass scaling of clutch mass may also explain why the scaling exponent (slope) for clutch size is about 1, when variation in egg size is statistically controlled (see General Patterns section). Similarly, the scaling exponent for clutch size approaches 1 when egg size naturally varies weakly or not at all with body mass (as in decapods and euphausiids: compare Fig. 3.4 and Fig. 3.5 in the next section (Taxonomic Variation; also see Hines 1982, Mauchline 1985, Ross and Quetin 2000). Likewise, the scaling exponent for egg mass approaches 1 when clutch size varies weakly or not at all with body mass (as in copepods: compare Fig. 3.4 and Fig. 3.5; also see Mauchline 1988). Spatial limits of the brood pouch may also explain why (1) egg size is positively correlated with maternal body size within various species of cladocerans (Robertson 1988, Glazier 1992, Bartosiewicz et al. 2015) and *Ligia* isopods (Tsai and Chen 1997); (2) egg size-versus-number trade-offs are most pronounced in relatively small mothers of the freshwater amphipod *Gammarus minus* (Glazier 2000); and (3) thicker species of subterranean amphipods tend to carry larger eggs than thinner species (Fišer et al. 2013).

Resource availability and demand (rate of embryonic metabolism) may importantly influence egg size and number, as well. Numerous studies have shown that food quantity and quality affect egg size, composition, and number in various crustaceans (Glazier 1992, Ramirez Llodra 2002, Wacker and Martin-Creuzburg 2007). Metabolic costs to produce and maintain eggs and embryos with different sizes may also be critical. For example, if one assumes that embryonic metabolism scales with embryo mass to a power less than 1 (similarly to that of adult animals), then the mass-specific metabolic rate of large embryos will be lower, and their energetic efficiency (biomass maintained per metabolic energy used) higher than that of small embryos. Based on this assumption, Goulden et al. (1987) argued that natural selection should often favor the production of larger offspring, especially in environments with relatively scarce food (also see Pettersen et al.

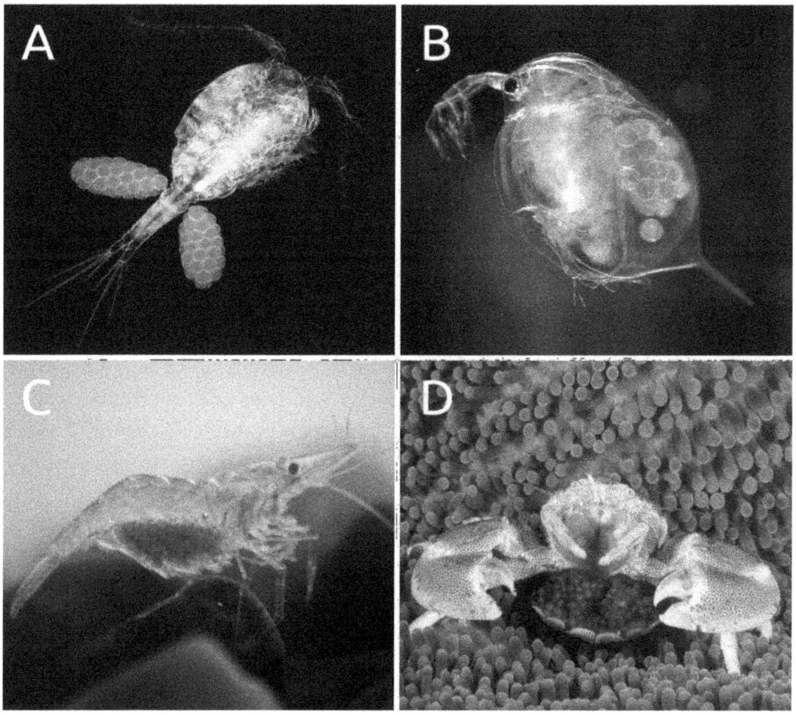

Fig. 3.2.
Photographs of egg-bearing crustaceans in the classes Maxillipoda (A), Branchiopoda (B), and Malacostraca (C, D). (A) Copepod with egg sacs in a sample from Surrey Bend Regional Park (British Columbia, Canada). Photograph by Waldo Nell ©, https://www.flickr.com/photos/pwnell/16119409481/. (B) Female cladoceran (*Daphnia magna*) with eggs. Photograph by Hajime Watanabe, under Creative Commons license (BY), doi:10.1371/image.pgen.v07.i03. (C) Decapod shrimp (*Palaemonetes pugio*) carrying eggs. Photograph by Brian Gratwicke, under Creative Commons license (BY), https://commons.wikimedia.org/wiki/File:Palaemonetes_pugio.jpg. (D) Berried porcelain crab (Decapoda; *Neopetrolisthes maculatus*). Photograph by Klaus Stiefel, under Creative Commons license (BY-NC), https://www.flickr.com/photos/pacificklaus/17369531096/. See color version of this figure in the centerfold.

2016). If so, egg number per clutch may be limited by, not only body space availability, but also an adaptive advantage of producing larger, fewer, energy-efficient eggs and embryos. However, it is unknown whether the above metabolic hypothesis applies generally to crustaceans. Across five *Daphnia* species, larger embryos use energy (triacylglycerol) reserves more slowly than do small embryos, as expected (Goulden et al. 1987). As a result, females can produce more offspring biomass at less cost (Visman et al. 1996, Pettersen et al. 2016). However, this advantage may occur only in food-limited or otherwise stressful environments (cf. Goulden et al. 1987). Whether large, newly released offspring have proportionately larger storage reserves than smaller offspring, thus enabling them to resist starvation better, should depend on embryonic developmental time. If embryonic developmental time is positively correlated with egg size, as generally observed in crustaceans, both within and among species (Steele and Steele 1975a, Clarke 1982, Lonsdale and Levinton 1985), differences in relative storage content between large and small offspring may be relatively small or even absent. Furthermore, contrary to the metabolic hypothesis, the mass- or volume-specific metabolic rate is not consistently related to egg or neonate size in two *Daphnia* species (Barber et al. 1990, Glazier 1992, Boersma 1995); the copepod *Acartia tonsa* (Hammervold et al. 2015); or three barnacle species (Barnes and Barnes 1965).

In addition, the metabolic hypothesis, as formulated by Pettersen et al. (2016), focuses only on volume-related metabolic demand and ignores volume-related water conservation and surface area–related oxygen supply. Larger eggs and offspring may be beneficial in dry environments because their smaller surface area-to-volume ratio results in lower rates of water loss per unit mass (also see later Water to Land section). However, smaller eggs may be beneficial, especially in warm water with low oxygen concentrations, because they have a larger surface area-to-volume ratio, thus enabling more oxygen uptake and carbon dioxide release per unit cytoplasm (Wear 1974, Neuheimer et al. 2015, Thatje and Hall 2016). For example, in three barnacle species the surface area-to-volume ratio increased more than two-fold when comparing the smallest to largest nauplii (calculated using data in Table 2 of Barnes and Barnes 1965). This geometric advantage of smaller eggs may not only permit higher metabolic rates (Steele and Steele 1975a, 1991a) but also faster growth and development (McLaren 1966, Clarke 1982, Lonsdale and Levinton 1985), which might be advantageous in environments where high predatory mortality favors rapid maturation (Wellborn 1994, Glazier 1999, Glazier et al. 2011). In short, a holistic view of the energetics and ecology of offspring at various stages of development is required to assess the relative adaptive advantage of small versus large eggs.

TAXONOMIC VARIATION

Clutch Mass

Isometric or near-isometric scaling of clutch mass (volume) with maternal body mass (volume) is seen not only for crustaceans as a whole (Fig 3.1A), but also within most major taxa, including copepods (Fig. 3.3), amphipods (Glazier 1999), cumaceans (my analysis of data from Ivanova and Vassilenko 1987), mysids (Mauchline 1973, 1985, 1988) and other peracarids (Fig. 3.3), euphausiids (Mauchline 1985, 1988), brachyurans (Hines 1982, 1991, Fukui 1988), and other decapods (Fig. 3.3, Mauchline 1988), although negative allometry (log-log slope < 1) appears to occur in stomatopods (Reaka 1979) and perhaps branchiopods (scaling slope = 0.86, but is not significantly different from 1; see Fig. 3.3). Nelson (1980) and Corey (1981) reported negative allometry of clutch-volume scaling in amphipods and cumaceans, respectively, which differs from the near-isometric scaling of clutch mass that the author has observed in these taxa. However, at lower taxonomic levels, clutch-mass scaling can vary substantially. For example, the intraspecific scaling slope for clutch mass (volume) versus maternal body mass (volume) varies greatly among species of cumaceans (0.03 to 0.50 for arithmetic data; Corey 1981), isopods (0.492 to 1.254 for $\log_{10}$-transformed data; Glazier et al. 2003) and decapods (0.436 to 2.397 for $\log_{10}$-transformed data; Hines 1982, 1991, Mauchline 1988). Why this variation occurs is not known, but it suggests that clutch mass may not always be a simple function of body volume, and that other factors (e.g., food and oxygen availability) should be considered in some cases (also see section titled Structural and Resource Limits).

Clutch mass may also vary significantly among taxa, independently of body size. For example, pinnotherid crabs carry larger clutch masses than other brachyuran crabs of comparable size, apparently because they have been able to increase the amount of body space available for egg packaging by using both their thorax and abdomen and by having distensible exoskeletons (Hines 1992). Among crustacean species, clutch mass as a percentage of maternal body mass typically varies from 5 to 25%, but more extreme values are possible (e.g., Ivanova and Vassilenko 1987, Hartnoll 2006, Achouri et al. 2008). In addition, by using ANCOVA with body mass as the covariate (the preferred method of body size correction: Packard and Boardman 1999), I have found that cladocerans tend to have larger clutch masses than copepods ($F_{1, 27} = 12.39$, $P = 0.0016$; also see Fig. 3.3).

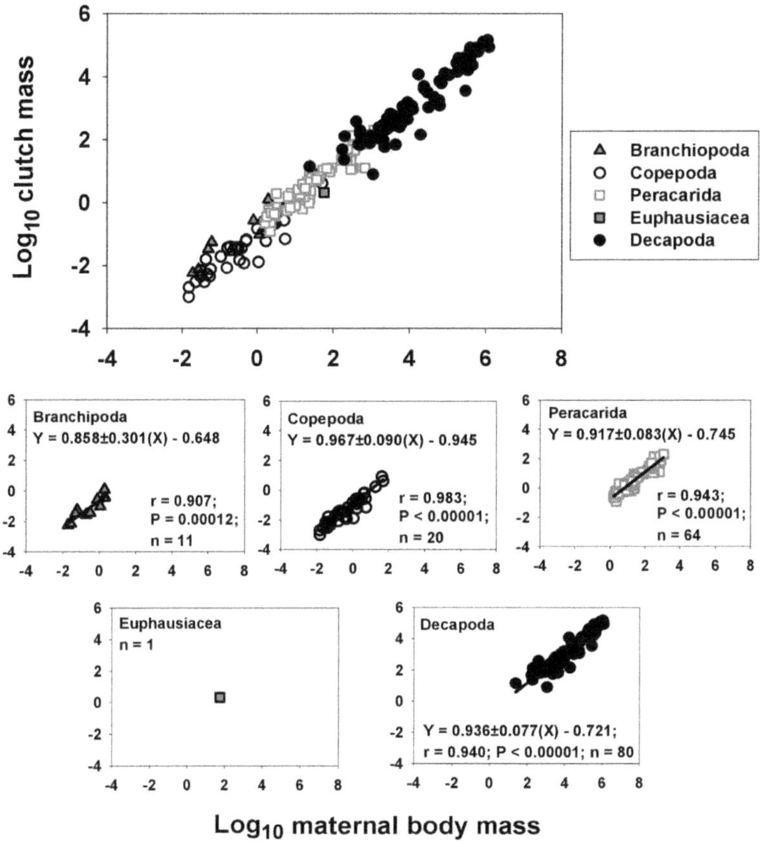

Fig. 3.3.
Log_{10} total wet mass of eggs (or embryos) in a clutch (mg) in relation to log_{10} maternal wet body mass (mg) of crustacean species in the taxa Branchipoda, Copepoda, Peracarida, Euphausiacea, and Decapoda. Least squares regression equations with 95% confidence intervals for the scaling slopes are shown. All slopes are significantly different from 0 but not from 1.

Egg (Offspring) Mass

Egg mass varies by more than 5 orders of magnitude among crustacean species, and more than half of this variation (about 57%) relates to maternal body size. However, the egg masses of various taxa may vary significantly, even after correcting for the effect of body size. For example, freshwater cladocerans produce significantly larger eggs than marine copepods of equal size (ANCOVA: $F_{1, 90} = 71.72$, $P < 0.00001$; also see Fig. 3.4), possibly because of environmental differences related to their different habitats, as will be seen. In the Decapoda, lobsters (Astacidea) have relatively large eggs, shrimps (Caridea) intermediate-sized eggs, and crabs (Brachyura) the smallest eggs (Rosa et al. 2007). Other taxonomic differences in egg size have apparently persisted over long geological periods, as seen in fossil eggs (Shen and Huang 2008, Gueriau et al. 2016).

The body mass scaling of egg mass also varies significantly among major taxa of crustaceans: copepods show relatively steep scaling (slope = 0.84), peracarids moderately steep scaling (slope = 0.64), branchiopods shallow scaling (slopes = 0.38), and decapods essentially zero scaling (nonsignificant slope = 0.094; Fig. 3.4). Similarly, Mauchline (1988) reported significant heterogeneity of the scaling of egg volume versus body volume (slopes = 0.301–1.003) in various

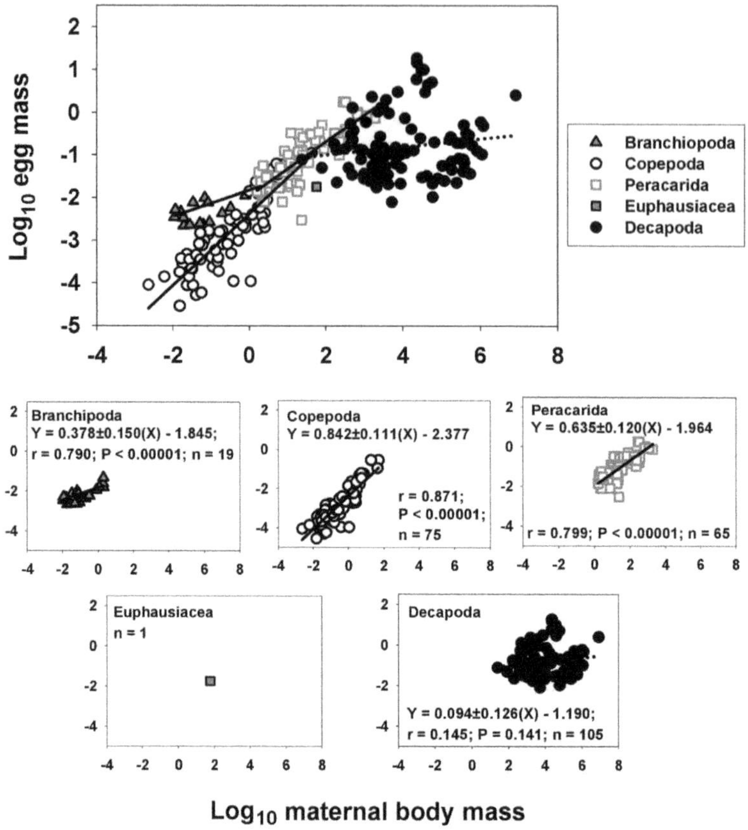

Fig. 3.4.
Log_{10} egg (or embryo) wet mass (mg) in relation to $\log_{10}$ maternal wet body mass (mg) of crustacean species in the taxa Branchipoda, Copepoda, Peracarida, Euphausiacea, and Decapoda. Least squares regression equations with 95% confidence intervals for the scaling slopes are shown. All slopes are significantly different from 0, except for the slope for the Decapoda. All slopes are also significantly less than 1, although the slope for the Copepoda is near 1.

taxa of pelagic crustaceans (calanoids, mysids, euphausiids, and decapods). Clearly, crustaceans do not display a single pattern of egg size scaling (contrary to Neuheimer et al. 2015). Investigators should consider this scaling heterogeneity when adjusting for the effects of body size on egg size.

Egg (Offspring) Number Per Clutch

Clutch size varies by more than 6 orders of magnitude among crustacean species, and nearly two-thirds of this variation (~64%) relates to maternal body size. However, the clutch sizes of various taxa may vary significantly, even after correcting for the effect of body size. For example, copepods produce significantly larger clutches than cladocerans of equal size (ANCOVA: $F_{1,51} = 7.57$, $P = 0.008$; also see Fig. 3.5), which may be at least partly a result of a trade-off with egg mass, which is significantly larger in cladocerans (see previous section). The sessile barnacles (Cirripedia) also produce clutch sizes more than 10 times larger than most crustaceans of similar mass (Fig. 3.5).

Furthermore, clutch size scaling varies significantly among crustacean taxa: decapods show relatively steep scaling (slope = 0.91), cirripedes moderately steep scaling (slope = 0.70), branchiopods and peracarids shallow scaling (slopes = 0.30 and 0.18, respectively), and copepods essentially zero

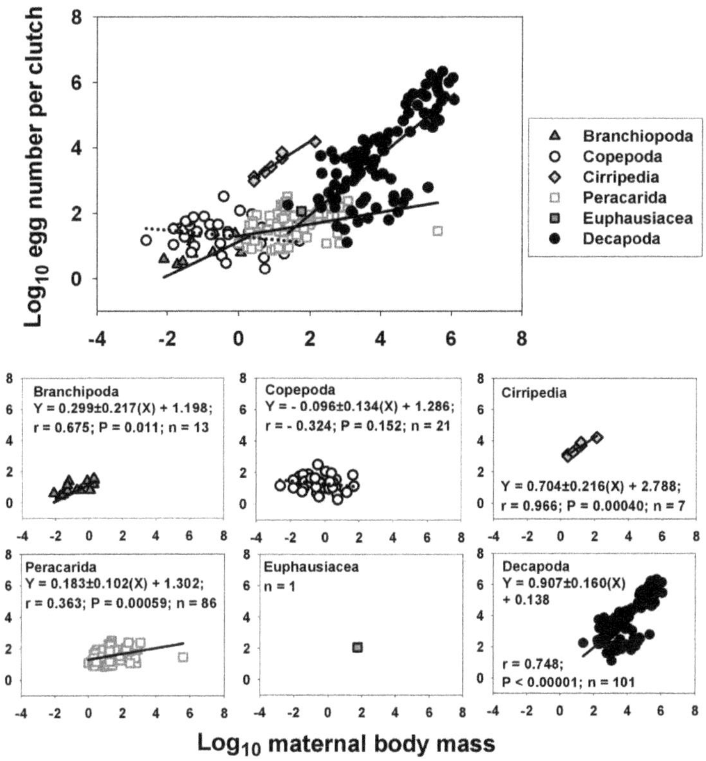

Fig. 3.5.
Log_{10} egg number in a clutch in relation to log_{10} maternal wet body mass (mg) of crustacean species in the taxa Branchipoda, Copepoda, Cirripedia, Peracarida, Euphausiacea, and Decapoda. Least squares regression equations with 95% confidence intervals for the scaling slopes are shown. All slopes are significantly different from 0, except for the slope for the Copepoda. All slopes are also significantly less than 1, except for the slope for the Decapoda.

scaling (nonsignificant slope = 0.096; Fig. 3.5). The difference in scaling between the decapods and peracarids results in the largest species of these taxa having strikingly different clutch sizes, such as more than 1 million in some relatively large crabs (Hines 1982, 1991), versus only 2–3 dozen in the giant deep sea isopod *Bathynomus giganteus* (Briones-Fourzán and Lozano-Alvarez 1991, Johnson et al. 2001). Species differences in clutch size scaling also occur within specific taxa, such as *Daphnia* (Green 1956), *Gammarus* (Steele and Steele 1975b), Cumacea (Corey 1981), Isopoda (Sutton et al. 1984, Ma et al. 1991, Glazier et al. 2003), Mysidacea (Mauchline 1973), Anomura (Reid and Corey 1991), *Cancer* (Hines 1991), Brachyura (Hines 1982), and Caridea (Corey and Reid 1991). As previously noted for egg size, investigators should consider this scaling heterogeneity when adjusting for the effects of body size on clutch size.

ENVIRONMENTAL VARIATION

Although many kinds of environmental factors affect crustacean offspring investment, this chapter emphasizes differences between major environmental realms that are especially significant for crustaceans, namely (1) aquatic versus terrestrial environments and (2) marine versus freshwater environments. Crustaceans apparently originated in seawater and secondarily colonized freshwater and terrestrial environments several times.

Clutch Mass

Although the body mass scaling of clutch mass does not vary much among major taxa and environments, the scaling slope of clutch mass differs significantly among environments, being essentially isometric (0.99) in marine species and somewhat negatively allometric in freshwater (0.91) and terrestrial species (0.84) (Fig. 3.6A and legend). Furthermore, in Caribbean grapsid crabs relative clutch mass decreases from 11% in seawater to 9% in surface freshwater, 3% in subterranean freshwater and 2% on land (Diesel et al. 2000). Southwestern Atlantic grapsid crabs also tend to show higher relative clutch masses in lower intertidal and subtidal habitats than in upper intertidal habitats (Spivak et al. 2012). In addition, isopods in relatively dry terrestrial microhabitats produce smaller clutch masses (adjusted to maternal body mass by ANCOVA) than do those in relatively moist soil or aquatic habitats (Glazier et al. 2003). Perhaps decreased resource acquisition and increased stress-related maintenance (survival) costs have reduced the amount of resources available for reproduction in freshwater and desiccating habitats (Glazier et al. 2003). Reduced offspring production may also be adaptive for conserving water in relatively dry terrestrial environments (Glazier et al. 2003).

Offspring (Egg) Mass and Number Per Clutch

Water to Land

A general analysis shows that terrestrial and semiterrestrial crustaceans tend to produce fewer, larger eggs than marine species of equal body size (Fig. 3.6B,C; also see Richardson and Araujo 2015). This trend occurs within specific taxa, including amphipods (Hurley 1968, Friend and Richardson 1986); decapods (e.g., Rabalais and Gore 1985, Anger 1995, Diesel et al. 2000, Cannicci et al. 2011); and isopods (Glazier et al. 2003). Presumably, drier conditions on land have selected for larger eggs and juveniles that can better resist desiccation than smaller ones. Given that clutch mass is subject to strong physical constraints and may be relatively low in some terrestrial crustaceans, it follows that terrestrial species should also brood smaller clutches than marine species. These trends also occur at smaller local and taxonomic scales involving closely related species or conspecific populations occupying different tidal zones with different degrees of desiccation. For example, more inland populations of the same or different species in the isopod genus *Ligia* tend to produce fewer, larger eggs relative to their body size than do littoral populations (Tsai and Chen 1997, Tsai and Dai 2001). Land crabs that produce large numbers of small eggs that are released into seawater are an "exception that proves the rule" (Richardson and Araujo 2015).

Seawater to Freshwater

Another significant ecological shift for crustaceans occurs between seawater and freshwater environments. In general, freshwater crustaceans tend to produce fewer, larger eggs than marine species of equal body size (Fig. 3.6B,C). This trend is similar to the water-to-land transition, except that the seawater versus freshwater difference in reproductive strategy is greater at larger adult sizes, and the egg size and number scaling relationships for freshwater species are linear, rather than curvilinear, as seen in the marine species (Fig. 3.6B,C). Similar salinity-related differences in offspring size and number occur between related species of gammaridean amphipods (Nelson 1980, Steele and Steele 1991a, Mirzajani et al. 2011; but see Sainte-Marie 1991, and below) and decapods (Rabalais and Gore 1985, Anger 1995, Cannicci et al. 2011, Vogt 2013), including palaeomonid shrimps (Magalhães and Walker 1988, Jalihal et al. 1993; also see Fig. 3.2C) and brachyuran crabs (Diesel et al. 2000, Devi and Smija 2013, Swetha et al. 2015). Remarkably, intraspecific associations between decreasing salinity and fewer, larger eggs or offspring even occur, including in the mysid *Mesopodopsis orientalis*

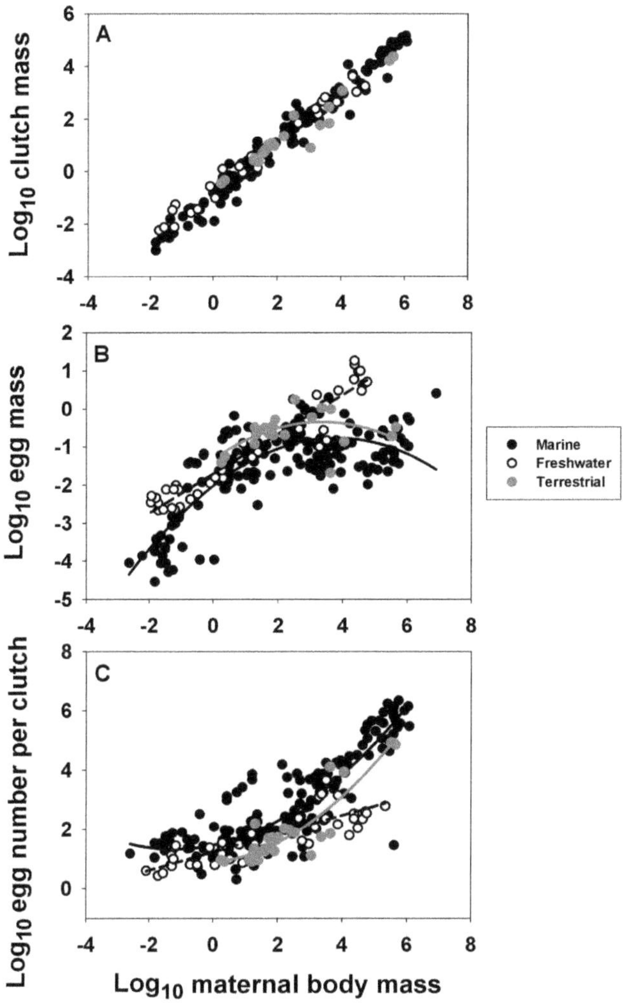

Fig. 3.6.

Scaling of 3 reproductive traits in relation to $\log_{10}$ maternal wet body mass (mg) of crustacean species from marine (salt and brackish water), freshwater, and terrestrial environments (the last category includes semiterrestrial and fully terrestrial species). (A) Log_{10} total wet mass of eggs (or embryos) in a clutch (mg): least squares regression equations are (marine) Y = 0.992 ± 0.026(X) – 0.926 [r = 0.988; P < 0.00001; n = 146]; (freshwater) Y = 0.907 ± 0.054(X) – 0.658 [r = 0.989; P < 0.00001; n = 28]; (terrestrial) Y = 0.844 ± 0.108(X) – 0.640 [r = 0.964; P < 0.00001; n = 22]. (B) Log_{10} egg (or embryo) wet mass (mg): least squares linear or polynomial (quadratic) regression equations are (marine) Y = 0.657(X) – 0.086(X^2) – 2.019 [r = 0.787; P < 0.00001 for both X and X^2 terms; n = 180], (freshwater) Y = 0.518 ± 0.051(X) – 1.704 [r = 0.959; P < 0.00001; n = 39], and (terrestrial) Y = 0.582(X) – 0.087(X^2) – 1.325 [r = 0.506; P = 0.021 for X term, and 0.032 for X^2 term; n = 22]. (C) Log_{10} egg number in a clutch: least squares linear or polynomial (quadratic) regression equations are (marine) Y = 0.224(X) + 0.092(X^2) + 1.458 [r = 0.890; P = 0.00017 for X term, and < 0.00001 for X^2 term; n = 174], (freshwater) Y = 0.310 ± 0.066(X) + 1.227 [r = 0.830; P < 0.00001; n = 43], and (terrestrial) Y = 0.120(X) + 0.110(X^2) + 0.866 [r = 0.905; P = 0.646 for X term, and 0.015 for X^2 term; n = 33]. Polynomial lines are shown only if they explain significantly more of the variation in egg mass or egg number (clutch size) than corresponding linear regressions.

(Hanamura et al. 2008) and several decapod species (Walsh 1993, Urzúa and Anger 2011, Meireles et al. 2013). In the prawn *Macrobrachium nipponense,* salinity-related population differences in egg size are under genetic control (Mashiko 1992) and have apparently evolved recently in the Holocene (Vogt 2013). A shift to fewer, larger eggs has even been reported for a population of the grass prawn *Palaemon elegans* that has invaded the low-salinity southern Baltic Sea only since the year 2000 (Janas and Mańkucka 2010). Reproductive responses to salinity may depend on the native habitat of a species. Freshwater species often show decreases in offspring production when exposed to increased salinity (Grzesiuk and Mikulski 2006), whereas marine species frequently show the opposite response (Steele and Steele 1991b, Subida et al. 2005; but see Kevrekidis et al. 2009).

Why freshwater crustaceans typically produce fewer, larger eggs than marine species is unknown. However, this life history strategy may relate to lower or more variable food supplies for juveniles (Magalhães and Walker 1988, Rabalais and Gore 1985, Anger 1995, Vogt 2013) and higher ionic and osmotic regulatory demands in freshwater (Glazier and Sparks 1997) that favor larger yolky eggs enabling higher rates of growth and survival under these stressful conditions (cf. Diesel et al. 2000). However, changes in egg size (volume) observed in laboratory manipulations of salinity may not be adaptive. For example, in the brackish-water amphipod *Gammarus lawrencianus,* increases in egg size at low salinities appear to be due to osmosis (water absorption; Steele and Steele 1991b). This observation calls attention to the need to examine egg mass and composition in salinity studies. In *G. salinus* egg mass decreases, despite increases in egg volume, with decreasing salinity (Skadsheim 1989). By contrast, the egg mass of the intertidal marine grapsoid crab *Neohelice granulata* increases with decreasing salinity (Gimenez and Anger 2001). The biochemical composition of prawn eggs or larvae also differs between brackish and freshwater populations (Urzúa and Anger 2011, Meireles et al. 2013).

Body size differences and other environmental factors (e.g., climate) can confound intra- and interspecific relationships between salinity and offspring size and number (Alon and Stancyk 1982, Anger et al. 2002, Ituarte et al. 2007). For example, in the polar isopod *Mesidotea entomon,* marine populations produce larger eggs and clutch sizes than freshwater populations (Korcznski 1991), but these differences may be merely incidental effects of larger female body sizes in the marine populations. After body size correction, reproductive differences between marine and freshwater populations may disappear. For example, using data for 66 populations of 27 *Gammarus* species inhabiting saline, brackish, and fresh waters (Subida et al. 2005), the author found that neither brood size nor egg diameter (both corrected for total body length, which is significantly higher in marine versus brackish/freshwater populations: ANOVA: $F_{2,63} = 4.10$, $P = 0.021$) varied significantly with habitat salinity (ANCOVA: $F_{2,60} = 0.25$, $P = 0.78$; $F_{2,35} = 1.82$, $P = 0.18$, respectively). This finding supports the analysis of gammaridean amphipods by Sainte-Marie (1991), but not those of Nelson (1980) and Steele and Steele (1991a). A similar lack of significance was found for brood size within *Gammarus* species for which data on at least two marine and two brackish water populations were available (*G. duebeni*: $F_{1,5} = 5.22$, $P = 0.071$; *G. locusta*: $F_{1,3} = 0.085$, $P = 0.79$; *G. salinus*: $F_{1,3} = 0.10$, $P = 0.77$), although brackish water populations tended to produce larger clutches than marine populations in all three species (also see Sastry 1983). Given these conflicting results, there is a need for further studies of salinity effects on egg size and number that distinguish the relative influence of phenotypic acclimation versus genotypic adaptation, and that tease out effects of body size and other potentially confounding intrinsic and extrinsic factors.

THEORETICAL VIEWS

Body Mass Scaling

The linear, isometric scaling of clutch mass with body mass is likely due to morphological volumetric constraints (see Structural and Resource Limits section), but why are the scaling of egg mass and number per clutch both negatively allometric (slope < 1) and curvilinear? Given that clutch mass (= egg mass × number) scales isometrically or nearly so (Figs. 3.1A, 3.3, 3.7), it follows logically that the linear scaling exponents for egg mass and number should add up to 1, which they nearly do (0.390 + 0.581 = 0.971: see Fig. 3.1). However surprisingly, both egg mass and number also scale curvilinearly. Egg mass scales concave downward (Fig. 3.1B), whereas egg number scales concave upward (Fig. 3.1C). I suggest that the opposite curvilinear relationships of egg size and number indicate not only a fundamental trade-off, but also that the sensitivity of these traits to body size is differently related to offspring versus parental fitness. At small body sizes, maximizing offspring fitness prevails, thus resulting in large increases in egg size, but relatively little change in egg number, with increasing maternal body mass. However, at large body sizes, maximizing parental fitness predominates, thus resulting in large increases in clutch size, but relatively little change in egg size, with increasing maternal body mass (Fig. 3.7). This size-related shift in the relative priority of offspring versus parental fitness may in turn relate to a change in the ratio of offspring (juvenile) to parental (adult) mortality from being low at small body sizes but high at large sizes. The concave downward scaling curve for egg mass makes sense if the relationship between offspring size and fitness is asymptotic (Fig. 3.7), as typically assumed by life history theory (reviewed by Rollinson and Hutchings 2013). The concave upward scaling curve for egg number would then follow as a simple trade-off. In addition, and nonexclusively, the curvilinear scaling relationships for egg mass and number may result from size-related variation in scaling relationships among taxa with different life history strategies (also see next section). The curvilinearity of the scaling of metabolic rate in mammals and terrestrial invertebrates has been explained in a similar way (Capellini et al. 2010, Ehnes et al. 2011, Müller et al. 2012).

Visman et al. (1996) suggested that the negative allometry of clutch size in many kinds of animals and plants has resulted from selection favoring more, but smaller size-specific clutches in larger individuals during their longer lifetimes. This size-related shift in reproductive strategy may

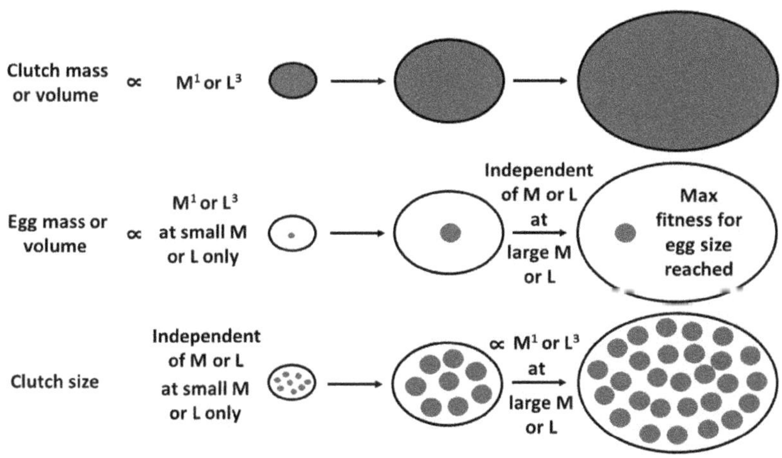

Fig. 3.7.
Schematic diagram showing general trends in how clutch mass (volume), individual egg mass (volume), and clutch size (egg number per clutch) change with increasing maternal body mass (M) or length (L) across crustacean species from tiny copepods to large crabs. Clutch mass tends to increase proportionately (isometrically) with increasing body mass (or to the cubic power with body length) because of body-volume constraints. However, individual egg mass tends to increase proportionately (isometrically), or nearly so, with increasing body mass, only at small body sizes, and varies independently of (or relatively weakly with) body mass at large body sizes, perhaps because offspring fitness is no longer increased by enlarging egg size beyond a certain point. By contrast, clutch size tends to vary independently of (or relatively weakly with) body mass at small body sizes, but approaches an isometric relationship with body mass at large body sizes, perhaps because relatively large females can increase their fitness more by increasing the number rather than size of their eggs.

in turn relate to increases in the ratio of juvenile to adult mortality in larger animals (following the classical life history model of Charnov and Schaffer 1973; also see Chapter 4 in this volume). If so, patterns of juvenile versus adult mortality may underlie both the negative allometry and curvilinearity of clutch size scaling in crustaceans. The above hypothesis is consistent with the common occurrence of inverse correlations between clutch size and breeding frequency (i.e., the number of clutches produced per breeding season or in a lifetime) in crustaceans, as reported for copepods (Niehoff 2007), amphipods (Nelson 1980, Sainte Marie 1991), ocypodid crabs (Henmi 2003), and stomatopods (Reaka 1979), though not in mysids (Hartnoll 1985).

Taxonomic Variation

In crustaceans, the taxonomic variation of egg mass and number per clutch and their scaling with body mass is considerably greater than that of clutch mass (compare Fig. 3.3 with Fig. 3.4 and Fig. 3.5). Much of this variation may relate to intrinsic and extrinsic effects on the relative rates of growth and mortality in juveniles versus adults, as predicted by life history theory (Sibly and Calow 1986, Stearns 1992, Roff 2002, Falster et al. 2008, Neuheimer et al. 2015). For example, barnacles (Cirripedia) tend to produce many more eggs per clutch than other crustaceans of equivalent body mass (Fig. 3.5), perhaps in part because of the high uncertainty of survival of their offspring (Hurley 1973, Barnes 1989). Peracarids with similar body masses produce far fewer eggs probably because their offspring develop in a protective brood pouch, thus increasing their probability of survival.

Taxonomic differences in the body mass scaling of egg mass and number may be explained using life history theory, as well. For example, interspecifically, copepods show a strong nearly isometric correlation between egg mass and body mass (and thus a relatively high, nearly constant ratio between offspring and maternal body sizes), whereas decapods show no such correlation (and thus a relatively variable ratio between offspring and maternal sizes that declines greatly with maternal size; see Fig. 3.4). Apparently, evolutionary interspecific increases in adult body size and clutch mass are associated with increases in egg mass in copepods, but mainly increases in egg number in decapods. These distinctly different size-related reproductive strategies may relate to different patterns of juvenile growth and mortality. Perhaps the ratio of offspring mass to maternal mass must be relatively high and invariant in copepods (ranging mostly from ~0.01 to 0.001), regardless of their adult size, because there is strong selection to reach maturity and breed quickly before being eaten, thus placing a greater premium on egg size versus number. As evidence, growth and developmental rates are nearly invariant in copepods with different adult sizes (Kiørboe and Sabatini 1995). However, the ratio of offspring mass to maternal mass is usually much smaller and more variable in decapods (ranging mostly from ~0.001 to 0.0000001), perhaps because high larval mortality (McConaugha 1992) has resulted in stronger selection for egg number than size. In addition, larval decapods may often experience high food levels, thus allowing them to grow rapidly from small hatchling sizes. By contrast, low food levels and low offspring growth rates tend to favor relatively large offspring (Hutchinson 1967, Glazier 1992, Ramirez Llodra 2002).

Another hypothesis requiring testing is that steep, relatively tight scaling of offspring size, as observed in copepods, should be associated with density-dependent larval growth and mortality, whereas weakly shallow or absent scaling of offspring size, as observed in decapods, should be associated with density-independent larval growth and mortality (following Neuheimer et al. 2015). This hypothesis assumes that relatively large offspring that are proportional to adult size are favored when high intraspecific competition associated with high population densities reduces larval growth and survival (as may occur in many copepods whose pelagic offspring and adults coexist, and thus may compete for the same food). By contrast, relatively small offspring (with little or no relation to adult size) should be favored when environmental factors other than population density inhibit larval growth or survival (as may occur in many decapods whose pelagic larvae avoid competition with benthic adults). The related theoretical model of Falster et al. (2008) also predicts shallower scaling of offspring size when density-independent juvenile mortality increases, which helps explain the decapod pattern. However, its prediction that the scaling of offspring size should be relatively steep when larger females have larger clutch sizes, thus increasing sibling competition that favors larger offspring, does not apply to copepods, which show no significant interspecific increase in clutch size with maternal size (see Fig. 3.5). The copepod pattern better fits the minimal offspring-size model of Meiri et al. (2015), which predicts steeper offspring size scaling when clutch size does not covary with body size. According to this model, when clutch size is invariant, egg size can increase more nearly in proportion to maternal body volume, than is possible when clutch size increases with maternal body size.

However, when testing theoretical models, one should realize that egg size is a relatively simplistic measure of offspring quality: the energy and nutrient contents of eggs are also important. Egg biochemical composition (independent of egg size) relates to maternal age and size in some decapods (e.g., Swiney et al. 2013), to food quantity/quality in cladocerans (e.g., Tessier and Consolatti 1991, Guisande and Gliwicz 1992) and copepods (e.g., Acheampong et al. 2011), and to other environmental factors (e.g., salinity; see Seawater to Freshwater section).

Environmental Variation

Life history theory may also help explain environmental variation in egg size and number. For example, according to triangular habitat-templet models (e.g., Grime and Pierce 2012), adversity selection in unfavorable environments should favor the production of fewer, larger, more resistant offspring in relatively dry land and ion-poor freshwater habitats than in seawater, as is generally observed in crustaceans (Figs. 3.6 and 3.8). Relatively low offspring growth and survival rates in harsh environments are also predicted to favor fewer, larger offspring (Sibly and Calow 1986), as occurs in polar, deep sea, and subterranean habitats (Sastry 1983, Sainte-Marie 1991, Clarke 1992, Kosobokova et al. 2007, Fišer et al. 2013, Thatje and Hall 2016).

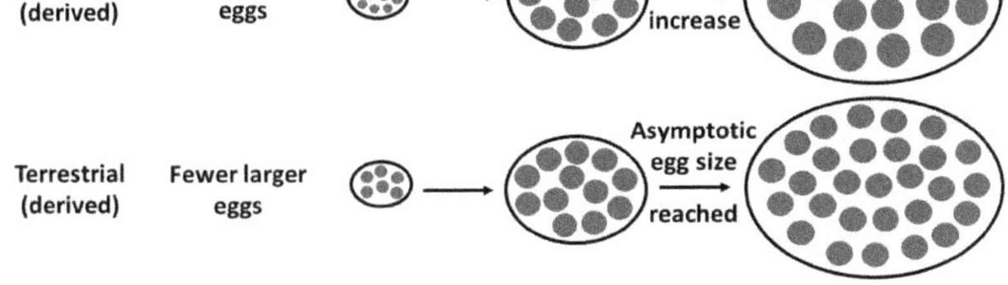

Fig. 3.8.
Schematic diagram showing general trends in how egg mass and number per clutch change with increasing maternal body mass across crustacean species in ancestral marine and derived freshwater and terrestrial environments (based on data in Fig. 3.6). For a given body mass, freshwater and terrestrial species tend to produce fewer, larger eggs than marine species. Furthermore, although freshwater species show linear increases in egg size with increasing body mass in log-log space, marine and terrestrial species exhibit curvilinear relationships, such that egg size reaches an asymptote at intermediate body masses.

The body mass scaling of offspring investment varies with environment as well. The scaling of clutch mass is essentially isometric in marine crustaceans but negatively allometric in freshwater and terrestrial species. These different scaling relationships may result, at least in part, from size-specific, taxon-related differences in offspring investment between environments. At the low end of the body mass spectrum, freshwater cladocerans have larger clutch masses than marine copepods of similar size (Fig. 3.3; also see Clutch Mass section within previous main section titled Taxonomic Variation), perhaps because cladocerans tend to have higher efficiencies of food acquisition and (or) more volatile population dynamics, thus causing them to be more r-selected than copepods (cf. Allan 1976). At the high end of the body mass spectrum, terrestrial decapods have smaller clutch masses than marine decapods (Fig. 3.5), perhaps because resource acquisition is lower and (or) resource allocation to body maintenance is higher in dry environments (also see Clutch Mass section in earlier main section titled Environmental Variation).

By contrast, the body mass scaling of egg size is steeper in freshwater versus marine environments, not only for crustaceans as a whole (Fig. 3.6), but also for specific taxa (e.g., amphipods; Steele and Steele 1991a). Recent models predict that steeper offspring-size scaling should occur when density-independent juvenile mortality is lower (Falster et al. 2008, Neuheimer et al. 2015). This theory may explain why offspring size scaling is steeper in fresh water versus salt water; freshwater crustaceans usually show relatively high per offspring investment, which is typically associated with better juvenile survival (Tessier and Consolatti 1989, Gliwicz and Guisande 1992, Cleuvers et al. 1997). However, it is a mystery why egg mass scales less steeply in freshwater cladocerans (branchiopods) than in marine copepods (see Fig. 3.4), despite the former having relatively larger eggs (see previous section titled Egg (Offspring) Mass).

In addition, why do freshwater crustaceans show linear scaling relationships for egg size and number, whereas marine and terrestrial species show curvilinear relationships (Figs. 3.6 and 3.8)? Perhaps the egg mass scaling relationship for freshwater crustaceans does not plateau at large body sizes, because lower or more unpredictable food supplies and (or) higher metabolic costs of ionic regulation in freshwater versus seawater favor relatively large offspring with high amounts of energy reserves and low surface area-to-volume ratios at all body sizes (also see earlier discussion on Seawater to Freshwater). These scaling differences do not appear to be a simple result of taxonomic differences because all sample species are decapods at the large end of the body size distribution, where environmental differences in egg size and number are most pronounced (compare Fig. 3.4 with Fig. 3.6B and Fig. 3.5 with Fig. 3.6C).

COMPARISONS WITH OTHER TAXA

Several patterns of crustacean offspring investment described in this chapter occur in other taxa of animals and plants. For example, the strong, nearly isometric scaling of clutch mass with body mass shown by crustaceans (log-log slope ≈ 0.97: Fig. 3.1A; 1.16: Blueweiss et al. 1978; 0.92: Visman et al. 1996) also occurs in fishes (0.90: Visman et al. 1996; 0.86–1.08: Hendriks and Mulder 2008), ectothermic animals generally (0.92: Blueweiss et al. 1978), and even plants (0.98: Visman et al. 1996; 0.67–1.08: Hendriks and Mulder 2008). The similarly steep scaling of clutch mass in crustaceans and spiders (1.09: Marshall and Gittleman 1994) and of clutch size and volume in insects (1.04: Honěk 1993; 0.79–1.00: Berrigan 1991, Hendriks and Mulder 2008) is not surprising because these arthropods share hard exoskeletons that exert physical volume constraints on egg mass packaging and storage of supporting energy reserves. Fishes also show volume-related clutch mass scaling, perhaps because they have a confining scaly exterior whose shape diversity is typically limited by hydrodynamic constraints. However, tetrapod vertebrates (with internal skeletons) tend to show lower scaling exponents for clutch mass (0.60–0.88: Blueweiss et al. 1978, Visman et al. 1996,

Hendriks and Mulder 2008). The lower scaling exponents observed in terrestrial versus aquatic vertebrates, parallel the lower exponent shown by terrestrial versus marine crustaceans (Fig. 3.6 legend). These similar environmental differences suggest a common cause or causes. Perhaps buoyancy in water allows large aquatic animals to carry proportionately larger clutch masses than can terrestrial animals of equal size. Another possibility is that desiccating environments on land select for proportionately larger offspring, thus increasing total clutch mass, especially in small animals with relatively high rates of water loss because of their large surface area-to-volume ratios, compared to aquatic animals of equal size. I consider other possible factors in the section Environmental Variation (in the main section titled Theoretical Views).

Another general life history pattern shown by crustaceans and other animals and plants is that the body mass scaling slopes and elevations for offspring size and number are much more variable among taxa than are those for clutch mass (compare Figs. 3.1, 3.3, 3.4 and 3.5 with Figures 2–4 in Hendriks and Mulder 2008; also see Blueweiss et al. 1978, Peters 1983, Visman et al. 1996). In addition, the residual variation tends to be greater for the scaling of egg size and clutch size than for clutch mass, as observed in both fishes (Duarte and Alcaraz 1989) and crustaceans (Fig. 3.1), as well as in other nonamniote animals and plants (Visman et al. 1996, Hendriks and Mulder 2008). These patterns suggest that offspring size and number are freer to vary than is total offspring investment. For example, although clutch (litter) mass (volume) is similar in crustaceans and tetrapod vertebrates at a given adult mass, neonate mass is considerably (2–3 orders of magnitude) smaller in crustaceans (Blueweiss et al. 1978). Furthermore, the greater interspecific variation of egg (offspring) size and number observed in large versus small crustaceans (Figs. 3.4 and 3.5) is a pattern shared by fishes (Duarte and Alcaraz 1989) and other animals (Visman et al. 1996). This pattern suggests that the evolution of offspring size and number is more constrained in smaller animals (Duarte and Alcaraz 1989, Visman et al. 1996), perhaps because their small eggs are closer to the minimal viable offspring size than are the larger eggs of larger animals (also see Mauchline 1988, Hendriks and Mulder 2008).

Trade-offs between egg size and number often occur, both within and among species, not only in crustaceans (Glazier 1992, 1999, Clarke 1993, Guisande et al. 1996, Veríssimo et al. 2011; also see section General Patterns above), but also in other animals (Roff 1992, Bernardo 1996, Visman et al. 1996, Fox and Czesak 2000). Remarkably, the inverse scaling relationships shown for egg mass (concave downward) and clutch size (concave upward) in crustaceans (Fig. 3.1B,C) are also shown for egg volume and clutch size in salamanders (Salthe 1969). In both taxa, increasing body size and offspring investment is associated mainly with increases in offspring (egg) size in small animals, but increases in clutch size in larger animals. Perhaps beyond a certain point increasing egg size does not significantly improve offspring fitness (an asymptotic curve of diminishing returns is empirically supported in Figure 9 of Glazier 1992; also see Rollinson and Hutchings 2013) and ultimately parental fitness, and thus increases in clutch size are increasingly favored at larger body sizes (also see Salthe 1969). This hypothesis is supported by the observation that in both decapods and teleost fishes, which are relatively large compared with many other crustaceans, clutch size, but not egg size is related to body size (Figs. 3.4 and 3.5; Elgar 1990).

In addition, crustaceans show life history responses to habitat differences that are similar to those of other animals. For example, terrestrial crustaceans tend to produce larger offspring than aquatic crustaceans of equal body size (Figs. 3.6 and 3.8), a pattern also shown by other animals (Hendriks and Mulder 2008). Similarly both fishes and crustaceans tend to produce larger eggs in freshwater than in seawater (Figs. 3.6 and 3.8; Duarte and Alcaraz 1989, Wootton 1991). In the sea, both fishes and decapods produce many tiny offspring: a strategy that may be favored by natural selection when food occurs densely in isolated patches (Winemiller and Rose 1993) and (or) for other reasons discussed in the section Environmental Variation (in the main section titled Theoretical Views).

FUTURE DIRECTIONS

Although useful, the scaling analyses presented in this chapter should be considered preliminary, because they do not use phylogenetically informed methods, and suitable data for many crustacean taxa are not yet available. Future research should (1) expand our life history knowledge of little studied taxa, (2) use robust standard methodologies that permit broad comparisons across crustacean taxa and with other animals in an ecological and evolutionary context, (3) take a holistic perspective on offspring size and production in the context of the entire life histories and ecological lifestyles (niches) of various taxa, and (4) use empirical findings not only to test hypotheses and models but also to formulate new hypotheses and theories.

CONCLUSIONS

This chapter has focused on how crustacean clutch mass, egg (offspring) mass, and clutch size vary with adult body mass, taxonomic affiliation, and major environmental differences. These analyses reveal only a few of the many intrinsic and extrinsic factors that affect crustacean offspring investment. Major patterns that emerge are that (1) clutch mass is strongly and nearly isometrically related to maternal body mass (and thus to the cubic power of body length), probably because of body-volume constraints (Figs. 3.1A and 3.7); (2) egg mass exhibits a positive, concave downward relationship with maternal body mass, possibly because of an asymptotic relationship between egg size and offspring fitness (Figs. 3.1B and 3.7); (3) egg number per clutch exhibits a positive, concave upward relationship with maternal body mass, and thus at larger body masses, maternal fitness appears to be enhanced more by increases in egg number than egg size (Figs. 3.1C and 3.7); (4) there is a strong, nearly proportional trade-off between egg mass and number per clutch; (5) egg mass and clutch size, and their scaling with maternal body mass, appear to be more ecologically sensitive and evolutionarily malleable than that of clutch mass, (6) small crustaceans (e.g., copepods) have evolved larger clutch masses with increasing body size mainly by increasing egg size, whereas large crustaceans (e.g., decapods) have done so mainly by increasing clutch size (Figs. 3.4 and 3.5): these trends may relate to a size-related shift in the relative priority of offspring versus maternal fitness, which is hypothesized to be the result of an increase in the ratio of juvenile to adult mortality with increasing body size, and (7) crustaceans in freshwater and terrestrial environments tend to produce larger, fewer eggs per clutch than those in marine environments (Figs. 3.6 and 3.8).

Undertaking further studies of crustacean offspring investment is important for both theoretical and practical reasons. Many kinds of crustaceans are useful for answering fundamental, theoretical core questions in evolutionary ecology, such as "Why do egg size and number vary?" In addition, a better understanding of offspring production in crustaceans would have many practical uses, given that many of these animals are important food organisms for humans or other organisms that humans eat, and thus require careful management. Also, they play major roles in the functioning of many kinds of aquatic and terrestrial ecosystems, including energy flow and nutrient cycling, and they are excellent bioindicators of environmental changes, including pollution.

ACKNOWLEDGMENTS

I thank Gary Wellborn and Martin Thiel for inviting me to write this chapter. I also thank them and Mika Tan and Tim Kiessling for their helpful ideas, comments, and encouragement.

REFERENCES

Acheampong, E., R.W. Campbell, A.B.S. Diekmann, and M.A. St John. 2011. Food availability effects on reproductive strategy: the case of *Acartia tonsa* (Copepoda: Calanoida). Marine Ecology Progress Series 428:151–159.

Achouri, M.S., F. Charfi-Cheikhrouha, and M. Zimmer. 2008. Reproductive patterns in syntopic terrestrial isopod species (Crustacea, Isopoda, Oniscidea) from Morocco. Pedobiologia 52:127–137.

Allan, J.D. 1976. Life history patterns in zooplankton. American Naturalist 110:165–180.

Alon, N., and N.E. Stancyk. 1982. Variation in life-history patterns of the grass shrimp *Palaemonetes pugio* in two South Carolina estuarine systems. Marine Biology 68:265–276.

Amin, S.M.N., A. Arshad, J.S. Bujang, S.S. Siraj, and S. Goddard. 2009. Reproductive biology of the sergestid shrimp *Acetes indicus* (Decapoda: Sergestidae) in coastal waters of Malacca, Peninsular Malaysia. Zoological Studies 48:753–760.

Andersen, K.H., T. Berge, R.J. Gonçalves, M. Hartvig, J. Heuschele, S. Hylander, N.S. Jacobsen, C. Lindemann, E.A. Martens, A.B. Neuheimer, K. Olsson, A. Palacz, A.E.F. Prowe, J. Sainmont, S.J. Traving, A.W. Visser, N. Wadhwa, and T. Kiørboe. 2016. Characteristic sizes of life in the oceans, from bacteria to whales. Annual Review of Marine Science 8:217–241.

Anger, K. 1995. The conquest of freshwater and land by marine crabs: adaptations in life-history patterns and larval bioenergetics. Journal of Experimental Marine Biology and Ecology 193:119–145.

Anger, K., and G.S. Moreira. 1998. Morphometric and reproductive traits of tropical caridean shrimps. Journal of Crustacean Biology 18:823–838.

Anger, K., G.S. Moreira, and D. Ismael. 2002. Comparative size, biomass, elemental composition (C, N, H), and energy concentration of caridean shrimp eggs. Invertebrate Reproduction and Development 42:83–93.

Arcos, F.G., I.S. Racotta, and A.M. Ibarra. 2004. Genetic parameter estimates for reproductive traits and egg composition in Pacific white shrimp *Penaeus* (*Litopenaeus*) *vannamei*. Aquaculture 236:151–165.

Austin, C.M. 1998. A comparison of clutch and brood size in the Red Claw, *Cherax quadricarinatus* (von Martens) and the Yabby, *C. destructor* Clark (Decapoda: Parastacidae). Aquaculture 167:135–145.

Baeza, J.A., L. Simpson, L.J. Ambrosio, R. Guéron, N. Mora, and D. Owen. 2015. Reproductive investment in a phyletic giant, the Caribbean King Crab *Damithrax spinosissimus*: exploring egg production costs in large brooding marine invertebrates. Journal of Shellfish Research 34:1049–1056.

Barber, I., D.J. Baird, and P. Calow. 1990. Clonal variation in general responses of *Daphnia magna* Straus to toxic stress. II. Physiological effects. Functional Ecology 4:409–414.

Barnes, M. 1989. Egg production in cirripedes. Oceanography and Marine Biology—An Annual Review 27:91–166.

Barnes, H., and M. Barnes. 1965. Egg size, nauplius size, and their variation with local, geographical, and specific factors in some common cirripedes. Journal of Animal Ecology 34:391–402.

Barnes, H., and M. Barnes. 1968. Egg numbers, metabolic efficiency of egg production and fecundity; local and regional variations in a number of common cirripedes. Journal of Experimental Marine Biology and Ecology 2:135–153.

Bartosiewicz, M., J. Jabłoński, J. Kozłowski, and P. Maszczyk. 2015. Brood space limitation of reproduction may explain growth after maturity in differently sized *Daphnia* species. Journal of Plankton Research 37:417–428.

Béguer, M., J. Bergé, M. Girardin, and P. Boët. 2010. Reproductive biology of *Palaemon longirostris* (Decapoda: Palaemonidae) from Gironde estuary (France), with a comparison with other European populations. Journal of Crustacean Biology 30:175–185.

Bernardo, J. 1996. The particular maternal effect of propagule size, especially egg size: patterns, models, quality of evidence and interpretations. American Zoologist 36:216–236.

Berrigan, D. 1991. The allometry of egg size and number in insects. Oikos 60:313–321.

Blueweiss, L., H. Fox, V. Kudzma, D. Nakashima, R. Peters, and S. Sams. 1978. Relationships between body size and some life history parameters. Oecologia 37:257–272.

Boersma, M. 1995. The allocation of resources to reproduction in *Daphnia galeata*: against the odds? Ecology 76:1251–1261.

Brante, A., C. Cifuentes, H.O. Pörtner, W. Arntz, and M. Fernandez. 2004. Latitudinal comparisons of reproductive traits in five Brachyuran species along the Chilean coast. Revista Chilena de Historia Natural 77:15–27.

Brillon, S., Y. Lambert, and J. Dodson. 2005. Egg survival, embryonic development, and larval characteristics of northern shrimp (*Pandalus borealis*) females subject to different temperature and feeding conditions. Marine Biology 147:895–911.

Briones-Fourzán, P., and E. Lozano-Alvarez. 1991. Aspects of the biology of the giant isopod *Bathynomus giganteus* A. Milne Edwards, 1879 (Flabellifera: Cirolanidae), off the Yucatan Peninsula. Journal of Crustacean Biology 11:375–385.

Briones-Fourzán, P., C. Barradas-Ortíz, F. Negrete-Soto, and E. Lozano-Álvarez. 2010. Reproductive traits of tropical deep-water pandalid shrimps (*Heterocarpus ensifer*) from the SW Gulf of Mexico. Deep Sea Research Part I: Oceanographic Research Papers 57:978–987.

Cannicci, S., R. Simoni, and F. Giomi. 2011. Role of the embryo in crab terrestrialisation: an ontogenetic approach. Marine Ecology Progress Series 430:121–131.

Capellini, I., C. Venditti, and R.A. Barton. 2010. Phylogeny and metabolic scaling in mammals. Ecology 91:2783–2793.

Charnov, E.L., and W.M. Schaffer. 1973. Life history consequences of natural selection: Cole's result revisited. American Naturalist 107:791–793.

Clarke, A. 1982. Temperature and embryonic development in polar marine invertebrates. International Journal of Invertebrate Reproduction 5:71–82.

Clarke, A. 1992. Reproduction in the cold: Thorson revisited. Invertebrate Reproduction and Development 22:175–183.

Clarke, A. 1993. Reproductive trade-offs in caridean shrimps. Functional Ecology 7:411–419.

Cleuvers, M., B. Goser, and H.T. Ratte. 1997. Life-strategy shift by intraspecific interaction in *Daphnia magna*: change in reproduction from quantity to quality. Oecologia 110:337–345.

Clutter, R.I., and G.H. Theilacker. 1971. Ecological efficiency of a pelagic mysid shrimp, estimates from growth, energy budget, and mortality studies. Fishery Bulletin 69:93–115.

Corey, S. 1981. Comparative fecundity and reproductive strategies in seventeen species of the Cumacea (Crustacea: Peracarida). Marine Biology 62:65–72.

Corey, S., and D.M. Reid. 1991. Comparative fecundity of decapod crustaceans. I. The fecundity of thirty-three species of nine families of caridean shrimp. Crustaceana 60:270–294.

Dagg, M.J. 1976. Complete carbon and nitrogen budgets for the carnivorous amphipod, *Calliopius laeviusculus* (Krøyer). Internationale Revue der gesamten Hydrobiologie und Hydrographie 61:297–357.

Devi, A.S., and M.K. Smija. 2013. Reproductive biology of the freshwater crab, *Travancoriana schirnerae* Bott, 1969 (Brachyura: Gecarcinucidae). Indian Journal of Fisheries 60:13–21.

Diesel, R., C.D. Schubart, and M. Schuh. 2000. A reconstruction of the invasion of land by Jamaican crabs (Grapsidae: Sesarminae). Journal of Zoology 250:141–160.

Duarte, C.M., and M. Alcaraz. 1989. To produce many small or few large eggs: a size-independent reproductive tactic of fish. Oecologia 80:401–404.

Dudycha, J.L., and M. Lynch. 2005. Conserved ontogeny and allometric scaling of resource acquisition and allocation in the Daphniidae. Evolution 59:565–576.

Echeverría-Sáenz, S., and I.S. Wehrtmann. 2011. Egg production of the commercially exploited deepwater shrimp, *Heterocarpus vicarius* (Decapoda: Pandalidae), Pacific Costa Rica. Journal of Crustacean Biology 31:434–440.

Ehnes, R.B., B.C. Rall, and U. Brose. 2011. Phylogenetic grouping, curvature and metabolic scaling in terrestrial invertebrates. Ecology Letters 14:993–1000.

Elgar, M.A. 1990. Evolutionary compromise between a few large and many small eggs: comparative evidence in teleost fish. Oikos 59:283–287.

Falster, D.S., A.T. Moles, and M. Westoby. 2008. A general model for the scaling of offspring size and adult size. American Naturalist 172:299–317.

Fernández, M., R. Calderón, M. Cifuentes, and P. Pappalardo. 2006. Brooding behaviour and cost of brooding in small body size brachyuran crabs. Marine Ecology Progress Series 309:213–220.

Fišer, C., M. Zagmajster, and V. Zakšek. 2013. Coevolution of life history traits and morphology in female subterranean amphipods. Oikos 122:770–778.

Forster, J., A.G. Hirst, and D. Atkinson. 2011. How do organisms change size with changing temperature? The importance of reproductive method and ontogenetic timing. Functional Ecology 25:1024–1031.

Fox, C.W., and M.E. Czesak. 2000. Evolutionary ecology of progeny size in arthropods. Annual Review of Entomology 45:341–369.

Friend, J.A., and A.M.M. Richardson. 1986. Biology of terrestrial amphipods. Annual Review of Entomology 31:25–48.

Fukui, Y. 1988. Comparative studies on the life history of the Grapsid crabs (Crustacea, Brachyura) inhabiting intertidal cobble and boulder shores. Publications of the Seto Marine Biological Laboratory 33:121–162.

Gimenez, L., and K. Anger. 2001. Relationships among salinity, egg size, embryonic development, and larval biomass in the estuarine crab *Chasmagnathus granulata* Dana, 1851. Journal of Experimental Marine Biology and Ecology 260:241–257.

Glazier, D.S. 1992. Effects of food, genotype, and maternal size and age on offspring investment in *Daphnia magna*. Ecology 73:910–926.

Glazier, D.S. 1999. Variation in offspring investment within and among populations of *Gammarus minus* Say (Crustacea: Amphipoda) in ten mid-Appalachian springs (U.S.A.) Archiv für Hydrobiologie 146:257–284.

Glazier, D.S. 2000. Smaller amphipod mothers show stronger trade-offs between offspring size and number. Ecology Letters 3:142–149.

Glazier, D.S. 2010. A unifying explanation for diverse metabolic scaling in animals and plants. Biological Reviews 85:111–138.

Glazier, D.S., and B.L. Sparks. 1997. Energetics of amphipods in ion-poor waters: stress resistance is not invariably linked to low metabolic rates. Functional Ecology 11:126–128.

Glazier, D.S., J. Wolf, and C. Kelly. 2003. Reproductive investment in aquatic and terrestrial isopods of central Pennsylvania. Crustaceana Monographs 2:151–179.

Glazier, D.S., E.M. Butler, S.A. Lombardi, T.J. Deptola, A.J. Reese, and E.V. Satterthwaite. 2011. Ecological effects on metabolic scaling: amphipod responses to fish predators in freshwater springs. Ecological Monographs 81:599–618.

Gliwicz, Z.M., and C. Guisande. 1992. Family planning in *Daphnia*: resistance to starvation in offspring born to mothers grown at different food levels. Oecologia 91:463–467.

Goulden, C.E., L. Henry, and D. Berrigan. 1987. Egg size, postembryonic yolk, and survival ability. Oecologia 72:28–31.

Green, J. 1956. Growth, size and reproduction in *Daphnia* (Crustacea: Cladocera). Proceedings of the Zoological Society of London 126:173–204.

Grime, J.P., and S. Pierce. 2012. The evolutionary strategies that shape ecosystems. Wiley-Blackwell, Hoboken, New Jersey.

Grzesiuk, M., and A. Mikulski. 2006. The effect of salinity on freshwater crustaceans. Polish Journal of Ecology 54:669–674.

Gueriau, P., N. Rabet, G. Clément, L. Lagebro, J. Vannier, D.E. Briggs, S. Charbonnier, S. Olive, and O. Béthoux. 2016. A 365-million-year-old freshwater community reveals morphological and ecological stasis in branchiopod crustaceans. Current Biology 26:383–390.

Guisande, C., and Z.M. Gliwicz. 1992. Egg size and clutch size in two *Daphnia* species grown at different food levels. Journal of Plankton Research 14:997–1007.

Guisande, C., J. Sanchez, I. Maneiro, and A. Miranda. 1996. Trade-off between offspring number and offspring size in the marine copepod *Euterpina acutifrons* at different food concentrations. Marine Ecology Progress Series 143:37–44.

Hammervold, S.H., R.N. Glud, J.O. Evjemo, A. Hagemann, and B.W. Hansen. 2015. A new large egg type from the marine live feed calanoid copepod *Acartia tonsa* (Dana)—Perspectives for selective breeding of designer feed for hatcheries. Aquaculture 436:114–120.

Hanamura, Y., R. Siow, and P.E. Chee. 2008. Reproductive biology and seasonality of the Indo-Australasian mysid *Mesopodopsis orientalis* (Crustacea: Mysida) in a tropical mangrove estuary, Malaysia. Estuarine, Coastal and Shelf Science 77:467–474.

Harlıoğlu, M.M., and Ö. Barım. 2004. The effect of dietary vitamin E on the pleopodal egg and stage-1 juvenile numbers of freshwater crayfish *Astacus leptodactylus* (Eschscholtz, 1823). Aquaculture 236:267–276.

Hartnoll, R.G. 1985. Growth, sexual maturity and reproductive output. Pages 101–128 *in* A.M. Wenner, editor. Crustacean issues 3: Factors in adult growth. Balkema, Rotterdam, The Netherlands.
Hartnoll, R.G. 2006. Reproductive investment in Brachyura. Hydrobiologia 557:31–40.
Hartnoll, R.G., M.S. Baine, A. Britton, Y. Grandas, J. James, A. Velasco, and M.G. Richmond. 2007. Reproduction of the black land crab, *Gecarcinus ruricola,* in the San Andres Archipelago, Western Caribbean. Journal of Crustacean Biology 27:425–436.
Hartnoll, R.G., A.C. Broderick, B.J. Godley, S. Musick, M. Pearson, S.A. Stroud, and K.E. Saunders. 2010. Reproduction in the land crab *Johngarthia lagostoma* on Ascension Island. Journal of Crustacean Biology 30:83–92.
Hendriks, A.J., and C. Mulder. 2008. Scaling of offspring number and mass to plant and animal size: model and meta-analysis. Oecologia 155:705–716.
Henmi, Y. 2003. Trade-off between brood size and brood interval and the evolution of underground incubation in three fiddler crabs (*Uca perplexa, U. vocans,* and *U. dussumieri*). Journal of Crustacean Biology 23:46–54.
Hernáez, P., Palma, S., and I. S. Wehrtmann. 2008. Egg production of the burrowing shrimp *Callichirus seilacheri* (Bott 1955) (Decapoda, Callianassidae) in northern Chile. Helgoland Marine Research 62:351–356.
Hildrew, A.G. 1985. A quantitative study of the life history of a fairy shrimp (Branchiopoda: Anostraca) in relation to the temporary nature of its habitat, a Kenyan rainpool. Journal of Animal Ecology 54:99–110.
Hines, A.H. 1982. Allometric constraints and variables of reproductive effort in brachyuran crabs. Marine Biology 69:309–320.
Hines, A.H. 1986. Larval patterns in the life histories of brachyuran crabs (Crustacea, Decapoda, Brachyura). Bulletin of Marine Science 39:444–466.
Hines, A.H. 1988. Fecundity and reproductive output in two species of deep-sea crabs, *Geryon fenneri* and *G. quinquedens* (Decapoda: Brachyura). Journal of Crustacean Biology 8:557–562.
Hines, A.H. 1991. Fecundity and reproductive output in nine species of *Cancer* crabs (Crustacea, Brachyura, Cancridae). Canadian Journal of Fisheries and Aquatic Science 48:267–275.
Hines, A.H. 1992. Constraint on reproductive output in brachyuran crabs: pinnotherids test the rule. American Zoologist 32:503–511.
Honěk, A. 1993. Intraspecific variation in body size and fecundity in insects: a general relationship. Oikos 66:483–492.
Hornung, E., and M.R. Warburg. 1993. Breeding patterns in the oniscid isopod, *Porcellio ficulneus* Verh., at high temperature and under different photophases. Invertebrate Reproduction and Development 23:151–158.
Hornung, E., and M.R. Warburg. 1994. Oosorption and oocyte loss in a terrestrial isopod under stressful conditions. Tissue and Cell 26:277–284.
Hurley, A.C. 1973. Fecundity of the acorn barnacle *Balanus pacificus* Pilsbry: a fugitive species. Limnology and Oceanography 18:386–393.
Hurley, D.E. 1968. Transition from water to land in amphipod crustaceans. American Zoologist 8:327–353.
Hutchinson, G.E. 1967. A treatise on limnology, volume II: introduction to lake biology and the limnoplankton. John Wiley, New York.
Iguchi, N., and T. Ikeda. 1994. Experimental study on brood size, egg hatchability and early development of a euphausiid *Euphausia pacifica* from Toyama Bay, southern Japan Sea. Bulletin of the Japan Sea National Fisheries Research Institute 44:49–57.
Ituarte, R.B., E.D. Spivak, and K. Anger. 2007. Intraspecific variability in life-history traits of a "freshwater shrimp," *Palaemonetes argentinus.* International Journal of Limnology 43:293–302.
Ivanova, M.B., and S.V. Vassilenko. 1987. Relationship between number of eggs, brood weight, and female body weight in Crustacea. Internationale Revue der gesamten Hydrobiologie und Hydrographie 72:147–169.
Jalihal, D.R., K.N. Sankolli, and S. Shenoy. 1993. Evolution of larval developmental patterns and the process of freshwaterization in the prawn genus *Macrobrachium* Bate, 1868 (Decapoda, Palaemonidae). Crustaceana 65:365–376.
Janas, U., and A. Mańkucka. 2010. Body size and reproductive traits of *Palaemon elegans* Rathke, 1837 (Crustacea, Decapoda), a recent colonizer of the Baltic Sea. Oceanological and Hydrobiological Studies 39:3–24.

Johnson, D.D., C.A. Gray, and W.G. Macbeth. 2010. Reproductive biology of *Portunus pelagicus* in a south-east Australian estuary. Journal of Crustacean Biology 30:200–205.
Johnson, W.S., M. Stevens, and L. Watling. 2001. Reproduction and development of marine peracaridans. Advances in Marine Biology 39:105–260.
Kevrekidis, T., G. Kourakos, and T. Boubonari. 2009. Life history, reproduction, growth, population dynamics and production of *Gammarus aequicauda* (Crustacea: Amphipoda) at extremely low salinities in a Mediterranean lagoon. International Review of Hydrobiology 94:308–325.
Kiørboe, T., and M. Sabatini. 1995. Scaling of fecundity, growth and development in marine planktonic copepods. Marine Ecology Progress Series 120:285–298.
Korcznski, R.E. 1991. Fecundity and age structure of freshwater and marine populations of the polar isopod *Mesidotea entomon* (L.). Invertebrate Reproduction and Development 20:45–48.
Kosobokova, K.N., H.-J. Hirche, and R.R. Hopcroft. 2007. Reproductive biology of deep-water calanoid copepods from the Arctic Ocean. Marine Biology 151:919–934.
Kotb, M.M., and R.G. Hartnoll. 2002. Aspects of the growth and reproduction of the coral gall crab *Hapalocarcinus marsupialis*. Journal of Crustacean Biology 22:558–566.
Lardies, M.A., M.J. Carter, and F. Bozinovic. 2004a. Dietary effects on life history traits in a terrestrial isopod: the importance of evaluating maternal effects and trade-offs. Oecologia 138:387–395.
Lardies, M.A., T.P. Catalán, and F. Bozinovic. 2004b. Metabolism and life-history correlates in a lowland and highland population of a terrestrial isopod. Canadian Journal of Zoology 82:677–687.
Lease, H.M., and B.O. Wolf. 2011. Lipid content of terrestrial arthropods in relation to body size, phylogeny, ontogeny and sex. Physiological Entomology 36: 29–38.
Lee, H.W., S. Ban, T. Ikeda, and T. Matsuishi. 2003. Effect of temperature on development, growth and reproduction in the marine copepod *Pseudocalanus newmani* at satiating food condition. Journal of Plankton Research 25:261–271.
Leme, M.H.A. 2006. Seasonal changes in reproductive traits of the crab *Sesarma rectum* (Grapsoidea: Sesarmidae) on the northern coast of São Paulo State, Brazil. Journal of Crustacean Biology 26:141–147.
Litulo, C. 2004. Breeding patterns of a tropical population of the fiddler crab, Uca inversa (Hoffmann, 1874) (Decapoda, Brachyura, Ocypodidae). Crustaceana 77:1045–1054.
Lonsdale, D.J., and J.S. Levinton. 1985. Latitudinal differentiation in embryonic duration, egg size, and newborn survival in a harpacticoid copepod. Biological Bulletin 168:419–431.
López-Sánchez, B., and E. Quintero-Torres. 2015. Inversión reproductiva de *Aratus pisonii* (Decapoda: Sesarmidae): diferencias entre hábitats y análisis de rutas. Revista de Biología Tropical 63:385–399.
Lynch, M. 1980. The evolution of cladoceran life histories. Quarterly Review of Biology 55:23–42.
Ma, H.H.T., P.K.S Lam, and D. Dudgeon. 1991. Inter- and intraspecific variation in the life histories of three sympatric isopods in a Hong Kong forest. Journal of Zoology 224:677–687.
Magalhães, C., and I. Walker, I. 1988. Larval development and ecological distribution of Central Amazonian palaemonid shrimps (Decapoda, Caridea). Crustaceana 55:279–292.
Maguire, I., G.I.V. Klobučar, and R. Erben. 2005. The relationship between female size and egg size in the freshwater crayfish *Austropotamobius torrentium*. Bulletin Français de la Pêche et de la Pisciculture 376–377:777–785.
Marshall, S., and J. Gittleman, J. 1994. Clutch size in spiders: is more better? Functional Ecology 8:118–124.
Mashiko, K. 1992. Genetic egg and clutch size variations in freshwater prawn populations. Oikos 63:454–458.
Mauchline, J. 1973. The broods of British Mysidacea (Crustacea). Journal of the Marine Biological Association of the United Kingdom 53:801–817.
Mauchline, J. 1985. Growth in mysids and euphasiids. Pages 337–354 *in* A.M. Wenner, editor. Crustacean issues 3: Factors in adult growth. Balkema, Rotterdam, The Netherlands.
Mauchline, J. 1988. Egg and brood sizes of oceanic pelagic crustaceans. Marine Ecology Progress Series 43:251–258.
McConaugha, J. R. 1992. Decapod larvae: dispersal, mortality, and ecology. A working hypothesis. American Zoologist 32:512–523.
McLaren, I.A. 1966. Predicting development rate of copepod eggs. Biological Bulletin 131:457–469.

Meireles, A.L., W.C. Valenti, and F.L. Mantelatto. 2013. Reproductive variability of the Amazon River prawn, *Macrobrachium amazonicum* (Caridea, Palaemonidae): influence of life cycle on egg production. Latin American Journal of Aquatic Research 41:718–731.

Meiri, S., A. Feldman, and L. Kratochvíl. 2015. Squamate hatchling size and the evolutionary causes of negative offspring size allometry. Journal of Evolutionary Biology 28:438–446.

Mirzajani, A., M. Sayadrahim, and A. Sari. 2011. Reproductive traits of some amphipods (Crustacea: Peracarida) in different habitats of Iran and Southern Caspian Sea. International Journal of Zoology 2011:598504

Müller, D.W., D. Codron, J. Werner, J. Fritz, J. Hummel, E.M. Griebeler, and M. Clauss. 2012. Dichotomy of eutherian reproduction and metabolism. Oikos 121:102–115.

Müller, Y., D. Ammar, and E. Nazari. 2004. Embryonic development of four species of palaemonid prawns (Crustacea, Decapoda): pre-naupliar, naupliar and post-naupliar periods. Revista Brasileira de Zoologia 21:27–32.

Nelson W.G. 1980. Reproductive patterns of gammaridean amphipods. Sarsia 65: 61–71.

Neuheimer, A.B., M. Hartvig, J. Heuschele, S. Hylander, T. Kiørboe, K.H. Olsson, J. Sainmont, and K.H. Andersen. 2015. Adult and offspring size in the ocean over 17 orders of magnitude follows two life history strategies. Ecology 96:3303–3311.

Niehoff, B. 2007. Life history strategies in zooplankton communities: The significance of female gonad morphology and maturation types for the reproductive biology of marine calanoid copepods. Progress in Oceanography 74:1–47.

Packard, G.C., and T.J. Boardman 1999. The use of percentages and size-specific indices to normalize physiological data for variation in body size: wasted time, wasted effort? Comparative Biochemistry and Physiology Part A: Molecular and Integrative Physiology 122: 37–44.

Paul, A.J., and A. Fuji. 1989. Bioenergetics of the Alaskan crab *Chionoecetes bairdi* (Decapoda: Majidae). Journal of Crustacean Biology 9:25–36.

Pérez-González, R., L.M. Valadez, G. Rodríguez-Domínguez, and E.A. Aragón-Noriega. 2012. Seasonal variation in brood size of the spiny lobster *Panulirus gracilis* (Decapoda: Palinuridae) in Mexican waters of the Gulf of California. Journal of Shellfish Research 31:935–940.

Peters, R.H. 1983. The ecological implications of body size. Cambridge University Press, Cambridge, England.

Pettersen, A.K., C.R. White, and D.J. Marshall. 2016. Why does offspring size affect performance? Integrating metabolic scaling with life-history theory. Proceedings of the Royal Society London B 282: 20151946.

Phillipson, J., and J. Watson. 1965. Respiratory metabolism of the terrestrial isopod *Oniscus asellus* L. Oikos 16:78–87.

Pinheiro, M.A., and O.S. Terceiro. 2000. Fecundity and reproductive output of the speckled swimming crab *Arenaeus cribrarius* (Lamarck, 1818) (Brachyura, Portunidae). Crustaceana 73:1121–1137.

Poltermann, M., H. Hop, and S. Falk-Petersen. 2000. Life under Arctic sea ice–reproduction strategies of two sympagic (ice-associated) amphipod species, *Gammarus wilkitzkii* and *Apherusa glacialis*. Marine Biology 136:913–920.

Poulin, R. 1995. Clutch size and egg size in free-living and parasitic copepods: a comparative analysis. Evolution 49: 325–336.

Poulin, R., and W.J. Hamilton. 1995. Ecological determinants of body size and clutch size in amphipods: a comparative approach. Functional Ecology 9:364–370.

Powell, M.L., and S.A. Watts. 2010. Energy allocation for egg production during starvation in sympatric crayfish, *Procambarus clarkii* and *Procambarus zonangulus*. Journal of the Alabama Academy of Science 81:226–234.

Przemysław, C., and D.G. Marcello. 2013. Realized fecundity in the first brood and size of eggs of Chinese mitten crab (*Eriocheir sinensis*)-laboratory studies. International Research Journal of Biological Science 2:1–6.

Quadros, A.F., P.B. Araujo, and C.C. Sokolowicz. 2008. Reproduction of neotropical isopods (Crustacea, Oniscidea) in southern Brazil: similarities and differences relative to temperate and tropical species. Pages 81–90 *in* M. Zimmer, F. Charfi-Cheikhrouha, and S. Taiti, editors. Proceedings of the International Symposium of Terrestrial Isopod Biology ISTIB 07. Shaker Verlag, Aachen, Germany.

Rabalais, N.N., and R.H. Gore. 1985. Abbreviated development in decapods. Pages 67–126 *in* A.M. Wenner, editor. Crustacean issues 2: Larval growth. Balkema, Rotterdam, The Netherlands.

Ramirez Llodra, E. 2002. Fecundity and life-history strategies in marine invertebrates. Advances in Marine Biology 43:88–170.

Ramirez Llodra, E., P.A. Tyler, and J.T. Copley. 2000. Reproductive biology of three caridean shrimp, *Rimicaris exoculata, Chorocaris chacei* and *Mirocaris fortunata* (Caridea: Decapoda), from hydrothermal vents. Journal of the Marine Biological Association of the UK 80:473–484.

Rashid, M.A., R.M. Shahjahan, R.A. Begum, M.S. Alam, Z. Ferdous, and M. Kamruzzaman. 2013. Fecundity and embryonic development in three *Macrobrachium* species. Journal of Entomology and Zoological Studies 1:3–11.

Reaka, M.L. 1979. The evolutionary ecology of life history patterns in stomatopod Crustacea. Pages 235–260 *in* S.E. Stancyk, editor. Reproductive ecology of marine invertebrates. University of South Carolina Press, Columbia, South Carolina.

Reid, D.M., and S. Corey. 1991. Comparative fecundity of decapod crustaceans, II. The fecundity of fifteen species of anomuran and brachyuran crabs. Crustaceana 61:175–189.

Reiss, M.J. 1989. The allometry of growth and reproduction. Cambridge University Press, New York.

Richardson, A., and P.B. Araujo. 2015. Lifestyles of terrestrial crustaceans. Pages 299–336 *in* M. Thiel and L Watling, editors. The natural history of the Crustacea, volume 2: Lifestyles and feeding biology. Oxford University Press, Oxford, England.

Robertson, A.L. 1988. Life histories of some species of Chydoridae (Cladocera: Crustacea). Freshwater Biology 20:75–84.

Roff, D.A. 1992. The evolution of life histories: theory and analysis. Chapman & Hall, New York.

Roff, D.A. 2002. Life history evolution. Sinauer, Sunderland, Massachusetts.

Rollinson, N., and J.A. Hutchings. 2013. The relationship between offspring size and fitness: integrating theory and empiricism. Ecology 94:315–324.

Rosa, R., R. Calado, L. Narciso, and M.L. Nunes. 2007. Embryogenesis of decapod crustaceans with different life history traits, feeding ecologies and habitats: a fatty acid approach. Marine Biology 151:935–947.

Ross, R.M., and L.B. Quetin. 2000. Reproduction in Euphausiacea. Pages 150–181 *in* I. Everson, editor. Krill: biology, ecology and fisheries. Blackwell Science, Oxford, England.

Sainte-Marie, B. 1991. A review of the reproductive bionomics of aquatic gammaridean amphipods: variation of life history traits with latitude, depth, salinity and superfamily. Hydrobiologia 223:189–227.

Saito, S. 1969. Energetics of isopod populations in a forest of central Japan. Researches on Population Ecology 11:229–258.

Salthe, S.N. 1969. Reproductive modes and the number and sizes of ova in the urodeles. American Midland Naturalist 81:467–490.

Sastry, A.N. 1983. Ecological aspects of reproduction. Pages 179–270 *in* F.J. Vernberg and W.B. Vernberg, editors. The biology of Crustacea, volume 8: Environmental adaptations. Academic Press, New York.

Sazhina, L.I. 1985. Fecundity and growth rate of copepods in different dynamic zones of equatorial countercurrent of the Indian Ocean. Polskie Archiwum Hydrobiologii 32:491–505.

Shakuntala, K. 1977. The relation between body size and number of eggs in the freshwater prawn, *Macrobrachium lamarrei* (H. Milne Edwards) (Decapoda, Caridea). Crustaceana 33:17–22.

Shen, Y., and D. Huang. 2008. Extant clam shrimp egg morphology: taxonomy and comparison with other fossil branchiopod eggs. Journal of Crustacean Biology 28:352–360.

Sibly, R. M., and P. Calow. 1986. Physiological ecology of animals. Blackwell Scientific Publications, Oxford, England.

Skadsheim, A. 1989. Regional variation in amphipod life history: effects of temperature and salinity on breeding. Journal of Experimental Marine Biology and Ecology 127:25–42.

Spivak, E.D., P.V. Silva, and T.A. Luppi. 2012. Habitat related variation in reproductive traits among intertidal crabs from the southwestern Atlantic. Journal of Crustacean Biology 32:57–66.

Stearns, S.C. 1992. The evolution of life histories. Oxford University Press, Oxford, England.

Steele, D.H., and V.J. Steele. 1975a. Egg size and duration of embryonic development in Crustacea. Internationale Revue der Gesamten Hydrobiologie und Hydrographie 60:711–715.

Steele, D.H., and V.J. Steele. 1975b. The biology of *Gammarus* (Crustacea, Amphipoda) in the northwestern Atlantic. XI. Comparison and discussion. Canadian Journal of Zoology 53:1116–1126.

Steele, D.H., and V.J. Steele. 1991a. Morphological and environmental restraints on egg production in amphipods. Pages 157–170 *in* A. Wenner and A. Kuris, editors. Crustacean issues 7: Crustacean egg production. Balkema, Rotterdam, The Netherlands.

Steele, D.H., and V.J. Steele. 1991b. Effects of salinity on the survival, growth rate, and reproductive output of *Gammarus lawrencianus* (Crustacea, Amphipoda). Marine Ecology Progress Series 78:49–56.

Strathmann, R.R., and M.F. Strathmann. 1995. Oxygen supply and limits on aggregation of embryos. Journal of the Marine Biological Association of the United Kingdom 75:413–428.

Subida, M.D., M.R. Cunha, and M.H. Moreira 2005. Life history, reproduction, and production of *Gammarus chevreuxi* (Amphipoda: Gammaridae) in the Ria de Aveiro, northwestern Portugal. Journal of the North American Benthological Society 24:82–100.

Sutcliffe, D.W. 1993. Reproduction in *Gammarus* (Crustacea, Amphipoda): female strategies. Freshwater Forum 3:26–64.

Sutton, S.L., M. Hassall, R. Willows, R.C. Davis, A. Grundy, and K.D. Sunderland. 1984. Life histories of terrestrial isopods: a study of intra- and interspecific variation. Symposia of the Zoological Society of London 53:269–294.

Swetha, C.H., B.P. Girish, and P.S. Reddy. 2015. Reproductive cycle and fecundity in natural population of edible freshwater crab, *Oziothelphusa senex senex* (Fabricius, 1798) (Decapoda: Brachyura). Journal of Aquaculture Research & Development 6:7.

Swiney, K.M., G.L. Eckert, and G.H. Kruse. 2013. Does maternal size affect red king crab, *Paralithodes camtschaticus*, embryo and larval quality? Journal of Crustacean Biology 33:470–480.

Tessier, A.J., and N.L. Consolatti. 1989. Variation in offspring size in *Daphnia* and consequences for individual fitness. Oikos 56:269–276.

Tessier, A.J., and N.L. Consolatti. 1991. Resource quantity and offspring quality in *Daphnia*. Ecology 72:468–478.

Thatje, S., and S. Hall. 2016. The effect of temperature on the evolution of per offspring investment in a globally distributed family of marine invertebrates (Crustacea: Decapoda: Lithodidae). Marine Biology 163:1–9.

Thessalou-Legaki, M. 1992. Reproductive variability of *Parapandalus narval* (Crustacea: Decapoda) along a depth gradient. Estuarine, Coastal and Shelf Science 35:593–603.

Thessalou-Legaki, M., and V. Kiortsis. 1997. Estimation of the reproductive output of the burrowing shrimp *Callianassa tyrrhena*: a comparison of three different biometrical approaches. Marine Biology 127:435–442.

Tollrian, R. 1995. Predator-induced morphological defenses: costs, life history shifts, and maternal effects in *Daphnia pulex*. Ecology 76:1691–1705.

Tropea, C., M. Arias, N.S. Calvo, and L.S.L. Greco. 2012. Influence of female size on offspring quality of the freshwater crayfish *Cherax quadricarinatus* (Parastacidae: Decapoda). Journal of Crustacean Biology 32:883–890.

Tsai, M.L., and H.C. Chen. 1997. Variations of reproductive traits of two semi-terrestrial isopods, *Ligia exotica* and *L. taiwanensis* (Crustacea: Ligiidae), in southern Taiwan. Zoological Studies 36:33–41.

Tsai, M.L., and C.F. Dai. 2001. Life history plasticity and reproductive strategy enabling the invasion of *Ligia exotica* (Crustacea: Isopoda) from the littoral zone to an inland creek. Marine Ecology Progress Series 210:175–184.

Urzúa, Á., and K. Anger. 2011. Larval biomass and chemical composition at hatching in two geographically isolated clades of the shrimp *Macrobrachium amazonicum*: intra-or interspecific variation? Invertebrate Reproduction and Development 55:236–246.

Verísimo, P., C. Bernárdez, E. González-Gurriarán, J. Freire, R. Muiño, and L. Fernández. 2011. Changes between consecutive broods in the fecundity of the spider crab, *Maja brachydactyla*. ICES Journal of Marine Science 68:472–478.

Visman, V., S. Pesant, J. Dion, B. Shipley, and R.H. Peters. 1996. Joint effects of maternal and offspring sizes on clutch mass and fecundity in plants and animals. Ecoscience 3:173–182.

Vogt, G. 2013. Abbreviation of larval development and extension of brood care as key features of the evolution of freshwater Decapoda. Biological Reviews 88:81–116.

Wacker, A., and D. Martin-Creuzburg. 2007. Allocation of essential lipids in *Daphnia magna* during exposure to poor food quality. Functional Ecology 21:738–747.

Walsh, C.J. 1993. Larval development of *Paratya australiensis* Kemp, 1917 (Decapoda: Caridea: Atyidae), reared in the laboratory, with comparisons of fecundity and egg and larval size between estuarine and riverine environments. Journal of Crustacean Biology 13:456–480.

Warburg, M.R., N. Cohen, D. Weinstein, and M. Rosenberg. 1993. Life history of a semelparous oniscid isopod, *Schizidium tiberianum* Verhoeff, inhabiting the Mediterranean region of northern Israel. Israel Journal of Zoology 39:79–93.

Watanabe, H. 1980. A study of the three species of Isopods in an evergreen broad-leaved forest in southwestern Japan. Revue d'Ecologie et de Biologie du Sol 17:229–239.

Wear R.G. 1974. Incubation in British decapod crustacean, and the effect of temperature on the rate of success of embryonic development. Journal of the Marine Biological Association of the United Kingdom 54:745–762.

Wellborn, G.A. 1994. Size-biased predation and prey life histories: a comparative study of freshwater amphipod populations. Ecology 75:2104–2117.

Węsławski, J.M., and J. Legeżyńska. 2002. Life cycles of some Arctic amphipods. Polish Polar Research 23:253–264.

Willows, R.I. 1987. Intrapopulation variation in the reproductive characteristics of two populations of *Ligia oceanica* (Crustacea: Oniscidea). Journal of Animal Ecology 56:331–340.

Winemiller, K.O., and K.A. Rose. 1993. Why do most fish produce so many tiny offspring? American Naturalist 142:585–603.

Wootton, R.J. 1991. Fish ecology. Chapman and Hall, New York.

Zamora-Terol, S., and E. Saiz. 2013. Effects of food concentration on egg production and feeding rates of the cyclopoid copepod *Oithona davisae*. Limnology and Oceanography 58:376–387.

Zairion, Y.W., M. Boer, and A. Fahrudin. 2015. Reproductive biology of the blue swimming crab *Portunus pelagicus* (Brachyura: Portunidae) in east Lampung waters, Indonesia: fecundity and reproductive potential. Tropical Life Sciences Research 26:67–85.

Žganec, K., P. Đurić, and S. Gottstein. 2011. Life history traits of the endangered endemic amphipod *Echinogammarus cari* (Crustacea, Gammaridae) from the Dinaric Karst. International Review of Hydrobiology 96:686–708.

4

SEMELPARITY AND ITEROPARITY

Øystein Varpe and Maciej J. Ejsmond

Abstract

Diversity in reproduction schedules is a central component of life history variability, with life span and age at maturity as key traits. Closely linked is the number of reproductive attempts and if organisms reproduce only once followed by death (semelparity) or spread reproduction over multiple and separated episodes during the reproductive lifespan (iteroparity). Amphipoda and Isopoda are two crustacean groups with many semelparous species, but semelparity is also part of other groups such as Decapoda, Copepoda, and Lepostraca. We briefly review theories posited for the evolution of semelparity and iteroparity, covering models on demography in both deterministic and fluctuating environments, and examine models on optimal resource allocation. We provide predictions of these theories, a guide on how to test them in crustaceans, and illustrate how theory can help us understand the diversity within this major taxon. We also point out a few shortcomings of these theories. One is that immediate recruitment is usually assumed in studies of semelparity, which is a poor assumption for the many crustaceans that form egg banks with prolonged recruitment. Another is the lack of models where iteroparity versus semelparity emerge as a consequence of life history trade-offs, rather than the more common approach that assumes demographic parameters. Furthermore, we argue that treating semelparity and iteroparity as a dichotomy is sometimes problematic and that viewing these strategies as a continuum can be useful. We discuss life history correlates and the particularly relevant links between the semelparity-iteroparity axis and capital breeding and seasonality, parental care, and terminal molts. We also discuss some of the indirect methods used to conclude if a crustacean is semelparous or not, such as a rapid drop in adult abundance after reproduction or signs of growth or storage after reproduction. A central message in the chapter is the high value of life history theory as a guide when formulating explanations and projecting evolutionary changes in reproductive lifespan of crustaceans.

Life Histories. Edited by Gary A. Wellborn and Martin Thiel.

INTRODUCTION

Life history theory aims to explain the diversity of life schedules or life cycles as well as the evolution of life history traits and combinations of traits (Stearns 1989, Roff 2002). At the core of this theory is the analysis of how limited resources are best allocated to fundamental processes such as growth, reproduction, and storage of energy and nutrient reserves in order to maximize fitness (Williams 1966, Stearns 1992). Understanding life history trade-offs is therefore a central dimension of life history theory. It is intuitive that a limited amount of resources cannot simultaneously allow both maximum growth and reproduction or maximum survival and reproduction. The challenge, therefore, is to predict how resources should be divided between competing needs. Different solutions to life history trade-offs are optimal under different conditions and in different environments. For instance, reproductive effort models predict that organisms experiencing high adult mortality should favor reproduction over survival, whereas those in safer environments should favor growth and longevity (Williams 1966, Reznick et al. 1990). Crustaceans, with their diverse life histories (Fig. 4.1), offer great opportunities for investigations of these predictions. For example, short-lived cladocerans in freshwater ponds shift sooner from asexual to sexual reproduction

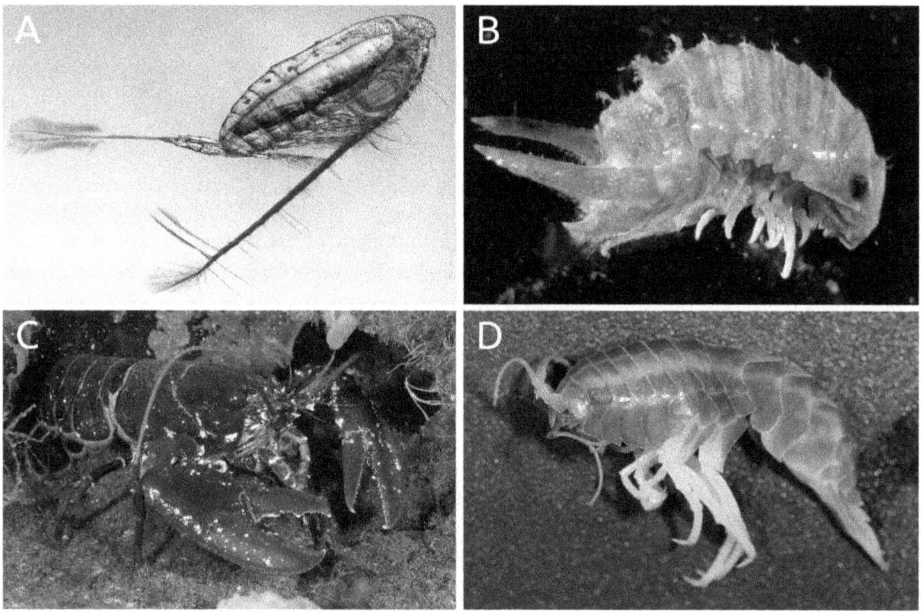

Fig. 4.1.
Photographs of crustacean representatives with life history strategies ranging from strict semelparity to long-lived species with iteroparity. See Fig. 4.3 for a schematic representation of the parity continuum. (A) Semelparous copepods of the genus *Neocalanus* spp., here represented by *Neocalanus plumchrus*. These copepods do not develop feeding appendices in their mature stage and rely on energy reserves for reproduction, with death following reproduction (see also Fig. 4.4). Note the well-filled oil sac, a large energy reserve. Photograph by Ross Hopcroft ©. (B) In the isopod *Paracerceis sculpta*, the females are strictly semelparous and die after one reproductive event, whereas males mate with multiple females and live longer. Depicted here is an "alpha" male known to attract and guard females from other males. Photograph by Alice Lodola ©. (C) Many lobsters and crabs, here represented by *Homarus gammarus*, are iteroparous. In these taxa iteroparity is often combined with indeterminate growth and long lifespans. Fecundity then typically increases with age and body size. Photograph by Erling Svensen/UWPhoto ©. (D) Many amphipods are semelparous, but some also represent the very iteroparous side of the continuum. *Eurythenes gryllus* is such an example of a long-lived, indeterminately growing, and iteroparous amphipod. Photograph by Armin Rose ©. See color version of this figure in the centerfold.

in riskier compared to safer environments (Ślusarczyk et al. 2012). In contrast, longer-lived shrimp species, such as *Palaemon adspersus*, may skip reproduction in response to adverse conditions (Berglund and Rosenqvist 1986), and in hermit crabs of the genus *Pagurus*, females continue to grow larger after first reproduction, and thereby invest in future reproductive success at the expense of current offspring production (Wada et al. 2008).

The diversity of life history strategies includes variability in how reproduction is scheduled through life. Some organisms have a short reproductive life span and reproduce only once followed by death (semelparity), whereas others spread reproduction over multiple, iterative attempts during life (iteroparity; Stearns 1992). Classic examples of the big-bang type (sensu Gadgil and Bossert 1970) of reproduction (i.e., semelparity) include salmonids migrating long distances from oceanic feeding grounds to rivers and lakes where they reproduce and afterward die (Crespi and Teo 2002, Quinn 2011), and *Lobelia* spp. or *Agave* spp. that live for many years before engaging in a single but massive reproductive episode (Young and Augspurger 1991). Examples of semelparity in Crustacea can be found in several groups, including marine (Shuster 1991) and terrestrial isopods (Warburg and Cohen 1991, Linsenmair 2007), amphipods (Sainte-Marie 1991), decapods (Kobayashi and Matsuura 1995), and marine copepods (Miller et al. 1984).

In this chapter, we (1) provide an introduction to the theories explaining the evolution of semelparity and iteroparity; (2) discuss how predictions from these theories can be tested and how Crustacea and the trait distribution within this group can serve as test cases; (3) argue for a need to view semelparity versus iteroparity more as a continuum and gradient than a dichotomy; (4) discuss some life history correlates, evolutionary relationships, and ecological consequences of semelparity in Crustacea, sometimes also with comparisons with other taxa; and (5) discuss the challenges and difficulties involved in determining semelparity and iteroparity from field data, the most common source of data for descriptions of life history diversity in Crustacea. Few in-depth analyses exist of the extent and adaptive value of semelparity and iteroparity within crustaceans. We start some of that work and add to more taxon-specific work provided earlier, such as the substantial review and analyses on amphipods (Sainte-Marie 1991), work on isopods (Harrison 1984), and perspectives provided for calanoid copepods (Hairston and Bohonak 1998, Varpe 2012).

Documenting and understanding the reproduction schedule is a central aim of empiricists. Quotes from studies of copepods can illustrate and inspire. As an example of death after reproduction, Miller et al. (1984) describe the spent stage of reproducing female copepods of the species *Neocalanus plumchrus* by writing that "[the] last clutch of eggs remain in the oviducts, the body tissue is entirely gone, and there is no visible ovary," and he further describes the females as "basically just exoskeletons." Kosobokova (1999), on the other hand, observes essential properties of iteroparity when describing the female reproductive biology of the copepod *Calanus glacialis*: "Many of them survive for several months after reproduction and are able to overwinter again."

THEORIES AND PREDICTIONS OF SEMELPARITY AND ITEROPARITY

Lamont Cole is often acknowledged for coining the terms semelparity and iteroparity (Cole 1954) in the first model to form a general, but overly simplistic, explanation of the adaptive value of semelparity versus iteroparity. Cole (1954) contrasted the population growth of annual and perennial organisms. An analysis of intrinsic population growth rates led him to conclude that any population of an annual organism able to produce just one additional individual offspring by sacrificing itself will grow in number at the same rate as a population of perennials. Because size of parents is usually much larger than the size of offspring, it seemed likely that by sacrificing itself any organism should be able to produce several more offspring than only one individual, and therefore an annual life history should be widely favored. The model by Cole was formulated with no explicit consideration of mortality (Gadgil and Bossert 1970), but the same result is achieved when mortality is included (Bryant 1971). The conclusion that the world should be filled with annuals dying after reproduction (later termed *Cole's paradox*) triggered a long-lasting debate in evolutionary biology on why iteroparity and a perennial lifestyle is so common. This debate led to a resolution of Cole's paradox but more importantly also brought forth three main and somewhat independent scientific threads that investigate the evolutionary drivers of survival and reproductive strategies, including evolution of parity. Subsequently, we provide a brief review of this theoretical development, highlighting the key characteristics and predictions derived under each approach, and then we link these to the diversity observed in Crustacea. Emerging from this review we also show that the routinely adopted dichotomous classification of life histories into semelparous or iteroparous, a consequence of decades dedicated to discussions of Coles' paradox, is crude and a potentially misleading simplification.

The Demographic Approach: Bet-Hedging in Fluctuating Environments

One of the first attempts to solve Cole's paradox focused on year-to-year variability. Using a computer model, Murphy (1968) showed that year-to-year stochastic variation in the number of offspring that recruit to the population promotes long-lived iteroparous forms. In contrast to annuals, reproduction in perennials can serve as a bet-hedging strategy (Slatkin 1974), because reproduction over several years increases the chances for favorable conditions for reproduction and recruitment. Schaffer (1974a) later showed that stochastic variation in adult survival, in contrast to uncertainty in fecundity, is expected to act in the opposite direction and promote semelparity. Several polemic articles followed, aiming at generalizations of earlier conclusions (e.g., Orzack and Tuljapurkar 1989). Important advancements were achieved by relaxing previous assumptions, such as inclusion of density-dependent juvenile recruitment (Bulmer 1985). The overall conclusion from these studies is that iteroparous reproduction is usually a response to a stochastically fluctuating environment, but general rules have proven difficult to formulate (Orzack and Tuljapurkar 1989, Benton and Grant 1999) and the topic is much discussed.

Among more recent and detailed studies, often with complex predictions, are those that introduced spatial structure (Ranta et al. 2000), combinations of gradual and stochastic change in environmental conditions (Zeineddine and Jansen 2009), and density-dependent vital rates in juveniles and adults (Benton and Grant 1999). How density dependence shapes costs and benefits of semelparity, and whether the density dependence operates via birth or death rates (Benton and Grant 1999), are central questions in the demographic approach. Sessile crustaceans such as barnacles could serve as a model group. Settlement probability in barnacles is usually negatively related to density, but growth rate and mortality of settled individuals are sometimes independent of density (e.g., Raimondi 1990). This circumstance allows separation of the density-dependent effects into components driven by recruitment and vital rates in adults and juveniles. More work on how density dependence operates across crustacean functional groups and taxa would be useful for our understanding of semelparity and iteroparity.

While theory provides complex predictions, rough generalizations are possible (e.g., Benton and Grant 1999). Before we stress some generalizations, it is important to note that the reviewed theory assumes immediate recruitment of offspring (e.g., no resting eggs). One generalization is that stochastic variation in fecundity selects for iteroparity, whereas uncertainty in adult survival promotes short life cycles with semelparous life history as one of the outcomes (Schaffer 1974a, Orzack and Tuljapurkar 1989). Barnacles illustrate these principles. In this group, selection towards iteroparity can be driven by the consequences of sedentary adults. Successful recruitment of larvae is highly dependent on whether a suitable habitat can be reached when ready to settle. As shown in *Semibalanus balanoides*, the year-to-year variability in transportation of planktonic larvae has much

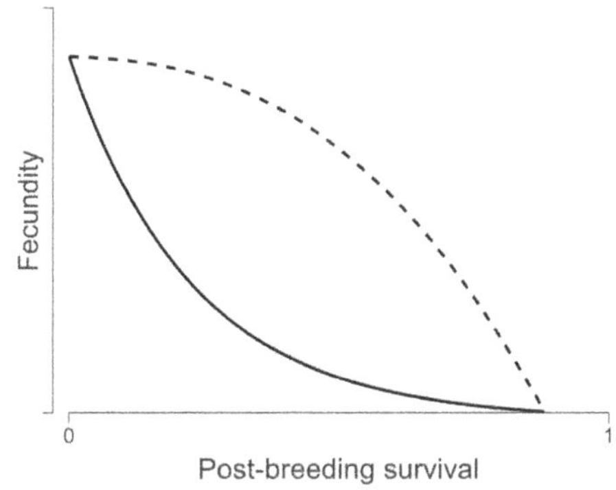

Fig. 4.2.
Schematic presentation of a central trade-off driving the evolution of reproductive life span. Faster-than-linear (convex, solid line) decline in fecundity with survival promotes semelparity. Under slower-than-linear (concave, dashed line) decrease in postbreeding survival, iteroparity or semelparity can evolve but iteroparity is more likely to be selected for (for details see Schaffer 1974b, Takada 1995).

impact on settlement probability and recruitment (Gaines and Bertness 1992); this stochastic variation in recruitment should select for iteroparity. Another generalization is related to the stability of reproductive success in iteroparous organisms. Whereas an iteroparous perennial lifestyle is a bet-hedging strategy that buffers the negative impact of poor years, it also comes with a price. The rule of thumb, supported by several intraspecific empirical estimates (e.g., see Fig. 4.20 in Roff 2002), is that offspring quality or number, or both, in a single reproductive event are greater in semelparous than in iteroparous organisms of similar size. This result suggests that extending reproduction to several years by iteroparous perennials comes with reduced maximum reproductive effort in years with favorable conditions. Hence, under stochastic variability, both strict semelparity and iteroparity with a particularly long reproductive life span are selected against (Benton and Grant 1999).

Here is a summary of the emerging predictions for semelparity versus iteroparity in fluctuating environments: (1) the greater the stochastic fluctuations in adult survival the stronger the selection toward semelparity; (2) the greater the stochastic fluctuations in fecundity the more iteroparous the lifestyle (however, semelparous organisms may buffer stochastic variation in recruitment by production of offspring that enter the population in different years, e.g., resting eggs, see the section "Dormant Eggs and Seed Bank Implications" below); (3) in a highly unpredictable environment, where strict semelparous and long-lived iteroparous organisms are selected against, intermediate strategies are promoted, including short-lived perennials or annuals with prolonged diapause or dormant eggs that hatch in different years.

The Demographic Approach: Deterministic Environments

Iteroparity as a bet-hedging strategy is only a partial solution to Cole's paradox. Given that the paradox emerged from a naively simple model, it might seem surprising that it took two decades to propose a solution based on a demographic process in a deterministic environment. Such a solution was offered when Charnov and Schaffer (1973) introduced differential juvenile and adult mortality to Cole's model. They concluded that evolution toward semelparity should indeed be expected when juveniles and adults suffer the same mortality risk. However, as juvenile survival, and so chances for recruitment, is usually much lower than adult survival, iteroparity should not be surprising. The higher the adult survival relative to juvenile survival, the more natural selection is expected to promote iteroparity. Semelparity is also expected to evolve when fecundity negatively affects adult postbreeding survival and an increase in offspring production results in a disproportionally higher decrease in the postbreeding survival (Fig. 4.2; Schaffer 1974b, Takada 1995). For instance, in semelparous species from the isopod family Sphaeromatidae, females perform a reproductive molt that highly affects their morphology and risk of mortality (Harrison 1984, Shuster 1991). The marsupium restricts female mobility and likely increases predation risk. Selection toward semelparity would be expected if production of a few more offspring (increased reproductive effort) disproportionally increases the mortality component related to the increased marsupium size and restricted mobility.

Other additions to the influential work by Charnov and Schaffer (1973) include flexible age at maturity (Young 1981) and density-dependent fecundity or juvenile survival (Charlesworth 1994). Overall, these additions confirmed the original conclusions that high adult survival and low juvenile survival should characterize iteroparous species. Semelparous organisms on the other hand are expected to have low postbreeding survival and relatively high chances of offspring recruitment. This is in contrast to conclusions derived under stochastically variable environments (see the previous section), which are sensitive to the form of density dependence in vital rates (e.g. Benton and Grant 1999).

The three emerging predictions in deterministic environments are: (1) increasing adult mortality should result in more semelparity, and increased juvenile mortality should select for iteroparity and a perennial lifestyle; (2) assuming positive scaling of fecundity with reproductive effort, semelparity is favored if adult survival decreases faster than linearly with an increase in fecundity (Fig. 4.2); and (3) iteroparity is a more likely outcome if postbreeding survival decreases slower than linearly with an increase in fecundity (Fig. 4.2).

The Optimal Resource Allocation Perspective

The demographic approach described above focuses on the differences between adult and juvenile prospects, but it ignores the underlying life history trade-offs that mediate these differences. For instance, the risk of death is unavoidably related to the duration of the juvenile period. A shortened juvenile phase, a potential driver of semelparity, would increase survival to maturity (Young 1981) but would also generate life history costs. One such cost is reduced body size, as the juvenile period is typically dedicated to growth. Because body size determines fecundity (Hines 1982, Corey and Reid 1991), smaller size would lead to fewer offspring, and a shortened juvenile period would not necessarily mean selection for semelparity. Similarly, the trade-off between costs of juvenile growth and future fecundity deserves closer attention. Theory predicts that adult size, and therefore future fecundity, is affected because growing juveniles compromise between safety and foraging (Werner and Anholt 1993, Abrams et al. 1996). In the salt-marsh grass shrimp (*Palaemonetes pugio*) for instance, juveniles migrate to shallow parts of the intertidal zone. These aquatic microhabitats, reached during high tide, are relatively safe because predatory fish occupy deeper parts (Kneib 1987). Safe habitats for juveniles affect the evolution of body size and, in line with this, the abundance of fish predators correlates negatively with the size attained by grass shrimps (Bass et al. 2001). Again, the effect of increased juvenile survival, expected to promote semelparity, is balanced by decreased fecundity. Growth strategies for juveniles may not only affect fecundity but also adult survival because in many crustaceans the juvenile and adult mortality schedules are dependent on size. *Hyalella azteca*, a freshwater amphipod, widely distributed and with an iteroparous but annual life cycle, is one example. Low growth rate in these amphipods, caused for example by food deprivation, extends the juvenile period, reduces body size at maturity, and as above, reduces fecundity (Moore and Farrar 1996). But it also affects mortality rate, and in fishless ponds, in contrast to ponds with fish predators, juvenile mortality exceeds adult mortality rate (Wellborn 1994). Hence, fishless lakes are inhabited by larger species of *Hyalella* than lakes with fish (Wellborn and Broughton 2008). Similarly, predation was suggested to explain spatial variability in reproductive investment in the amphipod *Gammarus minus* (Glazier 1999). In general, all the above examples show that the costs of growth, growth rate, and relations between survival of juveniles and adults in many cases are interdependent and cannot be understood in isolation. Understanding the combination of these processes is very valuable when predicting whether selection would favor semelparity or iteroparity. Life history trade-offs provide a mechanistic link between adult and juvenile survival, reproductive effort, and consequences of life history traits such as body size, size and age at maturity, or rate of aging. In a modeling perspective, iteroparity versus semelparity should therefore emerge as a consequence of life history trade-offs rather than assumed demographic parameters.

Growth-Reproduction Trade-off

Several life history trade-offs influence lifetime reproductive effort and schedules of juvenile versus adult survival. These trade-offs include current reproduction versus adult survival, mortality related to external parasites versus mortality and energetic costs of molting, reproductive effort versus rate of ageing, and foraging rate versus mortality rate. Allocation of resources to growth at the expense of reproduction deserves special attention here as larger size means higher fecundity (Hines 1982, Corey and Reid 1991). This trade-off should also include allocation to storage because energy reserves and other stores could often be viewed as an investment in future reproduction, although typically at a shorter time scale than growth (Varpe 2017, Ejsmond et al. 2015). In tanner crabs (*Chionoecetes bairdi*), primiparous females (i.e., breeding for the first time) are less fecund than females of the same size that have reproduced before. This was attributed to the fact that primiparous females grow and allocate resources to maturing oocytes (Somerton and Meyers 1983). The growth-reproduction trade-off is a significant factor in the evolution of parity in crustaceans also because numerous members such as decapods or many amphipods have indeterminate growth (i.e., they continue to grow after maturation). Indeterminate growth affects future fecundity and is age-specific as older, and thus larger, animals produce more offspring (see, for example, studies of the littoral prawn *Palaemon adspersus*; Berglund and Rosenqvist 1986). This effect of indeterminate growth changes the expectations of the demographic approach because higher fecundity due to increased size is an additional factor selecting for iteroparity. On the contrary, some crustacean species have determinate growth with a terminal molt, which we expect to select for a shorter reproductive phase and therefore potentially semelparity, a topic discussed further below.

In theoretical life history work that explores allocation trade-offs, most studies assume semelparity (Cohen 1971, Perrin and Sibly 1993) or that allocation to reproduction does not affect postbreeding survival (Kozłowski and Wiegert 1987, Kozłowski and Teriokhin 1999). However, life history models for plants have examined the link between evolution of parity and allocation of resources to growth and reproduction (Iwasa and Cohen 1989, Klinkhamer et al. 1997). In plants, it is relatively easy to determine the trade-off between current reproduction and future survival; a species is semelparous if no storage is left after reproduction and regrowth of vegetative parts after a harsh (e.g., winter) period is impossible (Iwasa and Cohen 1989, Klinkhamer et al. 1997). These models predict that semelparous annuals are favored when the growth season is short, when chances of surviving to the next breeding season decreases, or when large parts of storage are lost (Iwasa and Cohen 1989). Semelparity also brings fitness gains when reproduction attracts herbivores that cause higher mortality risk for adults (Klinkhamer et al. 1997). These predictions could be of value also to animal studies, including work on crustaceans. Many empirical studies on survival costs induced by reproduction in crustaceans show that multiple mechanisms are typically involved (see Browne 1982, Winfield and Townsend 1983, Sarma et al. 2002).

Overhead Costs

In many organisms, significant portions of survival and energetic costs related to reproduction are relatively constant and do not scale with fecundity. These costs are sometimes called "overhead" costs of reproduction. Overhead costs in Crustacea include energetic expenses related to a prebreeding molt (if a certain molt is obligatory for breeding), costs of migration to the breeding grounds, and physiological changes necessary for the onset of egg or sperm production. One example could be the synthesis of pigments for claw coloration in male fiddler crabs (*Austruca mjoebergi*). Colored claws attract females (Detto 2007) but may also attract predators. Links between semelparity and high overhead costs of reproduction have recently been postulated by Bonnet (2011), who was inspired by work on ectothermic vertebrates and viviparous snakes such as Aspic vipers (*Vipera aspis*). In this species reproduction comes with substantial energetic costs and for the relatively rare cases of iteroparity, females need two years to restore body condition and breed again (Bonnet 2011). Breeding females suffer high predation risk related to pregnancy (because they need to bask in the sun and are slower in escaping predators) but also a high postbreeding mortality rate; females are exhausted and in poor condition after releasing offspring. In a verbal model, Bonnet (2011) concluded that semelparity evolved because reproducing females pay high overhead costs (independent of the number of offspring produced). However, it is important to note that demographic costs of reproduction (high postbreeding mortality) and energetic costs of reproduction may influence adult and juvenile vital rates differently, therefore affecting the degree of selection for semelparity. Semelparity as a consequence of high overhead costs is therefore only partly supported by theory. High postbreeding mortality related to the poor condition of females is expected to select for shortened life and even semelparity.

The overhead cost of high postbreeding mortality applies to many crustaceans with determinate growth and a terminal molt, such as the snow crab *Chionoecetes opilio* (Sainte-Marie et al. 1995). Determinate growth with a terminal molt is possibly a constraint in some taxa, but it may also be an adaptation that increases current reproduction at the expense of life expectancy. A terminal molt restricts the ability to repair damage or remove external parasites via molting. Consequently, postbreeding mortality would rise and semelparity would be selected for. The role of energetic overhead, in selection for semelparity does however not seem to have theoretical support. In an early study, Schaffer (1974b) showed that fecundity should increase in an accelerating fashion (e.g., exponentially), with increasing reproductive allocation (including overhead costs) for semelparity to be optimal. To our knowledge, there are no examples of crustaceans in which offspring recruitment increases faster than linearly with reproductive allocation. This could perhaps be the case in species in which juveniles in a batch create a structure protecting themselves from predators, as in some insects, such that the more offspring in the batch, the higher the chance that the predator will not be able to break through the hypothetical shelter. However, Schaffer (1974b) showed that energetic overhead costs play a minor role in the evolution of semelparity because these do not scale with fecundity.

In sum, from the perspective of optimal resource allocation to reproduction or survival, theory predicts that (1) high mortality related to breeding accompanied by high costs of reproduction leads to semelparity, but the nature of reproductive costs is crucial; (2) life history trade-offs that increase juvenile survival at the expense of adult postbreeding survival create a selection gradient toward semelparity; and (3) the higher the overhead costs of reproduction in terms of high postbreeding mortality related to breeding, the more semelparity is favored (a terminal molt is a sign of selection for shortened life or even semelparity).

REVISITING DEFINITIONS AND CONCEPTS: DICHOTOMY OR CONTINUUM, ANNUAL AND PERENNIAL

The terms *semelparity* and *iteroparity* are not always easily applied. Traditionally, they are viewed as a dichotomy—as two contrasting solutions. Semelparity in the strict sense is death following *one* reproductive episode (Fig. 4.3A). A common challenge, however, is to evaluate what *one* reproductive episode is. Less strictly semelparous are those species that produce consecutive clutches

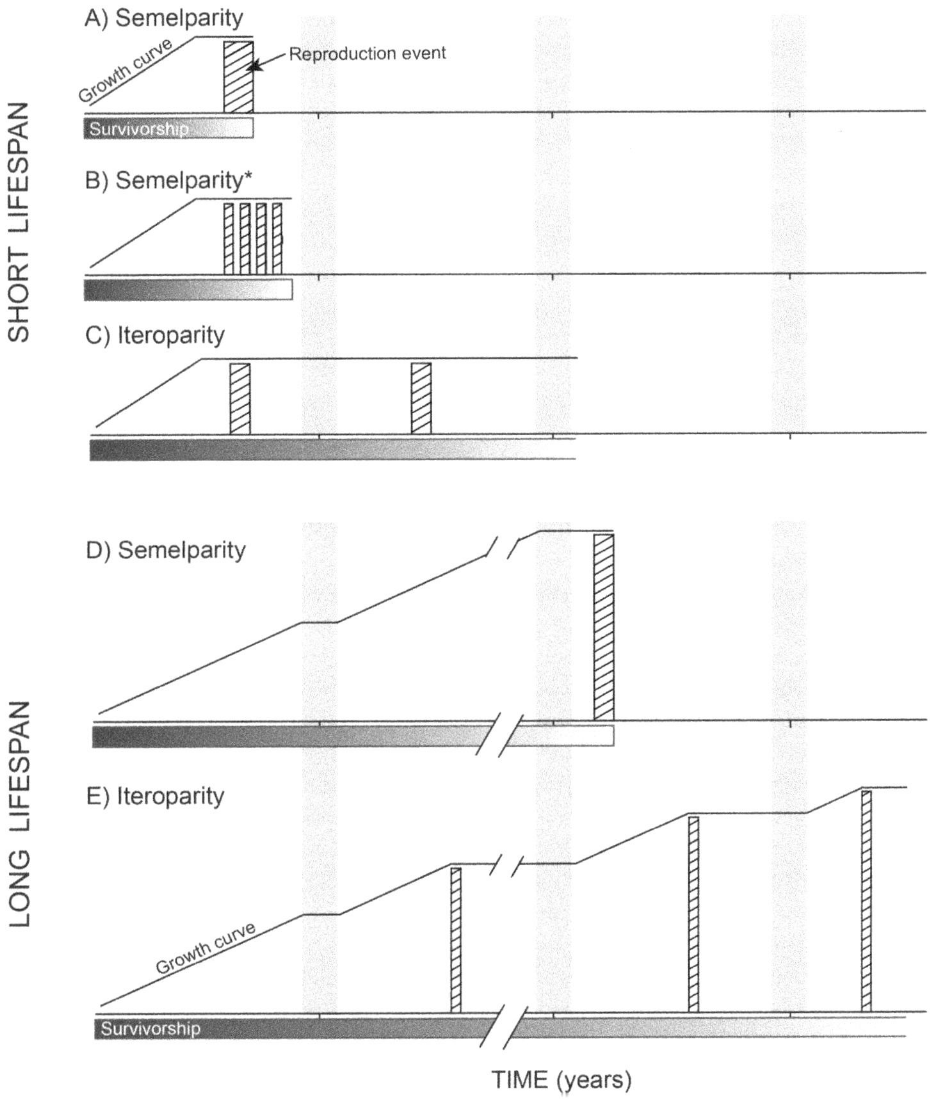

Fig. 4.3.
A schematic representation of life history variability and some of the forms that semelparity and iteroparity can take in an environment that to some degree is seasonal. The panels display a gradient in parity from (A) an annual and strictly semelparous strategy with immediate death following a big-bang reproductive event to (E) a long-lived iteroparous organism with indeterminate growth. (B) This represents the less clear situation where all reproduction falls within the same breeding season but is distributed in more or less separate clutches, which are here referred to as semelparity. Others have preferred terms such as "iteroparous annuals" for this lifestyle. (C) This represents iteroparity combined with a short life span. (D) This shows semelparity in species with a long life span but obviously short reproductive life span. The black line is a simplified and schematic growth profile. Hatched bars represent reproduction events. A nonfeeding season is conceptually marked with gray vertical bars (the duration of the nonfeeding period can differ in real cases). Survivorship is illustrated within the black- to gray-filled horizontal bar, kept closed to the right where death is certain and open where that is not the case. Time is assumed to represent years, although other timescales could be assumed.

of young over some time within the same breeding season, but still die afterward (Fig. 4.3B). Timescales, in particular reproducing during one year compared to multiple years, have sometimes been used as criteria for semelparity versus iteroparity, typically for work on mammals, birds, and plants (see Fritz et al. 1982 for discussion). In shorter-lived organisms this method works less well, and greater attention to how reproductive effort is distributed during the adult life span is needed, as is argued for insects (Fritz et al. 1982). This insect-based view would parallel cases seen in many crustaceans. Some have used "iteroparous annual" to describe the case of multiple sequential clutches within only one year or breeding season (e.g., Hairston and Bohonak 1998), whereas others have preferred to call this semelparity (e.g., Varpe 2012). In sum, timescales as described here are problematic and not a satisfying criterion for classification of semelparity and iteroparity. More fundamental properties are needed.

Begon et al. (1996) stressed that one should evaluate whether survival is prioritized or not between each reproductive event, and if it is indeed prioritized, one should consider it iteroparity. Recent and very interesting experimental work on plants suggests that there is variability within a semelparous strategy and that reproductive effort varies along a semelparous-iteroparous axis (Hughes and Simons 2014). Through experimental work modifying expected adult survival by manipulating expected season length, it was shown that this variability is generated by phenotypic plasticity (Hughes and Simons 2014).

Important insights on defining semelparity and iteroparity also come from theoretical models of optimal energy allocation in seasonal environments. In models that include capital breeding, it is optimal (i.e., brings the highest fitness gain) to allocate all stored energy within one day, which is physiologically impossible. Organisms are constrained with respect to the rate of transfer of resources from mother to eggs or offspring. If the rate of storage utilization is constrained, then the rate of reproduction per day decreases and the time necessary to utilize all reserves is much extended (Varpe et al. 2007, Varpe et al. 2009, Ejsmond et al. 2015). This allocation constraint gives resource acquisition and transfer from mother to offspring an important place in evaluating semelparity and iteroparity (see Houston et al. 2007). We would argue that if a female is not feeding between the clutches and dies afterward due to reproduction, it could belong in the semelparity category (or at least be closer to semelparity), compared to income breeders that feed again to build each new clutch (see also Capital Breeding and Seasonality section, below).

A continuum view is also evident in influential theoretical studies. For instance, in their work on the evolution of semelparity and iteroparity, Orzack and Tuljapurkar (1989) stated, "we regard this dichotomy as too simple and potentially misleading. Instead, we assume the temporal clumping and positioning of reproduction during life to be a continuous character." Extensive mapping of the reproduction and life history of gammaridean amphipods (Sainte-Marie 1991) also illustrated such a gradient. The copepod family Calanidae could be another case where semelparity (e.g., *Neocalanus* spp.; Miller et al. 1984) as well as perennial iteroparity (e.g., *Calanus hyperboreus*; Halvorsen 2015) are represented.

For strict semelparity, one may ask if death directly following reproduction is unavoidable. Semelparity is clearest when such a "programmed death" follows reproduction (Fig. 4.3A) and is caused by irreversible and dramatic changes in physiology as for instance in *Neocalanus* copepods (Miller 1984). Extreme cases include mothers that sacrifice themselves as a resource for their offspring, as in some semelparous *Schizidium* isopods in which offspring feed on the mother's body (Warburg and Cohen 1991). However, this type of semelparity can in some cases be "reversed" as in the desert spider (*Stegodyphus lineatus*) in which mothers separated from their offspring can produce a second clutch (Schneider and Lubin 1997). The desert spider example shows how a relatively simple behavioral decision about the timing of brood desertion is responsible for a species to be considered semelparous or iteroparous.

Death after reproduction is, however, more often an indirect effect due to heavy investment in reproduction and therefore less investment in maintenance, immune function, or predator avoidance. Death is in such cases less distinct. Reproductive investment would, for instance, drain resources that could be used to survive an inevitable and harsh period (e.g., the winter of seasonal environments). Thus, overwintering survival often trades off with reproduction. In this perspective, reproduction, instead of building up reserves, can be viewed as suicidal reproduction as offspring production decreases the chances to survive to the next reproductive event. But there are no clear-cut thresholds here, and therefore it seems more useful to consider semelparity and perennial iteroparity as endpoints of a continuum.

Furthermore, the extent that iteroparous organisms spread reproduction over independent breeding episodes can vary substantially (Fig. 4.3). Whereas semelparity covers a relatively narrow set of life histories, iteroparous strategies are diverse. Iteroparous crustaceans include short-lived species expected to breed for only two or three seasons (Fig. 4.3C), such as small isopods or large snow crabs, but also extremely long-lived species, such as the coconut crab (*Birgus latro*; Vogt 2012) that breed regularly for decades (Fig. 4.3D).

Reproductive Value

If we consider the evolution of parity with respect to demographic processes, fitness is closely linked to expected life duration and reproduction. If external mortality of adults is high, we would not observe long-lived iteroparous perennials. Life expectancy is a primary axis along which life histories are distributed (see Fig. 4.3 for this structure), but not a sufficient axis for a full understanding of the evolution of parity. We need to also consider reproductive value (Fisher 1930, Stearns 1992), a classic measure in evolutionary biology.

Reproductive value is the expected number of offspring an organism of a given age will produce throughout its remaining life. With respect to semelparity and iteroparity, it is also useful to refer to residual reproductive value (i.e., the reproductive value excluding the current reproductive event). In strictly semelparous species, the residual reproductive value at maturation is therefore extremely low. The duration of the juvenile phase can, however, vary considerably. Thus, we observe semelparity in short-lived as well as in long-lived species (Fig. 4.3A vs. Fig. 4.3D). If residual reproductive value decreases rapidly after reproduction (e.g., due to high postbreeding mortality), there would be strong selection to intensify early life reproductive allocation. Several mechanisms (reviewed above) can be responsible for a rapid decrease in residual reproductive value, and in turn evolution of semelparity. When using reproductive value as a fitness measure, we need to assume that the population is not in a long-term expansion or decline. In such cases, the reproductive success must be discounted according to trends in population numbers (Stearns 1992). Overall, considerations based on reproductive value are very useful starting points when attempting to position a given life history on the semelparity-iteroparity axis.

LIFE HISTORY TRADE-OFFS AND CORRELATES ON THE SEMELPARITY-ITEROPARITY AXIS

Can traits such as an exoskeleton or frequent molts preadapt crustaceans for either semelparity or iteroparity? Because of allocation trade-offs, life history theory predicts that certain combinations are more likely to occur than others (Stearns 1989, Roff 2002). We should ask if certain combinations of traits exist for crustaceans with semelparity and iteroparity as one of the focal traits. Here, we propose some candidate traits for such life history correlates, explain the logic behind them, and provide case studies of some crustacean groups as evidence. Questions related to these correlates could stimulate future studies.

Capital Breeding and Seasonality

Stores used for offspring production play a significant role in the life histories of crustaceans (Tessier and Goulden 1982, Varpe et al. 2009, Zeng et al. 2014). However, crustaceans are represented along the whole continuum from capital breeding to income breeding (Varpe et al. 2009, Griffen et al. 2012, Sainmont et al. 2014, Zeng et al. 2014). Capital breeders reproduce based on resources acquired in the past and stored until reproduction, whereas income breeders fuel reproduction with current acquisition. Semelparity often occurs in combination with capital breeding (Miller et al. 1984, Liu et al. 2011) as stored resources help maximize reproductive output when breeding takes place over a short time window (Tammaru and Haukioja 1996, Varpe et al. 2009). Pure capital breeders with lacking or impaired feeding appendages in adult stages, such as the copepod *Neocalanus* sp. (Miller et al. 1984; see Fig. 4.1A and Fig. 4.4) or females of the isopod *Paracerceis sculpta* (Shuster 1991; see Fig. 4.1B), are extreme examples of this case. There are several potential evolutionary drivers of capital breeding, with two deserving a closer look in relation to parity: the physical separation of feeding and breeding grounds (see Migration section, below), and seasonal variation in conditions that affect offspring prospects.

Seasonal variation in the environmental conditions that allow production of offspring is a fundamental constraint that affects breeding tactic and allocation strategy (McNamara and Houston 2008). Timing of reproduction affects reproductive value of released offspring, and this creates a life history dilemma for parents: when to grow, store, and reproduce over the season or year (Varpe 2017)? Capital

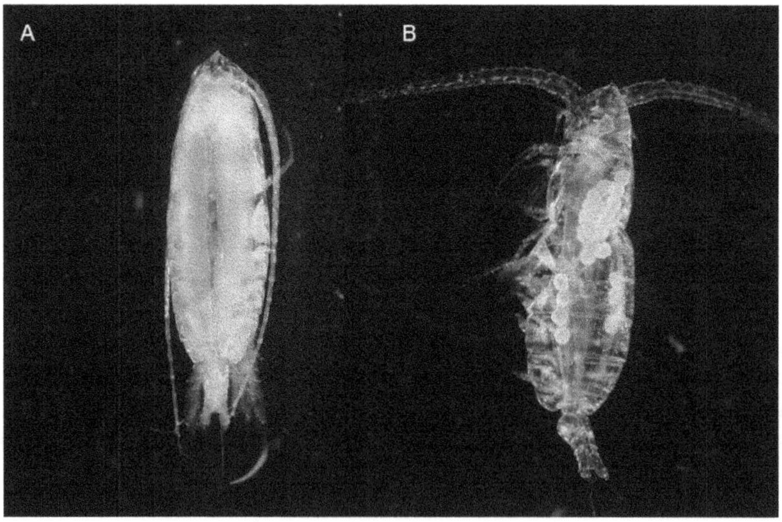

Fig. 4.4.
Two female individuals of the calanoid copepod *Neocalanus cristatus* in different reproductive stages. Individual (A) represents an early stage, full of energy reserves and with some of the early eggs seen. Individual (B) is almost fully spent, with only a few eggs left to be released, and little more than the exoskeleton remains (Miller et al. 1984). In this species, the mature female stage does not develop feeding appendages and is unable to feed. Egg production is therefore fully based on capital breeding through stores gathered near the surface the previous summer and brought to depth where the female later develops, releases eggs, and then dies. For a semelparous organism, it is adaptive to use all available resources for the single reproductive event, as illustrated by the spent stage in this copepod. In general, knowing that a given species cannot regain strength after reproduction (e.g., because of the inability to feed) represents strong evidence of semelparity. Photographs by Toru Kobari ©. See color version of this figure in the centerfold.

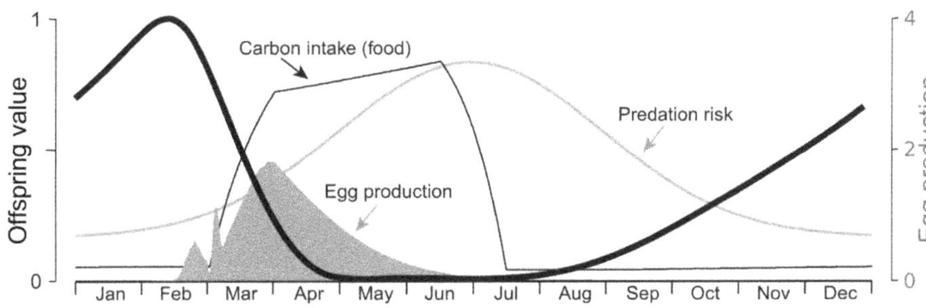

Fig. 4.5.
Semelparity and iteroparity have different consequences, for instance in seasonal environments where the windows of high offspring value are narrow, as illustrated by model predictions from a copepod life history model (Varpe et al. 2007, 2009). Seasonality in growth potential (thin black line) and predation risk (gray line) was assumed. The highly seasonal offspring value (thick black line) emerged, as possible in optimal annual routine models with fitness maximization by dynamic programming (McNamara and Houston, 2008). Note that peak offspring value is prior to the feeding season, selecting for some capital breeding. The population-level seasonality in egg production (filled gray area), as produced by individuals following the optimal state–dependent strategy, is mostly at times with intermediate and even low offspring value. These offspring are produced through income breeding. Because semelparity (only one breeding season per female) was assumed, egg production continued even if the late offspring had a very low probability to survive and contribute to future generations. If iteroparity had been allowed, it is more likely that the part of the feeding season with low offspring value would instead have been used for growth or storage, as preparation for future reproduction, as predicted by Ejsmond et al. (2015). For details on the copepod model and its outputs, see Varpe et al. (2007, 2009). Model predictions were originally for southern hemisphere seasonality, but are here shown for a northern hemisphere time axis.

breeding can be a response to seasonal environments because it allows high reproduction rate when prospects for offspring success are high (Varpe et al. 2009, Ejsmond et al. 2015), such as at times (or locations) when there is no food in the environment for income breeding. Calanoid copepods, which are abundant grazers in marine ecosystems, serve as an example. Capital breeding, based on energy reserves in well-defined lipid sacs (e.g. Vogedes et al. 2010), allows copepods to reproduce ahead of the phytoplankton bloom. This early reproduction allows offspring to exploit the phytoplankton growth in the same season (Niehoff et al. 2002, Daase et al. 2013). In a model by Varpe and colleagues (2007, 2009), with semelparity (or iteroparous annuals, sensu Hairston and Bohonak 1998) and deterministic environments assumed, some cases favored strategies that produced offspring in advance of the phytoplankton bloom, driven by a high fitness contribution by early-born young (Fig. 4.5). The model predicts that breeding (via capital breeding) should start when offspring prospects are maximal and continue (via income breeding) until death, even if the late produced offspring have a very low chance to survive and contribute to future generations (Varpe et al. 2007, 2009; Fig. 4.5). Instead, if iteroparity is assumed, we observe that mothers prioritize future survival late in the season by preparing for the next year (work in progress by the authors). In case of the copepods, this means rebuilding reserves preceding a long hibernation, instead of income breeding. We consequently believe that studies able to document breeding at times that are suboptimal from the offspring perspective would form valuable contributions to understanding evolution of semelparity in seasonal environments.

Migration

Spatially separated feeding and breeding grounds, and hence the associated breeding migrations, might select for semelparity and usually requires resources to be carried as stores (e.g., fat reserves) from the feeding to the breeding grounds (i.e., capital breeding). A famous crustacean example is the semelparous Japanese mitten crab (*Eriocheir japonica*) and Chinese mitten crab (*E. sinensis*), two closely related species (Tang et al. 2003). *Eriocheir japonica* and *E. sinensis* migrate from freshwater to the ocean to breed, and their egg production has a considerable capital breeding component (Wen et al. 2001, Dittel and Epifanio 2009). The catadromous lifestyle of the mitten crabs (Kobayashi and Matsuura 1995) resembles that of catadromous fish such as the European eel (*Anguilla anguilla*). Reproduction (which includes multiple broods) and migration in the mitten crab is exhaustive, and females die after one spawning season (Kobayashi and Matsuura 1995, Dittel and Epifanio 2009). We would refer to that as semelparity (Fig. 4.3B), but not in its strictest sense as multiple broods are possible. The first brood is usually the largest (Dittel and Epifanio 2009), and for the later broods there are indications of concurrent feeding (and therefore income breeding) because the hepatopancreas is not filled before egg production (Liu et al. 2011). Production of multiple broods may pay off better than a return migration and thus be optimal. These mitten crab examples show again that classification of a species into semelparous or iteroparous can be difficult and context dependent.

If a return migration to the feeding habitat is risky, then fitness may be maximized by semelparity (spawn and die) rather than preparations for another breeding in the future. In ornate rock lobsters (*Panulirus ornatus*), a large and commercially fished spiny lobster, juveniles travel about 500 km to spawn (MacFarlane and Moore 1986; Fig. 4.6), and a great majority of them die afterward. Mating and oviposition take place during the travel, a migration that typically starts in August. The breeding season is mostly from November to February. Up to three broods are produced, and there are several indications of high natural postspawning mortality (Moore and MacFarlane 1984, MacFarlane and Moore 1986). In sum, there is some support for evolution of crustaceans that combine long breeding migrations with semelparity (see Chapter 8 in this volume for a full chapter on migration in crustaceans).

Exoskeleton and Molting

An exoskeleton that is molted as the organism grows is one of the central and unifying characteristics of crustaceans and all arthropods. Some consequences of a molt are not related to an increase in body size but still have implications for semelparity and iteroparity. For instance, Tarling and Cuzin-Roudy (2008) have shown that frequent molts allow adult Antarctic krill (*Euphausia superba*) to remove ectoparasites. Molts are energetically costly and make individuals vulnerable, but in this case, they would potentially increase adult survival and reduce aging (mortality rate that increases with age, to which gradual accumulation of ectoparasites could contribute). These molts could then, independently of impacts on body size, select for iteroparity through increased survival. Other studies have found that under limited food conditions body size increments between molts are considerably reduced (Hartnoll 2001). This also suggests advantages of molts for future reproduction, beyond increased size and fecundity.

Many crustaceans are known for their ability to regenerate lost limbs (Savage and Sullivan 1978). In decapods, losing limbs mean decreased foraging ability or increased risk of being killed by a predator (reviewed in Juanes and Smith 1995) as well as loss of dominant status (Sekkelsten 1998). Consequently, molts that make regrowth of limbs possible (see Hopkins and Das 2015) increase survival and reproduction chances and thereby allow for iteroparity.

Many Decapoda, Ostracoda, and Copepoda do not grow after maturation and have a terminal molt (Hartnoll 1984). A terminal molt does not mean death, as for strict semelparity. Ectoparasite accumulation and poor regeneration is, however, expected to cause increased postbreeding mortality (Drouineau et al. 2013), which causes selection toward semelparity. The molt schedule in crustaceans is under endocrine control (Hartnoll 2001) and is consequently a potential proximate

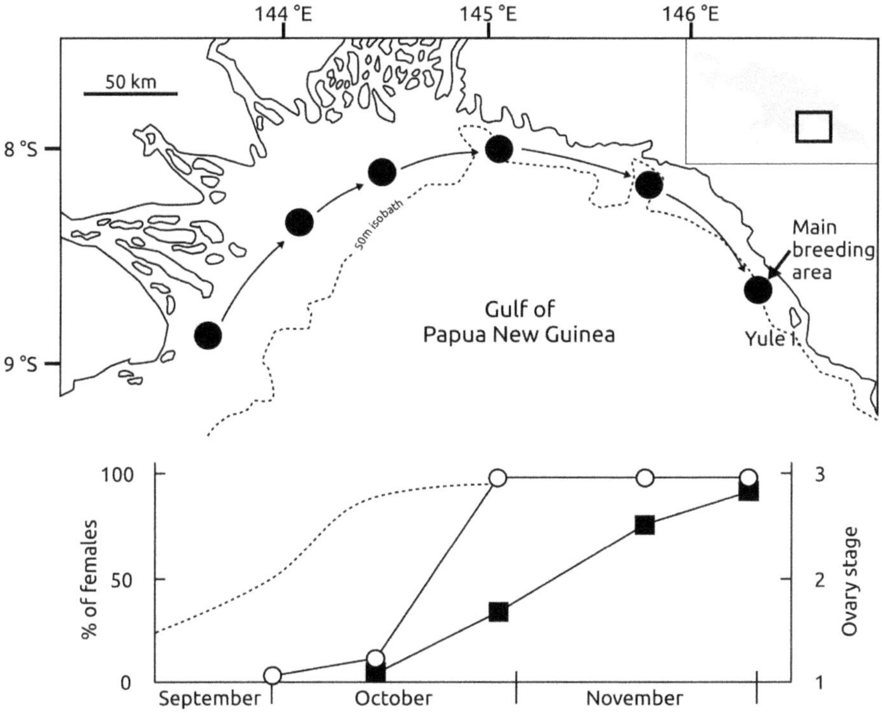

Fig. 4.6.
Annual breeding migration of the spiny lobster *Panulirus ornatus* in the Gulf of Papua (Papua New Guinea). Mating and oviposition takes place during the migration. Specimens of *P. ornatus* were collected at different sites (black dots in the map; locations are approximate), and the relationship between mating (circles) and oviposition (squares) was examined. The dotted line in the graph represents the mean stage of ovary development of females. During early September, about 50% of females had their ovaries developed to stage 2. Mating took place between mid to late October, and 80% had ovaries developed at stage 3 at that time. Oviposition occurred mostly during November. The breeding season mainly lasts from November to February, with up to three broods produced and large postspawning mortality in adults. There is variation between seasons and between different migratory waves of the same season (see MacFarlane and Moore 1986 for details). Modified from MacFarlane and Moore (1986), with permission from CSIRO Publishing.

mechanism for regulation of investment in future reproduction through growth (see Webster 2015 for a full chapter on the endocrinology of molting). From this perspective, the terminal molt, often seen as an internal developmental constraint, should perhaps rather be seen as an adaptation linked to favoring current reproduction over adult survival and growth.

Indeterminate Growth

Many crustaceans are indeterminate growers (i.e., species in which adults grow after maturation). The adaptive value of indeterminate growth is a central topic in life history theory (Heino and Kaitala 1999, Ejsmond et al. 2010) and much discussed in Crustacea (Hartnoll 2001, see also Chapter 2 in this volume).

Increased size translates to a range of ecological consequences, but the most general and important effect of a larger body is higher potential fecundity (Hines 1982, Corey and Reid 1991, Kiørboe and Sabatini 1995), with increased dominant status also being central in many cases (van der Meeren 1994, Duffy and Thiel 2007). To reach a large size, reproduction must be postponed. This is the classic evolutionary dilemma between benefits of investments in current versus future reproduction (Williams 1966). Investment in future reproduction through growth must be discounted by the chances that the organism dies before reproduction (Kozłowski 2006). Growth after maturation is therefore expected if organisms have relatively high chances of surviving to later reproductive events (Kozłowski and Teriokhin 1999, Ejsmond et al. 2010). Consequently, organisms with a short life span and a life history toward semelparity should not grow after reproduction but rather invest more in current reproductive efforts (e.g., Fig. 4.3B).

Importantly, growth after maturity is a strong sign of iteroparity. Gammaridean amphipods represent a full spectrum from semelparity to iteroparity (Sainte-Marie 1991). At the long-lived and iteroparous end, a good illustration is provided by the indeterminately growing *Eurythenes gryllus*, which matures at around 9 years of age and with an asymptotic size of about 14 cm and produces successive broods interspaced by molting and growth (Ingram and Hessler 1987). On the contrary, lack of growth after maturation cannot be used to conclude semelparity, as evident from the many iteroparous crustaceans that are determinate growers (e.g., the snow crab *Chionoecetes opilio* or the copepod *Calanus hyperboreus*).

Parental Care

Levels of parental care vary between and across crustacean taxa. For example, there are both egg-carrying and free-spawning copepods (Kiørboe and Sabatini 1994); there are long-lived decapods in which females carry the eggs for nearly a year, such as in the two commercially harvested lobsters *Homarus gammarus* and *H. americanus* (Phillips 2008); and there are crayfish where even the two first juvenile stages are carried by their mother (Vogt and Tolley 2004). Furthermore, broods are incubated in the marsupium by all (aquatic and terrestrial) amphipods (Sainte-Marie 1991, Thiel 1998). A link between parental care and semelparity has been suggested for insect species, where parental care has been predicted to be associated with reproduction that terminates life (Tallamy and Brown 1999). This link, however, has not been confirmed by comparative analyses. Only about 24% of the well-studied insect species with parental care were found to be semelparous (Trumbo 2013).

Parental care such as brood carrying can sometimes be viewed as an overhead cost of reproduction. For the evolution of semelparity, the nature of this cost is crucial. Importantly, it is not the brood carrying itself that triggers evolution of semelparity. With increasing parental care, juvenile survival increases and adult survival often decreases (Clutton-Brock 1991, Thiel 1999, Lewis and Loch-Mally 2010). Parental care therefore changes the relationship between adult and juvenile survival in a direction that favors a shorter adult phase with semelparity at the extreme. Semelparity is strongly selected for if chances for future reproduction (including mortality risk) are equal or lower than chances that a juvenile enters the adult phase (Charnov and Schaffer 1973). Even if various biological aspects might affect the threshold condition derived by Charnov and Schaffer (1973), semelparity will not be selected if survival of adults is orders of magnitude higher than juvenile survival. Therefore, intensive and long parental care can be expected to correlate with semelparity, but only for species with short life expectancy.

Semelparity is therefore relatively frequent in groups of short-lived crustaceans in which females care for the offspring by carrying embryos in the marsupium, a ventral brood pouch. In almost 200 species of gammaridean amphipods, a group with parental care (brood incubation in the marsupium), the great majority has short (< 2 years) life expectancy (see Appendix 1 in Sainte-Marie 1991) and almost 20% of the reviewed species were classified as semelparous (Sainte-Marie 1991). Even without control of the absolute measures of mortality in adults and juveniles, comparative studies show evidence consistent with this prediction. A clear example of extensive parental care and semelparity (suggested by death after one brood only) is the mud-dwelling amphipod *Casco bigelowi* (Thiel 1998). Within insects that care for offspring, semelparity was rare (12% of species) in species in which parental care is associated with creating structures that decrease mortality of juveniles but likely also adults (e.g., nests, burrows; Trumbo 2013). In contrast, for species that guard offspring in the open, semelparity was more common (45%; Trumbo 2013). What are the patterns for Crustacea? We are not aware of such an analysis, but observed variability in parental care provides a good starting point. Investigations should take into account the absolute differences between survival of adults and juveniles, and not only analyze the relative consequences of parental care.

Sex-Specific Strategies

Semelparity can be sex-specific, as observed and predicted for some fish species (Huse 1998). It is common that males display greater tendency toward semelparity than females. Crustacean examples of males, but not females, dying shortly after reproduction include some copepods (Miller et al. 1984, Boxshall et al. 1997) and amphipods such as *Monoporeia affinis* (Jacobson and Sundelin 2006). Loss of feeding appendages in males, but not females, is a clear sign of sex-specific semelparity, such as in the copepod *Euchaeta rimana* (Boxshall et al. 1997). Females may still be semelparous in the sense of breeding once or during one season, but live longer than males for reasons such as parental care (e.g., brood carrying in amphipods) or fertilization and egg production over some time using stored sperm (Titelman et al. 2007). In the copepod species *Eucalanus bungii*, there is evidence for true iteroparity in females, whereas males are semelparous and alive during a shorter time window than females (Miller et al. 1984). Similar observations exist for other calanoid copepods (Kosobokova 1999), as also discussed by Varpe (2012). In some species and breeding systems, however, it may be the females that are strictly semelparous and the males that breed for longer. An example is the alpha males of the isopod *Paracerceis sculpta* that mate with multiple females (where females reproduce only once) and also guard the females from other males (Fig. 4.1B; Shuster 1987). These alpha males can be active for a period corresponding to several female gestations (Shuster 1987).

For comparison, detailed studies on suicidal reproduction in mammals suggest that lethal effort of males is selected for under short and predictable conditions favorable for breeding, short and intensive period of mating, and intensive postcopulatory sexual selection (e.g., sperm competition; Fisher et al. 2013). We can predict that if the mating period is intensive and energetically costly, then males in particular are selected toward shorter life spans and semelparity.

Dormant Eggs and Seed Bank Implications

The theoretical studies on evolution of semelparity that we have reviewed previously assume (for simplicity) immediate recruitment of produced offspring. This is partly the reason that it is sometimes argued that annual reproduction (with nonoverlapping generations, and in many cases semelparous) is a strategy unable to persist over evolutionary time because any year with no offspring recruitment would lead to extinction. This argument is also used to explain why iteroparity is common. However, it ignores adaptations of many annual and semelparous organisms that make them resistant to year-to-year fluctuations in environmental conditions. A well-known example is annual plants with seeds produced during one season but germinating over several years (Cohen 1971). This "seed bank" strategy has also evolved among many annual or semelparous animals, including the resting egg solution of many crustaceans (Hairston 1996, Marcus 1996, Brendonck and de Meester 2003). The survival of resting eggs can be high, and emergence can take place over many years (Moritz 1987, Hairston et al. 1995, Hairston 1996). For instance, many freshwater cladocerans produce resting eggs that survive the winter and that form egg banks (Weider et al. 1997). Resting eggs are also the solution to other harsh periods, such as the dry stages of rock pools (e.g., in the fairy shrimp *Branchipodopsis wolfi*; Brendonck and Riddoch 2000). Resting eggs with differential emergence is a bet-hedging strategy similar to the way iteroparity is about "putting eggs in more than one basket." From a demographic point of view, the strategy of a univoltine (i.e., having one generation per year) crustacean (e.g., *Cyclops abyssorum* and *Leptodiaptomus minutus* in a high-latitude lake ecosystem; Antonsson 1992) dying at the end of the reproductive season is closer to an iteroparous perennial than a strictly annual and semelparous organism, because produced eggs recruit to the population over several years (cf. Orzack and Tuljapukar 1989; see Chapter 5 in this volume for a full discussion of voltinism). Another example is the "prolonged diapause" of larvae (i.e., diapause that in some individuals may last several years, as observed in several insect species; Menu and Debouzie 1993).

Egg bank strategies that allow recruitment to be prolonged for several years after reproduction show that recruitment does not have to follow the pulsed reproduction expected under semelparity. Several theoretical studies have investigated the adaptive value of prolonged diapause or seed banks (e.g. Cohen 1971, Menu et al. 2000). Interestingly, the model by Tachiki and Iwasa (2013) has suggested that prolonged diapause is an effect of coevolution with a fluctuating food resource. However, these models assumed either semelparity or iteroparity. For models on parity, however, it would be instructive if reproductive life span could emerge in parallel with the diapause and recruitment strategy. Little has been done to show when iteroparity (a bet-hedging strategy in fluctuating environments) can be outcompeted by semelparity with prolonged recruitment of juveniles.

THE EMPIRICAL CHALLENGE OF INFERRING SEMELPARITY AND ITEROPARITY

Several types of observations lead researchers to conclude that an organism is semelparous, but few are based on direct observations of adult mortality following a big-bang reproductive episode, as for instance in females of the isopod *Paracerceis sculpta* (Shuster 1991). Instead, several indirect methods are predominantly used when seeking evidence of semelparity versus iteroparity. Here we look into five categories of observations and discuss their strengths as well as potential pitfalls.

Oocytes

Among the strongest indirect evidence for iteroparity is the presence of small oocytes in spawning females or two generations of oocytes in gravid females. This criterion has, for instance, been applied in studies of shrimp (Lacoursiere-Roussel and Sainte-Marie 2009), similar to its use in determining if production of multiple clutches within the same breeding season is possible for squat lobsters (Dellatorre and Baron 2008). Kosobokova (1999) used gonad morphology to conclude that iteroparity is possible in the copepod *Calanus glacialis*.

Growth or Storage After Reproduction

Continuing growth or refilling reserves after a reproductive episode is strong additional evidence for investment in future reproduction and hence iteroparity. Growth can be viewed as a long-term investment in future reproduction and storage as preparation for future survival (such as diapause) or an upcoming breeding event (Varpe et al. 2009, Ejsmond et al. 2015). However, lack of growth after maturation does not necessarily equate to semelparity. Many iteroparous crustaceans are determinate growers. For copepods, sometimes thought of as semelparous, field studies have found some evidence of a switch from current reproduction to energy storage, which should be interpreted as a sign of iteroparity (Swalethorp et al. 2011). Interestingly, these signs of iteroparity were found in *Calanus hyperboreus* and *C. glacialis*, the most long-lived and capital breeding copepods, but not in the shorter-lived and income breeding *C. finmarchicus* (Swalethorp et al. 2011).

Rapid Drop in Adult Abundance

A rapid postreproductive decline in adult abundance is a clear sign of high postreproductive mortality, as long as migrations out of the study system can be excluded. It is reasonable to conclude semelparity in such cases. Examples include the leptostracan *Nebalia daytoni* (Vetter 1996), the observed massive male death shortly after mating in the amphipod *Monoporeia affinis* (Jacobson and Sundelin 2006), and investment in only one brood combined with male and female absence during parts of the year as in the amphipod *Casco bigelowi* (Thiel 1998; Fig. 4.7.). However, some degree of synchronized timing of breeding is needed for disappearance of a whole cohort to be a useful signal of semelparity. Such synchronization is typically pronounced in seasonal environments. Furthermore, iteroparity may for many species be physiologically possible and the strategy opted for, yet realized so rarely (because of high mortality) that semelparity is concluded. The possibility of iteroparity in the copepod *Calanus glacialis* (Kosobokova 1999) illustrates this point. It is stronger evidence of semelparity when rapid disappearance of adult individuals is accompanied with signs of programmed death following reproduction. *Schizidium* females that die during parturition provide such a contrasting example (Warburg and Cohen 1991).

Body Size and Abundance Data Combined

Size structure and analyses of cohorts are frequently used for evaluating basic life cycle features of crustaceans and have been used to suggest semelparity; examples include *Mysis mixta* (Richoux et al. 2004), *Onisimus litoralis* (Nygård et al. 2010), and *Uromunna naherba* (Esquete et al. 2014). The approach helps determine at which time of the year juveniles mature into adults and the time (age) needed to reach size of first reproduction. If adults continue to grow after first reproduction and form cohorts of different adult size, then iteroparity is the likely explanation. For instance, for *Onisimus litoralis*, for which Nygård et al. (2010) concluded semelparity, Węsławski et al. (2000) noted some particularly large females indicative of a second breeding season for a minority of the population. Similarly, Vetter (1996) concluded semelparity for *Nebalia daytoni* and iteroparity for *Nebalia hessleri*. However, if growth is deterministic, individuals may remain in the same size category and reproduce multiple times. That would be iteroparity (potentially combined with a terminal molt) but could be mistaken for semelparity, particularly if adult mortality is high and few would survive to their second breeding attempt.

Nonfeeding Adults

The absence of feeding appendages in the adult stage is a strong correlate of semelparity. It is then hardly possible (or beneficial) to prepare for a second breeding event. Crustacean examples include the copepods *Neocalanus* spp. (Miller et al. 1984, Fig. 4.4) and some sphaeromatid isopods (Harrison 1984, Shuster 1991). Interestingly, the lack of functional feeding appendages in adults is sparsely distributed among crustaceans when compared to insects where large taxonomic groups are characterized by no feeding in the adult stage. For instance, in mayflies (Ephemeroptera), adults do not feed and only live for hours or up to a few days (Brittain 1982), and similarly, a nonfeeding adult stage is common in several groups within Lepidoptera (Tammaru and Haukioja 1996).

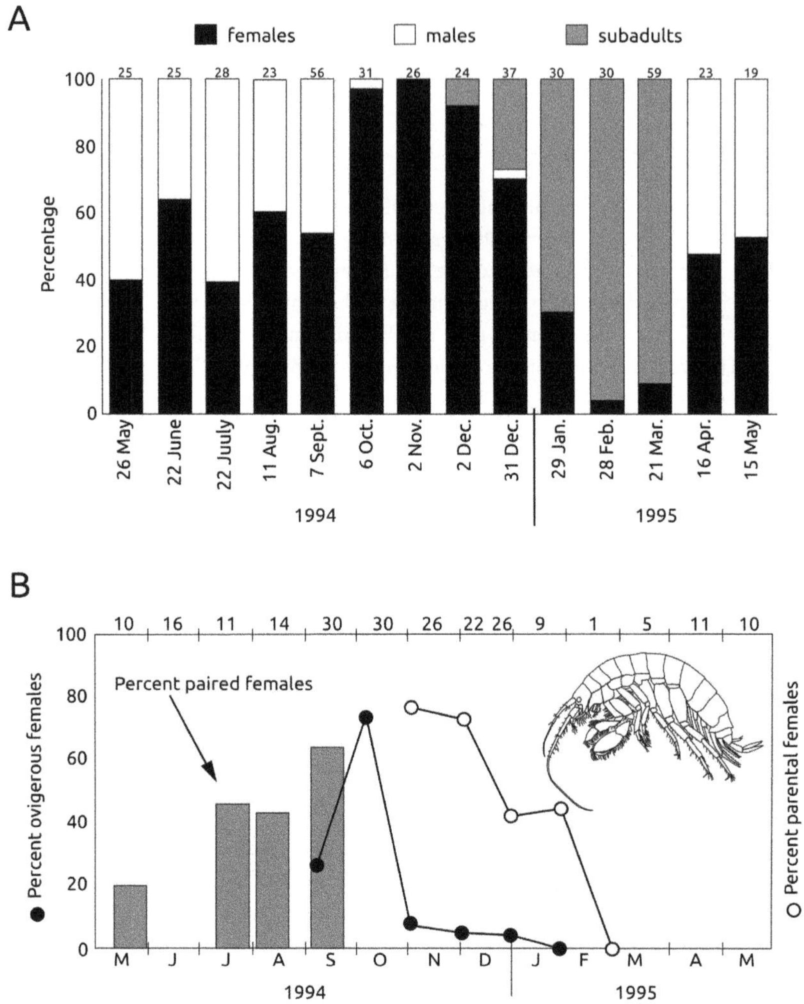

Fig. 4.7.
Inferring semelparity and iteroparity from empirical observations is often based on indirect evidence such as abundances and reproductive states of adults over time. The case of the deposit-feeding amphipod *Casco bigelowi*, studied by Thiel (1998), can illustrate this. Examination of (A) concurrent seasonal changes in relative abundance of females, males, and subadults (number of individuals collected on each sampling date is given on top of bars) and (B) reproductive status of females as seasonality of the percentage of paired females, ovigerous females, and parental females (number of adult females found on each sampling date is given on top of abscissa) shows that the proportion of females that are paired with males slowly builds up during the summer and the proportion of ovigerous females peaks, at the same time that males disappear from the population. Females care for offspring in their burrows for several months before they also disappear, probably dying. Consequently, subadults dominate the population during winter. These seasonal population dynamics illustrate that this species follows an annual life cycle, and in this case is also strictly semelparous, as it can be inferred from the population demographics that *C. bigelowi* produces only one brood during its lifetime (Thiel 1998). Other species in the same habitat, *Leptocheirus pinguis* and *Dyopedos monacanthus*, have multiple broods within the one breeding season (Thiel 1998). Modified from Thiel (1998), with permission from Springer. Inset of *C. bigelowi* modified from Bousfield (1973), with permission from NRC Research Press.

CONCLUSIONS AND FUTURE DIRECTIONS

Evolution of semelparity is directly related to the very central trade-off of current versus future reproduction (Williams 1966, Stearns 1992). Before asking where in the landscape of semelparity and iteroparity a species belongs, it is wise to first think about juvenile and adult reproductive value by mapping adaptations, behavior, and traits of a species on two life history axes: current versus future reproduction and juvenile versus adult survival. The theory of life history evolution, reviewed above, provides several hints and predictions regarding how the results of that mapping would relate to semelparity versus iteroparity, and the theory helps formulate ultimate explanations and project further evolutionary changes in reproductive life span.

In this chapter, we divided theoretical studies on evolution of semelparity into three groups: (1) bet-hedging in fluctuating environments, (2) age-specific birth and death rates in deterministic environments, and (3) fitness consequences of explicitly formulated life history trade-offs (optimal allocation of resources). Each provides important insights and suggests future directions for studies on evolution of semelparity. Models of fluctuating environments show that iteroparity is a bet-hedging strategy preventing complete recruitment failure in adverse years. However, semelparous crustaceans with resting eggs that recruit in several following years are also well adapted to fluctuating conditions. An important future direction would be to determine the critical conditions that select toward either iteroparous or semelparous reproduction with delayed recruitment. We would further argue that the demographic approach focused on deterministic environments, in which birth and death rates are assumed, should develop toward a more mechanistic approach, where birth and death rates emerge from life history trade-offs that are explicitly incorporated and linked to fitness. A very important future direction is to determine mechanisms underlying the trade-off between rate of reproduction and postbreeding survival in animals. Without this step, generalizations about the evolution of semelparity, similar to those invented for plant life histories, would be difficult to formulate.

Several ecological and evolutionary correlates discussed in this chapter generate interesting directions for future studies. For instance, signs of growth or renewing of storage after reproduction is a clear sign of iteroparity. Determinate growth, on the other hand, could be optimal also in long-lived perennial iteroparous species, and there is no reason to assume that determinate growth is a sign of semelparity. In both theoretical and empirical scientific literature, relatively little attention is given to the evolutionary connection between capital breeding and semelparous versus iteroparous reproduction. Whereas semelparity in a majority of cases should be combined with capital breeding, many perennial iteroparous species in Crustacea and other taxa are capital breeders. We see this variability as a fruitful direction for further investigation. Also, from the perspective of optimal resource allocation, we see a great potential in investigations of the adaptive value of the terminal molt in Crustacea. The terminal molt is routinely seen as a constraining factor and its adaptive value should be investigated more often. Finally, and inspired by recent work on plant reproduction (Hughes and Simons 2014), we would recommend greater attention to within-species variability in semelparity and iteroparity, particularly toward the semelparity end of the continuum where phenotypic plasticity would be expected to cause observed variability in reproductive life span.

ACKNOWLEDGMENTS

We are most thankful to Gary Wellborn, Martin Thiel, Mika Tan, and Tim Kiessling for very helpful and insightful suggestions on several versions of this text and for substantial and valuable help with the figures, particularly Figs. 4.6 and 4.7. We are very thankful to those who contributed photographs to the figures, and we thank Gro van der Meeren and Jan Kozłowski for valuable comments on a previous version of this chapter. Funding from the National Science Centre in Poland for project number 2014/15/B/NZ8/00236 supported our work on the chapter.

REFERENCES

Abrams, P.A., O. Leimar, S. Nylin, and C. Wiklund. 1996. The effect of flexible growth rates on optimal sizes and development times in a seasonal environment. The American Naturalist 147:381–395.

Antonsson, U. 1992. The structure and function of zooplankton in Thingvallavatn, Iceland. Oikos 64:188–221.

Bass, C.S., S. Bhan, G.M. Smith, and J.S. Weis. 2001. Some factors affecting size distribution and density of grass shrimp (*Palaemonetes pugio*) populations in two New Jersey estuaries. Hydrobiologia 450:231–241.

Begon, M., J.L. Harper, and C.R. Townsend. 1996. Ecology. Blackwell Science, Oxford, England.

Benton, T.G., and A. Grant. 1999. Optimal reproductive effort in stochastic, density-dependent environments. Evolution 53:677–688.

Berglund, A., and G. Rosenqvist. 1986. Reproductive costs in the prawn *Palaemon adspersus*: effects on growth and predator vulnerability. Oikos 46:349–354.

Bonnet, X. 2011. The evolution of semelparity. Pages 645–672 *in* R.D. Aldridge, and D.M. Sever, editors. Reproductive biology and phylogeny of snakes. Science Publishers, Enfield, New Hampshire.

Bousfield, E.L. 1973. Shallow-water Gammaridean amphipoda of New England. Cornell University Press, Ithaca, New York.

Boxshall, G.A., J. Yen, and J.R. Strickler. 1997. Functional significance of the sexual dimorphism in the cephalic appendages of *Euchaeta rimana* Bradford. Bulletin of Marine Science 61:387–398.

Brendonck, L., and L. de Meester. 2003. Egg banks in freshwater zooplankton: evolutionary and ecological archives in the sediment. Hydrobiologia 491:65–84.

Brendonck, L., and B.J. Riddoch. 2000. Egg bank dynamics in anostracan desert rock pool populations (Crustacea: Branchiopoda). Archiv Für Hydrobiologie 148:71–84.

Brittain, J.E. 1982. Biology of mayflies. Annual Review of Entomology 27:119–147.

Browne, R.A. 1982. The costs of reproduction in brine shrimp. Ecology 63:43–47.

Bryant, E.H. 1971. Life history consequences of natural selection: Cole's result. The American Naturalist 105:791–793.

Bulmer, M.G. 1985. Selection for iteroparity in a variable environment. The American Naturalist 126:63–71.

Charlesworth, B. 1994. Evolution in age-structured populations. Cambridge University Press, Cambridge, England.

Charnov, E.L., and W.M. Schaffer. 1973. Life-history consequences of natural selection: Cole's result revisited. The American Naturalist 107:791–793.

Clutton-Brock, T.H. 1991. The evolution of parental care. Princeton University Press, Princeton, New Jersey.

Cohen, D. 1971. Maximizing final yield when growth is limited by time or by limiting resources. Journal of Theoretical Biology 33:299–307.

Cole, L.C. 1954. The population consequences of life history phenomena. Quarterly Review of Biology 29:103–137.

Corey, S., and D. Reid. 1991. Comparative fecundity of decapod crustaceans. I. The fecundity of thirty-three species of nine families of caridean shrimp. Crustaceana 60:270–294.

Crespi, B.J., and R. Teo. 2002. Comparative phylogenetic analysis of the evolution of semelparity and life history in salmonid fishes. Evolution 56:1008–1020.

Daase, M., S. Falk-Petersen, Ø. Varpe, G. Darnis, J.E. Søreide, A. Wold, E. Leu, J. Berge, B. Philippe, and L. Fortier. 2013. Timing of reproductive events in the marine copepod *Calanus glacialis*: a pan-arctic perspective. Canadian Journal of Fisheries and Aquatic Sciences 70:871–884.

Dellatorre, F.G., and P.J. Barón. 2008. Multiple spawning and length of embryonic development of *Munida gregaria* in northern Patagonia (Argentina). Journal of the Marine Biological Association of the United Kingdom 88:975–981.

Detto, T. 2007. The fiddler crab *Uca mjoebergi* uses colour vision in mate choice. Proceedings of the Royal Society B: Biological Sciences 274:2785–2790.

Dittel, A.I., and C.E. Epifanio. 2009. Invasion biology of the Chinese mitten crab *Eriochier sinensis*: a brief review. Journal of Experimental Marine Biology and Ecology 374:79–92.

Drouineau, H., B. Sainte-Marie, and D. Duplisea. 2013. Estimating natural mortality and egg production of snow crab *Chionoecetes opilio* adult females. Aquatic Biology 18:261–270.

Duffy, J.E., and M. Thiel, editors. 2007. Evolutionary ecology of social and sexual systems: crustaceans as model organisms. Oxford University Press, New York.

Ejsmond, M.J., M. Czarnoleski, F. Kapustka, and J. Kozlowski. 2010. How to time growth and reproduction during the vegetative season: an evolutionary choice for indeterminate growers in seasonal environments. The American Naturalist 175:551–563.

Ejsmond, M.J., Ø. Varpe, M. Czarnoleski, and J. Kozlowski. 2015. Seasonality in offspring value and trade-offs with growth explain capital breeding. The American Naturalist 186:E111–E125.

Esquete, P., G.D.F. Wilson, and J.S. Troncoso. 2014. Ecology and systematics of a new species of Uromunna (Crustacea: Isopoda) from Spanish eelgrass beds. Helgoland Marine Research 68:329–339.

Fisher, D.O., C.R. Dickman, M.E. Jones, and S.P. Blomberg. 2013. Sperm competition drives the evolution of suicidal reproduction in mammals. Proceedings of the National Academy of Sciences of the USA 110:17910–17914.

Fisher, R.A. 1930. The genetical theory of natural selection. Oxford University Press, Oxford, England.

Fritz, R.S., N.E. Stamp, and T.G. Halverson. 1982. Iteroparity and semelparity in insects. The American Naturalist 120:264–268.

Gadgil, M., and W.H. Bossert. 1970. Life historical consequences of natural selection. The American Naturalist 104:1–24.

Gaines, S.D., and M.D. Bertness. 1992. Dispersal of juveniles and variable recruitment in sessile marine species. Nature 360:579–580.

Glazier, D.S. 1999. Variation in offspring investment within and among populations of *Gammarus minus* Say (Crustacea: Amphipoda) in ten mid-Appalachian springs (USA). Archiv Für Hydrobiologie 146:257–283.

Griffen, B.D., I. Altman, B.M. Bess, J. Hurley, and A. Penfield. 2012. The role of foraging in the success of invasive Asian shore crabs in New England. Biological Invasions 14:2545–2558.

Hairston, N.G. 1996. Zooplankton egg banks as biotic reservoirs in changing environments. Limnology and Oceanography 41:1087–1092.

Hairston, N.G., and A.J. Bohonak. 1998. Copepod reproductive strategies: life-history theory, phylogenetic pattern and invasion of inland waters. Journal of Marine Systems 15:23–34.

Hairston, N.G., R.A. Vanbrunt, C.M. Kearns, and D.R. Engstrom. 1995. Age and survivorship of diapausing eggs in a sediment bank. Ecology 76:1706–1711.

Halvorsen, E. 2015. Significance of lipid storage levels for reproductive output in the arctic copepod *Calanus hyperboreus*. Marine Ecology Progress Series 540:259–265.

Harrison, K. 1984. The morphology of the sphaeromatid brood pouch (Crustacea: Isopoda: Sphaeromatidae). Zoological Journal of the Linnean Society 82:363–407.

Hartnoll, R.G. 1984. Strategies of crustacean growth. Australian Museum Memoir 18:121–131.

Hartnoll, R.G. 2001. Growth in crustacean—twenty years on. Hydrobiologia 449:111–122.

Heino, M., and V. Kaitala. 1999. Evolution of resource allocation between growth and reproduction in animals with indeterminate growth. Journal of Evolutionary Biology 12:423–429.

Hines, A.H. 1982. Allometric constraints and variables of reproductive effort in brachyuran crabs. Marine Biology 69:309–320.

Hopkins, P.M., and S. Das. 2015. Regeneration in crustaceans. Pages 168–198 *in* E.S. Chang, and M. Thiel, editors. The natural history of the Crustacea, volume 4: Physiology. Oxford University Press, New York.

Houston, A.I., P.A. Stephens, I.L. Boyd, K.C. Harding, and J.M. McNamara. 2007. Capital or income breeding? A theoretical model of female reproductive strategies. Behavioral Ecology 18:241–250.

Hughes, P.W., and A.M. Simons. 2014. The continuum between semelparity and iteroparity: plastic expression of parity in response to season length manipulation in *Lobelia inflata*. BMC Evolutionary Biology 14:1–11.

Huse, G. 1998. Sex-specific life history strategies in capelin (*Mallotus villosus*)? Canadian Journal of Fisheries and Aquatic Sciences 55:631–638.

Ingram, C.L., and R.R. Hessler. 1987. Population biology of the deep-sea amphipod *Eurythenes gryllus*: inferences from instar analyses. Deep Sea Research Part A: Oceanographic Research Papers 34:1889–1910.
Iwasa, Y., and D. Cohen. 1989. Optimal growth schedule of a perennial plant. The American Naturalist 120:480–505.
Jacobson, T., and B. Sundelin. 2006. Reproductive effects of the endocrine disruptor fenarimol on a Baltic amphipod *Monoporeia affinis*. Environmental Toxicology and Chemistry 25:1126–1131.
Juanes, F., and L.D. Smith. 1995. The ecological consequences of limb damage and loss in decapod crustaceans: a review and prospectus. Journal of Experimental Marine Biology and Ecology 193:197–223.
Kiørboe, T., and M. Sabatini. 1994. Reproductive and life cycle strategies in egg-carrying cyclopoid and free-spawning calanoid copepods. Journal of Plankton Research 16:1353–1366.
Kiørboe, T., and M. Sabatini. 1995. Scaling of fecundity, growth and development in marine planktonic copepods. Marine Ecology Progress Series 120:285–298.
Klinkhamer, P., T. Kubo, and Y. Iwasa. 1997. Herbivores and the evolution of the semelparous perennial life-history of plants. Journal of Evolutionary Biology 10:529–550.
Kneib, R.T. 1987. Predation risk and use of intertidal habitats by young fishes and shrimp. Ecology 68:379–386.
Kobayashi, S., and S. Matsuura. 1995. Reproductive ecology of the Japanese mitten crab *Eriocheir japonicus* (De Haan) in its marine phase. Benthos Research 49:15–28.
Kosobokova, K.N. 1999. The reproductive cycle and life history of the arctic copepod *Calanus glacialis* in the White Sea. Polar Biology 22:254–263.
Kozłowski, J. 2006. Why life histories are diverse. Polish Journal of Ecology 54:585–605.
Kozłowski, J., and A.T. Teriokhin. 1999. Allocation of energy between growth and reproduction: the pontryagin maximum principle solution for the case of age-and season-dependent mortality. Evolutionary Ecology Research 1:423–441.
Kozłowski, J., and R.G. Wiegert. 1987. Optimal age and size at maturity in annuals and perennials with determinate growth. Evolutionary Ecology 1:231–244.
Lacoursiere-Roussel, A., and B. Sainte-Marie. 2009. Sexual system and female spawning frequency in the sculptured shrimp *Sclerocrangon boreas* (Decapoda: Caridea: Crangonidae). Journal of Crustacean Biology 29:192–200.
Lewis, S.E., and A.M. Loch-Mally. 2010. Ovigerous female amphipods (*Gammarus pseudolimnaeus*) face increased risks from vertebrate and invertebrate predators. Journal of Freshwater Ecology 25:395–402.
Linsenmair, K.E. 2007. Sociobiology of terrestrial isopods. Pages 339–364 *in* J.E. Duffy, and M. Thiel, editors. Evolutionary ecology of social and sexual systems: crustaceans as model organisms. Oxford University Press, New York.
Liu, Z., X. Wu, Y. Cheng, C. Zeng, and X. Yang. 2011. Ovarian re-maturation following the first spawning in the Chinese mitten crab, *Eriocheir sinensis* (H. Milne-Edwards). Aquaculture Research 42:417–426.
MacFarlane, J., and R. Moore. 1986. Reproduction of the ornate rock lobster, *Panulirus ornatus* (Fabricius), in Papua New Guinea. Australian Journal of Marine and Freshwater Research 37:55–65.
Marcus, N.H. 1996. Ecological and evolutionary significance of resting eggs in marine copepods: past, present, and future studies. Hydrobiologia 320:141–152.
McNamara, J.M., and A.I. Houston. 2008. Optimal annual routines: behaviour in the context of physiology and ecology. Philosophical Transactions of the Royal Society B: Biological Sciences 363:301–319.
Menu, F., and D. Debouzie. 1993. Coin-flipping plasticity and prolonged diapause in insects: example of the chestnut weevil *Curculio elephas* (Coleoptera: Curculionidae). Oecologia 93:367–373.
Menu, F., J.P. Roebuck, and M. Viala. 2000. Bet-hedging diapause strategies in stochastic environments. American Naturalist 155:724–734.
Miller, C.B., B.W. Frost, H.P. Batchelder, M.J. Clemons, and R.E. Conway. 1984. Life histories of large, grazing copepods in a subarctic ocean gyre: *Neocalanus plumchrus, Neocalanus cristatus*, and *Eucalanus bungii* in the northeast pacific. Progress in Oceanography 13:201–243.
Moore, D.W., and J.D. Farrar. 1996. Effect of growth on reproduction in the freshwater amphipod, *Hyalella azteca* (Saussure). Hydrobiologia 328:127–134.
Moore, R., and MacFarlane, J.W. 1984. Migration of the Ornate Rock Lobster, *Panulirus ornatus* (Fabricius), in Papua New Guinea. Australian Journal of Marine and Freshwater Research 35:197–212.

Moritz, C. 1987. A note on the hatching and viability of *Ceriodaphnia ephippia* collected from lake sediment. Hydrobiologia 145:309–314.
Murphy, G.I. 1968. Pattern in life history and environment. The American Naturalist 102:391–403.
Niehoff, B., S.D. Madsen, B.W. Hansen, and T.G. Nielsen. 2002. Reproductive cycles of three dominant *Calanus* species in Disko Bay, west Greenland. Marine Biology 140:567–576.
Nygård, H., J. Wallenschus, L. Camus, Ø. Varpe, and J. Berge. 2010. Annual routines and life history of the amphipod *Onisimus litoralis*: seasonal growth, body composition and energy budget. Marine Ecology Progress Series 417:115–126.
Orzack, S.H., and S. Tuljapurkar. 1989. Population dynamics in variable environments. VII. The demography and evolution of iteroparity. The American Naturalist 133:901–923.
Perrin, N., and R.M. Sibly. 1993. Dynamic models of energy allocation and investment. Annual Review of Ecology and Systematics 24:379–410.
Phillips, B. 2008. Lobsters: biology, management, aquaculture and fisheries. Wiley-Blackwell, Hoboken, New Jersey.
Quinn, T.P. 2011. The behavior and ecology of pacific salmon and trout. University of British Columbia Press, Vancouver, Canada.
Raimondi, P.T. 1990. Patterns, mechanisms, consequences of variability in settlement and recruitment of an intertidal barnacle. Ecological Monographs 60:283–309.
Ranta, E., D. Tesar, S. Alaja, and V. Kaitala. 2000. Does evolution of iteroparous and semelparous reproduction call for spatially structured systems? Evolution 54:145–150.
Reznick, D.A., H. Bryga, and J.A. Endler. 1990. Experimentally induced life-history evolution in a natural population. Nature 346:357–359.
Richoux, N.B., D. Deibel, and R.J. Thompson. 2004. Population biology of hyperbenthic crustaceans in a cold water environment (Conception Bay, Newfoundland). I. *Mysis mixta* (Mysidacea). Marine Biology 144:881–894.
Roff, D.A. 2002. Life history evolution. Sinauer Associates, Sunderland.
Sainmont, J., K.H. Andersen, Ø. Varpe, and A.W. Visser. 2014. Capital versus income breeding in a seasonal environment. The American Naturalist 184:466–476.
Sainte-Marie, B. 1991. A review of the reproductive bionomics of aquatic gammaridean amphipods: variation of life-history traits with latitude, depth, salinity and superfamily. Hydrobiologia 223:189–227.
Sainte-Marie, B., S. Raymond, and J.C. Brethes. 1995. Growth and maturation of the benthic stages of male snow crab, *Chionoecetes opilio* (Brachyura: Majidae). Canadian Journal of Fisheries and Aquatic Sciences 52:903–924.
Sarma, S.S.S., S. Nandini, and R.D. Gulati. 2002. Cost of reproduction in selected species of zooplankton (Rotifers and Cladocerans). Hydrobiologia 481:89–99.
Savage, T., and J.R. Sullivan. 1978. Growth and claw regeneration of the stone crab, *Menippe mercenaria*. Florida Marine Research Publications No. 32. Marine Research Laboratory, Florida Department of Natural Resources, St. Petersburg, Florida.
Schaffer, W.M. 1974a. Optimal reproductive effort in fluctuating environments. The American Naturalist 108:783–790.
Schaffer, W.M. 1974b. Selection for optimal life histories: effects of age structure. Ecology 55:291–303.
Schneider, J.M., and Y. Lubin. 1997. Does high adult mortality explain semelparity in the spider *Stegodyphus lineatus* (Eresidae)? Oikos 79:92–100.
Sekkelsten, G.I. 1988. Effect of handicap on mating success in male shore crabs *Carcinus maenas*. Oikos 51:131–134.
Shuster, S.M. 1987. Alternative reproductive behaviors: three discrete male morphs in *Paracerceis sculpta*, an intertidal isopod from the northern Gulf of California. Journal of Crustacean Biology 7:318–327.
Shuster, S.M. 1991. Changes in female anatomy associated with the reproductive moult in *Paracerceis sculpta*, a semelparous isopod crustacean. Journal of Zoology 225:365–379.
Slatkin, M. 1974. Hedging ones evolutionary bets. Nature 250:704–705.
Ślusarczyk, M., A. Ochocka, and D. Cichocka. 2012. The prevalence of diapause response to risk of size-selective predation in small- and large-bodied prey species. Aquatic Ecology 46:1–8.

Somerton, D.A., and W.S. Meyers. 1983. Fecundity differences between primiparous and multiparous female Alaskan tanner crab (*Chionoecetes bairdi*). Journal of Crustacean Biology 3:183–186.
Stearns, S.C. 1989. Trade-offs in life-history evolution. Functional Ecology 3:259–268.
Stearns, S.C. 1992. The evolution of life histories. Oxford University Press, Oxford, England.
Swalethorp, R., S. Kjellerup, M. Dünweber, T.G. Nielsen, E.F. Møller, S. Rysgaard, and B.W. Hansen. 2011. Grazing, egg production, and biochemical evidence of differences in the life strategies of *Calanus finmarchicus, C. glacialis* and *C. hyperboreus* in Disko Bay, western Greenland. Marine Ecology Progress Series 429:125–144.
Tachiki, Y., and Y. Iwasa. 2013. Coevolution of mast seeding in trees and extended diapause of seed predators. Journal of Theoretical Biology 339:129–139.
Takada, T. 1995. Evolution of semelparous and iteroparous perennial plants: comparison between the density-independent and density-dependent dynamics. Journal of Theoretical Biology 173:51–60.
Tallamy, D.W., and W.P. Brown. 1999. Semelparity and the evolution of maternal care in insects. Animal Behaviour 57:727–730.
Tang, B.P., K.Y. Zhou, D.X. Song, G. Yang, and A.Y. Dai. 2003. Molecular systematics of the Asian mitten crabs, genus *Eriocheir* (Crustacea: Brachyura). Molecular Phylogenetics and Evolution 29:309–316.
Tammaru, T., and E. Haukioja. 1996. Capital breeders and income breeders among Lepidoptera: consequences to population dynamics. Oikos 77:561–564.
Tarling, G.A., and J. Cuzin-Roudy. 2008. External parasite infestation depends on moult-frequency and age in Antarctic krill (*Euphausia superba*). Polar Biology 31:121–130.
Tessier, A.J., and C.E. Goulden. 1982. Estimating food limitation in cladoceran populations. Limnology and Oceanography 27:707–717.
Thiel, M. 1998. Reproductive biology of a deposit-feeding amphipod, *Casco bigelowi* with extended parental care. Marine Biology 132:107–116.
Thiel, M. 1999. Extended parental care in marine amphipods—II. Maternal protection of juveniles from predation. Journal of Experimental Marine Biology and Ecology 234:235–253.
Titelman, J., Ø. Varpe, S. Eliassen, and Ø. Fiksen. 2007. Copepod mating: chance or choice? Journal of Plankton Research 29:1023–1030.
Trumbo, S.T. 2013. Maternal care, iteroparity and the evolution of social behavior: a critique of the semelparity hypothesis. Evolutionary Biology 40:613–626.
van der Meeren, G.I. 1994. Sex-dependnet and size-dependent mating tactics in a natural population of shore crabs *Carcinus maenas*. Journal of Animal Ecology 63:307–314.
Varpe, Ø. 2012. Fitness and phenology: annual routines and zooplankton adaptations to seasonal cycles. Journal of Plankton Research 34:267–276.
Varpe, Ø. 2017. Life history adaptations to seasonality. Integrative and Comparative Biology 57:943–960.
Varpe, Ø., C. Jørgensen, G.A. Tarling, and Ø. Fiksen. 2007. Early is better: seasonal egg fitness and timing of reproduction in a zooplankton life-history model. Oikos 116:1331–1342.
Varpe, Ø., C. Jørgensen, G.A. Tarling, and Ø. Fiksen. 2009. The adaptive value of energy storage and capital breeding in seasonal environments. Oikos 118:363–370.
Vetter, E.W. 1996. Life-history patterns of two southern California *Nebalia* species (Crustacea: Leptostraca): The failure of form to predict function. Marine Biology 127:131–141.
Vogedes, D., Ø. Varpe, J.E. Søreide, M. Graeve, J. Berge, and S. Falk-Petersen. 2010. Lipid sac area as a proxy for individual lipid content of arctic calanoid copepods. Journal of Plankton Research 32:1471–1477.
Vogt, G. 2012. Ageing and longevity in the Decapoda (Crustacea): a review. Zoologischer Anzeiger 251:1–25.
Vogt, G., and L. Tolley. 2004. Brood care in freshwater crayfish and relationship with the offspring's sensory deficiencies. Journal of Morphology 262:566–582.
Wada, S., T. Oba, K. Nakata, and A. Ito. 2008. Factors affecting the interval between clutches in the hermit crab *Pagurus nigrivittatus*. Marine Biology 154:501–507.
Warburg, M., and N. Cohen. 1991. Reproductive pattern, allocation, and potential in a semelparous isopod from the Mediterranean region of Israel. Journal of Crustacean Biology 11:368–374.
Webster, S.G. 2015. Endocrinology of molting. Pages 1–35 *in* E.S. Chang, and M. Thiel, editors. The natural history of the Crustacea, volume 4: Physiology. Oxford University Press, New York.

Weider, L.J., W. Lampert, M. Wessels, J.K. Colbourne, and P. Limburg. 1997. Long-term genetic shifts in a microcrustacean egg bank associated with anthropogenic changes in the Lake Constance ecosystem. Proceedings of the Royal Society B: Biological Sciences 264:1613–1618.
Wellborn, G.A. 1994. Size-biased predation and prey life histories: a comparative study of freshwater amphipod populations. Ecology 75:2104–2117.
Wellborn, G.A., and R.E. Broughton. 2008. Diversification on an ecologically constrained adaptive landscape. Molecular Ecology 17:2927–2936.
Wen, X., L. Chen, C. Ai, Z. Zhou, and H. Jiang. 2001. Variation in lipid composition of Chinese mitten-handed crab, *Eriocheir sinensis* during ovarian maturation. Comparative Biochemistry and Physiology Part B: Biochemistry and Molecular Biology 130:95–104.
Werner, E.E., and B.R. Anholt. 1993. Ecological consequences of the trade-off between growth and mortality rates mediated by foraging activity. The American Naturalist 142:242–272.
Węsławski, J.M., K. Opaliński, and J. Legeżyńska. 2000. Life cycle and production of *Onisimus litoralis* (Crustacea Amphipoda): the key species in the Arctic soft sediment littoral. Polskie Archiwum Hydrobiologii 47:585–596.
Williams, G.C. 1966. Natural selection, the costs of reproduction, and a refinement of Lack's principle. The American Naturalist 100:687–690.
Winfield, I.J., and C.R. Townsend. 1983. The cost of copepod reproduction: increased susceptibility to fish predation. Oecologia 60:406–411.
Young, T.P. 1981. A general model of comparative fecundity for semelparous and iteroparous life histories. The American Naturalist 118:27–36.
Young, T.P., and C.K. Augspurger. 1991. Ecology and evolution of long-lived semelparous plants. Trends in Ecology and Evolution 6:285–289.
Zeineddine, M., and V.A.A. Jansen. 2009. To age, to die: parity, evolutionary tracking and Cole's paradox. Evolution 63:1498–1507.
Zeng, Y., C. McLay, and D.C.J. Yeo. 2014. Capital or income breeding crabs: who are the better invaders? Crustaceana 87:1648–1656.

5

LIFE HISTORY PERSPECTIVES ON VOLTINISM

Carlos San Vicente

Abstract

Patterns of voltinism are well documented in many crustaceans and other invetebrates, and these studies provide diverse insights into species biology, population ecology, and drivers of the evolution of voltinism. This chapter examines voltinism across crustacean taxa, with a focus on mysids as an informative model taxon that exhibits a broad range of life pattern diversity. Voltinism, which describes the number of generations per year for population or species, can be measured as generation time and is shaped by multiple environmental factors, including temperature, latitude, salinity, and depth. Generation time also varies with important biological traits, such as body size, life span, and maturation size and age. I discuss the relationships between voltinism and life history strategies, and the influence of voltinism on adaptative plasticity of species and their populations. Many factors shape evolution of voltinism, including fitness components such as survival, reproduction, and dispersal, as well as tradeoffs among age and size at maturity, reproductive investment, and lifespan. I highlight the importance of voltinism for population modeling in crustaceans, and for understanding regional differences in voltinism. Studies comparing and contrasting voltinism will be critical to better understand how climate change, strong habitat modifications, pollution, and invasive species will impact crustacean populations and their dependent communities.

INTRODUCTION

Voltinism (i.e., the number of generations per year for a species or population) has been widely studied because of its value as a descriptor of general biological strategies. An understanding of voltinism is important for (1) elucidating the adaptive value of life cycles across different ecological and evolutionary circumstances and (2) understanding dynamics of populations and communities (Hamilton 1969). In this sense, voltinism can be seen as a characteristic of a species or population

Life Histories. Edited by Gary A. Wellborn and Martin Thiel.

that affects its population dynamics, colonizing ability, resilience, and ultimately its evolutionary success. Natural populations are subject to multifaceted abiotic conditions such as temperature, pressure, and light. These populations are also exposed to elements of the biological environment, including prey, predators, and competitors, all of which affect the evolution and ecology of voltinism.

The present chapter reviews the diversity of voltinism in Crustacea, especially within mysids (order Mysida), because these crustaceans exhibit distinct patterns of voltinism across a spectrum of environmental conditions. Although a broad analysis of voltinism in mysids was presented by Mauchline (1980), this chapter provides more recent data, defines the main environmental drivers of voltinism evolution, and explores plasticity in voltinism. I draw conclusions that may be extrapolated to other taxonomic groups and finally highlight challenges and future directions in the study of voltinism.

APPROACHES TO ASSESS VOLTINISM

Although a species' voltinism may sometimes be simple to discern, for many species technical challenges in assessment of voltinism are substantial. Knowledge of certain biological features such as the general life cycle, mating and reproduction cycles, and life span or survivorship rates help better determine a population's voltinism (Fig. 5.1). When reproduction continues without interruption throughout the year, as in many temperate and tropical species, it will likely be difficult to infer voltinism in the field. Moreover, the practical difficulty of continuing a study for several consecutive years means that records for voltinism in species with long life cycles tend to be underrepresented and interannual changes are usually neglected. Lastly, most published studies are conducted in habitats with large populations because they are liable to yield clear results, but this limitation leads to less abundant populations being underrepresented.

One may be able to quantify voltinism by recognizing cohorts in population dynamics data (e.g., density and biomass, demographic structure, fecundity), but in practice it is often difficult to recognize distinct cohorts. A cohort is a group of same-aged individuals born more or less at the same time (Brey 2001), and in the simplest scenario, reproduction occurs simultaneously among all members of a population over a short period of time. In populations with identifiable cohorts, there is a clear separation of age groups or modes in population histograms, which implies that the cohorts are discrete and recognizable. Such populations are, in general, semelparous or occasionally biparous and usually breed intensively during a short period of the year. Under these conditions, the study of periodic samples along the annual cycle allows researchers to infer, in a simple way, the demographic parameters of the population.

In contrast to synchronous reproduction, a large number of crustaceans, including many mysids, have iteroparous females and extensive spawning without synchrony in egg production or offspring release, and later exhibit among-individual desynchronization in offspring growth rates. Such complex life history patterns make it very difficult to identify and track the overlapping cohorts produced during a year. In these populations, seasonal or regional differences may cause changes in adult morphology, different sizes at attainment of sexual maturity, and different seasonal growth curves (Wittmann 1992). Such growth relationships pose severe difficulties for assessing the number of cohorts and generations produced throughout the year.

The concept of generation can be applied when several offspring cohorts are produced by the same batch of brooding females; all individuals of these various cohorts define a daughter generation (Mauchline 1980). It corresponds to the intuitive notion of the time it takes for a generation to be replaced by the next. Generation time then is defined as the average time between 2

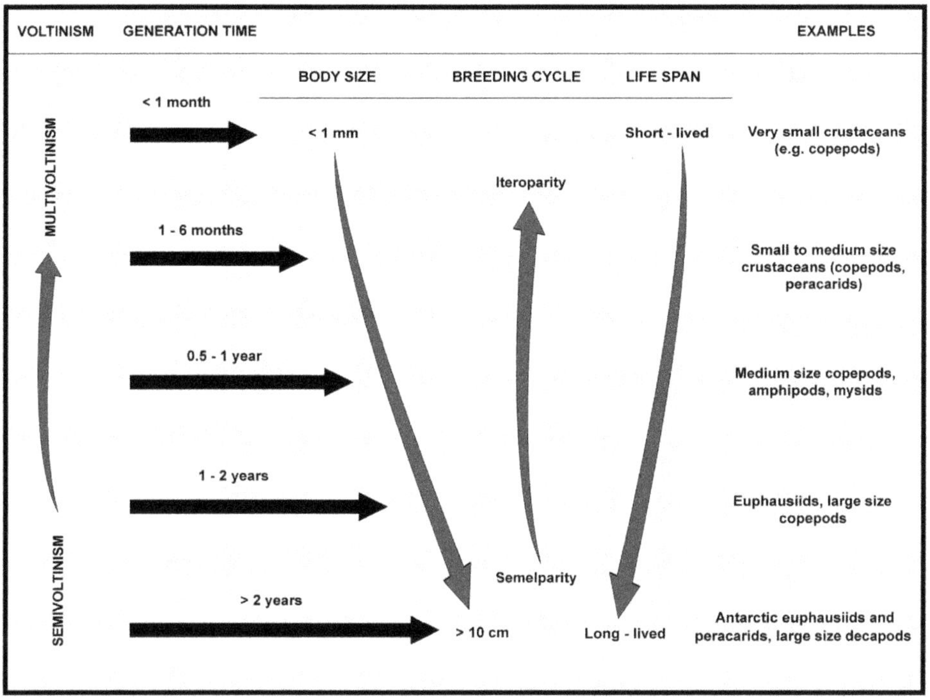

Fig. 5.1.
Schematic diagram showing voltinism and generation time of various crustaceans in relation to their size ranges, breeding cycles, and life spans. In general, species with a shorter life expectancy may maximize their lifetime fitness by investing more in reproduction, and individuals of species with longer life spans enhance their fitness by investing less in reproduction. Such trade-offs between reproductive investment and life span have strong implications for voltinism diversity. Based on Escribano and Riquelme-Bugueño (2015), with permission from Oxford University Press.

reproductive events in the genealogy of the population (Bienvenu and Legendre 2015). Individuals within any generation are inextricably linked to and shaped by their environment, and unpredictable environmental variation during one or more developmental periods can have effects that persist throughout their course of life. Furthermore, individuals of different cohorts can have different life histories and may vary greatly in abundance, depending on the number of births that occurred during a given period.

Determination of a population's voltinism depends on accurate sampling of a population throughout its life cycle, but abundance estimates for all life stages of a population may be confounded by population distributions that are highly dynamic across space and time. For this reason, accuracy in determination of a population's voltinism depends on the proportion of its generation time that it is available in its habitat for sampling. A number of crustacean species change their ecological niche at least once during their life cycle. Life stage-specific mass migration of aggregations of crustaceans occurs in some species. These migrations, which often involve long distances, allow life stages to occupy distinct niches (e.g., Delgado et al. 2013). Such abundance changes across the spatial distribution of individuals, often resulting from migrations, can result in apparent increases and decreases in densities (see Chapter 8 in this volume) and therefore distort inferences on voltinism in field populations (Fig. 5.2).

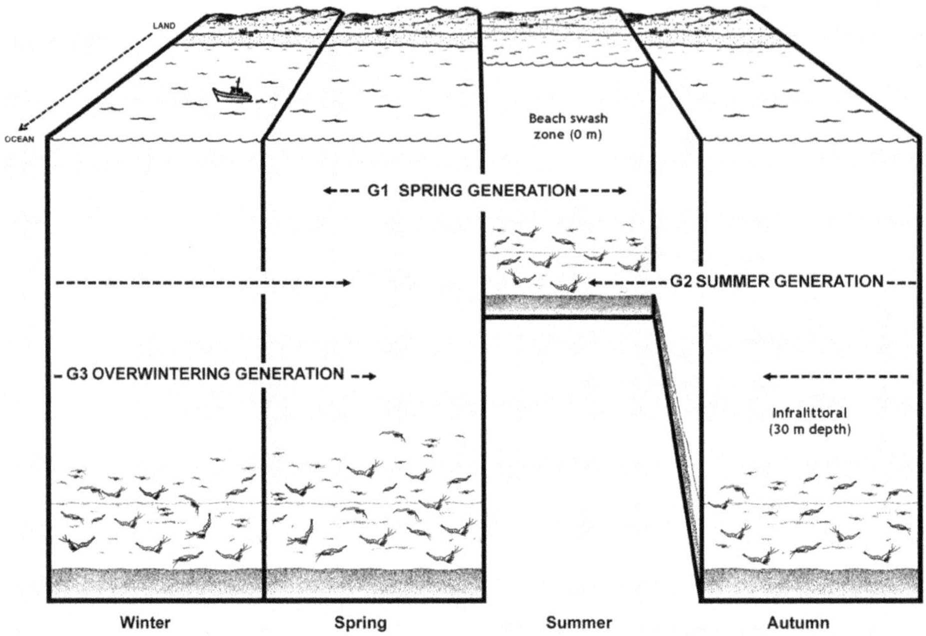

Fig. 5.2.
Schematic diagram of the seasonal cycle of the mysid *Schistomysis spiritus* in the southeasthern Bay of Biscay. G1, G2, and G3 refer to the spring generation, summer generation, and autumn generation, respectively. Every year the population disappears from infralittoral bottoms during summer and migrates to shallower coastal grounds in the Bay of Biscay. In such migrant populations, the study of only 1 area may significantly distort the number and timing of their generations and additionally underestimate mysid abundance. Modified from San Vicente and Sorbe (1995).

ECOLOGICAL DRIVERS OF VOLTINISM

Temperature

Temperature is the most important determinant of voltinism patterns, at least in poikilothermic organisms, and a large body of literature relates the effects of temperature to developmental rates and phenology (e.g., Powell and Jenkins 2000). Temperature has a marked influence on growth and reproduction of individuals because higher temperature provides a favorable condition for faster growth and development and may allow multiple generations within a year. Mysid voltinism is directly influenced by habitat temperature because the growth rate of individuals generally slows with lower temperatures (Fig. 5.3A) and the generation time increases, thereby reducing the number of generations that can be completed (Mauchline 1980).

The duration of breeding in any crustacean species depends on reproductive strategies, such as semelparity versus iteroparity, and on voltinism patterns. For example, in mysid populations there are essentially two fundamentally different reproductive strategies: one for cold-water breeders and another for warm-water breeders. Mysids in cold waters and meso- and bathypelagic environments tend to be semelparous or occasionally biparous and usually breed intensively during the cold season at 0°C to 7°C, whereas those in warm waters and epipelagic and coastal environments are usually iteroparous and breed intensively above 10°C (Wittmann 1984). Moreover, as suggested by Sudo et al. (2011), there may be an alternation between higher fecundity and iteroparity in

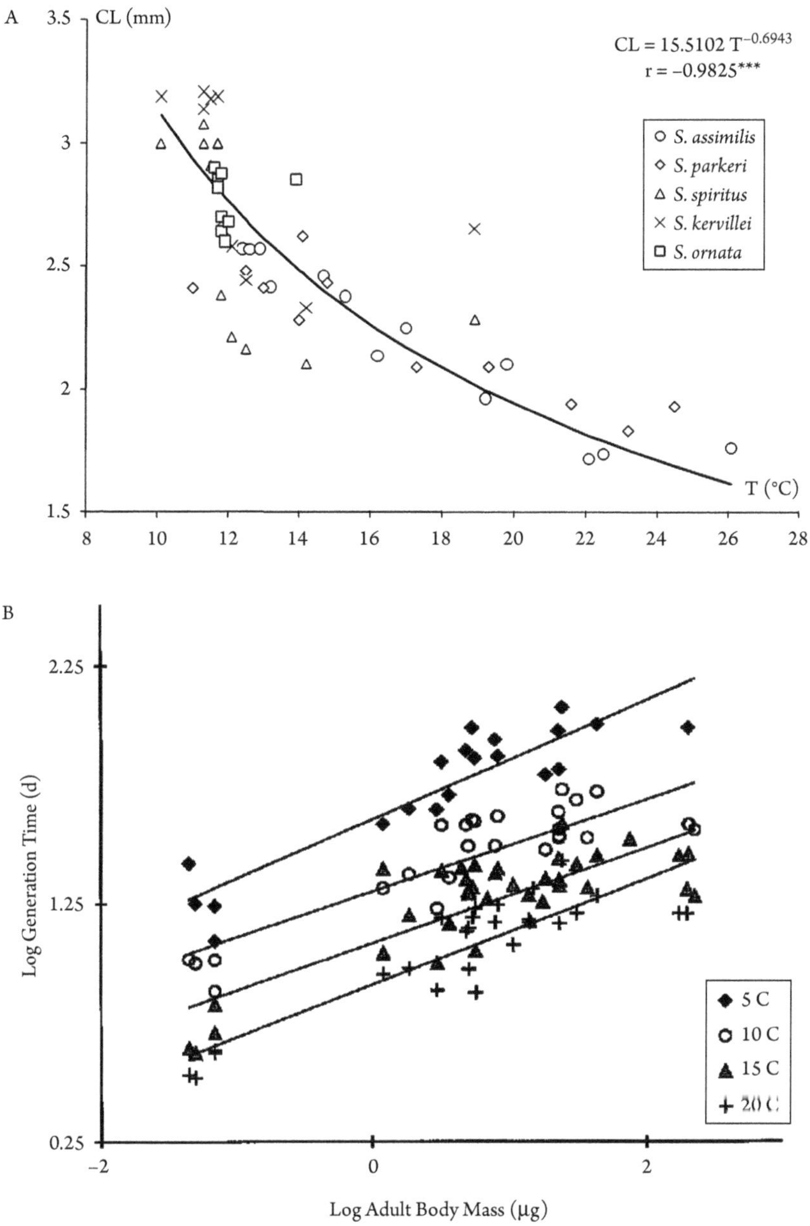

Fig. 5.3.
(A) Relationship between the carapace length of brooding females of the mysid *Schistomysis* spp. and the corresponding near-bottom temperatures showing how size is negatively correlated to temperature. From San Vicente and Sorbe (2013), with permission from Elsevier. (B) Zooplankton generation times (d) as a function of adult body mass at 5°C, 10°C, 15°C, and 20°C. Generation time increases with temperature and adult body mass. From Gillooly (2000), with permission from Oxford University Press.

autumn and winter cohorts and lower fecundity and semelparity in spring and summer cohorts of *Orientomysis robusta* in a shallow warm-temperate beach.

Although temperature is probably the main (and easier to measure) factor that determines voltinism in crustaceans, reproduction patterns are typically not determined by just one or a few factors but rather by complex coordination of endogenous and environmentally mediated events (Sastry 1983). Temperature, food availability, and seasonal time constraints regulate life cycle duration in copepods that can alternate between univoltine and multivoltine strategies (Chen and Folt 1996, Twombly et al. 1998, Gerten and Adrian 2002). For ectotherms, the value of scaling relationships of body mass and generation time is limited because they do not account for the influence of temperature on generation time. Generation times of ectotherms are generally thought to increase with body size but decrease with increasing temperature for a given body size. Thus, for a scaling relationship to be useful in the study of ectotherms, it must account for the effects of both body size and temperature (Gillooly 2000; Fig. 5.3B).

Latitude and Season Length

The number of generations per year typically decreases with latitude, as reproductive output becomes increasingly concentrated within the warmer months at higher latitudes. Boreal and polar mysid species are more likely to be large and have a single large, well-timed brood (Mauchline 1980). In contrast, warm temperate species tend to be small in size, produce many small broods, and have several generations in the year (San Vicente and Sorbe 2013). At lower latitudes, rapid growth rates and smaller size at first reproduction, coupled with short incubation times for the smaller eggs, all contribute to shorter summer generations. In an extensive review of reproductive patterns in amphipods, Sainte-Marie (1991) explored broad patterns in their voltinism arising from the influence of latitude and the associated temperature gradient. As a general rule, temperate species often have long overwintering generations and multiple shorter summer generations.

For ectotherms in seasonal environments, growth and reproduction are typically possible only during a limited period of the year (Kivelä et al. 2009), and length of the favorable season influences the number of generations per year (Stearns 1992). For example, polar and deep-sea peracarids have much longer generations, whereas tropical species may produce broods year-round in rapid succession (Johnson et al. 2001). Also, data available for deep-sea amphipods and isopods suggest rather long life spans with late maturity and lengthy incubation times (e.g. Ingram and Hessler 1987). In temperate regions, ectotherms, including freshwater crustaceans, have a characteristic regular summer-winter cycle with some interyear variation in length of the breeding period (Sainte-Marie 1991, Lewis 2009). Length of the favorable season also influences the number of generations per year (Stearns 1992).

Salinity

The effects of salinity alone, or in combination with temperature, on patterns of voltinism are evident in temporary ponds, tidal rock pools, estuaries, and other environments characterized by large fluctuations of temperature and salinity over hours, days, or seasons. As observed by Ganning (1971), salinity and temperature have strong influences on development and voltinism of ostracods occupying brackish water rock pools. For example, low salinity reduced hatching time, and reproduction almost doubled in comparison with high salinity. This earlier onset of development and increased voltinism may promote faster population growth and have major implications for ostracod egg bank accumulation (Rossi et al. 2013).

Interactions between temperature and salinity are often reported for estuarine invertebrates and can alter voltinism by shifting development rate and survival of crustaceans, including mysids, the brine shrimp *Artemia* spp., and the copepod *Acartia tonsa* (Fockedey et al. 2005, Browne and

Wanigasekera 2000, Peck et al. 2015). Generation times of the estuarine cladoceran *Latonopsis australis* vary from 5.2 to 16.7 days and are negatively correlated with salinity, with the generation time shorter in freshwater conditions than in high saline conditions (Haridevan et al. 2015).

Food Quality

Food quality affects voltinism through its effects on growth rate and egg provisioning. For example, food quality differences explain accelerated growth rates, increased voltinism, and high population densities of filter-feeding invertebrates in some lake outlet streams (Richardson 1984). Likewise, when food quality and primary production vary spatially, separate populations of the same species may differ in generation time because higher quality food is associated with shorter generation times (Ogonowski 2012). Within amphipod species, there is some plasticity in life cycles because of, in part, differences in food availability (Kolding and Fenchel 1981, Leonardsson et al. 1988).

Body Size

Voltinism data across many organisms suggest that, in general, smaller species undergo more generations per year than larger organisms. Bonner (2006) examined the relationship between size and duration of life cycles expressed as generation time for a wide variety of organisms, including planktonic crustaceans. Roughly, for every doubling in the length of an organism, the generation time doubles as well. Small size is associated with fast growth rates and short generation times, whereas organisms with large mass have slower growth rates and long generation times.

Allometric relationships between body size and other life history variables, such as voltinism, have been the basis for comparing the life history and physiological properties of taxa like birds, mammals, and insects (Demetrius et al. 2009). Inclusion of temperature in such relationships provides new insights for zooplankton ecology. For example, the generation time of the relatively large cladoceran, *Eurycercus lamellatus*, is 88 days longer than the smaller rotifer, *Notholca caudata*, at 5°C, but only 15 days longer at 20°C, a result that implies a significant advantage to population growth of smaller-bodied species in colder waters. In addition, it suggests that body size and temperature do constrain generation time and subsequent population growth of larger-bodied zooplankton in colder waters (Gillooly 2000).

VOLTINISM IN MYSIDA AND COMPARISON WITH OTHER TAXA

Mysids are a valuable taxon for investigating general principles of voltinism for several reasons. One, the order Mysida includes species from diverse habitats, including subterranean, freshwater, brackish, and coastal areas, as well as surface to deep-sea habitats. Two, mysids are ubiquitous crustaceans that play an important role in pelagobenthic flux exchanges and trophodynamics of aquatic ecosystems, and they are economically important for fisheries, especially in Asian countries. Three, the impact of invasive mysid species in some plankton communities can be remarkable: they reduce native species abundance drastically, and their invasion success may be determined by several population characteristics, such as short generation times (Borza 2014). Four, patterns of voltinism apparent in mysids can inform our understanding of voltinism across other crustacean taxa.

The general life cycle of mysids is relatively simple (Fig. 5.4). It begins with a period characterized by maternal care in which breeding females carry their embryos and lecithotrophic larvae within a marsupium. First instar juveniles are immediately able to swim and feed, and these free-living individuals grow by molting before and after reaching sexual maturity. Time spent in the different stages may change interspecifically and intraspecifically through adaptation to diverse environments.

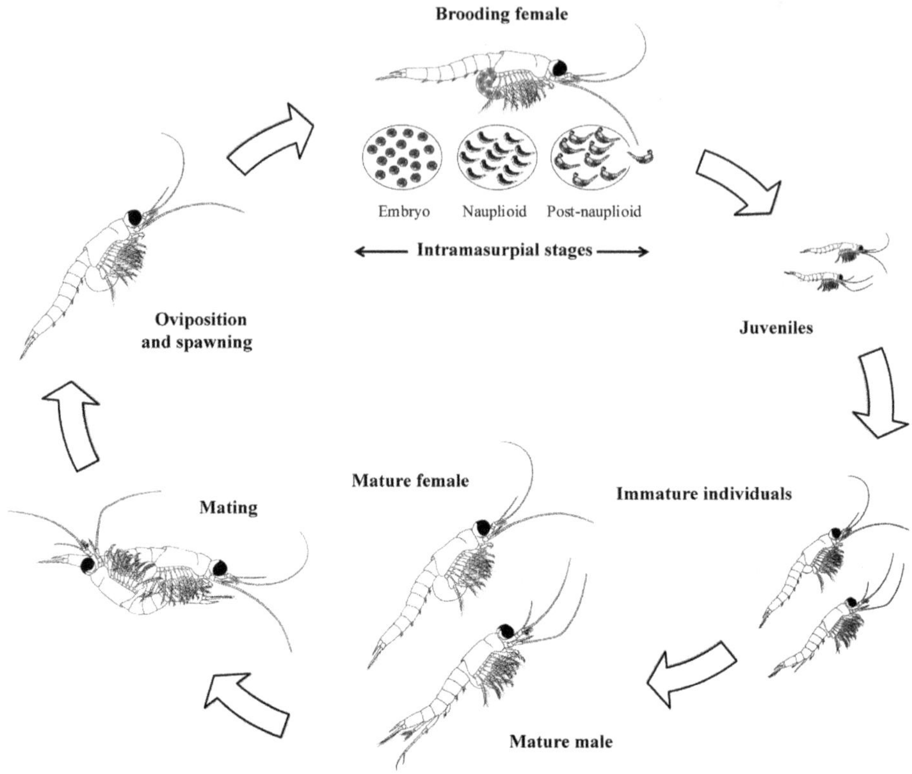

Fig. 5.4.
A generalized summary of the life cycle of a mysid. Time spent in stages may change interspecifically and intraspecifically through adaptation to environmental conditions. Furthermore, reproductive strategies range from semelparous to iteroparous; hence, the number of potential broods differs among species. Adaptive and phenotypically plastic changes in time spent within each life cycle phase determine potential voltinism for species and populations. Illustrated by Carlos San Vicente ©.

Furthermore, reproductive strategies range from semelparous to iteroparous; hence, the number of potential broods differs across species. Variation within the basic life cycle is common among species and populations as they adapt to the diverse environments inhabited by mysids.

Species with high risk of adult mortality should benefit from earlier reproduction and higher investment in fecundity (Johnson et al. 2001). Although predation risk is the most common explanation for higher fecundity and faster development times, other selective pressures may be involved. Sudden and abrupt variations in physical and chemical conditions of some environments are likely responsible for elevated death rates in resident populations (Mees et al. 1994, Winkler and Greve 2002, Hanamura et al. 2009). For example, temperature and salinity changes can modulate evolution of voltinism in populations because these often lead to high mortality among juveniles, affect growth rates and maturity, and influence brood size of gravid females (Bremer and Vijverberg 1982, McKenney and Celestial 1995, Winkler and Greve 2002, Fockedey et al. 2005).

VOLTINISM DIVERSITY

The diversity and plasticity of mysid life cycles vary from semi- to multivoltine among species and populations, and with distinct voltinisms predictably associated with specific environmental conditions. These patterns are also observed in a wide range of organisms, such as diverse orders of insects (e.g., Corbet et al. 2006, Altermatt 2010, Kivelä 2011) and vertebrates (e.g., Stearns 1977, Saether et al. 2005). The patterns of reproduction and succession of generations vary among different species and among populations of the same species that occupy different habitats. Multiple generations may occur within a single year, or completion of 1 breeding cycle may take a little more than a year, or even several years.

Semivoltine Populations

In mysids, populations with long generation times (producing less than 1 generation per year) typically occur in cold water (0–7°C) and are semelparous (Wittmann 1984). Spawning typically occurs in autumn or winter, and juveniles are released in spring when food is abundant (Fig. 5.5A). Semivoltine mysid populations are characteristic of those found in the Southern Ocean (Ward 1984, 1985, Siegel and Mühlenhardt-Siegel 1988, San Vicente et al. 2006), boreal waters (Geiger 1969, Mauchline 1980, Astthorsson 1984), bathypelagic environments (Mauchline 1988, Ikeda 1992), and oligotrophic lakes (Morgan 1980, Johannsson et al. 2009).

Semivoltine species are found in other crustacean taxa, exemplified by Antarctic krill, the euphausiid *Euphausia superba*, which is a long-lived crustacean with a life span, depending on its location, of 4 and 8 years, and with an age at maturity of about three to four years (Siegel 2000). Contingent on food conditions, an individual may reproduce each subsequent summer, or may delay reproduction and produce new a brood every two or three years (Pakhomov 2000). Semivoltine populations in these extreme environments are reported in a wide range of crustacean taxa as, for example, the mesopelagic and bathypelagic euphausiid species *Thysanopoda* spp. and *Bentheuphausia amblyops* (Mauchline 1980), the deep-sea/polar lysianassoid amphipod *Eurythenes gryllus* (Ingram and Hessler 1987) and other abyssal amphipod species (Sainte-Marie 1991), and the northern populations of the decapod *Nephrops norvegicus* (Sardà 1995).

Longer generation length may be associated with an extended period in sexually mature conditions, which may be important for finding mates in the low population density of the deep sea (Chikugo et al. 2013). Furthermore, minor changes in offspring survival rate can alter the abundance of an entire population, and the timing of offspring release is critical for semelparous breeders because if high juvenile mortality occurs the population has no potential for further recruitment until the following year (Richoux et al. 2004). The typical life history characteristics of semivoltine populations (long life, large size, and multiple spawning seasons) buffer populations from recruitment failures and allow persistence of populations across sequences of failures and successes in recruitment (Quetin and Ross 2003).

Univoltine Populations

When juveniles mature by the end of the same growth season in which they were released, the generation time becomes one year (Wittmann 1984). Life cycles extending over a full year often apply to mysid species in the latitudinal band of 40–60° and to neritic rather than littoral species, and breeding females may produce 1 or more broods during the breeding season (Mauchline 1980; Fig. 5.5B). Univoltinism is more frequent in species living in relatively extreme environments (e.g., high latitudes, deeper waters), where some environmental resources are limited during periods of the year. This voltinism pattern may be associated with seasonality in food availability (Saulich and Musolin 1996).

Mysis stenolepis has a conventional univoltine life cycle. Juveniles appear in early summer and attain sexual maturity and breed the following midwinter to spring. All mature individuals disappear before the second summer, with males disappearing long before the females (Smith 1879, Amaratunga and Corey 1975). *Mysis mixta* generally has a one-year life cycle in the northern Baltic

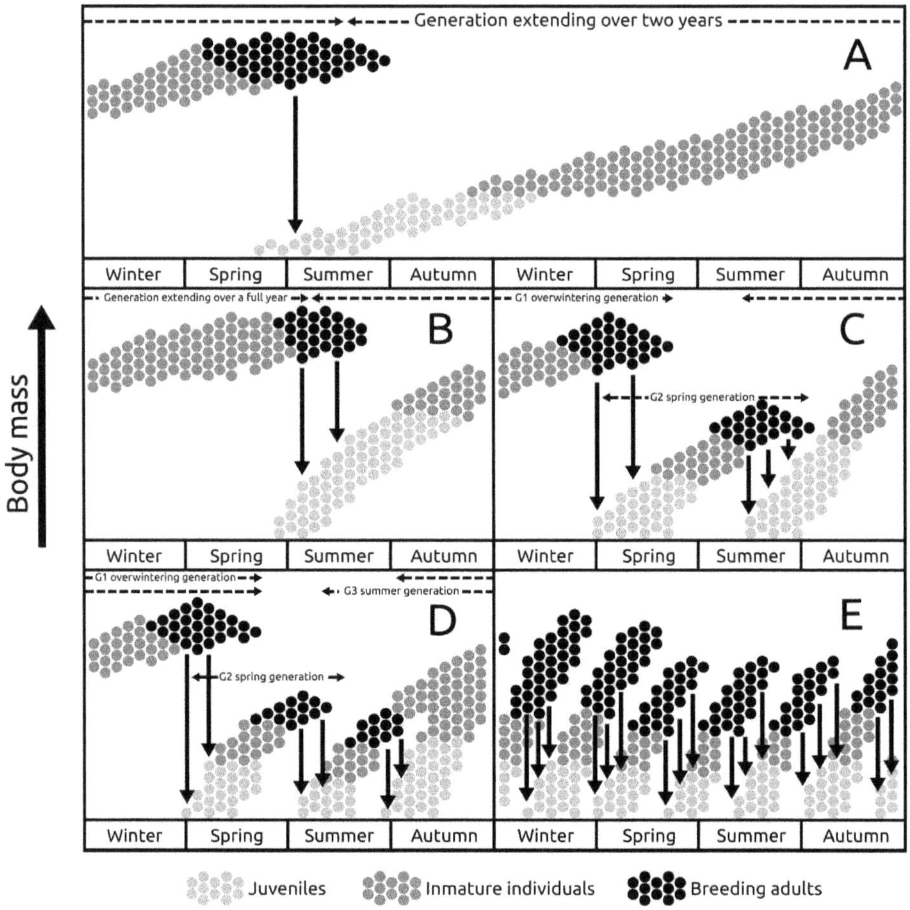

Fig. 5.5.
Schematic diagram showing the seasonal composition of a (A) semivoltine population with a 2-year life cycle, (B) univoltine population, (C) bivoltine population, (D) trivoltine population, and (E) multivoltine population. Arrows indicate production of the next generation by breeding adults. Redrawn from Mauchline (1980).

Sea. Its offspring are released in early spring after the ice breakup, and new juveniles grow during summer and autumn and initiate breeding in late autumn. Ovigerous females carry their brood for four to five months and begin to release them in early spring (Lehtiniemi et al. 2002). Some females may remain immature during their first winter and breed only in their second autumn (Salemaa et al. 1986). In the north and east of Iceland, evidence suggests that some *M. mixta* females that breed at the age of 1 year may live for another year to breed a second time (Astthorsson 1984).

Other records of univoltine annual populations are known for mysids *Mysidopsis didelphys, Schistomysis ornata* (Mauchline 1980), *Boreomysis megalops* and *Anchialina agilis* (Elizalde 1994, Sorbe 1984), euphausiids *Meganyctiphanes norvegica* and *Thysanoessa longicaudata* of the northern North Atlantic (Saunders et al. 2007), isopods *Ligia exotica* (Tsai and Dai 2001) and *Cyathura carinata* (Amanieu, 1970, Ferreira et al. 2004), diverse gammarid amphipods of the genus *Gammarus* inhabiting inland waters of Central Europe (Grabowski et al. 2007) and *Ampelisca sarsi* (Dauvin 1979), some cumaceans inhabiting high latitudes or deep-sea environments (e.g., Cartes and Sorbe 1996, Corbera et al. 2000), populations of *Nephrops norvegicus* at Mediterranean latitudes (Sardà 1995), and the deep-water rose shrimp *Parapenaeus longirostris* (Sobrino et al. 2005).

Bivoltine Populations

In mysid populations with two generations per year, often the first generation is released in the spring, and these offspring quickly become sexually mature to produce a second generation that reproduces in the summer. The summer cohort overwinters before reproducing again the following spring (Fig. 5.5C). In bivoltine populations, a lack of synchrony between females' broods, especially because females usually have several broods, often prevents simple empirical assessment of the number of generations produced. This problem is particularly prevalent around 40° latitude, where individuals of summer generations with greater longevity partially overlap those of the spring generation (Sorbe 1984).

Boreomysis arctica, a mysid, showed a near-continuous reproductive pattern in the deep western Mediterranean. This species was at least bivoltine, with peaks of recruits detected, particularly during late winter to early spring (Cartes and Sorbe 1998). Mysids *Tenagomysis chiltoni, T. macropsis,* and *Gastrosaccus australis* displayed bivoltine life cycles in the Taieri Estuary (New Zealand), with breeding peaks in December and February (Bierschenk 2014). The Mediterranean population of *Siriella clausii* is described as bivoltine (Barberá et al. 2013). *Limnomysis benedeni,* one of the most important pontocaspian invaders, was found to be bivoltine in Lake Constance (Hanselmann et al. 2011). For *Neomysis integer,* two generations per year were reported for the Baltic Sea population (e.g., Barz and Hirche 2009).

In the warmer waters of the eastern North Atlantic, the euphausiid *Thysanoessa longicaudata* produces discrete generations during two annual spawning pulses. The first generation is spawned during the spring, and these animals reach sexual maturity and breed by autumn, giving rise to a second generation that overwinters and breeds in the following spring (Lindley, 1978). Other records of bivoltine populations include the mysids *Neomysis americana, Praunus flexuosus,* and *Schistomysis ornata* (Pezzack and Corey 1979, Mauchline 1980, Sorbe 1991), the amphipods *Orchestia gammarella* (Amanieu 1970), *Gammarus salinus* (Kolding and Fenchel 1981), and *Ampelisca typica* (Dauvin 1988), diverse cumacean populations (e.g., Corbera et al. 2000), and the euphausiids *Thysanoessa raschi* and *Nyctiphanes couchi* (Mauchline 1980).

Multivoltine Populations

Species producing 3 or more generations per year exhibit more or less continuous breeding, and diverse periods of seasonal maxima in abundance throughout the year. This type of life history is most common among shallow living neritic and littoral mysid species occurring in temperate and warm waters at latitudes less than 40° (Mauchline 1980). Multivoltine populations with more than three generations in the annual cycle have been described in many species of mysids (Fager and Clutter 1968, Toda et al. 1982, Vilas 2005, Yamada et al. 2007), the isopod *Idotea viridis* (Amanieu 1970), and diverse species of amphipods (LaFrance and Ruber 1985, Sainte-Marie 1991).

Production of three generations in the annual cycle is typical of diverse mysid genera living in temperate estuaries, such as *Neomysis* (Mees et al. 1994, Vilas 2005, Yamada et al. 2007, Grzeszczyk-Kowalska et al. 2014), *Rhopalophtalmus* (Wooldridge 1986, Vilas 2005), *Mesopodopsis* (Azeiteiro et al. 1999), and *Tenagomysis* (Bierschenk 2014). Coastal trivoltine populations have been described in many mysid species (e.g., Matsudaira et al. 1952, Fager and Clutter 1968, Gaudy and Guerin 1979, Lejeusne and Chevaldonné 2005, Sudo et al. 2011, San Vicente and Sorbe 2013), the euphausiid *Nyctiphanes australis* (Mauchline 1980) or the cumaceans *Cumopsis goodsir* (Corbera et al. 2000) and *Claudicuma platense* (Roccatagliata 1991; Fig. 5.5D).

Species with more than three generations per year are characterized by extremely complex populations dynamics (Fig. 5.5E). Two main characteristics of these populations are nearly continuous reproduction and recruitment throughout much of the year and occasional episodes of enhanced reproductive intensity (Dean et al. 2005). Experimental studies conducted in the laboratory are therefore desirable for estimating rates of growth and reproduction of such mysids in the field, especially in areas where the water temperature exceeds 20°C for a significant portion of the annual cycle. Rearing experiments in the laboratory have demonstrated that various mysid species become sexually mature in fewer than 30 days at 20°C (e.g. Delgado et al. 2013). Such early maturity suggests mysids might have more than 3 generations per year in warm-water habitats (Sudo et al. 2011). A survey on *Mysidium columbiae* in a mangrove lagoon in Jamaica provided a rare instance of seasonality in abundance and reproductive characteristics of tropical coastal mysids, and it showed year-round reproduction in this species (Goodbody 1965). The relatively stable environmental conditions of these tropical regions throughout the year undoubtedly contribute greatly to reduced seasonality in reproductive traits (Hanamura et al. 2009).

Borza (2014) inferred as many as four to six generations per year from field data on nonindigenous freshwater populations of the mysids *Limnomysis benedeni, Hemimysis anomala,* and *Katamysis warpachowskyi* that invaded the Hungarian reach of the Danube River (Fig. 5.6). In all three species, juveniles were produced between spring and autumn, and the offspring born in the autumn form the overwintering generation and become fertile by the following spring. The number of generations revealed in this study can be regarded as extraordinarily high when considering the relatively large body size of the animals and the temperate climatic conditions. Such a high degree of voltinism might contribute to their invasion success.

Phenotypic Plasticity in Voltinism

As a result of temperature and other environmental differences between regions and habitats, voltinism plasticity within a species is common. For example, *Mysis relicta* may exhibit either a one-year or two-year life cycle (i.e., univoltine or semivoltine), depending on temperature, food availability, and lake productivity (Hakala 1978, Chess and Stanford 1998). *Schistomysis ornata* shows 1 generation in its northern populations and two generations in southern populations (Sorbe 1991). *Neomysis integer* populations have two generations in northernmost populations (> 53°N) and three or more in southern populations (Fockedey 2005). Ranges of two to four generations per year have also been identified in the mysid *Mesopodopsis slabberi* (Delgado et al. 1997, Azeiteiro et al. 2001, Vilas 2005). Numerous transplanted populations of *Mysis relicta* in lakes with different environmental conditions have also resulted in rapid changes in life history and growth, which further supports a strong phenotypically plastic component in their life cycle. Even in the large and stable environment, where *Mysis* has coexisted with its main food items and predators for 8,000–10,000 years, a flexible life history is maintained and is probably an important buffer against year-to-year fluctuations in food and predator abundances (Kjellberg et al. 1991).

Studies of marine mysids have found that most species undergo two or three generations per year. Such bivoltine-trivoltine life patterns are typical of peracarids from cold-temperate climates (Johnson et al. 2001). These two life patterns are probably a result of differences in duration of intermolt stages, with warmer temperatures in summer and autumn likely allowing shorter intermolt durations and colder temperatures leading to longer intermolt durations (Lejeusne and Chevaldonné 2005). The longer overwintering generation releases its young in the spring with rising temperature, increasing day length, and blooming phytoplankton. The shortest spring-autumn generations, in contrast, develop quickly from eggs to sexually mature adults. This strategy is particularly adaptive when high predation pressure occurs during the season (Toda et al. 1982).

Plasticity in voltinism has been described in many species and populations of crustaceans with different mesoscale and macroscale distributions. For example, the life cycles of the amphipod *Bathyporeia pilosa* from three sites of decreasing salinity show progressive restriction from (1) bivoltine cycles with continuous breeding and spring and autumn peaks in west Wales, to

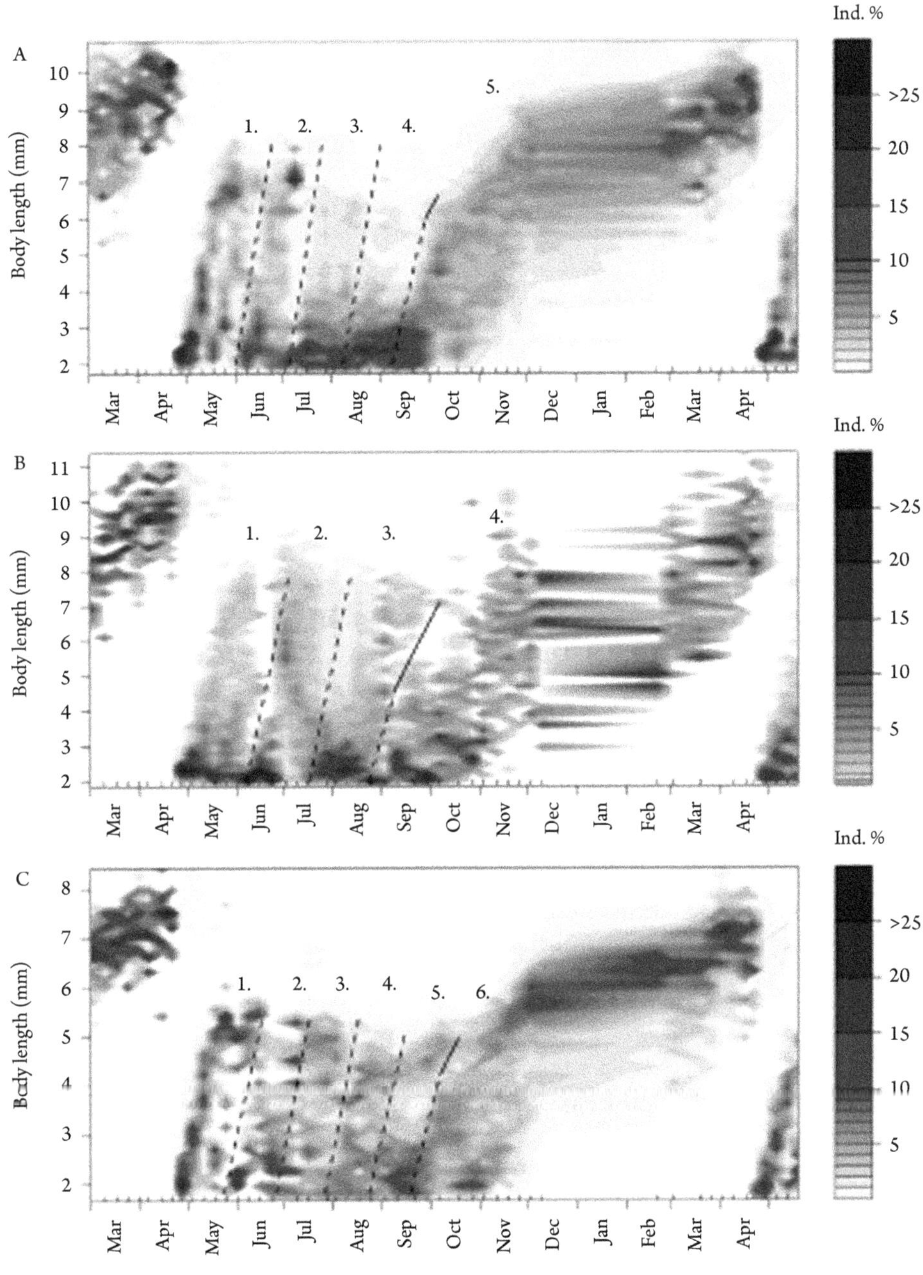

Fig. 5.6.
Length-frequency distributions of mysids (A) *Limnomysis benedeni*, (B) *Hemimysis anomala*, and (C) *Katamysis warpachowskyi*. Solid lines are borderlines between summer-type and overwintering-type individuals. Dashed lines are presumed borderlines between generations (numbered). Upward ticks on the horizontal axis show the sampling dates. From Borza (2014), with permission from Elsevier.

(2) discontinuous breeding with a dominant spring cohort and a minor autumn breeding peak at the mouth of the Severn estuary, to (3) a simple univoltine cycle with summer breeding in the lowest salinity region at the head of the Severn estuary (Mettam 1989).

In the northern waters of its distribution, the euphausiid *Thysanoessa longicaudata*, with individual life spans of 12 to 16 months, has a univoltine life cycle in which sexual maturity is reached around 1 year of age and a new generation is produced during a single spawning pulse in late April to late May (Saunders et al. 2007). In contrast, Lindley (1978) observed that in the warmer waters of the eastern North Atlantic, *T. longicaudata* produces discrete generations during annual spawning pulses.

In the freshwater crayfish *Procambarus clarkii*, Cuchol (2011) showed that there is a general trend from multivoltine life cycles with year-round breeding at tropical climates to a univoltine life cycle with seasonal recruitment at higher latitudes. There is also a latitudinal shift in the onset of the recruitment period in the annual cycle. At lower latitudes, the first recruitment wave usually begins in the spring, whereas at higher latitudes, recruitment does not occur before late summer.

Voltinism and Life Span

Information on longevity is available for a few species of Mysida, with a range of 19 days in *Orientomysis robusta* from shallow warm-temperate habitats (Sudo et al. 2011) to six to seven years in *Antarctomysis maxima* from the Southern Ocean (Siegel and Mühlenhardt-Siegel 1988, San Vicente et al. 2006). Polar and deep-sea mysids generally have the longest life spans, temperate species often have longer life spans for the overwintering generations interspersed with several summer generations with shorter life spans, and tropical species may be the shortest-lived. For example, *Neomysis intermedia* from warm-temperate (6–30°C) freshwater lakes of Japan shows life spans of only 1.5, two, two, and five months in summer, spring, autumn, and winter generations, respectively (Murano 1964). *Orientomysis robusta* from Igarashi Beach (9–27°C) shows life spans of 21–48 days for spring cohorts, 19–33 days for summer cohorts, and 69–138 days for autumn-winter cohorts (Sudo et al. 2011). In *Americamysis bahia*, at test conditions of 22°C and 20 psu, sexual characters were noticeable after the fourth molt (nine to 12 days). Mating occurred after the fifth or sixth molts (17–19 days), and females have frequent brood production (average of five to seven per female) over the full life span of 90 days (Verslycke et al. 2004).

In temperate climates, life span of coastal mysids generally varies in the range of two to 18 months, and rarely up to two years and the different generations have different life spans. For example, the summer generation may live only two to three months (Matsudaira et al. 1952, Delgado et al. 2013), whereas the overwintering generation may live two to three times longer (e.g., San Vicente and Sorbe 2013).

Long-lived species with life spans of years are found at high geographical latitudes, in the deep sea, and in freshwater oligotrophic lakes. Mauchline (1972) concluded that bathypelagic species of euphausiids, lophogastrids, and mysids probably live three to seven times as long as epipelagic species. Effects of the low temperature regime in the bathypelagic environment include decreased growth rates, increased longevity, larger size, and, in some cases, attainment of sexual maturity is prevented (Mauchline 1980). Gigantism of individuals is probably the extreme expression of these effects and may couple with aberrations of behavior (Mauchline 1972).

VOLTINISM AND SECONDARY PRODUCTION

Life span and voltinism are two of the most important life history features influencing secondary production estimation (Waters 1977). Environmental variables such temperature and food availability may influence the production and production/biomass (P/B) ratios, generally by altering some aspects of the life history of the population. Secondary production and P/B ratios increase with the number of generations produced per year, and low temperatures can slow the growth rates of animals and reduce their P/B ratio. Selective predation on adult individuals also alters production, and consequently P/B values, by changing the age distribution of the population; populations dominated by older individuals will have lower P/B ratios than those comprising younger individuals (Robertson 1979).

Several methods have been used to estimate the annual secondary production of crustacean populations. Menzies's (1980) formulation of the Hynes's size-frequency method assumes that the annual mean size-frequency distribution approximates the mortality curve of an average cohort (Benke 1979). The method is applied when it is not possible to identify and follow the growth and mortality of single cohorts through time, and therefore, it is suitable for short-lived and fast-growing species, such as those characteristic of many peracarid populations (Cartes et al. 2011). The formula has a correction factor, the cohort production interval, which requires calculation of a mean longevity of individuals estimated for the diverse annual generations.

Published annual P/B ratios in the literature (Table 5.1) appear to be quite variable. Nonetheless, multivoltine populations have higher rates of production than univoltine ones, and the modes of annual P/B ratios appear to be more or less as follows, from the highest to the lowest: P/B = 25 in multivoltine mysids (probably with four to five or more generations per year), P/B = 10 in bivoltine to trivoltine mysids (two to three generations per year), P/B = 5 in univoltine mysids (one generation per year) and P/B = 1–2 in semivoltine mysids (less than one generation per year and two to three or more years life span).

Mysid production estimates are strongly related to habitat, with coastal lagoons and brackish lakes, in general, having greater production than freshwater lakes and estuaries, whereas deep-ocean and shelf species have relatively low secondary production (Lill et al. 2011). Within species, values of the annual P/B ratio increase with decreases in the life span (Sudo et al. 2011), and high water temperatures result in high P/B ratios (e.g., Cartes et al. 2002).

It should be noted that the effect of voltinism on annual P/B ratios is not entirely deterministic but is broadly consistent among groups of animals with the same voltinism (Waters 1977). Faster growth rates in the littoral zone permit more generations per unit time and thus, multivoltinism may be an effect of high ecosystem production rates, not a cause of them (Downing and Rigler 1984). Estimates for *Tenagomysis chiltoni* suggest that the relatively stable hydrological and food conditions found in intermittently closed estuaries, which lead to dense, stable populations that are maintained through much of the year, implies relatively high production and turnover rate, and also that environmental factors are more important than the high turnover rate in driving the production of this population (Lill et al. 2011).

VOLTINISM AND EVOLUTION

Interspecific and intraspecific variations in crustacean life histories indicate that natural selection has favored different reproductive strategies that vary by life span, habitat use, environmental conditions, geographic location, and other factors. Determining the extent to which different factors have acted via natural selection to influence the evolution of life histories is difficult precisely because of the many factors involved. Nevertheless, one of the more significant factors is clearly voltinism and, in particular, its relationship with the plasticity in populations adapted to diverse and fluctuating environments.

One adaptive feature of any voltinism pattern would be to ensure (1) that the young are released into an optimal environment for individual growth, and (2) the reproduction success of brooding females (Wittmann 1984, Yamada et al. 2007). This is why some authors (e.g., Hébert et al. 2016) suggest that generation time appears to be an important trait with the potential to affect evolutionary processes. In general, species with a shorter life expectancy may maximize their lifetime fitness by investing more in reproduction than in survival, and individuals in species with longer

Table 5.1. Generation length (months), voltinism (number of generations per year) and mean annual production/biomass (P/B) ratio of some marine, brackish, and freshwater mysids. (—) indicates no data.

Species	Geographical zone	Latitude	Generation length	Voltinism	Annual P/B	References
Anchialina agilis	Bay of Biscay	44°N	12	1	4.3–3.9	Sorbe 1984
Anchialina agilis	Adriatic Sea	44°N	11	2	3.3–4.3	Ligas et al. 2009
Anchialina agilis	Catalan Sea	41°N	11	—	1.7–3.9	Cartes et al. 2009
Anisomysis mixta australis	Australia	43°S	7.5	1.5	7.5	Fenton 1996
Antarctomysis maxima	Southern Ocean	64°S	72	<0.5	1.0	Siegel and Mühlenhardt-Siegel 1988, San Vicente et al. 2006
Boreomysis arctica	Catalan Sea	40–41°N	6	2	6.9	Cartes and Sorbe 1999
Boreomysis megalops	Bay of Biscay	45°N	12	1	3.9	Elizalde 1994
Boreomysis californica	Japan	41°N	34	<2	0.3	Chikugo et al. 2013
Erythrops elegans	Adriatic Sea	44°N	6	2	3.4	Ligas et al. 2009
Erythrops elegans	Catalan Sea	44°N	6	—	7.1–7.2	Cartes and Sorbe 1999
Hemimysis margalefi	NW Mediterranean	43°N	6	4	—	Lejeusne and Chevaldonné 2005
Haplostylus normani	Catalan Sea	44°N	7	—	5.9–6.4	Cartes and Sorbe 1999
Leptomysis gracilis	Adriatic Sea	44°N	7	2	3.9–4.1	Ligas et al. 2009
Leptomysis gracilis	Catalan Sea	44°N	7	—	6.0–7.5	Cartes and Sorbe 1999
Mesopodopsis slabberi	NW Mediterranean	40°N	<8	3	—	Delgado et al. 1997
Mesopodopsis slabberi	Portugal	40°N	0.5	2–3?	9.3	Azeiteiro et al. 2001
Mesopodopsis slabberi	S Spain	36°N	2.9	4	38.2	Vilas 2005
Mesopodopsis wooldridgei	South Africa	33°S	—	4	8.0	Wooldridge 1983
Metamysidopsis neritica	South Brazil	25°S	4	3	—	Calil and Borzone 2008
Meterythrops microphtalma	Japan	41°N	>24	<2	0.6	Chikugo et al. 2013
Mysidopsis gibbosa	Adriatic Sea	44°N	7	2	2.9–3.0	Ligas et al. 2009

Mysis mixta	Baltic Sea	57°N	>12	2?	5. 0–7.2	Simm and Kotta 1992
Mysis mixta	Newfoundland	47°N	27	1–2	1.2–1.6	Richoux et al. 2004
Mysis relicta	Pääjarvi lake	61°N	12–24	0.5–1.0	3.0–3.8	Hakala 1978
Mysis relicta	Great Lakes	45°N	19	1	2.2–3.3	Sell 1982
Neomysis americana	Long Island Sound	41°N	12	2 (?)	3.7	Richards and Riley 1967
Neomysis awatschensis	N Japan	43°N	1–8	2	3.0–5.2	Yamada et al. 2007
Neomysis czerniawskii	N Japan	43°N	6?	2	4.1	Yamada et al. 2007
Neomysis integer	SW Netherlands	52°N	7	3	8.3	Mees 1994, Mees et al. 1994
Neomysis integer	S Spain	36°N	4	3	10.7	Vilas 2005
Neomysis mercedis	British Columbia	49°N	12	1	11.1	Cooper et al. 1992
Neomysis mirabilis	N Japan	43°N	2–8	2	1.8–3.5	Yamada et al. 2007
Orientomysis robusta	N Japan	38°N	8.6	>3	21.3	Sudo et al. 2011
Paramesopodopsis rufa	Australia	43°S	7.2	1.6	5.4	Fenton 1996
Parapseudomma calloplura	Catalan Sea	40–41°N	6.3	2	6.7	Cartes et al. 2001
Rhopalopthalmus tartessicus	S Spain	36°N	4	3	10.3	Vilas 2005
Rhopalopthalmus terranatalis	South Africa	33°S	6.7	3	7.9–8.7	Wooldridge 1983, 1986
Siriella clausi	NW Mediterranean	38°N	12	1	4.7	Barberá Cebrián 2002
Siriella clausi	NW Mediterranean	38°N	6	2	9.3	Barberá et al. 2013
Schistomysis assimilis	NW Mediterranean	41°N	6	3	9.1	San Vicente and Sorbe 2003
Schistomysis kervillei	Bay of Biscay	44°N	6	3	9.4	San Vicente and Sorbe 1990
Schistomysis ornata	Bay of Biscay	44°N	8	2	6.1	Sorbe 1991
Schistomysis parkeri	Bay of Biscay	43°N	6	3	9.7	San Vicente and Sorbe 1993
Schistomysis spiritus	Bay of Biscay	44°N	7	3	6.8	San Vicente and Sorbe 1995
Tenagomysis chiltoni	New Zealand	45°S	6	2	13.2	Lill et al. 2011, 2012
Tenagomysis novaezealandiae	New Zealand	45°S	4	3	6.8	Lill et al. 2011, 2012
Tenagomysis tasmaniae	Australia	43°S	9.2	1.3	5.4	Fenton 1996

life spans enhance their fitness by investing proportionately less in reproduction and more in survival. Such trade-offs between reproductive investment and life span have strong implications for voltinism diversity and plasticity.

Differences in generation time can affect rates of molecular evolution in invertebrates because species with faster generation turnover tend to have higher rates of molecular evolution, presumably because their genomes are copied more frequently and therefore collect more DNA replication errors per unit time, as described by Thomas et al. (2010). These authors also suggested that voltinism and rates of molecular evolution are significantly correlated in many invertebrates, including arthropods, and discussed the practical and theoretical implications of such relationships.

Finally, recent taxonomic analyses incorporating the molecular data have revealed high genetic diversification in several cosmopolitan species of mysids (e.g., Audzijonyte and Väinölä 2005, Remerie et al. 2006). These diversifications suggest that differences in life history features among local populations should be reassessed from the viewpoint of not only environmental adaptation but also their genetic diversity (Hanamura et al. 2009), and in particular in light of species that compose a cryptic species complex.

FUTURE DIRECTIONS: CHALLENGES AND OPPORTUNITIES

Some current challenges of aquatic sciences, including the effects of climate change and impacts of invasive species, have a close relationship with voltinism plasticity. Indeed, an increased number of generations in multivoltine arthropod species is expected as a result of shorter development times under changed climatic conditions, with the logical assumption that increases in surface temperatures will permit multivoltine species to increase the number of generations per year (Tobin et al. 2008). Moreover, multivoltinism and the greater reproductive capacity of some species in light of changing climate conditions have implications for the ability of a species to invade new habitats.

Increases in temperature can facilitate an expansion of species distributional range, and in short-lived species warmer temperatures and longer growing seasons may allow additional generations per year (e.g., Tobin et al. 2008). Both possibilities have major implications for populations because, from an ecological and evolutionary perspective, the impact of changes in the number of generations (voltinism) and investment into each generation may be even more important than seasonality, because an additional generation per unit time may accelerate population growth or adaptation (Altermatt 2010). However, not all species are capable of multiple broods or generations in a year, even when the length of the growing season is adequate; therefore, some species will be affected more than others from warming (Forrest and Miller-Rushing 2010). For example, multivoltine populations of the mysid *Hemimysis margalefi* have recently experienced range expansion as a result of warmer conditions in the northwestern Mediterranean, and they have colonized new areas and replaced its bivoltine-trivoltine congener *H. speluncola* in most of the latter's former range (Lejeusne and Chevaldonné 2005).

Climate-induced life history changes are likely to be complex because effects of climate warming can be mediated through other components of an ecosystem, such as resource availability. Population responses of the calanoid copepod *Leptodiaptomus ashlandi* to long-term changes in temperature in Lake Washington were observed between 1962 and 2005, coincident with a pronounced decline in *Leptodiaptomus* densities. However, production was maintained because of an increase in the P/B ratio and a life cycle shift in *Leptodiaptomus* from an annual to a 6-month cycle. This shift from one to two generations per year was most likely caused by a longer and warmer growing period combined with changing fluctuations in resource (phytoplankton) availability (Winder et al. 2009). Greater reproductive capacity of alien species compared to natives has also been proven for almost all alien peracarid populations in European fresh and brackish waters. These alien species have a large brood size, high fecundity, early maturation, and more generations per year (Grabowski et al. 2007, Borza 2014). Combined with predatory behavior, heavy pollution, and high tolerance to warm water temperatures, multivoltinism makes alien peracarids extremely successful invaders.

From an applied point of view, knowledge of voltinism is essential to optimize the most appropriate requirements for crustacean aquaculture (Herrera 2013). Such knowledge is also useful in ecotoxicological studies with multigenerational toxicity tests, as proposed by Verslycke et al. (2004), and to assess the risk of a species being affected by organic toxicants, as in monitoring to assess the ecological status of river basins (Liess and von der Ohe 2005, von der Ohe et al. 2007). Similarly, voltinism is considered an important indicator of the state of macroinvertebrate communities and their potential resilience (Kovalenko et al. 2014).

Voltinism knowledge also facilitates better understanding of population dynamics in metapopulations. Long-term persistence of classic metapopulations is essentially due to asynchrony in the dynamics of local populations, which reduces the variance in the intrinsic rate of metapopulation increase and hence reduces the risk of metapopulation extinction (Hanski 1999). In this sense, voltinism appears as a significant parameter to understand how past, present, and future environmental changes may affect crustacean species.

CONCLUSIONS

Voltinism in crustaceans is an important life history trait that shapes the dynamics of populations and their roles in aquatic ecosystems. Voltinism is measured by the number of generations produced by a population in a year, which can be equal to the number of cohorts for semelparous populations, or, in iteroparous populations, there may be multiple cohorts in a single generation.

Adaptive and phenotypically plastic changes in time spent within each life cycle phase determine the potential voltinism of species and populations. Plasticity in voltinism is common, and has been described in many species and populations with different mesoscale and macroscale distributions.

The number of generations per year typically decreases with latitude and depth and is closely linked to environmental temperature. Trade-offs between reproductive investment and life span have strong implications for voltinism diversity and its plasticity. Boreal and polar species are more likely to be large and have a single large, well-timed brood. In contrast, warm temperate species tend to be small in size, produce many small broods, and have several generations a year. Voltinism influences secondary production, with secondary production and P/B ratios increasing with the number of generations produced per year. The modes of annual P/B ratios analyzed in the order Mysida fluctuate between a maximum of 25 in multivoltine mysids and a minimum of 1–2 in semivoltine mysids. These values can be extrapolated to a great diversity of crustaceans.

Ramifications of some new environmental challenges facing aquatic ecosystems, including the effects of climate change and impacts of invasive species, are mediated by voltinism plasticity. These new threats suggest even more importance for studies of voltinism in crustacean populations.

REFERENCES

Altermatt, F. 2010. Climatic warming increases voltinism in European butterflies and moths. Proceedings of the Royal Society 277:1281–1287.

Amanieu, M. 1970. Cycles reproducteurs et fécondité de quelques crustacés péracarides littoraux. Déterminisme écologique et comparaisons biogéographiques. Bulletin de la Société d'Écologie (France) 1:223–232.

Amaratunga, T., and S. Corey. 1975. Life history of *Mysis stenolepis* Smith (Crustacea, Mysidacea). Canadian Journal of Zoology 53:942–952.

Astthorsson, O.S. 1984. The distribution and biology of mysids in Icelandic subarctic waters as demonstrated by analysis of cod stomach contents. Sarsia 69:107–116.

Audzijonyte, A., and R. Väinölä. 2005. Diversity and distribution of circumpolar fresh- and brackish-water *Mysis* (Crustacea: Mysida): descriptions of *M. relicta* Lovén, 1862, *M. salemaai* n. sp., *M. segerstralei* n. sp. and *M. diluviana* n. sp., based on molecular and morphological characters. Hydrobiologia 544:89–141.

Azeiteiro, U.M.M., L. Jesus, and J.C. Marques. 1999. Distribution, population dynamics, and production of the suprabenthic mysid *Mesopodopsis slabberi* in the Mondego Estuary, Portugal. Journal of Crustacean Biology 19:498–509.

Azeiteiro, U.M.M., J.C. Fonseca, and J.C. Marques. 2001. Biometry, estimates of production and seasonal variation in the biochemical composition of *Mesopodopsis slabberi* (Van Beneden, 1861) (Crustacea: Mysidacea). Boletín Instituto Español de Oceanografía 17:15–25.

Barberá C.C. 2002. Misidáceos (Crustacea, Peracarida) asociados a fanerógamas marinas en el sudeste ibérico. Ph.D. dissertation, Universitat d'Alacant, Spain:1–248.

Barberá, C., P. Sanchez-Jerez, and J.C. Sorbe. 2013. Population structure and secondary production of *Siriella clausii*, a dominant detritus feeding mysid in *Posidonia oceanica* meadows (W Mediterranean Sea). Estuarine, Coastal and Shelf Science 131:103–116.

Barz, K., and H.J. Hirche. 2009. Ecology of mysid shrimps in the Bornholm Basin (central Baltic Sea). Helgoland Marine Research 63:317–326.

Benke, A.C. 1979. A modification of the Hynes method for estimating secondary production with particular significance for multivoltine populations. Limnology and Oceanography 24:168–171.

Bienvenu, F., and S. Legendre. 2015. A new approach to the generation time in matrix population models. The American Naturalist 185:834–843.

Bierschenk, B.M. 2014. Life history and distribution of mysid species in a large open estuary. Ph.D dissertation, University of Otago, Dunedin, New Zealand:1–146.

Bonner, J.T. 2006. Matters of size. Natural History 115:54–59.

Borza, P. 2014. Life history of invasive Ponto-Caspian mysids (Crustacea: Mysida): a comparative study. Limnologica, Ecology and Management of Inland Waters 44:9–17.

Bremer, P., and J. Vijverberg. 1982. Production, population biology and diet of *Neomysis integer* (Leach) in a shallow Frisian lake (The Netherlands). Hydrobiologia 93:41–51.

Brey, T. 2001. Population dynamics in benthic invertebrates. A virtual handbook. Version 01.2. http://www.thomas-brey.de/science/virtualhandbook (accessed March 27, 2017).

Browne, R.A., and G. Wanigasekera. 2000. Combined effects of salinity and temperature on survival and reproduction of five species of *Artemia*. Journal of Experimental Marine Biology and Ecology 244:29–44.

Calil, P., and C.A. Borzone. 2008. Population structure and reproductive biology of *Metamysidopsis neritica* (Crustacea: Mysidacea) in a sand beach in south Brazil. Revista Brasileira de Zoologia 25:403–412.

Cartes, J.E., and J.C. Sorbe. 1996. Temporal population structure of deep-water cumaceans from the western Mediterranean slope. Deep-Sea Research 43:l23–1438.

Cartes, J.E., and J.C. Sorbe. 1998. Aspects of population structure and feeding ecology of the deep-water mysid *Boreomysis arctica*, a dominant species in western Mediterranean slope assemblages. Journal of Plankton Research 20:2273–2290.

Cartes, J.E., and J.C. Sorbe. 1999. Estimating secondary production in bathyal suprabenthic peracarid crustaceans from the Catalan Sea slope (western Mediterranean; 391–1255 m). Journal of Experimental Marine Biology and Ecology 239:195–210.

Cartes, J.E., M. Elizalde, and J.C. Sorbe, 2001. Contrasting life-histories, secondary production, and trophic structure of peracarid assemblages of the bathyal suprabenthos from the Bay of Biscay (NE Atlantic) and the Catalan Sea (NW Mediterranean). Deep-Sea Research Part I 48:2209–2232.

Cartes, J.E., T. Brey, J.C. Sorbe, and F. Maynou. 2002. Comparing production–biomass ratios of benthos and suprabenthos in macrofaunal marine crustaceans. Canadian Journal of Fisheries and Aquatic Sciences 59:1616–1625.

Cartes, J.E., A. Ligas, A.M. De Biasi, L. Pacciardi, and P. Sartor. 2009. Small-spatial scale changes in productivity of suprabenthic and infaunal crustaceans at the continental shelf of Ebro Delta (western Mediterranean). Journal of Experimental Marine Biology and Ecology 378:40–49.

Cartes, J.E., V. Mamouridis, and E. Fanelli. 2011. Deep-sea suprabenthos assemblages (Crustacea) off the Balearic Islands (western Mediterranean): Mesoscale variability in diversity and production. Journal of Sea Research 65:340–354.
Chen, C.Y., and C.L. Folt. 1996. Consequences of fall warming for zooplankton overwintering success. Limnology and Oceanography 41:1077–1086.
Chess, D.W., and J.A. Stanford. 1998. Comparative energetics and life cycle of the opossum shrimp (*Mysis relicta*) in native and non-native environments. Freshwater Biology 40:783–794.
Chikugo, K., A. Yamaguchi, K. Matsuno, R. Saito, and I. Imai. 2013. Life history and production of pelagic mysids and decapods in the Oyashio region, Japan. Crustaceana 86:449–474.
Cooper, K.L., K.D. Hyatt, and D.P. Rankin. 1992. Life-history and production of *Neomysis mercedis* in two British Columbia coastal lakes. Hydrobiologia 230:9–30.
Corbera, J., C. San Vicente, and J.C. Sorbe. 2000. Small-scale distribution, life cycle and production of *Cumopsis goodsir* in Creixell Beach (western Mediterranean). Journal of the Biological Association of the United Kingdom 80:271–282.
Corbet, P.S., F. Suhling, and D. Soendgerath. 2006. Voltinism of Odonata: a review. International Journal of Odonatology 9:1–44.
Cuchol, C. 2011. Population ecology of an alien "warm water" crayfish (*Procambarus clarkii*) in a new cold habitat. Knowledge and Management of Aquatic Ecosystems 401:1–21.
Dauvin, J.C. 1979. Life cycle, dynamics and productivity of Crustacea-Amphipoda from the western English Channel. 5. *Ampelisca sarsi* Chevreux. Journal of Experimental Marine Biology and Ecology 128:31–56.
Dauvin, J.C. 1988. Biologie, dynamique, et production de populations de crustacés amphipodes de la Manche occidentale. 3. *Ampelisca typica* (Bate). Journal of Experimental Marine Biology and Ecology 121:1–22.
Dean, A.F., S.M. Bollens, C. Simenstad, and J. Cordell. 2005. Marshes as sources or sinks of an estuarine mysid: demographic patterns and tidal flux of *Neomysis kadiakensis* at China Camp marsh, San Francisco estuary. Estuarine, Coastal and Shelf Science 63:1–11.
Delgado, L., G. Guerao, and C. Ribera. 1997. Biology of the mysid *Mesopodopsis slabberi* (van Beneden, 1861) (Crustacea, Mysidacea) in a coastal lagoon of the Ebro delta (NW Mediterranean). Hydrobiologia 357:27–35.
Delgado, L., G. Guerao, C. San Vicente, and C. Ribera. 2013. Population structure and life history of *Hemimysis lamornae mediterranea* (Malacostraca: Mysida) in the Ebro Delta (NW Mediterranean). Journal of Sea Research 83:137–145.
Demetrius, L., S. Legendre, and P. Harremoës. 2009. Evolutionary entropy: a predictor of body size, metabolic rate and maximal life span. Bulletin of Mathematical Biology 71:800–818.
Downing, J.A., and F.H. Rigler. 1984. A manual on methods for the assessment of secondary productivity in fresh waters. Blackwell Scientific Publications, Oxford, England.
Elizalde, M. 1994. Les communautés suprabenthiques bathyales de la marge sud du canyon du Cap-Ferret (golfe de Gascogne). Ph.D. dissertation, Université de Bordeaux, France:1–216.
Escribano, R., and R. Riquelme-Bugueño. 2015. Planktonic crustaceans: lifestyles in the water column. Pages 262–278. *in* M. Thiel, and L. Watling, editors. The natural history of the Crustacea, volume 2: Lifestyles and feeding biology. Oxford University Press, New York.
Fager, E.W., and R.I. Clutter. 1968. Parameters of a natural population of a hypopelagic marine mysid, *Metamysidopsis elongata* (Holmes). Physiological Zoology 41:257–267.
Fenton, G.E., 1996. Production and biomass of *Tenagomysis tasmaniae* Fenton, *Anisomysis mixta australis* (Zimmer) and *Paramesopodopsis rufa* Fenton from south-eastern Tasmania (Crustacea: Mysidacea). Hydrobiologia 323:23–30.
Ferreira, S.M., M.A. Pardal, A.I. Lillebø, P.G. Cardoso, and J.C. Marques. 2004. Population dynamics of *Cyathura carinata* (Isopoda) in a eutrophic temperate estuary. Estuarine, Coastal and Shelf Science 61:669–677.
Fockedey, N. 2005. Diet and growth of *Neomysis integer* (Leach, 1814) (Crustacea, Mysidacea). Ph.D. dissertation, University of Gent, Belgium:1–297.
Fockedey, N., J. Mees, M. Vangheluwe, T. Verslycke, C.R. Janssen, and M. Vincx. 2005. Temperature and salinity effects on post-marsupial growth of *Neomysis integer* (Crustacea: Mysidacea). Journal of Experimental Marine Biology and Ecology 326:27–47.

Forrest, J., and A.J. Miller-Rushing. 2010. Toward a synthetic understanding of the role of phenology in ecology and evolution. Philosophical Transactions of the Royal Society B 365:3101–3112.

Ganning, B.1971. On the ecology of *Heterocypris salinus, H. incongruens* and *Cypridopsis aculeata* (Crustacea, Ostracoda) from Baltic brackish-water rockpools. Marine Biology 8:271–279.

Gaudy, R., and J.P. Guerin. 1979. Écophysiologie comparée des mysidacés *Hemimysis speluncola* Ledoyer (cavernicole) et *Leptomysis lingvura* G.O. Sars (non cavernicole). Action de la température sur la croissance en élevage. Journal of Experimental Marine Biology and Ecology 38:101–119.

Geiger, S.R. 1969. Distribution and development of mysids (Crustacea, Mysidacea) from the Arctic Ocean and confluent seas. Bulletin of the Southern California Academy of Sciences 68:103–111.

Gerten, D., and R. Adrian. 2002. Species specific response of freshwater copepods to recent summer warming. Freshwater Biology 47:2163–2173.

Gillooly, J.F. 2000. Effect of body size and temperature on generation time in zooplankton. Journal of Plankton Research 22:241–251.

Goodbody, I. 1965. Continuous breeding in populations of two tropical crustaceans, *Mysidium columbiae* (Zimmer) and *Emerita portoricensis* Schmidt. Ecology 46:195–197.

Grabowski. M., K. Bacela, and A. Konopacka. 2007. How to be an invasive gammarid (Amphipoda: Gammaroidea)—comparison of life history traits. Hydrobiologia 590:75–84.

Grzeszczyk-Kowalska, A., J.C. Chojnacki, M. Raczyńska, and M. Raczyński. 2014. Biological characterization of *Neomysis integer* (Leach, 1815) from the Pomeranian bay in 2006–2007. Journal of Ecological Engineering 15:40–52.

Hakala, I. 1978. Distribution, population dynamics and production of *Mysis relicta* (Lovén) in southern Finland. Annales Zoologici Fennici 15:243–258.

Hamilton, A.L. 1969. On estimating annual production. Limnology and Oceanography 14:771–782.

Hanamura Y., R. Siow, P.E. Chee, and F.M. Kassim. 2009. Seasonality and biological characteristics of the shallow water mysid *Mesopodopsis orientalis* (Crustacea: Mysida) on a tropical sandy beach, Malaysia. Plankton Benthos Research 4:53–61.

Hanselmann, A.J., R. Gergs, and K.O. Rothhaupt. 2011. Seasonal shifts in the life cycle of the ponto-caspian invader *Limnomysis benedeni* (Crustacea: Mysida): a physiological adaptation? Hydrobiologia 673:193–204.

Hanski, I. 1999. Metapopulation ecology. Oxford University Press, New York.

Haridevan, G., R. Jyothibabu, N. Arunpandi, L. Jagadeesan, and A. Biju. 2015. Influence of salinity on the life table demography of a rare Cladocera *Latonopsis australis*. Environmental Monitoring and Assessment 187:1–15.

Hébert, J.B., J.S.J. Scheffer, and D.J. Hawthorne. 2016. Evidence for ecological speciation via a host shift in the holly leaf miner, *Phytomyza glabricola* (Diptera: Agromyzidae). Ecology and Evolution 6:6565–6577.

Herrera, A. 2013. Identification, abundance and rearing of Mysids from Gran Canaria: application to laboratory and oceanographic respiratory metabolism studies. Tésis Doctoral, Universidad de las Palmas de Gran Canaria, Spain:1–208.

Ikeda, T. 1992. Growth and life history of the mesopelagic mysd *Meterythrops microphthalma* in the southern Japan Sea. Journal of Plankton Research 14:1767–1779.

Ingram, C. L., and R.R. Hessler. 1987. Population biology of the deep-sea amphipod *Eurythenes gryllus*: inferences from instar analysis. Deep-Sea Research 34A:1889–1910.

Johannsson, O.E., K.L. Bowenı, M.T. Arts, and R.W. Smith. 2009. Field assessment of condition indices (nucleic acid and protein) in *Mysis diluviana*. Aquatic Biology 5:249–262.

Johnson, W.S., M. Stevens, and L. Watling. 2001. Reproduction and development of marine Peracaridans. Advances in Marine Biology 39:105–260.

Kivelä, S.M., 2011. Evolution of insect life histories in relation to time constraints in seasonal environments. Polymorphism and clinal variation. Acta Universitatis Ouluensis A 569:1–71.

Kivelä, S.M., P. Välimäki, J. Oksanen, A. Kaitala, and V. Kaitala. 2009. Seasonal clines of evolutionarily stable reproductive effort in insects. The American Naturalist 174:526–536.

Kjellberg, G., D.O. Hessen, and J.P. Nilssen. 1991. Life history, growth and reproduction of *Mysis relicta* in the large fjordtype Lake Mjosa, Norway. Freshwater Biology 26:165–173.

Kolding, S., and T.M. Fenchel. 1981. Patterns of reproduction in different populations of five species of the amphipod genus *Gammarus*. Oikos 37:167–172.
Kovalenko, K.E., V.J. Brady, J.J.H. Ciborowski, S. Ilyushkin, and L.B. Johnson. 2014. Functional changes in littoral macroinvertebrate communities in response to watershed-level anthropogenic stress. PLoS ONE 9:e101499.
LaFrance, K., and E. Ruber. 1985. The life cycle and productivity of the amphipod *Gammarus mucronatus* on a northern Massachusetts salt marsh. Limnology and Oceanography 30:1067–1077.
Lehtiniemi, M., M. Viitasalo, and H. Kuosa. 2002. Diet composition influences the growth of the pelagic mysid shrimp, *Mysis mixta* (Mysidacea). Boreal Environment Research 7:121–128.
Lejeusne, C., and P. Chevaldonné. 2005. Population structure and life history of *Hemimysis margalefi* (Crustacea: Mysidacea), a 'thermophilic' cave-dwelling species benefiting from the warming of the NW Mediterranean. Marine Ecology Progress Series 287:189–199.
Leonardsson, K., T. Sörlin, and L. Samberg. 1988. Does *Pontoporeia affinis* (Amphipoda) optimize age at reproduction in the Gulf of Bothnia? Oikos 52:328–336.
Lewis, J.J. 2009. Isopoda (aquatic sowbugs). Pages 346–355 *in* G.E. Likens, editor. Encyclopedia of inland waters, volume 2. Elsevier, Oxford, England.
Liess, M., and P.C. von der Ohe. 2005. Analyzing effects of pesticides on invertebrate communities in streams. Environmental Toxicology and Chemistry 24:954–965.
Ligas, A., A.M. de Biasi, M. Demestre, L. Pacciardi, P. Sartor, and J.E. Cartes. 2009. Effects of chronic trawling disturbance on the secondary production of suprabenthic and infaunal crustacean communities in the Adriatic Sea (NW Mediterranean). Ciencias Marinas 35:195–207.
Lill, A.W.T., G.P. Closs, C. Savage, and M. Schallenberg. 2011. Annual secondary production of two estuarine mysid species (Mysidacea: Mysidae) inhabiting an intermittently closed estuary, south-eastern New Zealand. Marine and Freshwater Research 62:23–834.
Lill, A.W.T., G.P. Closs, C. Savage, and M. Schallenberg. 2012. Annual secondary production of two estuarine mysid species (Mysidacea: Mysidae) inhabiting an intermittently closed estuary, south-eastern New Zealand. Corrigendum. Marine and Freshwater Research 63:385–386.
Lindley, J.A., 1978. Population dynamics and production of euphausiids. I. *Thysanoessa longicaudata* in the North Atlantic. Marine Biology 46:121–130.
Matsudaira, C., T. Kariya, and T. Tsuda. 1952. The study of the biology of a mysid *Gastrosaccus vulgaris* Nakasawa. Tohoku Journal of Agriculture Research 3:155–174.
Mauchline, J. 1972. The biology of bathypelagic organisms, especially Crustacea. Deep-Sea Research 19:753–780.
Mauchline, J., 1980. The biology of mysids and euphausiids. Advances in Marine Biology 18:1–680.
Mauchline, J. 1988. Egg and brood sizes of oceanic pelagic crustaceans. Marine Ecology Progress Series 43:51–258.
McKenney, C.L., Jr., and D.M. Celestial. 1995. Interactions among salinity, temperature, and age on growth of the estuarine mysid *Mysidopsis bahia* reared in the laboratory through a complete life cycle. I. Body mass and age-specific growth rate. Journal of Crustacean Biology 15:169–178.
Mees, J., 1994. The hyperbenthos of shallow coastal waters and estuaries: community structure and biology of the dominant species. Ph.D. dissertation, University of Ghent, Belgium:1–212.
Mees, J., Z. Abdulkerim, and O. Hamerlynck. 1994. Life-history, growth and production of *Neomysis integer* in the Westerschelde Estuary (SW Netherlands). Marine Ecology Progress Series 109:43–57.
Menzies, C.A., 1980. A note on the Hynes method of estimating secondary production. Limnology and Oceanography 25:770–773.
Mettam, C. 1989. The life cycle of *Bathyporeia pilosa* Lindström (Amphipoda) in a stressful, low salinity environment. Scientia Marina 53:543–550.
Morgan, M.D. 1980. Life history characteristics of two introduced populations of *Mysis relicta*. Ecology 61:551–561.
Murano, M. 1964. Fisheries biology of a marine relict mysid, *Neomysis intermedia* Czerniavsky. III. Lifecycle, with special reference to the reproduction of the mysid. Aquaculture Science 12:19–30.
Ogonowski, M. 2012. Studies on Baltic Sea mysids. Ph.D. dissertation, Stockholm University, Sweden:1–31.

Pakhomov, E.A. 2000. Demography and life cycle of Antarctic krill, *Euphausia superba*, in the Indian sector of the Southern Ocean: long-term comparison between coastal and open-ocean regions. Canadian Journal of Fisheries and Aquatic Sciences 57:68–90.

Peck, N., J. Peters, R. Deikmann, S. Laakmann, and J. Renz. 2015. Interactive effects of temperature and salinity on population dynamics of the calanoid copepod *Acartia tonsa*. Journal of Plankton Research 37:197–210.

Pezzack, D.S., and S. Corey. 1979. The life history and distribution of *Neomysis americana* (Smith) (Crustacea, Mysidacea) in Passamaquody Bay. Canadian Journal of Zoology 57:785–793.

Powell, J.A., and J.L. Jenkins. 2000. Seasonal temperature alone can synchronize life cycles. Bulletin of Mathematical Biology 62:977–998.

Quetin, L.B., and R.M. Ross. 2003. Episodic recruitment in Antarctic krill *Euphausia superba* in the Palmer LTER study regi6n. Marine Ecology Progress Series 259:185–200.

Remerie, T., T. Bourgois, D. Peelaers, A. Vierstraete, J. Vanfleteren, and A. Vanreusel. 2006. Phylogeographic patterns of the mysid *Mesopodopsis slabberi* (Crustacea, Mysida) in western Europe: evidence for high molecular diversity and cryptic speciation. Marine Biology 149:465–481.

Richards, S.W., and G.A. Riley. 1967. The benthic epifauna of Long Island Sound. Bulletin of the Bingham Oceanographic Collection 19:89–135.

Richardson, J.S. 1984. Effects of seston quality on the growth of a lake-outlet filter feeder. Oikos 43:386–390.

Richoux, N.B., D. Deibel, and R.J. Thompson. 2004. Population biology of hyperbenthic crustaceans in a cold water environment (Conception Bay, Newfoundland). I. *Mysis mixta* (Mysidacea). Marine Biology 144:881–894.

Robertson, A. I. 1979. The relationship between annual production: biomass rations and lifespans for marine macrobenthos. Oecologia 38:193–202.

Roccatagliata, D. 1991. *Claudicuma platense* Roccatagliata 1981 (Cumacea) a new reproductive pattern. Journal of Crustacean Biology 11:l13–122.

Rossi, V., A. Martorella, and P. Menozzi. 2013. Hatching phenology and voltinism of *Heterocypris barbara* (Crustacea: Ostracoda) from Lampedusa (Sicily, Italy). Journal of Limnology 72:227–237.

Saether, B-E., R. Lande, S. Engen, H. Weimerskirch, M. Lillegard, R. Altwegg, P.H. Becker, T. Bregnballe, J.E. Brommer, R.H. McCleery, J. Merilä, E. Nyholm, W. Rendell, R.R. Robertson, P. Tryjanowski, and M.E. Visser. 2005. Generation time and temporal scaling of bird population dynamics. Nature 436:99–102.

Sainte-Marie, B. 1991. A review of the reproductive binomics of aquatic gammaridean amphipods: variation of life history traits with latitude, depth, salinity and superfamily. Hydrobiologia 223:189–227.

Salemaa, H., K. Tyystjärvi-Muuronen, and E. Aro. 1986. Life histories, distribution and abundance of Mysis mixta and *Mysis relicta* in the northern Baltic Sea. Ophelia Supplementum 4:239–247.

San Vicente, C., and J.C. Sorbe. 1990. Biología del misidáceo suprabentónico *Schistomysis kervillei* (Sars, 1885) en la plataforma continental aquitana (suroeste de Francia). Actas del VI Simposio Ibérico de Estudios del Bentos Marino. Ed. Bilbilis, Palma de Mallorca, Spain:245–267.

San Vicente, C., and J.C. Sorbe. 1993. Biologie du mysidacé suprabenthique *Schistomysis parkeri* Norman, 1892, dans la zone sud du Golfe de Gascogne (plage d'Hendaye). Crustaceana 65:222–252.

San Vicente, C., and J.C. Sorbe. 1995. Biology of the suprabenthic mysid *Schistomysis spiritus* (Norman, 1860) in the southeastern part of the Bay of Biscay. Scientia Marina 59:71–86.

San Vicente, C., and J.C. Sorbe. 2003. Biology of the suprabenthic mysid *Schistomysis assimilis* (Sars, 1877) on Creixell beach, Tarragona (northwestern Mediterranean). Boletín del Instituto Español de Oceanografía 19:391–120.

San Vicente, C., and J.C. Sorbe. 2013. Comparative life-histories, population dynamics and productivity of *Schistomysis* populations (Crustacea, Mysida) in European shelf environments. Journal of Sea Research 81:13–32.

San Vicente, C., A. Ramos, and J.C. Sorbe. 2006. Suprabenthic euphausiids and mysids from the South Shetland Islands and the Bransfield Strait, Southern Ocean (BENTART-95 cruise). Polar Biology 29:211–222.

Sardà, F. 1995. A review (1967–1990) of some aspects of the life history of *Nephrops norvegicus*. ICES Marine Science Symposium 199:78–88.

Sastry, A.N. 1983. Ecological aspects of reproduction. Pages 179–270 *in* W.B. Vernberg, editor. The biology of Crustacea: environment adaptations. Academic Press, New York.
Saulich, A.K., and D.L. Musolin. 1996. Univoltinism and its regulation in some temperate true bugs (Heteroptera). European Journal of Entomology 93:07–518.
Saunders, R.A., A. Ingvarsdóttir, J. Rasmussen, S.J. Hay, and A.S. Brierley. 2007. Regional variation in distribution pattern, population structure and growth rates of *Meganyctiphanes norvegica* and *Thysanoessa longicaudata* in the Irminger Sea, North Atlantic. Progress in Oceanography 72:313–342.
Sell, D.W., 1982. Size-frequency estimates of secondary production by *Mysis relicta* in Lakes Michigan and Huron. Hydrobiologia 93:69–78.
Siegel, V., 2000. Krill (Euphausiacea) life history and aspects of population dynamics. Canadian Journal of Fisheries and Aquatic Sciences 57:130–150.
Siegel, V., and Mühlenhardt-Siegel U. 1988. On the occurrence and biology of some Antarctic Mysidacea (Crustacea). Polar Biology 8:181–190.
Simm, M., and I. Kotta. 1992. The life cycle and production of *Mysis mixta* in the Gulf of Riga. Pages 45–54 *in* J. Köhn, M.B. Jones, and A. Moffat, editors. Taxonomy, biology and ecology of (Baltic) mysids (Mysidacea: Crustacea). International Expert Conference, September 1991, Hiddensee, Rostock University, Germany.
Smith, S.I. 1879. II. The stalk-eyed crustaceans of the Atlantic coast of North America north of Cape Cod. Transactions of the Connecticut Academy of Arts and Sciences 5:27–138.
Sobrino, I., C. Silva, M. Sbrana, and K. Kapiris. 2005. Review of the biology and fisheries of the deep water rose shrimp *Parapenaeus longirostris* in European Atlantic and Mediterranean waters (Decapoda, Dendrobranchiata, Penaeidae). Crustaceana 78:1153–1184.
Sorbe, J.C., 1984. Contribution à la connaissance des peuplements suprabenthiques néritiques sud-Gascogne. Ph.D. dissertation, Université de Bordeaux I, France:1–265.
Sorbe, J.C., 1991. Biología del misidáceo suprabentónico *Schistomysis ornata* (Sars, 1864) en la plataforma continental aquitana (suroeste de Francia). Actas del V Simposio Ibérico de Estudios del Bentos Marino 24:273–298.
Stearns, S.C. 1977. The evolution of life history traits: a critique of the theory and review of the data. Annual Review of Ecology and Systematics 8:145–171.
Stearns, S.C. 1992. The evolution of life history. Oxford University Press, London, England.
Sudo, H., N. Kajihara, and M. Noguchi. 2011. Life history and production of the mysid *Orientomysis robusta*: high P/B ratio in a shallow warm-temperate habitat of the Sea of Japan. Marine Biology 158:1537–1549.
Thomas, J.A., J.J. Welch, R. Lanfear, and L. Bromham. 2010. A generation time effect on the rate of molecular evolution in invertebrates. Molecular Biology and Evolution 27:1173–1180.
Tobin, P.C., S. Nagarkatti, G. Loeb and M.C. Saunders. 2008. Historical and projected interactions between climate change and insect voltinism in a multivoltine species. Global Change Biology 14:951–957.
Toda, H., M. Takahashi, and S. Ichimura. 1982. Abundance and life history of *Neomysis intermedia* Czerniavsky in Lake Kasumigaura. Hydrobiologia 93:31–39.
Tsai, M.L., and C.F. Dai, 2001. Life history plasticity and reproductive strategy enabling the invasion of *Ligia exotica* (Crustacea: Isopoda) from the littoral zone to an inland creek. Marine Ecology Progress Series 210:175–184.
Twombly, S., N. Clancy, and C.W. Burns. 1998. Life history consequences of food quality in the freshwater copepod *Boeckella triarticulata*. Ecology 79:1711–1724.
Verslycke, T., N. Fockedey, C.L. McKenney, Jr., S.D. Roast, M.B. Jones, J. Mees, and C.R. Janssen. 2004. Mysids as potential test organisms for the evaluation of environmental endocrine disruption: a review. Environmental Toxicology and Chemistry 23:1219–1234.
Vilas, C., 2005. Estructura y dinámica de las poblaciones de misidáceos del estuario del Guadalquivir. Importancia ecológica. Ph.D. dissertation, University of Cádiz, Spain:1–204.
von der Ohe, P.C., A. Prüf, R.B. Schäfer, M. Liess, E. de Deckere, and W. Brack. 2007. Water quality indices across Europe—a comparison of the good ecological status of five river basins. Journal of Environmental Monitoring 9:970–978.
Ward, P. 1984. Aspects of the biology of *Antarctomysis maxima* (Crustacea: Mysidacea). Polar Biology 3:85–92.

Ward, P. 1985. On the biology of *Antarctomysis ohlini* (Crustacea: Mysidacea) at South Georgia. Bulletin British Antarctic Survey 67:13–23.

Waters, T.F., 1977. Secondary production in inland waters. Advances Ecological Research 24:91–164.

Winder, M., D.E. Schindler, T.E. Essington, and A.H. Litt. 2009. Disrupted seasonal clockwork in the population dynamics of a freshwater copepod by climate warming. Limnology and Oceanography 54:2493–2505.

Winkler, G., and W. Greve. 2002. Laboratory studies of the effect of temperature on growth, molting and reproduction in the co-occurring mysids *Neomysis integer* and *Praunus flexuosus*. Marine Ecology Progress Series 235:177–188.

Wittmann, K.J. 1984. Ecophysiology of marsupial development and reproduction in Mysidacea (Crustacea). Oceanography and Marine Biology: an Annual Review 22:293–428.

Wittmann, K.J. 1992. Cyclomorphosis in the temperate zone Mysidacea: evidence and possible adaptive taxonomical significance. Pages 25–32 *in* J. Köhn, M.B. Jones, and A. Moffat, editors. Taxonomy, biology and ecology of (Baltic) mysids (Mysidacea: Crustacea). International Expert Conference, September 1991, Hiddensee, Rostock University, Germany.

Wooldridge, T.H., 1983. Ecology of beach and surf-zone mysid shrimps in the Eastern Cape, South Africa. Pages 449–460 *in* A. McLachlan, and T. Erasmus, editors. Sandy beaches as ecosystems. Dr. W. Junk Publishers, The Hague, The Netherlands.

Wooldridge, T.H., 1986. Distribution, population dynamics and estimates of production for the estuarine mysid, *Rhopalophthalmus terranatalis*. Estuarine, Coastal and Shelf Science 23:205–223.

Yamada, K., K. Takahashi, C. Vallet, C., S. Taguchi, and T. Toda. 2007. Distribution, life history, and production of three species of *Neomysis* in Akkeshi-ko estuary, northern Japan. Marine Biology 150:905–917.

6

LARVAE AND DIRECT DEVELOPMENT

Richard R. Strathmann

Abstract

Modes of development of marine crustaceans and other marine invertebrates include presence or absence of a larval stage, of larval feeding, and of maternal protection of offspring. These different developmental modes impose different compromises (trade-offs) between the number of offspring and their size or the extent of maternal protection. Crustaceans differ from many marine animals in not shedding eggs prior to fertilization, which eliminates the complication of selection on size of eggs as a target for sperm. Features shared with marine invertebrates of several phyla include rare and ancient origins of feeding larvae, irreversible losses of a feeding larval stage, a constraint on brooding imposed by embryos' need for oxygen, and possible benefits from slower development of protected embryos. Crustaceans differ, however, in having a diverse exoskeletal tool kit that has provided unusual capabilities. Nauplii and zoeae are diverse in form, behavior, and habitat, despite each being nominally one type of larva. Nauplii, as feeding larvae, have adapted to both the benthos and plankton. Settling stages (cyprids and decapodids) with enhanced speed have evolved twice. Some very large adults can supply their large broods with oxygen. Capacity for defense of offspring and home has led a few times to eusociality. The need to molt to grow and change form imposes episodic risk and growth and, in some cases, links evolution of egg size and size at metamorphosis. Crustaceans' diverse life histories enable comparisons with broad implications for marine invertebrates: opportunity for dispersal is similar for larvae and adults of some crustaceans, demonstrating that marine larvae need not be adaptations for dispersal; development from very small eggs is enabled by less equipment needed for first larval feeding and also by postlarval stages being parasites; eggs shed into the water suffer greater mortality than planktonic larvae or brooded eggs, yet some planktonic crustaceans depend on benthic resting eggs for persistence of populations; larvae escape predation in diverse ways, and bigger larvae are not consistently safer; predation near the seafloor makes settlement a risky stage. Parallels with other taxa are

Life Histories. Edited by Gary A. Wellborn and Martin Thiel.

numerous, but the crustacean exoskeletal tool kit has conferred unusual evolutionary opportunities and constraints. Even among marine crustaceans, however, evolutionary options for life histories differ among clades because of rare evolutionary origins of traits of larvae and mothers and biased evolutionary transitions in those traits.

INTRODUCTION

The complexity of crustacean modes of development is illustrated by three features that are combined among taxa in different ways: the presence and extent of metamorphosis, parental care, and larval feeding (Fig. 6.1). Evolutionary changes in crustaceans' development have included (1) more abrupt transitions in form and function; (2) losses of the ancestral feeding larval form; (3) origin of new feeding larval stages; (4) origin of larval stages adapted for settlement; (5) origins of parental protection of embryos; (6) embryonization of protected larval stages; (7) abbreviation of larval development, even to hatching as an adult-like juvenile; (8) juvenile offspring remaining with their mother; and (9) in a few cases adult offspring forgoing reproduction while aiding their mother with successive broods (eusociality).

Possible trade-offs (compromises) associated with these traits are numerous. Some in the following list are common, some occasional. With more extended feeding and greater growth of the larva, initial size can be smaller, enabling greater fecundity; or initial juvenile size can be greater, increasing juvenile growth and survival. Potential costs are exposure of the larva to risk for a longer time, with a smaller fraction surviving. A planktonic larval life may sometimes provide greater growth or greater survival than a benthic one, but potential costs are transport away from favorable benthic habitat, risk of predation during the transition to benthic habitat, and special equipment for settlement. Larval dispersal can confer both benefits and costs. Maternal protection can increase survival of offspring. Extended protection can be associated with release of offspring that are larger or at a more advanced stage of development, with perhaps greater survival after their release, earlier growth to reproduction, retention at a favorable parental site, or other benefits of larger size or advanced stage. Extended protection is often associated with slower development (independent of

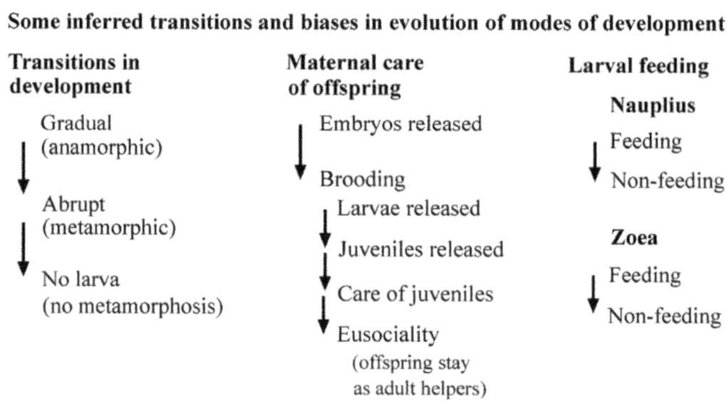

Fig. 6.1.
Modes of development that occur in crustaceans in differing combinations. The arrows indicate inferred predominate directions of evolutionary transitions. Evolutionary reversals (not shown) are a possibility in some cases but in others may be precluded by extensive changes in development and capabilities. Some steps in sequences may be skipped. Loss of a larval stage without prior evolution of a metamorphosis is a possibility, and release of larvae does not preclude evolution of care of juveniles or evolution of eusociality if the larvae do not disperse away from the parent.

size or stage at hatching), with several potential benefits. Costs of maternal protection can be fewer offspring produced or lower maternal survival or growth. Embryos' need for oxygen can impose costs or limits on brooding.

Models for developmental adaptations have especially focused on ways that environmental features affecting an offspring's growth (e.g., food, temperature) and survival (e.g., predators, dispersal relative to favorable habitat) influence optimal maternal investment per egg and in protection during embryonic development. Models that were intended to be simple and applicable to diverse animals omitted numerous challenges, compromises, and constraints. However, constraints and trade-offs affecting evolution of larvae and maternal protection differ among taxa even within the same environment. This review emphasizes constraints and opportunities in the evolution of developmental adaptations of marine crustaceans that are not addressed in general models of size-number trade-offs or differ from common assumptions on costs. It discusses (1) some modeled trade-offs in the evolution of larvae and brooding and then (2) ways that the crustaceans' exoskeletal tool kit, with diverse appendages, has enabled or constrained adaptations of larvae and brooding mothers. It next examines two features of marine life histories for which crustaceans have been unusually informative: (3) larvae as adaptations for dispersal and (4) mortality of embryos and larvae.

Use of terms varies among authors. Here, the egg stage includes unhatched embryos. Embryonic development ceases when the offspring hatch or are liberated from a brood. In this sense, an egg nauplius is an embryo; a free nauplius is a larva. The term *nauplius* refers to both ortho- and metanauplius stages. An animal becomes a juvenile when its form becomes adult-like.

MODELS OF SIZE-DEPENDENT TRADE-OFFS

Attempts to understand modes of development of disparate kinds of animals have focused on how natural selection might maximize the number of surviving offspring. The best compromises between number and mortality of offspring depend on both the environment and the traits of the animals. Adaptationist hypotheses on how modes of development suit an environment do not necessarily assume that the traits have changed to fit the animals to where they live. Instead, animals with a given set of traits live where they can.

Size and Number of Eggs

Size is an approximate and easily measured indicator of the organic content of eggs or of similar larvae or juveniles. More data exist for size of offspring than other measures of maternal investment; however, although useful for the following discussion, size is an imperfect indication of quantity or quality of material in offspring. Even the early stage embryos on different pleopods within a lobster's clutch can differ in fatty acid profiles, a difference expected to affect both energy reserves and other aspects of an offspring's quality (Leal et al. 2013).

In Smith and Fretwell's (1974) model the number of offspring that a mother can produce is inversely proportional to the provisioning for each; the probability that an offspring will survive to reproduce increases with provisioning by the parent but with diminishing returns (Fig. 6.2A). Larger offspring are expected to benefit from a briefer exposure to risk and from greater rates of growth or lower rates of mortality. Their model predicts a single optimal investment per offspring that maximizes the number that survive to a given later size and stage. It does not explicitly address several features of development that affect size-number trade-offs: presence or absence of parental protection, larval stages, larval feeding, change in habitat, or metamorphosis.

Vance's (1973) model includes the extent to which larval feeding and growth bridge the gap between materials provided in the egg and materials present at metamorphosis, thus linking parental

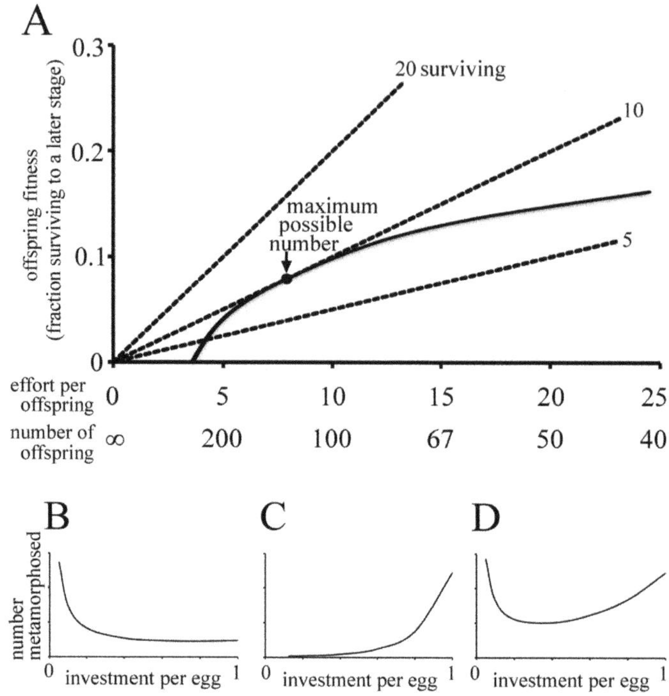

Fig. 6.2.
Models of maternal investment per offspring and offspring fitness. (A) Model of the optimal trade-off between number and size of offspring (after Smith and Fretwell 1974). For 1,000 units devoted to reproduction, different combinations of effort per offspring and number of offspring are possible (*x*-axis). Each dashed straight line shows a number of surviving offspring that would result from combinations of size and number of offspring produced (*x*-axis) and fraction surviving (*y*-axis). The curved line is the fraction that would actually survive for any given number of offspring and effort per offspring. The number of survivors is maximal for the effort per offspring at which the curved line meets the straight line of greatest slope. In this plot, returns diminish as investment per offspring is increased, and no offspring survive below a minimum investment per offspring. (B–D) Vance's (1973) model in which investment per egg ranges from 0 to 1, with 1 being the investment sufficient for development to metamorphosis without larval feeding. Depending on constant rates of growth and mortality, survival to metamorphosis is maximal with the smallest possible egg (B), with an egg large enough for development without larval feeding (C), or with local optima at either extreme of egg size (D).

investment per offspring to mode of development. Size at metamorphosis, rates of growth (weight added per time), and mortality rates are assumed to be constant; this is mathematically convenient but biologically unrealistic. The model predicts that, depending on conditions for larval growth and mortality, the number of offspring reaching metamorphic competence is maximized by (1) growth to metamorphosis that started with minimal material (Fig. 6.2B), (2) no growth required before metamorphosis (Fig. 6.2C), or (3) either extreme as a local optimum (Fig. 6.2D). An appealing feature is prediction of more than 1 mode of development as possible simultaneous optima, but the prediction of extremes as optima is inconsistent with the variation in egg sizes among and within species.

Subsequent models, starting with Christiansen and Fenchel (1979), predict intermediate egg sizes and degrees of larval feeding by assuming that growth rate increases or mortality rate decreases with increasing larval size. A decrease in egg size presumably results in a disproportionately longer time for a given amount of growth because if weight added per body weight per time changes little during growth, then weight added per time increases with increasing weight. With a constant instantaneous growth rate (g in exponential growth $W_t = W_0 e^{gt}$) and constant increase in weight from egg to metamorphosis (W_t–W_0), the period required for growth (t) is longer when initial weight (W_0) is less. Differences in assumed size-dependent growth or mortality of larvae can result in different evolutionarily stable egg sizes greater than zero and less than the size at metamorphosis. Optimal egg sizes and durations of larval feeding can vary over a wide range if one assumes that different environments or larval traits result in different size-dependent rates of growth or mortality (Strathmann 1985). Inclusion of facultative feeding, in which a larva can subsist without food for a time after it is first able to feed, also yields intermediate optimal egg sizes (McEdward 1997). Other complications have been modeled (Havenhand 1995, Marshall and Keough 2008). For example, unpredictable variation in rates of mortality or growth could favor bet hedging. Adult traits, such as reproductive effort or longevity of mothers, could be associated with selection on egg size or protection of embryos.

Size at Metamorphosis

The assumption of a constant size at metamorphosis is convenient but often incorrect. Selection for a briefer larval duration might increase size of eggs, reduce size at metamorphosis, or both (Strathmann 1977). Selection could act separately on size at metamorphosis if the optimal form differs with size or habitat. Appealing for its simplicity is the hypothesis that an animal should change its form or habitat at the size that minimizes the ratio between the instantaneous mortality rate and the instantaneous growth rate (Werner and Gilliam 1984, Werner 1988). This criterion requires the assumption of a population growth of zero.

Parental Protection

A mother may devote resources to protecting offspring that could otherwise be devoted to producing more embryos. One resource is time, as when a second brood cannot be produced until the preceding one is released. Another resource is materials, as when the mother reduces feeding or expends more energy while brooding. Protection of offspring may therefore increase survival but restrict production of eggs (Vance 1973, Grant 1983). Under that assumption, selection for protection depends on mortality rates of protected and unprotected early stages and the duration of their development. Pechenik (1979) suggested that greater vulnerability of early stages is the cause of mixed development, with parental protection followed by a free larval stage. The model of Sargent et al. (1987) predicts that the reduction in mortality due to parental protection results in a greater optimal egg size. The model includes greater duration of development to hatching for protected offspring, in part because greater provisions in the egg extend duration of development to hatching and also because larger eggs produce larger hatchlings, possibly with faster growth, lower mortality, and less time to become adults.

Experimental Demonstration of Correlations

Models that examine costs and benefits of developmental traits and comparisons of traits among taxa can be complemented by microevolutionary experiments that examine heritability, genetic correlations, and response to selection. Some small crustaceans, as in the copepod genus *Tigriopus*, can be reared and bred in petri dishes on a desk top through numerous generations with controlled matings (e.g., Barreto and Burton 2013). Experimental selection on egg size and size at metamorphosis of such species could test for correlations in their evolutionary change. Also, exoskeletal parts are easily measured, and examining genetic correlations of dimensions of appendages through nauplius and copepodid stages could test the hypothesis that metamorphosis decouples selection on larval and adult forms (Ebenman 1992, Jordan West, personal communication, 1996, Aguirre et al. 2014).

EQUIPMENT FOR CRUSTACEANS' MODES OF DEVELOPMENT

Part of the usefulness of the previously discussed models is that they call attention to what they omit. The models examine trade-offs that result from effects of investment per offspring on performance in growth and survival. That performance depends, however, on the particular equipment that larvae use to swim and feed and that parents use to protect embryos. That equipment provides some flexibility but also imposes some constraints on the evolution of larvae and the protection of embryos. The following examples illustrate how equipment has affected the evolution of modes of development in clades of marine crustaceans.

The Nauplius: A Head for Swimming and Feeding

Crustaceans, like many bilaterians, develop with additions to the body near the posterior end. Anterior appendages differentiate first and can potentially function early, but are few; hence head larvae that swim or crawl early in development do so with anterior appendages that differ from the adults. The nauplius is a head larva (Fig. 6.3A,B). The 3 most anterior pairs of appendages are sufficient for locomotion and feeding. Equipment for feeding includes spines near the base of the antenna and mandible, as well as a labrum (Sanders 1963). The taxonomic distribution of nauplii indicates that a feeding nauplius was the larval form of the ancestral crustacean (Dahms 2000, Martin et al. 2014, Akther et al. 2015), an inference supported by fossils. A metanauplius from the Lower Cambrian has a labrum and basal medially directed spines on the second antenna and mandible, indicating that it fed (Zhang et al. 2010). Larvae of phosphatocopines from the Upper Cambrian

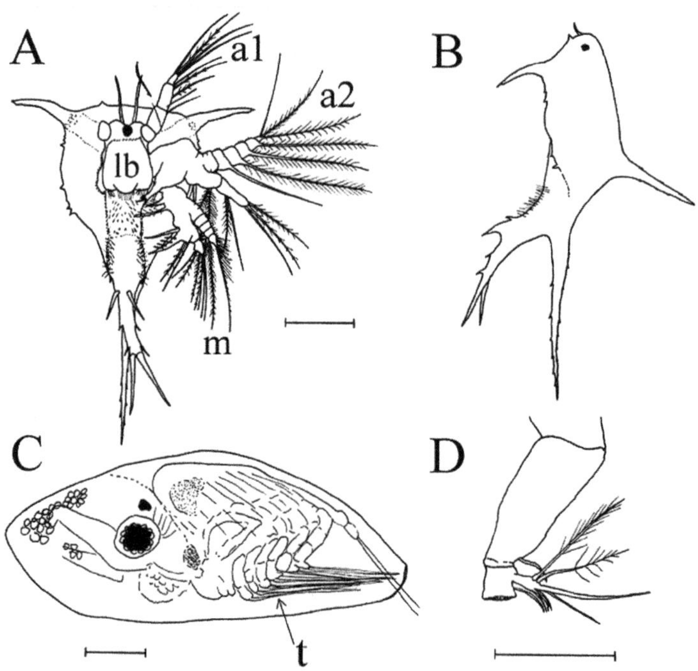

Fig. 6.3.
Larvae of thoracican barnacles. Stage II nauplius in (A) ventral and (B) lateral views. (C) Cyprid in lateral view, and (D) detail of cyprid antennule. a1 = first antenna; a2 = second antenna; lb = labrum; m = mandible; t = thoracic appendages. Scale bars = 0.1 mm. From Lang (1979).

were similarly equipped, indicating similar feeding capabilities, and because the phosphatocopines are known from the Lower Cambrian and are the inferred sister group of eucrustaceans, the equipment for naupliar feeding is inferred to predate the common ancestor of existing crustaceans (Maas and Waloszek 2005, Zhang et al. 2012). Although feeding naupliar stages may have evolved once and broadly share equipment such as the labrum and spines at the base of second antennae and mandibles, they are diverse in the methods of feeding. Even among planktonic copepods, the nauplii differ in methods of food capture (Bruno et al. 2012). Planktonic nauplii of other taxa and benthic nauplii further extend the diversity of methods of feeding within the naupliar body plan.

Capabilities Conferred by Nauplii

With a head larva, the vulnerable embryonic period can be short. Embryos released into the water can hatch as nauplii in less than a day, as in many copepods, euphausiaceans, and dendrobranchiate shrimps. Beginning locomotion early may be safer than remaining an unhatched embryo. Hatching is not delayed until development of capacity for larval feeding. In euphausiaceans, dendrobranchiate shrimp, and many calanoid copepods, the first several naupliar stages swim but do not feed. Rapid development to hatching is not always advantageous, however, because risk for embryos may be relatively low in some cases. Some barnacles brood embryos for weeks during the winter before releasing nauplii (Branscomb et al. 2014), and some copepods produce resting eggs that hatch much later as nauplii (Lindley 1997).

With only three pairs of appendages, a nauplius can be small, which enables evolutionary miniaturization at hatching, with the advantage of smaller and thus more numerous eggs. Posterior parts of the body develop as the larva grows. There is much overlap in sizes of eggs of crustaceans that hatch as feeding nauplii or at later stages, but many crustaceans with feeding nauplii have eggs smaller than the smallest eggs of crustaceans that first feed when they have more appendages. Many copepods and some barnacles that hatch as nauplii have eggs 100 μm or less in diameter, whereas the smallest eggs of malacostracans, which lack feeding naupliar stages, are nearly all greater than 100 μm (Fig. 6.4). Although Fig. 6.4 illustrates general trends in miniaturization, limited sampling of taxa influences extreme reported values; Johnson et al. (2001) mentioned a tanaidacean with an egg less than 100 μm but did not give the diameter of that small egg.

Parasitism following a nonfeeding larval stage has also enabled extreme miniaturization of eggs. Rhizocephalans (Kentrogonida and Akentrogonida; Fig. 6.4) are parasites that hatch as nonfeeding nauplii or cyprids and never feed with appendages. Many have eggs smaller than 100 μm (Høeg and Lützen 1993, 1995). Some rhizocephalan cyprids are less than 200 μm in carapace length yet possess thoracic limbs (Glenner et al. 1989); however, an extreme miniaturization of a rhizocephalan cyprid is associated with absence of some body parts. Bocquet-Védrine (1961) describes eggs of 35 to 45 μm diameter and cyprids of approximately 70 μm in carapace length for the akentrogonid *Chthamalophilus delagei*. This cyprid lacks thoracic limbs for swimming. Drifting is a possibility; one pair of appendages (antennules) is sufficient for walking and attachment; its target host is a sessile barnacle.

Bias in Loss of Naupliar Feeding

The labrum and medially directed spines at the base of antenna and mandible are commonly reduced or absent in nonfeeding nauplii (Martin et al. 2014, Akther et al. 2015). For crustaceans with nonfeeding nauplii (e.g., rhizocephalans, remipedes, euphausiaceans, dendrobranchiate shrimp), evolutionary restoration of naupliar feeding might be precluded because with no equipment for capturing food, there could be no feeding capability to be enhanced by selection. Existing evidence points to a single origin of naupliar feeding and numerous irreversible losses.

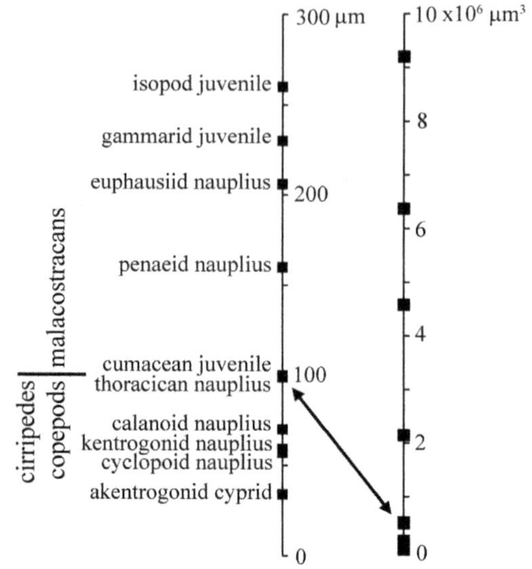

Fig. 6.4.
Smallest egg diameters and volumes for taxa differing in stage at hatching. Points plotted on line to the left are egg diameters and are labeled with taxon name and stage at hatching. Points plotted on line to the right are volumes for the same eggs, to indicate relative differences in organic material per egg. The arrow points to eggs of 100 μm diameter. Diameters for nonspherical eggs were calculated as diameters of a sphere of a volume equivalent to a spheroid of the reported length and width. Smallest eggs are as recorded in reviews of several crustacean taxa (Barnes 1989, Dall et al. 1990, Kiørboe and Sabatini 1995, Johnson et al. 2001, Gómez-Gutiérrez et al. 2010). Diameters for copepod eggs are from papers cited by Kiørboe and Sabatini (1995) for small eggs.

The taxonomic distribution of nauplii within Crustacea raises questions, however, about the evolution of a nonfeeding but swimming nauplius from an embryonized nauplius. Most malacostracans lack a free nauplius stage. Euphausiaceans and dendrobranchiate shrimp, which hatch as nonfeeding nauplii, are the exceptions. Inferred phylogenies of malacostracans imply that either the ancestor lacked a free nauplius and free nauplii subsequently evolved again, or that egg nauplii evolved multiple times (Strathmann and Eernisse 1994). Scholtz's (2000) hypothesis of evolutionary restoration of a free but nonfeeding nauplius from an egg nauplius gains plausibility because little is required of a swimming but nonfeeding nauplius (Fig. 6.5). These nonfeeding nauplii lack a labrum, midgut, articulation of the limbs, masticatory spine on the second antenna, and mandibular gnathobase, among other structures. Jirikowski et al. (2015) analyzed the shifts in timing of development that would be necessary for restoration of a free nauplius from an embryonized nauplius. Reduction of structures for feeding is expected, however, whether derivation is from a feeding nauplius or egg nauplius (Akther et al. 2015). The free-swimming but nonfeeding larva of the branchiopod *Leptodora kindtii* suggests similar hypotheses: either the larva evolved from ancestral brooded embryonized stages or the embryonized brooded stages evolved twice (Olesen 2004). In any case, there is as yet no evidence for restoration of a feeding nauplius after loss of the free naupliar stage.

Larvae with More than a Head for Swimming and Feeding

Other kinds of crustacean larvae are more taxonomically restricted than the nauplius (Martin et al. 2014) and therefore are inferred to have evolved later. These larvae use more posterior appendages in swimming or feeding and cannot provide as early a head start as a nauplius. For the most rapid

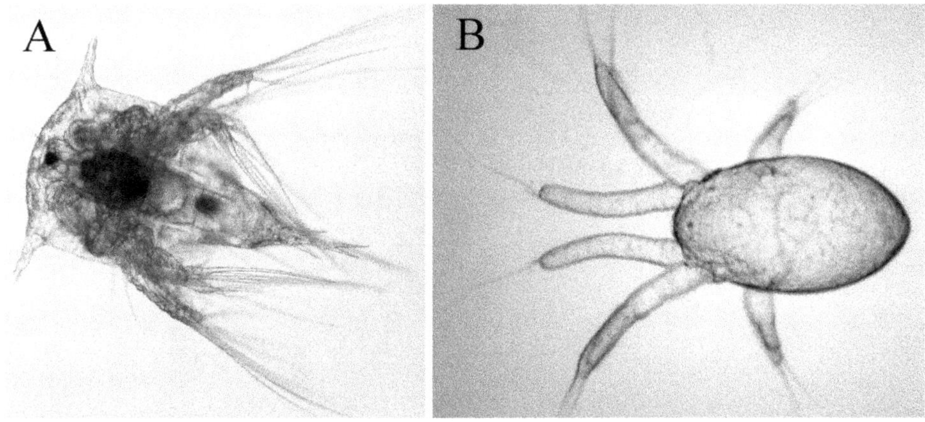

Fig. 6.5.
(A) Feeding nauplius (Thoracica) and (B) nonfeeding nauplius (Euphausiacea). Photographs by Richard R. Strathmann ©.

development, time to first larval feeding is greater. For the smallest offspring, eggs and first feeding stages are larger.

Nonfeeding naupliar stages are followed in development by feeding calyptopis and furcilia stages in euphausiaceans and by feeding protozoea and zoea stages in dendrobranchiate decapods. Similarities between euphausiacean and dendrobranchiate larvae suggest a single origin of this sequence of stages (Akther et al. 2015). These malacostracans have retained (or reevolved) a free nauplius with advantages of early hatching, although the first feeding stages (calyptopis and protozoea) use more posterior appendages (maxillae and maxillipeds) in feeding and swimming. As development proceeds, thoracic appendages first supplement and then replace antennal function in swimming. The earliest feeding larval stage of the remaining decapods (Pleocyemata) is a zoea. Zoeae capture a wide range of prey, from phytoplankton (Fileman et al. 2014) to cannibalizing zoeae about as large as themselves (Epelbaum and Borisov 2006). Zoeae use thoracic exopods for swimming and thoracic endopods for grasping prey (Anger 2001, Epelbaum and Borisov 2006; Fig. 6.6). Maxillules have also been observed to grasp and manipulate prey (Crain 1999), and at least some zoeae use mouthparts in suspension feeding on smaller particulate food. Almost all zoeae appear to be planktonic and to feed on plankton, but some, such as *Ebalia tuberosa*, may reside and feed at the sea floor (Schembri 1982).

Stomatopods hatch as nonfeeding larvae that remain with the mother. In lysiosquilloids these nonfeeding benthic larvae have biramous thoracic appendages and no abdominal appendages, whereas in other stomatopods they have uniramous thoracic appendages and biramous pleopods (Martin et al. 2014). Offspring leave the mother as feeding planktonic larvae, with the second maxillipeds as large raptorial claws for grasping prey and pleopods for swimming.

Fossils of feeding larval stages of decapods and stomatopods are described from the Mesozoic (Martin et al. 2014). Fossils described as (nonlarval) stomatopods and decapods date from the Paleozoic (Schram 1982). Hence, a Paleozoic origin for these feeding larvae is possible.

Settling Larvae

The special requirements for changing from the plankton to the benthos or to a host were not included in models emphasizing size-dependent trade-offs but have resulted in evolution of distinctive transitional larval forms. For example, a nonfeeding cyprid stage follows the nauplius stages

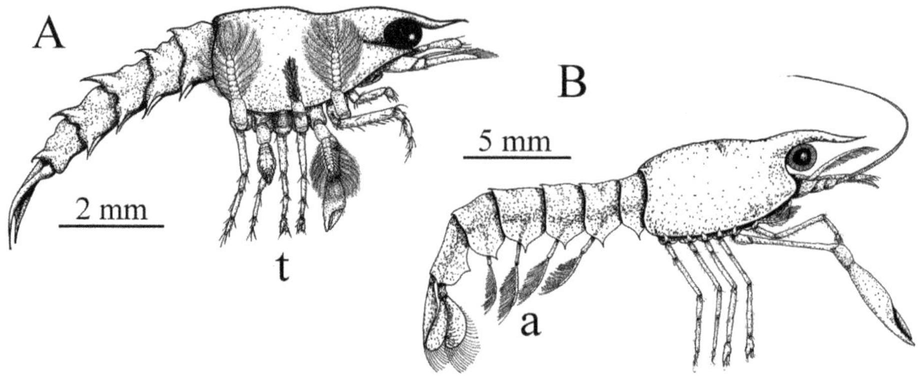

Fig. 6.6.
Larvae of the lobster *Homarus gammarus*. (A) The zoea (stage I) swims with the setose exopodites of thoracic limbs and seizes prey with endopodites of thoracic limbs (t). (B) The decapodid (stage IV, settling stage) walks with thoracic limbs and swims with setose abdominal pleopods (a). From Neil et al. (1976), with permission from The Royal Society of London.

in cirripedes (Fig. 6.3). A decapodid stage (a term including megalopa, glaucothoe, and puerulus) follows the zoeal stage in many decapods (Fig. 6.6); many decapodids feed, some with versatility (McConaugha 2002), but the stage can also be nonfeeding, as in some pagurids, lithodids, and palinurids (Nishida et al. 1990, Harvey 1993, Lemmens and Knott 1994, Epelbaum and Borisov 2006). A function of these late larval stages is the selection of a benthic habitat for continued life as a juvenile and adult. These transitional stages are capable of benthic attachment and walking, by antennules in barnacle cyprids and by thoracic legs in decapodids.

Another feature, less studied, is their greater swimming speed. The greater speed of settling stages is not entirely explained by increased size (Walker 2004). Form and movement of appendages also affect speed of small crustaceans (Yen 2013). More posterior appendages are used for swimming than in the previous larval stage: thoracic appendages instead of cephalic in the cyprid (Fig. 6.3C) and abdominal instead of thoracic in the decapodid (Fig. 6.6B). Some attain increased swimming speeds, which could be adaptive because the requirements for leaving planktonic life and finding a favorable benthic site may be more challenging than the requirements for swimming while remaining planktonic (Eckman et al. 1994, Phillips and Pearce 1997, Koehl and Cooper 2015).

Adaptations for speed differ among nauplii and cypridiform larvae, but greater speed is commonly an attribute of settling stages in thoracican and rhizocephalan barnacles. Estimated swimming speeds are less than 0.5 cm/s for nauplii and 1–5 cm/s or more for cyprids (Walker 2004, DiBacco et al. 2011). Some reported speeds for cirripede cyprids are slow, perhaps because less active or directed cyprids were observed, but the slower swimming of cypridiform larvae of related groups, the Ascothoracida and Facetoecta, is inferred to represent the ancestral condition. It is therefore thought that evolution of settling cypridiform stages preceded evolution of fast swimming by cirripede cyprids (Høeg et al. 2004). Speeds of nauplii of several other taxa are similar to those of cirripede nauplii (Williams 1994, Yen 2013), but some nauplii are capable of speeds approaching those of cirripede cyprids. During jumps, copepod nauplii attain speeds often greater than 1 cm/s and in some cases greater than 2 cm/s (Titelman and Kiørboe 2003a, Wu et al. 2011, Borg et al. 2012, Bradley et al. 2013, Gemmell et al. 2013, Wadhwa et al. 2014).

A similar increase in speed may occur from zoeal to decapodid stages. Jacoby (1982) observed speeds of 1–1.5 cm/s for zoeae and 4 cm/s for the decapodid of the brachyuran *Cancer magister*. In reviews by Mileikovsky (1973) and Chia et al. (1984), most speeds for zoeal and decapodid stages were not from the same study, but estimated speeds of zoeae were up to 2 cm/s and (with one exception) estimated speeds for decapodids were greater than 3 cm/s. Zoeal speeds in reports cited by Queiroga and Blanton (2005) were similarly low, whereas mean speeds reported for settling stages of lobsters were 8.5 cm/s and 15 cm/s for the decapodids of *Panulirus* spp. (Phillips and Olsen 1975, Calinski and Lyons 1983) and 13 cm/s for the decapodid of *Homarus americanus* (Cobb et al. 1989). In a few cases, however, metamorphosis and settlement is initiated by late zoeal stages responding to benthic cues, as in the anomuran *Emerita talpoida* (Harvey 1993) and some thalassinid shrimp (*Callichirus* spp.; Strasser and Felder 1999). Also, the antennules in the last nauplius stage of the barnacle *Newmaniverruca albatrossiana* suggest capacity for attachment (Watanabe et al. 2008).

In facetotectans and kentrogonid rhizocephalans, additional larval stages evolved. In the kentrogonids, the cyprid attaches to the host, becomes first a kentrogon, and then a slug-like vermigon that enters the host. The vermigon is unsegmented and has few cell types but is motile. The ypsigon of facetotectans appears to have the same function. Inferred directional biases in evolutionary transformations and relationships within the Thecostraca indicate that these remarkably unarthropod-like larval stages evolved convergently (Pérez-Losada et al. 2009).

Metamorphosis

Development proceeds through successive molts with gradual or abrupt changes in form (Sanders 1963). Metamorphic changes have evolved several times. Even in the gradual development of cephalocarids there is a molt in which the antenna loses its enditic process and the mandible its exopod and endopod: that is, a loss of naupliar equipment in the transition from metanauplius to juvenile (Sanders 1963; Fig. 6.7). If all crustaceans developed so gradually, however, we might not designate nauplii and metanauplii as larvae, not call that transition a metamorphosis, and not call that development indirect. Conspicuous metamorphoses can occur without a change in habitat, as in the striking change in form and function from planktonic nauplius to planktonic copepodid, or a change in habitat can involve multiple metamorphoses, as in nauplius to cyprid to juvenile in barnacles or zoea to megalopa (decapodid) to juvenile in crabs.

One evolutionary trend is toward increasingly pronounced metamorphosis and thus more indirect development. In the Thecostraca, cypridoid larvae occur in both the Ascothoracida and Rhizocephala, but some ascothoracidan adults remain cypridoid-like (a nearly direct development from settlement to adult; Pérez-Losada et al. 2009), whereas rhizocephalan adults no longer resemble arthropods (a very indirect development). The opposite evolutionary trend is from indirect to direct development, with abbreviation of free larval stages and embryonization of brooded larval stages, and in the extreme case release of adult-like juveniles from broods, as in peracarids and some decapods.

Some species that are similar as larvae and adults differ in size at metamorphosis. I know of no evidence that these differences evolved because of differences in the ratio of instantaneous mortality to instantaneous growth as they change in size or habitat, as in the model of Werner and Gilliam (1984). In some crustacean clades, the sizes of eggs and sizes at metamorphosis may not be free to evolve independently (Strathmann 1977). A change in one size without a change in the other would require a compensatory change in number of larval molts or in size increase per molt, and in some clades there is variation in egg size but little variation in number of larval stages and in enlargement per molt. Also, stresses on embryos or larvae affect size at metamorphosis and postlarval survival and growth (Emlet and Sadro 2006, Giménez 2006, Thiyagajaran et al. 2007). Effects carrying through metamorphosis potentially connect selection on embryonic, larval, and postlarval traits. Presumably, therefore, interdependence of egg size and size at metamorphosis differs among crustacean clades, an evolutionary complication not addressed in the life history models reviewed above.

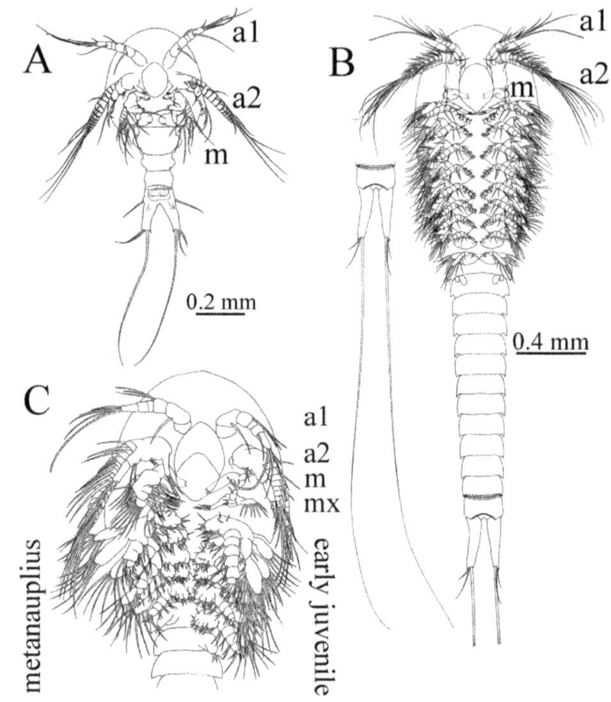

Fig. 6.7.
Cephalocarid stages. (A) Nauplius. (B) Adult. (C) Comparison of last metanauplius stage on one side and early juvenile on the other. a1 = first antenna; a2 = second antenna; m = mandible; mx = maxillule. Metamorphosis is marked by changes in second antenna, mandible, and maxillule. From Sanders (1963), with permission from the Connecticut Academy of Arts and Sciences.

Brooding and Other Parental Protection

Parental protection of embryos, such as diverse ways of brooding, had evolved numerous times among Cambrian arthropods (Caron and Vannier 2016). In crustaceans, brooding on the mother's body is common, benthic egg masses not on the mother's body are less common, and unguarded aggregations of embryos are rare (Lindley 1997, Richter and Scholtz 2001, Martin et al. 2014). Some large and diverse clades are characterized by a single method of brooding. The Pleocyemata (decapods minus the dendrobranchiate shrimps) brood on the abdomen with embryos attached to the pleopods by filaments. The peracarids retain unconnected embryos in a ventral thoracic marsupium that is bordered by oostegites. If each group evolved its distinctive mode of brooding only once, the origins are ancient and the methods persistent. Exceptional variations have evolved, however. Internalized brooding in pouches evolved several times in peracarids (Johnson et al. 2001). Although most dendrobranchiate shrimps and euphausiaceans shed fertilized eggs into the water, luciferids brood embryos on posterior thoracic limbs, and some euphausiaceans retain an ovigerous sac on the last pair of thoracic limbs; thermosbaenaceans brood in a dorsal thoracic chamber (Olesen et al. 2015). A stomatopod mother shapes and cleans a mass of adhering embryos without attaching them to herself, although in some species she attaches the embryos to the side of her burrow (Dingle and Caldwell 1972, Caldwell 1991). Thoracican barnacles brood two sacs of embryos (lamellae) in a mantle cavity inside their shell; rhizocephalans brood in the mantle cavity of an externa that extends outside their host (Høeg and Lützen 1995). Some copepods shed fertilized eggs into the water while others carry eggs, usually in sacs, in a variety of positions. Branchiurans and some copepods deposit eggs on substrata (Lindley 1997, Martin et al. 2014).

Transfer of nutrients from mother to developing embryos is rare in crustaceans. However, in brooding isopods, loss of maternal tissue suggests transfer of nutrients to embryos of some sphaeromatids (Shuster 1991), and maternal epithelial structure suggests transfer of nutrients in some oniscideans (Hoese and Janssen 1989). Numbers and lengths of brooded young suggest adelphophagy (sibling cannibalism) in a giant chaetiliid isopod in the Antarctic (Janssen and Hoese 1993).

Brooding is often the precursor of abbreviation or loss of free larval stages and embryonization of larval stages, but mixed development commonly persists, with brooding followed by free larval stages. Some clades vary in degrees of abbreviation, and some are fixed at an extreme. A large range of hatching stages exists in the Pleocyemata: species may have numerous, few, or no feeding zoeal stages, have zoeae that do not feed, release decapodids, or release adult-like juveniles (Rabalais and Gore 1985, Anger 2001, Vogt 2013). The number of zoeal stages tends to be more variable in species with more zoeal stages; scarce food or other stresses that slow growth result in more zoeal stages (Knowlton 1974, Anger 2001). This plasticity suggests a basis for evolutionary changes in the number of stages. The term *abbreviated development* implies a bias toward reduction and loss, but reversal of abbreviation seems a possibility until structures for larval feeding are lost (see Bias in Loss of Naupliar Feeding section in this chapter).

Clades differ greatly in frequency and range of abbreviation. The peracarids have lost ancestral larval stages entirely. Many brooding copepods retain six naupliar stages. Thoracican barnacles brood, but most have 6 naupliar stages; only a few species lack free nauplii and release cyprids. In contrast, all akentrogonid rhizocephalans lack free nauplii and release cyprids (Høeg and Lützen 1995, Høeg and Møller 2006). Species that differ in abbreviation coexist in the same marine habitats, but abbreviation is more common in some environments (Anger 1995, 2013, Vogt 2013, Pappalardo and Fernández 2014). In marine Pleocyemata, abbreviation is more common where temperature is low or larval food is scarce. Abbreviation is also common in Pleocyemata in freshwater and on land, but some terrestrial and freshwater adults migrate to place larvae in the sea, and some freshwater larvae migrate to the sea (Rabalais and Gore 1985, Anger 2001, Bauer 2004, Vogt 2013).

Not hatching as a dispersing larva can lead to extended cohabitation with the mother (Thiel 2007). Here are some examples: stomatopod hatchlings remain in the maternal burrow or cavity as nonfeeding antizoeae or pseudozoeae before they leave as feeding larvae (Martin et al. 2014), juveniles of some boring isopods remain with the mother (Thiel 2003), some phronimid amphipods remain with their mother in a home derived from gelatinous zooplankton (Aoki et al. 2013), and female bromeliad crabs release larvae into water in the axil of a plant and then protect them from predatory insect nymphs and spiders (Diesel and Schubart 2007). Where generations overlap at a site, evolution sometimes proceeds toward eusociality, with some sibling offspring of the resident mother foregoing reproduction and cooperating in protecting her offspring (Thiel 2007, Diesel and Schubart 2007, Duffy 2007, Vogt 2013).

Parental Costs of Brooding

Parental protection of broods comes with costs and limitations beyond time and materials. Brooding may increase risk of predation, such as in copepods that are more visible when brooding (Winfield and Townsend 1983, Svensson 1995, Logerwell and Ohman 1999). Brooders that remain at greater depth reduce risk from visual predators but obtain less food (Vuorinen et al. 1983, Bollens and Frost 1991). Moreover, in the presence of fish, an amphipod reduced its body flexing that would normally ventilate its brood, perhaps because flexing attracted the predators' attention (Arundell et al. 2014).

Brooding is constrained by supply of oxygen, which is a surface area-to-mass problem, such that in many marine invertebrate taxa, brooding is limited to species with small adults (Strathmann and Strathmann 1982). Large crabs and lobsters are a striking exception. Tethering embryos by filaments provides a flexible interstitial space and greater flow through the brood (Strathmann 1985). Brooding crabs adjust ventilation by abdominal flapping in response to the oxygen concentration within the brood (Baeza and Fernández 2002, Fernández et al. 2002). Even with this arrangement, embryos at the center of the brood receive less oxygen and develop more slowly than those at the periphery (Fernández et al. 2003), and small crabs expend less energy in ventilating broods than large ones (Fernández et al. 2006). Embryos of peracarids are not tethered but are held unconnected, often loose, in a marsupial chamber. Some amphipod mothers flex their bodies to ventilate the brood, and in contrast to crabs that increased abdominal flapping as embryos developed (Baeza and Fernández 2002), an amphipod flexed less as development advanced, perhaps because thinning of the egg envelope, the embryo's heart, and eventually the hatchlings' movements enhanced oxygen supply (Dick et al. 2002). In contrast, thoracican barnacles brood 2 lamellae, masses of adherent embryos surrounded by a sac, that are presumably supplied oxygen by diffusion. Crisp (1959) calculated limits on oxygen supply to such broods. The thinness of barnacles' egg masses should aid diffusive supply. Many rhizocephalan barnacles ventilate broods by contractions of the mantle (Høeg and Lützen 1995). Ohman and Townsend (1998) calculated that the linear strings of eggs of some brooding copepods could be an adaptation to environments low in oxygen. Crustaceans differ in use of their exoskeletal tool kit in ventilating broods and may therefore differ in limits on brood size and adult size.

Risk and Rate of Development

Embryos protected by a parent tend not only to hatch at a more advanced stage but also to develop more slowly than embryos shed into the water (Strathmann et al. 2002b, Hirst and López-Urrutia 2006). The slower development is evident even in duration of cell cycles in early development. For example, at similar egg diameters and temperature, brooded embryos of the barnacle *Balanus glandula* take 3.45 hours to develop from 2 to 4 cells, whereas shed embryos of the copepod *Calanus finmarchicus* take only 1.25 hours to develop from 2 to 8 cells (Strathmann et al. 2002b). Possible benefits of longer early cell cycles include less maternal investment in rate-limiting materials, more or different early transcription, correction of errors, and less synchronous early cell divisions that alter interactions between cell lineages. Possible trade-offs for developmental rates are influenced by protection but include traits unrelated to size, thus differing from the trade-offs in the model of Sargent et al. (1987).

LARVAE AS ADAPTATIONS FOR DISPERSAL AND MIGRATION

Planktonic larval development provides much of the dispersal of animals that have a sedentary adult stage, and differences in larval dispersal can result in differences in speciation and extinction rates of crustaceans (Vogt 2013). However, the persistence of nauplii and zoeae in life histories does not depend on advantages of larval dispersal. Many crustaceans with planktonic adults have planktonic larvae, as with many copepods, brine shrimp, euphausiaceans, and dendrobranchiate shrimp. The adults can disperse at least as well as the larvae. Similarly, many crustaceans with benthic adults have benthic nauplii or metanauplii, as with cephalocarids, mystacocarids, and many harpacticoid copepods (Sanders 1963, Dahms 2000). Benthic nauplii are adapted to benthic feeding (Hicks and Coull 1983, Dahms 2000). A feeding larval stage need not be an adaptation for dispersal or maintained by selection for dispersal. For benthic animals, a planktonic larval stage may be an ontogenetic migration to the plankton for a better habitat for early stages (Strathmann et al. 2002a, Burgess et al. 2015).

Furthermore, larval behavior can result in less dispersal than would occur for a passively drifting particle (Shanks 2009). Barnacle and decapod larvae can remain near shore even along open coasts (Tapia and Pineda 2007, Fisher et al. 2014). Some crustacean larvae remain in estuaries by timing their vertical swimming to tides (Forward and Tankersley 2001). In estuaries of eastern North America, vertical migration of larvae of the crab *Rhithropanopeus harrisii* results in little net motion over tidal cycles in estuarine channels (Cronin and Forward 1982), whereas larvae of the crab *Callinectes sapidus* initially swim upward and are flushed out of the estuary (Sulkin and van Heukelem 1982). Export from estuaries is not universal or inevitable. It appears to be a migration to avoid estuarine predators (Morgan and Christy 1997, Christy 2011) rather than an adaptation for dispersal. The harpacticoid copepod *Tigriopus californicus* dwells in high intertidal pools and can swim as nauplius and adult, but it disperses so little that mating of adults from different sites from central to southern California results in outbreeding depression from intergenic incompatibilities (Barreto and Burton 2013). For a full discussion of larval migration, see Chapter 8 in this volume.

Other reasons for doubting that extended larval durations are adaptations for dispersal are that the larval dispersal is commonly an obligate part of reproduction rather than a response to deteriorating local conditions, and that spreading offspring as part of bet-hedging or as a means of reducing competition or inbreeding with relatives confers diminishing returns as spread is increased (Strathmann et al. 2002a, Burgess et al. 2015). Planktonic development risks export to unfavorable habitats. Mobile benthic crustaceans can disperse as juveniles or adults either as a response to deteriorating conditions or fortuitously. Rafting on floating objects also results in extensive dispersal of benthic animals in some environments (Thiel and Haye 2006).

Dispersal to a suitable new site appears to be the function of nonfeeding larvae that are released from broods when nearly competent to settle. Larvae released at or near competence commonly swim upward for minutes to hours initially, then swim downward to encounter benthic sites. A brief period of swimming before competence could be optimal for dispersal even when larval growth does not compensate for mortality. Some nonfeeding larvae, however, are released well before they become competent to settle, suggesting either an advantage from the prolonged dispersal or disadvantages of longer brooding. For example, some rhizocephalans hatch as cyprids that swim for two or three days before competence, but that interval may be required for preparation for the molt to the next stage (Høeg and Ritchie 1987). Some other rhizocephalans release nonfeeding nauplii that are planktonic for days to weeks before metamorphosing to a cyprid (Høeg and Lützen 1995). The advantages of obligate dispersal of such duration are not obvious. An alternative hypothesis is that a longer precompetent planktonic development allows a shorter period of brooding thereby reducing costs or constraints of brooding (see Parental Costs of Brooding section in this chapter).

DEPENDENCE OF RATES OF GROWTH AND MORTALITY ON SIZE AND STAGE

Effects of Egg Size on Larval Growth Rate

In intraspecific comparisons, effects of egg or hatchling size on larval growth rate are more varied than assumed in most life history models. For the crab *Neohelice granulata* and the shrimp *Palaemonetes varians*, greater initial biomass is associated with a greater larval growth rate and also longer survival during starvation (Giménez and Torres 2002, Oliphant and Thatje 2013). Development to decapodid or juvenile stages is either briefer or size at metamorphosis is greater, depending on the number of larval instars (either 4 or 5 in both species) (Giménez and Torres 2002, Giménez et al. 2004, Oliphant and Thatje 2013).

As usually assumed, larger feeding larvae generally grow more in weight per time than smaller larvae. In crustaceans, this is manifested in approximately similar intervals between molts and increases in length from one stage to the next, with organic content increasing as more than length squared (Landry 1978, Rabalais and Gore 1985, Anger 2001). A weight-specific growth rate (weight gained per body weight per time) often decreases with increasing body weight (Anger 2001, Hirst et al. 2003), but that does not imply less weight gained per time.

Also, crustacean larval growth is punctuated by molting cycles and is thus more episodic than in general life history models. With a continuing food supply, most of the growth occurs in the first half of the molt cycle; later in the molt cycle larvae reach a point of reserve saturation, proceeding to molt even without further food (Anger 2001, Miller 2008). Scarce food for a late stage nauplius or zoea commits it to becoming a smaller cyprid or decapodid, which restricts the animal to a smaller size as it begins benthic life (Hentschel and Emlet 2000, Howard and Hentschel 2005).

Effects of Size, Stage, Behavior, Location, and Protection on Mortality Rates During Development

The head start from a bigger egg can increase survival by reducing time to grow to a safer stage, but mortality rates (deaths per time) also affect survival. As discussed below, estimates of mortality rates have demonstrated greater risk for planktonic embryos and larvae than for protected embryos; different mortality rates for larvae of different species in the same habitat; and greater risk near the sea floor. However, the estimates have not indicated consistently greater safety at greater larval size or later stage. Differences in size-dependent vulnerability presumably influence selection on size of eggs and size at metamorphosis, thereby contributing to the wide range of sizes at egg and metamorphosis for co-occurring species.

Mortality of Planktonic Larvae

Bigger larvae are not consistently safer. In earlier reviews, most estimated mortality rates for nauplii of copepods are above 0.1/d and most estimates for zoeae were below 0.1/d (Strathmann 1985, Rumrill 1990, Morgan 1995), a result consistent with the assumption that larger larvae with more equipment are usually safer. Estimates of larval mortality in planktonic populations are complicated, however, by the moving and mixing habitat and by vertical swimming that changes both vertical and horizontal distributions. Most early estimates were by the equation $N_t = N_0e^{-dt}$, where N_0 and N_t are numbers at two stages, the time t between stages is from laboratory rearing, and d is the mortality rate. Attempts to improve estimates have used vertical life tables (Aksnes and Ohman 1996, Ohman 2012, Tapia and Pineda 2007) and corrections for spatial variation (White et al. 2014). Tapia and Pineda (2007) estimated high mortality rates (0.23/d and 0.33/d) for nauplii of 2 barnacles off southern California, a result consistent with previous high estimates for nauplii. However, White et al. (2014), comparing larvae at a single location near the coast of northern California and with a correction for spatial variation, estimated overlapping vulnerability of nauplii and zoeae. Estimated mortality rates for zoeae of pagurid hermit crabs, pinnotherid crabs, and porcellanid crabs were between 0.06/d and 0.1/d, less than the estimate for nauplii of the barnacle *Balanus glandula* (0.14/d), but greater than the estimates for nauplii of the barnacle *Balanus crenatus* (~0.01/d). The zoeae were larger and had more appendages than the nauplii (Gonor and Gonor 1973, Shanks 2001) but were apparently as vulnerable.

Planktonic copepod nauplii are diverse in swimming and escape (Titelman and Kiørboe 2003b), and behavior may affect mortality rates more than size. At Georges Bank (northwest Atlantic), estimated mortality rates at late naupliar stages were greater for *C. finmarchicus* with the larger nauplius (about 0.14/d) than for *Pseudocalanus* spp. with a smaller nauplius (about 0.06/d; Ohman et al. 2002). Similarly, in a comparison of early naupliar stages at the Fladen Ground (North Sea, northeast Atlantic), mortality rates were greatest for *C. finmarchicus* (largest), less for *Pseudocalanus elongatus* (smaller), and lowest for *Oithona similis* (smallest), at about 0.25/d, 0.08/d, and 0.03/d, respectively (Eiane and Ohman 2004). These great differences between species were attributed to a better escape response by *P. elongatus* and the quiescent swimming by *O. similis*.

The effect of naupliar stage on mortality rate differs among species and locations. At Georges Bank, naupliar mortality rates for *C. finmarchicus* decreased through early stages but then increased through later stages. At the same location, for *Pseudocalanus* spp., there was little difference in mortality rates through the naupliar stages (Ohman et al. 2002). Although at Georges Bank, mortality rates for *C. finmarchicus* increased at later naupliar stages (Ohman et al. 2002), at Fladen Ground they did not (Eiane and Ohman 2004). Higher mortality at later stages in some circumstances is not surprising. In the laboratory, vulnerability to a variety of predators was greater for later stage nauplii of *C. pacificus* (Greene 1986).

Tethered baits lack morphological and behavioral defenses, isolating the effect of size. In experiments with tethered nauplii and adults of *Artemia franciscana* deployed from docks at tropical locations, adult mortality was greater or about equal to naupliar mortality (Kerr et al. 2014, 2015). The results suggested the presence of risk from visual predators such as fish, with risk to the less visible nauplii greater by day than by night. The nauplii were about the size of some early zoeae and the adults about the size of some decapodids.

In species with benthic adults, settling stages must traverse the space just above the seafloor, which appears to be especially risky. Mortality was greater near the sea floor than near the sea surface for tethered decapodids of the crab *Cancer magister* in a bay in the San Juan Islands, northeast Pacific (Allen and McAlister 2007), tethered decapodids of the spiny lobster *Panulirus argus* in a bay off south Florida (Acosta and Butler 1999), and tethered adult brine shrimp over a coral reef in the Red Sea (Motro et al. 2005). Similarly, Gaines and Roughgarden (1987) found fewer cyprids of *Balanus glandula* inshore than offshore of kelp beds and attributed the difference to predation by fish as water moved through the kelp.

The diverse means by which larvae reduce risk from predators include spines, vertical migration, and escape responses. Developmental plasticity in response to presence of predators can reduce risk for larvae. When brooded embryos of a shrimp were exposed to stimuli from predatory fish, larvae developed with fewer instars (abbreviated development) and possibly with an enhanced defense from the larval rostrum (Ituarte et al. 2014). For further discussion, see Chapter 12 in this volume. By moving vertically, some crustacean larvae occupy safer depths or take advantage of currents for transport to safer waters. Some mothers time release of larvae to increase larval safety. For example, many estuarine brachyuran crabs release zoeae at nocturnal ebb tides so that the zoeae avoid visual predation by estuarine fish while being exported for further development outside the estuary (Christy 2011, Rasmuson and Morgan 2013). The larvae return to the estuary as decapodids to settle in benthic habitats. Some brachyuran zoeae complete larval development within estuaries and are equipped with more formidable spines (Morgan 1989) or are more cryptic (Morgan and Christy 1997) as means of avoiding predation by fish. Larval movements may reflect compromises based on risk of predation, abundance of food, and probability of return to habitats in which the life cycle can be completed. Evidence of chemical defense is lacking for crustacean larvae (Vaughn and Allen 2010). The horns (with glands) of cirripede nauplii have been suspected of providing a chemical defense, but tests with predators have not supported that hypothesis (Taylor 1970, Bullard et al. 1999). Molting may affect vulnerability of larvae and therefore affect trade-offs associated with more or fewer larval stages. Molting at night, as observed for copepodites of *C. pacificus* (Miller et al. 1984), could reduce vulnerability while molting.

Mortality of Benthic Larvae

There are few estimates of mortality for benthic larvae. Annual production of eggs and of copepodids together with seasonally varying naupliar development of 5 to more than 40 days for *Microarthridion littorale* in a western Atlantic estuary suggest mortality rates between 0.1/d and 0.8/d (Morris and Coull 1992). Counts of nauplii of *Nannopus palustris* in the Gironde estuary, eastern Atlantic, indicated a mean naupliar mortality rate of about 0.4/d (Santos et al. 2003). Estimated mortality rates for harpacticoid nauplii adapted to the benthos can be at least as high as mortality rates of nauplii adapted to the plankton but presumably depend on habitat. For example, nauplii burrowing in algae (Shimono et al. 2007) could experience mortality very different from nauplii associated with estuarine sediments.

Mortality of Eggs Shed into the Water and of Benthic Resting Eggs

Eggs shed into the water appear to be at greater risk than brooded embryos or hatched motile larvae. For calanoid copepods, estimated mortality rates for shed eggs were greater than mortality rates in subsequent larval stages and greater than rates for adult females of species that brood (Peterson and Kimmerer 1994, Hirst and Kiørboe 2002, Ohman et al. 2002, Hirst et al. 2007, Bi et al. 2011). A trade-off is greater production of shed eggs than of brooded eggs but lower survivorship to hatching (Kiørboe and Sabatini 1994, Ohman et al. 2002). Losses of shed eggs can be from nondevelopment, sinking, advection, or predation.

Although copepod eggs shed into the water suffer high mortality, mortality of benthic resting eggs may be another matter. Many short-lived crustaceans have resting eggs (Hairston and Cáceres 1996). Some cladocerans and copepods survive as resting eggs at the sea floor when planktonic stages are rare or absent (Lindley 1997). Mortality rates estimated from hatching of eggs from sediments dated for depth ranged from 0.07/y to 0.53/y. Those extraordinarily low mortality rates are from several studies and locations (Sichlau et al. 2011) but are all for eggs buried in undisturbed sediments. Resting eggs must survive their sinking to the sea floor and predation before burial and then be resuspended for hatching. Hence, benthic fauna affect abundance of resting eggs by predation, burial, and resuspension. In laboratory experiments, pectinariid and maldanid polychaetes buried resting eggs of a copepod and also returned them to the sediment surface (Marcus and Schmidt-Gegenbach 1986), and predation by surface deposit feeders, a tellinid bivalve and a gammarid amphipod, decreased emergence of cladocerans and often decreased emergence of calanoid copepods, although emergence of one calanoid was sometimes increased, perhaps by resuspension and oxygenation (Viitasalo et al. 2007). Lastly, some benthic habitats are more conducive to deposition and later resuspension of eggs than others (Glippa et al. 2011, Scheef and Marcus 2011).

Mortality of Brooded Embryos

Brooded embryos are commonly safer than shed embryos or planktonic larvae but still suffer losses. The percent loss during brooding is often estimated from counts of broods at early or late stages of embryonic development, adjusted for size of the brooding mother. Kuris (1991) reviewed losses from the abdominal broods of Pleocyemata and marsupial broods of mysids, isopods, and amphipods. Causes of loss included predation, microbial infection, failure to attach or loss from brood chamber, ventilation by the mother, and abrasion against substrata. Mean losses reported for species ranged from 10% to 71% in the Pleocyemata and 5% to 57% in the Peracarida. These losses are mostly lower than estimates for planktonic larvae. For context, a planktonic mortality rate of 0.1/d over 10 days yields 37% survival (63% loss).

To assess the safety of broods, one must also consider their durations, which vary with temperature within species and with stage at release among taxa. For example, brood duration of the lobster *Nephrops norvegicus* is about six months to one year in the North Atlantic (Eiríksson 2014). Estimated losses during incubation range up to about 50%, or to 65% with inclusion of losses from failure to attach to pleopods during oviposition (McQuaid et al. 2009). Loss of 65% over 180 days is equivalent to a low mortality rate of 0.006/d. In contrast, for the riverine-estuarine shrimp *Macrobrachium nobilii* with a 12- to 14-day brood period, Balasundaram and Pandian (1982) estimated losses in the first 3 days to be 36% in the laboratory when brooders were kept with 6 other shrimp and to be 33% in the field. In the laboratory with other shrimp, losses were 53% in 11 days, equivalent to a mortality rate of about 0.07/d, which is similar to rates estimated for planktonic zoeae. These estimates exclude mortality of brooding mothers, which are expected to be much lower than mortality for embryos.

Hatching plasticity reduces risk when the mother herself is attacked. In some taxa, embryos at sufficiently advanced stages hatch when the brood is removed from the mother and dispersed. When crabs break and consume barnacles, some barnacle nauplii escape (Branscomb et al. 2014).

COMPARISONS

The exoskeletal tool kit has resulted in unusual features in crustacean larvae and maternal care, as discussed above, but numerous features of development are shared with other marine invertebrates. Like other marine invertebrates, crustaceans vary in protection of embryos, feeding and nonfeeding larvae, and degrees of metamorphosis. In many free-spawning invertebrates, the size of eggs as a target for sperm can affect selection on egg size, whereas with few exceptions (Barazandeh et al. 2013), crustacean males transfer sperm by contacting females so that eggs that are released to drift are already fertilized. The same result is achieved by other means in bryozoans, phoronids, and some annelids, in which females receive drifting spermatophores or bundles of sperm before releasing fertilized eggs. As in other phyla, feeding larval forms originated rarely and were lost more frequently. In addition to the nauplius, head larvae in arthropods include the protaspid stage of trilobites (Chatterton and Speyer 1997) and the protonymphon of pycnogonids (Vilpoux and Waloszek 2003). Examples in other phyla include the trochophore of annelids and tornaria of hemichordates (Lacalli 2005). Thus, head larvae with entirely different forms and feeding mechanisms evolved in ecdysozoans, lophotrochozoans, and deuterostomes. Only a few crustaceans deposit aggregations of eggs on substrata, perhaps because of opportunities for brooding on the mother's body, but even among other aquatic arthropods, horseshoe crabs bury clutches and caddis flies deposit gelatinous egg masses. An exoskeletal tool kit may account for more benthic larvae in the Arthropoda (some crustacean nauplii, some trilobite protaspids, and pycnogonid protonymphon) than in other phyla, but priapulids have benthic larvae despite lack of appendages, and benthic juveniles in other phyla, though not called larvae, often differ from adults. Muscle-powered limbs enable fast swimming, including modifications specifically for speed while settling, as with cyprids and decapodids, but larvae of holothuroids and some echinoids develop transverse ciliated bands that increase swimming speed as they become competent to settle. Muscle-powered equipment necessitates a high dry mass per wet mass that contrasts with gelatinous larval forms, such as larvae of echinoderms and hemichordates, but is similar to fish larvae (Kiørboe 2013).

CONCLUSIONS

The crustacean exoskeletal tool kit has allowed evolution of unusual capabilities from the beginning to the end of their development. A cost is that growth and change of the tool kit requires molting, which imposes episodic growth and vulnerability. Environmental challenges for diverse marine animals can be inferred from the adaptations available to crustaceans. Larval exoskeletons that differ among species and among instars within species have enabled unusually good estimates of mortality in the plankton.

ACKNOWLEDGMENTS

I am grateful for help from the University of Washington library at the Friday Harbor Laboratories; for advice from Jens Høeg, Jørgen Olesen, Martin Thiel, and Gary Wellborn on crustacean modes of development; and for comments on drafts from Colette Feehan, Lisbeth Francis, Shawn Luttrell, Florence McAlary, Megumi Strathmann, and Mika Tan.

REFERENCES

Acosta, C.A., and M.J. Butler. 1999. Adaptive strategies that reduce predation on Caribbean spiny lobster postlarvae during onshore transport. Limnology and Oceanography 44:494–501.

Aguirre, J.D., M.W. Blows, and D.J. Marshall. 2014. The genetic covariance between life cycle stages separated by metamorphosis. Proceedings of the Royal Society of London B: Biological Sciences 281:20141091.

Aksnes, D.L., and M.D. Ohman. 1996. A vertical life table approach to zooplankton mortality estimation. Limnology and Oceanography 44:494–501.

Akther, S., M.D. Agersted, and J. Olesen. 2015. Naupliar and metanaupliar development of *Thysanoessa raschii* (Malacostraca, Euphausiacea) from Godthåbsfjord, Greenland, with a reinstatement of the ancestral status of the free-living nauplius in malacostracan evolution. Plos One 10:e0141955.

Allen, J.D., and J.S. McAlister. 2007. Testing rates of planktonic versus benthic predation in the field. Journal of Experimental Marine Biology and Ecology 347:77–87.

Anger, K. 1995. The conquest of freshwater and land by marine crabs: adaptations in life-history patterns and larval bioenergetics. Journal of Experimental Marine Biology and Ecology 193:119–145.

Anger, K. 2001. The biology of decapod crustacean larvae. Crustacean Issues 14:1–419.

Arundell, K.L., N. Wedell, and A.M. Dunn. 2014. The impact of predation risk and of parasitic infection on parental care in brooding crustaceans. Animal Behaviour 96:97–105.

Aoki, K.N., C.N. Matsumoto-Ohshima, E. Hirose, and J. Nishikawa. 2013. Mother-young cohabitation in *Phronimella elongata* and *Phronima* spp. (Amphipoda, Hyperiidea, Phronimidae). Journal of the Marine Biological Association of the UK 93:1553–1556.

Baeza, J.A., and M. Fernández. 2002. Active brood care in *Cancer setosus* (Crustacea: Decapoda): the relationship between female behaviour, embryo oxygen consumption and the cost of brooding. Functional Ecology 16:241–251.

Balasundaram, C., and T.J. Pandian. 1982. Egg loss during incubation in *Macrobrachium nobilii*. Journal of Experimental Marine Biology and Ecology 59:289–299.

Barazandeh, M., C.S. Davis, C.J. Neufeld, D.W. Coltman, and A.R. Palmer. 2013. Something Darwin didn't know about barnacles: spermcast mating in a common stalked species. Proceeding of the Royal Society B: Biological Sciences 280:20122919.

Barnes, M. 1989. Egg production in cirripedes. Oceanography and Marine Biology: An Annual Review 27:91–166.

Barreto, F.S., and R.S. Burton. 2013. Elevated oxidative damage is correlated with reduced fitness in interpopulation hybrids of a marine copepod. Proceedings of the Royal Society of London B: Biological Sciences 280: 20131521.

Bauer, R.T. 2004. Remarkable shrimps: adaptations and natural history of the carideans. University of Oklahoma Press, Norman, Oklahoma.

Bi, H., K.A. Rose, and M.C. Benfield. 2011. Estimating copepod stage-specific mortality rates in open ocean waters: a case study from the northern Gulf of Mexico, USA. Marine Ecology Progress Series 427:145–159.

Bocquet-Védrine, J. 1961. Monographie de *Chthamalophilus delagei* J. Bocquet-Vedrine, rhizocéphale parasite de *Chthamalus stellatus* (Poli). Cahiers de Biologie Marine 2:455–593.

Bollens, S.M., and B.W. Frost. 1991. Ovigerity, selective predation, and variable diel vertical migration in *Euchaeta elongata* (Copepoda: Calanoida). Oecologia 87:155–161.

Borg, C.M.A., E. Bruno, and T. Kiørboe. 2012. The kinematics of swimming and relocation jumps in copepod nauplii. PLoS ONE 7:e47486.

Bradley, C.J., J.R. Strickler, E.J. Buskey, and P.H. Lenz. 2013. Swimming and escape behavior in two species of calanoid copepods from nauplius to adult. Journal of Plankton Research 35:49–65.

Branscomb, E.S., K. Vedder, and R.R. Strathmann. 2014. Risks for larvae mediated by plasticity in hatching. Invertebrate Biology 133:158–169.

Bruno E., C.M.A. Borg, and T. Kiørboe. 2012. Prey detection and prey capture in copepod nauplii. PLoS ONE 7:e47906.

Bullard, S.G., N.L. Lindquist, and M.E. Hay. 1999. Susceptibility of invertebrate larvae to predators: how common are post-capture larval defenses? Marine Ecology Progress Series 191:153–161.

Burgess, S.C., M.L. Baskett, R.K. Grosberg, S.G. Morgan, and R.R. Strathmann. 2015. When is dispersal for dispersal? Unifying marine and terrestrial perspectives. Biological Reviews 91:867–882.

Caldwell, R.L. 1991. Variation in reproductive behavior in stomatopod Crustacea. Pages 67–90 *in* R.T. Bauer, and J.W. Martin, editors. Crustacean sexual biology. Columbia University Press, New York.

Calinski, M.D., and W.G. Lyons. 1983. Swimming behavior of the puerulus of the spiny lobster *Panulirus argus* (Latreille, 1804) (Crustacea: Palinuridae). Journal of Crustacean Biology 3:329–335.

Caron, J.B., and J. Vannier. 2016. *Waptia* and the diversification of brood care in early arthropods. Current Biology 26:69–74.

Chatterton, B.D.E., and S.E. Speyer. 1997. Ontogeny. Pages 173–247 *in* R.L. Kaesler, editor. Treatise on invertebrate paleontology. Volume 1, part O. Arthropoda 1. Trilobita, revised. Geological Society of America, Boulder, Colorado.

Chia, F.S., J. Buckland-Nicks, and C.M. Young. 1984. Locomotion of marine invertebrate larvae: a review. Canadian Journal of Zoology 62:1205–1222.

Christiansen, F.B., and T.M. Fenchel, 1979. Evolution of marine invertebrate reproductive patterns. Theoretical Population Biology 16:267–282.

Christy, J.H. 2011. Timing of hatching and release of larvae by brachyuran crabs: patterns, adaptive significance and control. Integrative and Comparative Biology 51:62–72.

Cobb, J.S., D. Wang, D.B. Campbell, and P. Rooney. 1989. Speed and direction of swimming by postlarvae of the American lobster. Transactions of the American Fisheries Society 118:82–86.

Crain, J.A. 1999. Functional morphology of prey ingestion by *Placetron wosnessenskii* Schalfeew zoeae (Crustacea: Anomura: Lithodidae). Biological Bulletin 197: 207–218.

Crisp, D.J. 1959. The rate of development of *Balanus balanoides* (L.) embryos in vitro. Journal of Animal Ecology 28:119–132.

Cronin, T.W., and R.B. Forward, Jr. 1982. Tidally timed behavior: effects on larval distributions in estuaries. Pages 505–520 *in* V.S. Kennedy, editor. Estuarine comparisons. Academic Press, New York.

Dahms, H.-U. 2000. Phylogenetic implications of the Crustacean nauplius. Hydrobiologia 417:91–99.

Dall, W., B.J. Hill, P.C. Rothlisberg, and D.J. Staples. 1990. The biology of the Penaeidae. Advances in Marine Biology 27:1–484.

DiBacco, C., H.L. Fuchs, J. Pineda, and K. Helfrich. 2011. Swimming behavior and velocities of barnacle cyprids in a downwelling flume. Marine Ecology Progress Series 433:131–148.

Dick, J.T.A., R.J. Bailey, and R.W. Elwood. 2002. Maternal care in the rockpool amphipod *Apherusa jurinei*: developmental and environmental cues. Animal Behaviour 63:707–713.

Diesel, R., and C.D. Schubart. 2007. The social breeding system of the Jamaican bromeliad crab *Metopaulias depressus*. Pages 365–386 *in* J.E. Duffy, and M. Thiel, editors. Evolutionary ecology of social and sexual systems: crustaceans as model organisms. Oxford University Press, New York.

Dingle, H., and R.L. Caldwell. 1972. Reproductive and maternal behavior of the mantis shrimp *Gonodactylus bredini* Manning (Crustacea: Stomatopoda). Biological Bulletin 142:417–426.

Duffy, J.E. 2007. Ecology and evolution of eusociality in sponge-dwelling shrimp. Pages 387–409 *in* J.E. Duffy, and M. Thiel, editors. Evolutionary ecology of social and sexual systems: crustaceans as model organisms. Oxford University Press, New York.

Ebenman, B. 1992. Evolution in organisms that change their niches during the life cycle. American Naturalist 139:990–1021.
Eckman, J.E., F.E. Werner, and T.F. Gross. 1994. Modeling some effects of behavior on larval settlement in a turbulent boundary layer. Deep-Sea Research II 41:185–208.
Eiane, K., and M.D. Ohman. 2004. Stage-specific mortality of *Calanus finmarchicus, Pseudocalanus elongatus* and *Oithona similis* on Fladen Ground, North Sea, during a spring bloom. Marine Ecology Progress Series 268:183–193.
Emlet, R.B., and S.S. Sadro. 2006. Linking stages of life history: how larval quality translates into juvenile performance for an intertidal barnacle (*Balanus glandula*). Integrative and Comparative Biology 46: 334–346.
Eiríksson, H. 2014. Reproductive biology of female Norway lobster, *Nephrops novegicus* (Linnaeus, 1758) Leach, in Icelandic waters during the period 1960–2010: comparative overview of distribution areas in the Northeast Atlantic and the Mediterranean. Advances in Marine Biology 68:65–210.
Epelbaum, A., and R. Borisov. 2006. Feeding behaviour and functional morphology of the feeding appendages of red king crab *Paralithodes camtschaticus* larvae. Marine Biology Resesarch 2:77–88.
Fernández, M., L.M. Pardo, and J.A. Baeza. 2002. Patterns of oxygen supply in embryo masses of brachyuran crabs throughout development: the effect of oxygen availability and chemical cues. Marine Ecology Progress Series 245:181–190.
Fernández, M., N. Ruiz-Tagle, S. Cifuentes, H.O. Pörtner, and W. Arntz. 2003. Oxygen-dependent asynchrony of embryonic development in embryo masses of brachyuran crabs. Marine Biology 142:559–565.
Fernández, M., R. Calderón, M. Cifuentes, and P. Pappalardo. 2006. Brooding behaviour and cost of brooding in small body size brachyuran crabs. Marine Ecology Progress Series 309:213–220.
Fileman, E.S., P.K. Lindeque, R.A. Harmer, C. Halsband, and A. Atkinson. 2014. Feeding rates and prey selectivity of planktonic decapod larvae in the Western English Channel. Marine Biology 161:2479–2494.
Fisher, J.L., W.T. Peterson, and S.G. Morgan. 2014. Does larval advection explain latitudinal differences in recruitment across upwelling regimes? Marine Ecology Progress Series 503:123–137.
Forward, R.B., Jr., and R.A. Tankersley. 2001. Selective tidal-stream transport of marine animals. Oceanography and Marine Biology: An Annual Review 39:305–353.
Gaines, S.D., and J. Roughgarden. 1987. Fish in offshore kelp forests affect recruitment to intertidal barnacle populations. Science 235:479–481.
Gemmell, B.J., J. Sheng, and E.J. Buskey. 2013. Compensatory escape mechanism at low Reynolds number. Proceedings of the National Academy of the USA 110:4661–4666.
Giménez, L. 2006. Phenotypic links in complex life cycles: conclusions from studies with decapod crustaceans. Integrative and Comparative Biology 46:615–622.
Giménez, L., and G. Torres. 2002. Larval growth in the estuarine crab *Chasmagnathus granulata*: the importance of salinity experienced during embryonic development, and the initial larval biomass. Marine Biology 141:877–885.
Giménez, L., K. Anger, and G. Torres. 2004. Linking life history traits in successive phases of a complex life cycle: effects of larval biomass on early juvenile development in an estuarine crab, *Chasmagnathus granulata*. Oikos 104:570–580.
Glenner, H., J.T. Høeg, A. Klysner, and B.B. Larsen. 1989. Cypris ultrastructure metamorphosis and sex in seven families of parasitic barnacles (Crustacea, Cirripedia, Rhizocephala). Acta Zoologica 70:229–242.
Glippa, O., S. Souissi, L. Denis, and S. Lesourd. 2011. Calanoid copepod resting egg abundance and hatching success in the sediment of the Seine estuary (France). Estuarine, Coastal and Shelf Science 92:255–262.
Gómez-Gutiérrez, J., W.T. Peterson, and C.B. Miller. 2010. Embryo biometry of three broadcast spawning euphausiid species applied to identify cross-shelf and seasonal spawning patterns along the Oregon coast. Journal of Plankton Research 32:739–760.
Gonor, S.L., and J.J. Gonor. 1973. Descriptions of the larvae of four North Pacific Porcellanidae (Crustacea: Anomura). Fishery Bulletin 71:189–223.
Grant, A. 1983. On the evolution of brood protection in marine benthic invertebrates. The American Naturalist 122:549–555.

Greene, C.H. 1986. Patterns of prey selection: implications of predator foraging tactics. The American Naturalist 128:824–839.

Hairston, N.G., Jr., and C.E. Cáceres. 1996. Distribution of crustacean diapause: micro- and macroevolutionary pattern and process. Hydrobiologia 320:27–44.

Harvey, A.W. 1993. Larval settlement and metamorphosis in the sand crab *Emerita talpoida* (Crustacea, Decapoda, Anomura). Marine Biology 117:575–581.

Havenhand, J.N. 1995. Evolutionary ecology of larval types. Pages 79–122 *in* L. McEdward, editor. Ecology of marine invertebrate larvae. CRC Press, Boca Raton, Florida.

Hentschel, B.T., and R.B. Emlet. 2000. Metamorphosis of barnacle nauplii: effect of food variability and a comparison with amphibian models. Ecology 81:3495–3508.

Hicks, G.R.F., and B.C. Coull. 1983. The ecology of marine meiobenthic harpacticoid copepods. Oceanography and Marine Biology: An Annual Review 21:67–175.

Hirst, A.G., and T. Kiørboe. 2002. Mortality in marine planktonic copepods: global rates and patterns. Marine Ecology Progress Series 230:195–209.

Hirst, A.G., and A. López-Urrutia. 2006. Effects of evolution on egg development time. Marine Ecology Progress Series 326:29–35.

Hirst, A.G., J.C. Roff, and R.S. Lampitt. 2003. A synthesis of growth rates in marine epipelagic invertebrate zooplankton. Advances in Marine Biology 44:1–142.

Hirst, A.G., D. Bonnet, and R.P. Harris. 2007. Seasonal dynamics and mortality rates of *Calanus helgolandicus* over two years at a station in the English Channel. Marine Ecology Progress Series 340:189–205.

Høeg, J.T., and J. Lützen. 1993. Comparative morphology and phylogeny of the family Thompsoniidae (Cirripedia, Rhizocephala, Akentrogonida) with descriptions of three new genera and seven new species. Zoologica Scripta 22:363–386.

Høeg, J.T., and J. Lützen. 1995. Life cycle and reproduction in the Cirripedia Rhizocephala. Oceanography and Marine Biology: An Annual Review 33:427–485.

Høeg, J.T., and O.S. Møller. 2006. When similar beginnings lead to different ends: constraints and diversity in cirripede larval development. Invertebrate Reproduction and Development 49:125–142.

Høeg, J.T., and L.E. Ritchie. 1987. Correlation between cypris age, settlement rate and anatomical development in *Lernaeodiscus porcellanae* (Cirripedia: Rhizocephala). Journal of the Marine Biological Association of the UK 67:65–75.

Høeg, J.T., N.C. Lagersson, and H. Glenner. 2004. The complete cypris larva and its significance in thecostracan phylogeny. Pages 197–215 *in* G. Scholtz, editor. Evolutionary developmental biology of Crustacea. Crustacean issues, volume 15. Balkema Publishers, A.A./Taylor & Francis, The Netherlands.

Hoese, B., and H.H. Janssen. 1989. Morphological and physiological studies on the marsupium in terrestrial isopods. Monitore Zoologico Italiano (N.S.) Monografia 4:153–173.

Howard, S.C., and B.T. Hentschel. 2005. Effects of short-term food variability on the plasticity of age and size at metamorphosis of porcelain crab larvae. Limnology Oceanography 50:1960–1971.

Ituarte, R.B., M.G. Vázquez, M. de los Á. González-Sagrario, and E.D. Spivak. 2014. Carryover effects of predation risk on postembryonic life-history stages in a freshwater shrimp. Zoology 117:139–145.

Jacoby, C.A. 1982. Behavioral responses of the larvae of *Cancer magister* Dana (1852) to light, pressure, and gravity. Marine Behavior and Physiology 8:267–283.

Janssen, H., and B. Hoese. 1993. Marsupium morphology and brooding biology of the Antarctic giant isopod, *Glyptonotus antarcticus*. Eights 1853 (Crustacea, Isopoda, Chaetiliidae). Polar Biology 13:145–149.

Jirikowski, G.J., C. Wolff, and S. Richter. 2015. Evolution of eumalacostracan development—new insights into loss and reacquisition of larval stages revealed by heterochrony analysis. EvoDevo 6:4.

Johnson, W.S, M. Stevens, and L. Watling. 2001. Reproduction and development of marine peracaridans. Advances in Marine Biology 39:105–260.

Kerr, K.A., A. Cornejo, F. Guichard, and R. Collin. 2014. Planktonic predation risk varies with prey life history stage and diurnal phase. Marine Ecology Progress Series 503:99–109.

Kerr, K.A., A. Cornejo, F. Guichard, A.C. Crespi Abril, and R. Collin. 2015. Planktonic predation risk: effects of diel state, season and prey life history stage. Journal of Plankton Research 37:452–461.

Kiørboe, T. 2013. Zooplankton body composition. Limnology and Oceanography 58:1843–1850.

Kiørboe, T., and M. Sabatini. 1994. Reproductive and life cycle strategies in egg-carrying cyclopoid and free-spawning calanoid copepods. Journal of Plankton Research 16:1353–1366.
Kiørboe, T., and M. Sabatini. 1995. Scaling of fecundity, growth and development in marine copepods. Marine Ecology Progress Series 120:285–298.
Knowlton, R.E. 1974. Larval developmental processes and controlling factors in decapod Crustacea, with emphasis on Caridea. Thalassia Jugoslavica 10:139–158.
Koehl, M.A.R., and T. Cooper. 2015. Swimming in an unsteady world. Integrative and Comparative Biology 55:683–697.
Kuris, A.M. 1991. A review of patterns and causes of crustacean brood mortality. Pages 117–141 *in* A. Wenner, and A. Kuris, editors. Crustacean egg production. Crustacean issues, volume 7. Balkema Publishers, A.A., Rotterdam, The Netherlands.
Lacalli, T.C. 2005. Protochordate body plan and the evolutionary role of larvae: old controversies resolved? Canadian Journal of Zoology 83:216–224.
Landry, M.R. 1978. Population dynamics and production of a planktonic marine copepod, *Acartia clausii*, in a small temperate lagoon on San Juan Island, Washington. Internationale Revue der gesamten Hydrobiologie 63:77–119.
Lang, W.H. 1979. Larval development of shallow water barnacles of the Carolinas (Cirripedia: Thoracica) with keys to naupliar stages. NOAA Technical Report NMFS Circular 421. US Department of Commerce, Washington, D.C.
Leal, M.C., P.N. Pochelon, P. Lopes da Silva, A. Reis, R. Rosa, and R. Calado. 2013. Variable within-brood maternal provisioning in newly extruded embryos of *Homarus gammarus*. Marine Biology 160:763–772.
Lemmens, J.W.T.J., and B. Knott. 1994. Morphological changes in external and internal feeding structures during the transition phyllosoma-puerulus-juvenile in the western rock lobster (*Panulirus cygnus*: Decapoda: Palinuridae). Journal of Morphology 220:271–280.
Lindley, J.A. 1997. Eggs and their incubation as factors in the ecology of planktonic crustacea. Journal of Crustacean Biology 17:569–576.
Logerwell, E.A., and M.D. Ohman. 1999. Egg-brooding, body size and predation risk in planktonic marine copepods. Oecologia 121:426–431.
Maas, A., and D. Waloszek. 2005. Phosphatocopina—ostracode-like sister group of Eucrustacea. Hydrobiologia 538:139–152.
Marcus, N.H., and J. Schmidt-Gengenbach. 1986. Recruitment of individuals into the plankton: the importance of bioturbation. Limnology and Oceanography 31:206–210.
Marshall, D.J., and M.J. Keough. 2008. The evolutionary ecology of offspring size in marine invertebrates. Advances in Marine Biology 53:1–60.
Martin, J.W., J. Olesen, and J.T. Høeg. 2014. Atlas of crustacean larvae. Johns Hopkins University Press, Baltimore, Maryland.
McConaugha, J. 2002. Alternative feeding mechanisms in megalopae of the blue crab *Callinectes sapidus*. Marine Biology 140:1227–1233.
McEdward, L. 1997. Reproductive strategies of marine benthic invertebrates revisited: facultative feeding by planktotrophic larvae. The American Naturalist 150:48–72.
McQuaid, N., R.P. Briggs, and D. Roberts. 2009. Fecundity of *Nephrops norvegicus* from the Irish Sea. Journal of the Marine Biological Association of the UK 89:1181–1188.
Mileikovsky, S.A. 1973. Speed of active movement of pelagic larvae of marine bottom invertebrates and their ability to regulate their vertical position. Marine Biology 23:11–17.
Miller, C.B. 2008. Copepod growth in detail: pattern similarity to decapod larvae. ICES Journal of Marine Science 65:332–338.
Miller, C.B., M.E. Huntley, and E.R. Brooks. 1984. Post-collection molting rates of planktonic, marine copepods: measurement, applications, and problems. Limnology and Oceanography 29:1274–1289.
Morgan, S.G. 1989. Adaptive significance of spination in estuarine crab zoeae. Ecology 70:464–482.
Morgan, S.G. 1995. Life and death in the plankton: larval mortality and adaptation. Pages 279–321 *in* L. McEdward, editor. Marine invertebrate larvae. CRC Press, Boca Raton, Florida.
Morgan, S.G., and J.H. Christy. 1997. Planktivorous fishes as selective agents for reproductive synchrony. Journal of Experimental Marine Biology and Ecology 209:89–101.

Morris, J.T., and B.C. Coull. 1992. Population dynamics, numerical production, and potential predation impact on a meiobenthic copepod. Canadian Journal of Fisheries and Aquatic Sciences 49:609–616.

Motro, R., I. Ayalon, and A. Genin. 2005. Near-bottom depletion of zooplankton over coral reefs III: vertical gradient of predation pressure. Coral Reefs 24:95–98.

Neil, D.M., D.L. MacMillan, R.M. Robertson, and M.S. Laverack. 1976. The structure and function of the thoracic exopodites in the larvae of the lobster *Homarus gammarus* (L.). Philosophical Transactions of the Royal Society B: Biological Sciences 274:53–68.

Nishida, S., B.D. Quigley, J.D. Booth, T. Nemoto, and J. Kittaka. 1990. Comparative morphology of the mouthparts and foregut of the final-stage phyllosoma, puerulus, and postpuerulus of the rock lobster *Jasus edwardsii* (Decapoda: Palinuridae). Journal of Crustacean Biology 10:293–305.

Ohman, M.D. 2012. Estimation of mortality for stage-structured zooplankton populations: What is to be done? Journal of Marine Systems 93:4–10.

Ohman, M.D., and A.W. Townsend. 1998. Egg strings in *Euchirella pseudopulchra* (Aetideidae) and comments on constraints on egg brooding in planktonic marine copepods. Journal of Marine Systems 15:61–69.

Ohman, M.D., J.A. Runge, E.G. Durbin, D.B. Field, and B. Niehoff. 2002. On birth and death in the sea. Hydrobiologia 480:55–68.

Olesen, J. 2004. On the ontogeny of the Branchiopoda (Crustacea): contribution of development to phylogeny and classification. Pages 217–269 *in* G. Scholtz, editor. Evolutionary developmental biology of Crustacea. Crustacean issues, volume 15. Balkema Publishers, A.A./Taylor & Francis, The Netherlands.

Olesen, J., T. Boesgaard, and T.M. Iliffe. 2015. The unique dorsal brood pouch of Thermosbaenacea (Crustacea, Malacostraca) and description of an advanced developmental stage of *Tulumella unidens* from the Yucatan Peninsula (Mexico), with a discussion of mouth part homologies to other Malacostraca. PLoS ONE 10:e0122463.

Oliphant, A., and S. Thatje. 2013. Per offspring investment implications for crustacean larval development: evolutionary insights into endotrophy and abbreviated development. Marine Ecology Progress Series 493:207–217.

Pappalardo, P., and M. Fernández. 2014. Mode of larval development as a key factor to explain contrasting effects of temperature on species richness across oceans. Global Ecology and Biogeography 23:12–23.

Pechenik, J.A. 1979. The role of encapsulation in invertebrate life histories. The American Naturalist 114:859–870.

Pérez-Losada, M., J.T. Høeg, and K.A. Crandall. 2009. Remarkable convergent evolution in specialized parasitic Thecostraca (Crustacea). BMC Biology 7:15.

Peterson, W.T., and W.J. Kimmerer. 1994. Processes controlling recruitment of the marine calanoid copepod *Temora longicornis* in Long Island Sound: egg production, egg mortality, and cohort survival rates. Limnology and Oceanography 39:1594–1605.

Phillips, B.F., and L. Olsen. 1975. Swimming behaviour of the puerulus larvae of the western rock lobster. Australian Journal of Marine and Freshwater Research 26:415–417.

Phillips, B.F., and A.F. Pearce. 1997. Spiny lobster recruitment off western Australia. Bulletin of Marine Science 61:21–41.

Queiroga, H., and J. Blanton. 2005. Interactions between behaviour and physical forcing in the control of horizontal transport of decapod crustacean larvae. Advances in Marine Biology 47:108–214.

Rabalais, N.N., and R.H. Gore. 1985. Abbreviated development in decapods. Pages 67–126 *in* A.M. Wenner, editor. Larval growth. Crustacean issues, volume 2. Balkema Publishers, A.A., Rotterdam, The Netherlands.

Rasmuson, L.K., and S.G. Morgan. 2013. Fish predation after weakly synchronized larval release in a coastal upwelling system. Marine Ecology Progress Series 490:185–198.

Richter, S., and G. Scholtz. 2001. Phylogenetic analysis of the Malacostraca (Crustacea). Journal of Zoological Systematics and Evolutionary Research 39:113–136.

Rumrill, S.S. 1990. Natural mortality of marine invertebrate larvae. Ophelia 32:163–198.

Sanders, H.L. 1963. The Cephalocarida. Functional morphology, larval development, comparative external anatomy. Memoirs of the Connecticut Academy of Arts and Sciences 15:3–79.

Santos, P.J.P., L.P. Souza-Santos, and J. Castel. 2003. Population dynamics of *Nannopus palustris* (Copepoda, Harpacticoida) in the oligo-mesohaline area of the Gironde estuary (France). Cahiers de Biologie Marine 44:329–338.

Sargent, R.C., P.D. Taylor, and M.R. Gross. 1987. Parental care and the evolution of egg size in fishes. The American Naturalist 129:32–46.
Scheef, L.P., and N.H. Marcus. 2011. Mechanisms for copepod resting egg accumulation in seagrass sediments. Limnology and Oceanography 56:363–370.
Schembri, P.J. 1982. Locomotion, feeding, grooming and the behavioural responses to gravity, light and hydrostatic pressure in the stage I zoea larvae of *Ebalia tuberosa* (Crustacea: Decapoda: Leucosidae). Marine Biology 72:125–134.
Scholtz, G. 2000. Evolution of the nauplius stage in malacostracan crustaceans. Journal of Zoological Systematics and Evolutionary Research 38:175–187.
Schram, F.R. 1982. The fossil record and evolution of Crustacea. Pages 93–147 *in* L.G. Abele, editor. The biology of Crustacea, volume 1. Systematics, the fossil record, and biogeography. Academic Press, New York.
Shanks, A.L. 2001. An identification guide to the larval marine invertebrates of the Pacific Northwest. Oregon State University Press, Corvallis, Oregon.
Shanks, A.L. 2009. Pelagic larval duration and dispersal distance revisited. Biological Bulletin 216:373–385.
Shimono, T., N. Iwasaki, and H. Kawai. 2007. A new species of *Dactylopusioides* (Copepoda: Harpacticoida: Thalestridae) infesting brown algae, and its life history. Zootaxa 1582:59–68.
Shuster, S.M. 1991. Changes in female anatomy associated with the reproductive molt in *Paracereis sculpta*, a semelparous isopod crustacean. Journal of Zoology 225:365–379.
Sichlau, M.H., J.L.S. Hansen, T.J. Andersen, and B.W. Hansen. 2011. Distribution and mortality of diapause eggs from calanoid copepods in relation to sedimentation regimes. Marine Biology 158:665–676.
Smith, C.C., and S.D. Fretwell. 1974. The optimal balance between size and number of offspring. The American Naturalist 108:499–506.
Strasser, K.M., and D.L. Felder. 1999. Settlement cues in an Atlantic coast population of the ghost shrimp *Callichirus major* (Crustacea: Decapoda: Thalassinidea). Marine Ecology Progress Series 183:217–225.
Strathmann, R.R. 1977. Egg size, larval development, and juvenile size in benthic marine invertebrates. The American Naturalist 111:373–376.
Strathmann, R.R. 1985. Feeding and nonfeeding larval development and life-history evolution in marine invertebrates. Annual Review of Ecology and Systematics 16:339–361.
Strathmann, R.R., and D.J. Eernisse. 1994. What molecular phylogenies tell us about the evolution of larval forms. American Zoologist 34:502–512.
Strathmann, R.R., and M.F. Strathmann. 1982. The relation between adult size and brooding in marine invertebrates. The American Naturalist 119:91–101.
Strathmann, R.R., T.P. Hughes, A.M. Kuris, K.C. Lindeman, S.G. Morgan, J.M. Pandolfi, and R.R. Warner. 2002a. Evolution of local-recruitment and its consequences for marine populations. Bulletin of Marine Science 70:377–396.
Strathmann, R.R., J.M. Staver, and J.R. Hoffman. 2002b. Risk and the evolution of cell cycle durations of embryos. Evolution 56:708–720.
Sulkin, S.D., and W. van Heukelem. 1982. Larval recruitment in the crab *Callinectes sapidus* Rathbun: an amendment to the concept of larval retention in estuaries. Pages 459–473 *in* V.S. Kennedy, editor. Estuarine comparisons. Academic Press, New York.
Svensson, J.-E. 1995. Predation risk increases with clutch size in a copepod. Functional Ecology 9:774–777.
Tapia, F.J., and J. Pineda. 2007. Stage-specific distribution of barnacle larvae in nearshore waters: potential for limited dispersal and high mortality rates. Marine Ecology Progress Series 342:177–190.
Thiel, M. 2003. Reproductive biology of *Limnoria chilensis*: another boring peracarid species with extended parental care. Journal of Natural History 37:1713–1726.
Thiel, M. 2007. Social behavior of parent-offspring groups in crustaceans. Pages 294–318 *in* J.E. Duffy, and M. Thiel, editors. Evolutionary ecology of social and sexual systems: crustaceans as model organisms. Oxford University Press, New York.
Thiel, M., and P.A. Haye. 2006. The ecology of rafting in the marine environment. III. Biogeographical and evolutionary consequences. Oceanography and Marine Biology: An Annual Review 44:323–429.
Thiyagarajan, V., J.A. Pechenik, L.A. Gosselin, and P.Y. Qian. 2007. Juvenile growth in barnacles: combined effect of delayed metamorphosis and sublethal exposure of cyprids to low salinity stress. Marine Ecoogy Progress Series 344:173–184.

Titelman, J., and T. Kiørboe. 2003a. Motility of copepod nauplii and implications for food encounter. Marine Ecology Progress Series 247:123–135.

Titelman J., and T. Kiørboe. 2003b. Predator avoidance by nauplii. Marine Ecology Progress Series 247:137–149.

Taylor, P.B. 1970. Observations on the function of the frontolateral horns and horn glands of barnacle nauplii (Cirripedia). Biological Bulletin 138:211–218.

Vance, R.R. 1973. On reproductive strategies in marine benthic invertebrates. American Naturalist 107:339–352.

Vaughn, D., and J.D. Allen. 2010. The peril of the plankton. Integrative and Comparative Biology 50:552–570.

Viitasalo, S., T. Katajisto, and M. Viitasalo. 2007. Bioturbation changes the patterns of benthic emergence in zooplankton. Limnology and Oceanography 52:2325–2339.

Vilpoux, K., and D. Waloszek. 2003. Larval development and morphogenesis of the sea spider *Pycnogonum litorale* (Ström, 1762) and the tagmosis of the body of Pantopoda. Arthropod Structure & Development 32:349–383.

Vogt, G. 2013. Abbreviation of larval development and extension of brood care as key features of the evolution of freshwater Decapoda. Biological Reviews 88:81–116.

Vuorinen, I., M. Rajasilta, and J. Salo. 1983. Selective predation and habitat shift in a copepod species—support for the predation hypothesis. Oecologia 59:62–64.

Wadhwa, N., A. Andersen, and T. Kiørboe. 2014. Hydrodynamics and energetics of jumping copepod nauplii and copepodids. Journal of Experimental Biology 217:3085–3094.

Walker, G. 2004. Swimming speeds of the larval stages of the parasitic barnacle, *Heterosaccus lunatus* (Crustacea: Cirripedia: Rhizocephala). Journal of the Marine Biological Association of the UK 84:737–742.

Watanabe, H., J.T. Høeg, B.K.K. Chan, R. Kado, S. Kojima, and A. Sari. 2008. First report of antennular attachment organs in a barnacle nauplius larva. Journal of Zoology 274:284–291.

Werner, E.E. 1988. Size, scaling, and the evolution of complex life cycles. Pages 60–81 *in* B. Ebenman, and L. Persson, editors. Size-structured populations: ecology and evolution. Springer-Verlag, Berlin, Germany.

Werner, E.E., and J.F. Gilliam. 1984. The ontogenetic niche and species interactions in size- structured populations. Annual Review of Ecology and Systematics 15:393–425.

White, J.W., S.G. Morgan, and J.L. Fisher. 2014. Planktonic larval mortality rates are lower than widely expected. Ecology 95:3344–3353.

Williams, T.A. 1994. Locomotion in developing *Artemia* larvae: mechanical analysis of antennal propulsors based on large-scale physical models. Biological Bulletin 187:156–163.

Winfield, I.J., and C.R. Townsend. 1983. The cost of copepod reproduction: increased susceptibility to fish predation. Oecologia 60:406–411.

Wu, C.-H., H.-U. Dahms, S.-H. Cheng, and J.-S. Hwang. 2011. Effects of food and light on naupliar swimming behavior of *Apocyclops royi* and *Pseudodiaptomus annandalei* (Crustacea, Copepoda). Hydrobiologia 666:167–178.

Yen, J. 2013. Appendage diversity and mode of locomotion: swimming at intermediate Reynolds numbers. Pages 296–318 *in* L. Watling, and M. Thiel, editors. The natural history of the Crustacea, volume 1: Functional morphology and diversity. Oxford University Press, New York.

Zhang, X.-G., A. Maas, J.T. Haug, D.J. Siveter, and D. Waloszek. 2010. A eucrustacean metanauplius from the Lower Cambrian. Current Biology 20:1075–1079.

Zhang, H., X.-P. Dong, and S. Xiao. 2012. Three head-larvae of *Hesslandona angustata* (Phosphatocopida, Crustacea) from the Upper Cambrian of western Hunan, South China and the phylogeny of Crustacea. Gondwana Research 21:1115–1127.

7

GROWING OLD: AGING IN CRUSTACEA

Günter Vogt

Abstract

This chapter gives an overview on aging and longevity in the Crustacea. Crustaceans have life spans that range from a few weeks to several decades, depending on taxonomic affiliation, size, lifestyle, and environment. Some short-lived crustaceans can produce dormant eggs that remain viable for decades and centuries. Longevity differences of populations between habitats and geographical regions can be explained by life history theory. Crustaceans possess a variety of highly effective antiaging mechanisms, including molting, adult stem cell activity, removal of cellular waste products, detoxification of environmental toxicants, and immune response against pathogens. The longer-lived, indeterminately growing species show negligible senescence until old age and would thus provide valuable models for biogerontology.

INTRODUCTION

Crustacea is the animal group with the broadest range of morphological diversity on earth (Martin and Davis 2001). Similarly diverse are their physiology, life histories, and ecological adaptations. The extant members of the Crustacea include six classes, 51 orders, 1,003 families, 9,522 genera, and approximately 66,914 described species (Ahyong et al. 2011). Maximum body length of the adults ranges from about 0.1 mm in a tantulocarid ectoparasite to approximately 60 cm in the American lobster, representing a 600-fold difference. Crustaceans occur in marine, limnic, and terrestrial habitats—from polar regions to the tropics and from lowlands to high mountain areas. They exhibit life histories from extended larval development to direct development and brood care, from semelparity to iteroparity, and from determinate to indeterminate growth. Most crustaceans are free living, but some are sessile or parasitic. Many species, notably decapods, are exploited for human

Life Histories. Edited by Gary A. Wellborn and Martin Thiel.

consumption, reared in aquaculture, kept as pets, or used as research models. Because of their particularly high diversity and comprehensive research database, crustaceans should be among the prime animals for aging research, but this is not at all the case.

Research on aging in the Crustacea is severely hampered by the lack of permanent age information-bearing structures that other organisms possess, such as the growth rings in the shells of bivalves and the otoliths of fishes. Precise data on longevity can be obtained only by rearing in captivity from hatching to death and by long-term marking with internal tags. In practice, most life span data are calculated from growth models based on length-frequency distribution, mark and recapture, and the analysis of molt increment, intermolt duration, and reproduction parameters. An alternative approach is measurement of the age pigment lipofuscin, which is continuously accumulated in some areas of the brain. These indirect aging techniques have a small probability of error at younger ages but a large one at older ages. Therefore, in long-lived species, they give only a rough estimate of life span (Hartnoll 2001, Vogt 2012a).

This chapter summarizes the present knowledge on aging and longevity in the Crustacea, focusing on the Decapoda. It starts with a compilation of life span data from different crustacean groups and continues with the relationships between longevity and genetic background and environment. The differences observed in longevity between habitats are explained by life history theory. The following sections discuss the role of resource allocation in aging, the relationship between life history strategy and mortality, the link between growth format and senescence, and mechanisms that counteract aging and age-related diseases. At the end, I will compare aging in crustaceans with aging in other animal taxa and emphasize the relevance of aging research in crustaceans for biology and biogerontology.

LIFE SPANS IN CRUSTACEA

Life span data given in the literature do not follow a common standard. They either indicate the age of the oldest specimen, the age of the oldest cohort, or the maximum age estimated by growth models. Sometimes, longevity is also expressed as the mean age of the reproducing adults or the mean age of the oldest 10% of the population. Of course, these different life span values vary considerably for a given species. As an example, in my laboratory population of marbled crayfish *Procambarus virginalis*, mean life span of the reproducing adults was 718.8 days, mean life span of the longest-lived 10% was 1,154 days, and maximum individual life span was 1,610 days (Vogt 2010).

I have collected life span data for 211 Decapoda, among them 98 marine, 85 freshwater, and 28 semiterrestrial and terrestrial species. Some examples are given in Table 7.1. These data suggest that most decapods live between 1 and 10 years: 4.7% have life spans less than 1 year, 77.7% have life spans of 1–10 years, and 17.6% have life spans greater than 10 years. The shortest-lived decapod known is the planktonic shrimp *Lucifer faxoni*, which has a life span of only 30–40 days (Lee et al. 1992). The longest-lived decapod reliably aged is the European lobster *Homarus gammarus*, which can live as long as approximately 72 years (Sheehy et al. 1999). The American lobster *Homarus americanus* is often assumed to live for 100 years and more, but there is yet no proof of this speculation.

The Euphausiacea, the closest relatives of the Decapoda, live between two years (*Euphausia frigida*) and six years (*Euphausia superba*), as shown for five Antarctic species by Siegel (1987). Isopoda have life spans between one and ten years. Warburg (2011) compiled longevity data for 14 terrestrial species and concluded that most of them live for less than three years. Only three species exceeded an age of five years, among them *Armadillo officinalis* with a life span of about ten years. Amphipoda have longevities between 11 and 156 months (as estimated for the deep-sea species *Eurythenes gryllus*) but most species live for one to two years only (Sainte-Marie 1991). The sessile Cirripedia also vary considerably in life span. *Semibalanus balanoides* lives for two years and *Tetraclita squamosa* for approximately

Table 7.1. Life spans of decapod crustaceans.

Life span (years)	Species	Infraorder, family
Marine		
0.1	*Lucifer faxoni*	Dendrobranchiata, Luciferidae
0.5	*Thor manningi*	Caridea, Hippolytidae
1	*Acetes chinensis*	Dendrobranchiata, Sergestidae
1.7	*Litopenaeus stylirostris*	Dendrobranchiata, Penaeidae
2	*Pagurus brevidactylus*	Anomura, Paguridae
3.3	*Crangon crangon*	Caridea, Crangonidae
4.5	*Uca pugnax*	Brachyura, Ocypodidae
6	*Carcinus maenas*	Brachyura, Portunidae
8	*Callinectes sapidus*	Brachyura, Portunidae
9	*Aristeus antennuatus*	Dendrobranchiata, Aristeidae
11	*Pandalus borealis*	Caridea, Pandalidae
15	*Nephrops norvegicus*	Astacidea, Nephropidae
25	*Chaceon maritae*	Brachyura, Geryonidae
30	*Panulirus argus*	Achelata, Palinuridae
72	*Homarus gammarus*	Astacidea, Nephropidae
Freshwater		
0.7	*Elamenopsis kempi*	Brachyura, Hymenosomatidae
1	*Caridina pristis*	Caridea, Atyidae
1.5	*Cambarellus shufeldtii*	Astacidea, Cambaridae
2	*Aegla jarai*	Anomura, Aeglidae
3	*Macrobrachium rosenbergii*	Caridea, Palaemonidae
4.4	*Procambarus virginalis*	Astacidea, Cambaridae
5	*Eriocheir sinensis*	Brachyura, Varunidae
6	*Liberonautes latidactylus*	Brachyura, Potamonautidae
7	*Astacus leptodactylus*	Astacidea, Astacidae
11	*Cambaroides japonicus*	Astacidea, Cambaridae
15	*Potamon fluviatile*	Brachyura, Potamidae
16	*Pacifastacus leniusculus*	Astacidea, Astacidae
18	*Xiphocaris elongata*	Caridea, Atyidae
25	*Astacoides granulimanus*	Astacidea, Parastacidae
29	*Paranephrops zealandicus*	Astacidea, Parastacidae
Freshwater caves		
15	*Palaemonias ganteri*	Caridea, Atyidae
16	*Procambarus erythrops*	Astacidea, Cambaridae
38	*Orconectes australis australis*	Astacidea, Cambaridae
Terrestrial		
13	*Gecarcoidea natalis*	Brachyura, Gecarcinidae
20	*Cardisoma guanhumi*	Brachyura, Gecarcinidae
40	*Birgus latro*	Anomura, Coenobitidae

Note: Compiled from Vogt (2012a, 2014, 2015), aging methods and references cited therein.

14 years (Knox 2001). The planktonic Cladocera and Copepoda usually have longevities of a few weeks or months. For example, cultured *Daphnia pulex* and *Daphnia pulicaria* reached a maximum age of 57 and 98 days, respectively (Dudycha and Tessier 1999), and cultured copepods had maximum life spans of 32–75 days as demonstrated for five species (Hopp et al. 1997).

Several small and short-lived crustaceans, usually planktonic or from temporary waters, are able to produce dormant eggs that remain viable for years, decades, and even centuries (Radzikowski 2013). For instance, under laboratory conditions, dormant eggs of the anostracan *Branchinecta packardi*, the notostracan *Triops longicaudatus*, and the conchostracan *Caenestheriella gynecia* remained viable for 16, 14, and eight years, respectively. Eggs of the cladoceran *Daphnia pulicaria* and the copepod *Diaptomus sanguineus* isolated from lake sediments that were reliably aged with the ^{210}Pb-method even hatched after 700 and 332 years, respectively (Radzikowski 2013, Frisch et al. 2014).

LONGEVITY AND GENETICS

Because of their broad systematic diversity, the Decapoda seem particularly suitable for investigating the relationship between longevity and taxonomic affiliation or genetic background. The importance of genetic makeup in longevity, even within the same higher taxon, becomes particularly evident when the Penaeidae are compared with the Astacidea. Penaeid shrimps, which are considered as basic decapods, are rather short-lived with life spans of only one to two years, whereas the derived clawed lobsters and freshwater crayfish are typically longer-lived, including species with life spans of decades (Table 7.1; Vogt 2012a). Differences in life spans in higher taxa have also been demonstrated for the Amphipoda (Sainte-Marie 1991). The mean longevity ranges from 11 months in the Talitroidea to 31 months in the Eusiroidea.

Small, isolated populations are often characterized by low genetic variation, which can result in distinct phenotypic differences from their parental population, including longevity. A well-known founder effect is inbreeding depression, which usually results in shortening of longevity. This phenomenon may explain the shorter life span of only 14 months in a Taiwanese landlocked population of the shrimp *Caridina gracilipes* when compared to the 22 months in neighboring amphidromous populations (Han et al. 2011). On the other hand, if longevity promoting genes are coincidentally frequent in the founder population, longevity can be prolonged in comparison to the parent population. An example is the ancient and isolated population of freshwater crab *Potamon fluviatile* in the center of Rome, Italy, which shows gigantism and a life span of 15 years compared to nine to ten years in rural populations elsewhere in Italy (Scalici et al. 2008).

LONGEVITY AND THE ENVIRONMENT

Crustaceans occur in almost all aquatic habitats and geographical regions and are therefore particularly suitable to investigate the relationship between longevity and the environment. The evolution of different life spans in different environmental contexts can be explained by life history theory, which deals with fitness, adaptation, natural selection, and constraints. It explains the evolution of life history features and the trade-offs that bind them together (Stearns 1992). Significant evolutionary changes of life history traits such as age at maturity, brood size, interval between broods, and probably also longevity can be genetically established in fewer than 30 generations as shown for the guppy *Poecilia reticulata* (Reznick et al. 1990).

Populations and species from higher latitudes have generally longer life spans when compared to their closest lower latitude relatives. A good example is the intertidal mole crab *Emerita brasiliensis* that occurs along a 2,700 km range from Rio de Janeiro, Brazil, to Arachania, Uruguay.

Life span increases along this coastline from 6.1 to 16 months in females and from 7.1 to 11 months in males (Defeo and Cardoso 2002). The same tendency becomes evident when Arctic and Antarctic caridean shrimps are compared with their relatives in temperate regions. For instance, *Notocrangon antarcticus* from the Weddell Sea has a life span of 8–10 years whereas *Crangon crangon* from the Irish Sea lives for only 3.3 years (Oh et al. 1999, Bluhm and Brey 2001). This life span "extension" with latitude is also reported for freshwater decapods. For example, the red swamp crayfish *Procambarus clarkii*, which usually lives for 12–18 months in its native range in subtropical Louisiana, reaches greater ages in temperate Europe, increasing from 3.5 years at 40° N to 6.5 years at 49° N (Huner 2002, Chucholl 2011).

This phenomenon is also observed in the deep sea and high altitudes. For example, the dendrobranchiate shrimp *Aristeus antennuatus*, which occurs at a depth of 200–1,440 m in the Mediterranean Sea, lives for about nine years, whereas its surface water relative *Melicertus kerathurus* has a life span of only two years (Orsi Relini and Relini 1998, Vitale et al. 2010). In the case of high altitudes, the signal crayfish *Pacifastacus leniusculus* in California reaches an age of approximately six years in lowland rivers and 12 years in a mountain lake at 1,900 m altitude (Flint 1975, Shimizu and Goldman 1983).

High geographical latitude and high altitude both have cooler water temperatures, which usually results in slower growth, later onset of maturation, reduced molting and reproduction frequencies, and prolonged longevity (Vogt 2012a). Although these effects are certainly triggered by low temperature, they cannot simply be explained by a temperature-dependent reduction of the metabolic rate. They are rather the effect of life history changes that include alterations in activity and reproduction. For example, if a crayfish species is transferred to cooler water, its breeding period is unavoidably prolonged because a certain amount of degree days (sum of heat) is required to complete embryonic development (Skurdal et al. 2011). Hence, annually reproducing species from temperate regions often reproduce only every second or third year in high latitudes as shown for the noble crayfish *Astacus astacus* (Zimmerman 2009). The reduced allocation of resources toward growth and reproduction probably enables higher investments in maintenance and repair, which would explain prolongation of longevity in cool waters. Evidence for such a trade-off comes from the longest-lived bilaterian animal, the ocean quahog *Arctica islandica*, which has a maximum life span of 375 years. In this species slow growth, late onset of reproduction, and long life span are correlated with very efficient maintenance of body proteins when compared to faster growing, faster reproducing, and faster aging bivalves from more southern regions (Strahl et al. 2007).

Differences in longevity were also found between populations of the same species in the same geographical area or even in the same water body. For instance, in a 3.4 km long lowland river in Normandy, France, upstream and downstream populations of the white-clawed crayfish *Austropotamobius pallipes* had life spans of four to five years and five to six years, respectively (Neveu 2000). The New Zealand crayfish *Paranephrops planifrons* had a maximum age of seven years in native forest streams but of only four years in neighboring pasture streams (Parkyn et al. 2002). Such differences can be elicited by habitat complexity, food resource availability, and predator pressure and disease. Extrinsic mortality by predation and disease seems of particular relevance. It is theorized that the life span of a population is gradually prolonged if juveniles are preferably predated but instead shortened if adults are preferably predated. Indeed, experiments with guppies verified that predation of juveniles can result in selection for inherently longer-lived organisms (Reznick et al. 2004). This finding may explain the observed prolongation of longevity in shrimp *Xiphocaris elongata* living in streams with predator fish when compared to conspecifics living in streams without predator fish (Cross et al. 2008), which appears counterintuitive at first glance.

Particularly great differences in longevity were found between cave-dwelling crayfish and their epigean relatives. For example, the Alabama cave crayfish *Orconectes australis australis* has a life span of about 22 years, with some individuals reaching an age of 38 years, as revealed by mark-recapture–based growth models (Venarsky et al. 2012). In contrast, its relative *Orconectes placidus* from Alabamian surface waters lives for only two to three years (Taylor 2003). A similar relationship was found between the Santa Fe cave crayfish *Procambarus erythrops* and its epigean relative *Procambarus clarkii* in the southern United States, which have life spans of 16 years and less than two years, respectively (Streever 1996, Huner 2002). Cave animals usually experience low and erratic food supplies, as well as constantly low temperature and low oxygen. These factors were shown to result in the reduction of metabolism, motility and growth rate, later onset of maturity, and irregular reproduction when compared to epigean relatives (Streever 1996, Venarsky et al. 2012). A positive trade-off between these energy-saving mechanisms and maintenance and repair is probably causative for the particularly long life spans of subterranean decapods.

Differences in life span among environments have also been demonstrated for other crustacean classes and orders. For example, the mean longevity values for cold and warm water amphipods were 19.6 and 11.8 months, for deep and shallow water amphipods 156 (only one species in dataset) and 16.8 months, and for marine, brackish water, and freshwater amphipods 19.5, 12.6, and 16.5 months, respectively (Sainte-Marie 1991). In the intertidal barnacle *Chamaesipho tasmanica*, localization from the midshore level to the more air-exposed higher areas was positively correlated with longevity (Jeffery and Underwood 2001). Curiously, in *Daphnia magna*, dissolved humic substances resulted in gender-specific life span modulation (Euent et al. 2008). The mean life span of females was reduced in a concentration-dependent manner by more than 25%, whereas the mean life span of males was increased by 20%.

LONGEVITY IN SYMBIOTIC ASSOCIATIONS

Longevity of the partners in a symbiotic relationship is a main determinant of the dynamics of the association, as illustrated by pea crabs and their hosts. The Pinnotheridae, which live in the mantle cavity of bivalves, have life spans of 1.5 to three years. Their hosts have shorter, equal or longer life spans. If the life span of the crab is much shorter than that of the bivalve such as in the case of *Zaops ostreus* and *Crassostrea virginica* (two to three years vs. 20 years), the parasitizing female crab can stay in the host until death and the same bivalve can be infested by several generations of crabs (Christensen and McDermott 1958). However, if the life span of the crab is longer than that of its host as in the pair *Pinnotheres taichungae* and *Laternula marilina* (two years vs. nine months), the crab has to change its host repeatedly (Hsueh 2003). An interesting situation develops if host and parasite have similar life spans. There is evidence from the symbiosis of *Pinnotheres tsingtaoensis* and *Soletellina acuta*, which both have life spans of approximately two years, that the crabs do not leave their hosts after infestation, even under conditions of severe stress, and die together with the host (Soong 1997). An adjustment of longevity between partners was also found in the symbiosis of the shrimp *Palaemonetes paludosus* and the isopod *Probopyrus pandalicola*, of which the latter parasitizes in the branchial chamber of the former. Both animals have life spans of approximately one year, and the parasite rarely dies earlier than the host (Beck 1980).

An interesting behavior was observed in the rhizocephalan *Sacculina senta*, which parasitizes in the Japanese shore crab *Hemigrapsus sanguineus*. After infestation, the cirriped inhibits molting of the host, but after releasing several broods of larvae, the external reproductive system of the parasite detaches and induces molting of the crab (Takahashi and Matsuura 1994). This measure results in renewal of the cuticle and associated masticatory and sensory structures of the host and the removal of fouling organisms. It thus counteracts mechanical senescence of the crab, probably promoting longevity of the parasitic association.

LONGEVITY AND THE ALLOCATION OF METABOLIC RESOURCES

The allocation of limited resources between competing activities such as growth, reproduction, and maintenance and repair is central to the disposable soma theory of aging (Kirkwood 2008, Jones et al. 2014). Life span is expected to be extended when high proportions of energy are channeled into maintenance and repair. The optimal allocation, as determined by natural selection, is that which maximizes the production of progeny in the long-term. Most Crustacea reproduce throughout their entire adult life span, and there is a positive correlation between body size (which itself is positively correlated with age) and clutch size. Therefore, perpetuating investment into maintenance and repair until old age should be favored in crustaceans. This is not the case in animals with a long postreproductive life period such as mammals (Kirkwood and Austad 2000).

The relationship between the allocation of metabolic resources and longevity has not yet been investigated in detail for crustaceans. However, in various decapod species, limbs have been shown to regenerate at the expense of growth or reproduction (Juanes and Smith 1995). Interestingly, in damaged freshwater crabs *Spiralothelphusa hydrodroma*, energy was allocated to regeneration in the nonbreeding season but not in the breeding season (Maginnis 2006).

I have used the parthenogenetic marbled crayfish *Procambarus virginalis* to study the allocation-longevity issue in more detail. This crayfish has the advantage of producing genetically identical offspring, which can be raised either communally or individually in simple, highly standardized housing systems (Vogt 2008a). Surprisingly, life span varied considerably among batch-mates despite identical genetic makeup, identical feeding, and exposure to the same environmental conditions (Fig. 7.1; specimens B1–B8). These differences are interpreted as the result of stochastic developmental variation (Vogt 2015). The results further suggest that (1) there is no simple relationship between life span and investments in growth and reproduction (Fig. 7.1; compare the relative position of individuals in left and right panels), (2) individual rearing promotes life span extension (Fig. 7.1; compare specimens A, A1, $B3_1$, and $B3_{1-1}$ with the rest), and (3) damage and repair does not necessarily impair longevity if food is available in excess (Fig. 7.1; see specimens $B3_1$ and $B3_{1-1}$).

Pietrzak et al. (2013) presented an interesting example of life span extension in a subpopulation of the cladoceran *Daphnia pulicaria* that expressed a concomitant change of behavior and allocation strategy. Instead of growing and reproducing during summer as their conspecifics in the same lake, these individuals migrated into colder waters at increased water depth, reduced metabolic rates, and saved the metabolic expenses usually used for reproduction. In the next spring they migrated closer to the surface and started to reproduce as their short-lived relatives. By these means, they extended life span to more than a year, which is several times the usual life span.

MORTALITY AND AGE STRUCTURE

Individual life expectancy and the age structure of a population are dependent on the life history strategy of the species and limiting environmental influences such as food availability and predation. The Decapoda are particularly suitable for examining this relationship because they have evolved different life history strategies ranging from extended larval development to direct development and posthatching brood care.

Extended development is predominant in the sea while direct development is predominant in fresh water. Species with extended development have high egg numbers and develop in a series of planktonic larvae before settling to the ground (Anger 2001). For example, lobsters have egg numbers per clutch in the ten thousands, penaeid shrimps in the hundred thousands, and some brachyuran crabs even in the millions. In contrast, directly developing species have egg numbers in the

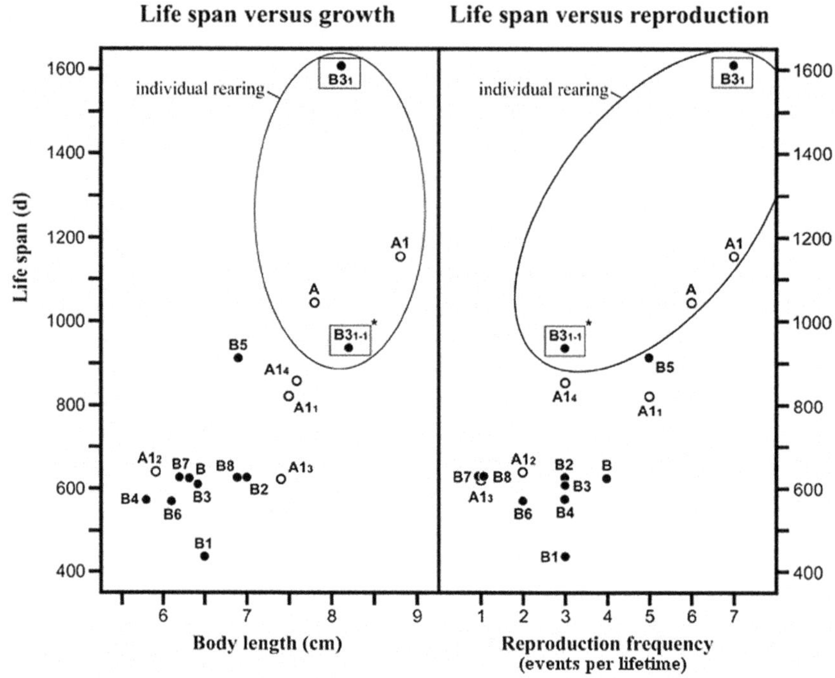

Fig. 7.1.
Relationship between life span and allocation of resources in genetically identical marbled crayfish, *Procambarus virginalis*. Specimens A and B are the founders of 2 laboratory lineages indicated by white and black dots. A1 and B1–B8 are F1-progeny, $B3_1$ and $A1_1$–$A1_4$ are F2-progeny, and $B3_{1\text{-}1}$ is F3-progeny. Female B and batch-mates B1–B8 were communally housed in a 60 × 30 × 30 cm aquarium at room temperature and fed daily ad libitum with TetraMin Wafer Mix pellets. The same holds for $A1_1$–$A1_4$. Specimens A, A1, $B3_1$, and $B3_{1\text{-}1}$ were individually kept in 30 × 20 × 20 cm aquaria during their adolescent and adult life periods. Water and feeding conditions were identical to communal rearing. Rectangular frames indicate specimens with regeneration of several appendages. The asterisk denotes the most heavily damaged individual (still alive at end of experiment). The data indicate that life span varies considerably among isogenic batch-mates, that there is no simple relationship between life span and investments in growth and reproduction, that individual rearing promotes life span extension, and that damage and repair does not impair longevity when food is available in excess. See text for details. From Vogt (2010), with permission from Springer.

tens and hundreds. Embryos of such direct developers are carried on the pleopods until hatching of the first juvenile stage, which is an effective means to reduce mortality of the early life stages (Thiel 2000, 2003, Anger 2001, Vogt 2013). The reduction of egg number per clutch in species with abbreviated development is accompanied by an increase in egg size due to extensive accumulation of nutrient reserves, enabling independence of the hatched stages from fluctuating food sources. This switch from planktotrophy to lecithotrophy certainly contributes to survival of the first free life stages (Rabalais and Gore 1985, Anger 2001).

A further means to reduce mortality of the early offspring is posthatching brood care. This feature is typical of most freshwater decapods but also occurs in several other crustacean taxa (Thiel 2000, 2003, Vogt 2013). In some groups such as the Peracarida, posthatching care is obligatory. In others such as the Decapoda and Euphausiacea, it is restricted to a fraction of the group. In the directly developing and brood-caring freshwater crayfish, average mortality from spawning to the release of juveniles from the pleopods is estimated to be about 50% (Reynolds 2002). In terrestrial isopods, the average brood pouch mortality is even below 10% (Warburg et al. 1984). A particularly intense mode of brood care was observed in sponge-dwelling *Synalpheus* species. These eusocial marine shrimps have very low fecundity, for instance, with a mean of only 4.5 eggs per clutch in *S. filidigitus,* and they protect all juvenile stages by a coordinated group response against intruders (Duffy 2007).

The production of yolkier eggs and prolongation of brood care in decapods has certainly reduced mortality of early life stages and enhanced mean individual life expectancy from days to weeks and months. But has this evolutionary shift also influenced maximum life span of the species? Comparison of life span data between marine (predominantly extended development) and freshwater decapods (predominantly direct development) suggests that this is not the case. In the 109 freshwater species with a known age, 0.9% had life spans less than 1 year, 63.3% between one and five years, 20.2% between five and ten years, and 15.6% greater than 10 years. These figures do not differ fundamentally from those of 98 marine species, which are 9.2%, 56.1%, 17.3%, and 17.3%, respectively (Vogt 2012a, 2014), particularly if the short-lived planktonic penaeids are removed from the data set.

The age structure of a population is determined by the balance between recruitment and mortality. An example of the age structure of a longer-lived decapod crustacean is given in Fig. 7.2. In most natural populations of the Decapoda, much less than 1% of the offspring reach adult age. The vast majority of the planktonic larvae and early juvenile stages are predated or fall victim to food shortage and adverse abiotic conditions (Anger 2001). Without these limiting environmental factors, survival would be much higher as can be seen from the survival rate in aquaculture facilities. For example, in cultured *Penaeus monodon*, which produce a mean of 460,000 eggs per spawning, hatching rate is 77%; survival from nauplius to 20-day-old postlarva, the seed stage for grow-out ponds, is 43%; and survival in the grow-out ponds is 92%–97% (Vogt 2012a).

Estimation of the mortality in a population is relatively easy in captivity, as shown in Fig. 7.3A for daphnids, but it is difficult in the wild. In longer-lived decapods, most of the total mortality of a cohort occurs during the first year of life (Fig. 7.3B). Thereafter, mortality rate slows, depending on maximum life span of species, availability of food, population density, predators, diseases, and fishery intensity. For example, in a population of the signal crayfish *Pacifastacus leniusculus* (life span, eight years) in Lake Billy Chinook, Oregon, the instantaneous mortality rate (natural mortality plus fisheries mortality) was Z = 0.67, which means that about 50% of the population dies on an annual basis (Lewis 1997). This rate was twice the mortality rate (Z = 0.32) estimated for *P. leniusculus*

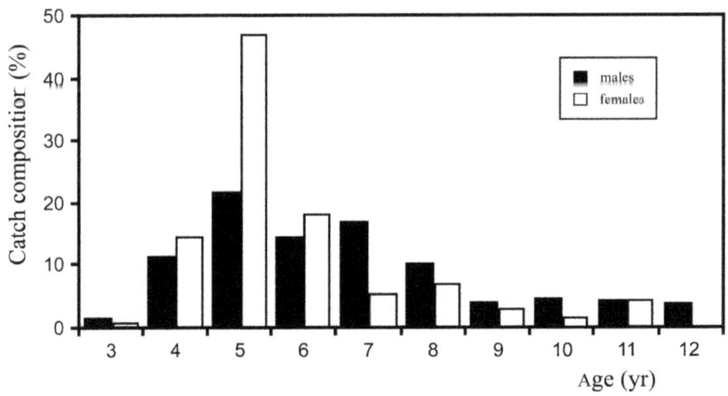

Fig. 7.2.
Age structure of population of the squat lobster, *Cervimunida johni* in a fishing area off northern-central Chile. Ages of males and females were calculated from series of length-frequency distributions collected monthly from commercial catches between September 1996 and September 1997. Redrawn after Arancibia et al. (2005), under Creative Commons license 3.0 (BY-NC).

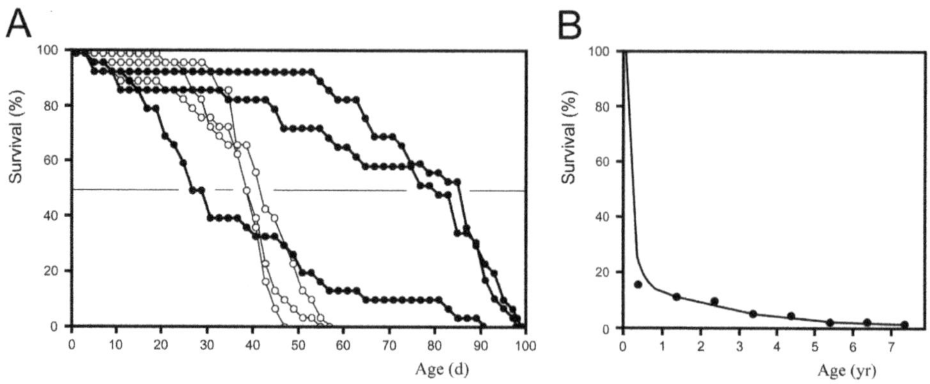

Fig. 7.3.
Mortality in water flea and crayfish. (A) Survivorship curve of 3 cultured populations of *Daphnia pulex* (open circles) and 3 cultured populations of *D. pulicaria* (closed circles). Mean life span is the age where survivorship equals 50%. Mortality is highest at older age. Redrawn after Dudycha and Tessier (1999). (B) Survivorship curve of signal crayfish, *Pacifastacus leniusculus* in Lake Billy Chinook, Oregon. Mortality is highest in the first months of free life. Redrawn after Lewis (1997).

in the colder Lake Tahoe (life span, 12 years; Flint 1975). In the shorter-lived African freshwater prawn *Macrobrachium macrobrachion* (life span, 1.5 years) in Luubara Creek, Nigeria, the instantaneous mortality rate was $Z = 2.75$, which corresponds to an annual mortality rate of roughly 90% of the population (Deekae and Abowei 2010).

GROWTH FORMAT AND SENESCENCE

Senescence refers to age-related changes in structure, physiology, reproduction, behavior, and health that have adverse effects on an organism and that increase the likelihood of its death. According to the speed and pattern of degenerative changes in the adult life period, senescence can be broadly categorized into rapid, gradual, and negligible (Finch 1990, Jones et al. 2014). This classification makes little sense in annual and semelparous species but is useful in longer-lived and iteroparous species.

In the Decapoda there are examples of all categories of senescence. The mitten crab *Eriocheir sinensis,* which reaches an age of 5 years, can be assigned to the rapid senescence type because both genders die soon after their first breeding season (Veilleux and de Lafontaine 2007). Examples of gradual senescence are spider crabs of the families Majidae and Oregoniidae. These determinately growing species stop growth and molting at the onset of maturation or around midage and suffer thereafter from mechanical senescence. For example, in the snow crab *Chionoecetes opilio,* which continues life after the last molt for another five to seven years, old specimens often have worn through carapaces and an incomplete set of limbs (Fonseca et al. 2008). Moreover, in the final anecdysial stage of such determinate growers, epifauna can develop excessively and obstruct proper functioning of the eyes and mouthparts as observed in the Caribbean spider crab *Stenorhynchus seticornis* (Hartnoll 1965).

Negligible senescence means that structural and functional decay is either undetectable or restricted to a very short final life period. This type of senescence is typical for most of the longer-lived decapods, which grow, molt, and reproduce with no fixed limits (Vogt 2012a). Indeterminate growers show no pronounced mechanical, cellular, reproductive, and behavioral senescence. Reproduction in the marbled crayfish may serve as an example. In our laboratory population, the reproducing adults had a mean life span of 718.8 ± 221.6 days and a clutch size of 7.41 ± 2.35% of body weight (Vogt 2010). Females younger than six months and older than two years had comparable relative clutch sizes of 7.45% and 7.97%, respectively (Fig. 7.4). Moreover, clutch sizes in the same individual at first and last spawning were often similar, corroborating the absence of reproductive senescence. Negligible senescence was also recorded with respect to the histology and cytology of the hepatopancreas, heart, and ovary (the main organs of metabolism, circulation, and reproduction, respectively), as revealed by comparison of 6-month-old and 3-year-old marbled crayfish (Vogt 2010).

In the wild, animals mainly die of extrinsic causes such as predation, accidents, lack of food, extreme abiotic conditions, and infectious diseases. In captivity, most of these risk factors are under control, and therefore, average longevity is usually extended. Under these conditions the internal causes of death become predominant, particularly in determinate growers. For example, in humans and pet rodents, age-related diseases such as cardiovascular diseases, Alzheimer's disease, diabetes, and cancer are prime causes of mortality (Hoyert and Xu 2012). In contrast, indeterminately growing decapods kept in a predator- and disease-free environment mainly die of cannibalism, accidents, random failure of organs and, most importantly, molting problems. For example, in a laboratory population of marbled crayfish, more than 85% of the adults died during ecdysis (Vogt 2010).

The scarcity of age-related diseases in decapods also concerns cancer. So far, almost 200 different diseases have been described for the Decapoda, most of them being infectious diseases (references in Vogt 2008b, Stentiford 2011). There are, however, only 16 incidences of tumor-like growths known, some of them being of dubious validity (Vogt 2008b, Shields and Small 2013). Most of these tumors were spontaneous, but two cases were apparently related to environmental carcinogens. Among the few identified tumors of decapods, there is only a single one that could be categorized as age-related: a growth in the stomach wall of a large American lobster (Vogt 2008b).

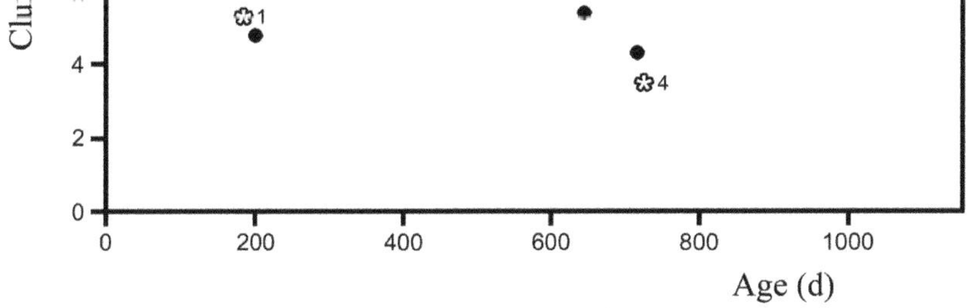

Fig. 7.4.
Negligible reproductive senescence in indeterminately growing marbled crayfish, *Procambarus virginalis*. Clutch size is expressed as percentage of body weight (bw). Rectangles denote the first and third (last) clutches of a specimen that had reached an age of 437 days, and asterisks denote the first, fourth, and seventh (last) clutches of a specimen that had lived for 1,154 days. The graph illustrates that clutch size varies considerably among individuals and among successive clutches of the same individual but, on average, it is not markedly reduced with age. From Vogt (2012a), with permission from Elsevier.

In determinately growing mammals, prevalence of cancer increases rapidly in the second half of life. For instance, in the United States, cancer rates for humans aged 50 to 64 years are seven- to sixteenfold higher than rates for younger persons, and rates for persons aged 65 to 74 years are two- to threefold higher than rates for persons aged 50 to 64 years (Edwards et al. 2002). This tendency is also typical for other determinately growing mammals and fish. For example, in laboratory rodents, more than 30% have tumors at the end of their two-year to three-year lives (Chandra and Frith 1992), and in zebrafish *Danio rerio*, which has a mean life span of 3.5 years, the prevalence of seminomas can amount to 50% by two years of age (Spitsbergen and Kent 2003).

Potential reasons for the high resistance of decapods to cancer are discussed in detail in Vogt (2008b, 2016). Among them are specific detoxification pathways for environmental carcinogens and lifelong telomere protection without adverse side effects such as uncontrolled cell proliferation.

ANTIAGING MECHANISMS

Crustaceans have evolved many effective antiaging mechanisms that may explain the scarcity of age-related disorders. These are molting, regeneration of lost appendages, lifelong renewal of tissues by stem cells, detoxification of free radicals, lysosomal removal of cellular waste, detoxification of environmental pollutants, and isolation and destruction of pathogens and diseased tissue areas by the immune system.

The most obvious antiaging mechanism is molting, which serves not only for regular renewal of the cuticle but also for replacement of mouthparts, the grinding and filtering structures of the stomach, the olfactory, gustatory, tactile and hydrodynamic sense organs, the statocysts, and the external reproductive structures (Fig. 7.5A–C; Vogt 2010, 2014). Fouling organisms that may impair motility and gas exchange at the gills are stripped by molting as well. Many decapods molt more than 20 times in their lifetime, and the Murray crayfish *Euastacus armatus* even molts up to 80 times in its 28 years of life (Gilligan et al. 2007).

Regeneration of lost limbs and replacement of old tissue cells by new ones preserve the integrity of the body and thus contribute to longevity as well. Autotomized limbs are regenerated in strict temporal relationship to the molting cycle. New appendages are formed during proecdysis in regeneration papillae and emerge in the following ecdysis. Regeneration is initiated by the formation of a blastema consisting of mitotically active epidermis cells and undifferentiated immigrant cells (Hopkins et al. 1999). This blastema produces different tissues such as epidermis, musculature, connective tissue, and nervous tissue. (For a full review of limb regeneration in crustaceans, see Hopkins and Das 2015.) In indeterminately growing decapods the ability to regenerate limbs persists until old age, whereas in determinately growing species it stops around midlife with the terminal molt. Hartnoll (1965) therefore assumed that the inability to regenerate lost appendages may be a life span–limiting factor in these species.

The replacement of damaged and old cells in the soft organs occurs through the activity of adult stem cells (Vogt 2012b, 2016). In indeterminately growing decapods, these cells remain active until old age, as shown for the marbled crayfish (Vogt 2010). Among the better-known examples are the E-cells of the hepatopancreas (Fig. 7.5D), the stem cells of the hematopoietic tissue (Fig. 7.5E), the progenitor cells of the neurogenic niche of the olfactory midbrain (Fig. 7.5F), and the satellite cells of the heart and skeletal musculature (Söderhäll 2013, Benton et al. 2014, Vogt 2016). These stem cells have quite different activity patterns. Some are more or less continuously active, such as the neurogenic and hematopoietic stem cells. Others become active in relatively short pulses and are silent in between. For example, the E-cells of the hepatopancreas are activated at the end of each food-digesting cycle, and the myogenic stem cells in the heart and skeletal musculature are activated during molting (Vogt 2016).

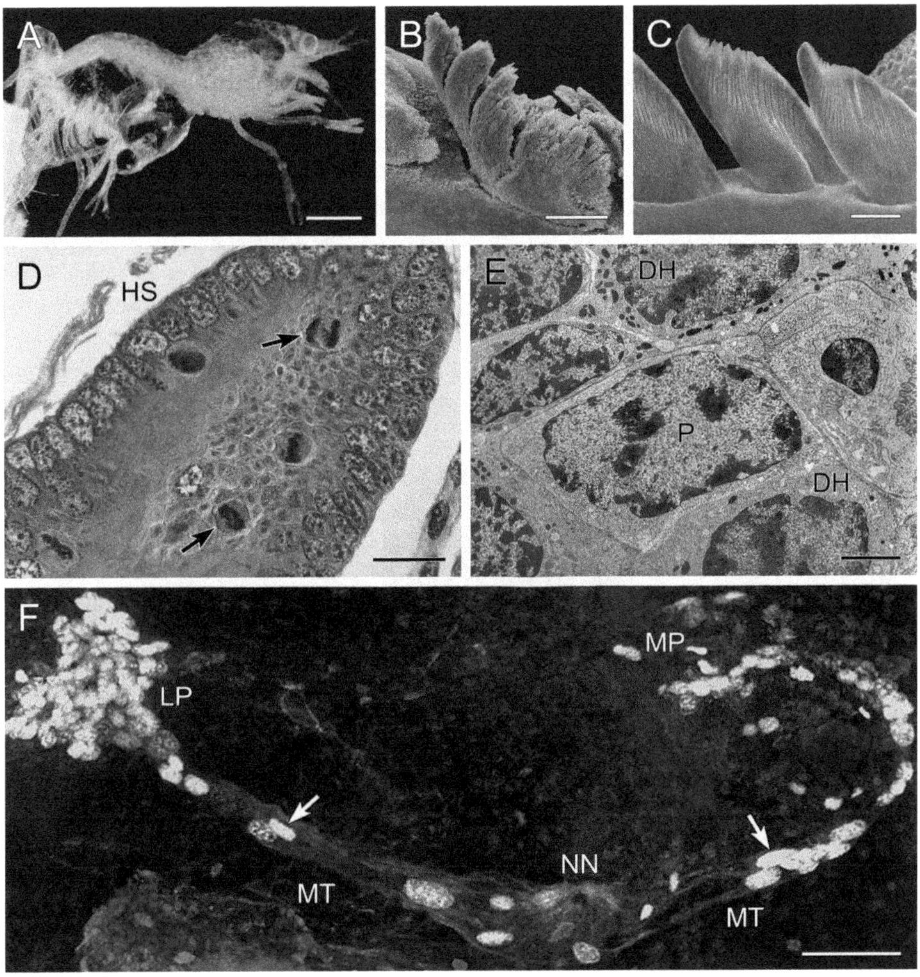

Fig. 7.5.
Prevention of aging by molting and stem cell activity. (A) Juvenile marbled crayfish *Procambarus virginalis* shedding its old exoskeleton. Scale bar = 1 mm. From Vogt (2008c), with permission from Wiley-Liss. (B–C) Replacement of (B) worn gustatory seta on pereopod by (C) new ones in marbled crayfish through molting. Scale bars = 20 μm. From Vogt (2012a), with permission from Elsevier. (D) Stem cell niche at blind end of hepatopancreas tubule of marbled crayfish. Arrows denote dividing stem cells. HS, hemolymph sinus. Scale bar = 20 μm. From Vogt (2008a), with permission from the Zoological Society of London. (E) Electron micrograph of hematopoietic tissue of signal crayfish *Pacifastacus leniusculus* showing undifferentiated progenitor cell (P) and surrounding differentiating hemocytes (DH). Scale bar = 5 μm. From Chaga et al. (1995). (F) Neurogenic system in hemideutocerebrum of red swamp crayfish, *Procambarus clarkii*, consisting of neurogenic niche (NN), migratory tracts (MT) and lateral (LP) and medial (MP) proliferation areas. Arrows denote neural precursors that migrate from the neurogenic niche to the proliferation areas where they divide and differentiate. Scale bar = 40 μm. From Song et al. (2007), with permission from the Company of Biologists. See color version of figure parts 7.5A, D, and F in the centerfold.

Renewal of cells in the organs of decapod crustaceans is only poorly understood, perhaps with the exception of the nervous system and the hepatopancreas. Neurogenesis occurs throughout adult life in decapod crustaceans within the peripheral nervous system and in the visual and olfactory centers of the central nervous system (for a full review, see Schmidt 2014). The hepatopancreas is apparently the organ with the fastest cellular turnover. It serves for the absorption of nutrients, synthesis of digestive enzymes, storage of metabolic reserves, and detoxification of xenobiotics (Gherardi et al. 2010). In adult decapods, it is composed of several hundred blindly ending tubules (Fig. 7.6D) that fuse together to form collecting ducts, which finally terminate in the stomach. This tubular system displays a distinct age polarity because all stem cells are confined to the blind ends of the tubules (Fig. 7.5D). They give rise to three mature cell types with different functions (Gherardi et al. 2010, Vogt 2016). Mitotic pulses of the E-cells push the older epithelial cells downstream, resulting in the establishment of a distinct age gradient along the tubules (Fig. 7.6D). In adult crayfish *Procambarus acutus acutus,* cell migration from the E-cell zone to the end of the tubules lasted approximately one week as shown by radioactive labeling (Davis and Burnett 1964), suggesting that the entire organ inclusive of the collecting ducts may be renewed in about two weeks.

Reactive oxygen species and free radicals, which are generated as byproducts of energy production in the mitochondria and by other intracellular processes, are detoxified by the enzymes

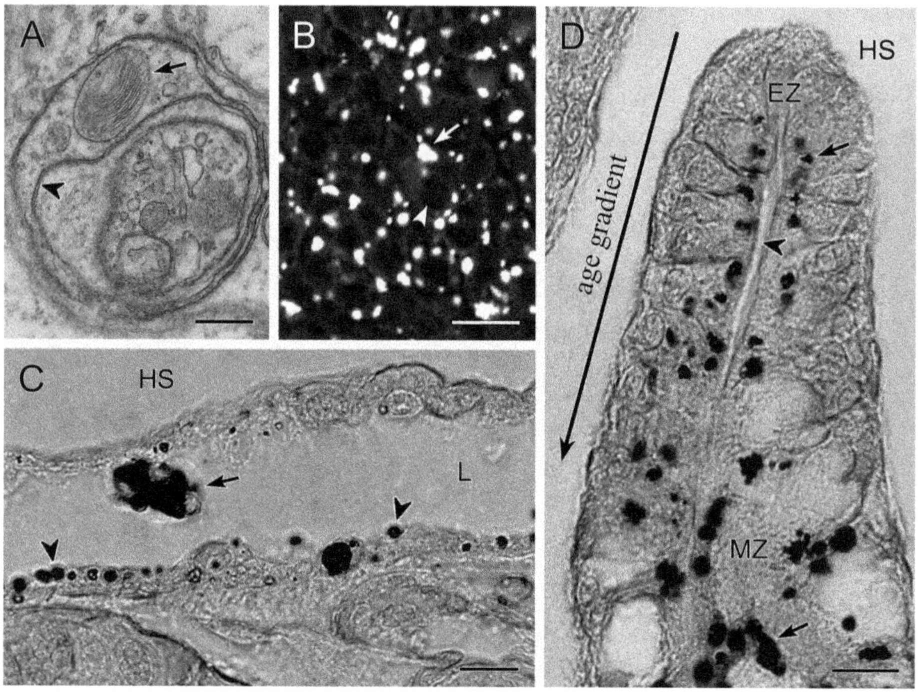

Fig. 7.6.
Prevention of aging by the lysosomal system. (A) Autophagosome in neuron of kuruma prawn *Penaeus japonicus* including decaying mitochondrion (arrow) and membrane remnants (arrowhead). Scale bar = 1 μm. From Medina et al. (2000), with permission from The Crustacean Society. (B) Permanent deposition of cellular waste as lipofuscin granules (arrow) in neurons of old edible crab, *Cancer pagurus*. Arrowhead denotes cell nucleus of neuron. Scale bar = 20 μm. From Sheehy and Prior (2008), with permission from Inter-Research. (C) Constitutive secretion of lead-sulfite granules in antennal gland of giant tiger prawn *Penaeus monodon* exposed to waterborne lead. The metal is absorbed from the hemolymph, detoxified and accumulated in lysosomes of the epithelium cells (arrowheads), and subsequently discharged into the gland lumen (L) and excreted as larger aggregates (arrow) together with the urine. HS, hemolymph sinus. Scale bar = 20 μm. From Vogt and Quinitio (1994), with permission from Elsevier. (D) Storage excretion of copper-sulfite granules in hepatopancreas tubule of giant tiger prawn exposed to waterborne copper. Both quantity and size of metal-detoxifying lysosomes (arrows) increase from the embryonic zone (EZ) toward the mature zone (MZ) of the tubule (i.e., with age of the cells). The metal granules leave the tissue by holocrine secretion of the oldest cells at the proximal end of the tubules. Arrowhead, tubular lumen. Scale bar = 20 μm. From Vogt and Quinitio (1994), with permission from Elsevier. See color version of figure part 7.6D in the centerfold.

superoxide dismutase, catalase, and glutathione peroxidase, as well as the free radical scavenger glutathione. These antioxidative molecules are widespread in the animal kingdom and have also been measured in several decapods, including shrimps, crayfish, and crabs (Elia et al. 2006). Many decapods seem to enhance their antioxidative capacity additionally by astaxanthin, one of the best free radical scavengers in the animal kingdom (Angeles et al. 2009).

Cellular waste resulting from defective cell organelles is degraded in lysosomes, usually resulting in myelinated residual bodies (Fig. 7.6A). These residual bodies can either be discharged as they are formed or accumulated in the cells for certain periods of time. Only in few areas of the crustacean body, such as the olfactory lobe of the deutocerebrum and the terminal medulla of the eyestalk, does cellular waste persist until death as lipofuscin granules (Fig. 7.6B). In mammals, such deposits are often suspected to be involved in age-related malfunctioning of the brain (Jung et al. 2007), but there is no evidence for such an effect in aged crustaceans with intense lipofuscin deposits (Sheehy and Prior 2008). Because of its continuous accumulation throughout life, lipofuscin can be used as an age marker in long-lived decapods (Sheehy et al. 1999).

Detoxification of environmental heavy metals can be quite variable depending on the species, metal, and route of uptake as shown in the tiger prawn *Penaeus monodon*. This prawn detoxifies waterborne copper in the hepatopancreas, the main metabolic organ of decapods, but waterborne lead accumulates in the antennal gland, the main excretory organ (Vogt and Quinitio 1994). In both organs, the metals are channeled into the lysosomes of the detoxifying cells and demobilized as inert sulfites. In the antennal gland, such metal-loaded lysosomes are individually discharged as they are formed and excreted together with the urine (Fig. 7.6C). The metal detoxifying cells of the hepatopancreas obviously lack the capability of apocrine secretion, and therefore, the lysosomes increase in number and size with the age of the cells (Fig. 7.6D). They leave the tissue at the end of the hepatopancreas tubules, where the oldest cells are regularly expelled by holocrine secretion. Organic pollutants are catabolized by the microsomal cytochrome P450 system (James and Boyle 1998).

Another important survival and antiaging strategy is the removal and destruction of pathogens and malignant cells. Decapods combat viruses, bacteria, fungi, and metazoan parasites by antimicrobial peptides, phagocytosis, and melanization and encapsulation (Cerenius et al. 2008). A particularly curious example of pathogen collecting and destroying cells are fixed phagocytes (Johnson 1987). These cells secrete a sievelike membrane that is spanned around the phagocyte like a tent-roof construction (Fig. 7.7A). Viruses and bacteria are rapidly attracted from the hemolymph underneath this tent and accumulated (Fig. 7.7B). Thereafter, they are phagocytosed one after the other and degraded by the lysosomal system (Fig. 7.7A).

Melanization and encapsulation is the most important immune defense mechanism of decapods (Cerenius et al. 2008). It is elicited by binding of recognition proteins of the host to pathogen-associated molecules such as the lipopolysaccharides of bacteria or ß-1,3-glucan of fungi. Recognition proteins are continuously secreted from the hepatopancreas. The reaction product binds to receptors of the hemocytes and triggers their degranulation. The released substances, among them prophenoloxidase, and further substances from the hemolymph provide a cascade of proteases and inhibitors (proPO-system) that finally generates melanin. The result of the melanization and encapsulation reaction is the complete isolation of individual pathogens or diseased tissue areas by an amorphous melanin layer and a surrounding cellular layer of hemocytes (Fig. 7.7C). The enclosed pathogens are destroyed by cytotoxic action of quinones (melanin precursors) and the physical shielding effect of the rigid melanin layer (Fig. 7.7D).

Phagocytotic hemocytes and the melanization and encapsulation response may also eliminate emerging tumors. Evidence comes from Sparks and Morado (1987), who illustrated phagocytosis of a neoplastic cell by a granulocyte in the hindgut tumor of red king crab *Paralithodes camtschaticus* (Fig. 7.7E). The authors also reported on histological evidence for attempts of hemocytes to encapsulate larger groups of tumor cells.

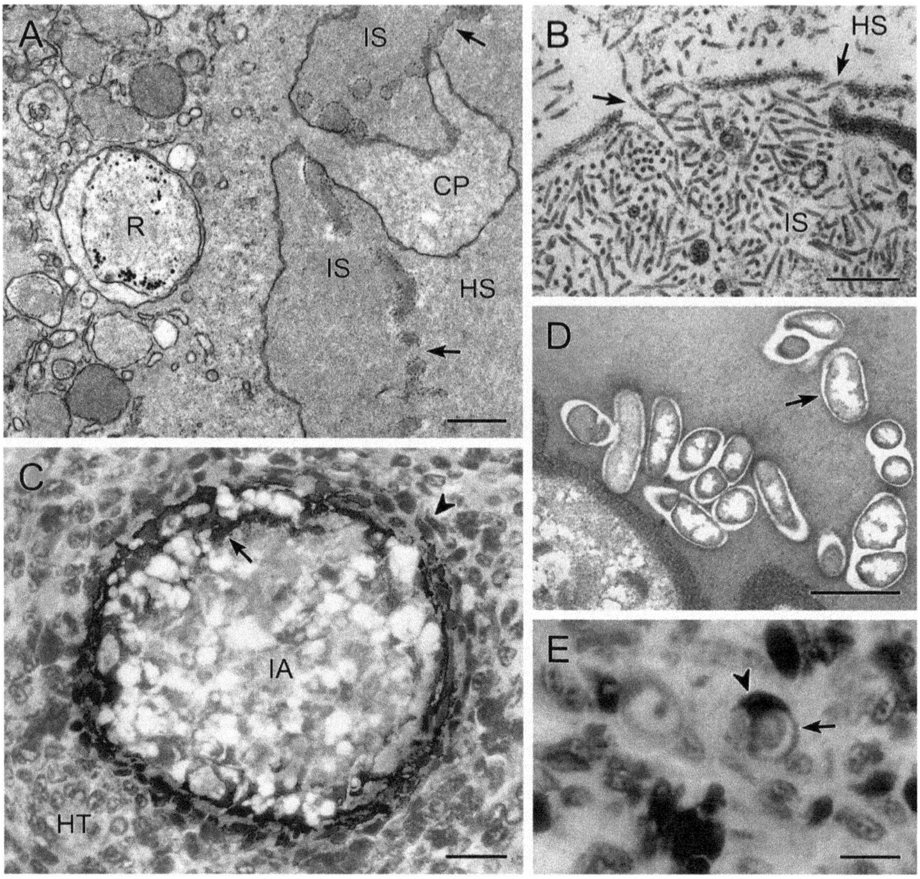

Fig. 7.7.
Prevention of aging by the immune system. (A) Peripheral area of fixed phagocyte in rickettsia-infected rock pool shrimp, *Palaemon elegans*. An interim storage space (IS) for pathogens is formed under a sieve-like net (arrows) that is spanned by cytoplasmic protrusions (CP). HS, hemolymph sinus; R, internalized rickettsia being degraded in phagosome. Scale bar = 500 nm. Photograph by G. Vogt ©. (B) Clearance of virus particles from hemolymph by fixed phagocyte in blue crab, *Callinectes sapidus*. Viruses enter the interim storage space through holes in the net (arrows) and are accumulated there in large amounts. Scale bar = 500 nm. From Johnson (1987), with permission from Elsevier. (C) Isolation of bacteria-infected hepatopancreas area in noble crayfish, *Astacus astacus*. The infected area (IA) is completely separated from the healthy tissue (HT) by a capsule composed of an amorphous melanin layer (arrow) and a surrounding cellular layer (arrowhead). Scale bar = 50 µm. From Vogt (2008b), with permission from Wiley-Liss. (D) Disintegrating bacteria (arrow) in polymerizing melanin mass in hepatopancreas of noble crayfish. Scale bar = 2 µm. From Vogt (2008b), with permission from Wiley-Liss. (E) Phagocytosis of neoplastic cell with hypertrophied nucleus (arrow) by granulocyte (arrowhead) in hindgut tumor of red king crab, *Paralithodes camtschaticus*. Scale bar = 20 µm. From Sparks and Morado (1987), with permission from Academic Press. See color version of figure part 7.7C in the centerfold.

COMPARISON OF AGING IN CRUSTACEANS AND OTHER ANIMAL PHYLA

Data on aging and longevity across higher taxa are available only for the vertebrate classes (Carey and Judge 2000) and the decapod crustaceans (Vogt 2012a). I had also expected such a compilation for the Bivalvia due to the ease of age estimation by the annual growth rings of the shells and the abundance of aging information for individual species (references in Philipp and

Abele 2010, Philipp et al. 2012), but there is yet no such compilation. In the decapods, longevity data are available for 1.43% of the extant 14,750 species, which is similar to fish (1.6% of the 24,000 extant species) but less than in birds (8% of the 9,000 species) and mammals (20% of the 4,500 species; Carey and Judge 2000).

In the Decapoda, there is a 650-fold difference between the shortest-lived (40 days) and the longest-lived species (72 years), which may even increase once the biggest-sized decapods are reliably aged. This range is greater than in most other higher animal taxa. In insects, the closest relatives of crustaceans, life span varies from one month in the fruit fly to about two decades in queens of termites (Thorne et al. 2002). In bivalves, the range of maximum life spans is one to 374 years (Abele et al. 2009), in fishes one to 152 years, in amphibians 1.8 to 55 years, in reptiles one to 153 years, in birds 1.5 to 73 years, and in mammals one to 122 years (Carey and Judge 2000). Higher maximum life span values sometimes quoted in the popular press, such as 211 years in whales, tend not to be based on sound scientific evidence. Nonetheless, particularly long life spans have been determined for cold water corals and sponges. The age of the Hawaiian coral *Leiopathe* sp. collected from 500 m water depth was estimated to be 4,265 years by the ^{14}C-radiocarbon method (Roark et al. 2009), but it has to be emphasized that this value holds for the colony and not for an individual. Large individuals of Antarctic sponges were assumed to be 15,000 years old (Gatti 2012), but the aging method used is unreliable. Animals with large numbers of pluripotent stem cells and high regeneration potential such as *Hydra* polyps and flatworms were even suspected to be immortal (Petralia et al. 2014). Further information on life spans of animals can be found in AnAge, the Animal Ageing and Longevity Database (http://genomics.senescence.info/species/. Last updated on January 27, 2015, it contains 3,769 records, mainly from vertebrates.

In mammals and other vertebrates with determinate growth, decline of tissue structure and function, behavior, and reproductive performance steadily rises from midlife on. These changes often lead to age-related diseases and are explained as an evolutionary consequence of determinate growth (Finch 1990, Ricklefs 2010a, 2010b). Differences in senescence between determinately and indeterminately growing decapods seem to support this hypothesis. However, there must be additional factors aside from growth format that influence the aging process, because indeterminately growing fish have higher rates of age-related deteriorations than indeterminately growing decapods and are significantly more susceptible to age-related cancer (Bailey et al. 1996, Almaida-Pagán et al. 2012, Vogt 2016). An obvious example of such a taxon-specific difference is molting, which allows lifelong renewal of the cuticle, chewing structures, and the external sense organs in crustaceans.

CRUSTACEANS AND BIOGERONTOLOGY—AN OUTLOOK

Research on aging in crustaceans serves primarily for a better understanding of crustacean biology. Crustacean fisheries, aquaculture, ecology, and conservation would certainly benefit from better knowledge in this field. In fisheries, data on aging and longevity are required to correctly estimate the fishery potential of natural populations and to manage stocks properly, and in aquaculture such information is helpful to optimize grow-out and reproduction strategies. In ecology, knowledge of mortality and longevity of populations and species is required to optimize models on natural dynamics and to determine their role in the ecosystem. In conservation biology, information on aging and longevity is helpful to model the extinction threat of populations and species.

Estimation of chronological age in crustaceans has been severely hampered by the absence of permanent annual growth structures, which were thought to be completely removed at each molting. However, Kilada et al. (2012), Leland et al. (2015), and Sheridan et al. (2015) have recently detected stratified structures in calcified areas of the eyestalk and ossicles of the gastric mill of crayfish, shrimps, crabs, and lobsters that are not shed during molting (Fig. 7.8). It is not yet

Fig. 7.8.
Possibility of direct age determination in crustaceans. Section of unmolted zygocardiac ossicle from gastric mill of edible crab *Cancer pagurus* of 175 mm carapace width, showing periodic growth rings (indicated by white dots). Scale bar = 100 μm. From Sheridan et al. (2015), with permission from The Crustacean Society.

unambiguously proven that these layers represent annual growth rings, but if this relationship should be verified in laboratory-raised specimens of known age, then this new technique would revolutionize age determination in longer-lived decapods. Periodical growth increments have also been detected in calcified structures of Cirripedia, but these growth rings correspond to molting frequency. Thus, only in older cold-water species that molt only once a year, they reflect an annual periodicity (Pohle 1992, Varfolomeeva et al. 2008).

Another future perspective of aging research in crustaceans is the use of obligately parthenogenetic species such as some daphnid lineages and marbled crayfish to investigate the nongenetic influences on aging in more detail. In these models, the influence of genetic variation of aging is close to zero, facilitating a more precise estimation of environmental and stochastic developmental impacts on aging and longevity (Vogt 2010, 2016).

Another interesting application of research in crustaceans relates to the long-term viability of dormant eggs of planktonic daphnids and copepods. Eggs that have been deposited in sediments decades or centuries ago can be harvested and subjected to further analysis, allowing for the reconstruction of plankton communities in earlier times. They are particularly suitable for investigating the impact of human activities on freshwater biota (Frisch et al. 2014, Ohtsuki et al. 2015).

Research on aging in crustaceans may also advance topical and controversially discussed issues of general biogerontology such as life prolongation by antioxidants, cellular waste treatment, caloric restriction, stem cell activity, and longevity genes. For example, particularly long-lived lobsters are suitable in investigations of long-term cell protection by antioxidants and the lysosomal system and to search for longevity genes. Species from caves, the deep sea, and polar waters may contribute to the exploration of life extension by reduced metabolism and caloric restriction. The stem cell systems of indeterminately growing decapods, which appear to function until old age without detrimental side effects, may even stimulate new ideas on how to reduce age-related diseases and cancer in humans.

CONCLUSIONS

Crustaceans have maximum life spans of a few weeks to about 70 years. The smaller, planktonic species are usually short lived, whereas larger benthic species live mostly between 1 and 5 years. Life spans of more than 2 decades are known only for the larger decapod species. Some short-lived crustaceans such as daphnids and copepods can produce dormant eggs that remain viable for decades and centuries. Longevity of a crustacean is basically determined by its genetic makeup but can be considerably modulated by abiotic factors, food availability, and the allocation of metabolic resources. In cold or nutrient-poor environments such as polar waters, the deep sea, and freshwater caves, natural selection favors longer life spans, which can be explained by trade-offs between growth, reproduction, and maintenance and repair. During the conquest of freshwater habitats, most decapods evolved a different life history strategy from the production of high numbers of eggs and low investment in the offspring to the production of smaller numbers of yolkier eggs and extension of brood care. This shift has dramatically raised mean individual life expectancy and markedly altered the mortality patterns of populations. However, maximum life spans of species remained largely unaffected. Senescence in the adult life period is obviously dependent on growth format. Indeterminate growers, the vast majority of crustaceans, only rarely develop age-related disorders, whereas determinate growers suffer at least from mechanical senescence. Crustaceans can counteract aging by a variety of antiaging mechanisms, including molting, regeneration of limbs, adult stem cell activity, detoxification of free radicals by antioxidants, cellular waste removal by lysosomes, detoxification of environmental pollutants by lysosomes and the cytochrome P450 system, and combating of pathogens and neoplastic cells by the immune system. Crustacea have evolved a particularly broad range of body plans, life histories, growth formats, environmental adaptations, types of senescence, and life spans. Therefore, they should be particularly promising for studying general biogerontological questions.

ACKNOWLEDGMENTS

Many thanks to Gary Wellborn, Martin Thiel, Mika Mei Jia Tan, and Tim Kiessling for valuable comments to improve the manuscript and careful editorial work.

REFERENCES

Abele, D., T. Brey, and E. Philipp. 2009. Bivalve models of aging and the determination of molluscan lifespans. Experimental Gerontology 44:307–315.

Ahyong, S.T., J.K. Lowry, M. Alonso, R.N. Bamber, G.A. Boxshall, P. Castro, S. Gerken, G.S. Karaman, J.W. Goy, D.S. Jones, K. Meland, D.C. Rogers, and J. Svavarsson. 2011. Subphylum Crustacea Brünnich, 1772. Zootaxa 3148:165–191.

Almaida-Pagán, P.F., J. de Costa, P. Mendiola, and D.R. Tocher. 2012. Age-related changes in mitochondrial membrane composition of rainbow trout (*Oncorhynchus mykiss*) heart and brain. Comparative Biochemistry and Physiology B 163:129–137.

Angeles, I.P., Jr., Y.-H. Chien, and A.V. Yambot. 2009. Effect of injected astaxanthin on survival, antioxidant capacity, and immune response of the giant freshwater prawn *Macrobrachium rosenbergii* (De Man, 1879) challenged with *Lactococcus garvieae*. Journal of Shellfish Research 28:931–937.

Anger, K. 2001. The biology of decapod crustacean larvae. Crustacean issues 14. Balkema, Lisse, The Netherlands.

Arancibia, H., L.A. Cubillos, and E. Acuña. 2005. Annual growth and age composition of the squat lobster *Cervimunida johni* off northern-central Chile (1996–97). Scientia Marina 69:113–122.

Bailey, G.S., D.E. Williams, and J.D. Hendricks. 1996. Fish models for environmental carcinogenesis: the rainbow trout. Environmental Health Perspectives 104(Suppl 1):5–21.

Beck, J.T. 1980. Life history relationships between the bopyrid isopod *Probopyrus pandalicola* and one of its freshwater shrimp hosts *Palaemonetes paludosus*. American Midland Naturalist 104:135–154.

Benton, J.L., R. Kery, J. Li, C. Noonin, I. Söderhäll, and B.S. Beltz. 2014. Cells from the immune system generate adult-born neurons in crayfish. Developmental Cell 30:322–333.

Bluhm, B.A., and T. Brey. 2001. Age determination in the Antarctic shrimp *Notocrangon antarcticus* (Crustacea: Decapoda), using the autofluorescent pigment lipofuscin. Marine Biology 138:247–257.

Carey, J.R., and D.S. Judge. 2000. Longevity records: life spans of mammals, birds, amphibians, reptiles, and fish. Monographs on Population Aging 8. University Press of Southern Denmark, Odense, Denmark.

Cerenius, L., B.L. Lee, and K. Söderhäll. 2008. The proPO-system: pros and cons for its role in invertebrate immunity. Trends in Immunology 29:263–271.

Chaga, O., M. Lignell, and K. Söderhäll. 1995. The haemopoietic cells of freshwater crayfish, *Pacifastacus leniusculus*. Animal Biology 4:59–70.

Chandra, M., and C.H. Frith. 1992. Spontaneous neoplasms in aged CD-1 mice. Toxicology Letters 61:67–74.

Christensen, A.M., and J.J. McDermott. 1958. Life-history and biology of the oyster crab, *Pinnotheres ostreum* Say. Biological Bulletin 114:146–179.

Chucholl, C. 2011. Population ecology of an alien "warm water" crayfish (*Procambarus clarkii*) in a new cold habitat. Knowledge and Management of Aquatic Ecosystems 401:29.

Cross, W.F., A.P. Covich, T.A. Crowl, J.P. Benstead, and A. Ramírez. 2008. Secondary production, longevity and resource consumption rates of freshwater shrimps in two tropical streams with contrasting geomorphology and food web structure. Freshwater Biology 53:2504–2519.

Davis, L.E., and A.L. Burnett. 1964. A study of growth and cell differentiation in the hepatopancreas of the crayfish. Developmental Biology 10:122–153.

Deekae, S.N., and J.F.N. Abowei. 2010. The mortality, exploitation rate and longevity of *Macrobrachium macrobrachion* (Herklots, 1851) in Luubara creek, Ogoni land, Niger delta, Nigeria. Current Research Journal of Biological Sciences 2:323–327.

Defeo, O., and R.S. Cardoso. 2002. Macroecology of population dynamics and life history traits of the mole crab *Emerita brasiliensis* in Atlantic sandy beaches of South America. Marine Ecology Progress Series 239:169–179.

Dudycha, J.L., and A.J. Tessier. 1999. Natural genetic variation of life span, reproduction, and juvenile growth in *Daphnia*. Evolution 53:1744–1756.

Duffy, J.E. 2007. Ecology and evolution of eusociality in sponge-dwelling shrimp. Pages 387–409 *in* J.E. Duffy, and M. Thiel, editors. Evolutionary ecology of social and sexual systems: crustaceans as model organisms. Oxford University Press, New York.

Edwards, B.K., H.L. Howe, L.A.G. Ries, M.J. Thun, H.M. Rosenberg, R. Yancik, P.A. Wingo, A. Jemal, and E.G. Feigal. 2002. Annual report to the nation on the status of cancer, 1973–1999, featuring implications of age and aging on U.S. cancer burden. Cancer 94:2766–2792.

Elia, A.C., A.J.M. Dörr, C. Mastrangelo, M. Prearo, and M.C. Abete. 2006. Glutathione and antioxidant enzymes in the hepatopancreas of crayfish *Procambarus clarkii* (Girard 1852) of Lake Trasimeno (Italy). Bulletin Français de la Pêche et de la Pisciculture 380/381:1351–1361.

Euent, S., R. Menzel, and C.E.W. Steinberg. 2008. Gender-specific lifespan modulation in *Daphnia magna* by a dissolved humic substances preparation. Annals of Environmental Science 2:7–10.

Finch, C.E. 1990. Longevity, senescence, and the genome. University of Chicago Press, Chicago, Illinois.

Flint, R.W. 1975. Growth in a population of the crayfish *Pacifastacus leniusculus* from a subalpine lacustrine environment. Journal of the Fisheries Research Board of Canada 32:2433–2440.

Fonseca, D.B., B. Sainte-Marie, and F. Hazel. 2008. Longevity and change in shell condition of adult male snow crab *Chionoecetes opilio* inferred from dactyl wear and mark-recapture data. Transactions of the American Fisheries Society 137:1029–1043.

Frisch, D., P.K. Morton, P.R. Chowdhury, B.W. Culver, J.K. Colbourne, L.J. Weider, and P.D. Jeyasingh. 2014. A millennial-scale chronicle of evolutionary responses to cultural eutrophication in *Daphnia*. Ecology Letters 17:360–368.

Gatti, S. 2012. The role of sponges in high-Antarctic carbon and silicon cycling—a modelling approach. Berichte zur Polar- und Meeresforschung 434:1–124.

Gherardi, F., C. Souty-Grosset, G. Vogt, J. Diéguez-Uribeondo, and K.A. Crandall. 2010. Infraorder Astacidea Latreille, 1802 p.p.: the freshwater crayfish. Pages 269–423 *in* F.R. Schram, and J.C. von Vaupel Klein, editors. Treatise on zoology—anatomy, taxonomy, biology. The Crustacea, volume 9A. Eucarida: Euphausiacea, Amphionidacea, and Decapoda (partim). Brill, Leiden, The Netherlands.

Gilligan, D., R. Rolls, J. Merrick, M. Lintermans, P. Duncan, and J. Kohen. 2007. Scoping the knowledge requirements for Murray Crayfish (*Euastacus armatus*). Fisheries Final Report Series 89. Narrandera Fisheries Centre, Narrandera, Australia.

Han, C.-C., C.-S. Chang, I.-M. Cheng, L.-S. Fang, and K.S. Tew. 2011. Population dynamics of a landlocked and amphidromous freshwater shrimp, *Caridina gracilipes* (Decapoda: Caridea) in subtropical waters. Journal of Crustacean Biology 31:278–285.

Hartnoll, R.G. 1965. The biology of spider crabs: a comparison of British and Jamaican species. Crustaceana 9:1–16.

Hartnoll, R.G. 2001. Growth in Crustacea—twenty years on. Hydrobiologia 449:111–122.

Hopkins, P.M., and S. Das. 2015. Regeneration in crustaceans. Pages 168–198 *in* E.S. Chang, and M. Thiel, editors. The natural history of the Crustacea, volume 4: Physiology. Oxford University Press, New York.

Hopkins, P.M., A.C.-K. Chung, and D.S. Durica. 1999. Limb regeneration in the fiddler crab, *Uca pugilator*: histological, physiological and molecular considerations. American Zoologist 39:513–526.

Hopp, U., G. Maier, and R. Bleher. 1997. Reproduction and adult longevity of five species of planktonic cyclopoid copepods reared on different diets: a comparative study. Freshwater Biology 38:289–300.

Hoyert, D.L., and J. Xu. 2012. Deaths: preliminary data for 2011. National Vital Statistics Reports 61(6). National Center for Health Statistics, Hyattsville, Maryland.

Hsueh, P.-W. 2003. Responses of the pea crab *Pinnotheres taichungae* to the life history patterns of its primary bivalve host *Laternula marilina*. Journal of Natural History 37:1453–1462.

Huner, J.V. 2002. *Procambarus*. Pages 541–584 *in* D.M. Holdich, editor. Biology of freshwater crayfish. Blackwell, Oxford, England.

James, M.O., and S.M. Boyle. 1998. Cytochromes P450 in Crustacea. Comparative Biochemistry and Physiology C 121:157–172.

Jeffery, C.J., and A.J. Underwood. 2001. Longevity determines sizes of an adult intertidal barnacle. Journal of Experimental Marine Biology and Ecology 256:85–97.

Johnson, P.T. 1987. A review of fixed phagocytic and pinocytotic cells of decapod crustaceans, with remarks on hemocytes. Developmental and Comparative Immunology 11:679–704.

Jones, O.R., A. Scheuerlein, R. Salguero-Gómez, C.G. Camarda, R. Schaible, B.B. Casper, J.P. Dahlgren, J. Ehrlén, M.B. García, E.S. Menges, P.F. Quintana-Ascencio, H. Caswell, A. Baudisch, and J.W. Vaupel. 2014. Diversity of ageing across the tree of life. Nature 505:169–173.

Juanes, F., and L.D. Smith. 1995. The ecological consequences of limb damage and loss in decapod crustaceans: a review and prospectus. Journal of Experimental Marine Biology and Ecology 193:197–223.

Jung, T., N. Bader, and T. Grune. 2007. Lipofuscin: formation, distribution, and metabolic consequences. Annals of the New York Academy of Sciences 1119:97–111.

Kilada, R., B. Sainte-Marie, R. Rochette, N. Davis, C. Vanier, and S. Campana. 2012. Direct determination of age in shrimps, crabs, and lobsters. Canadian Journal of Fisheries and Aquatic Sciences 69:1728–1733.

Kirkwood, T.B.L. 2008. Understanding ageing from an evolutionary perspective. Journal of Internal Medicine 263:117–127.

Kirkwood, T.B.L., and S.N. Austad. 2000. Why do we age? Nature 408:233–238.

Knox, G.A. 2001. The ecology of seashores. CRC Press, Boca Raton, Florida.

Lee, W.Y., M. Omori, and R.W. Peck. 1992. Growth, reproduction and feeding behavior of the planktonic shrimp, *Lucifer faxoni* Borradaile, off the Texas coast. Journal of Plankton Research 14:61–69.

Leland, J.C., D.J. Bucher, and J. Coughran. 2015. Direct age determination of a subtropical freshwater crayfish (redclaw, *Cherax quadricarinatus*) using ossicular growth marks. PLoS ONE 10:e0134966.

Lewis, S.D. 1997. Life history, population dynamics, and management of signal crayfish in Lake Billy Chinook, Oregon. Master's thesis. Department of Fisheries and Wildlife, Oregon State University, Corvallis, Oregon.

Maginnis, T.L. 2006. The costs of autotomy and regeneration in animals: a review and framework for future research. Behavioral Ecology 17:857–872.

Martin, J.W., and G.E. Davis. 2001. An updated classification of the recent Crustacea. Science Series 39. Natural History Museum of Los Angeles County, Los Angeles, California.

Medina, A., Y. Vila, C. Megina, I. Sobrino, and F. Ramos. 2000. A histological study of the age-pigment, lipofuscin, in dendrobranchiate shrimp brains. Journal of Crustacean Biology 20:423–430.

Neveu, A. 2000. Étude des populations d'*Austropotamobius pallipes* (Crustacea, Astacidae) dans un ruisseau forestier de Normandie. I. Structures démographiques et croissance: stabilité et variabilité au cours de six années. Bulletin Français de la Pêche et de la Pisciculture 356:71–98.

Oh, C.-W., R.G. Hartnoll, and R.D.M. Nash. 1999. Population dynamics of the common shrimp, *Crangon crangon* (L.), in Port Erin Bay, Isle of Man, Irish Sea. ICES Journal of Marine Science 56:718–733.

Ohtsuki, H., T. Awano, N.K. Tsugeki, S. Ishida, H. Oda, W. Makino, and J. Urabe. 2015. Historical changes in the ecosystem condition of a small mountain lake over the past 60 years as revealed by plankton remains and *Daphnia* ephippial carapaces stored in lake sediments. PLoS ONE 10:e0119767.

Orsi Relini, L., and G. Relini. 1998. Seventeen instars of adult life in female *Aristeus antennatus* (Crustacea: Decapoda: Aristeidae). A new interpretation of life span and growth. Journal of Natural History 32:1719–1734.

Parkyn, S.M., K.J. Collier, and B.J. Hicks. 2002. Growth and population dynamics of crayfish *Paranephrops planifrons* in streams within native forest and pastoral land uses. New Zealand Journal of Marine and Freshwater Research 36:847–861.

Petralia, R.S., M.P. Mattson, and P.J. Yao. 2014. Aging and longevity in the simplest animals and the quest for immortality. Ageing Research Reviews 16:66–82.

Philipp, E.E.R., and D. Abele. 2010. Masters of longevity: lessons from long-lived bivalves—a mini review. Gerontology 56:55–65.

Philipp, E.E.R., J. Strahl, and A.A. Sukhotin. 2012. Aging in marine animals. Pages 193–207 *in* D. Abele, J.P. Vazquez-Medina, and T. Zenteno-Savin, editors. Oxidative stress in aquatic ecosytems. Wiley-Blackwell, Chichester, England.

Pietrzak, B., A. Bednarska, M. Markowska, M. Rojek, E. Szymanska, and M. Slusarczyk. 2013. Behavioural and physiological mechanisms behind extreme longevity in *Daphnia*. Hydrobiologia 715:125–134.

Pohle, G.W. 1992. First Canadian record of *Paralomis bouvieri* Hansen, 1908 (Decapoda: Anomura: Lithodidae), infected by the rhizocephalan *Briarosaccus callosus* (Cirripedia: Peltogastridae) and carrying a hyperparasitic cryptoniscinid isopod (Epicaridea). Canadian Journal of Zoology 70:1625–1629.

Rabalais, N.N., and R.H. Gore. 1985. Abbreviated development in decapods. Pages 67–126 *in* A.M. Wenner, editor. Larval growth. Crustacean issues 2. Balkema, Rotterdam, The Netherlands.

Radzikowski, J. 2013. Resistance of dormant stages of planktonic invertebrates to adverse environmental conditions. Journal of Plankton Research 35:707–723.

Reynolds, J.D. 2002. Growth and reproduction. Pages 152–191 *in* D.M. Holdich, editor. Biology of freshwater crayfish. Blackwell, Oxford, England.

Reznick, D.A., H. Bryga, and J.A. Endler. 1990. Experimentally induced life-history evolution in a natural population. Nature 346:357–359.

Reznick, D.N., M.J. Bryant, D. Roff, C.K. Ghalambor, and D.E. Ghalambor. 2004. Effect of extrinsic mortality on the evolution of senescence in guppies. Nature 431:1095–1099.

Ricklefs, R.E. 2010a. Life-history connections to rates of aging in terrestrial vertebrates. Proceedings of the National Academy of Sciences of the USA 107:10314–10319.

Ricklefs, R.E. 2010b. Insights from comparative analyses of aging in birds and mammals. Aging Cell 9:273–284.

Roark, E.B., T.P. Guilderson, R.B. Dunbar, S.J. Fallon, and D.A. Mucciarone. 2009. Extreme longevity in proteinaceous deep-sea corals. Proceedings of the National Academy of Sciences of the USA 106:5204–5208.

Sainte-Marie, B. 1991. A review of the reproductive bionomics of aquatic gammaridean amphipods: variation of life history traits with latitude, depth, salinity and superfamily. Hydrobiologia 223:189–227.

Scalici, M., D. Macale, F. Schiavone, F. Gherardi, and G. Gibertini. 2008. Effect of urban isolation on the dynamics of river crabs. Fundamental and Applied Limnology 172:167–174.

Schmidt, M. 2014. Adult neurogenesis in crustaceans. Pages 175–205 *in* C. Derby, and M. Thiel, editors. The natural history of the Crustacea, volume 3: Nervous systems and control of behavior. Oxford University Press, New York.

Sheehy, M.R.J., and A.E. Prior. 2008. Progress on an old question for stock assessment of the edible crab *Cancer pagurus*. Marine Ecology Progress Series 353:191–202.

Sheehy, M.R.J., R.C.A. Bannister, J.F. Wickins, and P.M.J. Shelton. 1999. New perspectives on the growth and longevity of the European lobster (*Homarus gammarus*). Canadian Journal of Fisheries and Aquatic Sciences 56:1904–1915.

Sheridan, M., R. Officer, I. O'Connor, and C. Lordan. 2015. Investigating the feasibility of using growth increments for age determination of Norway lobster (*Nephrops norvegicus*) and brown crab (*Cancer pagurus*). Journal of Crustacean Biology 35:495–498.

Shields, J.D., and H.J. Small. 2013. An unusual cuticular tumor-like growth on the abdomen of a lobster, *Homarus americanus*. Journal of Invertebrate Pathology 114:245–249.

Shimizu, S.J., and C.R. Goldman. 1983. *Pacifastacus leniusculus* (Dana) production in the Sacramento River. Freshwater Crayfish 5:210–228.

Siegel, V. 1987. Age and growth of Antarctic Euphausiacea (Crustacea) under natural conditions. Marine Biology 96:483–495.

Skurdal, J., D.O. Hessen, E. Garnås, and L.A. Vøllestad. 2011. Fluctuating fecundity parameters and reproductive investment in crayfish: driven by climate or chaos? Freshwater Biology 56:335–341.

Söderhäll, I. 2013. Recent advances in crayfish hematopoietic stem cell culture: a model for studies of hemocyte differentiation and immunity. Cytotechnology 65:691–695.

Song, C.-K., L.M. Johnstone, M. Schmidt, C.D. Derby, and D.H. Edwards. 2007. Social domination increases neuronal survival in the brain of juvenile crayfish *Procambarus clarkii*. Journal of Experimental Biology 210:1311–1324.

Soong, K. 1997. Some life history observations on the pea crab, *Pinnotheres tsingtaoensis*, symbiotic with the bivalve mollusk, *Sanguinolaria acuta*. Crustaceana 70:855–866.

Sparks, A.K., and J.F. Morado. 1987. A putative carcinoma-like neoplasm in the hindgut of a red king crab, *Paralithodes camtschatica*. Journal of Invertebrate Pathology 50:45–52.

Spitsbergen, J.M., and M.L. Kent. 2003. The state of the art of the zebrafish model for toxicology and toxicologic pathology research—advantages and current limitations. Toxicologic Pathology 31(Suppl):62–87.

Stearns, S.C. 1992. The evolution of life histories. Oxford University Press, Oxford, England.

Stentiford, G.D., editor. 2011. Diseases of edible crustaceans. Journal of Invertebrate Pathology 106(Special Issue):1–164.

Strahl, J., E. Philipp, T. Brey, K. Broeg, and D. Abele. 2007. Physiological aging in the Icelandic population of the ocean quahog *Arctica islandica*. Aquatic Biology 1:77–83.

Streever, W.J. 1996. Energy economy hypothesis and the troglobitic crayfish *Procambarus erythrops* in Sim's Sink Cave, Florida. American Midland Naturalist 135:357–366.

Takahashi, T., and S. Matsuura. 1994. Laboratory studies on molting and growth of the shore crab, *Hemigrapsus sanguineus* de Haan, parasitized by a rhizocephalan barnacle. Biological Bulletin 186:300–308.

Taylor, C.A. 2003. Conservation assessment for a crayfish (*Orconectes placidus*). USDA Forest Service, Eastern Region, Champaign, Illinois.

Thiel, M. 2000. Extended parental care behavior in crustaceans—a comparative overview. Pages 211–226 *in* J.C. von Vaupel Klein, and F.R. Schram, editors. Crustacean issues 12. The biodiversity crisis and Crustacea. Balkema, Rotterdam, The Netherlands.

Thiel, M. 2003. Extended parental care in crustaceans—an update. Revista Chilena de Historia Natural 76:205–218.

Thorne, B.L., N.L. Breisch, and M.I. Haverty. 2002. Longevity of kings and queens and first time of production of fertile progeny in dampwood termite (Isoptera; Termopsidae; *Zootermopsis*) colonies with different reproductive structures. Journal of Animal Ecology 71:1030–1041.

Varfolomeeva, M., A. Artemieva, N. Shunatova, and E. Yakovis. 2008. Growth and survival of barnacles in presence of co-dominating solitary ascidians: growth ring analysis. Journal of Experimental Marine Biology and Ecology 363:42–47.

Veilleux, É., and Y. de Lafontaine. 2007. Biological synopsis of the Chinese mitten crab (*Eriocheir sinensis*). Canadian Manuscript Report of Fisheries and Aquatic Sciences 2812:1–45.

Venarsky, M.P., A.D. Huryn, and J.P. Benstead. 2012. Re-examining extreme longevity of the cave crayfish *Orconectes australis* using new mark-recapture data: a lesson on the limitations of iterative size-at-age models. Freshwater Biology 57:1471–1481.

Vitale, S., L. Cannizzaro, L. Lumare, and S. Mazzola. 2010. Population parameters of *Melicertus kerathurus* (Decapoda, Penaeidae) in southwest Sicilian shallow waters (Mediterranean Sea) using length-frequency analysis. Crustaceana 83:997–1007.

Vogt, G. 2008a. The marbled crayfish: a new model organism for research on development, epigenetics and evolutionary biology. Journal of Zoology 276:1–13.

Vogt, G. 2008b. How to minimize formation and growth of tumours: potential benefits of decapod crustaceans for cancer research. International Journal of Cancer 123:2727–2734.

Vogt, G. 2008c. Investigation of hatching and early post-embryonic life of freshwater crayfish by in vitro culture, behavioral analysis, and light and electron microscopy. Journal of Morphology 269:790–811.

Vogt, G. 2010. Suitability of the clonal marbled crayfish for biogerontological research: a review and perspective, with remarks on some further crustaceans. Biogerontology 11:643–669.

Vogt, G. 2012a. Ageing and longevity in the Decapoda (Crustacea): a review. Zoologischer Anzeiger 251:1–25.

Vogt, G. 2012b. Hidden treasures in stem cells of indeterminately growing bilaterian invertebrates. Stem Cell Reviews and Reports 8:305–317.

Vogt, G. 2013. Abbreviation of larval development and extension of brood care as key features of the evolution of freshwater Decapoda. Biological Reviews 88:81–116.

Vogt, G. 2014. Life span, early life stage protection, mortality, and senescence in freshwater Decapoda. Pages 17–51 *in* D.C.J. Yeo, N. Cumberlidge, and S. Klaus, editors. Advances in freshwater decapod systematics and biology. Crustaceana Monographs 19. Brill, Leiden, The Netherlands.

Vogt, G. 2015. Stochastic developmental variation, an epigenetic source of phenotypic diversity with far-reaching biological consequences. Journal of Biosciences 40:159–204.

Vogt, G. 2016. Research on stem cells, aging, cancer resistance, and epigenetics in marbled crayfish and relatives. Potential benefits for human biology and medicine. Pages 115–157 *in* T. Kawai, Z. Faulkes, and G. Scholtz, editors. Freshwater crayfish: global overview. CRC Press, Boca Raton, Florida.

Vogt, G., and E.T. Quinitio. 1994. Accumulation and excretion of metal granules in the prawn, *Penaeus monodon*, exposed to water-borne copper, lead, iron and calcium. Aquatic Toxicology 28:223–241.

Warburg, M.R. 2011. Cost of breeding in oniscid isopods: a partial review. Crustaceana 84:1561–1580.

Warburg, M.R., K.E. Linsenmair, and K. Bercovitz. 1984. The effect of climate on the distribution and abundance of isopods. Symposium of the Zoological Society of London 53:339–367.

Zimmerman, J.K.M. 2009. Conditions for sustainable crayfish populations in northern Sweden. Rapport. Mid-Sweden University, Sundsvall, Sweden.

8

LIFE CYCLE AND SEASONAL MIGRATIONS

Raymond T. Bauer

Abstract

Seasonal and life cycle migrations are mass movements in which individuals move horizontally for long distances to encounter favorable conditions for reproduction and development. Such migrations have been best studied in larger mobile decapod crustaceans, many of which are commercially important. Some decapod shrimps and brachyuran crabs are dependent on productive estuaries for completion of life cycles. In these species, planktonic larvae develop in oceanic waters. Postlarval stages utilize currents and appropriate behaviors to enter estuaries via selective tidal stream transport (STST). After growth, juveniles and subadults leave for the adult oceanic habitats, again using STST. Many subtropical and temperate zone neritic species make seasonal offshore migrations into deeper waters during the winter, with return nearshore in the spring; some high latitude species make these migrations but with seasons reversed. Numerous freshwater shrimps are amphidromic, that is, they live and reproduce in streams and rivers, but their planktonic larvae drift or are released directly into the sea for development and dispersal. Postlarvae find the mouths of streams, and then make spectacular mass migrations as juveniles back upstream to the adult habitat. Adults of terrestrial crabs live inland, but brooding females move into the littoral zone during new or full moon periods to hatch out larvae into high amplitude tides that carry the larvae out to sea for development and dispersal.

Environmental cues and behaviors used in selective tidal stream transport by postlarvae and juveniles to migrate into and from estuaries include salinity, temperature, hydrostatic pressure, and turbulence associated with tides. In addition, endogenous lunar or tidal rhythms may trigger appropriate behaviors. Orientation and navigational cues used in crustacean migrations are still poorly understood, although a magnetic map sense has been shown in spiny lobsters and indicated in terrestrial crabs. Light reflected from the sea and vibrations from surf may guide these crabs to coastal hatching sites. The mechanisms that have evolved in migrating crustaceans to reach their destinations will be a productive field of research in the years to come.

Life Histories. Edited by Gary A. Wellborn and Martin Thiel.

INTRODUCTION

Migration is defined here as horizontal movements of some duration by all or part of a population from one habitat to another for purposes of reproduction, growth, or to reach areas with more favorable environmental conditions. The familiar long-distance and often spectacular movements of many bird species from spring-summer breeding grounds to distant overwintering sites, sometimes in another hemisphere (e.g., Arctic terns; Alerstam 1985), are typical examples for migrations of individuals in distinct habitats or geographic locations. The mass movement of anadromous Pacific salmon (*Oncorhynchus* spp.), in which adults move from the sea into freshwater streams for spawning, with a return migration of juveniles into the sea for growth to maturation (Hasler et al. 1978), is an example of a life cycle migration between habitats and generations. Dingle (1996) provided a working consensus definition of migration as a persistent or focused movement between or among (but not within) habitats, in which the normal responses to the home range are stopped or inhibited. This definition distinguishes true migration from other movements such as "commuting," in which an organism shuttles back and forth on a daily basis among different areas of the same habitat (e.g., the vertical diel "migrations" of many planktonic crustacean adults and larvae). Dingle also distinguishes "station keeping" as involving movements that return organisms to their normal habitat on being displaced (e.g., tidal "migrations" of some estuarine shrimps) (e.g., Allen et al. 2015). As with most classifications of biological activity, differences among these categories of movements are not always distinct.

Migrations are ecologically important because migrating organisms are temporally variable components of different ecosystems, affecting habitat, productivity, and trophic relationships. Migrations promote exchange of biomass and energy between habitats, and the presence or absence of migrations has an important impact on the dispersal and population genetics of a species.

PENAEID SHRIMPS

Life cycle migrations occur in many penaeid shrimps (Decapoda, Dendrobranchiata, Penaeoidea), in which larvae, juveniles, and adults each may occur in distinctly different habitats (Dall et al. 1990; Fig. 8.1). Adults mate and spawn fertilized eggs directly into high-salinity coastal oceanic waters; these eggs hatch as nauplii and develop in the plankton before the postlarval (settling) stage. Postlarvae are transported by coastal and tidal currents toward "nurseries" (i.e., coastal shallows and estuaries). The postlarvae molt to benthic juveniles, which utilize the abundant food supplies of estuaries, mangroves, seagrass meadows, and other productive shallows to grow rapidly to subadults, which then leave the nursery grounds for offshore adult habitats (Garcia and Le Reste 1986). The degree of dependence on estuaries for completion of the life cycle, and thus the need for and distance of migrations, varies greatly among penaeid species (Dall et al. 1990).

The Gulf of Mexico, with its broad continental shelf and numerous riverine estuaries, mangroves, and seagrass meadows, is the home of several penaeid species, three of which are of considerable importance to both local and world fisheries (Bourgeois et al. 2015, Criales et al. 2015): *Litopenaeus setiferus, Farfantepenaeus aztecus,* and *F. duorarum*. All have been well studied and will serve as case studies for penaeid shrimp migrations.

Reproductive adults of the white shrimp *L. setiferus* are common in nearshore waters along the coasts of the southeastern Atlantic and Gulf of Mexico. Their biology and migrations were first studied in detail using mark-recapture methods by Linder and Anderson (1956), with subsequent work summarized by Bourgeois et al. (2015). Linder and Anderson used the term *outside waters* for subtidal waters beyond the outer beach line and *inside waters* for the salt marshes (*Spartina* species), bays, sounds, rivers, creeks, and bayous landward of the beach line. With increasing temperatures

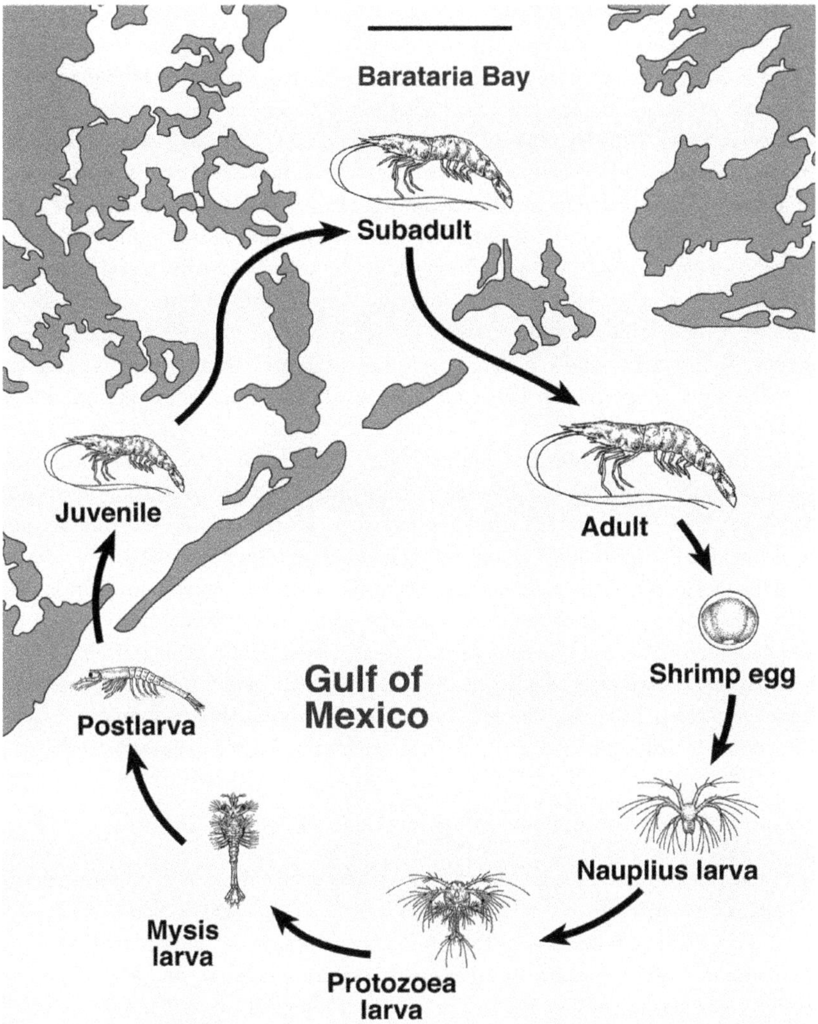

Fig. 8.1.
Life cycle and migrations of an estuary dependent penaeid shrimp from the north central Gulf of Mexico (Louisiana). Scale bar within Barataria Bay represents 6 km. Shrimps and distance offshore are not drawn to scale. Illustration courtesy of the Louisiana Sea Grant College Program.

in early spring, spawning of demersal eggs mainly occurs in the outside waters from March to November. Larvae develop in the water column and migrate as postlarvae, with the aid of coastal currents and tides, passes into inshore estuaries and bays mainly from May through November. White shrimp postlarvae and juveniles penetrate farther into inshore, lower salinity waters than any other Gulf penaeid, growing rapidly on the rich organic matter of the vegetated muddy bottoms they prefer. Within a few months, the large (total length, 9–12 cm) juveniles and subadults move from the inner marshes and estuary into the deeper, more saline waters nearer the beach line. As water temperatures decline in late August and September, especially with advancing cold weather fronts between September and November, the subadults migrate from the inside to outside waters and continue growth and maturation until spawning begins in the spring.

Besides these life history migrations, adult *L. setiferus* also may undertake seasonal migrations of hundreds of kilometers. As a basically shallow-water subtropical species, white shrimps are negatively affected by the colder winter temperatures of nearshore waters. Linder and Anderson (1956) found that tagged Atlantic coastal populations were essentially stationary during spring and summer, but with the advent of colder temperatures in the autumn, the shrimps migrated south parallel to the coast (e.g., from southern North Carolina to southern Florida), into deeper, warmer waters offshore and returned in the spring. However, in the northern Gulf of Mexico, weather fronts accompanied by cold northerly winds push warmer water offshore and quickly chill both the shallow outside waters and especially the very shallow bays and estuaries of the northern Gulf. As a result, adults in the northern Gulf move from shallow outside waters to farther offshore. Juveniles and subadults passing the fall and winter seasons in estuaries and inner bays temporarily move to outside waters. After the passage of a front, warmer southerly winds return, raising water temperatures, which allows both adults and juveniles to return to their previous locations.

Further east in the Gulf of Mexico (south Florida), *F. duorarum* (pink shrimp) is the dominant commercial penaeid, occurring over more sandy sediments than the white shrimp. Juveniles tagged by Costello and Allen (1966) in the high-salinity nursery grounds (mangroves, seagrass meadows, estuaries) along the Florida coast migrated during maturation at least 250 km to the adult spawning grounds at the Dry Tortugas and Sanibel island fishing grounds off southwestern Florida. The study found that although migration routes were broad, juveniles from different nursery grounds went to different adult grounds, and consequently there was relatively little mixing of local populations. Criales et al. (2015) further showed this movement was not passive dispersal, but a directed movement (migration). Favorable currents during the period of larval development, coupled with larval behaviors (described below), transported larvae from spawning to nursery grounds.

Mechanisms of Postlarval and Juvenile Migrations in Penaeid Shrimps

The aggregation of reproductive adults in specific offshore spawning areas may have evolved so that hatching larvae are carried by prevailing currents near to coastal shallows and estuaries (Criales et al. 2015). For example, in the brown shrimp *F. aztecus,* peak spawning occurs from September to November and postlarvae primarily enter estuaries during February and March. Westerly and northwesterly currents carry postlarvae near passes leading into coastal estuaries. During the passage of northerly cold fronts at this time, water is first pushed out of estuaries by north winds and ebb tides. Postlarvae encountering this cold, low-salinity water move down to the bottom, concentrating them in a favorable position to enter the estuaries. When the front passes, southerly winds produce a return flow into the estuary and the postlarvae enter, primarily on flood tides (Rogers et al. 1993). Modeling studies and results of previous studies on larval migrations of the pink shrimp *F. duorarum* show that larvae migrate from the south Florida Dry Tortugas and Marquesas into the Florida Bay nursery grounds by taking best advantage of coastal and tidal currents using similar behaviors (Criales et al. 2015).

The basic pattern of penaeid postlarval and juvenile migrations is that postlarvae migrate into coastal nursery grounds on flood tides and late-stage juveniles (subadults) leave on ebb tides (Fig. 8.2). In both migrations, movement is generally greatest at night during spring tides associated with the full and new moon (Dall et al. 1990, Criales et al. 2015). This use of tidal currents in larval and juvenile migrations is termed *selective tidal stream transport* (STST) (Forward et al. 2003). STST is a type of migratory behavior in which animals ascend in the water column during one tidal phase and are inactive near the bottom during the opposing tidal phase (Fig. 8.2). In this way, one-way horizontal movement is achieved with the aid of tidal currents, which is particularly helpful to weakly swimming larval stages.

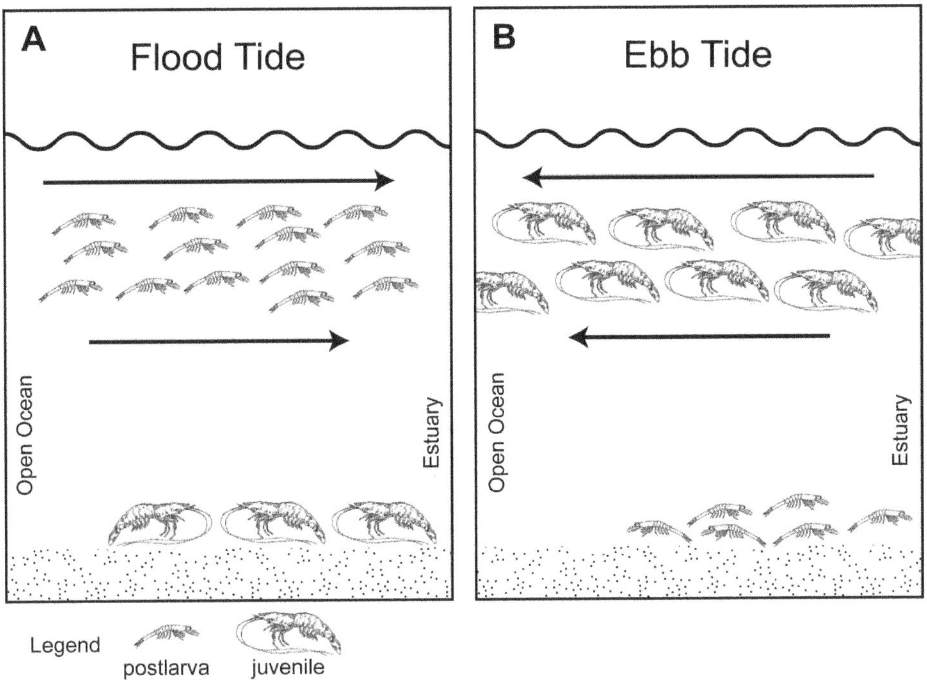

Fig. 8.2.
Diagram of selective tidal stream transport (STST) by penaeid life stages. (A) Postlarvae from the open sea use flood tidal flow to move into an estuary; younger juveniles maintain their position by being near or burrowed within the bottom. (B) Older juveniles/subadults use ebb tidal flow to leave the estuary for the open sea; recently arrived postlarvae maintain their position by remaining near the bottom where tidal flow is lowest.

In the pink shrimp *F. duorarum*, postlarvae become concentrated near the surface layer as they approach the entrance to Florida Bay, their nursery grounds, and they develop a semidiurnal tidal rhythm that keeps postlarvae near the surface during flood tides, especially nocturnal ones. To ensure their retention in the nursery grounds, postlarvae of *F. aztecus* and *F. duorarum* move to the bottom during ebb tides, perhaps using lower salinity as a primary cue (Rogers et al. 1993, Criales et al. 2011).

Postlarvae molt to the first of several benthic juvenile stages, which feed and grow on the nursery grounds in the entrance to the estuaries. Responses of juveniles to tides change during their ontogeny, with behavior of early stages the opposite of the later (subadult) stages. These subadults eventually migrate out of the estuaries and then to adult habitats offshore (Hughes 1969, Rogers et al. 1993, Criales et al. 2011). Early juveniles, which need to remain within the estuary, tend to move to or burrow in the bottom during ebb tides, stimulated by small salinity decreases typical of brackish water nursery areas and perhaps as well by hydrostatic pressure. During flood tides (cued by increases in salinity or hydrostatic pressure), they may ascend and swim with the flood to maintain any displacement or to enter deeper into the nursery. Behavioral responses to environmental cues change as juveniles reach subadult stages, when migration out of the estuary is adaptive. Such individuals ascend during ebb tides and swim or drift with the flow to move out of the estuary.

What are some of the factors that influence migratory behavior of postlarvae and juveniles? Small salinity changes associated with ebb and flood tides have been strongly indicated by observational and experimental evidence (Hughes 1969, Rogers et al. 1993). The postlarval and juvenile migratory responses to salinity described above are from studies done in brackish water nursery grounds. However, when coastal nursery areas are hypersaline in arid areas of the world, the responses to salinity may be opposite to those described above (Garcia and Le Reste 1986, Ogburn et al. 2013). Migrations in and out of nursery grounds are most intense during favorable night tides, especially those associated with the new moon, and this may be an adaptation to avoid visual predators (Kerr et al. 2014). Thus, diel responses to light are very important in migratory behaviors. Changes in pressure due to tidal height (hydrostatic cues) may also be important in STST (Tielmann et al. 2015).

Dall et al. (1990) pointed out that although postlarvae and juveniles could probably migrate using only environmental cues, circadian and tidal rhythms aid the migrators in making the correct behavioral response. Ogburn et al. (2013) used both field collections and laboratory experiments to study endogenous rhythms to compare migratory behavior in postlarval and juvenile white (*L. setiferus*), brown (*F. aztecus*), and pink (*F. duorarum*) shrimps. They concluded that brown and pink shrimp postlarvae showed circatidal rhythms (12.4 h) in vertical swimming activity that coincided with the early flood tide, whereas white shrimp postlarvae did not, depending only on environmental cues to migrate during the flood tides. Juveniles of brown and white shrimp had weak circatidal activity rhythms with peak vertical swimming during ebb to early flood tide, whereas pink shrimp juveniles switched from the postlarval circatidal to a circadian rhythm with nocturnal swimming. It is likely that similar mixes of environmental cues, supported by endogenous behavioral rhythms, are used by other estuary-dependent penaeid larvae and postlarvae.

MARINE CARIDEAN SHRIMPS

The infraorder Caridea is a diverse and speciose group of decapod shrimps, occurring in nearly all marine and freshwater habitats of the world (Bauer 2004, De Grave and Fransen 2011). As in penaeid shrimps, migrations of marine carideans are best known from schooling species important in fisheries, particularly those of the boreal genera *Pandalus* and *Crangon*. The migrations are mainly a response to seasonal or water-mass changes of temperature and life cycle requirements, such as breeding and juvenile growth. The general pattern in the temperate-boreal *Pandalus* species is a migration of ovigerous females into warmer shallow waters during spring to hatch larvae. The larvae and resultant juveniles are thus delivered to a habitat that is more favorable to growth (warmer, greater food supply). Individuals move offshore in autumn, usually in size- and age-specific waves, when cooling, shallow waters stimulate their movement into warmer deeper waters (Shumway et al. 1985, Bergström 2008). Higher wave action and turbulence during winter storms may also be factors stimulating migration to deeper, more stable areas.

In *Crangon crangon*, a commercially important North Atlantic species, offshore females hatch larvae that are carried by coastal currents and STST (flood tides) into coastal nurseries, including intertidal mud flats and marshes. Juveniles can grow rapidly into subadults and adults in these areas because of the higher temperatures, rich food supplies, and at least some protection from predators in these nursery areas (Hufnagl et al. 2014). Most of the subadult population migrates offshore during the autumn (mating season) to the adult spawning grounds with the help of ebb tides, completing the life cycle. The offshore migration of incubating females to deeper warmer waters may be an adaptation that promotes embryonic development, the speed of which is positively correlated with temperature (Bauer 2004). The degree of offshore migration depends on the severity of the winter; in mild winters, there may not be a migration at all (Bodekke 1976).

Studies on some other temperate and high-latitude coastal carideans indicate migration, with the same general pattern of movement offshore with seasonal cooling of shallows in autumn to deeper warmer waters in winter, especially by the reproductive adults. In some species, there is a return migration in the spring to hatch larvae in warmer coastal shallows (Höglund 1943, Allen 1966,

Bilgin et al. 2008). In other species, the migration may be one way: larvae are transported toward the shore by coastal currents and STST, juveniles grow in coastal shallows, and as individuals age and mature reproductively, they move out into deeper waters where they remain as adults (Allen 1966).

AMPHIDROMOUS FRESHWATER SHRIMPS

Approximately 25% of the nearly 3,500 caridean shrimp species live in freshwater habitats of subtropical and tropical areas of the world (De Grave et al. 2008, De Grave and Fransen 2011). In many species, especially those in landlocked habitats, females produce large yolky eggs that sustain direct or abbreviated larval development (ALD). Embryos hatch either as small juveniles or nonfeeding (lecithotrophic) larvae, with sufficient yolk reserves to complete development in the adult habitat (Bauer 2004). In contrast, females of many other freshwater species produce small eggs with much less yolk that hatch into nonfeeding stage 1 larvae that require saline water to molt to the first of several planktotrophic larval stages (extended larval development [ELD]). These species live in rivers and streams that empty into the sea, with reproductive adult populations found from several to many kilometers upstream. The larvae must be transported to the sea and, as postlarvae after development, must find the mouths of rivers and streams and migrate as young juveniles back upstream to the adult habitat (Fig. 8.3A). McDowall (2007) has defined this diadromous life history cycle as amphidromy, in which there are recurring migrations between freshwater and the sea for purposes other than reproduction. Amphidromy is found in *Xiphocaris elongata* (Xiphocarididae), many species of Atyidae, and some Palaemonidae (especially *Macrobrachium* species). Life cycles and other aspects of the biology of amphidromous shrimps have been reviewed in Bauer (2013).

Larvae may reach the sea with release (hatching) by upstream females and downstream drift with stream flow, or females may migrate downstream to hatch incubated stage 1 larvae in estuaries or open marine habitats (Fig. 8.3A). Species using larval drift inhabit higher-gradient, rapidly

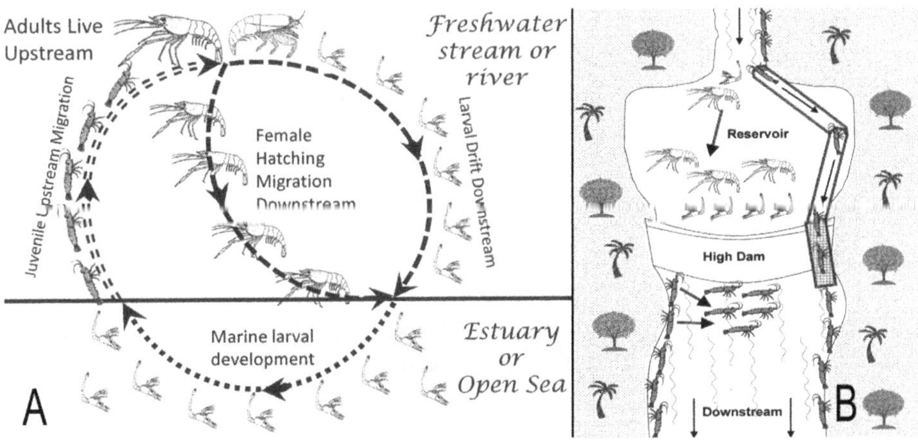

Fig. 8.3.
Amphidromy in caridean shrimps. (A) Life cycle: adults live upstream in fresh water. Larvae are either released upstream and drift to marine waters downstream, or females migrate downstream to release them. After larval development in the sea, postlarvae enter a stream or river and migrate upstream to the adult habitat. From Bauer (2013). (B) Human impacts of a high dam: blockage of larval drift downstream to and juvenile migration upstream from the sea. Shrimp ramps (right side of dam) may allow juveniles to move up and past the dam. From Bauer (2011), with permission from Brill.

flowing rivers and streams that are only several to a few hundred kilometers, at most, from saline waters (March et al. 1998, Benstead et al. 2000). Such species are common on small mountainous oceanic islands (e.g., Caribbean Sea, Indo-West Pacific; McDowall 2007) and continental streams close to the sea (e.g., Bertini et al. 2014). In these streams, flow is rapid enough to carry stage 1 larvae to the sea within a few days before residual yolk reserves are depleted (Benstead et al. 2000, March et al. 2003). In most amphidromous species, hatching or release of larvae occur during the wet or flood season with resultant high water flow, which facilitates rapid larval drift to downstream estuaries or the open sea (Bauer 2013).

At the other extreme are amphidromic species from low-gradient rivers in which adult populations may occur from a few hundred to several hundred kilometers upstream. In this case, larval drift may not be rapid enough for the stage 1 larvae to reach the sea before yolk supplies are depleted and starvation occurs. Females migrate down to estuaries or the sea coast and hatch out stage 1 larvae directly into saline water where larval development can be completed (e.g., *Cryphiops caementarius, Macrobrachium rosenbergii, M. ohione*; Bauer 2013; Fig. 8.3A).

Larval and female reproductive migrations to the sea have been inferred by changes in abundances from plankton sampling and trap or net sampling, respectively. However, the upstream migrations of juveniles back from the sea after larval development are rather easily observed directly in streams and rivers (Bauer 2013). Many thousands of small postlarval and juvenile individuals crawl and swim along the banks of streams, particularly at night (Fig. 8.4). When encountering an obstacle such as a cascade, juveniles may move to the wetted splash zones on surrounding banks and climb vertically past the barrier. They can similarly overcome a low natural or man-made dam, as long as there is some directional flow over the barrier.

Juvenile migrations take place principally at night (Bauer 2013), as does hatching and release of larvae to begin downstream drift to the sea (Benstead et al. 2000, March et al. 2003). Presumably, this is an adaptation for avoidance of visually hunting predators (Kikkert et al. 2009). The directional stimulus for upstream movement of juveniles is a positive rheotaxis (i.e., movement into and against the current of downstream flow). The concentration of migrations in shallow water along

Fig. 8.4.
Atyid shrimp postlarvae and young juveniles migrating upstream along and just above the wetted bank of a tropical stream. Original photograph courtesy of Alan Covich ©.

the bank allows the juveniles to migrate (and conserve energy) where the current is slowest but still sufficient to supply the upstream directional signal. In still water or along the bank with no flow, the juveniles become confused and mill about (Benstead et al. 1999). In most species, the juvenile migration occurs at low or decreasing flow periods, which reduces the stress of moving upstream against a current (Bauer 2013).

The construction of dams, the extraction of water above dams for urban use or irrigation, and other human impacts (e.g., bright nighttime lighting along migration routes) interrupt the downstream migrations of larvae and the upstream migrations of juveniles returning from the sea (Fig. 8.3B). Such impacts have severely damaged species diversity and ecological function in affected tropical streams (e.g., Holmquist et al. 1998, March et al. 2003) but can be reduced by an understanding of the shrimp's life cycle and behavior. March et al. (2003) suggested that mortality of the nocturnally drifting larvae could be mitigated by withdrawing water during the day, when few larvae are up in the water column. The inclusion or construction of inclined spillways or "shrimp ramps/ladders" can also successfully allow migrating juveniles to climb and bypass low dams (Fig. 8.3B; Bauer 2013, Olivier et al. 2013).

Adaptive Significance of Amphidromy in Shrimps

Why do some freshwater species retain the long larval series of the supposed marine ancestor? As McDowall (2007) asked, "why bother" with amphidromy? An examination of costs and benefits gives the probable answers. Freshwater habitats, both lentic and lotic, provided marine ancestors with a new, relatively empty habitat, rich in food supplies. However, freshwater species that became landlocked had to evolve ALD or become extinct. To retain the ELD of their marine ancestors, freshwater shrimps must live in stable, plankton-rich lentic habitats where developing larvae would find sufficient food, but relatively few species do so (Bauer 2013). Consequently, those species from plankton-poor lentic environments have evolved ALD where adults supply eggs and embryos with sufficient energy for development (Magaelhães and Walker 1988). On the other hand, the larvae of shrimps living in rivers and streams are swept downstream when hatched, away from the adult habitat. These species retain the larval development of their ancestors in the plankton-rich estuarine or marine environment but must pay the price of a long arduous migration to the sea as larvae and back up the rivers as newly settled postlarvae or juveniles. Shrimps may face a lower predation risk in freshwater streams, which house a lower diversity of fish predators, especially on oceanic islands where amphidromous shrimps are common (McDowall 2007). Moreover, migrating juveniles of many amphidromous shrimps are good climbers, allowing them to ascend steep gradients and waterfalls to upstream habitats that most fishes and large predatory shrimps cannot reach (Covich 2009).

Another advantage of larval development in the sea is dispersal, which allows invasion of new streams, high gene flow among populations, and a wide geographic distribution compared with landlocked ALD species (e.g., Dennenmoser et al. 2010, Cook et al. 2012). Interestingly, all freshwater crabs have direct development (Hartnoll 1988) and thus do not enjoy these benefits of amphidromy.

PALINUROID LOBSTERS

This group of achelate lobsters is composed of two families, the Palinuridae and Scyllaridae. Migrations are much better known and studied in the Palinuridae, with much research stimulated by the impressive single-file mass migrations of some species, especially that of *Panulirus argus,* documented by William Herrnkind and collaborators (Herrnkind 1969, Bill and Herrnkind 1976,

Herrnkind et al. 2001, Lavalli and Herrnkind 2009, Eggleston et al. 2013). Palinurid (spiny lobsters) are large decapods characterized by a defensive morphology consisting of a hard, highly calcified exoskeleton; long, stout, and spiny second antennae; and sharp projections or horns curving over the eyes. The antennae can be lashed about or thrust at attacking predators. The life cycle begins as embryos incubated by females hatch into transparent leaflike larvae, termed *phyllosoma*. These larvae are very long-lived in the plankton, drifting from several months up to 2 years during development. They settle out of the plankton as a small lobster-like puerulus into shallow benthic nursery areas (e.g., seagrass meadows, mangrove roots, algae, coral, and rocky crevices). As the juveniles grow, they move out to other deeper habitats and are gregarious in rocky or coral shelters or large seaweeds in some species (Butler et al. 2006, Hinojosa et al. 2014). Reproductive adults, especially ovigerous females, move out to areas where larvae can be released into oceanic water masses and currents for dispersal.

Herrnkind (1980) synthesized research on movements of spiny lobsters and categorized migrations as ontogenetic, reproductive, or mass movements. In *Panulirus interruptus*, larvae settle out as pueruli into surf grass (*Phyllospadix* spp.) in the wash zone of rocky shores, growing for 2 years before moving out to deeper rocky caves and crevices 10 to 30 m deep. The movement of juveniles from these shallow nursery grounds to deeper more seaward habitats may involve only several hundred meters. The overall distance is short because the continental shelf is quite narrow along the west coast of the United States and Mexico where *P. interruptus* occurs. On the other hand, the distance from *P. argus* nursery areas to outer deeper reefs may be several to many kilometers off the Bahamas and Florida Keys. In *P. argus* and *P. cygnus*, juveniles have been observed at night moving in queues (single-file formation) across open areas during this ontogenetic emigration.

In several spiny lobster species, females migrate out to the edges of coastal reefs and banks to release larvae into oceanic currents (e.g., *P. argus, P. interruptus, Jasus edwardsii*; Herrnkind 1980). The presence or extent of migrations related to reproduction may vary within genus, with little or no migration in some species and extensive movements in others. Females of the deep-water species *P. delagoae* off southeast Africa migrate in mass up from deep waters to the shallow extreme of their depth range to release larvae near the edge of the Agulhas current. In *P. argus* and presumably in other reproductive migrators, the females disperse back into their prereproductive habitats after spawning (Herrnkind 1980). Boles and Lohmann (2003) showed experimentally that *P. argus* is capable of true navigation based on a magnetic map sense, by which they can determine their position relative to another location (i.e., premigratory position) without relying on landmarks, cues from the destination, or memory of the outward journey. This capability may permit long-distance migration and homing in *P. argus* and other palinurids.

Spiny lobsters may also make long-distance migrations for nonreproductive purposes. Direct field observations and scuba studies of these migrations in *P. argus* (Fig. 8.5) have garnered a great deal of attention from both scientists and the general public alike (Herrnkind 1969). For *P. argus*, the adaptive value of these migrations from shallow sublittoral waters is escape from poor winter conditions (cold water temperatures, wave action, turbulence) to more stable and optimal conditions in deeper waters (Eggleston et al. 2013). Apparent hormonal changes stimulated by the fall photoperiod, as well as a marked increase in current speeds due to the passage of polar weather fronts, stimulate nonreproductive lobsters to move into deeper water. In doing so, they must pass over long stretches of open bottom that expose them to attack by a variety of predators. The lobsters form queues (single-file formations) when crossing such areas (Fig. 8.5A,C) and form defensive formations when attacked (Fig. 8.5B).

In other palinurid species, mass migrations appear to serve another purpose, that of counteracting displacement of larvae from up-current sites of larval release (Herrnkind 1980). When long-lived phyllosoma larvae are released into oceanic currents for dispersal, they may drift

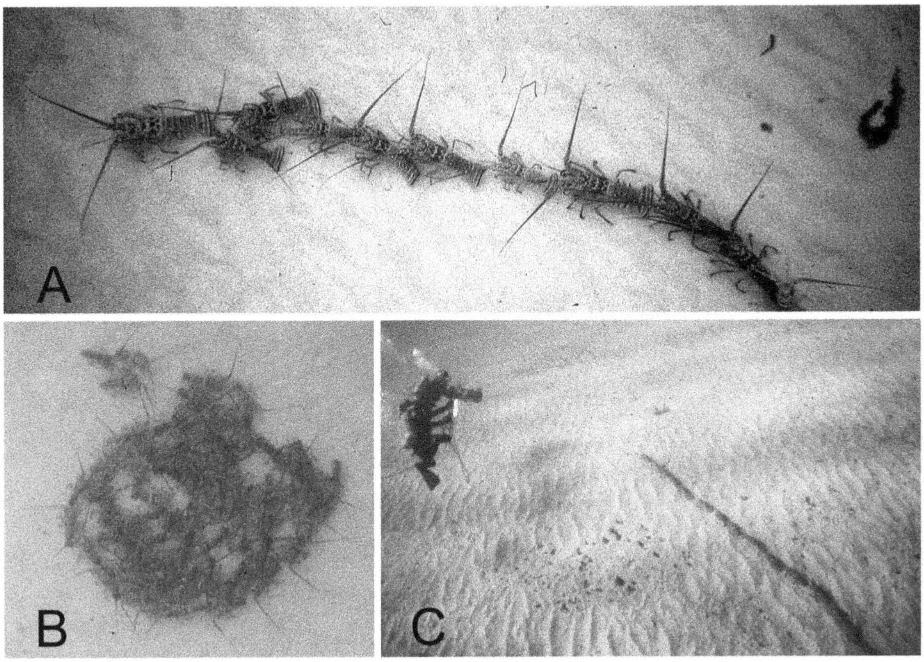

Fig. 8.5.
Spiny lobster (*Panulirus argus*) migration queues. (A) Closeup of a queue showing the head-to-tail single-file formation of the migrating lobsters. (B) Defensive rosette or spiral formed by a queue when threatened by a predator. (C) Migratory queue moving over a soft bottom; diver included for perspective. Original photographs courtesy of William Herrnkind ©. See color version of this figure in the centerfold.

long distances from their site of origin. Immature and young adult lobsters may make long distance migrations against the direction of current back to areas inhabited by reproductive adults. Tagging studies by Groeneveld and Branch (2002) with *Palinurus gilchristi* off the south coast of South Africa demonstrated such a countercurrent migration of immature and young lobsters. The Agulhas current flows southwesterly along the coast, and the average direction of movement of migrating lobsters was always easterly against the direction of current. Tagged lobsters migrated at least 50 km and as much as 450 km over one to three years in the correct direction to bring them back from settlement sites downstream to various up-current adult spawning areas.

Queuing (Fig. 8.5A,C) is characteristic of these mass migrations by spiny lobsters and is best studied in *P. argus*. Single-file queues can also form in nonmigratory lobsters without access to shelter, but the tendency to queue is much stronger in migratory lobsters (Herrnkind 1969). In migrating *P. argus*, queues from about 3–60 individuals form, traveling at speeds averaging 28 cm/s (~1 km/h); these queues can be observed by divers during the day (Fig. 8.5C). Parallel single-file lines of migrating individuals can be observed across an area as much as 1.6 km wide (Bill and Herrnkind 1976). In queuing, the cephalothorax of 1 lobster overlaps the abdomen of the next forward one, often hooking its anterior walking legs around the telson of the other individual (Fig. 8.5A). The inner branch of the chemotactile first antenna (antennule) of the following individual makes frequent contact with the body of the forward individual. Experimental observations showed that both visual and tactile cues were used in forming and maintaining queues. However, tactile cues were sufficient, so that queuing and migration could occur at night as well as during the day (Herrnkind 1969).

Hypotheses on the adaptive value of queuing during mass migrations over shelterless substratum were proposed and tested by Bill and Herrnkind (1976) and Herrnkind et al. (2001). Reduction of drag and defense against predators are the major benefits. Experiments on the effects of queuing on drag were performed in laboratory tow tanks using preserved lobsters. During queuing, a following lobster encounters a moving envelope of water that reduces drag by half in queues of five or more individuals, presumably decreasing the energy expended while migrating.

Spiny lobsters moving over open substratum are subject to attacks by predators such as triggerfish, large sharks, and octopi (Butler et al. 2006). When attacked by a predator in the field, queues form defensive rosettes (Fig. 8.5B), with the lead lobster pirouetting in circles, followed by others in the queue, which spirals to form a rosette of individuals with anterior ends facing outward. Predators are repulsed by the lashing and pointing of the large spiny antennae of the lobsters, which make short forward lunges. In experiments performed in large laboratory tanks, mortality on solitary lobsters by various numbers of triggerfish predators was nearly 50% compared to grouped lobsters, which only received minor bites (Lavalli and Herrnkind 2009).

CLAWED LOBSTERS AND CRAYFISHES

In the clawed lobsters and crayfishes (decapod infraorder Astacidea), migratory behavior has been shown in the commercially important American lobster, *Homarus americanus*, a generally solitary shelter dweller (Cooper and Uzman 1971, 1980, Lawton and Lavalli 1995). However, mark-recapture studies on the European lobster, *H. gammarus*, showed only very limited local movement (Agnalt et al. 2007), and no significant migratory behavior has been reported in *Nephrops* species (Chapman 1980).

An extensive mark-recapture study on *H. americanus* in New England showed that inshore populations exhibited little movement, but seasonal inshore-offshore migrations were found in lobsters living at deep continental platform or slope depths (Cooper and Uzman 1971). The movement was inshore in the spring and summer and then back offshore in fall and winter. Distances covered in these migrations were 16 to 160 km, with travel speeds of up to 10 km/d. In the spring and summer, the higher temperatures in shallow water are hypothesized to favor reproductive processes such as molting and the subsequent mating and spawning in females as well as more rapid growth (Cooper and Uzman 1971). Migration back to deeper, more stable waters in the fall and winter may be an escape from the low temperatures and wave action associated with those seasons in shallower water.

Mass movements of some freshwater cambarid crayfishes have been reported, and Walls (2009) asked the question: do crayfishes migrate? In the classical sense used in this chapter (Dingle 1996), Walls' answer was no. The often spectacular movement of hundreds of thousands of crayfishes across fields and highways, which may continue for days, occurs irregularly and uncommonly, usually involving the adults of a single species in deteriorated condition. Movements are usually from a marsh, canal, or swamp to another similar body of water on the other side of a road or levee that separates the two bodies of water. These movements have been interpreted as a search for more suitable habitat. Especially in late summer-autumn, crayfishes sporadically relocate in mass from waters with low oxygen to more aerated ones or from a drying habitat to a wetter one with better conditions for burrowing.

MARINE BRACHYURAN CRABS

Migrations have been reported in several brachyuran crabs, particularly those that have commercial or economic importance. Productive estuarine habitats may sustain life history stages of brachyuran crabs, such as the blue crab *Callinectes sapidus* (Portunidae), which has been particularly well studied (Kennedy and Cronin 2007). Like all portunids, the juveniles and adults are excellent swimmers using the paddle-like fifth pereopods. The blue crab supports important fisheries along the US southeast Atlantic coast and Gulf of Mexico. Unlike most estuary-dependent shrimps, all phases of the life cycle take place within the estuary except for the extended larval development, which takes place in euhaline oceanic waters (Fig. 8.6).

The blue crab life cycle involves various migrations, mainly within but also outside of the estuarine habitat, which have been modeled and summarized by Forward et al. (2003; Fig. 8.7). This model is derived from earlier studies on populations in the Chesapeake Bay estuary, later reviewed and added to by Kennedy and Cronin (2007). Juveniles and adults spend much of their life cycle in the low-salinity upper parts of the estuary. Mating and insemination of females occur in upper

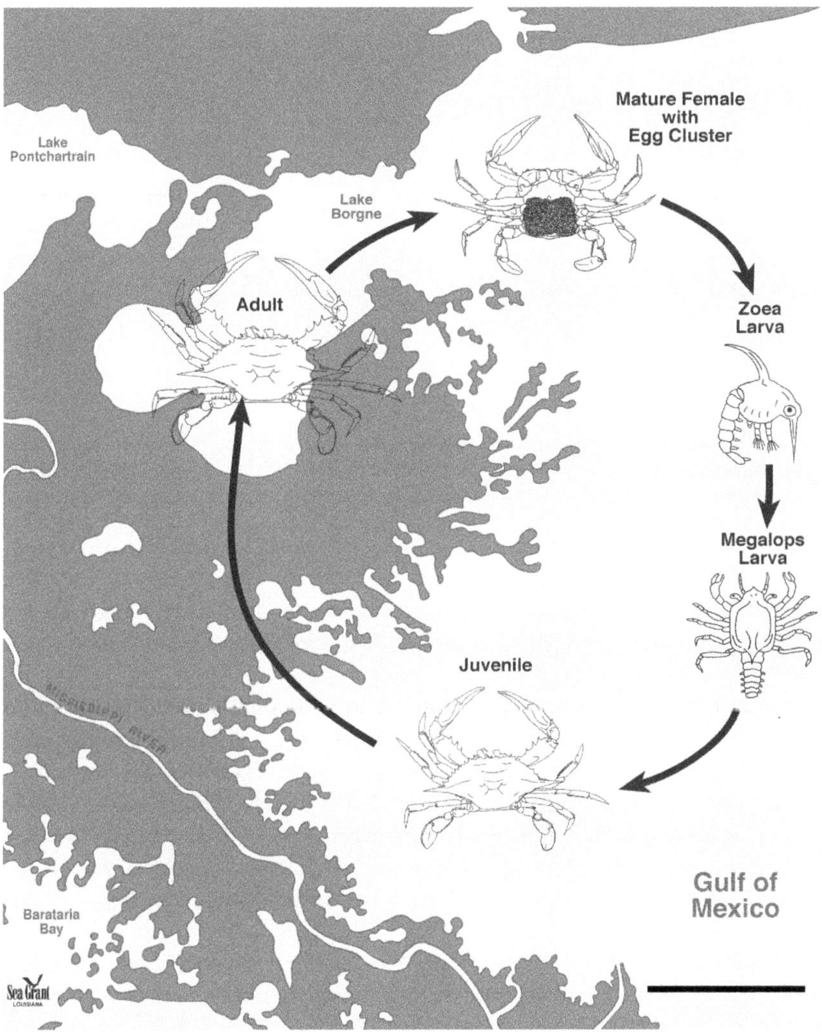

Fig. 8.6.
Life cycle of the estuary-dependent blue crab *Callinectes sapidus* in the north central Gulf of Mexico (Louisiana). Scale bar represents 20 km. Crabs are not drawn to scale. Illustration courtesy of the Louisiana Sea Grant College Program.

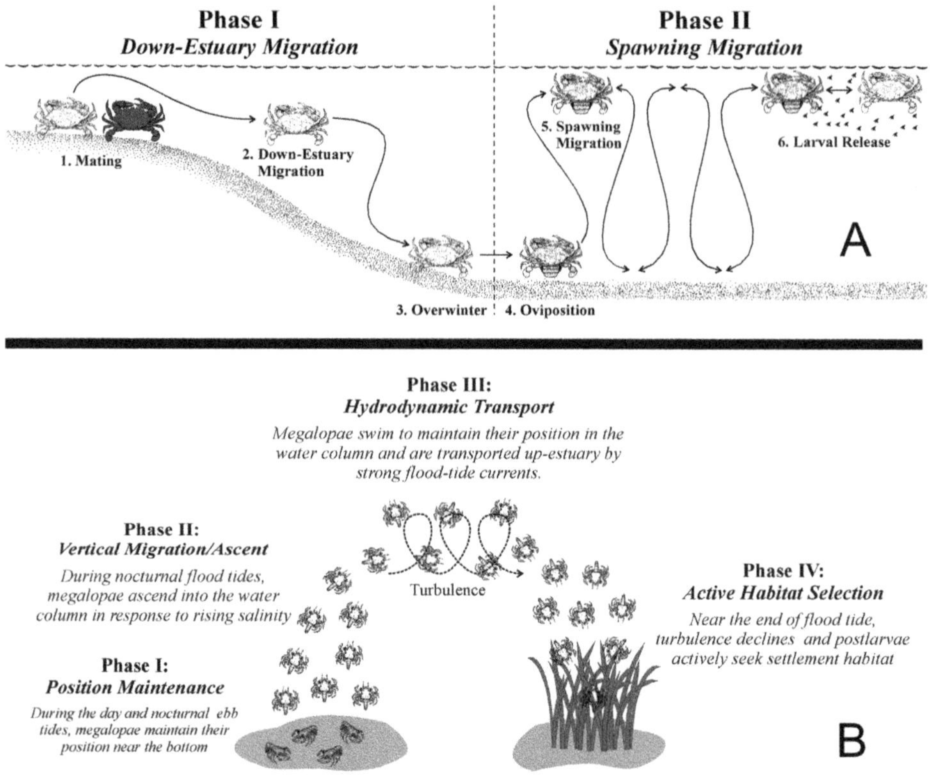

Fig. 8.7.
Blue crab (*Callinectes sapidus*) migrations, Chesapeake Bay, United States. (A) Down-estuary migrations of reproductive female for spawning of embryos and hatching of larvae. (B) Selective tidal stream transport (STST) migration into and settlement of megalopa larvae. From Forward et al. (2003), with permission from the Bulletin of Marine Science.

reaches of the estuary in late spring to early fall, and it is only the inseminated females which, during the fall, make a down-estuary migration taking three to four months and covering 150–300 km (phase I; Fig. 8.7A). In the lower estuary, the females overwinter in "staging areas" close to the estuary mouth. In the spring, females begin spawning (extruding and attaching fertilized eggs) and embryo incubation, which takes about two weeks (phase II; Fig. 8.7A). Females with embryos near hatching will then migrate, aided by nocturnal ebb tide transport, to the mouth or just outside of the estuary to release the larvae into high-salinity coastal waters. After hatching, females may reenter the lower estuary, assisted by nocturnal flood tide transport, to oviposit one or more additional broods during the same reproductive season, again with seaward migration for hatching larvae.

The zoeal larval stages develop in oceanic water into the settling stage, the megalopa. Megalopae are transported by wind-driven currents (onshore and along-shore currents with Ekman flow currents) toward the mouth of estuaries (Ogburn et al. 2012). Apparently influenced by chemical cues emanating from the estuary, their previous circadian activity rhythm of diel vertical migration changes. The increase in salinity associated with flood tides stimulates them to move up into the water column and swim with the flood, but only at night (Fig. 8.7B; Forward et al. 2003). Turbulence in the flood tide promotes sustained swimming before they move to the bottom during slack water just after the flood tide. This behavior allows them to enter and move up into the estuary using nocturnal flood-tide transport. Once encountering suitable nursery habitats such as seagrass meadows or other structured habitats (Kennedy and Cronin 2007), they settle and juvenile growth begins. With maturation, the juveniles move up into the less saline upper estuary to continue growth to adulthood.

Another group of migratory crabs that are estuary-dependent but much more closely tied to freshwater are the Chinese and Japanese mitten crabs (Grapsoidea: Varunidae), *Eriocheir sinensis* (Korea to south China), and *E. japonicus* (Japan and Taiwan). Most work has focused on *E. sinensis,* which has invaded rivers far outside its native range to Europe, western Asia, and the US west coast (Dittel and Epifanio 2009). Besides possible displacement of native fauna, this species' major deleterious effects on invaded river systems includes damage to stream banks, earthen dikes, and levees caused by its burrowing activities. The mass migrations of adults moving downstream to reproduce in estuaries affect fishing activities by clogging nets and traps.

E. sinensis is catadromous, with older juveniles and adults living in rivers for most of their lives. Depending on geographic location, the juvenile and adult phases in freshwater may last 2–5 years. The synchronized mass migration of adults down to estuaries occurs in late summer and fall after the upriver female puberty molt, with subsequent mating and spawning of embryos (Fig. 8.8). This migration may cover hundreds of kilometers over a period of some months. Once in the estuary, females incubate the embryos for about two months, releasing larvae in higher salinity areas of the estuary, which then move out to sea, presumably by STST. The resulting megalopae, unlike the second to fifth zoeal stages, are euryhaline. They reenter estuaries, presumably by processes similar to those described above for blue crabs. The megalopae settle in lower salinity areas of the estuary, where the young juvenile crabs grow until they begin their migration upstream (Fig. 8.8). The juvenile migration is not well described but appears to be a dilute dispersal upstream rather than a mass movement as in downstream migrating females. Although primarily aquatic, mitten crabs can and do leave the river to migrate over and around obstacles in the path of their downstream or upstream migrations.

Very different types of seasonal and reproductive migrations occur in the fully marine snow crab (*Chionoecetes opilio*), a stenohaline and stenothermal brachyuran (Majoidea: Oregoniidae). It inhabits boreal and arctic seas, supporting important fisheries in the Bering Sea and along the eastern Canadian coast. The basic reproductive biology is important in understanding the crab's migrations (Sainte-Marie et al. 2008). Males and females become fully mature after a terminal (puberty) molt. "Pubescent females" near the terminal molt are attractive to males but do not mate

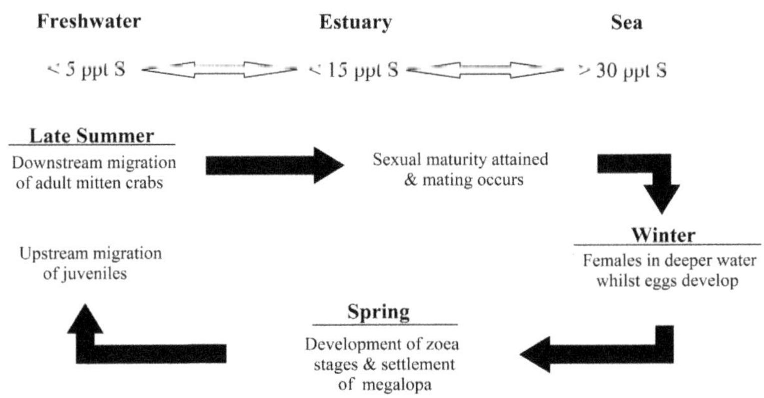

Fig. 8.8.
Life cycle migration of *Eriocheir sinensis,* a catadromous and estuary-dependent brachyuran crab. Modified from Crosier and Molloy (2004).

until after they molt. "Primiparous" females are those that have spawned their first brood; those with subsequent broods are termed *multiparous*. The distinction is important, because both females and males of different age and reproductive ages may have distinct migration patterns, resulting in a complex seasonal distribution over the continental shelf. Movements during these migrations have been traced using tagging of adults together with estimates of relative age, using a shell condition index (hardness, fouling) calibrated with radiochemical methods (Ernst et al. 2005, Nichol and Somerton 2015). Adults are long-lived, perhaps with a life span of 15 years or more (Sainte-Marie et al. 2008).

Both males and females undergo an ontogenetic migration over a period of years, with older individuals moving deeper and further out on the continental shelf. For females, this appears to be a one-way movement beginning after the puberty molt. Mating occurs during the winter (February-March) and incubation of embryos follows for about one year, with hatching of larvae over the shelf beginning in April of the next year. The older and larger males, occurring in the deepest shelf depths (> 100 m), may seasonally return to somewhat shallower depths (50–100 m) to mate with the older multiparous females. Smaller mature males in shallower depths may make more extensive inshore migrations to mate with pubescent females (after the puberty molt) and primiparous females, which occur at lesser depths (10–50 m) than multiparous females. Distances up to 100 km may be traveled during such migrations. Migrating individuals appear to track the narrow range of preferred bottom temperatures (−1°C–4.5°C), which move further inshore during the winter and offshore (and deeper) during the spring-summer months.

In addition to the ontogenetic and reproductive (mating) migrations of adult snow crabs, immature crabs make massive inshore migrations to shallow subtidal areas in the winter to molt, grow, and mature (Sainte-Marie and Hazel 1992, Lovrich et al. 1995). Other benefits of this inshore migration may include escape from higher predation pressures from offshore predators and cannibalism from larger crabs that occur at greater shelf depths. Additionally, smaller adult males may mate for the first time. This group of snow crabs moves offshore in late spring and summer as water temperatures become too warm (> 4.5°C) for this cold-water species.

Seasonal migrations occur in some temperate-water brachyurans but with the seasons opposite those of snow crabs. In *Callinectes sapidus*, juveniles and adult males overwinter in deeper channels in Chesapeake Bay, whereas in *Maja brachydactyla* (occurring along the Atlantic coast of Spain), only adults move from shallow kelp forests to deeper (10–40 m) waters for the winter, with return in both species to shallows for the summer (Hines et al. 1995). Similarly, both juveniles and adults of *Carcinus maenas* (Wadden Sea) move from intertidal mud flats to sublittoral waters to overwinter (Thiel and Dernedde 1994).

MARINE ANOMURAN CRABS

Reports of migrations, as defined for this review, are not as extensive in anomurans as for brachyuran crabs, although this may due to less study about them. However, the seasonal migrations of the red king crab *Paralithodes camtschaticus* have been described by Stone et al. (1992). This cold-water crab occurs in the Bering Sea (North Pacific), where it is the target of a commercial fishery. During the summer and fall, adults stay in feeding grounds in cooler, deeper (40–60 m) waters. There is a synchronous mass movement of individuals to shallower depths (20–30 m) during the winter where the water cools to preferred water temperatures (3°C–8°C). The population makes a gradual movement toward deeper water but returns to shallow water between March and May for molting, mating, and egg oviposition by females. In late spring, there is a gradual movement back to deeper waters to spend the summer and early fall as shallow waters warm. In this species, the adaptive value of the seasonal migration to and from shallow waters appears to allow king crabs to maintain themselves in different feeding grounds of optimal temperature during different parts of the year, as well as to take advantage of spring warming in shallow water to molt for growth and, in females, to mate.

Anomurans in the superfamilies Galatheoidea and Chirostyloidea are dorsoventrally flattened with the abdomen curled below the abdomen ("squat lobsters"). Although basically benthic in both deep-sea and continental shelf habitats, their mass aggregations and, in some species, vertical migrations as pelagic swarms, have attracted much attention. Lovrich and Thiel (2011) reported ontogenetic migrations in some species. In estuarine and fjord species, STST is utilized to carry newly hatched larvae out to sea (ebb tide) for development and the return of postlarvae (flood tide) for recruitment. In continental shelf species such as the pelagic red crab *Pleuroncodes planipes*, however, megalopae recruit after planktonic development to specific shelf nursery grounds in which sulfide microbial communities associated with the oxygen minimum zone provide food for the growing juveniles. With maturity, juveniles move away from these areas to more offshore locations or parallel to the coast, depending on the local topography.

Seasonal reproductive migrations have been reported in some squat lobsters such as *P. planipes*, *P. monodon*, and *Galathea squamifera* (Thiel and Lovrich 2011). In these continental shelf species, ovarian maturation occurs in the fall, with subsequent mating and then incubation of embryos during the winter, with hatching in the early spring. These species occupy deeper waters during the summer, migrate to shallow waters during the fall, and stay during the winter when females are incubating embryos. Larvae are released in the spring during the seasonal upwelling and spring bloom. The onshore movement of preferred bottom water temperatures may be the proximate factor stimulating the move onshore, but the adaptive value of the migration is that incubating females can hatch larvae when and where optimal feeding conditions will occur.

TERRESTRIAL CRABS

Terrestrial crabs are those active in air and more or less independent of aquatic habitats, and they include brachyuran and anomuran species (Hartnoll 1988). Members of the brachyuran family Gecarcinidae and the anomuran family Coenobitidae (land hermits) are the most terrestrial by this definition, and these species make dramatic mass migrations of up to several kilometers. The adaptive value of these migrations is the hatching (release) of larvae into the sea, where extended larval development takes place, as in their marine ancestors (Wolcott 1988). Hatching migrations are synchronized to lunar cycles, with hatching around new or full moons, during which the highest and lowest tides occur, promoting larval dispersal out to sea. Gene flow among populations is thus enhanced in these species, several of which occur on isolated oceanic islands.

These migrations do impose high risks and costs, as adults are exposed to increased natural predation and human impacts such as fishing pressure in some species, vehicular mortality, and human barriers to overland migration. Desiccation is always a threat to these crabs (Bliss 1979), which live in tropical and subtropical areas. When not migrating, the danger of desiccation to brachyuran land crabs is reduced by their burrow-living habits, and the land hermits avoid it by living in shells in which water is carried. Hatching migrations generally occur during the rainy season in these species, so that the threat of desiccation is reduced or avoided (Bliss 1979). The actual hatching event takes place at night or predawn hours, perhaps to reduce desiccation and mortality from visual predators.

Herrnkind (1983) pointed out that hatching migrations in land crabs may have three phases: (1) assembly of reproductive adults at sites near the shore or at more inland sites away from the surf zone (termed *staging areas*) where mating and oviposition may occur; (2) movement to the water's edge at night to hatch embryos, releasing larvae into the sea; (3) return movement back to inland habitats. The first phase, however, is omitted in some species, with mating and oviposition occurring at inland areas before migration of ovigerous females.

One of the best-studied and most spectacular land crab migrations is that of the red crab *Gecarcoidea natalis*, endemic to Christmas Island (Indian Ocean; Fig. 8.9C). Its population density on this small island is quite high, so that its mass movements and interactions with the human population during migration have attracted much attention (Fig. 8.9A–B). Adamczewska and Morris (2001) analyzed reproductive and migratory activities using radio-tracking and mark-recapture techniques to determine the routes, destinations, and travel speeds of migrating crabs. The crabs live in the forested part of the island, with the larger males higher up on inland plateaus than younger males and females, which live nearer the coast. The reproductive migration is directly stimulated by the arrival of monsoonal rains in November and December. If the rains do not come at all, as can occur in extreme El Niño Southern Oscillation, or ENSO, events, the migration may not take place in that year (Shaw and Kelly 2013).

Stimulated by the seasonal rains, adults move down during the early morning and late afternoon to the lowest coastal terraces (staging areas) near the shore, where burrowing, mating, oviposition, and incubation of embryos take place. Tagged crabs moved in straight lines, rather than following possibly easier but longer routes. The distances covered can be up to 4–5 km and take 1–2 weeks. However, if seasonal rains are late, the migration is rushed, as reproductive activities and larval release are tied to lunar cycles (Fig. 8.10). After long periods of relative inactivity during the

Fig. 8.9.
Reproductive migrations by the red crab *Gecarcoidea natalis* on Christmas Island (Indo-Pacific). (A) Mass migration of red crabs toward the coast. (B) Red crabs crossing a road during migration. (C) Red crab portrait. (D) Females descending rocky sea cliffs to hatch out larvae. Photographs courtesy of Allison Shaw ©, from a research project funded by the National Geographic Society/Waitt Grants program. See color version of this figure in the centerfold.

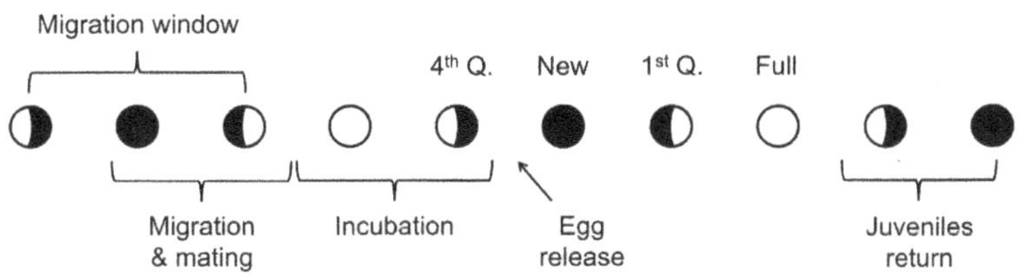

Fig. 8.10.
Schedule of *Gecarcoidea natalis* (red crab) reproductive migrations with respect to the lunar cycle: larvae are released a few days before the new moon; development takes about one month, with megalopae returning to land. Three to 5 weeks before the new moon, reproductive adults migrate to the coast, and females mate and then incubate their embryos for two weeks prior to hatching and larval release in the sea ("migration window"). From Shaw and Kelly (2013), with permission from John Wiley and Sons.

dry season, crabs must feed to meet the physiological demands of migration and reproduction. In hurried migrations, feeding time is reduced (Adamczewska and Morris 2001).

After mating, males move back inland, but females stay to incubate embryos in burrows for about 2 weeks. A few days before the new moon, they move down to the supralittoral rocky cliffs that surround the shore to release their embryos (Fig. 8.9D). During predawn hours, females descend the cliffs near the water's edge, clinging tightly to the rocky substratum, vigorously shaking the body and fanning the abdomen as waves wash over them, hatching and releasing larvae into the water. These activities occur a few days before the new moon as the high tide is turning, so that the following low tide will take larvae offshore toward open ocean waters. After hatching their larvae, the females return to their usual forest habitat. Larval development takes 3–4 weeks, after which the megalopae (settling stage) come ashore, molting and growing as juveniles, and within a few weeks are found in the adult forest habitat (Wolcott 1988).

Cardisoma guanhumi is a very large land crab inhabiting coastal areas of the tropical and subtropical West Atlantic and Caribbean. To date, the most extensive study on its ecology and reproductive biology is that by Gifford (1962). These burrow-living crabs inhabit mangrove fringes, swampy areas, riverbanks, and open fields as much as 5 to 8 km inland, provided that a source of surface or near-surface water is available. As in *G. natalis* and other land crabs, the reproductive and migratory activities occur during the rainy season (May to December in south Florida, where the Gifford study took place). Unlike *G. natalis*, there is no movement to staging areas next to the coast for mating and oviposition; these activities occur inland. Migrations and larval release are tied to monthly lunar cycles, with most hatching a few days before and after the full moon, with, curiously, much less around the new moon. Ovigerous females migrate toward the water within a few days of approaching the full (usually) or new moon to be at water's edge when the associated high tides favorable for larval release occur. After hatching, batches of females migrate back into inland areas.

The migration of the small land crab *Epigrapsus notatus* is similar to *C. guanhumi* in that there is no prehatching movement to shore staging areas, so mating and oviposition occur in the coastal forests of Taiwan where marked crabs were followed and studied (Liu and Jeng 2005). These crabs are quite cryptic and have very little activity outside of burrows until the rainy season. Unlike other land crabs, the hatching migration of the *E. notatus* occurs at the end of the rainy season. The late migration is thought to allow the crabs enough time to forage and accumulate energy stores as their activity during the rest of the year is very limited. The reproductive season is limited to only two months, September and October, and hatching is restricted to a few days after the full (but not new) moons of those months. The actual distance of migration from the coastal forest to near the shore by ovigerous females may be from only a few to as much as several hundred meters. At night the females move near the water and then wait for a few minutes to several hours to enter the water, as Liu and Jeng (2005) suggest, to wait for embryo readiness to hatch. Another alternative is that females are waiting for the best tidal stage for hatching and larval dispersal. Unlike other land crabs, *E. notatus* females move into and hatch larvae in the landward end or middle of surge channels that penetrate the fossil coral reef littoral of the region. Liu and Jeng (2005) hypothesized that these small crabs are not as well adapted for clinging to rocks during hatching as other land crabs and that at the bottom of these surge channels, risk of being swept away by surf is less. However, hatching in a somewhat confined space attracts fish predators of the larvae, so the reduction in mortality risk to adults is accompanied by increased predation on larvae before they are swept out of the channels into the open sea. After hatching, females return to their normal forest habitat.

The land hermit crabs have similar geographic and habitat distributions in tropical areas as their brachyuran counterparts and must also migrate to the sea to release their marine planktonic larvae. The two most important species in terms of abundance, distribution, and knowledge about their biology are *Coenobita clypeatus* ("land hermit" or "soldier" crab) in the tropical West Atlantic and *Birgus latro* ("coconut" or "robber" crab) of the Indo-West Pacific.

De Wilde (1973), using a variety of marking and tagging methods and experiments, described in detail the reproductive migrations of *C. clypeatus* on the Caribbean island of Curaçao. The inland populations may be found several kilometers from the coast in semiarid areas with shade and a source of surface water. These hermit crabs face similar problems with desiccation as do brachyuran land crabs but, instead of living in burrows, they live within a gastropod shell in which water is maintained to moisten the abdomen, respiratory surfaces, and incubated embryos. The reproductive season (summer through fall, especially September and October) overlaps with the relatively weak rainy season. In the Curaçao populations studied, males and females migrated in large numbers from the eastern part of the island to a few preferred staging areas (phase 1) near to but not adjacent to rocky shores along the south coast where larval release would later take place. These staging areas are characterized by the presence of daytime shelters (rock crevices, tree roots), drinking water, and a nearby sheltered and accessible shore. Such staging areas are near natural brackish water lagoons not connected to the sea, or near watering places for livestock.

Reproduction and the final hatching migration (phase 2) are synchronized by lunar cycles. Mating and oviposition take place around the full moon in the staging areas. Incubation takes 3–4 weeks; about a week before the next full moon, females with prehatching embryos migrate at night from staging areas down to the rocky shores to deposit and hatch out the embryos. Females carefully descend rocky cliffs to near the water line and attach masses of embryos to wetted areas below (dead coral, rock substrata, algae). After egg deposition, females return to inland habitats. The deposition of embryos onto the littoral substratum is quite distinct from brachyuran land crabs, which enter the water and use shaking and abdominal fanning to hatch and disperse larvae. To attach its embryos, female *C. clypeatus* reach back into the embryo mass and remove a portion with the small chelae of the fifth (last) pereopods. The female passes the small embryo mass to the maxillipeds, which mold the embryos into a ball that is then transferred to the tips of the large chelipeds. The female then flicks or tosses the embryo ball up to 20 cm onto the wet rocks below, where it adheres. The embryos hatch spontaneously as the rising tide immerses them in seawater. This unique behavior appears to be necessary to expose the embryos to seawater; otherwise the hermit crab would have to leave the shell to immerse the entire abdomen and embryo mass into the water. Additionally, the soft nonmuscular abdomen and reduced pleopod structure of hermit crabs are incapable of brachyuran hatching behaviors. After egg deposition, females make diffuse migrations back to inland areas, as do males after the mating and spawning season have terminated. Larval development in the sea is approximately one month, after which the settling stage (glaucothoe) arrive along the shore. Young growing juvenile crabs gradually move into the inland habitat.

The coconut or robber crab *Birgus latro*, although a paguroid hermit crab, only lives within a shell for a brief period after settlement and growth to about 1 cm (Drew et al. 2010). As it grows to a very large size (up to 11 cm thoracic length and 500 g, the largest terrestrial arthropod), it acquires the thickened and calcified exoskeleton, both cephalothorax and abdomen, which allows it to resist desiccation and live shell-free. Its reproductive migrations are still not well known (Drew et al. 2010) but are generally similar to those of *C. clypeatus* in that the movement of males and females from inland habitats to the coast (Krieger et al. 2012) and reproductive activities are synchronized with lunar cycles. Larval release is associated with the new, rather than full moon. As in *C. clypeatus*, mating and oviposition occur in staging areas near the shore, and females only produce one brood per year. Approximately 1 month after incubation, embryos are ready to hatch, and, quite unlike embryo deposition in *C. clypeatus*, females enter the shore wash zone, uncurl the abdomen, exposing the embryos to seawater, which stimulates them to hatch. Larval development, settlement, early growth, and movement to inland areas are similar to *C. clypeatus*, including the diffuse migration back to inland areas after females hatch their embryos, and when males no longer have mating opportunities.

The orientation and navigational cues and mechanisms used by land crabs are poorly known, as Wolcott (1988) reported, and little has changed since that time. The brighter seaward horizon from sea-surface reflection of moon and starlight is often cited as a cue for land crabs migrating at night. As they near the coast, these crabs may be guided by vibrations transmitted through the ground generated by crashing surf (e.g., Bliss 1979). The straight-line migrations of the Christmas Island crab *G. natalis* suggested some sort of magnetic compass migration to Adamczewska and Morris (2001). Other authors, such as de Wilde (1973) and Krieger et al. (2012), proposed that memory of migratory routes, as well as and learning by younger migrators from older experienced crabs, may play a role in navigation to staging and hatching areas.

PERACARIDA

Seasonal and life cycle migrations have been studied in some species of peracaridans (e.g., mysids, isopods, amphipods). The horizontal extent of peracaridan migrations is limited compared to the decapod cases described above. This disparity is partly a question of scale, because most peracarids have small body size compared to decapods, resulting in more limited capacity for horizontal movements. Additionally, all peracaridans have abbreviated development, resulting in the release of embryos of juveniles into the same habitat as the adult, without a series of planktonic stages, as in decapods. The most extensive peracarid migrations studied are made by mysids, which is not surprising given their shrimp-like body form with enhanced swimming abilities compared to more benthic peracarids. Some amphipods certainly have the locomotory power to make migrations, but studies to date indicate dispersal, often long range, is the function of such movements (Bringloe et al. 2013, Thiel 1998), not a true migration. However, in studying circadian and circatidal rhythms in *Corophium volutator*, Harris and Morgan (1986) found it feasible that these amphipods may make semilunar and seasonal migrations up and down the estuarine shore, in addition to their better-known tidal migrations.

San Vicente and Sorbe (2013) reviewed horizontal movements of mysid species. In some cases, they are limited to seasonal movement from shallower to deeper parts of the surf zone, whereas in others there is a more extensive movement from coastal and estuarine habitats to deeper water for more favorable temperature conditions or reproduction. In the latter case, the resulting juveniles move back to coastal areas during warmer months for growth and maturation. Suzuki et al. (2009) describe a life cycle migration in *Hyperacanthomysis longirostris* in the Chikugo River estuary in Japan. Brooding females release juveniles in the upper estuary, which grow as they drift downstream; they later migrate back up to the upper estuary for reproduction.

Although primarily benthic, free-living isopods may be active swimmers. The term *migratory* has been used to describe life cycle movements in some species. However, Eltringham and Hockley (1961) pointed out for wood-boring *Limnoria* species, such movements are for dispersal and not a true population migration. Likewise, kelp-holdfast *Limnoria* spp. abandon detached floating holdfasts and actively swim to young attached holdfasts to maintain themselves in the kelp forest; this behavior also serves as a dispersal mechanism (Miranda and Thiel 2008). On the other hand, in the littoral zone two species of *Dynamene* species do show a life cycle migration (Holdich 1976). Reproductive adults (a male and harem of several females) occupy empty barnacle shells or crevices in the upper intertidal zone. Females hatch juveniles into the plankton, and the latter settle on algae in the intertidal zone to live and grow. On reaching sexual maturity later in the late summer, they swim or crawl up to the reproductive habitat.

CONCLUSIONS AND FUTURE DIRECTIONS

As in other major groups of animals, both terrestrial and aquatic, many decapod and peracaridan crustaceans have evolved life cycle migrations in which young and adults move to and from spatially separated habitats, each advantageous to a particular ontogenetic stage for feeding and growth, reproduction, and dispersal. In many shrimps, lobsters, and crabs, the life cycle begins with adults hatching planktonic larvae for development into open ocean waters, with juveniles exploiting productive estuaries or sublittoral habitats. Behaviors have evolved in postlarvae and juveniles of such species to utilize STST to move into and out of, respectively, estuaries. In amphidromous freshwater shrimps, larvae are delivered to the sea by downstream migrating females or stream drift of larvae hatched out upstream. After development, the postlarvae must encounter a river mouth or stream and then make mass migrations upriver to adult habitats. Terrestrial crabs with planktonic larval development march overland or down river to release larvae into the sea and then return inland, as must newly metamorphosed juveniles arrive from the sea.

Seasonal migrations occur in the some subtropical and temperate near-shore crustaceans discussed above, with movement of benthic stages offshore in fall and winter to deeper waters of more stable or warmer temperature, with lower exposure to wave action from winter storms. These groups include some decapod shrimps, lobsters, crabs, and mysids (Fig. 8.11A). In some high-latitude species (snow, king crabs), the movement is the reverse (i.e., onshore in the winter, offshore in summer) to follow preferred colder temperatures in deeper waters (Fig. 8.11B). In some cases, these seasonal movements coincide with reproductive activities (e.g., molting and mating, larval release), so that they are not strictly distinct from life cycle migrations.

Studies on migrations, as defined in this review (exclusive of vertical and strictly tidal migrations), have been focused on decapods and peracarids. In smaller mobile species from a variety of crustacean taxa, life cycle and seasonal migrations may be discovered if the relative scale of species body size and horizontal distances traversed are considered. Direct observation and analysis of migratory movements and behavior of large, shallow water spiny lobsters was made possible by scuba studies. Similar in situ observations need to be made on other benthic crustaceans during migrations using scuba gear and the increasingly sophisticated monitoring equipment (e.g., surveillance video, remote underwater vehicles, radiotelemetry, tagging with global positioning systems).

There are still many questions about various processes utilized in the migrations considered here. Some of these are group-specific: after larval development in the open oceans, how do the postlarvae of amphidromous shrimps arrive or find the mouths of rivers and streams that they will need to enter to continue the life cycle? Aside from STST in larval and postlarval migrations in estuary-dependent species, virtually nothing is known about the navigational mechanisms used in crustacean migrations, with only a few enlightening studies (e.g., magnetic compass sense in spiny lobsters; Boles and Lohmann 2003). Many navigational cues and processes have been suggested, including polarized light, horizon brightness of the sea surface, magnetic compass sense, and vibrations from surf in land crabs, but both field and experimental studies are greatly needed to propose and test hypotheses such mechanisms.

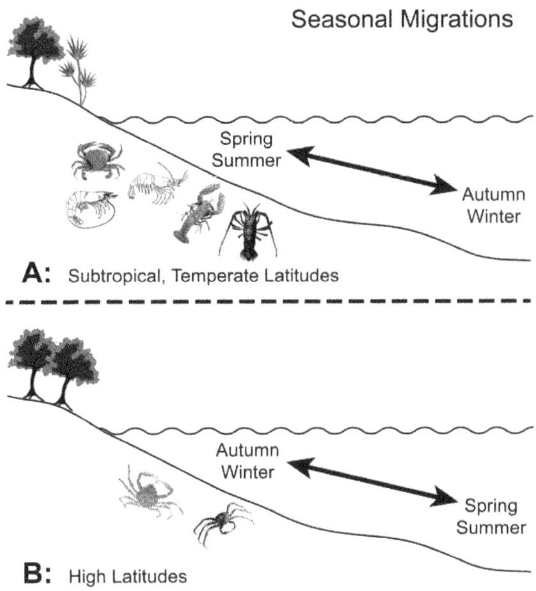

Fig. 8.11.
Seasonal migrations and latitude. (A) In neritic species from subtropical and temperate seas, all or part of the populations may move into deeper, more temperature-stable, calmer waters in the autumn and winter, returning to shallow, warmer depths in the spring and summer for growth and reproduction. (B) In some high-latitude crabs (e.g., snow; Brachyura: Majiidae) and king crabs (Anomura: Lithodidae), overwintering occurs in shallow waters where the crabs' preferred very low water temperatures occur, with the population moving with these temperatures into deeper waters of the continental shelf in late spring and summer.

ACKNOWLEDGMENTS

A number of colleagues have been very helpful with references and images for figures of migratory crustaceans, and their input is gratefully acknowledged. Special thanks are due to William Herrnkind (spiny lobsters), Alan Covich (amphidromous shrimps), Richard Forward (blue crabs), Allison Shaw and Mark Laidre (terrestrial crabs), and the Louisiana Sea Grant program (Roy E. Kron and Jessica A. Schexnayder). Finally, many thanks are due the editors, Gary Wellborn and Martin Thiel, both for the invitation to write this chapter and also for their patience and skill in editing it. This is Contribution No. 183 of the Laboratory for Crustacean Biology, University of Louisiana, Lafayette.

REFERENCES

Adamczewska, A.M., and S. Morris. 2001. Ecology and behavior of *Gecarcoidea natalis*, the Christmas Island red crab, during the annual breeding migration. Biological Bulletin 200:305–320.

Agnalt, A., T.S. Kristiansen, and K.E. Jorstad. 2007. Growth, reproductive cycle, and movement of berried European lobsters (*Homarus gammarus*) in a local stock off southwestern Norway. ICES Journal of Marine Science 64:288–297.

Alerstam, T. 1985. Strategies of migratory flight, as illustrated by arctic and common terns, *Sterna paradisea* and *Sterna hirudo*. Contributions to Marine Science 27(Suppl):580–603.

Allen, D.M., J.M Harding, K.B. Stroud, and K.L. Yozzo. 2015. Movements and site fidelity of grass shrimp (*Palaemonetes pugio* and *P. vulgaris*) in salt marsh intertidal creeks. Marine Biology 162:1275–1285.

Allen, J.A. 1966. The dynamics and interrelationships of mixed populations of Caridea found off the north-east coast of England. Pages 45–66 *in* H. Barnes, editor. Some contemporary studies in marine science. Allen and Unwin, London, England.

Bauer, R.T. 2004. Remarkable shrimps: adaptations and natural history of the carideans. University of Oklahoma Press, Norman, Oklahoma.
Bauer, R.T. 2011. Amphidromy and migrations of freshwater shrimps. II. Delivery of hatching larvae to the sea, return juvenile upstream migration, and human impacts. Pages 157–168 *in* A. Asakura, editor. New frontiers in crustacean biology. Brill, Leiden.
Bauer, R.T. 2013. Amphidromy in shrimps: a life history pattern between rivers and the sea. Latin American Journal of Aquatic Research 41:633–650.
Bergström, B. 2008. The biology of *Pandalus*. Advances in Marine Biology 38:1–245.
Benstead, J.P., J.G. March, C.M. Pringle, and F.N. Scatena. 1999. Effects of a low-head dam and water abstraction on migratory tropical stream biota. Ecological Applications 9:656–668.
Benstead, J.P., J.G. March, and C.M. Pringle. 2000. Estuarine larval development and upstream post-larval migration of freshwater shrimps in two tropical rivers of Puerto Rico. Biotropica 32:545–548.
Bertini, G., J.A. Baeza, and E. Perez. 2014. A test of large-scale reproductive migration in females of the amphidromous shrimp *Macrobrachium acanthurus* (Caridea: Palaemonidae) from south-eastern Brazil. Marine and Freshwater Research 65:81–93.
Bilgin, S., O. Ozen, and A.S. Ates. 2008. Spatial and temporal variation of *Palaemon adspersus, Palaemon elegans,* and *Crangon* (Decapoda: Caridea) in the southern Black Sea. Estuarine, Coastal and Shelf Science 79:671–678.
Bill, R.G., and W.F. Herrnkind. 1976. Drag reduction by formation movement in spiny lobsters. Science 193:1146–1148.
Bliss, D.E. 1979. From sea to tree: saga of a land crab. American Zoologist 19:385–410.
Bodekke, R. 1976. The seasonal migration of the brown shrimp *Crangon*. Netherlands Journal of Sea Research 10:103–130.
Boles, L.C., and K.J. Lohmann. 2003. True navigation and magnetic maps in spiny lobsters. Nature 421:60–63.
Bringloe, T.T., D, Drolet, M.A. Barbeau, M.R. Forbes, and T.G. Gerwing. 2013. Spatial variation in population structure and its relation to movement and the potential for dispersal in a model intertidal invertebrate. PLoS ONE 8:e69091.
Bourgeois, M., L. Landry, J. Lightner, J. Marx, and K. Semon. 2015. Louisiana shrimp: fishery management plan. Louisiana Department of Wildlife and Fisheries, Baton Rouge, Louisiana.
Butler, M.J., R.S. Steneck, and W.F. Herrnkind. 2006. Juvenile and adult ecology. Pages 263–309 *in* B. Phillips, editor. Lobsters: biology, management, aquaculture and fisheries. Blackwell Scientific Press, Oxford, England.
Chapman, C.J. 1980. Ecology of juvenile and adult *Nephrops*. Pages 143–175 *in* J.S. Cobb, and B.F. Phillips, editors. The biology and management of lobsters, volume II. Ecology and management. Academic Press, New York.
Cook, B.D., T.J. Page, and J.M. Hughes. 2012. Phylogeography of related diadromous species in continental and island settings, and a comparison of their potential and realized dispersal patterns. Journal of Biogeography 39:421–430.
Cooper, R.A., and J.R. Uzman. 1971. Migrations and growth of deep-sea lobsters, *Homarus americanus*. Science 171:288–290.
Cooper, R.A., and J.R. Uzman. 1980. Ecology of juvenile and adult *Homarus*. Pages 97–142 *in* J.S. Cobb, and B.F. Phillips, editors. The biology and management of lobsters, volume II. Ecology and management. Academic Press, New York.
Costello, T.J., and D.M. Allen. 1966. Migrations and geographic distribution of pink shrimp, *Penaeus duorarum,* of the Tortugas and Sanibel grounds, Florida. Fishery Bulletin 65:449–459.
Covich, A.P. 2009. Freshwater ecology. Pages 343–347 *in* R.G. Gillespie, and D.A. Clague, editors. Encyclopedia of islands. University of California Press, Berkeley, California.
Criales, M.M., M.B. Robblee, J.A. Browder, H. Cardenas, and T. Jackson. 2011. Field observations on selective tidal stream transport for postlarval and juvenile pink shrimp in Florida Bay. Journal of Crustacean Biology 31:26–33.
Criales, M.M., M. Laurent, L.M. Cherubin, and J.A. Browder. 2015. Modeling larval transport and settlement of pink shrimp in south Florida: dynamics of behavior and tides. Marine and Coastal Fisheries: Dynamics, Management, and Ecosystem Science 7:148–176.

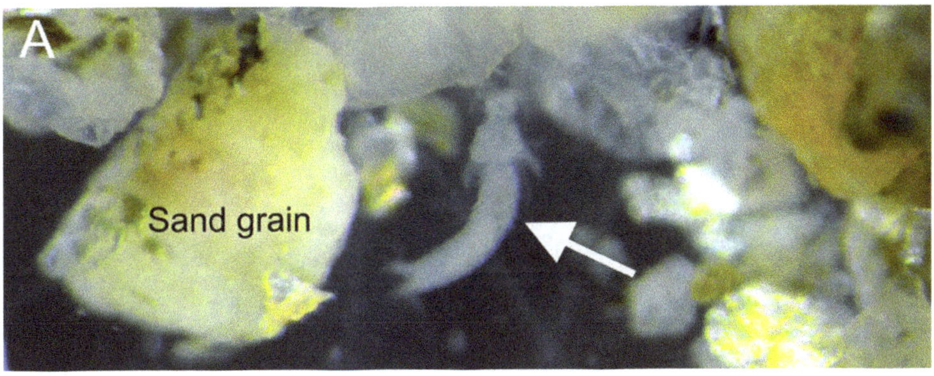

Fig. 1.3.
Life cycle and larval stages of Mystacocarida and Branchiopoda. (A) Adult *Derocheilocaris remanei* (Mystacocarida) among sand grains from beach at Canet Plage near Perpignan, France.

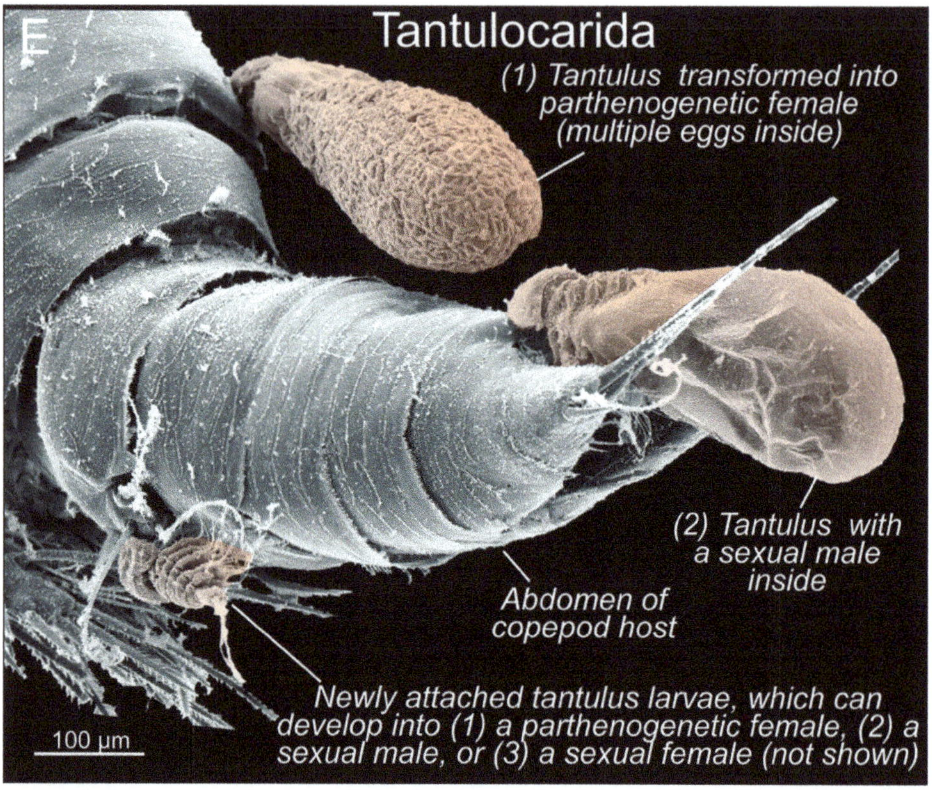

Fig. 1.9.

Life cycle and developmental stages of Cirripedia and Tantulocarida. (E) Three life stages of *Arcticotantulus kristensenii* (Tantulocarida) attached to hind body of copepod host (tantulocarids shaded in red).

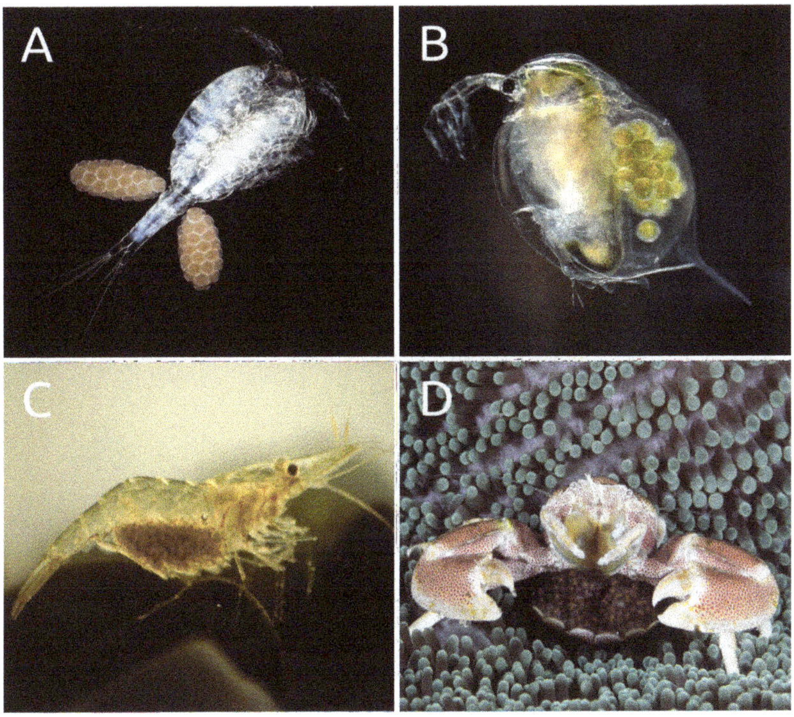

Fig. 3.2.

Photographs of egg-bearing crustaceans in the classes Maxillipoda (A), Branchiopoda (B) and Malacostraca (C, D). (A) Copepod with egg sacs in a sample from Surrey Bend Regional Park (British Columbia, Canada). Photograph by Waldo Nell ©, https://www.flickr.com/photos/pwnell/16119109181/. (B) Female cladoceran (*Daphnia magna*) with eggs. Photograph by Hajime Watanabe, under Creative Commons license (BY), doi:10.1371/image.pgen.v07.i03. (C) Decapod shrimp (*Palaemonetes pugio*) carrying eggs. Photograph by Brian Gratwicke, under Creative Commons license (BY), https://commons.wikimedia.org/wiki/File:Palaemonetes_pugio.jpg. (D) Berried porcelain crab (Decapoda: *Neopetrolisthes maculatus*). Photograph by Klaus Stiefel, under Creative Commons license (BY-NC), https://www.flickr.com/photos/pacificklaus/17369531096/.

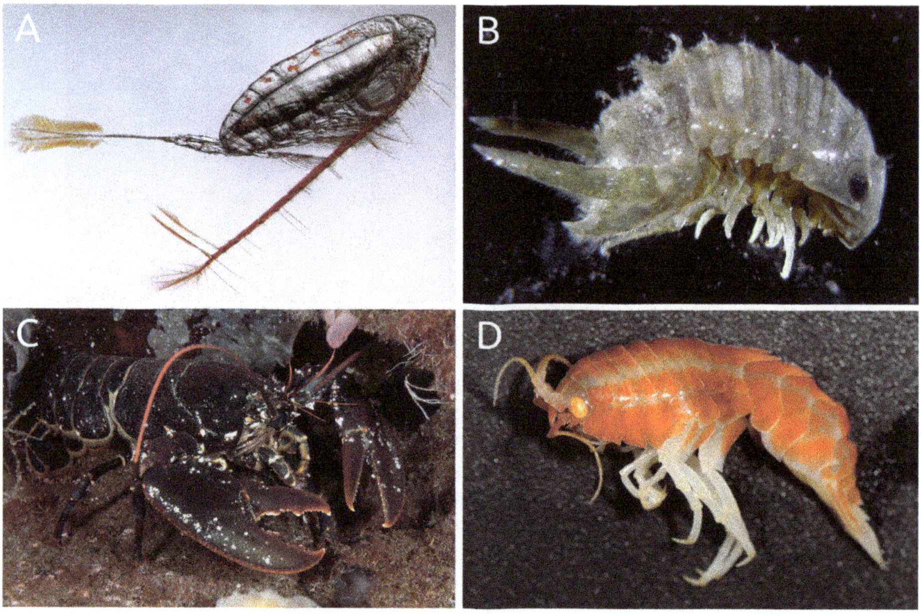

Fig. 4.1.
Photos of crustacean representatives with life history strategies ranging from strict semelparity to long-lived species with iteroparity. See Fig. 4.3 for a schematic representation of the parity continuum. (A) Semelparous copepods of the genus *Neocalanus* spp., here represented by *Neocalanus plumchrus*. These copepods do not develop feeding appendices in their mature stage and rely on energy reserves for reproduction, with death following reproduction (see also Fig. 4.4). Note the well filled oil sac, a large energy reserve. Photograph by Ross Hopcroft ©. (B) In the isopod *Paracerceis sculpta* the females are strictly semelparous and die after one reproductive event, whereas males mate with multiple females and live longer. Depicted here is an "alpha" male known to attract and guard females from other males. Photograph by Alice Lodola ©. (C) Many lobsters and crabs, here represented by *Homarus gammarus*, are iteroparous. In these taxa iteroparity is often combined with indeterminate growth and long lifespans. Fecundity then typically increases with age and body size. Photograph by Erling Svensen/UWPhoto ©. (D) Many amphipods are semelparous, but some also represent the very iteroparous side of the continuum. *Eurythenes gryllus* is such an example of a long-lived, indeterminately growing, and iteroparous amphipod. Photograph by Armin Rose ©.

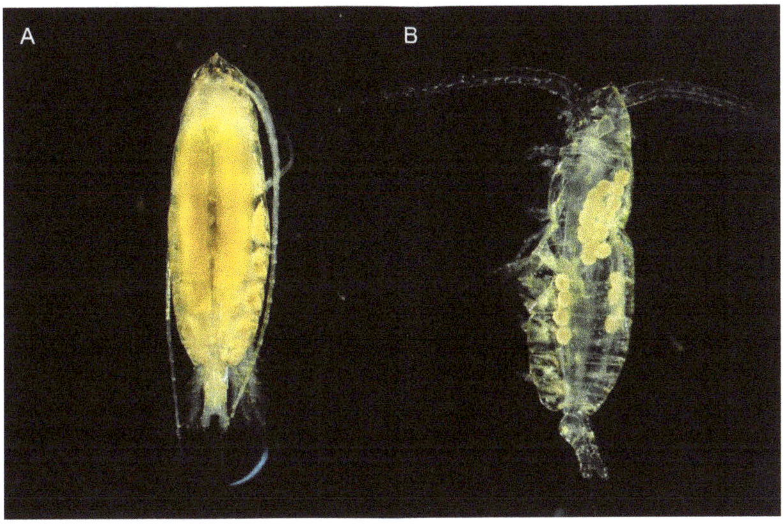

Fig. 4.4.
Two female individuals of the calanoid copepod *Neocalanus cristatus* in different reproductive stages. Individual (A) represents an early stage full of energy reserves and with some of the early eggs seen. Individual (B) is almost fully spent with only a few eggs left to be released, and little more than the exoskeleton remains (e.g., Miller et al. 1984). In this species the mature female stage does not develop feeding appendages and is unable to feed. Egg production is therefore fully based on capital breeding through stores gathered near the surface the previous summer and brought to depth where the female later develop and release eggs, and then die. For a semelparous organism it is adaptive to use all available resources for the single reproductive event, as illustrated by the spent stage in this copepod. In general, knowing that a given species cannot regain strength after reproduction (for instance because of the inability to feed) represents strong evidence for semelparity. Photographs by Toru Kobari ©.

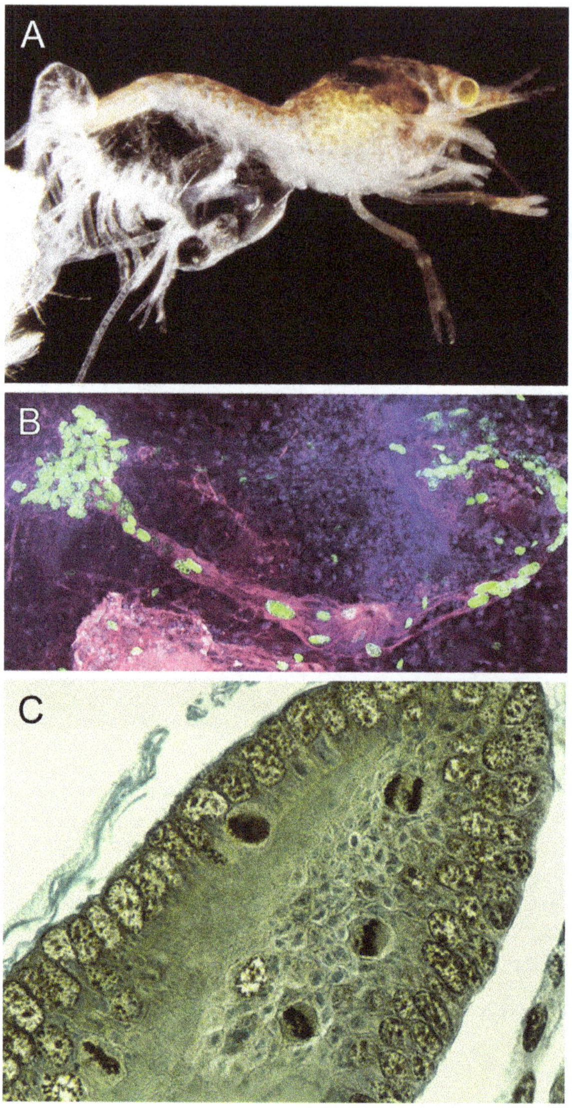

Fig. 7.5.
Anti- aging mechanisms in Decapoda. (A) Prevention of mechanical aging by molting in marbled crayfish, *Procambarus virginalis* (see Fig. 7.5A). (B– C) Activity of adult stem cells: (B) Neurogenic system in brain of red swamp crayfish, *Procambarus clarkii* (see Fig. 7.5F). (C) Stem cell niche in hepatopancreas of marbled crayfish (see Fig. 7.5D).

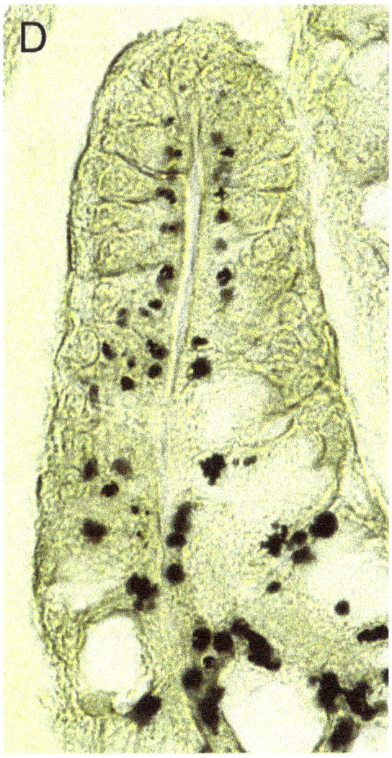

Fig. 7.6.
(D) Lysosomal detoxification of environmental copper in the hepatopancreas of giant tiger prawn, *Penaeus monodon.*

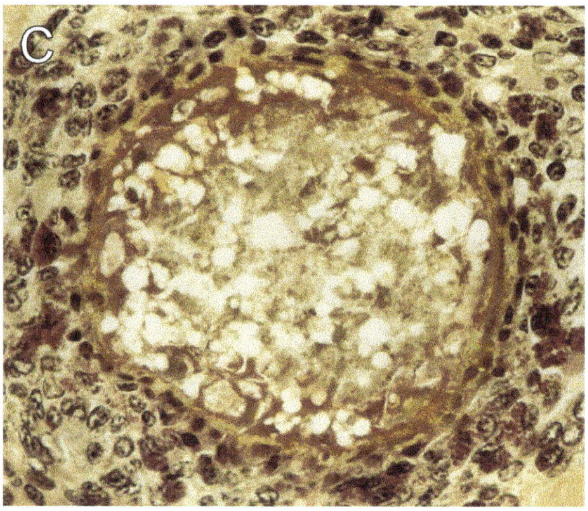

Fig. 7.7.
(C) Immune defense against bacterial infection in noble crayfish, *Astacus astacus.*

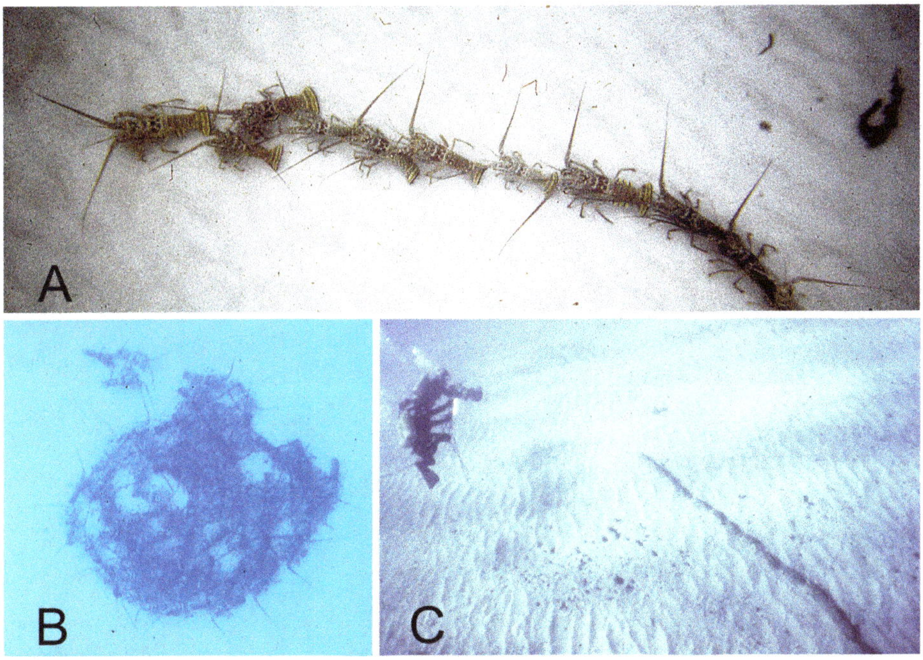

Fig 8.5.
Spiny lobster (*Panulirus argus*) migration queues. (A) Close up of a queue showing the head to tail single file formation of the migrating lobsters. (B) Defensive rosette or spiral formed by a queue when threatened by a predator. (C) Migratory queue moving over a soft bottom; diver included for perspective. Original photographs courtesy of William Herrnkind ©.

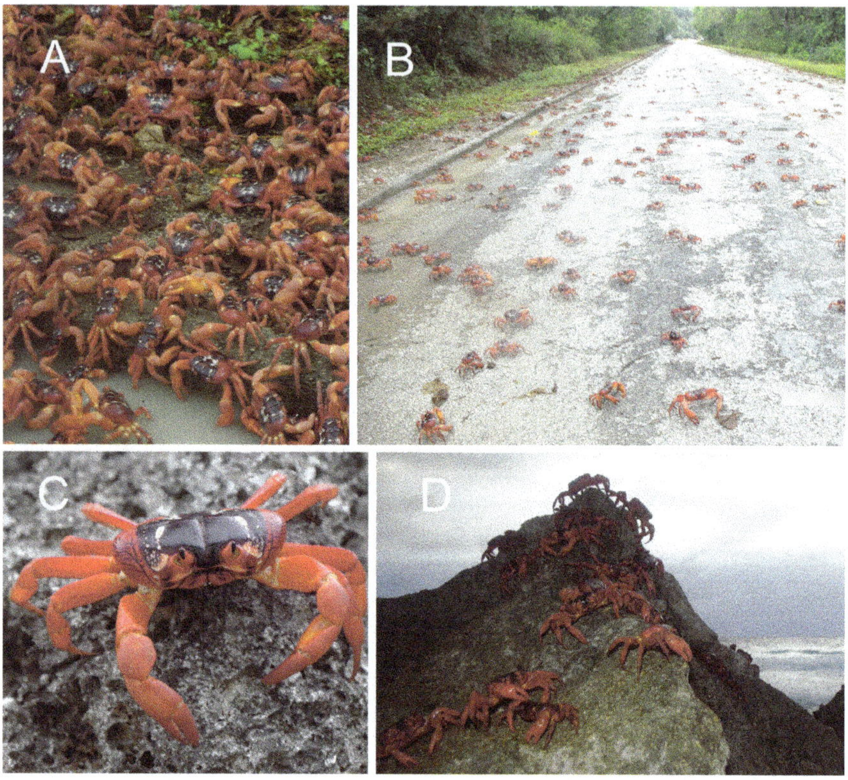

Fig 8.9.
Reproductive migrations by the red crab *Gecarcoidea natalis* on Christmas Island (Indo-Pacific). (A) Mass migration of red crabs towards the coast. (B) Red crabs crossing a road during migration. (C) Red crab portrait. (D) Females descending rocky sea cliffs to hatch out larvae. Photographs courtesy of Allison Shaw ©, from a research project funded by the National Geographic Society/Waitt Grants program.

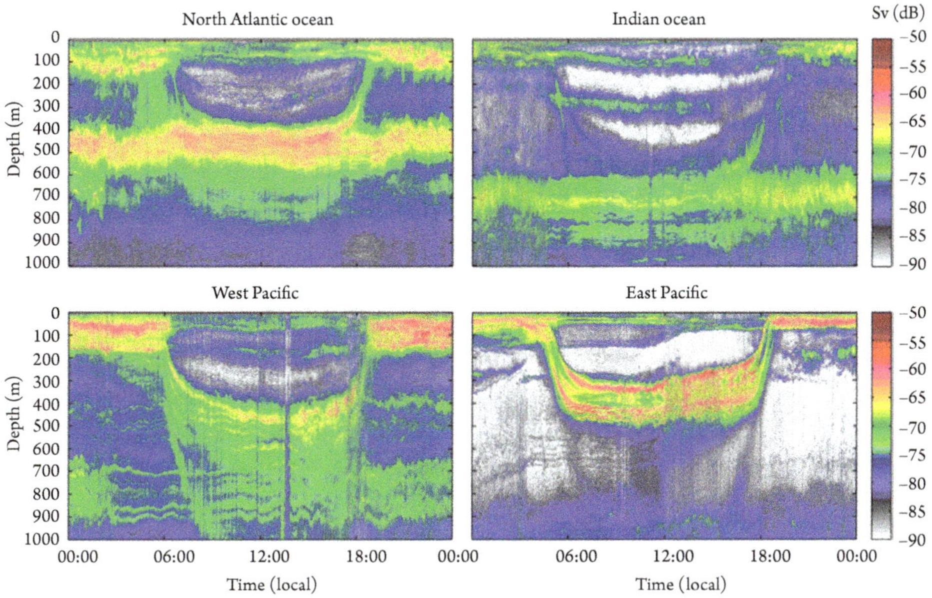

Fig. 9.1.
Examples of deep see echograms covering 24 hours periods, from different geographic regions in the mesopelagic zone of the North Atlantic ocean (upper left), Indian ocean (upper right), East Pacific (lower left), and West Pacific (lower right). Diel vertical displacements of pelagic animals forming sonic scattering layers (SSC) is evident in all areas, although proportion of migrating backscatter varies between ~20% in the Indian Ocean to ~90% in the Eastern Pacific. Most of this variability can be explained by physical properties of water masses in the localities, such as oxygen concentration, turbidity and temperature. From Klevjer (2016).

Fig. 11.1.
(A) Burrow of a coconut crab (*Birgus latro*) carved into the ground beneath the roots of a coconut tree (*Cocos nucifera*) in the Chagos Archipelago. Burrow owner is at entrance. Note substantial coconut husk located at mouth of burrow in front of owner.

Fig. 11.4.
Transportable burrow: a shell (*Nerita scabricosta*) carved out by terrestrial hermit crabs (*Coenobita compressus*). Left shows before excavation (unremodeled) and right shows after excavation (fully remodeled). The shell interior has been completely hollowed out by the crab, which has eliminated the shell columella.

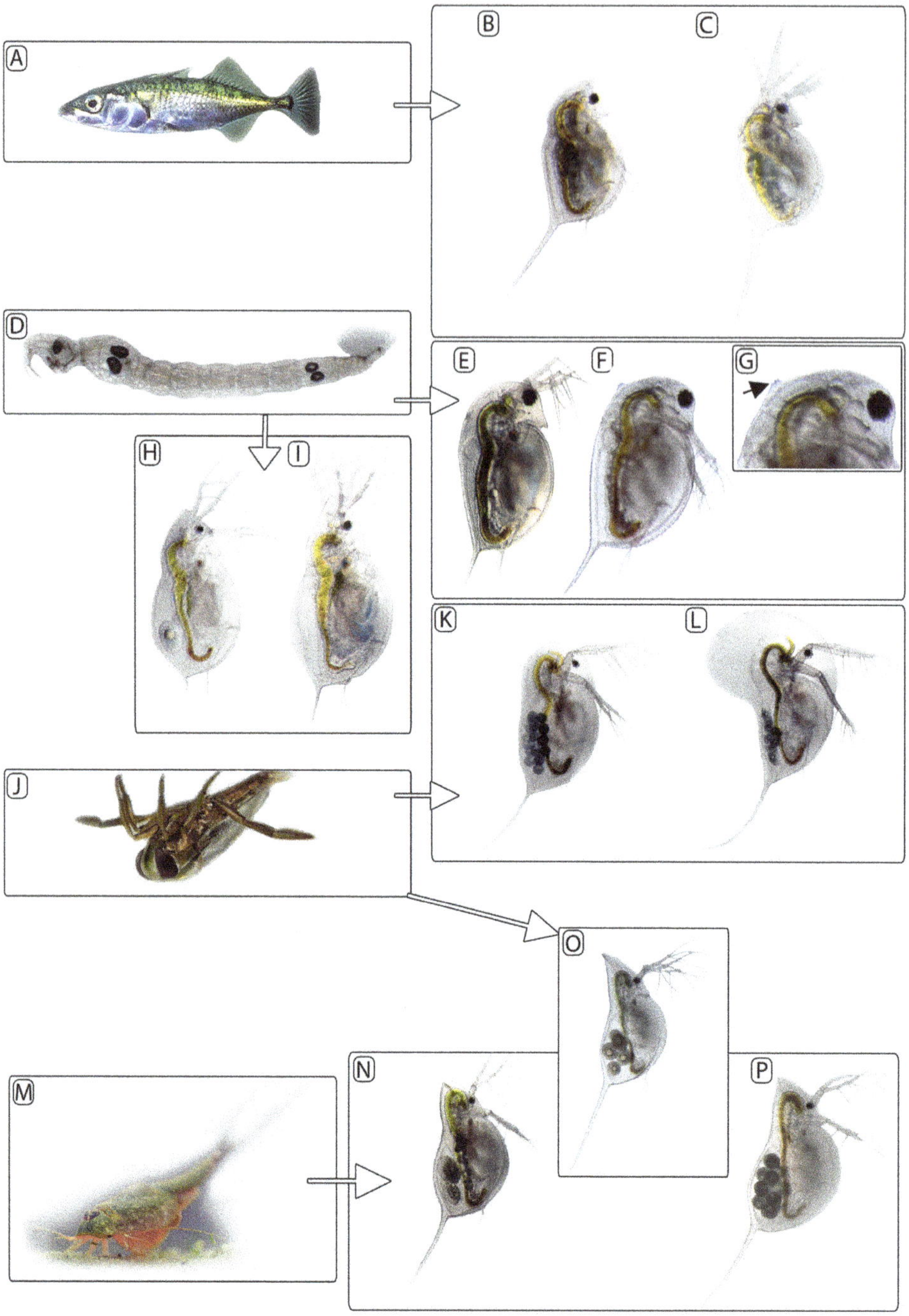

Fig. 12.2.
Predator induced defenses in different *Daphnia* species. (A) Predator *Gasterosteus aculeatus* (three-spined stickleback) that induces helmets in *D. lumholtzi*. (B) Undefended *D. lumholtzi*. (C) Defended *D. lumholtzi* with remarkably elongated head and tail-spines. (D) The invertebrate predator *Chaoborus obscuripes* commonly described to induce defenses in *D. pulex*. (E) Undefended *D. pulex* compared to (F) defended *D. pulex* carrying neckteeth in the dorsal head region. (G) Insert shows magnification of neckteeth displayed in (F). Likewise, *Chaoborus* induces helmet development in *D. cucullata*: (H) Undefended *D. cucullata* (I) Defended *D. cucullata* with helmet and elongated tail spine. (J) The backswimmer *Notonecta glauca* induces morphological defenses in *D. longicephala*. (K) The undefended *D. longicephala* morphotype is small and inconspicuous in comparison to the defended morphotype. (L) Defended *D. longicephala* grow large crests as well as elongated tail spines. (M) The ancient predator *Triops cancriformis* that induces defenses in *D. barbata*. (N) *D. barbata* (here: undefended form) develops defense modalities adapted to the predation regime. (O) *Notonecta*-defended *D. barbata* develop larger and straight helmets in comparison to (N) and to (P) the *Triops*-defended morphotype which has larger and backwards-bending helmets and tail-spines. Illustrated by Linda C. Weiss (2016) ©.

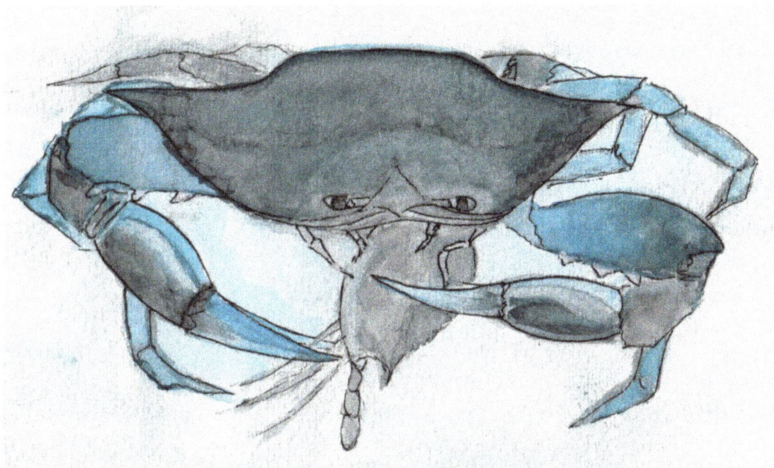

Fig. 14.1.
Intraspecific predation. An adult blue crab, *Callinectes sapidus*, kills and consumes a juvenile blue crab. Based on Smithsonian Environmental Research Center 2014. Illustrated by Bronwyn H. Bleakley ©.

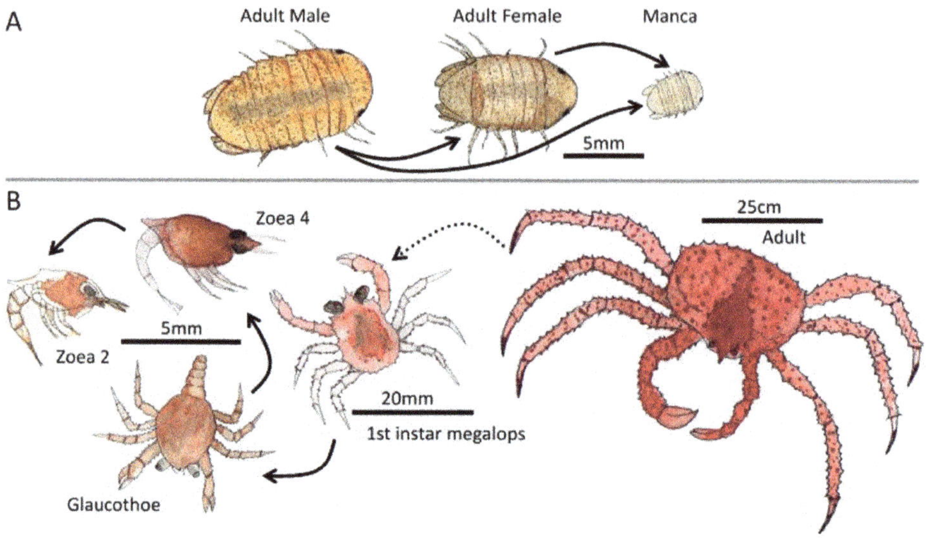

Fig. 14.3.
Size asymmetry and profitability of prey influence the incidence of cannibalism. (A) Cannibalism in Socorro isopods (*Thermosphaeroma thermophilum*), which display size-structured cannibalism. Any smaller individual is at risk, although manca (direct developing juveniles) and small females are at greatest risk of depredation. (B) Effects of size-asymmetry and profitability on cannibalism in red king crabs (*Paralithodes camtschaticus*). Smaller individuals within a developmental cohort and across close developmental classes are susceptible to cannibalism; however, adults likely do not eat larvae because extreme size differences reduce the profitability of cannibalism among those classes. Solid lines indicate common direction of attack. Dotted line indicates direction of attack when the size difference between the aggressor and victim is small enough to make the attack profitable. Illustrated by Bronwyn H. Bleakley ©.

Crosier, D.M., and D.P. Molloy. 2004. Species profiles: *Eriocheir sinensis*. Aquatic nuisance species research program. U.S. Army Corps of Engineers (Vicksburg, Mississippi), Albany, New York.

Dall, W., B.C. Hill, P.C. Rothlisberg, and D.J. Sharples. 1990. The biology of the Penaeidae. Advances in Marine Biology 27:1–489.

De Grave, S., and C.H.J.M. Fransen. 2011. *Carideorum catalogus*: the recent species of the dendrobranchiate, stenopodidean, procarididean and caridean shrimps. Zoologische Mededelingen 85:195–589.

De Grave, S., S. Cai, and A. Anker. 2008. Global diversity of shrimps (Decapoda: Caridea) in freshwater. Hydrobiologia 595:287–293.

Dennenmoser, S., M. Thiel, and C.D. Schubart. 2010. High genetic variability with no apparent geographic structuring in the mtDNA of the amphidromous river shrimp *Cryphiops caementarius* (Decapoda: Palaemonidae) in northern-central Chile. Journal of Crustacean Biology 30:762–766.

de Wilde, P.A.W.J. 1973. On the ecology of *Coenobita clypeatus* in Curaçao, with reference to reproduction, water economy and osmoregulation in terrestrial hermit crabs. Studies on the Fauna of Curaçao and other Caribbean Islands 44:1–138.

Dingle, H. 1996. Migration: the biology of life on the move. Oxford University Press, New York.

Dittel, A.I., and C.E. Epifanio. 2009. Invasion biology of the Chinese mitten crab *Eriocheir sinensis*: a brief review. Journal of Experimental Marine Biology and Ecology 374:79–92.

Drew, M., S. Harzsch, M. Stensmyr, S. Erland, and B.S. Hansson. 2010. A review of the biology and ecology of the robber crab, *Birgus latro* (Linnaeus, 1767) (Anomura: Coenobitidae). Zoologischer Anzeiger 249:45–67.

Eggleston, D.B., W.F. Herrnkind, and A.H. Hines. 2013. Behavior and ecology of mobile animals: insights from in situ observations. Pages 99–114 *in* M. Lang, editor. Research and discoveries: the revolution of science through scuba. Smithsonian Institution Press, Washington D.C.

Eltringham, S.K., and A.R. Hockley. 1961. Migration and reproduction of the wood-boring isopod, *Limnoria*, in Southampton water. Limnology and Oceanography 6:467–482.

Ernst, B., J.M. Orensanz, and D.A. Armstrong. 2005. Spatial dynamics of female snow crab (*Chionoecetes opilio*) in the eastern Bering Sea. Canadian Journal of Fisheries and Aquatic Sciences 62:250–268.

Forward, R.B., R.A. Tankersley, and J.M. Welch. 2003. Selective tidal stream transport of the blue crab *Callinectes sapidus*: an overview. Bulletin of Marine Science 72:347–365.

Garcia, S., and L. Le Reste. 1986. Life cycles, dynamics, exploitation and management of coastal penaeid shrimp stocks. FAO Fishery Technical Paper No. 203. FAO, Rome, Italy.

Gifford, C.A. 1962. Some observations on general biology of land crab, *Cardisoma guanhumi* (Latreille), in south Florida. Biological Bulletin 123:207–223.

Groeneveld, J.C., and G.M. Branch, 2002. Long distance migration of South African deep-water rock lobster, *Palinurus gilchristi*. Marine Ecology Progress Series 232:225–238.

Harris, G.J., and E. Morgan. 1986. Seasonal and semi-lunar modulation of the endogenous swimming rhythm in the estuarine amphipod *Corophium volutator* (Pallas). Marine and Freshwater Behaviour and Physiology 12:303–314.

Hartnoll, R.G. 1988. Evolution, systematics, and distribution. Pages 6–54 *in* W.W. Burggren, and B.R. McMahon, editors. Biology of land crabs. Cambridge University Press, Cambridge, England.

Hasler, A.D., A.T. Scholz, and R.M. Horrall. 1978. Olfactory imprinting and homing in salmon. American Scientist 66:347–355.

Herrnkind, W.F. 1969. Queuing behavior of spiny lobsters. Science 164:1425–1427.

Herrnkind, W.F. 1980. Spiny lobsters: patterns of movement. Pages 349–407 *in* J.S. Cobb, and B.F. Phillips, editors. The biology and management of lobsters, volume I. Physiology and behaviour. Academic Press, New York.

Herrnkind, W.F. 1983. Movement patterns and orientation. Pages 41–105 *in* F.J. Vernberg, and W.B. Vernberg, editors. Biology of Crustacea, volume 7. Behavior and ecology. Academic Press, New York.

Herrnkind, W.F., M.J. Childress, and K.L. Lavalli. 2001. Defense coordination and other benefits among exposed spiny lobsters: inferences from mass migratory and mesocosm studies of group size and behavior. Marine and Freshwater Research 52:1113–1124.

Hines, A.H., T.G. Wolcott, E. Gonzalez-Gurriaran, J.L. Gonzalez-Escalante, and J. Freire. 1995. Movement patterns and migrations in crabs: telemetry of juvenile and adult behavior in *Callinectes sapidus* and *Maja squinado*. Journal of the Marine Biological Association of the UK 75:27–42.

Hinojosa, I.A., B.S. Green, C. Gardner, and A. Jeffs. 2014. Settlement and early survival of southern rock lobster, *Jasus edwardsii*, under climate-driven decline of kelp habitats. ICES Journal of Marine Science 72(Suppl 1):i59–i68.

Holdich, D.M. 1976. A comparison of the ecology and life cycles of two species of littoral isopod. Journal of Experimental Marine Biology and Ecology 24:133–149.

Höglund, H. 1943. On the biology and larval development of *Leander squilla* (L.) forma typica De Man. Svenska Hydrografisk-Biologiska Kommissionens Skrifter Ny Serie Biologie. Band II, No. 6:1–44.

Holmquist, J.G., J.M. Schmidt-Gengenbach, and B. Buchanan-Yoshioka. 1998. High dams and marine freshwater linkages: effects on native and introduced fauna in the Caribbean. Conservation Biology 12:621–630.

Hufnagl, M., A. Temming, and T. Pohlmann. 2014. The missing link: tidal-influenced activity a likely candidate to close the migration triangle in brown shrimp *Crangon* (Crustacea, Decapoda). Fisheries Oceanography 23:242–257.

Hughes, D.A. 1969. Factors controlling emergence of pink shrimp (*Penaeus duorarum*) from the substrate. Biological Bulletin 136:43–53.

Kennedy, V.S., and L.E. Cronin, editors. 2007. The blue crab *Callinectes sapidus*. Maryland Sea Grant, College Park, Maryland.

Kerr, K.A., A. Cornejo, F. Guichard, and R. Collin. 2014. Planktonic predation risk varies with prey life history stage and diurnal phase. Marine Ecology Progress Series 503:99–109.

Kikkert, D.A., T.A. Crowl, and A.P. Covich. 2009. Upstream migration of amphidromous shrimp in the Luquillo Experimental Forest, Puerto Rico: temporal patterns and environmental cues. Journal of the North American Benthological Society 28:233–246.

Krieger, J., R. Grandy, M.M. Drew, S. Erland, M.C. Stensmyr, S. Harzsch, and B.S. Hansson. 2012. Giant robber crabs monitored from space: GPS-based telemetric studies on Christmas Island (Indian Ocean). PLoS ONE 7:e49809.

Lavalli, K., and W.F. Herrnkind. 2009. Collective defense by spiny lobster (*Panulirus argus* Latreille, 1804) against triggerfish (*Balistes capriscus* Gmelin, 1788): effects of number of attackers and defenders. New Zealand Journal of Marine and Freshwater Research 43:15–28.

Lawton, P., and K.L. Lavalli. 1995. Postlarval, adolescent, and adult ecology. Pages 47–88 *in* J.R. Factor, editor. Biology of the lobster, *Homarus americanus*. Academic Press, San Diego, California.

Linder, M.J., and W.W. Anderson. 1956. Growth, migrations, spawning and size distribution of shrimp, *Penaeus setiferus*. Fishery Bulletin 56:555–645.

Liu, H.C., and M.S. Jeng. 2005. Reproduction of *Epigrapsus notatus* (Brachyura: Gecarcinidae) in Taiwan. Journal of Crustacean Biology 25:135–140.

Lovrich, G.A., and M. Thiel. 2011. Ecology, physiology, feeding and trophic role of squat lobsters. Pages 183–222 *in* G.C.B. Poore, S.T. Ahyong, and J. Taylor, editors. The biology of squat lobsters. CSIRO Publishing, Victoria, Australia.

Lovrich, G.A., B. Sainte-Marie, and B.D. Smith. 1995. Depth distribution and seasonal movements of *Chionoecetes opilio* (Brachyura: Majidae) in Baie Sainte-Marguerite, Gulf of Saint Lawrence. Canadian Journal of Zoology 73:1712–1726.

Magaelhães, C., and I. Walker. 1988. Larval development and ecological distribution of central Amazonian palaemonid shrimps (Decapoda, Caridea). Crustaceana 55:279–292.

March, J.G., J.P. Benstead, C.M. Pringle, and F.N. Scatena. 1998. Migratory drift of larval freshwater shrimps in two tropical streams, Puerto Rico. Freshwater Biology 40:261–273.

March, J.G., J.P. Benstead, C.M. Pringle, and F.N. Scatena. 2003. Damming tropical island streams: problems, solutions, alternatives. Bioscience 53:1069–1078.

McDowall, R.M. 2007. On amphidromy, a distinct form of diadromy in aquatic organisms. Fish and Fisheries 8:1–13.

Miranda, L., and M. Thiel. 2008. Active and passive migration in boring isopods *Limnoria* spp. (Crustacea, Peracarida) from kelp holdfasts. Journal of Sea Research 60:176–183.

Nichol, D.G., and D.A. Somerton. 2015. Seasonal migrations of morphometrically mature male snow crab (*Chionoecetes opilio*) in the eastern Bering Sea in relation to mating dynamics. Fishery Bulletin 113:313–326.

Ogburn, M.B., M. Hall, and R.B Forward, Jr. 2012. Blue crab (*Callinectes sapidus*) larval settlement in North Carolina: environmental forcing, recruit-stock relationships, and numerical modeling. Fisheries Oceanography 21:20–32.

Ogburn, M.B., M.M. Criales, R.T. Thompson, and J.A. Browder. 2013. Endogenous swimming activity rhythms of postlarvae and juveniles of the penaeid shrimp *Farfantepenaeus aztecus, Farfantepenaeus duorarum*, and *Litopenaeus setiferus*. Journal of Experimental Marine Biology and Ecology 440:149–155.

Olivier, T.J., K.Q. Handy, and R.T. Bauer. 2013. Impacts of river control structures on the juvenile migration of the amphidromous river shrimp *Macrobrachium ohione* (Smith): possible solutions for the restoration of upstream populations. Freshwater Biology 58:1603–1613.

Rogers, B.D., R.F. Shaw, W.H. Herke, and R.H. Blanchet. 1993. Recruitment of postlarval and juvenile brown shrimp (*Penaeus aztecus* Ives) from offshore to estuarine waters of the northwestern Gulf of Mexico. Estuarine and Coastal Shelf Science 36:377–394.

Sainte-Marie, B., and F. Hazel. 1992. Moulting and mating of snow crabs, *Chionoecetes opilio* (O. Fabricius), in shallow waters of the northwestern Gulf of Saint Lawrence. Canadian Journal of Fisheries and Aquatic Sciences 49:1282–1293.

Sainte-Marie, B., T. Gosselin, J.M. Sévigny, and N. Urbani. 2008. The snow crab mating system: opportunity for natural and unnatural selection in a changing environment. Bulletin of Marine Science 83:131–161.

San Vicente, C., and J.C. Sorbe. 2013. Comparative life-histories, population dynamics and productivity of *Schistomysis populations* (Crustacea, Mysida) in European shelf environments. Journal of Sea Research 81:13–32.

Shumway, S.E., H.C. Perkins, D.F. Schick, and A.P. Stickney. 1985. Synopsis of biological data on the pink shrimp, *Pandalus borealis* Kröyer, 1838. NOAA Technical Report NMFS 30; FAO Fisheries Synopsis No. 144. U.S. Department of Commerce, Washington D.C.

Shaw, A.K., and K.A. Kelly. 2013. Linking El Niño, local rainfall, and migration timing in a tropical migratory species. Global Change Biology 19:3283–3290.

Stone, R.P., C.E. O'Clair, and T.C. Shirley. 1992. Seasonal migration and distribution of female red king crabs in a southeast Alaskan estuary. Journal of Crustacean Biology 12:546–560.

Suzuki, K.W., K. Nakayama, and M. Tanaka. 2009. *Hyperacanthomysis longirostris* along a temperate macrotidal estuary (Chikugo River estuary, Japan). Estuarine, Coastal and Shelf Science 83:516–528.

Thiel, M. 1998. Population biology of *Dyopedos monacanthus* (Crustacea: Amphipoda) on estuarine soft-bottoms: importance of extended parental care and pelagic movements. Marine Biology 132:209–221.

Thiel, M., and T. Dernedde. 1994. Recruitment of shore crabs *Carcinus maenas* on tidal flats: mussel clumps as an important refuge for juveniles. Helgoländer Meeresuntersuchungen 48:321–332.

Thiel, M., and G.A. Lovrich. 2011. Agonistic behaviour and reproductive biology of squat lobsters. Pages 223–247 *in* G.C.B. Poore, S.T. Ahyong, and J. Taylor, editors. The biology of squat lobsters. CSIRO Publishing, Victoria, Australia.

Tielmann, M., S. Reiser, M. Hufnagl, J.P. Herrmann, A. Eckardt, and A. Temming. 2015. Hydrostatic pressure affects selective tidal stream transport (STST) in the North Sea brown shrimp (*Crangon crangon*, L.). Journal of Experimental Biology 218:3241–3248.

Walls, J.G. 2009. Crawfishes of Louisiana. Louisiana State University Press, Baton Rouge, Louisiana.

Wolcott, T.G. 1988. Ecology. Pages 55–96 *in* W.W. Burggren, and B.R. McMahon, editors. The biology of land crabs. Cambridge University Press, Cambridge, England.

9

DIEL VERTICAL MIGRATION OF AQUATIC CRUSTACEANS—ADAPTIVE ROLE, UNDERLYING MECHANISMS, AND ECOSYSTEM CONSEQUENCES

Piotr Dawidowicz and Joanna Pijanowska

Abstract

The phenomenon of diel vertical migration (DVM) of planktonic crustaceans, recognized by biologists for at least 2 centuries, is a special case of habitat selection behavior by pelagic animals, with their depth preference changing over a diel cycle in a way that prevents encounters with visually oriented predators (mostly fish). Migrating populations usually move toward cold, dark deep-water strata deprived of algal food when there is sufficient ambient light and move back to food-rich and warm surface waters after dusk. DVM has been recognized in pelagic representatives of all aquatic phyla of the animal kingdom and is considered the most massive diel biomass displacement on Earth. DVM can be observed in nearly all lentic freshwater and marine environments. As zooplankton occupy the central position in pelagic food webs, their massive migrations dramatically affect ecological functioning of offshore biota, particularly the efficiency of primary production utilization, energy flow, and biogeochemical pathways of essential nutrients such as carbon fluxes. The phenomenon of DVM is perhaps the most suitable for quantitative description and the major environmental factors underlying the fitness consequences of DVM, including vertical gradients of light intensity (predation risk), temperature related metabolic rates, food concentration (growth and fecundity), and others, are easy to monitor track in the field and to manipulate in laboratory systems. DVM, as inducible behavior, can be experimentally manipulated, both in the field and in the laboratory, which, in turn, makes it possible to design experiments convenient for testing specific hypotheses on various proximate and ultimate factors underlying this behavior. These characteristics make DVM suitable for investigating the evolution of animal behavior, its adaptive value, and ecosystem consequences.

In the fondest memory of our friend Konrad Ciechomski with whom we made, years ago, our first steps into the world of plankton migrations.

Life Histories. Edited by Gary A. Wellborn and Martin Thiel.

INTRODUCTION: DIEL VERTICAL MIGRATION AS ADAPTIVE HABITAT CHANGE

For the purpose of this chapter, we define diel vertical migrations (DVM) after de Meester et al. (1999) as a special case of habitat selection behavior by pelagic animals, with their depth preference changing over a diel cycle in a way that prevents encounters with visually oriented predators (mostly fish) under light levels that cause high predation risk. This pattern of habitat selection translates into synchronized movements of migrating populations toward food-rich and warm surface waters after dusk and then back to cold, dark deep-water strata deprived of algal food when light in surface waters reaches levels that allow predators to visually locate their prey. Ascending to the surface at night is required due to the poor quality of deep refuges (low temperature, low food quantity and quality) as compared to that of surface strata in stratified lakes and marine habitats. If food availability does not decrease with depth, stable vertical distribution of planktonic animals may be observed, with no tendency to migrate (Pijanowska and Dawidowicz 1987).

The pattern described above is known as normal or nocturnal DVM. However, inverse migrations do also occur (i.e., planktonic animals stay closer to the surface during the day and migrate deeper at night); these are believed to evolve under selection by invertebrate tactile (nonvisual) predators, which themselves avoid fish predation by normal DVM (Ohman et al. 1983). In general, however, the pattern of DVM is more often related to pressure from planktivorous fish (Hays 2003).

The phenomenon of DVM of planktonic crustaceans has been recognized by biologists for at least two centuries. In his short resumé of the history of DVM studies, Bayly (1986) recalls the first probable descriptions of light-related migrations in freshwater *Daphnia* by Cuvier (1817) and in marine plankton by Bellingshausen (1823). Since then, the literature on vertical migrations has grown enormously, particularly in the second half of the twentieth century, and the phenomenon still attracts interest among marine and freshwater ecologists. This voluminous literature has been reviewed in several articles and books (Cushing 1951, Hutchinson 1967, Bayly 1986, Lampert 1989, 1993, Pijanowska 1993, de Meester et al. 1999, Pearre 2003, Cohen and Forward 2009, Ringelberg 2010).

DVM has been recognized in pelagic representatives of all phyla of the animal kingdom, from Cnidaria to Chordata, and is considered the most massive diel biomass displacement on Earth. In the oceans, aggregates of synchronously migrating animals form so-called sonic scattering layers (Fig. 9.1), which were falsely interpreted as the sea bottom by the operators of naval echo sounders during World War II. The universality of this behavior also has a geographical dimension; DVM can be observed in practically all freshwater and marine environments, from sufficiently deep pools to the oceans. This ubiquity makes DVM particularly suitable for investigations on the evolution of animal behavior and inspires questions about the adaptive value of migration. Research on DVM has also contributed extensively to the theory of evolutionary ecology, for example, in further developing the concepts of trade-offs (Lampert 1993) and habitat selection (Iwasa 1982). Furthermore, as zooplankton occupy the central position in pelagic food webs, their massive migrations dramatically affect the functioning of offshore biota, particularly the efficiency of primary production utilization, energy flow, and biogeochemical pathways of essential nutrients such as carbon fluxes. Last, but not least, the phenomenon of DVM is, by its nature, perhaps the most suitable out of all animal behaviors for quantitative, numerical description. In most cases, just two numbers sufficiently delineate the pattern: the weighted mean daytime (noon) and nighttime (midnight) population depths (Worthington 1931). Moreover, the major environmental factors underlying the fitness consequences of DVM, including vertical gradients of light intensity (hence predation risk), temperature (which translates into metabolic rate), food concentration (affecting growth and fecundity), and others, are all relatively easy to track in the field and to generate in laboratory systems. Consequently, DVM is amenable to mathematical modeling (Iwasa 1982, Gabriel and Thomas 1988, Fiksen 1997).

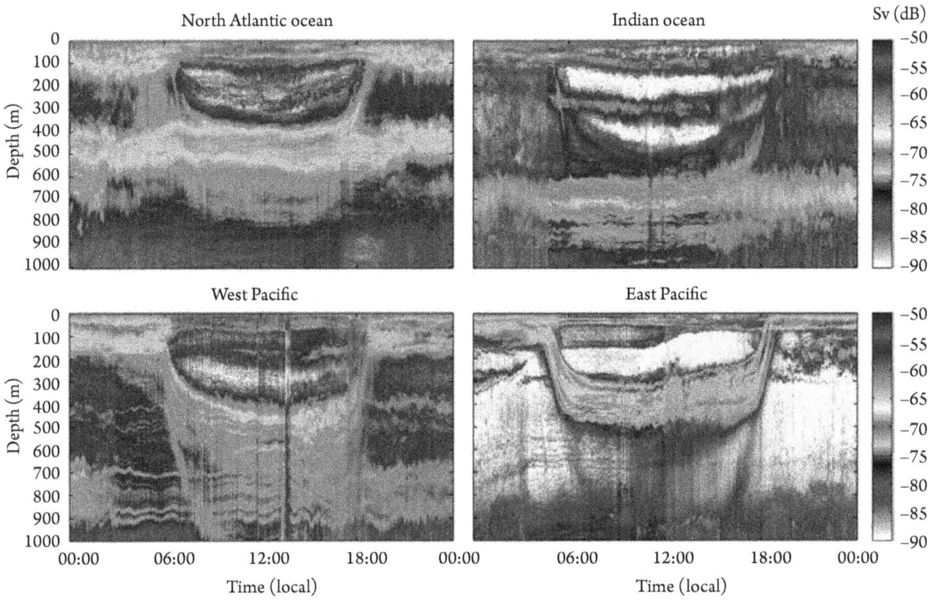

Fig. 9.1.
Examples of deep-sea echograms covering 24-hour periods, from different geographic regions in the mesopelagic zone of various oceans: the North Atlantic (upper left), Indian (upper right), East Pacific (lower left), and West Pacific (lower right). Diel vertical displacements of pelagic animals forming sonic scattering layers (SSLs) is evident in all areas, although the proportion of migrating backscatter varies between about 20% in the Indian Ocean to about 90% in the Eastern Pacific. Most of this variability can be explained by physical properties of water masses in the localities, such as oxygen concentration, turbidity, and temperature. From Klevjer (2016). See color version of this figure in the centerfold.

Several experimental studies conducted in the 1990s revealed that DVM in many crustacean zooplankton species is phenotypically plastic, and such inducible behavior (for reviews see Pijanowska 1993, de Meester et al. 1999, and Chapter 12 in this volume) can be experimentally manipulated, both in the field and in the laboratory. This, in turn, makes it possible to design experiments convenient for testing specific hypotheses on various proximate and ultimate factors underlying the behavior.

After introducing the notion of DVM as an adaptive habitat selection behavior, we present a short inventory of habitats and taxa where DVM is commonly observed. The history of DVM research is briefly presented, with a special emphasis given to hypotheses relating diel changes in depth preferences with avoidance of visual predators in surface waters. We discusse both field and laboratory evidence supporting the validity of DVM as predator-avoidance behavior. The benefits of DVM are compared with the measurable costs of such predator avoidance behavior, and documents the inducible character of this behavior is documented.

DIEL VERTICAL MIGRATION IN DIVERSE TAXA

DVM is widespread in pelagic representatives of all taxa in the animal kingdom and also occurs in planktonic protists, algae, and cyanobacteria (Pearre 2003). It occurs in the Crustacea that generally dominate freshwater and marine planktonic communities in numbers and biomass, as well as in other arthropods: Acarina (Bader 1980), Insecta (e.g., phantom midge larvae, Swift and Fedorenko 1975), as well as in Cnidaria (Muscatine and Marian 1982), Rotifera (Pennak 1944), Polychaeta

(Garland et al. 2002), Mollusca (bivalve larvae, Raby et al. 1994; pelagic gastropods, Seapy 1990), Echinodermata (larvae, Pennington and Emlet 1986), Chaetognatha (Pearre 1974), Tunicata (Madin et al. 1996), and Chordata (e.g., freshwater and marine fish, Neverman and Wurtsbaugh 1994, Clarke 1978). DVM does not depend on an animal's trophic status; migrations are common in herbivores (such as *Daphnia*), predators (such as fish), and even in ontogenetic stages that do not forage at all (such as *Chaoborus* pupae). This taxonomic and trophic variety among vertically migrating species does not imply taxon-specific diversity in the underlying proximate and ultimate mechanisms behind their particular migratory behaviors. On the contrary, examples of DVM in animals from different taxa have been invoked to test universal hypotheses, which can contribute to the general understanding of the phenomenon (e.g., Pearre 2003).

The majority of ecological and evolutionary ("eco-evo") studies on DVM were conducted using freshwater cladocerans, mainly *Daphnia*, as a model, whereas most evidence of community and ecosystem consequences of DVM derives from studies on marine taxa.

DIEL VERTICAL MIGRATION AS A PREDATOR AVOIDANCE BEHAVIOR

Mangel and Clark (1988) reviewed historical concepts concerning the evolution of DVM and listed a number of independent hypotheses proposed by various authors that explained DVM as (1) a mechanism of population self-regulation; (2) a way to prevent overexploitation of food resources; (3) a mechanism that maximizes the gene exchange rate within a population; (4) a way to avoid harmful ultraviolet radiation; (5) a way to maintain a relatively stable food intake rate throughout the year; (6) a response to algal toxins in surface waters; (7) a form of horizontal displacement (water masses at the surface and in the deep often flow in different directions); (8) a response to displacements of prey populations; (9) a way to avoid inter- and intraspecific competition; (10) a way to maximize energetic gain; (11) a response to physical conditions, without direct biological meaning; (12) a way to achieve optimal diel rhythm of ambient temperature change; and (13) a mechanism to avoid predation. Some of these hypotheses rely on group selection arguments (e.g., 1, 2, 3, 5, and 9), others fail to explain the timing (e.g., 6 and 7) and the depth range of DVM (e.g., 4), and some have been experimentally falsified (e.g., 10). The consensus among the majority of plankton ecologists, corroborated with evidence from the primary literature and our own research, is that the predator avoidance hypothesis (Zaret and Suffern 1976, Stich and Lampert 1981, Gliwicz 1986a) offers the most universal explanation of the phenomenon. Four categories of evidence support the validity of the predator avoidance hypothesis. First, a strong correlation is observed between the day depth of zooplankton and the strength of predation (van Gool and Ringelberg 2002) or the history of stocking (Gliwicz 1986a; Fig. 9.2). Second, a clear relationship has been documented between day depth and general visibility (body size, coloration, presence, or even number, of opaque eggs in brood sacs) of planktonic prey. Degree of this adaptive rhythmic migration behavior is correlated with the level of various size and visibility selective pressures experienced by different zooplankton species. Third, DVM can be experimentally induced by the presence of predators, and day depth can be manipulated by the strength of their predation intensity (Dawidowicz and Loose 1992a, Loose and Dawidowicz 1994; Fig. 9.3). Finally, evolutionary game theory models (Iwasa 1982, Gabriel and Thomas 1988, Hugie and Dill 1994) predict that in stratified pelagic systems the pay-offs of a nocturnal migration strategy for planktonic invertebrates can only surpass those of the nonmigration strategy under high predation pressure from visually oriented planktivores. In the absence of predation, the nonmigration strategy becomes evolutionary stable (Maynard Smith and Price 1973).

Under certain circumstances, however, the typical pattern of diel movements may be maladaptive. In the Cahora Bassa Reservoir on the lower Zambezi in southeastern Africa, Tanganyikan sardines *Limnothrissa miodon* crop crustacean zooplankton most efficiently on nights when the full

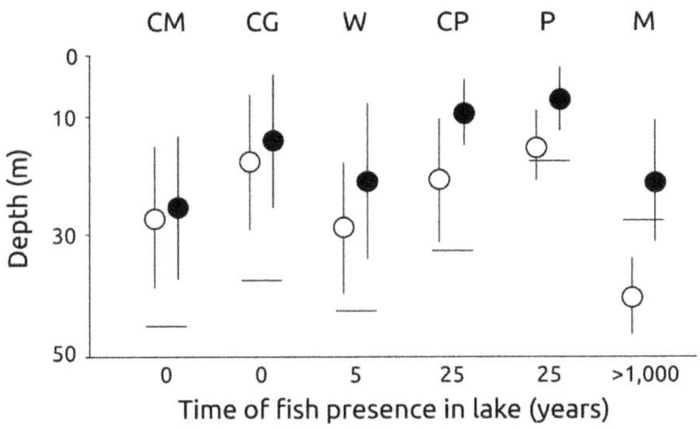

Fig. 9.2.
Noon (empty circles) and midnight (black circles) mean depths ± 1 standard deviation (SD) of the *Cyclops* populations in August 1985, in six Tatra mountains lakes in Poland. CM and CG = lakes with no fish; W, CP and P = recently stocked lakes; M = fish present for millennia; *x*-axis = length of period fish are present in the lake (names of the lakes abbreviated on the top of the panel, CM = Czarny nad Morskim, CG = Czarny Gąsienicowy, W = Wielki, CP = Czarny w Pięciu, P = Przedni, M = Morskie Oko). Horizontal lines represent 1% of solar irradiance. Whereas no differences between day and night population depths were observed in fishless lakes, the clear pattern of diel displacements is visible in lakes inhabited by fish. The larger the difference between day and night depth, the longer the history of fish presence in a lake. These results are one of the earliest and most convincing field studies showing the role of fish predation in the evolution of diel vertical migration. Modified from Gliwicz (1986a), with permission from Nature Publishing Group.

or nearly full moon rises after sunset (i.e., when zooplankton approach the surface during darkness and become vulnerable in the light of the rising moon; Gliwicz 1986b). This lunar trap may cause massive mortality among planktonic animals that migrate to the surface at night. In contrast, on a subtidal sand flat in the Gulf of California, the DVM patterns of demersal zooplankton (organisms inhabiting bottom substrata that periodically emerge to swim freely in the water column) were highly variable and did not correlate with the presence or absence of moonlight (Aldredge and King 1980). Cumaceans emerged from the benthos at dusk regardless of the phase of the moon, whereas species of amphipods and isopods exhibited significant avoidance of moonlight, delaying emergence until moonset or returning to the benthos at moonrise. In general, large crustaceans migrated less frequently into the water column during the moonlight phase than small forms, suggesting that nocturnal predation by visually hunting planktivorous fish may be an important selection pressure.

In some shallow lakes, crustaceans (mostly *Daphnia*) and copepods (*Onychodiaptomus sanguineus*) undergo diel horizontal migration (DHM) into macrophytes of the littoral zone (de Stasio and Bart 1993, Burks et al. 2002). It has been suggested that DHM, similarly to DVM, reduces predation by fish and invertebrate predators such as *Chaoborus* larvae (Kvam and Kleiven 1995). The costs and benefits of DHM are relatively unknown; they should be favored when macrophytes density in the littoral zone is sufficient to reduce planktivores (Burks et al. 2002).

SIZE AND SEX-RELATED PATTERNS OF DIEL VERTICAL MIGRATION

When planktonic animals are exposed to visual predators, the amplitude and timing of their migration may depend on body size, as smaller individuals may remain closer to the surface during the day compared with larger animals (Beklioglu et al. 2008). Larger organisms are more conspicuous and therefore initiate their migration earlier and descend deeper when exposed to fish cues (Beklioglu

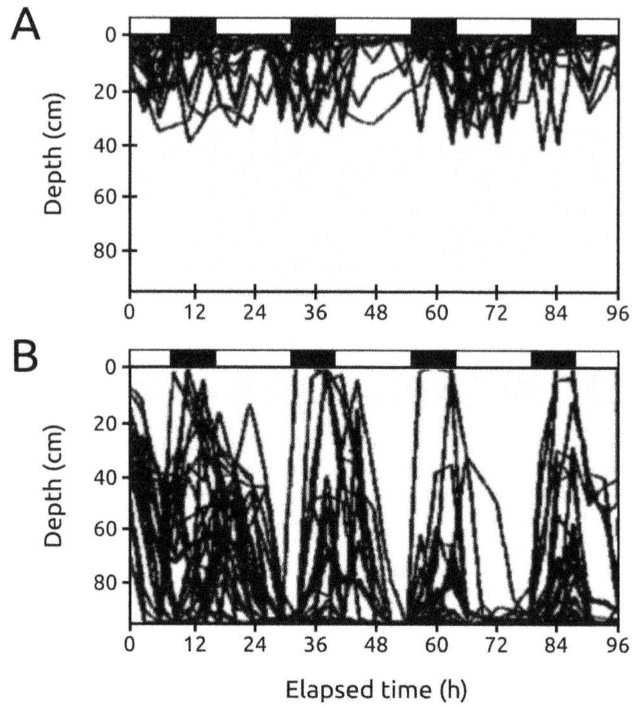

Fig. 9.3.
Swimming tracks of *Daphnia* individuals in the absence (A) and in the presence (B) of fish. Bars at the top of both panels indicate the day/night cycles. The experiment was conducted in a small-scale laboratory setup, called plankton organs, which is a system of glass tubes 1 m long and 1.5 cm in diameter designed by Dawidowicz (Dawidowicz 1993) to track individual migratory behaviors of different invertebrates under easily manipulated regimes of light, food, and temperature gradients. Whereas the typical pattern of diel vertical migration (descending toward the bottom during laboratory daylight and ascending back to the surface at nighttime) has been registered in the presence of fish kairomone, no significant differences between day and night depths were observed in the fishless treatment. From Dawidowicz and Loose (1992a), with permission from John Wiley and Sons.

et al. 2008). In addition to size-dependent differences in migratory behavior, sex-related differences are also observed within taxa. Male and female *Daphnia magna* adopt different strategies to maximize their fitness. Male *D. magna* permanently occupy deepwater strata, and thereby avoid the threat of predation in surface waters (Mikulski et al. 2011). Ontogenetic and sex-related changes in DVM of the planktonic copepod *Calanus sinicus* have also been observed in summer in the Inland Sea of Japan (Uye et al. 1990) as well as in *Calanus finmarchicus* in the Norwegian and Greenland Seas (Fig. 9.4; Dale and Kaartvedt 2000). Stage-specific differences in the DVM behavior of *C. sinicus* were found (Uye et al. 1990), where the onset of distinct DVM took place in the fourth copepodid stage and the amplitude of vertical migration increased with age, reaching its maximum in adult females; adult males, however, remained in the surface layer and did not migrate (Uye et al. 1990), like nonmigratory males of *C. finmarchicus* (Dale and Kaartvedt 2000).

CONSTITUTIVE VERSUS FACULTATIVE DEFENSE

Predators can cause differential mortality of different genetically fixed behavioral (migratory) types within a prey population, selecting for genotypes that efficiently avoid predators (Gliwicz 1986a,

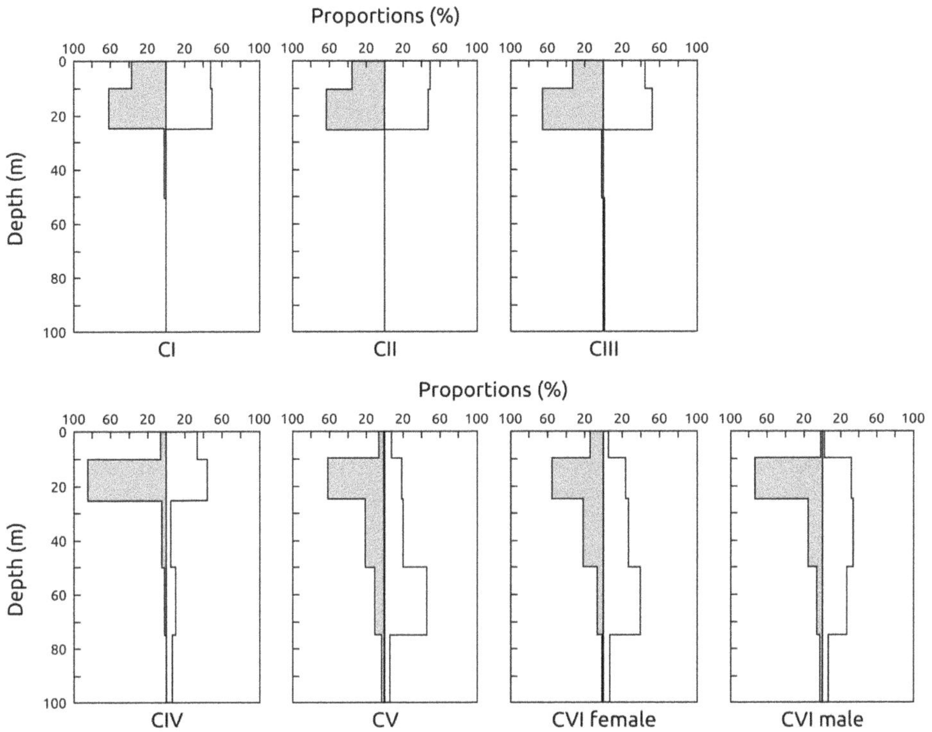

Fig. 9.4.
Diel patterns in stage-specific vertical distribution of *Calanus finmarchicus* in the North Atlantic. Copepodites from first to third instar (CI–CIV) stayed near surface day and night, whereas older stages moved to deeper strata, particularly during the daytime, thus showing typical pattern of diel vertical migrations. From Dale and Kaartvedt (2000), with permission from Oxford University Press.

Bollens and Frost 1989). Evidence for genotype-dependent depth preferences and DVM patterns are accumulating (Weider 1984, Mueller and Seitz 1993, King and Miracle 1995, de Meester et al. 1995) and genetic differences between the surface-dwelling and vertically migrating parts of lake populations are likely (Leibold and Tessier 1991). However, in the majority of cladocerans, the reaction norm is wide enough to explain the various patterns of habitat selection within a population. Many results unequivocally indicate phenotypic behavioral plasticity of individual prey as the mechanism underlying changes in the vertical distribution of planktonic animals (Bollens and Frost 1991; for review, see Pijanowska 1993). In fact, substantial variability in the patterns of crustacean migrations can be explained by phenotypically plastic responses cued by predator-borne infochemicals (i.e., kairomones) and light (Larsson and Dodson 1993, Pijanowska 1993). The inducibility of migratory behavior implies that certain costs are associated with the expression of the defense, which would decrease relative fitness of the prey in the absence of a predator. If there were no fitness costs involved, the DVM should instead have evolved as a constitutive defense (Harvell 1990; also see Chapter 12 in this volume). DVM has been proven to be inducible in many cases (reviewed by Pijanowska 1993 and de Meester et al. 1999), with light cues acting together with fish kairomones that serve as a direct signal of a predator's presence (Loose et al. 1993, von Elert and Loose 1996, von Elert and Pohnert 2000). Experimental and field studies on freshwater and marine invertebrates provide evidence that a fundamental environmental cue for DVM is a relative change in light intensity. An increase in light intensity can lead to a negatively phototactic response resulting in downward displacements, whereas a decrease in light intensity can induce a positive phototactic reaction that results in upward movements. The threshold rate of change in light level needed to induce such movements depends on species-specific adaptations to light intensity (Ringelberg et al. 1967, Forward 1988, Frank and Widder 1997). Some studies on DVM have been conducted under constant light intensity and stable vertical light gradient, and, consequently, day depth is considered as the primary phototactic response (de Meester 1991, 1993). Other studies have focused on the reaction of planktonic animals to the rate of change in light intensity, the so-called secondary phototaxis (Ringelberg 1964). The day depth, timing, and rate of vertical displacements can determine the eventual profit of DVM behavior. Thus, both the primary and secondary phototactic responses are subjected to natural selection (de Meester et al. 1999). Indeed, predation can contribute to a change in genotype frequencies in cladoceran populations, either via direct elimination of nonresponsive clones or indirectly, by favoring clones that effectively avoid the danger. If clones within a cladoceran population differ in their responsiveness to predation via differential migratory behavior, fish predators cause a rapid significant shift in clonal composition in favor of responsive clones (Pijanowska et al. 1993, de Meester et al. 1995).

In high-latitude marine environments with no solar illumination in winter, light-mediated patterns of vertical migration have long been considered nonexistent. However, DVM of zooplankton has been shown to occur even during the darkest part of the polar night, when illumination levels are exceptionally low (Berge et al. 2009). Recently, evidence of uniform migratory behavior of zooplankton driven by lunar illumination has been demonstrated across the entire Arctic in fjords, shelves, and open sea. Mass downward displacements of zooplankton from surface waters to a depth of 50 m occur every 29.5 days in winter, concurring with periods of the full moon. Moonlight may, like sunlight, enable predation of zooplankton by carnivorous zooplankters, fish, and birds known to feed during the polar night (Last et al. 2016).

Swimming in response to light intensity is the major mechanism underlying DVM. Indeed, the foraging efficiency of planktivorous fish is strictly dependent on ambient illumination, and, consequently, light intensity is a highly reliable measure of predation risk for a planktonic animal. However, other factors can also be directly involved or may modify the prey's response to light. In the presence of light, DVM in prey is triggered by kairomones (Fig. 9.5). Infochemicals released either from planktivorous or piscivorous fish induce DVM in *Daphnia magna* (Loose et al. 1993, von Elert and Loose 1996, von Elert and Pohnert 2000, von Elert and Stibor 2006). It had been suggested that trimethylamine may be the DVM-inducing infochemical (Boriss et al. 1999), but this was later falsified when trimethylamine was applied in ecologically relevant concentrations. Bioassays aimed to characterize the chemical nature of the DVM-inducing fish kairomone indicate that the cue is of low molecular weight (≤ 500 Da), lipophilic, thermally stable, immune to extreme values of pH, and resistant to digestion by proteinases (Loose et al. 1993). Activity of the compound depends on hydroxyl groups (von Elert and Pohnert 2000). The identical characteristics of the kairomone released by both planktivorous and piscivorous fish indicates that this nonspecific cue provides an evolutionarily general indicator of predation risk (von Elert et al. 2012). The properties of DVM-inducing cues from marine fish are reviewed by Cohen and Forward (2009). It seems that amino acids, as degradation products of fish mucus, may be essential constituents of infochemicals inducing DVM in marine zooplankton.

The response of planktonic crustaceans to kairomone-simulated predator presence is context-dependent (Effertz and von Elert 2014). As demonstrated by Brewer et al. (1999), increased evasiveness of *D. pulicaria* was induced by the chemical traces of fish presence (fish kairomone) in light, but in the relative safety of darkness the *D. pulicaria* escape behavior was not affected by the kairomone. Similarly, the DVM behavior (surface avoidance) was not induced by fish infochemicals under constant darkness (Loose 1993; Fig. 9.5). In stratified lakes, daytime residence of *Daphnia* in the dark hypolimnion may, in addition to avoiding light, be avoiding fish kairomones as well.

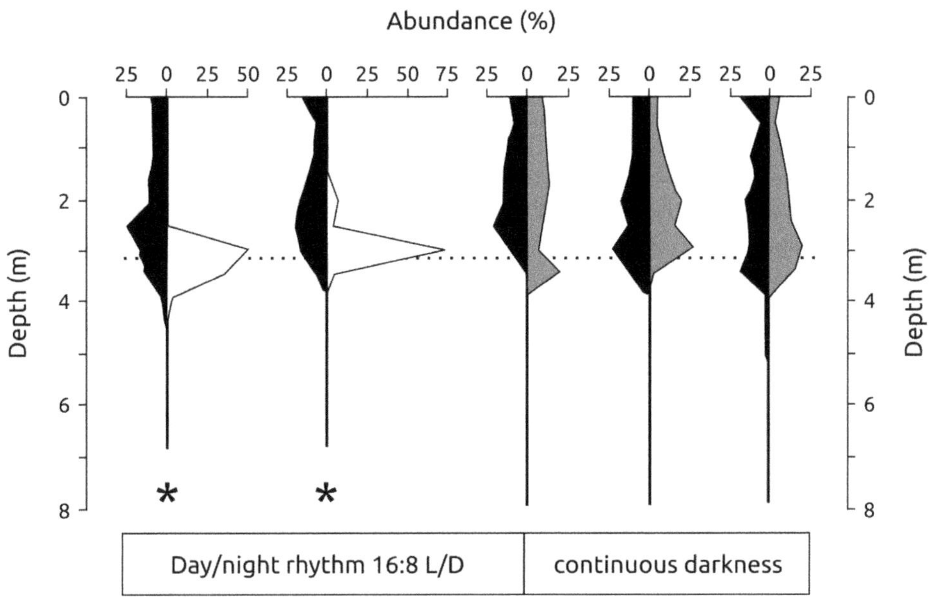

Fig. 9.5.
Diel changes in *Daphnia* vertical distribution in plankton towers in the presence of fish infochemicals. The first 2 profiles represent day (white surface) and night (black surface) distribution during a regular light/dark cycle in summer (16L:8D). The last 3 profiles represent day (white surface) and night (gray surface) distribution under continuous darkness. The dotted horizontal line shows depth of the thermocline. Plankton towers have been constructed in the Max-Planck Institute of Limnology (Plön, Germany) as a utility specifically dedicated to investigating the phenomenon of diel vertical migration (Lampert and Loose 1992). Although under continuous darkness, there was no difference between the day and night depth of *Daphnia*, under the summer photoperiod *Daphnia* moved to a deep refuge against fish during the day and ascended safely toward surface in the night. From Loose (1993), with permission from John Wiley and Sons.

Hence, both mechanisms, avoidance of kairomone and of high levels of light, may be involved in the induction of migratory behavior.

Daphnia can also respond with various behavioral reactions to chemical cues from freshly crushed conspecifics (i.e., indicating mortality risk), such as a shift in vertical distribution toward the bottom. The persistent tendency to occupy deeper strata indicates that *Daphnia* perceive the signal from crushed conspecifics as nonspecific information not necessarily associated with either vertebrate or invertebrate predators. The recognition of a signal originating from injured conspecifics can be particularly adaptive in encounters with unfamiliar predators and those that change their diet during ontogenesis. The combination of such a signal with a predator cue can reliably advertise the local predation risk (Pijanowska 1997).

Vertical migration behavior may also be triggered by mechanical stimuli generated by predator presence (e.g., predator-mediated mechanical or visual cues or a hierarchy of both). These cues were responsible for eliciting vertical migrations in adult females of the copepod *Acartia hudsonica* (Bollens et al. 1994). The specific stimuli eliciting vertical migration in zooplankton vary among species, in both marine and freshwater systems.

The unknown chemical nature of the DVM-inducing kairomone (and also of the signals originating from conspecifics) poses methodological problems, such as the necessity of using crude extracts or predator-conditioned water to obtain a medium imitating the predators' presence. Such media contain not only dissolved kairomones but also bacteria (depending on filtration efficiency) and a variety of other chemical compounds released by a predator. These compounds, as well as bacteria, may affect the target organism, either directly or indirectly, by supporting bacterial growth (Maszczyk and Bartosiewicz 2012). Their presence may conceal or interfere with the effects of the behavioral response to the kairomone of interest. Characterizing the predator-released compounds that induce defensive behavioral reactions in their prey is of high priority. It could provide insight into developmental, ecological, and evolutionary questions, as well the mechanisms of interspecific recognition.

Finally, the sensory systems involved in light and kairomone perception among crustaceans are also important in the discussion of predator detection, but these are beyond the scope of this chapter. These topics have been extensively reviewed by Mellon (2014), and additionally vision mechanisms and chemical senses are reviewed, respectively, in Glantz (2014) and Derby and Weissburg (2014).

PHYSIOLOGICAL AND GENETIC BASIS OF DIEL CYCLES

Not much is known about the genetic and physiological mechanisms underlying the DVM behavior, and the dispersed data originating from the existing few studies do not as yet provide clear conclusions. Adjustment of the migration rhythm of crustaceans to the change of day and night is probably mediated by the hormone melatonin. Melatonin in Cladocera is a stress signal inhibitor in molecular pathways associated with the response to predation threat, as was initially recognized by Bentkowski et al. (2010). In the presence of exogenous melatonin and under the threat of predation, *Daphnia* stayed closer to the surface and their distribution was more variable than that of individuals exposed solely to the kairomone (Bentkowski et al. 2010). The act of either remaining in or approaching surface waters in the presence of predation threat seems to be maladaptive.

In the process of melatonin synthesis, the rate-limiting enzyme is the aralkylamine N-transferase (AANAT). Three genes coding for insect-like AANATs in *Daphnia* were recently identified (Schwarzenberger and Wacker 2015). In *Daphnia,* AANAT gene expression oscillates daily, and the highest peak of expression after the onset of darkness is followed by a peak of melatonin production at midnight. In most organisms, melatonin synthesis is due to rhythmic expression of circadian clock genes. Putative clock genes and insect-like AANAT genes of *Daphnia* are expressed as well. Melatonin synthesis seems to be coupled to the expression of *Daphnia* clock genes, and insect-like AANATs of crustaceans have a similar function as AANATs in vertebrates (Schwarzenberger and Wacker 2015).

Daphnia exposed to fish kairomones show a significant decrease in actin protein concentration (Pijanowska and Kloc 2004). Whereas actin, as a major component of the cytoskeleton, can play an important role in decreasing size at first reproduction, its role in DVM behavior is unknown. A significant increase in the expression of one of the four actin paralogs was found after exposure to fish kairomones (Effertz and von Elert 2014). Apparently, different actin paralogs are differently affected by fish kairomones, probably depending on length of exposure to their presence. Also, the well-recognized role of highly conserved proteins, cyclophilines (peptidyl-prolyl *cis-trans* isomerases), in signal transduction of the optical system in *Drosophila* (Ferreira and Orry 2012) might indicate the involvement of cyclophiline in the response of *Daphnia* to fish kairomones as well (Effertz and von Elert 2014). Again, its possible role in the DVM response needs to be further elucidated.

PHYSIOLOGICAL AND LIFE HISTORY COSTS OF DIEL VERTICAL MIGRATION

DVM of cladocerans is probably the sole example of a massive animal displacement that relates to the selection of low-quality habitat. Other migration patterns, observed in birds, fish, or mammals (such as African ungulates) are generally associated with a search for better-quality habitat offering more favorable conditions for feeding, growth, and reproduction; this is not the case in plankton DVM. Considerable fitness costs are associated with DVM in stratified lakes, because migration to cold, deep water reduces metabolic and reproduction rates in planktonic animals (Loose and Dawidowicz 1994). These costs increase over the season as the steepness of the thermal gradient increases. Indeed, it has been observed that the amplitude of a migrating population in situ varies with thermocline depth (Marcogliese and Esch 1992). Although a large body size yields important fitness benefits in terms of foraging efficiency and fecundity in *Daphnia*, under the hazard of fish predation it also brings about a necessity to perform a long-range migration. At the peak of summer, when fish predation pressure is strong and metalimnetic gradients are steep, large migrating *Daphnia* may lose their competitive advantage over small-sized individuals (clones or species) (see Chapter 2 in this volume), which do not need to migrate as far to protect themselves against visually orienting planktivores. Hence, the ability to trade off 2 (or more) defensive strategies, namely shifts in major life history traits, such as size at maturity versus changes in diel patterns of habitat preferences, can be crucial in maximizing fitness as it enables to choose the optimal defense among those available, or the most beneficial combination of various defenses. For example, Sakwińska and Dawidowicz (2005) found a positive association between the day depth (i.e., illumination level and, as a consequence, predation risk) and size at first reproduction (SFR) among a set of 14 *D. hyalina* populations from northern Polish lakes of different trophic status.

Planktonic crustaceans migrate relatively long distances of several to a few hundred meters. The largest amplitude of migration by *Daphnia* is 60 m (Hutchinson 1967), whereas marine euphausiids and copepods migrate daily within a range of up to 300 m, or several thousand times their body length. Migrating animals usually experience dramatic changes in the quality of their environment en route. During the morning descent, they dive from the relatively warm and food-rich euphotic zone into dark, cold and usually food-deprived (and sometimes also oxygen-deprived; see Hidalgo et al. 2005) deep-water refugia, where they spend the daytime hours. This habitat shift unavoidably affects essential fitness components, such as food intake, metabolic and growth rates, resource allocation, and fecundity of migrants.

Several factors contribute to fitness costs of DVM. Upward and downward swimming requires energy. However, experimental estimates of the energetic costs of locomotion during DVM are rare, mostly due to the technical difficulty of their measurement. Torres and Childress (1983) found a strong positive relationship between swimming speed and respiration rate in *Euphausia pacifica* and applied it to published data on the velocity of movements of the sonic scattering layer formed by this species during DVM. They concluded that "... vertical migration is energetically expensive; its cost should be thoroughly considered in attempts to describe the energetics of vertically migrating species." This conclusion is in contradiction with a few earlier studies, which suggested that costs of swimming during DVM in planktonic copepods and mysids were negligible (Klyashtorin and Yarzhombek 1973, Foulds and Roff 1976). Nevertheless, all of the above estimates rely on the assumption that the velocity of ascending and descending of crustacean populations reflects the speed of the constituent individuals, which slow down their swimming when populations rest at their daytime and nighttime depths. However, this assumption is not necessarily true; as pointed out by Pearre (1979), it is incorrect to equate mean population displacements to the locomotory activity of individual animals if their migratory movements are not synchronized. Thus, the vertical migration of a population may, at least in part, result from synchronization of individual displacements (up or down) rather than from increased swimming speed of individuals, as was observed by Dawidowicz (1993) in migratory, noncrustacean phantom midge (*Chaoborus flavicans*) larvae. In such cases, DVM would not imply additional costs related to locomotion. Dawidowicz and Loose (1992b) applied a different approach to the assessment of metabolic costs of DVM in *Daphnia hyalina*. In a life table experiment, they determined individual life history parameters in a cohort of *Daphnia* forced to swim in vertical flow-through chambers against currents simulating diel migration with an amplitude of 60 m (which is the highest ever recorded in *Daphnia*; Hutchinson 1967) and 3 m, at high (1 mg organic Carbon L^{-1}) and at low food (0.1 mg organic Carbon L^{-1}) levels. None of the evaluated life history parameters (somatic growth rate, size and age at first reproduction, and fecundity) differed significantly between long- and short-distance migration treatments at the high food level, whereas in starving *Daphnia* only the growth rate was marginally lower in long-migrating individuals. This result suggests that energy expenditures on locomotion do not account for the metabolic costs of DVM, at least under favorable food conditions.

Although locomotion costs appear to be minimal, other costs may be more substantial. As a consequence of vertical migrations, the planktonic animals spend daytime hours in deep, nutrient-poor, cold waters, and this time is particularly long in temperate waters during summer. This exposure to cold, low-food waters may lead to a dramatic reduction in the growth rate of migrating individuals (by more than 50%; Dawidowicz and Loose 1992b) and in birth rate in migrating populations (Stich and Lampert 1981, Loose and Dawidowicz 1994). As estimated by Loose and Dawidowicz (1994), these costs in terms of birth rate reduction in *Daphnia* are generated mostly by a decrease of up to 70% in metabolic rate caused by low ambient temperatures in deep-water refugia; costs due to reduced food availability account for less than 10%. Similar predictions were generated by an individual-based demographic model developed by Vos et al. (2002), who argued that temperature effects dramatically surpass the effects of ambient food level on birth rates of migrating *Daphnia*.

Variation between ambient temperatures at daytime and nighttime depths may surpass 10°C in temperate lakes (e.g., Gliwicz and Pijanowska 1988) and 15°C in tropical marine waters (e.g., Elder and Seibel 2015); thus, migrating crustaceans experience a drastic temperature change during their passage through the thermocline. This abrupt change in the temperature is likely to induce thermal stress in the migrators. A physiological response to thermal stress includes the expression of heat shock proteins (HSPs), and indeed, such a response has been observed in vertically migrating marine copepods (*Calanus finmarchicus*; Voznesensky et al. 2004), hyperiid amphipods (*Phronima sedentaria*; Elder and Seibel 2015), and *Daphnia* (Mikulski et al. 2011). Nonmigrating, deep-dwelling males of *D. magna* reduce their molecular defenses against stress, such as altering the production of HSPs; they do not maintain the physiological machinery that elicits an increase in HSP levels in response to stress. In contrast, females of *D. magna* actively select between deep-water habitats that offer safety against predators during the day and surface strata that offer optimal conditions for growth and offspring production at night. Consequently, females are exposed to variable environmental conditions that may be associated with increased stress. To permit survival in both the surface and deep habitats, *D. magna* females require molecular mechanisms to protect their cells from rapid changes in stress levels. The different habitat selection strategies of male and female *D. magna* result in different patterns of HSP production, with females maintaining high constitutive levels of HSPs from HSP 60, 70, and 90 families (Mikulski et al. 2011). Additionally, exposure to extreme temperatures may drive the animals into oxidative stress due to excessive oxygen demand at high temperature, and reduced aerobic capacity of mitochondria at low temperatures (Pörtner 2002).

Low oxygen concentrations are common in the hypolimnion of stratified lakes; thus, the utilization of a low-oxygen refugium by migrating zooplankton may constitute an additional source of oxidative stress and enhance the metabolic costs of consequent protective responses (e.g., hemoglobin synthesis; Sell 1998). Different tolerance to oxygen deficiencies and various strategies are exhibited by migrating invertebrate species from the Eastern Tropical Pacific and the Red Sea (Seibel et al. 2016). Tolerance to low oxygen varies between migratory and nonmigratory species (Escribano and Riquelme-Bugueño 2015). The migrating species *Euphausia eximia* and *Nematoscelis gracilis* tolerate very low oxygen concentrations for at least 12 hours with no mortality risk, whereas *N. difficilis* is incapable of surviving prolonged oxygen deficiencies. Active avoidance of the hypoxic zone is an important behavioral mechanism, whereas the ability to reduce energy expenditure during daytime escapades into low oxygen layers is an essential physiological mechanism to cope with oxygen deficiencies. When exposed to hypoxia, individuals of some species may reduce aerobic metabolism by more than 50%. Growing evidence suggests that metabolic suppression is a widespread strategy among migrating zooplankton in oxygen minimum zones (Gonzalez and Quiñones 2002).

Other data suggest that hypoxic and low oxygen zones (up to 4 mL O_2/L) may still provide a refuge from visual predators for some marine crustaceans that can protect themselves from oxidative damage (Escribano et al. 2009, Webster et al. 2015). Such species include the mysid *Mysis relicta* and the copepod *Limnocalanus macrurus*, which show peaks of biomass in hypoxic (< 2 mL O_2/L) and low oxygen (2–4 mL O_2/L) strata below 75 meters. Mysids from areas with hypoxia had significantly amplified antioxidant potential as compared to conspecifics from oxygenated strata, and had no visible traces of oxidative damage. The interacting effects of oxygen and temperature on the metabolism of aquatic (both freshwater and oceanic) species give rise to predictions that patterns of vertical distribution may change along with global climate change.

DIEL VERTICAL MIGRATION AMONG OTHER ANTIPREDATOR RESPONSES

Among various adaptations to complex inland aquatic habitats (Covich 2015), planktonic crustaceans display a plethora of behavioral, morphological, and life history responses to predation threat. These responses are mostly examined separately and their adaptive value is discussed with no relation to other antipredator traits. DVM behavior provides perhaps the most powerful zooplankton defense against visually oriented predation (Lampert 1989, 1993). However powerful, DVM is not the only antipredator weapon in the repertoire of cladoceran defenses. The modification of SFR is also regarded as an adaptive response to size selective predation. SFR is a crucial animal life history trait (Stearns 1992). It is particularly important for planktonic cladocerans inhabiting an environment governed by size-dependent ultimate forces, where efficiency of food collection and predation risk are both functions of body size (Gliwicz 2003). Large *Daphnia* are competitively superior over small-bodied herbivorous zooplankton, because they can survive and reproduce at lower food concentrations (Gliwicz 1990, Gliwicz and Lampert 1990, Kreutzer and Lampert 1999). Moreover, both the quality (starvation resistance) and quantity of offspring in *Daphnia* increase with increasing mother body size. Consequently, *Daphnia* fitness is positively correlated with body size. However, large-bodied *Daphnia* suffer a high mortality risk imposed by planktivorous fish that hunt visually and prefer large, conspicuous prey (Hrbàček and Novotná-Dvořáková 1965). Thus, the relationship between body size and fitness turns negative in their presence (Brooks and Dodson 1965). A number of experimental studies indicate that in different *Daphnia* species and clones, size at first reproduction is also a flexible trait, being reduced in the presence of fish exudates, or kairomones (Macháček 1991, Dawidowicz and Loose 1992a, Stibor 1992; for a full discussion of size and life history trade-offs, see Chapter 2 in this volume).

Effertz and von Elert (2014) demonstrated that the kairomone-induced reduction in SFR is suppressed under permanent darkness (i.e., light conditions experienced by vertically migrating *Daphnia*) (Fig. 9.6). These experimental results are in line with field observations by Sakwińska and Dawidowicz (2005), who found a positive correlation between SFR and daytime residence depth among *D. longispina* populations from various Polish lakes that differed in the thickness of the anoxic zone (and thus in the availability of the dark refugium). Exposure to the kairomone also leads to the shortening of lifespan in cladocerans (Dawidowicz et al. 2010). However, the "physiological" longevity of vertically migrating, light-avoiding *Daphnia* was shown to surpass that of nonmigrators at the surface during daytime (Dawidowicz et al. 2013). Hence, DVM can be treated as a strategy conserving life history parameters unrelated to predation. As these studies demonstrate, a deeper

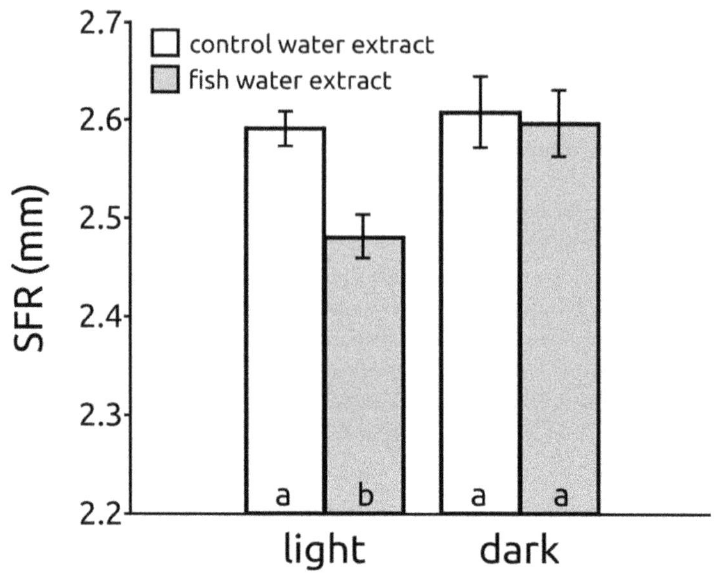

Fig. 9.6.
Size at first reproduction (SFR) of *Daphnia magna* grown in the presence (white bars) and absence of fish kairomones (gray bars), in light (left) and under permanent darkness (right). Different letters denote a significant difference ($P < 0.05$) between the treatments. The kairomone-induced reduction in SFR appeared only the illuminated *Daphnia*. Consequently, the light avoidance through diel vertical migration behavior may inhibit antipredator life history response in migrating cladocerans. Modified from Effertz and von Elert (2014), with permission from the Royal Society.

understanding of the degree to which specific antipredator responses are coupled is essential for elucidating the full scope of costs and benefits experienced by planktonic animals zooplankton in their complex natural environment. The consensus on whether demonstrating a full repertoire of antipredator defenses is adaptive or whether it may reduce fitness is still widely contested. De Meester and Pijanowska (1997) argued that the simultaneous expression of various antipredator defense mechanisms might be maladaptive. If single traits are effective enough in protecting animals from predation threat, then there is no need to pay the unnecessary costs of other defenses. Although the identification of costs associated with antipredator defenses is usually not easy (Tollrian and Harvell 1999), they have been quantified for some of the defenses. For example, in vertically migrating *Daphnia*, the intrinsic rate of increase (r) can be reduced by more than 50% as compared to nonmigrating conspecifics due to environmental costs such as low temperatures and low food concentrations in the hypolimnetic refuge (Dawidowicz and Loose 1992a, Loose and Dawidowicz 1994). Reduction in SFR, which also commonly occurs in *Daphnia* exposed to fish predation (Macháček 1991), results in the reduced quality of the undersized offspring produced by small-bodied females. Small *Daphnia* neonates are less resistant to starvation than larger ones (Tessier and Consolatti 1989, Gliwicz and Guisande 1992). Moreover, large females tend to produce more eggs (Ebert 1993); thus, maturing at a small size leads to a lower number of offspring in early clutches.

Because of their costs, antipredator defenses are seldom redundant. De Meester et al. (1995) showed that in the presence of fish, a small-bodied *D. hyalina/galeata* hybrid clone dwelled near the surface day and night, while clones with a larger size at maturity performed a long-range DVM. Behavioral defense (an extended DVM amplitude) occurred only if the life history defense (reduced SFR) did not.

A similar scenario of uncoupled defenses was further developed in the laboratory study on the responses of 12 potentially plastic traits in 16 *D. magna* clones originating from different habitats (Boersma et al. 1998). None of the studied clones responded to a fish kairomone with changes in all 12 examined defensive traits, although all clones exhibited a response in at least one trait. There was no (or weak) correlation between different possible responses to the kairomone, and it was concluded that reacting to the predation risk with more than 1 trait was of limited advantage to an individual. Despite this result, in the majority of the studied clones (9 of 16), at least 2 traits were simultaneously modified in response to the presence of kairomones. Hence, a simultaneous induction of various defensive traits within a single genotype (clone) appeared common, at least in the *D. magna* populations examined by Boersma et al. (1998).

This result raises the question of what the fitness benefit is of maintaining plasticity in multiple traits, especially given that the genetic and metabolic costs of plasticity (as defined by Tollrian and Harvell 1999) should increase with an increasing number of flexible traits. Nonetheless, having multiple flexible responses can be beneficial for an individual genotype exposed to a number of biotic agents that vary simultaneously (Tollrian and Harvell 1999). Zooplankton prey are often exposed to multiple predator species, which may differ in their predation strategies. Avoidance of visual predators through DVM may expose migrators to the pressure of deep-dwelling tactile predators, such as phantom midge larvae (e.g., Dawidowicz et al. 2002). Because not every antipredator response is efficient against all predators, exposure to a multiple-predator environment can cause conflicting prey responses, as was experimentally shown by Brett (1992), who exposed *Daphnia* to simultaneous threats from fish and *Chaoborus* larvae.

In lakes of the temperate zone, the entire water column is usually aerated in spring and thermal stratification is weak, but anoxia progresses with time; by midsummer in highly eutrophic lakes, only the upper few meters are available for aerobic organisms. This timing often coincides with peaks of fish density and foraging activity (Gliwicz and Pijanowska 1988). Thus, at the same time that predation pressure increases, availability of the hypolimnetic refuge shrinks. With no access to dark, deep asylum during daytime, DVM is no longer effective as a fish avoidance behavior. Consequently, a switch to another defense, such as reduction in size at maturity, appears essential for prey survival. Relying entirely on a single defensive trait, such as DVM, is adaptive only as long as the trait is effective under ambient environmental conditions (Sakwińska and Dawidowicz 2005). Inducible defenses may not simply be uncoupled so that they are independent of each other as was suggested by de Meester and Pijanowska (1997) and Boersma et al. (1998), but instead, they may be negatively coupled within a single genotype (Ślusarczyk and Pinel-Alloul 2010).

COMMUNITY AND ECOSYSTEM CONSEQUENCES

Due to its scale in terms of biomass and number of populations involved, and because of the central position occupied by herbivorous zooplankton in offshore pelagic food webs, DVM inevitably acts on several basic processes and interactions within offshore communities. First, variation in patterns of DVM among different taxa can directly or indirectly affect their competitive interactions. Second, DVM modulates the interactions of the migrating animals with those organisms from higher (predators) and lower (e.g., primary producers) trophic levels. Third, zooplankton migrations contribute to vertical fluxes of organic matter and energy and thereby affect biogeochemical cycles of many essential nutrients, including carbon, nitrogen, phosphorus, and others (Pearre 2003). Fourth, recent studies (Dewar et al. 2006, Wilhelmus and Dabiri 2014) have suggested that DVMs of zooplankton (or other migrating organisms) may have an impact on ocean mixing; however, the scale of the phenomenon seems to be questionable. Although biological mixing (so-called biomixing; Dean et al. 2016) may be of the same order of magnitude as mixing caused by winds and tides (e.g.,

Kunze 2011), other results show the negligible enhancement of vertical transport by zooplankton in comparison with the turbulent mixing typical in oceans and lakes (Noss and Lorke 2014).

The various DVM patterns of planktonic crustaceans that differ in body size and other traits linked with vulnerability to predation, such as body transparency or escape ability, result in spatial separation; consequently, there is reduced competition between taxa or ontogenetic stages within a taxon. For example, two of three sympatric calanoid species in Dabob Bay, *Calanus pacificus* and *Metridia lucens* that are similar in size but differ in escape ability migrated normally but occupied different daytime depths (Fig. 9.7), with the nonevasive *M. lucens* staying consistently deeper while the approximately twice smaller *Pseudocalanus newmani*, which migrated inversely, stayed above (Bollens et al. 1993). Consequently, their spatial niches were clearly separated.

As shown experimentally by Dawidowicz and Wielanier (2004), metabolic costs related to fish-induced vertical migration may handicap the otherwise competitively superior large-bodied *Daphnia* and indirectly favor its small-sized, inconspicuous competitor *Ceriodaphnia*, which, being ignored by planktivorous fish, does not migrate and safely exploits the resources of surface waters day and night. As a result, both species can coexist within a lake.

An obvious consequence of DVM is the daytime release of algae from grazing by herbivorous crustaceans in the euphotic zone, which in turn affects primary production by the phytoplankton (Bowers 1979). Reichwaldt and Stibor (2005) showed that higher phytoplankton biomass can develop in a lake when grazing cladocerans migrate vertically. Lampert and Taylor (1985) measured in situ day and night vertical community grazing profiles in a small eutrophic lake in Germany. They reported strong coupling between crustacean zooplankton DVM and grazing in the euphotic zone during summer (Fig. 9.8). Algae were harvested at night, but their growth was relatively unaffected by grazing during daytime. Because the migration patterns (large-sized *Daphnia* and *Eudiaptomus* did migrate, whereas small *Ceriodaphnia* did not) and food selectivity of major grazers differed substantially, the migrations led to diel variations in the relative grazing pressure on different size fractions of phytoplankton. Vertical and diel differences in grazing rates disappeared in the fall when

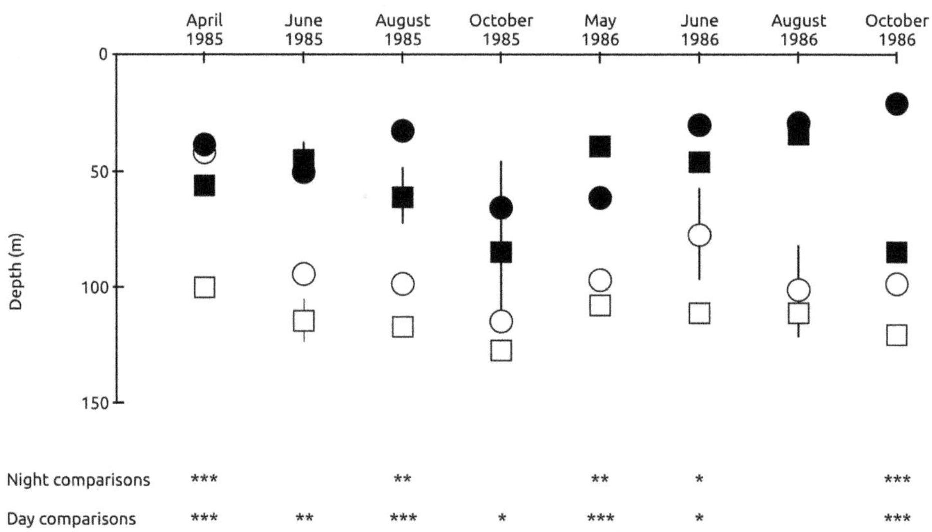

Fig. 9.7.
Weighted mean daytime (open) and nighttime (closed) depths of adult *Metridia lucens* (squares) and *Calanus pacificus* (circles) in the Dabob Bay. Asterisks indicate significant differences between species (t-test, *** $P < 0.001$, ** $P < 0.01$, * $P < 0.05$). Except in August 1986, the migrating populations of both species were clearly separated in space. Modified from Bollens et al. (1993), with permission from John Wiley and Sons.

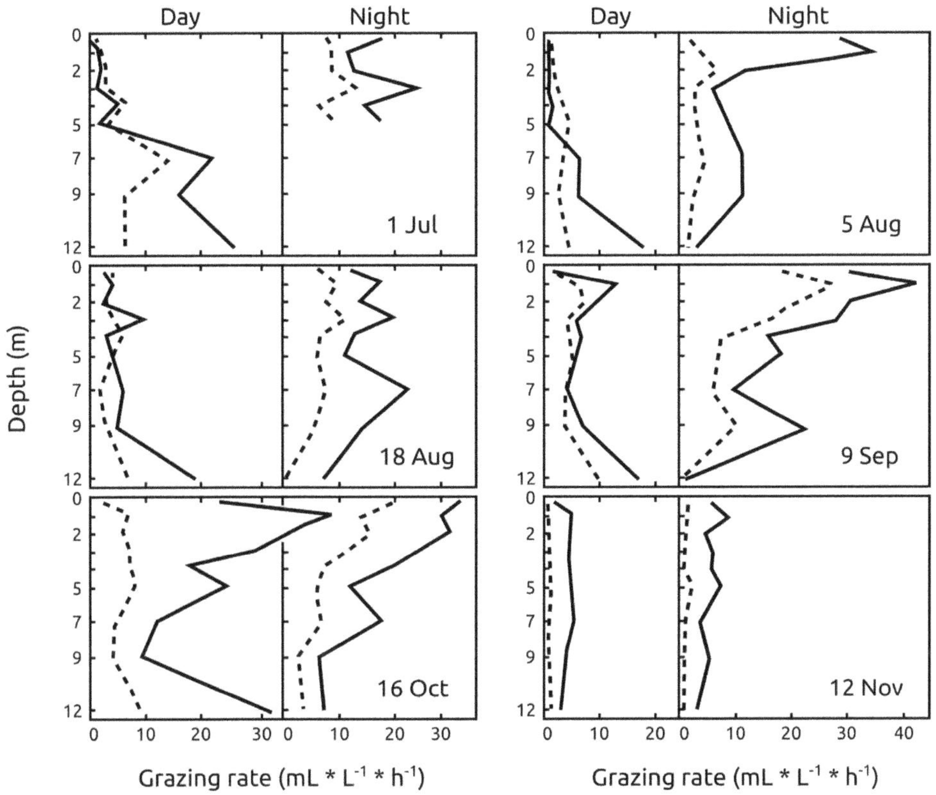

Fig. 9.8.
Seasonal variations in depth profiles of zooplankton grazing rates on green alga *Scenedesmus* (solid lines) and cyanobacterium *Synechococcus* (dashed lines) in eutrophic Schöhsee (Germany) during the day and the night. During long summer days, the epilimnetic phytoplankton are released from grazing pressure due to diel vertical migrations of planktonic herbivores. From Lampert and Taylor (1985), with permission from John Wiley and Sons.

vertical migrations ceased. According to the mechanistic model of Petzoldt et al. (2009), the reduction in daily grazing caused by DVM benefits edible small-sized algae during the summer stratification period and delays a shift toward poorly edible or nonedible phytoplankton in a lake. Using deep in situ enclosures with migrating and nonmigrating *Daphnia hyalina,* Haupt et al. (2009) experimentally confirmed that cladoceran DVM, and not just organism abundance, affects the composition of the phytoplankton community.

In extreme high-latitude marine environments deprived of solar illumination in winter, primary production is almost nil, but lunar vertical migration may still cause monthly pulses of carbon remineralization due to respiration of carnivorous and detritivorous zooplankton (Last et al. 2016).

Zooplankton DVM also affects higher trophic levels (i.e., plankton-feeding animals) by impeding their foraging efficiency. Numerous planktivores adjust their behavior to diel vertical displacements of their food resources. For example, some filter-feeding fish that do not need light to locate and capture planktonic prey tend to follow the migrating zooplankton (Nelson et al. 1997). Also, air-breathing planktivores (marine mammals, birds, and turtles) modify the depth of their dives according to diel patterns of prey migrations (see Hays 2003 and literature therein).

Massive vertical movements of zooplankton affect vertical fluxes of nutrients, organic matter, and energy, particularly when stratification of the water column prevents vertical mixing (see section

VIII in Pearre 2003 for thorough review). The downward direction of these zooplankton-mediated fluxes is a simple consequence of nocturnal foraging of the animals at the surface and the subsequent release of part of the ingested matter through egestion, respiration, and excretion in deep-water daytime refuges (e.g., Longhurst and Harrison 1989). According to various authors, cited by Pearre (2003), marine zooplankton can carry 13%–58% of the sinking particulate carbon, 8%–82% of the nitrogen, negligible amounts of the phosphorus (but in freshwater lakes, phosphorus export from the euphotic zone by migrating large-bodied *Daphnia* may be substantial; Wright and Shapiro 1984), and sometimes significant amounts of other materials such as radionuclides and metals (up to 80% of downward fluxes). The contribution of zooplankton to vertical fluxes is largely dependent on the taxonomic composition of local communities. The presence of large-bodied species, which are the most active migrants, such as euphausiids in the oceans, and *Daphnia*, and calanoid copepods in freshwaters, strengthens the zooplankton-mediated export of the materials from surface waters.

Anoxic bottom strata can be a barrier for vertical migrants. Downward migration depth may be constrained by an expanding hypoxic zone (e.g., due to eutrophication). Metabolic suppression resulting from oxygen deficiencies in deep-water strata may substantially reduce the contribution of migrating euphausids to the vertical flux of carbon and nitrogen by 49%–64% (Seibel et al. 2016). According to Honjo et al. (2008), the effects of zooplankton on vertical transport in the oceans become insignificant below a depth of 1.5 km due to reduced abundance of zooplankton. In lakes, however, active transport by zooplankton can directly supply the bottom sediments with material of surface origin (e.g., Madeira et al. 1982).

FUTURE DIRECTIONS

Planktonic animals live in a chemically complex habitat and are exposed to chemical signals from multiple sources. Part of this chemical information can be ignored, but detecting and interpreting some information may be crucial for fitness. Therefore, simultaneous exposure to kairomones from different predators, and to other information sources should be studied further, including information hierarchy, kairomone receptors, pathways, and processing of the information leading to the eventual migration response. Also, although examined in a few case studies (e.g., Pestana et al. 2013), the relative role of kairomones and alarm cues originating from injured conspecifics should be further elucidated. Furthermore, revealing the chemical origin of predator kairomones and alarm cues is an essential research challenge.

There is widespread variation in DVM behavior between individuals of the same species or even within clones of genetically identical individuals. The phenomenon itself has until now received relatively little attention (but see e.g., Dawidowicz 1993, Dawidowicz and Loose 1992a). It is not clear whether individual behaviors (so-called personalities; Dingemanse et al. 2010) are maintained as "conditional" strategies (phenotypically different individuals performing different migratory patterns to optimize fitness) or as "alternative" strategies (selection among several existing fitness peaks by an individual is flexible). Revealing the mechanisms that maintain such diversity is important because intraspecific behavioral diversity can affect population dynamics and the outcome of ecological interactions at the community level (see Community and Ecosystem Consequences section).

Similarly to vertebrates, melatonin synthesis in *Daphnia* (and probably also in other crustaceans) is coupled to expression of clock genes, and insect-like AANATs of crustaceans have a similar function in initiating melatonin synthesis as AANATs in vertebrates. Elucidation of the supposed coupling of cyclic clock gene expression in *Daphnia* (and, most probably, in other crustaceans) with melatonin production is very important for future ecological studies. This result may be supported by the daytime-dependent release of melatonin (and also corticosteroids; Möstl and Palme 2002), which, as was demonstrated by Foulkes et al. (1997), may alleviate stress responses. It also remains to be recognized whether the daytime-dependent synthesis of melatonin is responsible for *Daphnia*'s ability to behaviorally adjust to daily changes in the light regime. It has already been proposed that melatonin signaling plays a role in the circadian control of swimming to adjust the vertical position of marine annelids *Platynereis dumerilii* in response to ambient light (Tosches et al. 2014).

Although studies on the expression of clock genes in crustacean species at the mRNA and protein levels are very rare, they could serve to create a model of the crustacean molecular oscillator (Bernatowicz et al. 2016). Crustacean oscillators are considered adaptive mechanisms for living under a regimen of changing light conditions. More analyses of diel changes in the abundance of gene transcripts may reveal elements of the crustacean molecular oscillator (such a study has recently been completed for *Daphnia pulex* by Bernatowicz et al. 2016) and its input and output pathways.

One of the first consequences of ongoing and predicted climatic change is the extension of the stratification season, which can have major implications for water quality because a longer stratification period increases the risk of oxygen deficiencies in deep-water strata. Anoxic conditions at the bottom will enhance phosphorus release from the sediments to the lake in a process known as internal loading, which will support algal growth and bring about a subsequent decline in water transparency. Thus, the thickness of the surface layer penetrated by visual predators will shrink but, simultaneously, the availability of a deep-water refuge will also be reduced due to anoxia. Another concern is the frequency of deep mixing events in a lake. If deep mixing events occur less frequently or disappear, thermal stratification will be altered and migration behavior will change accordingly. This modification will impose ecosystem consequences, among them an altered efficiency of top-down control of algal biomass or nutrient transfer in the water column. The future consequences of climate warming may be mathematically modeled based on existing data or can be directly predicted via observations of lakes that are already undergoing related processes, such as lakes that are heated artificially (e.g., by power plants).

Widespread and common exposure to sky glow and artificial lighting can severely affect the foraging schedule of aquatic predators as well as the migratory behavior of animals (Navara and Nelson 2007 and references therein). Changes in ambient illumination drive migration patterns in many species. Exposure of *Daphnia* to urban light pollution in the wild was found to decrease the magnitude of migratory movements and the number of migrating individuals (Moore et al. 2001). Irregular light/dark patterns are now considered as endocrine disruptors for aquatic crustaceans. Many physiological effects of light pollution, most of which occur through endocrine pathways after exposure to extended periods of light, have been described (Navara and Nelson 2007). Increasing levels of urban sky glow and disturbances in the natural light cycle could bring about incidences of metabolic disorders, immunosuppression, or oxidative stress that can have further important ecological repercussions at the individual, population, and community levels. Circadian disruption accompanying exposure to light pollution and endocrine mediators that are potentially involved should be further investigated. A deeper understanding of the mechanisms by which exposure to an unnatural light regime may alter physiology and migratory behavior of aquatic crustaceans would be useful for modeling its consequences at the population and community levels (e.g., the outcome of primary producer-herbivore-predator interactions).

CONCLUSIONS

DVM can be considered the most powerful of all mechanisms enabling the existence of zooplankton in shelterless pelagic domains patrolled by relatively much larger and much faster swimming visually oriented fish predators.

Virtually all pelagic crustaceans, as well as representatives of many other planktonic taxa (including free-swimming ontogenetic stages of benthic animals), perform DVMs. The universality of DVM, combined with simplicity of its detection in the field and further mathematical description, makes it a convenient model behavior for developing the general ecoevolutionary concepts, such as trade-offs and habitat selection. Moreover, because zooplankton occupy the central position in pelagic food webs, their massive migrations dramatically affect the functioning of offshore biota, particularly the efficiency of primary production utilization, energy flow, and biogeochemical pathways of essential nutrients such as carbon fluxes.

Although the adaptive role, inducible character, and most ultimate and proximate aspects of DVM already have satisfactory explanations, there are still unresolved problems, including genetic and physiological mechanism underlying the evolution of DVM behavior. Other unresolved problems are linked to the complex impact of a fast-changing environment on DVM, including climate change and other factors that are now being recognized as new environmental threats generated by anthropopressure. The resulting community and ecosystem consequences, such as biomass and nutrients cycling in aquatic habitats, need further attention.

ACKNOWLEDGMENTS

Part of this study was subsidized by the Polish National Science Centre (NCN), grant number 2014/13/B/NZ8/04670. We also wish to express our gratitude to the editors of this volume of *The Natural History of Crustacea* for their great help in structuring the body of our chapter, as well as for many thoughtful comments to its content.

REFERENCES

Aldredge, A.L., and J.M. King. 1980. Effects of moonlight on the vertical migration patterns of demersal zooplankton. Ecology 44:133–156.

Bader, C. 1980. Some biological and ecological data on water-mites, mainly some significant data on the life-duration. International Journal of Acarology 6:239–243.

Bayly, I.A.E. 1986. Aspects of diel vertical migration in zooplankton, and its enigma variations. Limnology in Australia. Springer, The Netherlands.

Beklioglu, M., A.G., Gozen, F. Yildrim, P. Zorlu, and S. Onde. 2008. Impact of food concentration on diel vertical migration behavior of *Daphnia pulex* under fish predation risk. Hydrobiologia 614:321–327.

Bellingshausen, F.F. 1823. Doneseniye kapitana 2 ranga Bellinsgauzena iz Porta Zhaksona, o svoem plavanii [Dispatch from Captain Bellingshausen at Port Jackson about his voyage]. Zapiski izdavayemya gosudarstvennym admiralteiskim departamentom. Papers from the Russian Admiralty 5:201–219.

Bentkowski P., M. Markowska, and J. Pijanowska. 2010. Role of melatonin in the control of depth distribution of *Daphnia magna*. Hydrobiologia 643:43–50.

Berge, J., F. Cottier, K.S. Last, Ø. Varpe, E. Leu, J. Søreide, K. Eiane, S. Falk-Petersen, K. Willis, H. Nygård, D. Vogedes, C. Griffiths, G. Johnsen, D. Lorentzen, and A.S. Brierley. 2009. Diel vertical migration of Arctic zooplankton during the polar night. Biology Letters 5:69–72.

Bernatowicz, P., J. Kotwica-Rolińska, J., E. Joachimiak, A. Sikora, M.A. Polańska, J. Pijanowska, and P. Bębas. 2016. Temporal expression of the clock genes in the water flea *Daphnia pulex* (Crustacea: Cladocera). Journal of Experimental Zoology Part A: Ecological Genetics and Physiology 325:233–254.

Boersma, M., P. Spaak, and P.L. de Meester. 1998. Predator-mediated plasticity in morphology, life history, and behavior of *Daphnia*: the uncoupling of responses. The American Naturalist 152:237–248.

Bollens, S.M., and B.W. Frost. 1989. Predator-induced diel vertical migration in a planktonic copepod. Journal of Plankton Research 11:1047–1065.

Bollens, S.M., and B.W. Frost. 1991. Diel vertical migration in zooplankton: rapid individual response to predators. Journal of Plankton Research 13:1359–1365.
Bollens, S.M., K. Osgood, B.W. Frost, and S.D. Watts. 1993. Vertical distributions and susceptibilities to vertebrate predation of the marine copepods *Metridia lucens* and *Calanus pacificus*. Limnology and Oceanography 38:1827–1837.
Bollens, S.M., B.W. Frost, and J.R. Cordell. 1994. Chemical, mechanical and visual cues in the vertical migration behavior of the marine planktonic copepod *Acartia hudsonica*. Journal of Plankton Research 16:555–564.
Boriss, H., M. Boersma, and K.H. Wiltshire. 1999. Trimethylamine induces vertical migration in *Daphnia*. Nature 398.
Bowers, J.A. 1979. Zooplankton grazing in simulation models: the role of vertical migration. Pages 53–73 *in* D. Scavia, and A. Robertson, editors. Perspectives on lake ecosystem modelling. Ann Arbor Science, Ann Arbor, Michigan.
Brett, M.T. 1992. *Chaoborus* and fish-mediated influences on *Daphnia longispina* population structure, dynamics and life history strategies. Oecologia 89:69–77.
Brewer, M.C., P. Dawidowicz, and S.I. Dodson. 1999. Interactive effects of fish kairomone and light on *Daphnia* escape behavior. Journal of Plankton Research 21:1317–1335.
Brooks, J.L., and S.I. Dodson. 1965. Predation, body size and the composition of plankton. Science 150:26–35.
Burks, R.L., D. M. Lodge, E. Jeppesen, and T.L. Lauridsen. 2002. Diel horizontal migration of zooplankton: costs and benefits of inhabiting the littoral. Freshwater Biology 47:343–365.
Clarke, T.A. 1978. Diel feeding patterns of 16 species of mesopelagic fishes from Hawaiian waters. Fishery Bulletin 76:485–513.
Cohen, J.H., and R.B. Forward. 2009. Zooplankton diel vertical migration—a review of proximate control. Oceanography and Marine Biology—An Annual Review 47:77–111.
Covich, A.P. 2015. Freshwater crustaceans: Adaptations to complex inland habitats and species interactions. Pages 337–378 *in* M. Thiel and L. Watling, editors. The natural history of the crustacea, volume 2: Lifestyles and feeding biology. Oxford University Press, New York.
Cushing, D. 1951. The vertical migration of planktonic Crustacea. Biological Reviews 26:158–192.
Cuvier, G. 1817. Le Regne Animal. Vol. 17. Texte. Les Crustaces. Masson, Paris, France.
Dale, T., and S. Kaartvedt. 2000. Diel patterns in stage-specific vertical migration *of Calanus finmarchicus* in habitats with midnight sun. ICES Journal of Marine Science 57:1800–1818.
Dawidowicz, P. 1993. Diel vertical migration in *Chaoborus flavicans*: population patterns vs. individual tracks. Archiv für Hydrobiologie Beiheft Ergebnise der Limnologie 39:19–28.
Dawidowicz, P., and C.J. Loose. 1992a. Metabolic costs during predator-induced diel vertical migration of *Daphnia*. Limnology and Oceanography 37:1589–1595.
Dawidowicz, P., and C.J. Loose. 1992b. Cost of swimming by *Daphnia* during diel vertical migration. Limnology and Oceanography 37:665–669.
Dawidowicz, P., and M. Wielanier. 2004. Costs of predator avoidance reduce competitive ability of *Daphnia*. Hydrobiologia 526:105–109.
Dawidowicz, P., A. Prejs, A. Engelmayer, A. Martyniak, J. Kozłowski, L. Kufel, L., and M. Paradowska. 2002. Hypolimnetic anoxia hampers top–down food-web manipulation in a eutrophic lake. Freshwater Biology 47:2401–2409.
Dawidowicz, P., P. Prędki, and B. Pietrzak. 2010. Shortened lifespan: another cost of fish-predator avoidance in cladocerans? Hydrobiologia 643:27–32.
Dawidowicz, P., P. Prędki, and B. Pietrzak. 2013. Depth-selection behavior and longevity in *Daphnia*: an evolutionary test for the predation-avoidance hypothesis. Hydrobiologia 715:87–91.
Dean, C., A. Soloviev, A. Hirons, T. Frank, and J. Wood. 2016. Biomixing due to diel vertical migrations of zooplankton: comparison of computational fluid dynamics model with observations. Ocean Modelling 98:51–64.
Dingemanse, N.J., A.J. Kazem, D. Réale, and J. Wright. 2010. Behavioural reaction norms: animal personality meets individual plasticity. Trends in Ecology and Evolution 25:81–89.
de Meester, L. 1991. An analysis of the phototactic behaviour of *Daphnia magna* clones and their sexual descendants. Hydrobiologia 225:217–227.

de Meester, L. 1993. Genotype, fish-mediated chemical, and phototactic behavior in *Daphnia magna*. Ecology 74:1467–1474.

de Meester, L., and J. Pijanowska. 1997. On the trait-specificity of the response of *Daphnia* genotypes to the chemical presence of a predator. Pages 407–417 *in* P. Lenz, D.K. Hartline, J. Purcell, and D. Macmillian, editors. Zooplankton: sensory ecology and physiology. Gordon and Breach Publishers, Amsterdam, The Netherlands.

de Meester, L., L.J. Weider, and R. Tollrian. 1995. Alternative antipredator defenses and genetic-polymorphism in a pelagic predator-prey system. Nature 378:483–485.

de Meester, L., P. Dawidowicz, E. van Gool, and C. Loose. 1999. Ecology and evolution of predator induced behavior of zooplankton: depth selection behavior and diel vertical migration. Pages 160–176 *in* R. Tollrian, and C.D. Harvell, editors. The ecology and evolution of inducible defenses. Princeton University Press, New Jersey.

Derby, C.D., and M.J. Weissburg. 2014. The chemical senses and chemosensory ecology of crustaceans. Pages 263–292 *in* C.D. Derby, and M. Thiel, editors. The natural history of the Crustacea, volume 3: Nervous systems and control of behavior. Oxford University Press, New York.

de Stasio Jr., B.T. 1993. Diel vertical and horizontal migration by zooplankton: population budgets and the diurnal deficit. Bulletin of Marine Science 53:44–64.

Dewar, W.K., R.J. Bingham, R.L. Iverson, D.P. Nowacek, L.C. St Laurent, and P.H. Wiebe. 2006. Does the marine biosphere mix the ocean? Journal of Marine Research 64:541–561.

Ebert, D. 1993. The trade-off between offspring size and number in *Daphnia magna*: the influence of genetic, environmental and maternal effects. Archiv für Hydrobiologie, Suppl. 90:453–473.

Effertz, C., and E. von Elert. 2014. Light intensity controls anti-predator defenses in *Daphnia*: the suppression of life-history changes. Proceedings of the Royal Society of London B: Biological Sciences 281:20133250.

Elder, L.E., and B.A. Seibel. 2015. The thermal stress response to diel vertical migration in the hyperiid amphipod *Phronima sedentaria*. Comparative Biochemistry and Physiology Part A: Molecular and Integrative Physiology 187:20–26.

Escribano, R., and R. Riquelme-Bugueño. 2015. Planktonic crustaceans life styles in the water column. Pages 261–278 *in* M. Thiel, and L. Watling, editors. The natural history of the Crustacea, volume 2: Life styles and feeding biology. Oxford University Press, New York.

Escribano, R., P. Hidalgo, and C. Krautz. 2009. Zooplankton associated with the oxygen minimum zone system in the northern upwelling region of Chile during March 2000. Deep-Sea Research II 56:1083–1094.

Ferreira, P. A., and A. Orry. 2012. From *Drosophila* to humans: reflections on the roles of the prolyl isomerases and chaperones, cyclophilins, in cell function and disease. Journal of Neurogenetics 26:132–143.

Fiksen, Ø. 1997. Allocation patterns and diel vertical migration: modeling the optimal *Daphnia*. Ecology 78:1446–1456.

Forward, R.B. 1988. Diel vertical migration: zooplankton photobiology and behaviour. Oceanography and Marine Biology—An Annual Review 26:361–393.

Foulds, J.B., and J.C. Roff 1976. Oxygen consumption during simulated vertical migration in *Mysis relicta* (Crustacea, Mysidacea). Canadian Journal of Zoology 54:377–385.

Foulkes, N.S., J. Borjigin, and S.H. Snyder. 1997. Rhythmic transcription: the molecular basis of circadian melatonin synthesis. Trends in Neurosciences 20:487–492.

Frank, T. M., and E.A. Widder. 1997. The correlation of downwelling irradiance and staggered vertical migration patterns of zooplankton in Wilkinson Basin, Gulf of Maine. Journal of Plankton Research 19:1975–1991.

Gabriel, W., and B. Thomas. 1988. Vertical migration as evolutionary stable strategy. American Naturalist 132:199–216.

Garland, E.D., C.A. Zimmer, and S.J. Lentz. 2002. Larval distributions in inner-shelf waters: the roles of wind-driven cross-shelf currents and diel vertical migrations. Limnology and Oceanography 47:803–817.

Glantz, R. 2014. Visual systems of crustaceans. Pages 206–234 *in* C.D. Derby, and M. Thiel. The natural history of the Crustacea, volume 3: Nervous systems and control of behavior. Oxford University Press, New York.

Gliwicz, M.Z. 1986a. Predation and the evolution of vertical migration in zooplankton. Nature 320:746–748.

Gliwicz, Z.M. 1986b. A lunar cycle in zooplankton. Ecology 67:883–897.

Gliwicz, Z.M. 1990. Food thresholds and body size in cladocerans. Nature 343:638–640.

Gliwicz, Z.M. 2003. Between hazards of starvation and risk of predation: the ecology of offshore animals, volume 12. International Ecology Institute, Oldendorf/Luhe, Germany.

Gliwicz, Z.M., and C. Guisande. 1992. Family planning in *Daphnia*: resistance to starvation in offspring born to mothers grown at different food levels. Oecologia 91:463–467.

Gliwicz, Z.M., and W. Lampert. 1990. Food thresholds in *Daphnia* species in the absence and presence of blue-green filaments. Ecology 71:691–702.

Gliwicz, M.Z., and J. Pijanowska. 1988. Effect of predation and resource depth distribution on vertical migration of zooplankton. Bulletin of Marine Science 43:695–709.

Gonzalez, R.R., and R.A. Quiñones. 2002. Ldh activity in *Euphausia mucronata* and *Calanus chilensis*: implications for vertical migration behaviour. Journal of Plankton Research 24:1349–1356.

Harvell, C.D. 1990. The ecology and evolution of inducible defenses. Quarterly Review of Biology 65:323–340.

Haupt, F., M. Stockenreiter, M. Baumgartner, M. Boersma, and H. Stibor. 2009. *Daphnia* diel vertical migration: implications beyond zooplankton. Journal of Plankton Research 31:515–524.

Hays, G.C. 2003. A review of the adaptive significance and ecosystem consequences of zooplankton diel vertical migrations. Hydrobiologia 503:163–170.

Hidalgo, P., R. Escribano, and C.E. Morales. 2005. Ontogenetic vertical distribution and diel migration of the copepod *Eucalanus inermis* in the oxygen minimum zone off northern Chile (20–21 S). Journal of Plankton Research 27:519–529.

Honjo, S., S.J. Manganini, R.A. Krishfield, and R. Francois. 2008. Particulate organic carbon fluxes to the ocean interior and factors controlling the biological pump: a synthesis of global sediment trap programs since 1983. Progress in Oceanography 76:217–285.

Hrbáček, J., and M. Novotná-Dvořáková. 1965. Plankton of four backwaters related to their size and fish stock. Nakladatelství Československé akademie věd. 75:1–64.

Hugie, D.M., and L.M. Dill. 1994. Fish and game: a game theoretic approach to habitat selection by predators and prey. Journal of Fish Biology 45:151–169.

Hutchinson, G.E. 1967. A treatise on limnology. II. Introduction to lake biology and their limnoplankton. Wiley. New York.

Iwasa, Y. 1982. Vertical migration of zooplankton: a game between predator and prey. American Naturalist 120:171–180.

King, C.E., and M.R. Miracle. 1995. Diel vertical migration by *Daphnia longispina* in a Spanish lake: genetic sources of distributional variation. Limnology and Oceanography 40:226–231.

Klevjer, T.A., X. Irigoien, A. Røstad, E. Fraile-Nuez, V.M. Benítez-Barrios, and S. Kaartvedt. 2016. Large scale patterns in vertical distribution and behaviour of mesopelagic scattering layers. Scientific Reports 6: 19873; doi: 10.1038/srep19873.

Klyashtorin, L.B., and A.A Yarzhombek. 1973. Energy consumption in active movements of planktonic organisms. Oceanology 13:575–580.

Kreutzer, C., and W. Lampert. 1999. Exploitative competition in differently sized *Daphnia* species: a mechanistic explanation. Ecology 80:2348–2357.

Kunze E. 2011. Fluid mixing by swimming organisms in the low-Reynolds-number limit. Journal of Marine Research 69:591–60.

Kvam O.V., and O.T. Kleiven. 1995. Diel horizontal migration and swarm formation in *Daphnia* in the response to *Chaoborus*. Hydrobiologia 307:177–184.

Lampert, W. 1989. The adaptive significance of diel vertical migration of zooplankton. Functional Ecology 3:21–27.

Lampert, W. 1993. Ultimate causes of diel vertical migration of zooplankton: new evidence for the predator-avoidance hypothesis. Archiv für Hydrobiologie Beiheft Ergebnisse der Limnologie 39:79–88.

Lampert, W., and C.J. Loose. 1992. Plankton towers: bridging the gap between laboratory and field experiments. Archiv für Hydrobiologie 126:53–66.

Lampert, W., and B.E. Taylor. 1985. Zooplankton grazing in a eutrophic lake: implications of diel vertical migration. Ecology 66:68–82.

Larsson, P., and S. Dodson. 1993. Invited review: chemical communication in planktonic animals. Archiv für Hydrobiologie 129:129–155.

Last K.L., L. Hobbs, J. Berge, A.S. Brierley, and F. Cottier. 2016. Moonlight drives ocean-scale mass vertical migration of zooplankton during the Arctic winter. Current Biology 26:244–251.
Leibold, M., and A.J. Tessier 1991. Contrasting patterns of body size for *Daphnia* species that segregate by habitat. Oecologia 86:342–348.
Longhurst, A.R., and W.G. Harrison. 1989. The biological pump: profiles of plankton production and consumption in the upper ocean. Progress in Oceanography 22:47–123.
Loose, C.J. 1993. Lack of endogenous rhythmicity in *Daphnia* diel vertical migration. Limnology and Oceanography 38:1837–1841.
Loose, C.J., and P. Dawidowicz. 1994. Trade-offs in diel vertical migration by zooplankton: the costs of predator avoidance. Ecology 75:2255–2263.
Loose, C.J., E. von Elert, and P. Dawidowicz. 1993. Chemically-induced diel vertical migration in *Daphnia*: a new bioassay for kairomones exuded by fish. Archiv für Hydrobiologie 126:329–337.
Macháček, J. 1991. Indirect effect of planktivorous fish on the growth and reproduction of *Daphnia galeata*. Hydrobiologia 225:193–197.
Madeira, P.T., A.S. Brooks, and D.B. Seale. 1982. Excretion of total phosphorus, dissolved reactive phosphorus ammonia and urea by Lake Michigan *Mysis relicta*. Hydrobiologia 93:145–154.
Mangel, M., and C. Clark. 1988. Dynamic modelling in behavioral ecology. Princeton University Press, Princeton, New Jersey.
Madin, L.P., P. Kremer, and S. Hacker. 1996. Distribution and vertical migration of salps (Tunicata, Thaliacea) near Bermuda. Journal of Plankton Research 18:747–755.
Marcogliese, D.J., and G.W. Esch. 1992. Alterations of vertical distribution and migration of zooplankton in relation to temperature. American Midland Naturalist 128:139–155.
Maszczyk, P., and M. Bartosiewicz. 2012. Threat or treat: the role of fish exudates in the growth and life history of *Daphnia*. Ecosphere 3:1–19.
Maynard Smith, M., and G.R. Price. 1973. The logic of animal conflict. Nature 246:15–18.
Mellon, De F., Jr. 2014. Sensory systems of crustaceans. Pages 49–84 *in* N.C. Derby, and M. Thiel, editors. The natural history of the Crustacea, volume 3: Nervous systems and control of behavior. Oxford University Press, New York.
Mikulski, A., P. Bernatowicz, M. Grzesiuk, M. Kloc, and J. Pijanowska. 2011. Differential levels of stress proteins (HSPs) in male and female *Daphnia magna* in response to thermal stress: a consequence of sex-related behavioral differences? Journal of Chemical Ecology 37:670–676.
Moore, M.V., S.M. Pierce, H.M. Walsh, S.K. Kvalvik, and J.D. Lim. 2001. Urban light pollution alters the diel vertical migration of *Daphnia*. Internationale Vereinigung für Theoretische und Angewandte Limnologie Verhandlungen 27:779–782.
Möstl, E., and R. Palme. 2002. Hormones as indicators of stress. Domestic Animal Endocrinology 23:67–74.
Muscatine, L., and R.E. Marian. 1982. Dissolved inorganic nitrogen flux in symbiotic and nonsymbiotic medusae. Limnology and Oceanography 27:910–917.
Mueller, J.C., and A. Seitz. 1993. Habitat partitioning and differential vertical migration of some *Daphnia* genotypes in a lake. Archiv für Hydrobiologie Beiheft Ergebnisse der Limnologie 39:167–174.
Navara, K.J., and R.J. Nelson. 2007. The dark side of light at night: physiological, epidemiological, and ecological consequences. Journal of Pineal Research 43:215–224.
Neverman, D., and W.A. Wurtsbaugh. 1994. The thermoregulatory function of diel vertical migration for a juvenile fish, *Cottus extensus*. Oecologia 98:247–256.
Nelson, D.R., J.N. McKibben, W.R. Strong Jr, C.G. Lowe, J.A. Sisneros, D.M. Schroeder, and R.J. Lavenberg. 1997. An acoustic tracking of a megamouth shark, *Megachasma pelagios*: a crepuscular vertical migrator. Environmental Biology of Fishes 49:389–399.
Noss, C., and A. Lorke. 2014. Direct observation of biomixing by vertically migrating zooplankton. Limnology and Oceanography 59:724–732.
Ohman, M.D., B.W. Frost, and E.B.Cohen. 1983. Reverse diel vertical migration: an escape from invertebrate predators. Science 220:1404–1407.
Pearre, S. 1974. Ecological studies of three west-Mediterranean chaetognaths. Investigaciones Pesqueras 38:325–369.
Pearre, S. 1979. Problems of detection and interpretation of vertical migration. Journal of Plankton Research 1:29–44.

Pearre, S. 2003. Eat and run? The hunger/satiation hypothesis in vertical migration: history, evidence and consequences. Biological Reviews 78:1–79.

Pennak, R.W. 1944. Diurnal movements of zooplankton in some Colorado mountain lakes. Ecology 25:387–403.

Pennington, J.T., and R.B. Emlet. 1986. Ontogenetic and diel vertical migration of a planktonic echinoid larva, *Dendraster excentricus* (Eschscholtz): occurrence, causes, and probable consequences. Journal of Experimental Marine Biology and Ecology 104:69–95.

Pestana, J.L.T., D.J. Baird, and A.M.V.M. Soares. 2013. Predator threat assessment in *Daphnia magna*: the role of kairomones versus conspecific alarm cues. Marine and Freshwater Research 64:679–686.

Petzoldt, T., L. Rudolf, K. Rinke, and J. Benndorf. 2009. Effects of zooplankton diel vertical migration on a phytoplankton community: a scenario analysis of the underlying mechanisms. Ecological Modelling 220: 1358–1368.

Pijanowska, J. 1993. Diel vertical migration in zooplankton: fixed or inducible behavior? Archiv für Hydrobiologie Beiheft Ergebnisse der Limnologie 39:89–97.

Pijanowska, J. 1997. Alarm signals in *Daphnia*? Oecologia 112:12–16.

Pijanowska, J., and P. Dawidowicz. 1987. The lack of vertical migration in *Daphnia*: the effect of homogeneously distributed food. Hydrobiologia 148:175–181.

Pijanowska, J., and M. Kloc. 2004. *Daphnia* response to predation threat involves heat-shock proteins and the actin and tubulin cytoskeleton. Genesis 38:81–86.

Pijanowska, J., L.J. Weider, and W. Lampert. 1993. Predator-mediated genotypic shifts in a prey population: experimental evidence. Oecologia 96:40–42.

Pörtner, H.O. 2002. Climate variations and the physiological basis of temperature dependent biogeography: systemic to molecular hierarchy of thermal tolerance in animals. Comparative Biochemistry and Physiology Part A: Molecular and Integrative Physiology 132:739–761.

Raby, D., Y. Lagadeuc, J.J. Dodson, and M. Mingelbier.1994. Relationship between feeding and vertical distribution of bivalve larvae in stratified and mixed waters. Marine Ecology-Progress Series 103:275–275.

Reichwaldt, E.S., and H. Stibor. 2005. The impact of diel vertical migration of *Daphnia* on phytoplankton dynamics. Oecologia 146:50–56.

Ringelberg, J. 1964. The positively phototactic reaction of *Daphnia magna* Straus: a contribution to the understanding of diurnal vertical migration. Netherlands Journal of Sea Research 2:319–406.

Ringelberg, J. 2010. Diel vertical migration of zooplankton in lakes and oceans. Springer, Dordrecht, The Netherlands.

Ringelberg, J., J.V. Kasteel, and H. Servaas. 1967. The sensitivity of *Daphnia magna* Straus to changes in light intensity at various adaptation levels and its implication in diurnal vertical migration. Journal of Comparative Physiology A: Neuroethology, Sensory, Neural, and Behavioral Physiology 56:397–407.

Sakwińska O., and P. Dawidowicz. 2005. Life history strategy and depth selection behavior as alternative anitpredator defenses among natural *Daphnia hyalina* populations. Limnology and Oceanography 50:1284–1289.

Schwarzenberger, A., and A. Wacker. 2015. Melatonin synthesis follows a daily cycle in *Daphnia*. Journal of Plankton Research 37:636–644.

Seapy, R.R. 1990. Patterns of vertical distribution in epipelagic heteropod molluscs off Hawaii. Marine Ecology Progress Series 60:235–246.

Seibel, B.A., J.L. Schneider, S. Kaartvedt, K.F. Wishner, and K.L. Daly. 2016. Hypoxia tolerance and metabolic suppression in oxygen minimum zone euphausiids: implications for ocean deoxygenation and biogeochemical cycles. Integrative and Comparative Biology 56:510–523.

Sell, A.F. 1998. Adaptation to oxygen deficiency: contrasting patterns of haemoglobin synthesis in two coexisting Daphnia species. Comparative Biochemistry and Physiology Part A: Molecular and Integrative Physiology 120:119–125.

Stearns, S.C. 1992. The evolution of life histories. Oxford University Press, Oxford, England.

Stibor, H. 1992. Predator induced life-history shifts in a freshwater cladoceran. Oecologia 92:162–165.

Stich, H.B., and W. Lampert. 1981. Predator evasion as an explanation of diurnal vertical migration by zooplankton. Nature 293:396–398.

Swift, M.C., and A.Y. Fedorenko. 1975. Some aspects of prey capture by *Chaoborus* larvae. Limnology and Oceanography 20:418–425.

Ślusarczyk, M., and B. Pinel-Alloul. 2010. Depth selection and life history strategies as mutually exclusive responses to risk of fish predation in *Daphnia*. Hydrobiologia 643:33–41.

Tessier, A.J., and N.L. Consolatti. 1989. Variation in offspring size in *Daphnia* and consequences for individual fitness. Oikos 269–276.

Tollrian, R., and C.D. Harvell, editors. 1999. The ecology and evolution of inducible defenses. Princeton University Press, Princeton, New Jersey.

Torres, J.J., and J.J. Childress. 1983. Relationship of oxygen consumption to swimming speed in *Euphausia pacifica*. Marine Biology 74:79–86.

Tosches, M.A., D. Bucher, P. Vopalensky, and D. Arendt, 2014. Melatonin signaling controls circadian swimming behavior in marine zooplankton. Cell 159:46–57.

Uye S., C. Huang, and T. Onbe. 1990. Ontogenetic diel vertical migration of the planktonic copepod *Calanus sinicus* in the Inland Sea of Japan. Marine Biology 104:389–396.

van Gool, E. and J. Ringelberg. 2002. Relationship between fish kairomone concentration in a lake and phototactic swimming by *Daphnia*. Journal of Plankton Research 24:713–721.

von Elert, E., and C.J. Loose. 1996. Predator-induced diel vertical migration in *Daphnia*: enrichment and preliminary chemical characterization of a kairomone exuded by fish. Journal of Chemical Ecology 22:885–895.

von Elert, E., and G. Pohnert. 2000. Predator specificity of kairomones in diel vertical migration of *Daphnia*: a chemical approach. Oikos 88:119–128.

von Elert, E., and H. Stibor. 2006. Predator-mediated life history shifts in *Daphnia*: enrichment and preliminary chemical characterization of a kairomone exuded by fish. Archiv für Hydrobiologie 167:21–35.

von Elert, E., C. Brönmark, and L.A. Hansson. 2012. Information conveyed by chemical cues. Chemical Ecology in Aquatic Systems 19–38.

Vos, M., B.J.G. Flik, J. Vijverberg, J. Ringelberg, and W.M. Mooij, 2002. From inducible defences to population dynamics: modelling refuge use and life history changes in *Daphnia*. Oikos 99:386–396.

Voznesensky, M., P.H. Lenz, C. Spanings-Pierrot, and D.W. Towle. 2004. Genomic approaches to detecting thermal stress in *Calanus finmarchicus* (Copepoda: Calanoida). Journal of Experimental Marine Biology and Ecology 311:37–46.

Webster, C.N., C. Hansson, T. Didrikas, E. Gorokhova, H. Peltonen, A.S. Brierley, and M. Lehtiniemi. 2015. Stuck between a rock and a hard place: zooplankton vertical distribution and hypoxia in the Gulf of Finland, Baltic Sea. Marine Biology 162:1429–1440.

Weider, L.J. 1984. Spatial heterogeneity of *Daphnia* genotypes: vertical migration and habitat partitioning. Limnology and Oceanography 29:225–235.

Wilhelmus, M.M., and J.O. Dabiri. 2014. Observations of large-scale fluid transport by laser-guided plankton aggregations. Physics of Fluids 26:101302.

Wright, D.I., and J. Shapiro. 1984. Nutrient reduction by biomanipulation: An unexpected phenomenon and its possible cause. Verhandlung Internationale Vereinigung Limnologie 22:518–524.

Worthington, E.B. 1931. Vertical movements of fresh-water macroplankton. Internationale Revue der Gesamten Hydrobiologie 25:394–436.

Zaret, T.M., and J.S. Suffern. 1976. Vertical migration in zooplankton as a predator avoidance mechanism. Limnology and Oceanography 21:804–813.

10

UNCHARTED TERRITORIES: DEFENSE OF SPACE IN CRUSTACEA

Melissa Hughes and Whitney L. Heuring

Abstract

Territoriality is a special case of resource defense, in which space is actively defended for exclusive use. As active defense is likely to be costly, territoriality is expected only when the benefits of exclusivity outweigh these costs. In most territorial species of noncrustacean taxa, the defended space includes resources critical for reproduction or food. These resources are not only critical for reproductive success, but also are vulnerable to "looting", that is, the value of these resources may be reduced through short-term intrusions, even without loss of ownership, thus providing an advantage for active defense of exclusive space. Many crustaceans defend space, particularly burrows or other shelters that are refuges from predation or environmental stressors. While protection is obviously a critical resource, it is not a resource that necessarily requires exclusivity; indeed, many crustaceans that depend upon shelters for protection do not defend them for exclusive use. Nonetheless, many crustacean taxa aggressively defend exclusive access to their shelters. Crustaceans, then, may be especially suitable for testing alternative hypotheses of territoriality, including the potential benefits of interindividual spacing rather than defense of space per se. It is also worth considering a null hypothesis for territoriality: aggressive defense of space in crustaceans may be an artifact of relatively sedentary species with high intraspecific aggression favored in other contexts, rather than aggression favored for defense of particular resources. In addition to these questions, much remains to be learned about territorial behaviors in crustaceans. Most notably, the boundaries of defended space are unknown in many taxa. Understanding the boundaries of defended space is important for understanding the ecological consequences of territoriality, as well as aspects of territory acquisition and the roles of neighbor relationships and territorial advertisement signals in territory defense. Many crustacean territories appear to differ from those described for other animals, especially terrestrial species; it is not clear, however, whether these differences are due to differences in function or habitat, or rather result from our incomplete knowledge of crustacean territoriality.

Life Histories. Edited by Gary A. Wellborn and Martin Thiel.

INTRODUCTION

Competition over resources is widespread, if not ubiquitous, across animal taxa. Defense of space, then, can be seen as a special case of resource defense, whether that resource is space around an individual that moves with it (i.e., "personal space"), or a home range or territory, in which the defended space is a specific geographic location that is fixed for at least some period of time (Brown and Orians 1970). The distinction between a home range and a territory is not always clear, and usage of these terms differs across taxa (see review in Nice 1941, Maher and Lott 1995). In practice, this distinction is a false dichotomy, with species showing a range of behaviors intermediate in many characteristics typical of home ranges or territories (Kaufmann 1983). What is of biological interest is not the classification of spatial defense according to rigidly applied definitions, but rather understanding differences in *how* space is defended and the evolutionary consequences of spatial defense across taxa.

Traditionally (Mayr 1935, Nice 1941), territories are defined as spaces that are actively defended for exclusive use of the territory holders (i.e., individuals, pairs, or groups), whereas home ranges are defined as simply the space habitually used by the animal. This definition of territory contains two critical elements: active defense and exclusive use. The latter is self-explanatory. Whereas home ranges are not necessarily exclusive and may show considerable overlap, territories are expected to have little or no overlap between adjacent neighbors, and territory holders will chase out intruders when detected. In a territorial system, if the number of individuals exceeds available space, the exclusive defense of space commonly results in "nonterritorial" individuals (Nice 1941, Brown and Orians 1970), often termed *floaters*, who have either been unable to compete for space or have opted to delay territoriality rather than defend space in suboptimal habitats (Lee 2005). Floaters may reside along the edges of territorial space or range across multiple territories, waiting for an opportunity to overthrow a territory holder (Smith 1978). Home ranges, in contrast, are not a limited resource: by definition, all animals have a home range. There are no floaters or individuals excluded from having a home range, although lower quality individuals may be relegated to lower-quality habitats.

The role of "active defense" in territoriality is worth further consideration. Territories are defended against intruders when detected, as described above. Similarly, nonterritorial animals may engage in aggressive behavior when encountering each other in areas of home range overlap (Brown and Orians 1970). However, territoriality often involves behaviors that exclude intruders *before* intruders are detected—in effect, defense even in the absence of intruders. Such defense might involve regular patrolling behavior to search for potential intruders; active defense also includes the production of territorial advertisement (or "keep out") signals (Mayr 1935, Nice 1941), which are produced even when a receiver, such as a potential intruder, is not immediately evident. Because home ranges are not actively defended against potential intruders, they typically do not involve patrolling or advertisement signals. Thus, defense of space in a home range is similar to defense of any other resource, in that in most cases, energy is expended in defense only when the resource is immediately threatened by a competitor. The maintenance of exclusive territories, in contrast, includes not only expenditure of energy to evict intruders, but also energy expenditure to *prevent* intrusions through patrolling and territorial advertisement signals (Stamps 1994). Territoriality is thus expected only when the benefits of exclusive use of space exceed these costs (Brown and Orians 1970, Stamps 1994). Accordingly, many species are territorial only for limited times, often only during the breeding season; similarly, if the benefits of exclusivity are sex-specific, territories may also be sex-specific, such that one or both sexes defend exclusive space against members of the same but not opposite sex.

Territories and home ranges, then, differ in a number of characteristics, including degree of overlap, behaviors involved in active defense, and the presence of floaters (Table 10.1). All of these characteristics are functionally related to the degree to which the space defended is "exclusive" to the owner. Exclusivity can be difficult to measure in the wild, however, and implies a binary condition (i.e., space is either exclusive or it is not); in practice, territories may have different amounts

Table 10.1. Variation in defense of space. Rather than being discrete categories, home ranges and territories are better viewed as ends of a spectrum described by the characteristics below.

	More "home range-like"	More "territory-like"
Exclusivity	Low	High
Overlap with neighbors	High	Low
Compressibility	Probably low	May be high
Defense	When challenged; site-specific dominance	Active throughout space; patrol to prevent/evict intruders
Signals	Receiver (intruder)-directed; primarily in interior	Produced in the absence of intruders; primarily on boundaries
"Floaters"/individuals excluded from system	Few/none	May be common

of overlap and territory holders may allow different levels of intrusion (Kaufmann 1983). Because applying an arbitrary threshold as to the "degree of exclusivity" for a space to be classified as a territory versus home range seems counterproductive, some have argued for disposal of the distinction altogether and instead view competition for space in terms of site-specific dominance: competitive interactions in which an individual's likelihood of winning depends on location, such that individuals tend to be dominant within their own space (Kaufmann 1983). Site-specific dominance may be a productive model for understanding some variation across taxa in defense of space, but it does not address behaviors such as patrolling and advertisement signaling, which clearly play a substantial role, and at the cost of great energetic investment, in the defense of space in many taxa.

Hence, the central issue of territoriality is economics: the benefits of exclusive access to space (or the resources therein) must be sufficient to warrant the costs associated with active defense (Brown 1964, Brown and Orians 1970, Stamps 1994). Perhaps not surprisingly, then, territories are most commonly associated with reproductive resources (mating, nesting sites) or food (Stamps 1994). Indeed, early attempts to classify territories by function focused almost exclusively on reproductive and food resources (Table 10.2). Although these early descriptions of territorial functions were primarily based on terrestrial territories, the roles of reproductive and foraging resources in territoriality are widespread (Stamps 1994), including many aquatic taxa. For example, some fish aggressively exclude others from nesting and mating areas (e.g., Moyer and Sawyers 1973), whereas other species defend feeding territories (e.g., Robertson and Polunin 1981); similarly, limpets defend algal gardens as feeding territories (Stimson 1970, Branch 1975, Shanks 2002).

In addition to being critical to survival and reproductive success, food and reproductive resources share another characteristic that may be important to understanding territorial behavior: they are vulnerable to looting (i.e., loss in value resulting from brief incursions). Intruding neighbors or floaters can steal food resources. Nests can be damaged or parasitized by intruding females, eggs or young can be consumed, and intruding males may seek extra-pair or sneak fertilizations. Thus the value of food or reproductive resources within territorial space may be reduced by intrusions even if the territory holder's ownership of that space remains intact; correspondingly, the benefits associated with ownership of that space depend on exclusive access and preventing even brief incursions into the defended space.

Table 10.2. Classifications of territories according to the function of defended space. Note the emphasis on reproductive resources.

Function	Mayr (1935)	Nice (1941)	Hinde (1956)
All purpose (includes mating, nesting, foraging area)	I ("true territory")	A	A
Mating/nesting area	II(a)	B	B
Mating only	II(b)	C	D
Feeding only (between breeding seasons)	III	E ("winter territories")	
Nest only	IV	D	C
Roosting/refuges		F	

Defense of space differs from defense of other resources in additional interesting dimensions. First, until the habitat is fully saturated, space differs from other resources in being readily compressible or divisible (Stamps and Krishnan 1998). Indeed, even early discussions of territoriality note that territory holders are sometimes willing to cede part of their territory to intruders, suggesting that (1) they had initially defended more space than necessary (Nice 1941) or (2) the value of the additional space no longer exceeded the costs of defending it. If an intruder repeatedly challenges a territory holder in a particular subset of the territory and the costs to the territory holder of continued defense are higher than the advantage of maintaining that location, experience-based models of territoriality predict that territory holders may cease to defend that location even after winning all of the competitive interactions (Stamps and Krishnan 1999). In other words, to the extent that the defended space is subdividable or compressible, competitors for even relatively exclusive space may be able to acquire that resource without actually winning any competitive interactions (Stamps and Krishnan 1997). Second, defending a geographic location provides competitors with stable relationships; depending on degree of site fidelity, individuals may have long-term relationships with their territorial neighbors. It may be advantageous, then, to be able to distinguish between neighbors and strangers, or even between individual neighbors. Such recognition abilities have been demonstrated in a variety of territorial species (see review in Temeles 1994). Certainly, categorical or individual recognition are not restricted to species that defend space, but such recognition can have an important influence on territory defense behaviors. Because neighbors are competitors with whom a territory holder has established relationships, neighboring individuals may show reduced aggression toward territorial neighbors or even form alliances under some conditions to exclude new invaders (i.e., Dear Enemy Effect; Getty 1987, Temeles 1994).

As this chapter considers what is known about territoriality in crustaceans, we will use the term *territory* rather loosely, to refer to any defended space that does not move with the individual. That is, we exclude "individual space" and short-term defense of females (e.g., "neighborhoods of dominance," sensu Correa and Thiel, 2003). We do this not to ignore the differences between home ranges and territories described above but to focus on characteristics of defense rather than classification. The interesting questions with regard to crustacean defense of space lie not in whether crustaceans defend territories sensu stricto. Rather, these relate to the degree to which crustaceans' defense of space shares characteristics that might be more "territory-like" or "home range-like" (Table 10.1) and how a better understanding of crustacean defense of space might better inform understanding of variation in spatial defense.

For most crustaceans, dispersal occurs during early life history stages; perhaps not surprisingly then, we found no studies of territorial behavior in larval crustaceans. Indeed, most studies of territorial behavior focus on adults and so this chapter does as well, although we note when juveniles or younger stages are known to engage in territorial behavior. In this chapter, we first consider similarities and differences in function between crustacean territories and those described for other taxa and then explore behavioral aspects of territoriality in crustaceans.

TERRITORIALITY IN CRUSTACEANS

Many crustaceans possess formidable weaponry and notoriously aggressive dispositions. It is not surprising, then, that many benthic and terrestrial species defend space. Indeed, the pugnacious behavior typical of many crustaceans suggests an alternative hypothesis to territoriality: apparent defense of space could simply be a byproduct of high levels of aggression directed at conspecifics by relatively sedentary individuals, not defense of resources in that space. A direct test of this alternative hypothesis would require explicit comparisons of aggressive behavior both at and away from the presumed territory. Although some crustaceans that defend space also engage in aggressive behavior outside of their defended space (e.g., big-clawed snapping shrimp, *Alpheus heterochaelis*; Hughes 1996), explicit comparisons of aggressive behavior across spatial contexts are apparently lacking. The byproduct hypothesis also predicts low site fidelity, with the defended space moving with the defender, while territoriality predicts high and/or strategic site fidelity—defending the resource as a function of its value and moving when higher value resources can be obtained. Consistent with the byproduct hypothesis, some crustaceans defend shelters or refuges while using them but change refuges frequently. For example, Norway lobsters (*Nephrops norvegicus*) show low burrow fidelity, and the same burrow may be used by different individuals on different days (Chapman and Rice 1971). The freshwater crayfish *Orconectes virilis* shows considerable variation in site fidelity, with some individuals moving frequently and others remaining at the same location for months; females are more site faithful, but also show greater variation in burrow fidelity (Hazlett et al. 1974). In the burrowing crayfish *Parastacus pugnax*, variation in burrow fidelity is related to environmental constraints; burrow fidelity is high during the dry season, when soil conditions make burrow construction difficult, but lower during the wet season (Palaoro et al. 2016). Low site fidelity is unlikely to be the result of cognitive constraints; the red swamp crayfish *Procambarus clarkii* is able to learn and remember precise spatial locations, in spite of showing low burrow fidelity in the wild (Barbaresi and Gherardi 2006).

Low site fidelity, however, is not necessarily inconsistent with territoriality. In terrestrial vertebrates, variation in site fidelity among individuals of the same species may reflect individual differences in territorial defense strategies, including strategic movements to territorial locations of higher quality (Hughes and Hyman 2011 and references within). Understanding variation in site fidelity in crustaceans, then, requires an understanding of what characteristics contribute to variation in territory quality. The symbiotic crab *Allopetrolisthes spinifrons* aggressively defends its host anemones (Baeza et al. 2002), and site fidelity in the wild is longer for crabs inhabiting larger and more dispersed host species in the subtidal zone, with more frequent movements between hosts occurring for crabs inhabiting smaller, more abundant hosts in the intertidal zone (Thiel et al. 2003). Unfortunately, data on site fidelity (such as how long territory holders defend a particular location, or the likelihood of returning to the same shelter after forays to search for mates or food) and the consequences of relocation (such as whether individuals move to obtain a higher quality territory, or the extent to which site fidelity depends on territory quality) appear to be lacking in most territorial crustaceans.

What Space is Defended—and Why?

The overwhelming majority of crustacean territories are some form of shelter. Burrows, tubes, or, in the case of symbiotic crustaceans (Baeza 2015), shelter in the form of other organisms, are nearly ubiquitous as defended spaces across crustacean taxa. Indeed, the sole exceptions appear to be the amphipods *Dyopedos monacanthus* and *Dyopedos porrectus*, which build and aggressively defend masts (Mattson and Cedhagen 1989, Thiel 1997; Fig. 10.1), which appear to provide no protective function. In *D. monacanthus*, females build the masts (or "mud whips") and typically remain near their base, while their offspring reside along the length of the mast (Thiel 1997). These masts project into the water column facilitating food collection. In addition, they serve as a locus for mating and for female defense of juveniles, typically from intruding conspecifics (Thiel 1997). These amphipod masts, then, seem most similar to "all purpose" territories described in terrestrial systems, corresponding to Mayr's (1935) "true territories" (Table 10.2).

Symbiotic territorial crustaceans similarly defend territories that readily correspond to "true territories" (Mayr 1935) in function. These are a particularly interesting group of territorial crustaceans because the territory they defend is not a burrow but rather their host organism. For symbiotic territorial crustaceans, their territory (host) often provides food, shelter and a place to breed. Indeed, Ďuriš et al. (2011) contended that sponge-dwelling crustaceans are more likely to be parasitic than commensal. For example, the porcelain crab *Allopetrolisthes spinifrons* defends its host sea anemone *Phymactis clematis* from intruders (Baeza et al. 2002, Baeza and Thiel 2003); the anemone serves as both a refuge from predators, as well as a food source for the territorial resident crab (reviewed in Baeza et al. 2002). A similar relationship exists between xanthid crabs (genus *Trapezia*) and pocilloporid corals (Knudsen 1967). Pairs of xanthid crabs aggressively defend their coral colony from intruders, with each sex repelling same-sex intruders (Huber 1987). A less-studied species, the symbiotic shrimp *Vir euphyllius*, can be found in heterosexual pairs on small caryophyllid corals

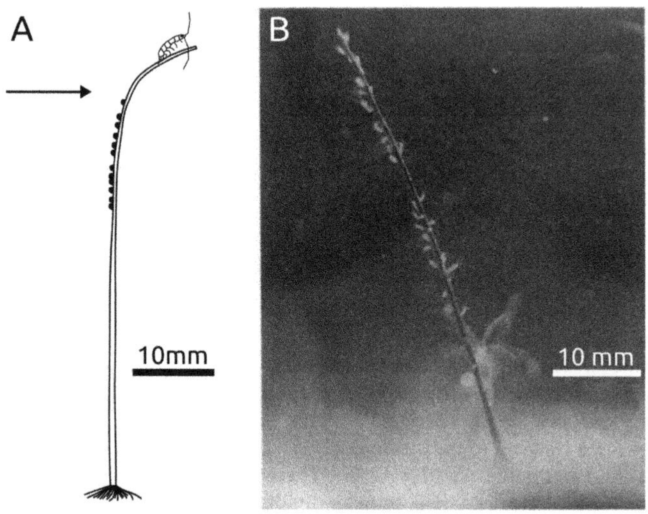

Fig. 10.1.
Unusual "mast" or "mud whip" territory of the amphipod *Dyopedos monacanthus*. These "all purpose" territories serve in foraging, mating, and protection of young. (A) Arrow indicates direction of water flow. Modified from Mattson and Cedhagen (1989), with permission from Elsevier. (B) Photograph showing female at base with young distributed along the upper portions. Modified from Thiel (1997), with permission from the Marine Biological Association of the United Kingdom.

(genus *Euphyllia*), suggesting that these shrimp may also be defending their coral from intruders (Marin 2007).

Given that symbiotic crustaceans defend all-purpose territories, when generations overlap or dispersal is limited, territories can become family affairs, resulting in interesting social dynamics. One of most terrestrial of all crabs, the Jamaican bromeliad crab (*Metopaulias depressus*), has an obligate relationship with bromeliads (Diesel 1989, Diesel and Schubart 2007). Bromeliad crab colonies, composed of an adult female and up to two generations of young, live in the freshwater-holding leaves of bromeliads; the bromeliad provides the crabs with food, shelter, and a place to mate (Diesel 1989, Diesel and Schubart 2007). The adult female crab will aggressively defend her bromeliad plant from intruders, securing her resource as well as protecting her offspring, while older offspring contribute both to defense of the bromeliad and to brood care of younger siblings (Diesel and Schubart 2007). Similarly, in the eusocial snapping shrimp (*Synalpheus* spp.), restricted dispersal abilities combined with patchy distributions of host sponges lead to large family groups cooperatively defending their hosts from intruders (Duffy et al. 2002). Sponges provide not only protection but also food resources for the defenders (Rützler 1976). Eusociality provides an advantage in defending these vital resources (Duffy and Macdonald 2010), with both male and female shrimp contributing to host defense (Tóth and Bauer 2007). The social structures within these colonies are varied and can be quite complex. For example, species differ in the degree to which 1 female monopolizes reproduction within the colony, suggesting variation in female-female competition within colonies (Chak et al. 2015b).

For symbiotic crustaceans, variation in territorial behavior and site fidelity may be explained by a combination of ecological and behavioral variables (Fig. 10.2). As modeled by Baeza and Thiel (2007), host guarding (i.e., territorial behavior) is likely to be favored in symbiotic crustaceans when potential hosts are rare or have a highly scattered distribution, and when the risk of predation is high; host switching (the inverse of site fidelity) is predicted to show an opposite response, higher when hosts are abundant and/or when the risk of predation is low. Territorial behavior in symbiotic crustaceans is also more likely when the host is small relative to the defender and has a simple structure making it economically defensible while site fidelity is not predicted to vary with host size and shape.

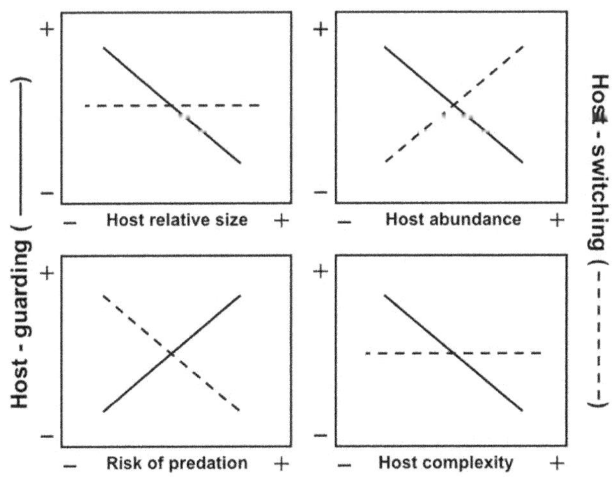

Fig. 10.2.
Territorial behavior (i.e., host guarding) and site fidelity (i.e., the inverse of host switching) in symbiotic crustaceans. From Baeza and Thiel (2007), with permission from Oxford University Press.

Generally speaking, however, the vast majority of crustacean territories consist of burrows or other shelters, frequently constructed by the crustaceans themselves, which typically function primarily as protection from predators or from environmental stressors (reviewed in Atkinson and Eastman 2015). In other words, most crustacean territories are inherently different in function from classical descriptions (Table 10.2); they are also more limited in function than the all-purpose territories of symbiotic crustaceans, and while the habitat suitable for burrow construction may be limited, the distribution of most crustacean territories is not limited by the distribution of a specific host.

Exclusivity of Defended Space and Relationship to Function

Many crustaceans aggressively exclude intruders from their burrows or shelters, resulting in high degrees of exclusivity. The degree to which exclusivity is necessary for or enhances the protective function of burrows, is unclear, and in some taxa, sharing of refuges is common if they are sufficiently large to accommodate multiple individuals (e.g., spotted spiny lobsters, *Panulirus guttatus*; Segura-Garcia et al. 2004). Indeed, proximity to conspecifics can often provide "safety in numbers," and predator protection is not a resource that can readily be "looted." Thus, an intriguing feature of crustacean territories is that they largely involve defense of a resource that, while obviously critical to fitness, does not appear to necessarily require exclusivity or active defense. Stamps (1994) proposed that predator search behavior (i.e., if predators increase search behavior in areas in which prey have already been detected) could be an overlooked possible advantage to territoriality; that is, if proximity to conspecifics increases the likelihood of being detected by a predator, the advantage of territorial aggression may not be in the benefits of space as a resource per se, but in the benefits of maintaining distance between nearest neighbors. Similarly, parasite avoidance could provide another advantage of maintaining exclusive use of space (Stamps 1994). In an aquatic environment other advantages of territoriality may include reduced costs associated with oxygen depletion and accumulation of wastes. These costs may be particularly important in intertidal organisms or those residing in microhabitats with reduced water flow (such as most burrows and tubes). In any case, crustaceans provide an excellent opportunity for exploring how diverse benefits associated with interindividual spacing may favor territoriality.

Advantages associated with interindividual spacing are likely to interact with the ecological and behavioral variables suggested to favor territorial defense in symbiotic crustaceans (Baeza and Thiel 2007). Here, we propose an adaptation of Baeza and Thiel's model to include costs that increase with the proximity of shelters for nonsymbiotic crustaceans—that is, territories that are not limited by the abundance and distribution of host organisms (Fig. 10.3). The *x*-axis combines ecological variables related to the value of maintaining ownership of a burrow. For burrows that function primarily for protection, value increases with increasing risk of exposure (to predation or environmental stressors) and with decreasing availability of refuges or suitable habitat in which to construct a burrow. Site fidelity is likely to increase with burrow value, independent of proximity costs. The relationship between burrow value and territorial behavior, however, may be more complex. If the costs associated with proximity to additional conspecifics are high, territorial behavior ensuring a minimal distance between individuals may be favored regardless of burrow value (i.e., regardless of the availability of suitable habitat or the risks of exposure). At lower proximity costs, however, territoriality may be favored only when ownership of a burrow is of intermediate value. If both the costs of proximity and the value of ownership are low, then the costs associated with exclusive defense exceed the value of exclusive ownership. On the other hand, if the value of ownership is high but the costs of proximity are low, it may be less costly to share the shelter than to invest in exclusive defense. In other words, variation across taxa in the likelihood of tolerating conspecifics within burrows versus behaving in a territorial fashion may often reflect differences in the costs of proximity rather than the value of the burrow per se.

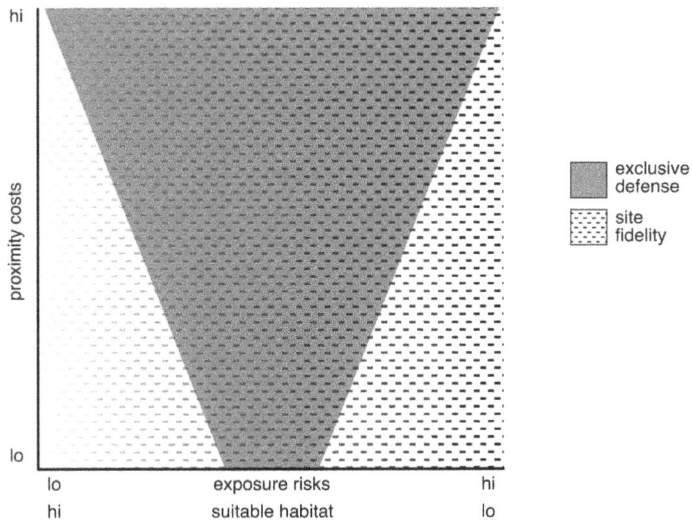

Fig. 10.3.
Model of territorial behavior for self-built burrows serving primarily as protection. Burrow value increases along the x-axis, as the risks of exposure (to predation or environmental stressors) increase and as the availability of habitat suitable for constructing new burrows decreases. Proximity costs may include risks associated with predator search behavior, parasite risk, or habitat degradation associated with higher densities of conspecifics.

Among vertebrate taxa, territories are often associated with providing exclusive access to mates and/or reproductive resources (Stamps 1994), and likewise, many crustacean burrows serve both as shelter and as a reproductive resource, providing protection from predation for the female before or during mating (especially in species that molt prior to mating), and subsequently for her eggs or larvae. In more terrestrial species, protection from desiccation may also be critical (e.g., desert isopods, Linsenmair 2007; semiterrestrial crayfish, Richardson 2007). The dual protection and reproduction function of such burrows is illustrated by sand bubbler crabs, in which the value of burrows is higher when individuals are under predation risk and when mated (Koga and Ikeda 2010). Many species mate in their burrows, including marsh crabs (*Sesarma reticulatum*; Seiple and Salmon 1982), American lobsters (Atema et al. 2007), and many species of fiddler crabs and stomatopods (Christy and Salmon 1991). When burrows are used for reproduction and defended by large males, alternative male strategies may be favored. For example, in the isopod *Paracerceis sculpta*, alpha males attempt to defend exclusive access to sponges for reproduction, whereas beta males use a strategy of female mimicry and the smaller gamma males attempt to access sponges by stealth (Shuster 1997). When burrows serve a reproductive function, it is obviously easier to hypothesize an adaptive function to exclusive defense. In most crustaceans, however, it is worth noting that the reproductive function of burrows is essentially an extension of their function in adult survival, and exclusivity is generally maintained even by nonreproductive individuals and outside the reproductive season.

In some taxa, males and females share burrows throughout the reproductive cycle; the sexes then may engage in joint territorial defense (snapping shrimp *Alpheus* spp., Mathews 2002, Hughes et al. 2014; desert isopod *Hemilepistus reaumuri*, Linsenmair 1984). If males and females do not share a territory throughout the reproductive cycle, females often join the male in his burrow (e.g., fiddler crabs, Christy and Salmon 1991; American lobsters, Atema et al. 2007). On the other hand, in stomatopods, males more commonly join the female in her burrow than vice versa (Christy and

Salmon 1991). Even if males and females typically share a territory, there may be differences in site fidelity. In the snapping shrimp *Alpheus armatus*, for example, males are more likely than females to leave their anemone in search of new mating opportunities (Knowlton 1980). In the snapping shrimp *Alpheus angulosus*, males are more likely to be evicted from their burrows than females, which may similarly suggest lower site fidelity in males (Mathews 2002). However, lower eviction rates in females may also be due to their higher levels of aggression (Hughes et al. 2014), and both sexes appear to have low site fidelity in this species (Heuring, unpublished data).

The protective function of crustacean territories may be extended to family groups as well, such as in desert isopods (Linsenmair 1984), as well as in some symbiotic crustaceans discussed earlier (i.e., snapping shrimp *Synalpheus* spp., Duffy et al. 2002; and the bromeliad crab, Diesel 1989, Diesel and Schubart 2007). Desert isopods provide an interesting example of how the availability of suitable habitat for burrow construction can affect territoriality and cooperative defense. In *Porcellio albinus*, burrows are readily constructed in loose sand and competitions for burrows are rare; in contrast, *Hemilepistus reaumuri* constructs burrows in densely packed soils. These burrows are essential for survival during the dry season but can be dug only during the wet season (Linsenmair 2007). Not surprisingly, competition for burrows in *H. reaumuri* can be intense. In *H. reaumuri*, but not *P. albinus*, both sexes care for brood and cooperatively defend their territory (Linsenmair 2007). Thus, as with symbiotic crustaceans, limited availability of suitable refuges (or the habitat in which to construct them) appears to favor increased site fidelity and cooperative defense.

Although territorial defense for exclusive access to foraging resources is common in other taxa, territories that function primarily for foraging appear to be rare among crustaceans. The unusual mast territories of *Dyopedos* amphipods are one exception: by extending into the water column, these mud whips likely facilitate feeding for juveniles (Thiel 1997). The round territories of tube-building amphipods (*Ericthonius brasiliensis*) are another exception; these territories contain a central burrow and space around the burrow with a radius approximately equal to the amphipods' body lengths (Connell 1963). The amphipods defend their space against wandering intruders, because the space around their burrows ensures access to food sources (Connell 1963). Here again, the burrow portion of the territory is likely to serve primarily a protective function, but the foraging area around the burrow is also defended for exclusive access. Tubes typically show a uniform distribution with nonoverlapping feeding areas, and although tubes may be established with overlapping feeding areas during colonization, individuals with overlapping feeding areas are eventually evicted (Connell 1963; Fig. 10.4).

Does Defense Extend Beyond the Shelter?

The previous example is unusual in that the boundary of defended space clearly extends beyond the burrow or shelter. Unless territory holders defend space around their shelter, the distribution of crustacean territories may simply be a function of the abundance and distribution of appropriate shelters, not a result of the behavior of the territory holders. Unfortunately, in most taxa, it is not clear whether the burrow alone is the entire territory or if the defended space extends beyond the burrow. There are at least some additional taxa, however, in which evidence suggests that territories may extend beyond the shelter itself. In the freshwater prawn *Macrobrachium lar*, individuals that had autotomized a cheliped defended smaller areas around their shelter than intact individuals (Seidel et al. 2007; Fig. 10.5). Territory size, in this study, was determined by staging competitive interactions for food at various locations within the experimental arena. In American lobsters (*Homarus americanus*), shelters adjacent to large males are more likely to be vacant, suggesting that large males may extend their defense to multiple shelters (Karnofsky and Price 1989). In some cases, defense of larger territories around the burrow appears to result in advantages for mating. Courtship burrows of the ghost crab *Ocypode jousseaumei* are overdispersed across a wide range of

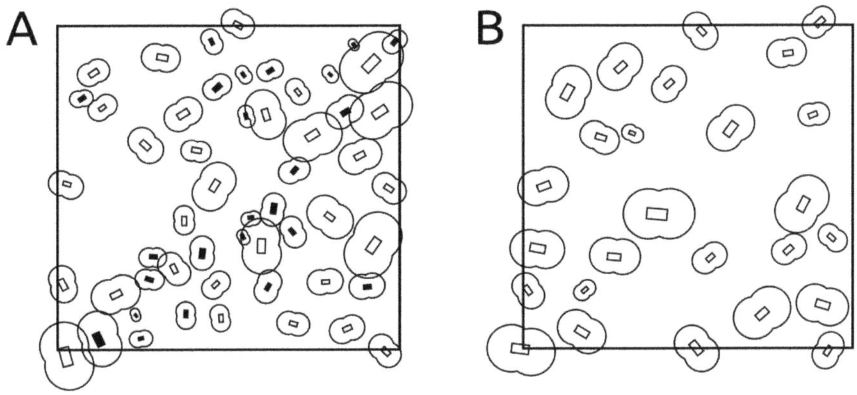

Fig. 10.4.
Feeding territories of tube-building amphipods (*Ericthonius brasiliensis*). Tubes are shown as rectangles, with feeding area in surrounding oval. (A) Location of feeding territories early in recolonization (following experimental removal of territories from a feeding surface); note that many feeding territories initially overlap. (B) Same feeding surface later in recolonization; note that the tubes with overlapping territories—shaded black in (A)—have disappeared, resulting in exclusive access for each territory holder to a feeding territory. Each square measures 5 × 5 cm. Modified from Connell (1963), with permission from Springer-Verlag.

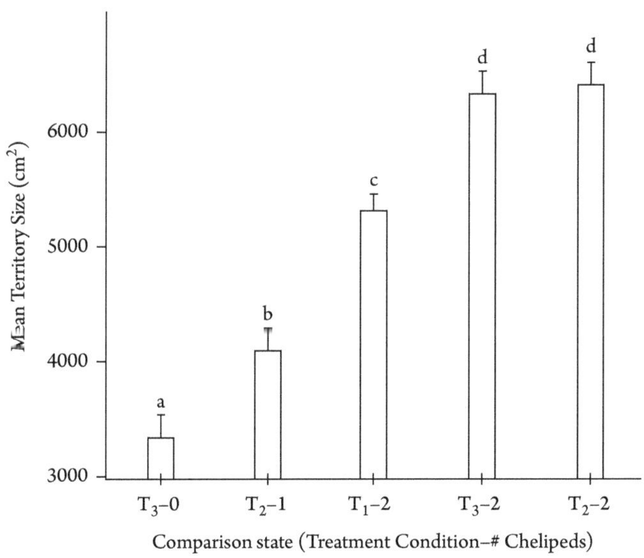

Fig. 10.5.
Territory size of freshwater prawns, *Macrobrachium lar*, by social environment and number of intact chelipeds. Treatments: T_1 = 4 intact prawns; T_2 = 2 intact prawns, 2 prawns with 1 intact cheliped (1 cheliped autotomized); T_3 = 2 intact prawns, 2 prawns with 0 chelipeds (both chelipeds autotomized). A different small-letter alphabet indicates significance. Compared to a social environment of all intact individuals (T_1), individuals with intact chelipeds defend larger territories when competing with individuals missing chelipeds (T_2–2, T_3–2), whereas the territories defended by individual missing chelipeds (T_2–1, T_3–0) are significantly smaller. From Seidel et al. (2007), with permission from The Crustacean Society.

crab densities (Clayton 2005), suggesting that larger territories may be advantageous for avoiding competition among males. In the fiddler crab *Tubuca capricornis*, some males may defend multiple burrows, allowing them to attract additional females (Mautz et al. 2011). Similarly, adult male grapsid crabs (*Pachygrapsus transversus*) defend crevice shelters and the surrounding area on intertidal rocks from other large males, whereas adult female crabs and smaller crabs occupy the other crevices in the males' territories, presumably leading to higher reproductive success for the males (Abele et al. 1986). The dotillid crab *Ilyoplax pusilla* defends a territory containing both its burrow and some space around its burrow (Wada 1993). Larger crabs are able to defend larger territories, presumably resulting in greater food access for both sexes and higher mating success for males (Wada 1993). Territorial individuals sometimes build barricades near the burrows of neighboring individuals; these barricades appear to demark a boundary of defended space, as the barricaded individual biases its home range away from the barricade, and the barricade builder responds aggressively if the barricaded neighbor trespasses beyond the barricade (Wada 1984). Clearly then, some crustaceans do defend space that extends beyond their shelters. For most taxa, however, the boundaries of defense remain largely unexplored.

The degree to which territories extend beyond a burrow or shelter has implications for territorial acquisition. Burrows differ from other forms of territories in that they are a discrete resource rather than compressible space; discrete burrows presumably must be won outright in a contest, whereas compressible space may be "won" through persistence (Stamps and Krishnan 1997). Similarly, territorial behavior in the owl limpet depends on recent site-specific agonistic experience and can be experimentally induced (Shanks 2002), suggesting intriguing lines of research that could be explored in crustacean systems as well.

Territory acquisition is also a function of the ontogeny of territorial behavior: how and when individuals begin to defend territories may determine the territory they acquire. In the symbiotic crab *Allopetrolisthes spinifrons*, recruits do not appear to engage in territorial behavior but juveniles and adults do (Baeza et al. 2002). Similarly, several species of spiny lobster share shelters as juveniles, but do not as adults, suggesting an ontogenetic shift in territorial behavior (Childress 2007). For most territorial crustaceans, however, territorial establishment has been studied primarily in adults. For crustaceans that defend territories as family groups, territories may sometimes be inherited. For example, in the bromeliad crab, although the life span of a bromeliad is roughly equivalent to the reproductive life span of a territorial female, her daughters may inherit clonal offshoots (Diesel and Schubart 2007). By and large, however, potential avenues for territory acquisition in crustaceans, other than winning access to refuges outright, have not been well explored.

Territorial Defense Behaviors: Patrolling and Advertisement Signals

Whether (or how far) territories extend beyond the entrances to shelters also has implications for territory defense, especially with regard to the need to patrol the territory or produce territorial advertisement signals, or both (Table 10.1); these behaviors are poorly studied in crustacean taxa. Rather than patrolling, defense of a burrow might involve guarding the entrance, as is seen in desert isopods (Linsenmair 2007). Similarly, female *Dyopedos* amphipods typically remain at the base of their mud whips, presumably to fend off conspecific intruders and protect the juveniles inhabiting the whips (Thiel 1997). If defended space does not extend beyond the burrow or shelter, then patrolling may be necessary only if burrows are much larger than defenders. Conversely, if the territory is too large to easily detect intruders throughout the defended space, active defense may be necessary. In eusocial snapping shrimp, for example, larger individuals are more active within the colony, potentially patrolling the colony to guard against intruders (Duffy et al. 2002). Similarly, in the bromeliad crab, experimentally introduced intruders are rapidly expelled (Diesel and Schubart 2007), suggesting they may be detected more rapidly than would be expected by chance movements around the territory. We have a poor understanding in most taxa of the degree to which the boundaries of territories extend beyond the defended shelter, and it is possible that we have overlooked patrolling behavior in other crustaceans as well.

The roles of territorial advertisement signals (i.e., signals produced even in the absence of an apparent intruder) in crustacean territorial defense is also underexplored. Waterborne chemical signals are produced by many crustaceans, and in many taxa these chemical signals may attract potential mates and/or repel potential intruders from the signaler's burrow (see Breithaupt and Thiel 2011 for reviews). Many of these chemical signals are likely produced regardless of whether an intruder is present and thus are excellent candidates for territorial advertisement signals. Whether these chemical signals are produced solely in the context of territory ownership (or change in composition with changes in territorial status), however, is not known; these signals may convey the same information regarding the signaler regardless of whether the signaler is on or off territory (or owns a territory or not). Nonetheless, further exploration of these waterborne chemical signals with regard to their potential functions as territorial advertisement signals seems warranted.

Chemical signals perceived by contact are also common among crustaceans, and it is possible that chemical signals deposited on the substratum around territorial boundaries could play a role in crustacean territorial defense. Although such "scent marking" may be unlikely in an aquatic environment (Thiel and Duffy 2007), desert isopods do appear to leave chemical signals around their burrow entrances. For example, *Porcellio* spp. surround their burrow entrances with sand scraped from inside their burrows, and the fecal rings of *Hemilepistus reaumuri* also provide a chemical signature (Linsenmair 2007). In *H. reaumuri,* returning foragers can orient to the vicinity of their burrows even if their fecal rings are removed, but the presence of these landmarks improves homing ability and reduces search time (Hoffmann 1985a). In addition to helping territory holders return to the correct burrows, these chemical marks could deter potential intruders. When returning foragers encounter the fecal rings of other family groups close to their own burrows, they do continue to search for the burrow within the ring; however, changes in their search behavior suggests they are able to distinguish the foreign landmark from their own, although not with complete certainty (Hoffmann 1985b). Such advertisement signals (i.e., signals produced in the absence of immediate territorial intruders) could be particularly advantageous for *H. reaumuri,* because their home ranges may overlap hundreds of conspecific family groups (Linsenmair 2007). The utility of these signals in territory defense, in other words, may depend on the likelihood of intruders intercepting such signals, as well as on physical constraints on signal transmission, which are likely to differ considerably between aquatic and terrestrial territories.

Landmarks can serve as visual boundaries to territories in both terrestrial and aquatic taxa, and when constructed by the territory holder, can be considered territorial advertisement signals as well (Heap et al. 2012). Some crustaceans "mark" their territories with structures, such as pillars or hoods adjacent to burrow entrances in fiddler crabs (Christy and Salmon 1991) or fecal rings in terrestrial isopods (Hoffmann 1985a). In some fiddler crabs, burrow hoods may serve to visually isolate displaying males, thus reducing aggression from territorial neighbors (Zucker 1981). In general, however, as visual cues these structures appear to assist territory holders (or in the case of fiddler crabs, mate-searching females) in orienting toward the burrow opening, not as "keep out" signals to intruders (Hoffmann 1985a, Christy and Salmon 1991). For fiddler crabs, these burrow structures may take advantage of the tendency to orient toward landmarks to avoid predators (Christy 2007), suggesting they serve an attractive function, rather than repelling intruders. Fiddler crabs also engage in claw-waving displays, but these appear to be directed at female receivers more so than males, thus functioning as mate attraction signals rather than advertisement signals for territorial defense (Christy and Salmon 1991, Pope 2000).

In contrast, the barricades constructed by the dotillid crab *I. pusilla,* described above (Does Defense Extend Beyond the Shelter? section), do appear to function as territorial advertisement signals. Produced primarily by larger males against smaller neighbors (both male and female), these structures alter space use by the barricaded neighbor, whose range is biased away from the barricade even if the barricade builder has been removed (Wada 1984). In other words, the signal defends the territory in the absence of the territory holder, in a manner similar to classic terrestrial territorial advertisement signals such as bird song (Krebs et al. 1978). When the barricade itself is removed, movements of the formerly barricaded neighbor extend toward the barricade builder, who responds aggressively, suggesting that the barricade served as a "keep out" signal to the barricaded neighbor (Wada 1984). Although the name *barricade* may suggest that these structures physically block the movements of neighbors, these barricades are relatively short and would appear to present only a minor obstacle to walk around (see Figure 1 in Wada 1984). These structures, then, appear to facilitate territory defense, marking a boundary of space defended by the barricade builder, and sending a "keep out" signal to neighbors even when those neighbors are not directly challenging the territory holder.

Many crustaceans produce mechanoacoustic signals in territorial defense interactions. For example, the ghost crab *O. jousseaumei* produces both a stridulation and longer sequences of substrate rapping during territory defense (Clayton 2005). Eusocial snapping shrimp (*Synalpheus rathbunae*) engage in coordinated snapping displays that appear to function as warning signals to intruders (Tóth and Duffy 2005). Although technically not advertisement signals (these displays are produced only after the intruder is detected), mechanoacoustic signals such as these are nonetheless intriguing in that they raise the possibility of similar signals used as advertisement signals in crustacean territorial defense. Indeed, *O. jousseaumei* also produces shorter sequences of substrate rapping beginning shortly after completing their burrows, and these crabs appear to do so in the absence of any direct territorial challenge. These vibrational signals may serve to advertise the burrow, although whether they function to attract females or repel male intruders (or both) remains unclear (Clayton 2005). Male aquatic isopods (*Cymodoce japonica*) produce a stridulatory sound that may serve in the defense of nesting cavities, although the function of this signal is yet to be resolved (Nakamachi et al. 2015). Interestingly, an unexplored hypothesis with regard to potential functions of burrows as crustacean territories is the possibility that burrows may serve as resonating cavities for mechanoacoustic signals (Atkinson and Eastman 2015).

Territoriality as Social Behavior: Neighbor Recognition and Alliances

Social recognition can take a variety of forms, ranging from discriminating between familiar and unfamiliar individuals, to discriminating between different classes of individuals (such as previously encountered dominants and previously encountered subordinates), to true individual recognition (Gherardi et al. 2012). The cognitive abilities necessary for such discrimination are well documented among crustaceans. For example, American lobsters can discriminate between familiar and unfamiliar opponents (Karavanich and Atema 1998); visual cues appear to play an important role in recognition (Gherardi et al. 2010). Visual cues also permit some fiddler crab males to distinguish between neighboring and unfamiliar females (Detto et al. 2006). Stomatopods can discriminate between previously encountered individuals who were dominant or subordinate to them by chemical signals, and also attend to differences among individuals in color patterns (Vetter and Caldwell 2015). Freshwater crayfish distinguish between familiar and unfamiliar individuals by chemical and possibly also visual cues (Patullo and Macmillan 2015), and chemical recognition of familiar individuals appears widespread among caridean shrimp (Chak et al. 2015a). The snapping shrimp *Alpheus heterochaelis* can discriminate between familiar dominant and unfamiliar individuals based on chemical cues; in addition, they can associate the chemical cues of different individuals with different spatial locations, suggesting the potential for true individual recognition (Ward et al. 2004). Crustaceans thus have social recognition abilities that would allow territorial

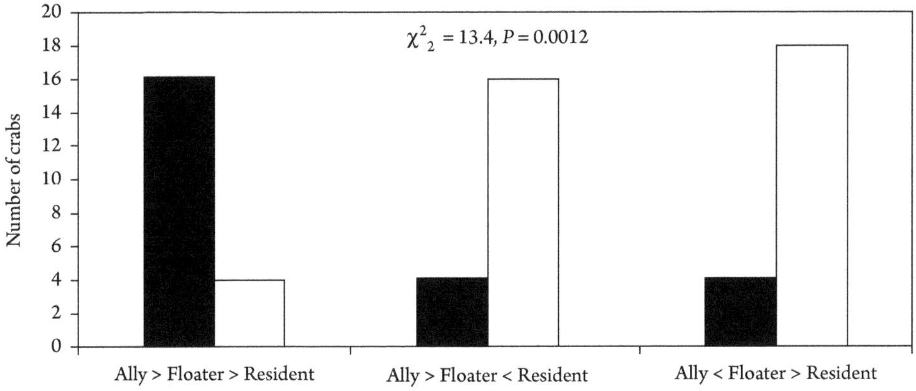

Fig. 10.6.
Territorial coalitions in the African fiddler crab, *Uca annulipes*, in response to experimentally introduced and tethered floaters. The likelihood of neighboring crabs (allies) assisting the resident against the floater depends on their relative sizes: allies are more likely to assist the resident (black bars) when that assistance is likely to be successful (ally larger than floater) and when the resident is at a disadvantage (floater larger than resident). From Detto et al. (2010), with permission from The University of Chicago Press.

systems to include complex social networks. The degree to which these social recognition abilities affect territorial dynamics, however, remains largely unexplored.

In terms of territorial social dynamics, fiddler crabs are perhaps the best studied crustaceans. Fiddler crabs clearly show the Dear Enemy Effect in their territorial defense, with lower levels of aggression directed at neighbors than at strangers (Pratt and McLain 2006), and in some species territorial individuals assist their neighbors in territorial defense (Backwell and Jennions 2004, Booksmythe et al. 2010, Detto et al. 2010). Specifically, larger male residents will help a smaller neighbor fight off intermediate-sized intruders, suggesting that these helpers benefit by keeping a smaller, potentially less threatening, male as a neighbor (Backwell and Jennions 2004, Detto et al. 2010; Fig. 10.6). Perhaps as a result, floater males seeking a territory will take neighbor size into account when determining which resident male to fight (Milner et al. 2011, Milner 2012). Such social complexity in territorial dynamics seems unlikely to be limited to fiddler crabs. Territorial crustaceans provide a largely underutilized opportunity for study of social cognition and network dynamics in territorial systems.

FUTURE DIRECTIONS

Critical to any understanding of territoriality, of course, is defining what space is defended, and determining the adaptive value of exclusive defense. Here much work remains to be done in crustaceans. First, although the adaptive value of a burrow for protection (from predation and/or environmental stressors) seems straightforward, the advantage of exclusivity for this resource remains unclear and indeed, many crustaceans share their shelters. Crustaceans, then, may provide excellent opportunities for testing alternative hypotheses for territoriality, particularly with regard to advantages associated with minimizing proximity to conspecifics. Second, the adaptive value of exclusive access to space is likely to vary temporally—territories may be advantageous at some life history stages and/or in some seasons but not in others. With few exceptions (e.g., dotillid crabs, Baeza et al. 2002; spiny lobsters, Childress 2007), ontogenetic or temporal variation in territorial behavior has rarely been explored in crustaceans. Third, the nature of these defended shelters range from other (host) organisms in territorial symbiotic crustaceans, to cavities in hard substratum that can be used and defended as refuges, to burrows excavated or constructed by the territorial holder. These different kinds of refuges are likely to differ not only in distribution, but also in the costs associated with refuge maintenance and the value of the refuge relative to the costs associated with finding or constructing a new one. How territoriality might differ between these types of territories has not been explicitly explored; indeed, with few exceptions (see, for example, burrowing crayfish [Palaoro et al. 2016] and terrestrial isopods [Linsenmair 2007]), how constraints on burrow construction affect territoriality has rarely been considered.

Although many taxa clearly defend burrows or other forms of shelter, the degree to which the defended space extends beyond that shelter is typically not well characterized. Such information is necessary to understand the ecological consequences of territoriality: is the distribution of animals determined by distribution of shelters or behavior of territory holders? Are some individuals excluded from the territorial system (floaters)? In addition, if space around burrows is defended, the extent to which that space is compressible leads to questions of territorial establishment: must competitors be dominant to establish a territory, or may subordinate individuals establish a territory through persistence? Crustaceans are potentially excellent subjects for study of territorial establishment, as there appears to be a range of variation in degree to which defended space is likely to be compressible.

Nearly all studies of territorial behavior in crustaceans focus on the outright aggressive behavior necessary to obtain or defend a territory; behaviors such as patrolling and territorial advertisement signals (i.e., "keep out" signals produced in the absence of an immediate intruder) are less well studied and may be less common, particularly in aquatic crustaceans. Although the apparent rarity of these behaviors may be a real difference between crustacean territoriality and that of other taxa, perhaps resulting from differences in territory function, it seems at least equally likely that the absence of these behaviors is an artifact of lack of study. Advantages of specialized territorial advertisement signals may also differ between terrestrial and aquatic territories, given differences in physical constraints on signal transmission. As crustaceans hold territories in environments ranging from entirely aquatic to entirely terrestrial, they may be particularly well suited for examining the benefits of such signals in territory defense as a function of environmental factors. Overall, a better understanding of behaviors involved in territorial maintenance in crustaceans would inform both our understanding of these taxa and of territoriality more generally.

Finally, with the notable exception of fiddler crabs, the role of social networks and social cognition in crustacean territoriality is largely unexplored, despite evidence of social recognition in a wide array of crustacean taxa. Given that there appears to be variation in site fidelity both across species and between sexes within species, crustaceans provide many possible comparisons of the degree to which neighbor recognition and coalition formation may be advantageous. Thus, crustaceans could also provide excellent model systems for study of these behaviors.

CONCLUSIONS

Most crustacean territories serve primarily to protect the territory holder, and hence most crustacean territories are functionally quite distinct from classical classifications that have been broadly applied in other systems (Table 10.2). Although the outright loss of a burrow or shelter would be costly, it is not clear why protection from predation requires exclusive access. It is not clear, in other words, how the safety provided by a burrow would be reduced by an intruder, because safety is not readily vulnerable to "looting"—the safety of a shelter is not necessarily reduced by sharing it with others. Indeed, in many contexts, increasing numbers provides added protection from predation (i.e., increases in detectability of a larger group are offset by decreases in per capita predation risk) and many crustaceans do share burrows or other shelters. For example, decreased risk of predation appears to be the primary advantage of gregarious behavior and shelter sharing in many species of spiny lobsters (Childress 2007). Nonetheless, many taxa defend exclusive space. Thus exclusivity may be favored for reasons that have not yet been explored; understanding the advantages of exclusive access to space is crucial for understanding crustacean territoriality. Crustacean territories may provide excellent opportunities to examine alternative hypotheses of territorial function (e.g., exclusivity as a means of maintaining optimal spacing rather than defending exclusive access to valuable resource). The optimal spacing hypothesis was originally proposed as a mechanism for limiting detection by predators or parasites (Stamps 1994). In crustaceans, if the costs associated with predator detection exceed any protective benefit of being in a group, territoriality may be favored to maintain dispersion. Risks of disease may also affect territorial behavior in crustaceans. In Caribbean spiny lobsters, for example, healthy individuals avoid sharing a shelter with individuals infected with a lethal virus even in the presence of a predator, a condition under which shelter sharing is usually preferred (Behringer and Butler 2010). Exclusive burrows may also benefit aquatic crustaceans due to improved water quality within the burrow; for intertidal crustaceans, increased spacing may result in fewer conspecifics sharing a tide pool and thus similarly result in better water quality. In any case, crustaceans seem likely to provide insight into the relationship between exclusive defense and the benefits of exclusive access to space.

Crustacean territories appear to share characteristics of both territories (sensu stricto) and home ranges (Table 10.1). Whether this mix of characteristics reflects the primary function of crustacean territories, however, or whether it is simply a result of insufficient study is unclear. Typical of territories, defense of space tends to be highly exclusive and, if only the burrows themselves are defended, there appears to be little overlap between neighbors (although the extent to which space around burrows is defended or overlaps with neighbors is unknown in most taxa). On the other hand, unless space surrounding burrows is also defended, compressibility of crustacean territories is likely to be low, given that they are defending discrete resources rather than open space. Similarly, behaviors such as patrolling and territorial advertisement signals have not been reported in the defense of most crustacean territories, and floaters appear to be similarly rare, most likely because survivorship is low without the protection of a shelter. These latter characteristics fall more on the home range end of the spatial defense spectrum; however, these characteristics are also the most poorly studied aspects of crustacean territoriality. Some of these apparent differences between crustacean territories and those described in classic terrestrial systems may be due to differences in function. Floaters, for example, may not survive long without the protection of a burrow in high-risk environments and therefore may be difficult to detect. In addition, some of the differences observed in crustacean territories may be due to differences inherent to the aquatic environment in which most crustaceans reside. Either hypothesis would be readily testable with crustacean systems. Given the ease with which many crustacean systems can be experimentally manipulated, crustaceans are vastly underutilized in studies of territoriality and have much to offer the future of the field.

ACKNOWLEDGMENTS

We thank John Christy, Denise Pope, Mika Tan, Martin Thiel, and Gary Wellborn, as well as the Behavioral Ecology Seminar for helpful comments on earlier drafts. Whitney Heuring was supported by the Graduate Program in Marine Biology Presidential Summer Research Award, College of Charleston.

REFERENCES

Abele, L.G., P.J. Campanella, and M. Salmon. 1986. Natural history and social organization of the semiterrestrial grapsid crab *Pachygrapsus transversus* (Gibbes). Journal of Experimental Marine Biology and Ecology 104:153–170.

Atema, J., and M.A. Steinbach. 2007. Chemical communication and social behavior of the lobster *Homarus americanus* and other decapod Crustacea. Pages 115–144 *in* J.E. Duffy, and M. Thiel, editors. Evolutionary ecology of social and sexual systems: crustaceans as model organisms. Oxford University Press, New York.

Atkinson, R.J.A., and L.B. Eastman. 2015. Burrow dwelling in Crustacea. Pages 78–117 *in* M. Thiel, and L. Watling, editors. The natural history of the Crustacea, volume 2: Lifestyles and feeding biology. Oxford University Press, New York.

Backwell, P.R.Y., and M.D. Jennions. 2004. Coalition among male fiddler crabs. Nature 430:417.

Baeza, J. A. 2015. Crustaceans as symbionts: an overview of their diversity, host use and life styles. Pages 163–189 *in* M. Thiel, and L. Watling, editors. The natural history of the Crustacea, volume 2: Lifestyles and feeding biology. Oxford University Press, New York.

Baeza, J.A., and M. Thiel. 2003. Predicting territorial behavior in symbiotic crabs using host characteristics: a comparative study and proposal of a model. Marine Biology 142:93–100.

Baeza, J.A., and M. Thiel. 2007. The mating system of symbiotic crustaceans: a conceptual model based on optimality and ecological constraints. Pages 249–267 *in* J.E. Duffy, and M. Thiel, editors. Evolutionary ecology of social and sexual systems: crustaceans as model organisms. Oxford University Press, New York.

Baeza, J.A., W. Stotz, and M. Thiel. 2002. Agonistic behaviour and development of territoriality during ontogeny of the sea anemone dwelling crab *Allopetrolisthes spinifrons* (H. Milne Edwards, 1837) (Decapoda: Anomura: Porcellanidae). Marine and Freshwater Behaviour and Physiology 35:189–202.

Barbaresi, S., and F. Gherardi. 2006. Experimental evidence for homing in the red swamp crayfish, *Procambarus clarkii*. Bulletin français de la pêche et de la pisciculture 380–381:1145–1154.

Booksmythe, I., M.D. Jennions, and P.R.Y. Backwell. 2010. Investigating the 'dear enemy' phenomenon in the territory defence of the fiddler crab, *Uca mjoebergi*. Animal Behaviour 79:419–423.

Branch, G.M. 1975. Mechanisms reducing intraspecific competition in *Patella* spp: migration, differentiation and territorial behaviour. Journal of Animal Ecology 44:575–600.

Breithaupt, T., and M. Thiel. 2011. Chemical communication in crustaceans. Springer-Verlag, New York.

Brown, J.L. 1964. The evolution of diversity in avian territorial systems. Wilson Bulletin 6:160–169.

Brown, J.L., and G.H. Orians. 1970. Spacing patterns in mobile animals. Annual Review of Ecology and Systematics 1:239–262.

Behringer, D.C. and M.J. Butler IV. 2010. Disease avoidance influences shelter use and predation in Caribbean spiny lobster. Behavioral Ecology and Sociobiology 64:747–755.

Chak, S.T.C, R. Bauer, and M. Thiel. 2015a. Social behaviour and recognition in decapod shrimps, with emphasis on the Caridea. Pages 57–84 *in* L. Aquiloni, and E. Tricarico, editors. Social recognition in invertebrates: the knowns and unknowns. Springer International Publishing, Switzerland.

Chak, S.T.C., J.E. Duffy, and D.R. Rubenstein. 2015b. Reproductive skew drives patterns of sexual dimorphism in sponge-dwelling snapping shrimp. Proceedings of the Royal Society B: Biological Sciences 282:20150342.

Chapman, C.J., and A.L. Rice. 1971. Some direct observations on the ecology and behaviour of the Norway lobster *Nephrops norvegicus*. Marine Biology 10:321–329.

Childress, M. J. 2007. Comparative sociobiology of spiny lobsters. Pages 271–293 *in* J.E. Duffy, and M. Thiel, editors. Evolutionary ecology of social and sexual systems: crustaceans as model organisms. Oxford University Press, New York.

Christy, J.H. 2007. Predation and the reproductive behavior of fiddler crabs (Genus *Uca*). Pages 211–231 *in* J.E. Duffy, and M. Thiel, editors. Evolutionary ecology of social and sexual systems: crustaceans as model organisms. Oxford University Press, New York.

Christy, J.H., and M. Salmon. 1991. Comparative studies of reproductive behavior in mantis shrimps and fiddler crabs. American Zoologist 31:329–337.
Clayton, D. 2005. Substrate (acoustic/vibrational) communication and ecology of the ghost crab *Ocypode jousseaumei* (Brachyura: Ocypodidae). Marine and Freshwater Behaviour and Physiology 38:53–70.
Connell, J.H. 1963. Territorial behavior and dispersion in some marine invertebrates. Researches on Population Ecology, Kyoto University 5:87–101.
Correa, C. and M. Thiel. 2003. Mating systems in caridean shrimp (Decapoda: Caridea) and their evolutionary consequences for sexual dimorphism and reproductive biology. Revista Chilena de Historia Natural 76:187–203.
Detto, T., P.R. Backwell, J. M. Hemmi, and J. Zeil. 2006. Visually mediated species and neighbour recognition in fiddler crabs (*Uca mjoebergi* and *Uca capricornis*). Proceedings of the Royal Society of London B: Biological Sciences, 273, 1661–1666.
Detto, T., M.D. Jennions, and P.R.Y. Backwell. 2010. When and why do territorial coalitions occur? Experimental evidence from a fiddler crab. The American Naturalist 175:E119–E125.
Diesel, R. 1989. Parental care in an unusual environment: *Metopaulias depressus* (Decapoda: Grapsidae), a crab that lives in epiphytic bromeliads. Animal Behaviour 38:561–575.
Diesel, R., and C.D. Schubart. 2007. The social breeding system of the Jamaican bromeliad crab *Metopaulias depressus*. Pages 365–386 *in* J.E. Duffy, and M. Thiel, editors. Evolutionary ecology of social and sexual systems: crustaceans as model organisms. Oxford University Press, New York.
Duffy, J.E., and K.S. Macdonald. 2010. Kin structure, ecology and the evolution of social organization in shrimp: a comparative analysis. Proceedings of the Royal Society B: Biological Sciences 277:575–584.
Duffy, J.E., C.L. Morrison, and K.S. Macdonald. 2002. Colony defense and behavioral differentiation in the eusocial shrimp *Synalpheus regalis*. Behavioral Ecology and Sociobiology 51:488–495.
Ďuriš Z., I. Horká, P.J. Juračka, A. Petrusek, and F. Sandford. 2011. These squatters are not innocent: the evidence of parasitism in sponge-inhabiting shrimps. PLoS ONE 6:e21987.
Getty, T. 1987. Dear enemies and the prisoner's dilemma: why should territorial neighbors form defensive coalitions? American Zoologist 27:327–336.
Gherardi, F., F. Cenni, G. Parisi, and L. Aquiloni. 2010. Visual recognition of conspecifics in the American lobster, *Homarus americanus*. Animal Behaviour 80:713–719.
Gherardi, F., L. Aquiloni, and E. Tricarico. 2012. Revisiting social recognition systems in invertebrates. Animal Cognition 15:745–762.
Hazlett, B., D. Rittschof, and D. Rubenstein. 1974. Behavioral biology of the crayfish *Orconectes virilis* I. Home range. American Midland Naturalist 92:301–319.
Heap, S., P. Byrne, and D. Stuart-Fox. 2012. The adoption of landmarks for territorial boundaries. Animal Behaviour 83:871–878.
Hinde, R.A. 1956. The biological significance of the territories of birds. Ibis 98:340–369.
Hoffmann, G. 1985a. The influence of landmarks on the systematic search behaviour of the desert isopod *Hemilepistus reaumuri*: I. Role of the landmark made by the animal. Behavioral Ecology and Sociobiology 17:325–334.
Hoffmann, G. 1985b. The influence of landmarks on the systematic search behaviour of the desert isopod *Hemilepistus reaumuri*. II. Problems with similar landmarks and their solution. Behavioral Ecology and Sociobiology 17:335–348.
Huber, M.E. 1987. Aggressive behavior of *Trapezia intermedia* Miers and *T. digitalis* Latreille (Brachyura: Xanthidae). Journal of Crustacean Biology 7:238–248.
Hughes, M. 1996. Size assessment via a visual signal in snapping shrimp. Behavioral Ecology and Sociobiology 38:51–57.
Hughes, M., and J. Hyman. 2011. Should I stay or should I go now: late establishment and low site fidelity as alternative territorial behaviors. Ethology 117:979–991.
Hughes, M., T. Williamson, K. Hollowell, and R. Vickery. 2014. Sex and weapons: contrasting sexual dimorphisms in weaponry and aggression in snapping shrimp. Ethology 120:982–994.
Karavanich, C., and J. Atema. 1998. Individual recognition and memory in lobster dominance. Animal Behaviour 56:1553–1560.

Karnofsky, E.B., and H.J. Price. 1989. Dominance, territoriality and mating in the lobster, *Homarus americanus*: a mesocosm study. Marine Behaviour and Physiology 15:101–121.

Kaufmann, J.H. 1983. On the definitions and functions of dominance and territoriality. Biological Reviews of the Cambridge Philosophical Society 58:1–20.

Knowlton, N. 1980. Sexual selection and dimorphism in two demes of a symbiotic, pair-bonding snapping shrimp. Evolution 34:161–173.

Knudsen, J.W. 1967. *Trapezia* and *Tetralia* (Decapoda, Brachyura, Xanthidae) as obligate ectoparasites of pocilloporid and acroporid corals. Pacific Science 21:51–57.

Koga, T., and S. Ikeda. 2010. Perceived predation risk and mate defense jointly alter the outcome of territorial fights. Behavioral Ecology and Sociobiology 64:827–833.

Krebs, J., R. Ashcroft, and M. Webber. 1978. Song repertoires and territory defense in the great tit. Nature 271:539–542.

Lee, J.S.E. 2005. Alternative reproductive tactics and status-dependent selection. Behavioral Ecology 16:566–570.

Linsenmair, K.E. 1984. Comparative studies on the social behaviour of the desert isopod *Hemilepistus reaumuri* and of a *Porcellio* species. Symposium, Zoological Society of London 53:423–453.

Linsenmair, K.E. 2007. Sociobiology of terrestrial isopods. Pages 339–364 *in* J.E. Duffy, and M. Thiel, editors. Evolutionary ecology of social and sexual systems: crustaceans as model organisms. Oxford University Press, New York.

Maher, C.R., and D.F. Lott. 1995. Definitions of territoriality used in the study of variation in vertebrate spacing systems. Animal Behaviour 49:1581–1597.

Marin, I.N. 2007. Notes on taxonomy and biology of the symbiotic shrimp *Vir euphyllius* Marin et Anker, 2005 (Decapoda: Palaemonidae: Pontoniinae), associated with scleractinian corals *Euphyllia* spp. (Cnidaria: Caryophyllidae). Invertebrate Zoology 4:15–23.

Mathews, L. 2002. Territorial cooperation and social monogamy: factors affecting intersexual behaviours in pair-living snapping shrimp. Animal Behaviour 63:767–777.

Mattson, S., and T. Cedhagen. 1989. Aspects of the behaviour and ecology of *Dyopedos monacanthus* (Metzger) and *D. porrectus* Bate, with comparative notes on *Dulichia tuberculata* Boeck (Crustacea: Amphipoda: Podoceridae). Journal of Experimental Marine Biology and Ecology 127:253–272.

Mautz, B., T. Detto, B.B.M. Wong, H. Kokko, M.D. Jennions, and P.R.Y. Backwell. 2011. Male fiddler crabs defend multiple burrows to attract additional females. Behavioral Ecology 22:261–267.

Mayr, E. 1935. Bernard Altum and the territory theory. Proceedings of the Linnaean Society of New York 45–46:24–38.

Milner, R.N.C. 2012. Everybody needs good neighbours: coalition formation influences floater fight choice. Ethology 118:373–376.

Milner, R.N.C., M.D. Jennions, and P.R.Y. Backwell. 2011. Know thine enemy's neighbor: neighbor size affects floaters' choice of whom to fight. Behavioral Ecology 22:947–950.

Moyer, J.T., and C.E. Sawyers. 1973. Territorial behavior of the anemonefish *Amphiprion xanthurus* with notes on the life history. Japanese Journal of Ichthyology 20:85–93.

Nakamachi, T., H. Ishida, and N. Hirohashi. 2015. Sound production in the aquatic isopod *Cymodoce japonica* (Crustacea: Peracarida). Biological Bulletin 229:167–172.

Nice, M.M. 1941. The role of territory in bird life. American Midland Naturalist 26:441–487.

Palaoro, A.V., E. del Valle, and M. Thiel. 2016. Life history patterns are correlated with predictable fluctuations in highly seasonal environments of semi-terrestrial burrowing crayfish. Hydrobiologica 767:51–63.

Patullo, B.W., and D.L. Macmillan. 2015. To what extent can freshwater crayfish recognise other crayfish? Pages 37–48 *in* L. Aquiloni, and E. Tricarico, editors. Social recognition in invertebrates: the knowns and unknowns. Springer International Publishing, Switzerland.

Pope, D.S. 2000. Testing the function of fiddler crab claw waving by manipulating social context. Behavioral Ecology and Sociobiology 47:432–437.

Pratt, A.E., and D.K. McLain. 2006. How dear is my enemy: intruder-resident and resident-resident encounters in male sand fiddler crabs (*Uca pugilator*). Behaviour 143:597–617.

Richardson, A.M.M. 2007. Behavioral ecology of semiterrestrial crayfish. Pages 319–338 *in* J.E. Duffy, and M. Thiel, editors. Evolutionary ecology of social and sexual systems: crustaceans as model organisms. Oxford University Press, New York.

Robertson, D.R., and N.V.C. Polunin. 1981. Coexistence: symbiotic sharing of feeding territories and algal food by some coral reef fishes from the western Indian Ocean. Marine Biology 62:185–195.

Rützler, K. 1976. Ecology of Tunisian commercial sponges. Tethys 7:249–264.

Segura-Garcia, I., E. Lozano-Alvarez, and P. Briones-Fourzan. 2004. Within-shelter behaviour of the spotted spiny lobster, *Panulirus guttatus* (Latreille) in simulated communal dens: an exploratory study. Marine and Freshwater Behavior and Physiology 37:17–30.

Seidel, R.A., R.L. Schaefer, and T.J. Donaldson. 2007. The role of cheliped autotomy in the territorial behavior of the freshwater prawn *Macrobrachium lar*. Journal of Crustacean Biology 27:197–201.

Seiple, W., and M. Salmon. 1982. Comparative social behavior of two grapsid crabs, *Sesarma reticulatum* (Say) and *S. cinereum* (Bosc). Journal of Experimental Marine Biology and Ecology 62:1–24.

Shanks, A.L. 2002. Previous agonistic experience determines both foraging behavior and territoriality in the limpet *Lottia gigantean* (Sowerby). Behavioral Ecology 13:467–471.

Shuster, S.M. 1997. Alternative reproductive behaviors: three discrete male morphs in *Paracerceis sculpta*, an intertidal isopod from the northern Gulf of California. Journal of Crustacean Biology 7:318–327.

Smith, S.M. 1978. The "underworld" in a territorial sparrow: adaptive strategy for floaters. The American Naturalist 112:571–582.

Stamps, J.A. 1994. Territorial behavior: testing the assumptions. Advances in the Study of Behavior 23:173–232.

Stamps, J.A., and V.V. Krishnan. 1997. Functions of fights in territory establishment. The American Naturalist 150:393–405.

Stamps, J.A., and V.V. Krishnan. 1998. Territory acquisition in lizards. IV. Obtaining high status and exclusive home ranges. Animal Behaviour 55:461–472.

Stamps, J.A., and V.V. Krishnan. 1999. A learning-based model of territory establishment. The Quarterly Review of Biology 74:291–318.

Stimson, J. 1970. Territorial behavior of the owl limpet, *Lottia gigantea*. Ecological Society of America 51:113–118.

Temeles, E.J. 1994. The role of neighbours in territorial systems: when are they 'dear enemies'? Animal Behaviour 47:339–350.

Thiel, M. 1997. Reproductive biology of an epibenthic amphipod (*Dyopedos monacanthus*) with extended parental care. Journal of Marine Biology 77:1059–1072.

Thiel, M., and J.E. Duffy. 2007. The behavioral ecology of crustaceans: a primer in taxonomy, morphology and biology. Pages 3–28 *in* J.E. Duffy, and M. Thiel, editors. Evolutionary ecology of social and sexual systems: crustaceans as model organisms. Oxford University Press, New York.

Thiel, M., A. Zander, N. Valdivia, J.A. Baeza, and C. Rueffler. 2003. Host fidelity of a symbiotic porcellanid crab: the importance of host characteristics. Journal of Zoology 261:353–362.

Tóth, E., and R.T. Bauer. 2007. Gonopore sexing technique allows determination of sex ratios and helper composition in eusocial shrimps. Marine Biology 151:1875–1886.

Tóth, E., and J.E. Duffy. 2005. Coordinated group response to nest intruders in social shrimp. Biology Letters 1:49–52.

Vetter, K.M., and R.L. Caldwell. 2015. Individual recognition in stomatopods. Pages 17–36 *in* L. Aquiloni, and E. Tricarico, editors. Social recognition in invertebrates: the knowns and unknowns. Springer International Publishing, Switzerland.

Wada, K. 1984. Barricade building in *Ilyoplax pusillus* (De Haan) (Crustacea: Brachyura). Journal of Experimental Marine Biology and Ecology 83:73–88.

Wada, K. 1993. Territorial behavior, and sizes of home range and territory, in relation to sex and body size in *Ilyoplax pusilla* (Crustacea: Brachyura: Ocypodidae). Marine Biology 115:47–52.

Ward, J., N. Saleh, D.W. Dunham, and N. Rahman. 2004. Individual discrimination in the big-clawed snapping shrimp, *Alpheus heterochelis*. Marine and Freshwater Behaviour and Physiology 37: 35–42.

Zucker, N. 1981. The role of hood-building in defining territories and limiting combat in fiddler crabs. Animal Behaviour 29:387–395.

11

EVOLUTIONARY ECOLOGY OF BURROW CONSTRUCTION AND SOCIAL LIFE

Mark E. Laidre

Abstract

Burrows represent a prominent example of animal architecture that fundamentally alters the surrounding physical environment, often with important consequences for social life. Crustaceans, in particular, offer a model system for understanding the adaptive functions of burrows, their ecological costs and benefits, and their long-term evolutionary impacts on sociality. In general, burrows are central to the life history of many species, functioning as protective dwellings against predators and environmental extremes. Within the refuge of a burrow, one or multiple inhabitants can feed, molt, grow, mate, and raise offspring in relative safety. Depending on the substratum, substantial construction costs can be incurred to excavate a burrow de novo or enlarge a preexisting natural crevice. This investment has been evolutionarily favored because the benefits afforded by the burrow outweigh these costs, making the burrow an "extended phenotype" of the architect itself. Yet even after a burrow is fully constructed, the architect must incur continued costs over its life history, both in maintenance and defense, if it is to reap further benefits of its burrow. Indeed, because burrows accumulate value based on the work involved in their construction, they can attract conspecific intruders who seek to shortcut the cost of construction by evicting an existing occupant and usurping its burrow. Consequently, a burrowing lifestyle can lead to escalating social competition, with many crustaceans evolving elaborate weapons and territorial signals to resolve conflicts over burrow ownership. Some burrows even outlast the original architect as an "ecological inheritance," serving as a legacy that impacts social evolution among subsequent generations of kin and nonkin. Comparative studies, using cutting-edge technology to dig deeper into the natural history of crustacean burrows, can provide powerful tests of general theoretical models of animal architecture and social evolution, especially the extended phenotype and niche construction.

Life Histories. Edited by Gary A. Wellborn and Martin Thiel.

INTRODUCTION

Many organisms can act as architects (von Frisch and von Frisch 1974, Gould and Gould 2007), fundamentally reshaping the surrounding physical environment during their lifetime and even bequeathing this altered environment as a legacy for subsequent generations of conspecifics (Vermeij 2010). Indeed, the building efforts of animal architects can lead to the modification of whole ecological landscapes, both above and below ground (Hansell 2005). Arguably, the bedrock of such animal architectural achievements are burrows, which are excavated down into a substratum, often with the adaptive function of creating protective homes within which the burrower lives (Fig. 11.1). Burrows underscore the dynamic feedback between organisms and their environment. This

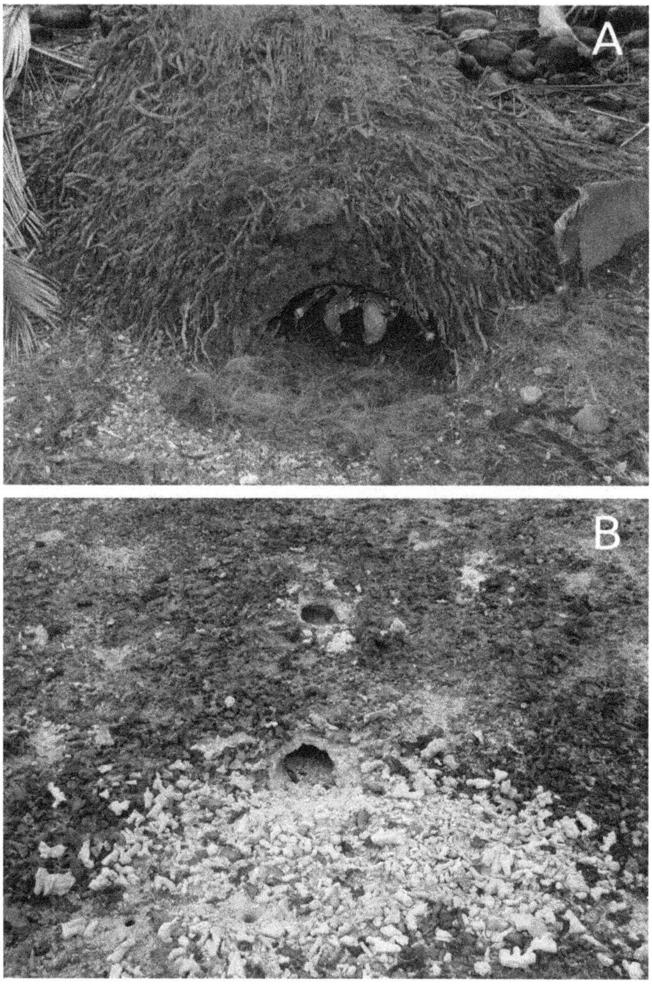

Fig. 11.1.
(A) Burrow of a coconut crab (*Birgus latro*) carved into the ground beneath the roots of a coconut tree (*Cocos nucifera*) in the Chagos Archipelago. Burrow owner is at entrance. Note substantial coconut husk located at mouth of burrow in front of owner. (B) Burrow (U-shaped with two entrances) excavated below ground in coral rubble by a brachyuran crab in the Chagos Archipelago. Burrow owner is in the tunnel between the entrances. Note pile of white coral rubble at mouth of nearest entrance. Photographs by Mark E. Laidre ©. See color version of figure part 11.1A in the centerfold.

chapter reviews the major variables (Table 11.1) that shape the evolutionary ecology of burrows in crustaceans, emphasizing how and why crustaceans construct burrows and the complex interactions among behavior, ecology, and evolution resulting from a burrowing lifestyle. Building on rich descriptive accounts of crustacean burrows (e.g., Atkinson and Eastman 2015, Mejaes et al. 2015), this chapter argues that burrows in crustaceans provide model empirical systems, which can be understood in the broader context of two overarching theoretical concepts: the extended phenotype (Dawkins 1999) and niche construction (Odling-Smee et al. 2003, Odling-Smee et al. 2013). Together these two concepts provide a unifying framework for predicting (1) when burrowing will be favored, based on its ecological costs and benefits, and (2) the long-term evolutionary impact of burrowing on social life.

Briefly, the extended phenotype concept (Dawkins 1999) emphasizes that genes create not only an organism's body (the standard conception of a phenotype), but also a farther-reaching phenotype, embodied by the alterations an organism makes to its physical surroundings. Critically, the extended phenotype concept predicts that genes will be evolutionarily favored if they cause organisms to modify environments in ways that directly favor the genes' own propagation; that is, they need to enhance the survival and reproduction of the organism within which the genes reside.

Table 11.1. Key variables shaping the evolutionary ecology of burrows, separated into (1) variables determining benefits and costs and (2) variables impacting social life. An assessment of the importance of these variables is provided: *** = highly important, ** = important, and * = potentially important but requires further research. More detailed considerations, including crustacean examples from the literature, are cited throughout the main text.

Key variables	Consequence
(i) Variables determining benefits and costs	
***Higher level of predation outside burrows ***Harsher environmental extremes outside burrow ***Kin co-residing in burrow	↑ Benefits of burrows
*Larger energy investment (***Substratum hardness is best proxy for energy investment) **More time investment (opportunity cost) **Greater burrow architectural complexity (length and size) **Elevated exposure to predators while creating burrow	↑ Construction costs of burrows
***Weaker substrata with less stability **No 'glue' (rare in crustaceans) to cement burrow together	↑ Maintenance costs of burrows
*More valuable burrow (linked to greater original construction cost) ***Greater conspecific density (generally due to ecological constraints on burrow availability)	↑ Defense costs of burrows
(ii) Variables impacting social life	
**Burrow size (bigger burrows allow more room for simultaneous cohabitation) *Burrow longevity (burrows in harder substrata last longer, allowing ecological inheritance across generations, and driving escalated social competition, eviction, and territorial signals) ***Burrow transportability (allow direct social exchange)	↑ Social life

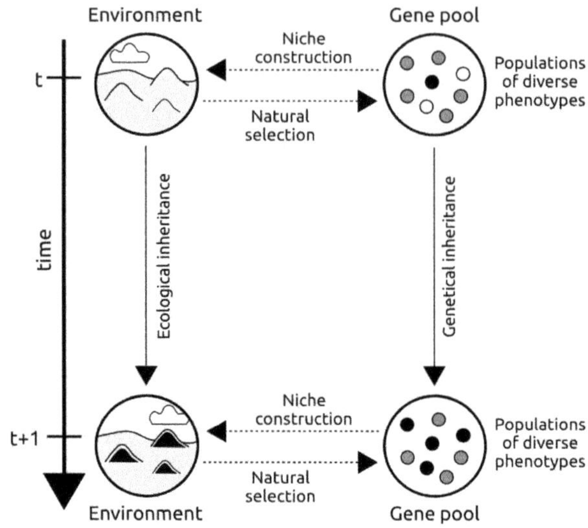

Fig. 11.2.
Conceptual framework of niche construction theory. In generation *t*, natural selection operates on the gene pool of a population of organisms. Simultaneously, this population of organisms constructs their niche by altering the surrounding environment. As a consequence of these environmental alterations, in the next generation ($t + 1$), an "ecological inheritance" is passed down alongside a genetic inheritance. The ecological inheritance is simply the difference in the environment between generations. Modified from Laland et al. (2000), with permission from Cambridge University Press.

Cases of animal architecture, like burrows, should therefore be evolutionarily shaped to maximize the adaptive benefits to the burrower, devalued by any costs the burrower incurs in constructing, maintaining, and defending its burrow. The extended phenotype concept thus sheds essential light on when burrowing will be evolutionarily favored.

Whereas the concept of an extended phenotype focuses on an organism's lifetime, the concept of niche construction goes further. Niche construction looks beyond an individual's own lifetime (Odling-Smee et al. 2003) to consider the impact an organism's environmental modifications may have on future generations (Odling-Smee et al. 2013), including those that share the organism's genes or are unrelated to the organism (Krakauer et al. 2009). The insight of niche construction is that organisms do not merely pass on a genetic inheritance. Rather, because organisms can actively construct the world around them, their burrows and other architectural constructs can outlast the original architect and persist as a so-called ecological inheritance (Fig. 11.2). When this altered ecology is passed down to others, including future generations, it can shift the natural selection pressures they experience, in some cases favoring greater levels of sociality or more intense territorial competition. Ultimately, both the extended phenotype and niche construction frameworks provide general principles to help make sense of the causes and consequences of a burrowing lifestyle, especially the fascinating diversity we find among crustacean burrowers.

FUNCTIONS OF BURROWS

In general, crustacean burrows function as protective dwellings against predators and environmental extremes, providing a safe haven for one or multiple inhabitants, including mated pairs, parent-offspring family groups, and congregations of unrelated conspecifics. Many essential activities can therefore take place within the refuge of a burrow. For instance, within a burrow an individual can feed, molt, grow, mate, brood eggs, and parent, all in relative safety compared to outside the burrow. A cost-benefit conceptual framework provides a means of tracing several key steps in the evolution of burrowing behavior, especially understanding the critical first step: how and why burrows become an extended phenotype that enhances the fitness of individuals that construct them. For a burrow to qualify as an extended phenotype, the burrow constructor must measurably benefit from its burrow and these benefits must exceed the costs involved in construction, because only then will the burrow enhance the burrower's fitness and be evolutionarily favored.

BENEFITS OF BURROWS

The benefits of burrow construction align with burrows' protective function. In a number of species, observational studies as well as manipulative experiments have directly measured the types of benefits that burrows bestow on their occupants. These studies have sought to compare response variables such as survival, foraging efficiency, and reproductive success of individuals in burrows versus out of burrows. Virtually all these studies provide strong support for a major adaptive benefit of burrows: an increased chance of survival and higher fitness within the burrow (Warner 1977, Rebach and Dunham 1983, Burggren and McMahon 1988). This benefit of burrows was well documented by Ball et al. (2001) in early benthic phase European lobsters (*Homarus gammarus*), where the survival of individuals placed on shelter-providing substrata (where individuals could burrow) was significantly higher than for unsheltered controls. Ultimately, it was by burrowing into the substratum that individuals evaded predation by fishes and crabs. Similar antipredator benefits of burrows have been found in American lobsters (*H. americanus*; Lavalli and Barshaw 1986, James-Pirri and Cobb 1999) as well as in fiddler crabs (Crane 1975, Macintosh 1979) and ghost crabs (Lucrezi and Schlacher 2014). In the fiddler crab *Leptuca beebei*, sexually selected traits, such as larger claws and brighter colors, make males more conspicuous to bird predators, including great-tailed grackles (*Quiscalus mexicanus*), which increases predation pressures on males compared to females when individuals are fleeing back into the burrow (Koga et al. 2001). Interestingly, ghost crabs, despite being the fastest crustaceans on land (Lucrezi and Schlacher 2014), still experience high degrees of predation outside their burrows, where they are eaten by diverse species of reptiles, birds, and mammals that forage at the land-sea interface (Lucrezi and Schlacher 2014). Ghost crabs thus underscore the critical importance of burrows as safe havens, even when species have the most extreme adaptations for speed outside the burrow.

In some cases, predators may be able to corner burrow occupants within their burrow, as when burrowing mud crabs (*Neopanope texana*) predate postlarval lobsters (*H. americanus*) by digging into the burrows (Lavalli and Barshaw 1986). Overall, however, the world is much more dangerous outside compared to inside the burrow. This is especially true when individuals are engaged in vital activities, such as mating or foraging, which prevent them from attending as effectively to external threats. Pursuing such activities within a burrow, and making only brief forays to the surface, can thus optimize the trade-off between protection and outside opportunities. For instance, sand bubbler crabs (*Scopimera inflata*) make only brief excursions out of their burrows to feed, often escaping back into their burrows to avoid intense predation pressure outside from red-capped plovers (*Charadrius ruficapillus*; Evans et al. 2010). Interestingly, in the New Zealand mud crab (*Austrohelice crassa*), individuals spend less time hiding in burrows if there are more conspecific neighbors around, which dilutes the predation risk (Guerra-Bobo and Brough 2011). Similarly, in fiddler crabs (*Leptuca pugilator*), if fellow conspecifics show flight behavior, then even when the burrow owner itself is not privy to this threat, it reacts to these cues and flees into its burrow (Wong et al. 2005). Hence, outside factors, such as the intensity of predation and the presence of conspecifics, can also modulate the inherent benefit of burrows as safe havens.

In addition to their antipredation benefits, burrows also constitute underground microhabitats, which can benefit occupants by keeping them insulated from extremes in temperature and humidity that they might otherwise be unable to endure. For instance, Allen et al. (2012) found that when individuals of the fiddler crab (*L. pugilator*) were excluded from their burrows those individuals with high surface area–to–volume ratios were particularly liable to desiccate and die in short periods because of the higher temperatures and lower humidity outside their burrows. Indeed, molting, one of the most vulnerable activities in which all crustaceans engage, may be especially challenging and dangerous for terrestrial and semiterrestrial crabs; there is a lack of constancy in temperature and humidity outside the burrow as well as the exposure to predators and conspecifics while the exoskeleton is soft. Molting may therefore have been one of the initial drivers of burrowing. Notably, because a burrow can eventually become a locus for vital social interactions, such as mate selection, parenting, and offspring development (Thiel 2007), the benefits of burrowing can extend not only to the burrow constructor itself but also to its mate and its kin, further enhancing the burrow constructor's overall inclusive fitness. Yet even with these rich benefits, spanning both self and family (Thiel 1999), the evolutionary equation still requires that benefits surpass costs, if burrowing is to evolve (Table 11.2).

Table 11.2. Simple cost-benefit parameters driving burrow construction and social evolution. B = benefit and C = cost, with currency in fitness.

Conceptual model	Criteria	Evolutionary interpretation
Extended phenotype	$B_{burrow} > C_{construction}$	Burrowing will evolve if the benefits of a burrow exceed the cost of constructing it.
Niche construction	$C_{construction} > C_{eviction\ (or\ waiting)}$	Individuals will circumvent burrow construction (by either evicting the burrow owner or waiting for it to die) if the cost of constructing their own burrow exceeds (1) the cost of evicting an owner or (2) the opportunity cost of simply waiting for the owner to die, leaving behind its burrow as an ecological inheritance.
	$B_{burrow} > C_{construction} + C_{defense}$	Burrow owners will repel evictors if the benefits of retaining ownership of their burrow exceeds the construction costs already invested plus the additional costs of defending/maintaining their burrow.
	$C_{defense} \overset{?}{=} C_{eviction}$	Balance between the cost to the burrow owner of defending its burrow versus the cost to the evictor of evicting determines which strategy wins. The question mark designates uncertainty in the relative magnitude of these costs, which vary based both on ecological constraints and individual differences, likely resulting in a frequency-dependent mix of constructor versus evictor strategies.

COSTS OF BURROWS

An essential part of the definition of burrowing is the excavation of a substratum (Hansell 2005). In contrast to other forms of animal architecture, in which organisms cobble material together to build something new, burrowing implies removal and hollowing out of an existing foundation. Typically, this involves carving out a space below ground (but see section on Transportable Burrows). Burrowing species may possess physiological specializations that distinguish them from nonburrowing species (Bridges 1986), especially because burrowing is an active construction process that involves work and hence energy spent by the burrower (Dorgan 2015). Simply occupying a preexisting hole or crevice—although it may constitute living in a burrow—does not constitute burrow construction, which requires active modification of the surroundings on the part of the organism.

Burrowing crustaceans undoubtedly incur costs in the construction process. For instance, crayfish use 2 key motor patterns, pushing and carrying, when excavating the sand from their underwater burrows, and both these actions require prolonged effort until the burrow is finished (Grow 1981). Although the benefits of burrows have been well measured (see earlier discussion), costs are notoriously more difficult for evolutionary biologists to pin down. Ideally, precise energetic or physiological measurements of the costs of burrow construction would be made and then be translated into the ultimate evolutionary currency: fitness impacts on survival and reproduction. However, the energetics of burrow construction have been quantified in only some vertebrate and invertebrate taxa (e.g., Luna and Antinuchi 2006, Suter et al. 2011), and only for a small number of crustacean species (e.g., Brown and Trueman 1996). Consequently, it is helpful to rely on various proxies, which provide at least a relative (if not absolute) basis for comparing costs and benefits, both within and across different species. In the next section, I consider several proxies for the construction costs of burrows and focus especially on a burrower's initial investment in creating its burrow (e.g., digging and excavating the burrow and then maintaining its architectural integrity). In subsequent sections, I consider additional costs, particularly defense costs, which arise once a burrow is fully established.

Proxies for Construction Costs

Several relevant proxies exist for the cost of burrow construction. Higher values of these proxies presumably translate to greater overall costs to the burrower. For instance, some crustaceans create longer or deeper burrows that can extend up to several meters into the substratum, with voluminous and elaborate interiors (Atkinson and Eastman 2015). Interestingly, in both *Lembos websteri* and *Crassicorophium bonnellii*, which are tubicolous, tube length is highly correlated with animal length (Shillaker and Moore 1978). The greater the length or space within the burrow, the greater the amount of material that must be excavated; hence, greater total effort is required on the part of burrower. In many species, individuals create complicated twists and extensions along a single long tunnel. The exact function of these extensions is not always clear, but the greater cost involved in making them may be offset by the larger number of protective hideaways contained within the burrow.

A second relevant proxy for the cost of burrow construction is the time spent burrowing, which itself often correlates with the length and internal space of a burrow. Burrowing shrimp (*Callianassa subterranea*), for instance, have been observed in the laboratory across their entire burrow construction process (Stamhuis et al. 1996). The more time these shrimp spend burrowing, the longer and more elaborate their burrow. However, the burrows in that study approached an asymptote after approximately 50–100 hours at 400 mm in tunnel length, apparently because of the size constraints of the containers in which the shrimp were tested. Interestingly, in mole crabs (*Emerita analoga*), Kolluru et al. (2011) found that forcing individuals to dig repeatedly made them slower on subsequent occasions (Fig. 11.3A). Mole crabs only dig, and they do not construct actual burrows; however, these data show that there are costs to moving sediment and thus indirectly suggest that time spent burrowing is energetically costly. The size of the burrower can also affect the time spent burrowing, with smaller individuals in *Sphaeroma terebrans* taking more time to build a burrow and ultimately having lower survival than medium-sized individuals (Thiel 2001). Essentially, both time spent burrowing and the final burrow length can be viewed as opportunity costs, because if a burrower is not engaged in the act of construction or in elongating its burrow, then it could devote this time and effort to foraging and other essential activities.

During the initial phase of burrow construction, burrowers may also face elevated risk from predation, which is yet a third proxy for the cost of burrow construction. Although this potential

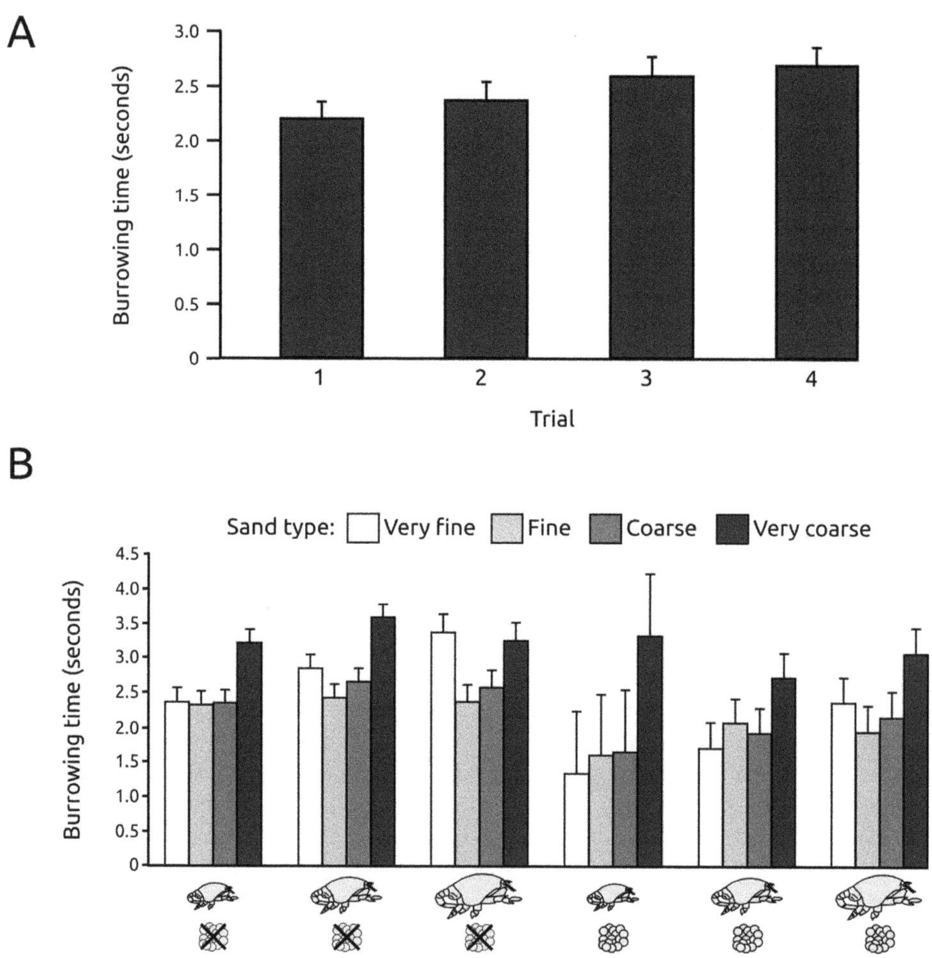

Fig. 11.3.
Construction costs based on burrowing time in the sand crab (*Emerita analoga*). (A) Burrowing time increased for individuals that were experimentally challenged to repeatedly dig, suggesting that burrowing is energetically costly. (B) Burrowing time as a function of sand type, crab size, and reproductive condition. Across nearly all crab sizes (small, medium, large) and reproductive conditions (nonovigerous, ovigerous), burrowing time is increased in very coarse sand, which represents a harder substrate for digging compared to finer sand. Bars in both panels show least-squares means + standard error. Modified from Kolluru et al. (2011), with permission from Elsevier. Drawing of *E. analoga* modified from Paul (1971), with permission from Springer.

cost has not been formally investigated, it is nevertheless clear that burrowing individuals must repeatedly foray between the burrow depth and the burrow opening to remove the material they are excavating. While engaged in this work, individuals may be less vigilant against predators, and their greater activity may also make them more conspicuous to predators. So even if a fully formed burrow offers a valuable antipredator resource, the early stages of making this burrow could expose the burrower to potentially deadly levels of predation risk. Notably, in sand crabs (*E. analoga*), more heavily parasitized individuals dig more slowly and are more susceptible to predation (Kolluru et al. 2011), suggesting it is advantageous to finish a burrow quickly.

Finally, a fourth proxy for the cost of burrow construction, and arguably the most valuable one for predicting behavior from a comparative perspective (Ghazoul 2001, Luna and Antinuchi 2006), is the type of substratum from which burrowers excavate (Fig. 11.3B). Broadly, crustaceans excavate burrows in an immense variety of substrata (Mejaes et al. 2015), which vary dramatically in degrees of hardness: from algae and loose-packed soft beach sand; to coarser sand with larger grain size; to wood; and ultimately to the hardest of solids, such as shell and rock, the last of which requires extraordinary construction costs to excavate. For example, some crustaceans excavate their burrows in soil and woody plant material, such as the burrows of coconut crabs, which excavate beneath the roots of coconut trees (Fig. 11.1A; Laidre 2018). Other crustaceans, such as many brachyuran crabs, excavate in pebble and coral rubble in the intertidal zone (Fig. 11.1B). And still others, such as the terrestrial hermit crabs *Coenobita* spp., hollow out marine shells that wash onto land by excavating directly into the calcium carbonate of the shell interior (Ball 1972, Laidre 2012a, Fig. 11.4). The ease of excavation into a substratum fundamentally depends on both the substratum material and the morphological specializations (in digging appendages and muscles) of the species (Faulkes 2013). Harder substrata translate to greater excavation effort, so these heightened costs must be offset by greater benefits to the burrower. Interestingly, the costs of burrow construction may not only vary among different habitats with variable degrees of substratum hardness, but may also vary within the

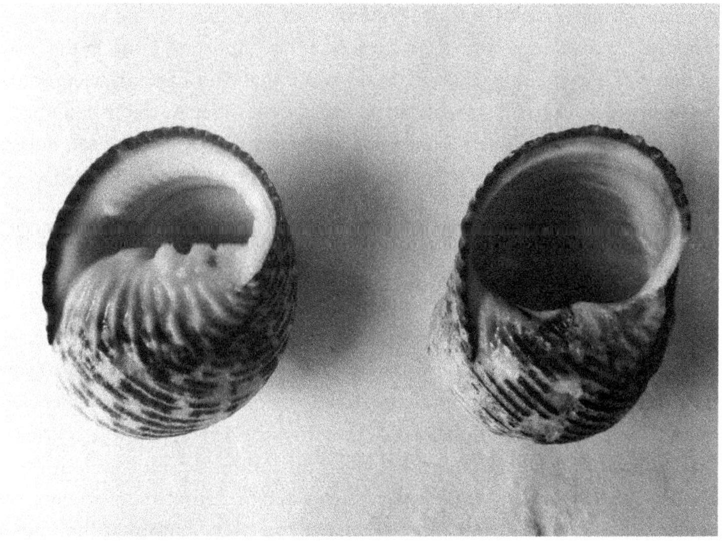

Fig. 11.4.
Transportable burrow: a shell (*Nerita scabricosta*) carved out by terrestrial hermit crabs (*Coenobita compressus*). The left image shows before excavation (unremodeled), and the right image shows after excavation (fully remodeled). The shell interior has been completely hollowed out by the crab, which has eliminated the shell columella. Photograph by Mark E. Laidre ©. See color version of this figure in the centerfold.

same habitat across seasons (e.g., as moisture content changes a substratum from harder to softer before and after rains). Ultimately, the greater investment in burrow construction by some species can have important consequences later for social life, as we will see in considering the impact of different architectural parameters on social evolution.

ARCHITECTURAL PARAMETERS THAT AFFECT SOCIAL EVOLUTION

Social evolution intersects with both the extended phenotype and the niche construction concepts. Social evolutionary theory asks why certain ecological and evolutionary circumstances favor greater or lesser degrees of social interaction (Bourke 2011). Several architectural parameters of crustacean burrows appear to strongly affect social evolution, providing a scaffolding for social life both within and across generations.

Burrow Size (Spatial Dimension)

Burrows can, literally, lay the foundation for social life by providing a central shared dwelling for interaction among conspecifics. However, if social interactions are to take place within a burrow, it is critical that the size of the burrow be spacious enough for individuals to coreside. Furthermore, these individuals must be tolerant of each other. In many crustaceans, large-enough burrows as well as social tolerance among conspecifics has enabled multiple cooccupants to be accommodated within the same burrow. The structure of these social assemblages include heterosexual mated pairs, parent-offspring family groups (sometimes including multiple generations of successive broods), and even congregations of unrelated conspecifics. For instance, desert isopods occupy burrows as family groups (Linsenmair 2007). And the breeding burrows of *L. pugilator*, which are constructed by courting males, accommodate up to three ovigerous females, each female being sequestered in a separate terminal chamber (Christy 1982). Thiel (2007) reviewed the literature on social behavior among crustaceans dwelling within the same burrow and describes current knowledge about their interactions, including cases of extended parental care. An important point in the present chapter is that even if burrows are exceedingly small, to the point that they can only accommodate a single individual, these burrows can still strongly affect social evolution. As described in a later section, burrows that do not allow individuals to coreside at the same point in time can nevertheless generate intense selection pressures for complex social interactions, particularly if they are passed down across generations.

Construction Investment (Original Cost to Construct)

The construction costs of a burrow can be viewed as an investment by the burrow constructor. This investment should be directly translatable to the value of the burrow as a dwelling. Burrows that required greater construction costs and effort to build thus are predicted to have higher value. Resource value may be socially recognized by individuals other than the original burrow constructor, because these other individuals might likewise benefit from inhabiting the burrow (Arnott and Elwood 2008). Consequently, conspecific intruders may be attracted to the burrow and may seek to evict the burrow owner and take over its burrow, reaping the benefits of the burrow without having paid any construction costs. Such evictions of burrow owners are common in many crustacean species (e.g., James-Pirri and Cobb 1999, Mathews 2002, Briffa and Elwood 2004, Koga and Ikeda 2010).

From a game-theoretical mathematical perspective (Maynard Smith 1982, Sigmund 1993), the value created by burrow constructors may lead to a producer-scrounger game (Barnard and Sibley 1981) in which a population initially composed exclusively of burrow constructors is subject to invasion by mutant evictors, ultimately leading to a frequency-dependent mix of both constructor and evictor "strategies" in the population (Table 11.2). The presence of evictors should in turn favor burrow constructors that further invest in defense of their burrow. In particular, there should be a correlation between a burrow owner's investment in construction and its investment in defense, such that the burrow owner is more willing to defend a more valuable burrow (Arnott and Elwood 2008), which required greater effort to build. In other words, if more investment was put in originally, there is more incentive to defend that investment (Ghazoul 2001). To my knowledge, data do not exist to undertake a broad cross-species test of this prediction across crustaceans. However, stomatopods provide an informative case study (Caldwell and Dingle 1975, Dingle and Caldwell 1978). In certain gonodactyloid stomatopods, which make their burrows in hard substrata that are limiting and require substantial construction costs, individuals are much more aggressive in defending their burrows compared with squilloid stomatopods, which build in softer substrata involving particulate sediments that require minimal construction costs. Further tests of the predicted correlation between investment in burrow construction and investment in burrow defense would be valuable.

Burrow Longevity (Temporal Dimension)

Greater construction costs not only raise the value of a burrow but also frequently correlate with increased longevity of the burrow, with some burrows persisting, even without maintenance, for extended periods. Of course, no burrows are permanent. Indeed, some are destroyed by the sheer act of excavation itself (this is especially true for burrows constructed in aquatic plants, where burrowing undermines the plant structure's integrity, causing it to fall apart; Mejaes et al. 2015). Also, the presence of too many individuals around or within a burrow may generate interference competition between overlapping species and may contribute to the demise of burrows (Aspey 1978). Yet burrows constructed within hard substrata can last for prolonged periods (see table 14.4 in Thiel 2007), potentially remaining durable for years or even decades. Burrows can also be reinforced along the walls with mucous secretions or with hard materials, such as coral or shell fragments, to help prevent collapse. Although "cementing glue" is common in various eusocial insects (Hansell 2005), it appears to be absent or at least rare in crustaceans. Nevertheless, relative to the generation time of various crustacean species, burrows can outlast their original constructor and be handed down as an ecological inheritance to subsequent generations. Niche construction models predict that such an ecological inheritance can have a profound influence on social evolution (Krakauer et al. 2009, Odling-Smee et al. 2013), because the inheritance itself offers a valuable resource, which can be reused across generations. Few empirical studies, however, have traced burrows from their original construction point to their inheritance and reuse by future generations. In the next section, I describe a particularly well-studied case of the inheritance of constructed burrows, which are themselves transportable.

Transportable Burrows

Unlike standard stationary burrows, which remain in one place, transportable burrows can be moved. Some transportable burrows move simply because they float around (e.g., wood pieces at the sea surface), but the most interesting are those that are actively transported by the burrow occupant itself. Individuals in such transportable burrows are no longer rooted to a central place foraging lifestyle and home site, so they can overcome a major constraint associated with life in burrows: reduced dispersal.

An exemplary case of transportable burrows (Laidre 2012a) is provided by hermit crabs, some of the most well-studied crustaceans (Hazlett 1981, McLaughlin 2015). Nearly all hermit crabs carry an external object (usually a gastropod shell) as their shelter (Laidre 2011), which Atkinson and Eastman (2015) have pointed out is "analogous to a burrow" (p. 80). But just as terrestrial crustaceans (Burggren and McMahon 1988, Greenaway 2003, Richardson and Araujo 2015) differ substantially in their lifestyle from marine ones, so terrestrial hermit crabs do something that no marine hermit crabs do: they actively carve out and remodel gastropod shells (Laidre 2012a, Laidre and Vermeij 2012), excavating them as constructed burrows (Fig. 11.4). This excavation of shells by terrestrial hermit crabs directly parallels the underground excavation of stationary burrows by coconut crabs (Fig. 11.1A). Indeed, it is perhaps no coincidence that coconut crabs (*Birgus latro*) are the closest evolutionary relatives of terrestrial hermit crabs (*Coenobita* spp.). As juveniles, coconut crabs likewise excavate shells, so their excavation of underground burrows after they outgrow shells may be behaviorally homologous to the act of excavating and hollowing out shells.

The act of excavating a shell for a terrestrial hermit crab is costly, requiring substantial time and energy (Laidre 2012a, Laidre 2012b, Laidre et al. 2012). Indeed, crabs sometimes even die in the process of trying to remodel shells (Laidre 2012a, 2012b). Yet because remodeled shells are so critical to life on land, providing individuals with a lighter load (Herreid and Full 1986) and more space for an internal water reservoir (de Wilde 1973), terrestrial hermit crabs have ultimately become specialized to live in these constructed shells. Indeed, after an early life stage, terrestrial hermit crabs cannot survive in anything other than a remodeled shell (Laidre 2012b). Remodeled shells, like the stationary burrows of other crustaceans, therefore represent an extended phenotype, one that is costly but highly beneficial for these crabs.

Given that the cost of remodeling a shell from scratch is so high, terrestrial hermit crabs seeking to move into bigger remodeled shells in the housing market frequently pursue an alternative strategy; instead of remodeling a shell themselves, they instead wait around in social groups (Laidre 2010) until another, larger individual either dies or is evicted (Laidre 2013a, 2013b). After the dead or evicted individual is removed, its remodeled shell is left behind, which others can then enter (Fig. 11.5). Precisely because remodeled shells last so long, potentially decades (Vermeij 2010, Laidre 2012a), and also because remodeled shells can be transported free of any movement constraints,

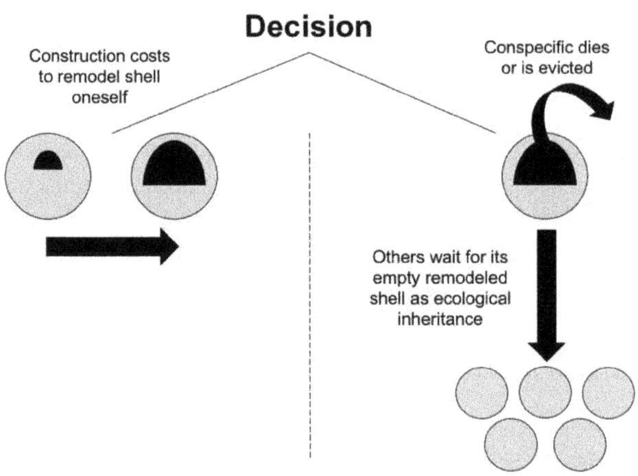

Fig. 11.5.
Cost-benefit decision tree for terrestrial hermit crabs (*Coenobita* spp.). When construction costs are high for remodeling a shell oneself, then individuals can wait until another larger conspecific either dies or is evicted, thereby leaving behind its remodeled shell as an ecological inheritance. By waiting in social gatherings, individuals can maximize the chances of inheriting another's shell and avoid the costly work of personally excavating a shell.

a social strategy is highly effective among terrestrial hermit crabs and has led to the formation of complex social groupings of non-kin (Laidre 2014). The oceanic larval dispersal stage of terrestrial hermit crabs also ensures that social interactions on land are between unrelated individuals (Greenaway 2003). Hence, the unusually high level of sociality among terrestrial hermit crabs, relative to their marine counterparts (Laidre and Trinh 2014, Laidre and Greggor 2015), is ultimately attributable to the construction individuals perform on shells and the persistence of these shells as an ecological inheritance.

ECOLOGICAL INHERITANCE AND LONG-TERM EVOLUTIONARY CONSEQUENCES OF BURROW LIFE

Niche construction models predict that ecological inheritance can have important evolutionary consequences for subsequent generations, both for kin and non-kin. Terrestrial hermit crabs provide a clear case of the impact of an ecological inheritance on social evolution among non-kin. Yet other burrow-constructing crustaceans offer support for the impact of an ecological inheritance on social evolution among kin. For example, in the marine isopods *Sphaeroma terebrans* (Thiel 2001) and *Limnoria annae* (Brearley and Walker 1995), juveniles are raised in the same burrow as their parents, later excavating their own burrow as branches off the parental burrow. Excavating from within the parental burrow is much less costly, because not only are juveniles in a safe environment while they excavate, but they can also excavate within the softer, interior tissue of an existing burrow, which is easier than excavating from scratch on the outer epidermal tissue of the host plant (Mejaes et al. 2015). Juveniles' burrowing from within a parental burrow constitutes a form of niche construction, which operates alongside extended parental care and kin selection, ultimately ensuring that the ecological inheritance of a preconstructed burrow passes directly to one's offspring.

Eviction and Escalated Social Competition

Among non-kin, which generally lack the shared interests of parent-offspring and other close family relatives, the ecological inheritance of constructed burrows can lead to severe competition and escalated social interactions immediately outside of the burrow. Particularly when burrows persist across generations, they may become a target for unrelated conspecifics seeking to evict the current owner or wait until it dies. Ironically, burrows, which originally evolved to protect against predators, may eventually lead to the need to protect against conspecifics. The longer burrows last, the greater the chance over the course of a burrow's existence that conspecific intruders will encounter it. And if the cost a potential evictor would incur in evicting the owner is less than its estimated future fitness gains from taking over ownership of the burrow, then conflicts between burrow owners and evictors may be frequent. For instance, in lobsters (*H. americanus*), even early benthic stage individuals, like their older juvenile and adult counterparts, will invade and evict conspecifics from burrows (James-Pirri and Cobb 1999). Indeed, as burrows become increasingly limiting resources, then ecological constraints may augment their inherent value further, making them even more enticing to potential intruders that are seeking to overtake ownership. Thus, even when burrows cannot be transported, they can nevertheless generate ample scope for complex social interaction once conspecifics are attracted to the site of these stationary resources.

If a burrow owner is to continue to reap the benefits of its burrow and avoid being evicted, it must now invest in defense, in addition to the substantial investment it may already have placed in burrow construction. Critically, for the burrow to be a worthwhile investment to the original owner, the benefits the burrow affords must exceed the total costs incurred, both in prior construction and ongoing defense (Maynard Smith 1982, Ghazoul 2001). As with construction costs, measuring defense costs is challenging, particularly because even if a burrow owner is successful in defense, the injuries or other harm it sustains during an escalated fight are not always visible. In only a few crustacean species have the metabolic costs of escalated fights been precisely quantified, revealing that energetic reserves in owners and evictors are key determinants of success (Briffa and Elwood 2004).

Ultimately, whether a burrow owner or an evictor is successful in a conflict will depend on the balance between the maximal defense costs the burrow owner is capable of suffering and the maximal eviction costs the evictor is willing to incur (Table 11.2). These parameters will undoubtedly vary across individuals, based on their strength and resource-holding potential. For instance, in the intertidal sand-dwelling crustacean *Leptochelia dubia,* males occupying burrows usually win fights against equal-sized intruders, but larger males always win fights with smaller males regardless of whether or not they are occupying a burrow (Highsmith 1983). Likewise, in the semiterrestrial crab *Neohelice granulata,* bigger males are more capable of defending burrows. Small males of this species, in contrast, do not even bother constructing burrows but instead pursue alternative mating strategies; they either try to intercept females on the surface, while females are searching for burrows, or they temporarily wait for females inside the burrows of large males, which they do not actually own (Moyano et al. 2016). In contests over burrows between male sand crabs (*Scopimera globosa*) the resource value of the burrow as well as the differential resource-holding potential (body size) of owners and intruders, jointly determine the conflict outcome (Koga and Ikeda 2010). When multiple individuals (such as a mated pair or a family) inhabit a burrow, their combined resource-holding potential can offer an advantage: territorial cooperation. For instance, experiments in snapping shrimp (*Alpheus angulatus*) have revealed that intruders were significantly less likely to evict individuals from their burrow if the burrow was occupied by a pair rather than a solitary individual (Mathews 2002). This result might be due to cooperation, or it might be due to the differential resource value of a mate, which could cause individuals to invest more in defense. For desert isopods *Hemilepistus reaumuri,* Linsenmair (2007) has provided strong evidence that cooperation in pairs is indeed critical to successful burrow defense.

Ecological constraints make burrows more limiting, either because burrows themselves, or the spaces needed to construct them, are in short supply. In such cases, the cost-benefit decision points will likely change (Rubenstein and Wrangham 1986). For instance, if burrowless roamers have few options other than evicting a current burrow owner, this may favor frequent and intense intraspecific aggression (Dalosto et al. 2013). In crustaceans in extremely burrow-limiting conditions, eviction from one's burrow may effectively be a death sentence for the burrow owner, meaning that owners will have no viable alternative but defending their burrow with all their available resources. This situation applies in many stomatopods whose raptorial appendages have evolved into dangerous weapons, the primary function of which is to keep intruding conspecifics away (Caldwell and Dingle 1975, Dingle and Caldwell 1978). The same may be true of coconut crabs (Fig. 11.1A), which have some of the most powerful crushing claws of any crustacean and frequently use these claws to defend ownership of burrows (Oka et al. 2016, Laidre 2017, 2018). In many other crustaceans, investments in defense are prominently represented in claw weaponry and escalated fighting at the entrances to their burrows.

Interestingly, if a burrow is successfully taken over by an evictor, then the evictor never has had to pay any construction costs. Evicting is thus equivalent to having another individual do the work and then stealing its property (Strassmann and Queller 2014). In its new role as burrow owner, the successful evictor's cost-benefit equation for investment in defense will require that the costs it expended on evicting the prior owner, plus the costs it will expend on future defense against other potential evictors, not exceed the benefits it reaps from the burrow. Several key predictions can be made based on the relationship among these costs and benefits (Table 11.2). However, it is critical to know the history of burrow construction and inheritance to correctly interpret behavior during burrow conflicts, because the baseline costs used as a reference point will differ between individuals that originally constructed a burrow as opposed to those that acquired the burrow through eviction. In most species, we still lack such detailed information, not to mention information on the relative density and limitation of burrows in the surrounding environment.

Territorial Signals, Sexual Selection, and Communicative Adaptations to Burrows

In many crustaceans, intense competition for burrows has led to territoriality (Nordhaus et al. 2009). In the ocypodid crab *Ilyoplax pusilla*, for instance, territorial behavior around burrows is most developed in large males compared to smaller individuals and females (Wada 1993). As the costs of defending burrows increases, requiring recurrent disputes with intruders, individuals may seek to minimize these costs using signals (Laidre and Johnstone 2013), which can dissuade intruders without the need to resort to escalated fighting or injury (Laidre 2007, Laidre and Elwood 2008, Laidre 2009). Some signals may even serve dual sexual selection functions, both attracting potential mates as well as keeping intruders at bay. Signals that advertise a burrow owner's size, strength, or motivation are the most relevant to intruders, because they correlate with the owner's claim over the burrow and hence its ability to defend this territory. Claw-waving displays, for instance, are common in some semiterrestrial burrow-dwelling species, such as fiddler crabs, because advertising claw size may be an important determinant of success during contests over burrows (Salmon and Atsaides 1968, but see Backwell et al. 2000). Acoustic signals have also evolved in some crustacean species, such as the ghost crabs *Ocypode ceratophthalmus* (Hughes 1966) and *Ocypode cordimana* (Horch 1975), which by stridulating can project sounds from the depth of their burrows, thereby conveying to intruders that the burrow is occupied, even if owner and intruder never visually confront one another. Interestingly, sound production may also convey aspects of the signaler's own body size, important information that can help resolve conflicts without recourse to actual physical fighting. And the burrow itself has been hypothesized to act as a resonating chamber for these sounds, amplifying and effectively broadcasting them more loudly to would-be intruders (Hughes 1966). The evolution of such signals can thus be linked back to the architectural foundation provided by constructed burrows as valuable, contestable dwellings, which can be inherited by others if an owner is removed.

In addition to signals that are direct behavioral acts or features of the burrow owner's own phenotype, some communicative adaptions to burrows may be represented by features of the burrow itself, remnants of earlier building and construction activity by the burrower. These so-called extended phenotype signals are common in noncrustacean taxa, such as birds, where built structures such as bowers can function as sexual selection advertisements, conveying information about the quality of the male builder (Hansell 2005, Gould and Gould 2007). In most crustaceans, extended phenotype signals have not been investigated, although fiddler crabs provide a notable exception. Fiddler crabs build structural ornaments near the openings to their burrows, including hoods, chimneys, pyramids, and pillars, all made from the sediment excavated from the burrow (Zucker 1981). Perhaps surprisingly, experiments by Christy et al. (2003) have revealed that these aboveground burrow "toppings" do not necessarily signal anything about the burrow owner itself and may not even be correlated with the owner's phenotypic quality or resource-holding potential. For instance, in *Leptuca terpsichores*, burrows offer a safe haven from avian predators, and both the sand hood built by a male and the male's claw waving have been found to act simply as beacon to these safe spots, rather than conveying information about male quality (Christy 2007, Perez et al. 2016). The same seems to apply in *L. beebei*, where under elevated predation risk females showed stronger preference for males who build pillars, apparently only because these pillars provided safety for the visiting female (Kim et al. 2009). Hence, many of the structures built by fiddler crabs near their burrows may simply be the equivalent of sensory traps, attracting females that burrow-owning males can then mate with. Females may still benefit by being attracted, given that these structures offer them a safe haven in predator-ridden landscapes. The structures can thus simultaneously benefit males that build them, by increasing their mating opportunities, and the females that are attracted to them, by providing a form of protection during mate-search. However, the structures are not without costs, because in addition to attracting receptive females, they may also attract interloping males and nonreceptive females (Muramatsu 2009).

No studies, to my knowledge, have experimentally investigated whether the physical features of burrows in other crustacean species might convey information about the burrow owner itself. Coconut crab burrows could provide an intriguing system for such experiments. As Darwin (2001, p. 414) observed, "these crabs inhabit deep burrows, which they hollow out beneath the roots of trees" and "they accumulate surprising quantities of the picked fibres of the cocoa-nut husk, on which they rest as on a bed." The accumulation of such large quantities of remnant husk, much of it placed outside at the mouth of the burrow (Fig. 11.1A), may not be simply a byproduct. Rather, this accumulation of husk might be a potential index signal (Laidre and Johnstone 2013), specifically advertising the burrow owner's strength (both to receptive females and to intruding males) based on the number of coconuts the burrow owner recently husked. Manipulative experiments testing this hypothesis would be constructive, not only in coconut crabs, but also in other crustaceans that accumulate byproducts that are potentially useful in advertising male quality.

COMPARISON WITH OTHER TAXA

In addition to crustaceans, a great many other taxa construct burrows, including mollusks, fishes, amphibians, reptiles, birds, mammals, prehistoric dinosaurs, and other arthropods. Like crustaceans, the burrows of these other taxa have in many cases strongly shaped their social evolution. Many mammals, for instance, excavate and live within burrows, ranging from Arctic climates (where polar bears, *Ursus maritimus*, carve out dens in the snow to raise their cubs) to hotter climates (where cooperatively breeding meerkats, *Suricata suricatta*, raise successive generations of offspring to remain in the parental territory burrow and help their parents reproduce). Meerkats, in particular, provide a compelling example of the cost of eviction (Clutton-Brock 2016), because subordinates who threaten a dominant pair's breeding status are ousted from the group's burrow and typically die as a consequence (Clutton-Brock 2009), not unlike eviction in terrestrial hermit crabs and other highly social crustaceans.

Construction costs have been found to be particularly influential in shaping social evolution in subterranean rodents (Luna and Antinuchi 2006), where the cost of digging in soils of different hardness is one of the most important factors influencing burrowing efficiency. Indeed, social burrowing, by digging together (Ebensperger and Bozinovic 2000), and group living in general (Ebensperger and Cofre 2001) may both have evolved in rodents specifically to reduce the costs of burrow construction. The African mole-rats (family Bathyergidae) provide perhaps the most compelling case study of construction costs driving social evolution. These subterranean rodents, which are endemic to sub-Saharan Africa, display levels of sociality ranging from solitary to eusocial. In the dry environments they inhabit, new burrow formation is extremely energetically expensive, with digging through soil being up to 3,600 times the cost of surface locomotion (reviewed in Faulkes and Bennett 2013). Independent phylogenetic contrasts have revealed that a few key ecological variables (greater aridity and the associated challenge of burrowing through the hard soil to feed on clumped underground tubers) are the most important predictors of larger social group size and increased cooperation (Faulkes et al. 1997). Indeed, construction costs and ecological constraints appear to have been the main drivers culminating in the pinnacle of social evolution, eusociality, which is found in the naked mole-rat (*Heterocephalus glaber*) and the Damaraland mole-rat (*Fukomys damarensis*). These results from mammals should motivate parallel studies in crustaceans, within and across species, to assess how economic trade-offs in construction costs relative to the benefits of burrows ultimately affect social evolution.

Strong arguments have been made for the central importance of underground burrows in the origin of social life not just in mammals but also in invertebrates, especially in social insects such as termites, which are specialized to burrow in wood. The so-called fortress hypothesis (Hansell 2005) posits that the monopolization of a central defensible nest (one that is rare, valuable, and long-lived as an ecological inheritance) is the critical preadaptation in the evolutionary road to eusociality, the most complex form of sociality known. Other than social insects, the only known eusocial invertebrates are snapping shrimps (Duffy 2007); this supports the fortress hypothesis, given that the sponges and coral rubble cavities where these crustaceans make their homes are limited resources that are extremely valuable and stable. More detailed data on the longevity of these dwellings as an ecological inheritance and more in-depth behavioral observations (of the extent to which certain subsocial snapping shrimps partially excavate burrows versus simply moving in to preexisting natural cavities) would help shed light on the hypothesized importance of construction in social evolution.

Among invertebrates, Suter and colleagues (2011) provided an exemplary case of quantifying the exact energetic costs of burrow construction in wolf spiders, using high-speed videography coupled with scanning electron microscopy and energetic calculations. Few other studies, in any taxa, have so carefully or thoroughly measured construction costs. The results of this study are especially intriguing because they reveal that the act of burrow construction costs spiders the equivalent of at least one and up to several eggs from their clutch. White (2001) estimated that, in scorpions, excavating a single burrow comprises approximately 2% of an adult's yearly energy budget. Clearly then, the construction costs of burrows are not negligible, and in some cases burrow construction may be expensive, involving major sacrifices in reproductive success, which only pay off once the burrow becomes a finished dwelling the burrower can live in.

Notably, despite the central importance of a gene-centered, extended phenotype concept for all burrowing animals, we have a strong grasp of the underlying genetic basis of burrowing, including the alternative extended phenotypes that can be produced for only one genus, *Peromyscus*. Hoekstra and colleagues (Weber and Hoekstra 2009, Weber et al. 2013) have revealed, through three-dimensional measurements of burrows and careful genetic crossing experiments between oldfield mice (*P. polionotus*) and their sister species deer mice (*P. maniculatus*), that just a few genetic changes are responsible for complex changes in burrow architecture. This includes longer burrows and the presence of an escape hatch near the surface, which allows mice with that extended phenotype to rapidly tunnel to the surface if cornered in their burrow by a predator. With increasing knowledge of the genomics of crustaceans, including the genetic basis of their behavior, it may eventually be possible to likewise elucidate genes coding for some of the astounding diversity in crustacean burrows.

FUTURE DIRECTIONS

Despite a rich body of empirical work on the evolutionary ecology of burrows in crustaceans, much still remains unknown. Indeed, if crustacean burrows are to realize their full potential as model systems for testing theoretical frameworks, such as the extended phenotype and niche construction frameworks, then it is essential that further critical observations and experiments be undertaken. Much of the required data can be collected with new, cutting-edge technologies, which measure energetic costs, record social interactions, and track movements. Once combined, these methodological innovations can ultimately catapult natural history studies to a new level of sophistication. The following discussion touches on some of the most promising directions for future research that can set the stage for major advances.

First, and most critically for understanding why burrowing evolved, we need more in-depth measures of the actual costs crustaceans incur in constructing their burrows, with both costs and benefits measured on similar scales, so that they can be compared. Ideally, cost measures will be energetic ones that quantify the work involved in burrow construction across different substrata, and these costs should then be linked back to benefits to elucidate what favored the evolution of burrowing behavior. Several recent studies in other behavioral domains have taken physiological blood lactate measures in crustaceans to quantify energetic costs (e.g., Briffa and Elwood 2004), and there now exist simple and widely available measurement devices that can make acquiring such data relatively easy. In principle, therefore, this same technology can revolutionize our understanding of the construction costs of burrows, enabling more rigorous tests of the extended phenotype model, especially if we are able to independently measure construction, maintenance, and defense costs.

Second, we still lack baseline data on the types and frequencies of social and territorial interactions taking place within and around crustacean burrows, especially rates of intrusion by conspecifics, attempted and successful eviction, and turnover in burrow ownership. Cutting-edge tracking technologies and remote recording techniques can be extremely helpful in providing answers, with miniature global positioning system (GPS) trackers, wildlife tracking (e.g., VHF tags), and networks of camera traps having the potential to generate detailed information about individuals' movements both to and from their own and others' burrows. Such data could ultimately test predictions about the frequency-dependent trade-offs between burrow constructors (who invest in building) versus evictors (who steal these valuable dwellings). Moreover, by using GPS tracking to follow the movements of crustaceans living in stationary versus transportable burrows, we can better understand how the absence of movement constraints in transportable burrow owners has fostered the evolution of complex social life among non-kin. Terrestrial hermit crabs, in particular, provide a model system for such studies (Laidre 2014).

Third, in addition to tracking the actual occupants living inside burrows, a major unanswered question is the longevity of the burrows themselves, both when they are actively maintained by the burrow owner and when they fall into disuse. Although some burrows may require near constant care to preserve their architectural integrity, and so collapse quickly if not maintained, others may remain in peak condition with little or no maintenance at all. Specifically, burrows that are excavated in hard substrata, which requires strong digging appendages to excavate, and burrows that are held together with powerful cementing compounds may both be key to burrow longevity. However, how often burrows exceed the lifetime of individual residents is poorly documented. Only once we can relate the relative durability of burrows to their construction costs and to the temporal turnover in burrow ownership will it be possible to test key predictions from niche construction models. Critically, if burrows across many species ultimately outlast their original constructors, providing homes for multiple successive occupants, then these burrows constitute an ecological inheritance that likely has major evolutionary impacts on subsequent generations.

Across all these focal topics, from energetic costs to social interactions to burrow longevity, it is essential not only to use cutting-edge technology for measurements, but to pair this technology with carefully designed experiments that build on a foundation of natural history and broader theoretical constructs.

SUMMARY AND CONCLUSIONS

Some of the most accomplished animal architects are invertebrates (Gould and Gould 2007). This chapter has reviewed a fascinating clade of invertebrate architects, the crustaceans, through the lens of two overarching theoretical concepts: the extended phenotype and niche construction. The first of these concepts predicts when burrowing will be favored, and the second of these concepts predicts the long-term impact burrows will have if they remain durable. This overview has examined both what we know and what still remains to be discovered in order to fully test these theoretical concepts. Across the phylogeny of crustaceans, many species excavate burrows. Generally, these burrows appear to enhance the burrower's fitness, maximizing the benefits the burrower reaps relative to the costs, as is predicted by the extended phenotype concept. Costs are paid both during the initial excavation and construction of the burrow, during the maintenance of the burrow, and during defense of the burrow against intruders. Some burrows are quite costly to construct, are durable across generations, and are in short supply. Burrows with these characteristics are the most relevant to the niche construction concept. As predicted by niche construction, such burrows frequently attract conspecific intruders who seek to take over ownership of the burrow and evict the current burrow owner. This escalation of social interaction, due ultimately to burrow construction, has been well studied in terrestrial hermit crabs, which excavate and occupy transportable burrows that they carry around on their backs. In other species, where burrows are likewise highly valuable and sought after, complex signals and weaponry have evolved as a mechanism of maintaining ownership of one's burrow. These results thus suggest that ecological inheritance of burrows has had important evolutionary impacts on social life in many burrowing crustaceans. However, in few species, crustacean or otherwise, have the full spectrum of costs and benefits relevant to burrowing been quantified. Furthermore, in most cases, the longevity of burrows is still unclear. Thus, to chart the path ahead, further measurements, experiments, and comparisons across taxa, are essential to testing and refining predictions of the extended phenotype and niche construction models. Ultimately, by combining natural history, experiments, and theory we can peer down the evolutionary "rabbit hole" to discover the full depth of burrows.

ACKNOWLEDGMENTS

This work was supported by Dartmouth College start-up funds, the National Geographic Society, the Smithsonian Institution, the Neukom Institute, the Miller Institute, and the National Science Foundation. I thank the editors, Martin Thiel and Gary Wellborn, for constructive and valuable feedback, and Tim Kiessling for help with creating figures. I am also grateful to Laurel Symes for generous discussion and to Ryan Calsbeek for a sentence of poetic-scientific insight.

REFERENCES

Allen, B.J., B. Rodgers, Y. Tuan, and J.S. Levinton. 2012. Size-dependent temperature and desiccation constraints on performance capacity: implications for sexual selection in a fiddler crab. Journal of Experimental Marine Biology and Ecology 438:93–99.

Arnott, G., and R.W. Elwood. 2008. Information gathering and decision making about resource value in animal contests. Animal Behaviour 76:529–542.

Aspey, W.P. 1978. Fiddler crab behavioral ecology: burrow density in *Uca pugnax* (Smith) and *Uca pugilator* (Bosc) (Decapoda Brachyura). Crustaceana 34:235–244.

Atkinson, R.J.A., and L.B. Eastman. 2015. Burrow dwelling in Crustacea. Pages 78–117 *in* M. Thiel and L. Watling, editors. The natural history of the Crustacea, volume 2: Lifestyles and feeding biology. Oxford University Press, New York.

Backwell, P.R., J.H. Christy, S.R. Telford, M.D. Jennions, and J. Passmore. 2000. Dishonest signalling in a fiddler crab. Proceedings of the Royal Society of London B, Biological Sciences 267:719–724.

Ball, B., A. Linnane, B. Munday, R. Browne, and J.P. Mercer. 2001. The effect of cover on in situ predation in early benthic phase European lobster *Homarus gammarus*. Journal of the Marine Biological Association of the United Kingdom 81:639–642.

Ball, E.E. 1972. Observations on the biology of the hermit crab, *Coenobita compressus* H. Milne Edwards (Decapoda; Anomura) on the west coast of the Americas. Revista de Biologia Tropical 20:265–273.

Barnard, C.J., and R.M. Sibley. 1981. Producers and scroungers: a general model and its application to feeding flocks in house sparrows. Animal Behaviour 29:543–550.

Bourke, A.F.G. 2011. Principles of social evolution. Oxford University Press, New York.

Brearley, A., and D.I. Walker. 1995. Isopod miners in the leaves of two Western Australian *Posidonia* species. Aquatic Botany 52:163–181.

Bridges, C.R. 1986. A comparative study of the respiratory properties and physiological function of haemocyanin in two burrowing and two non-burrowing crustaceans. Comparative Biochemistry and Physiology Part A: Physiology 83:261–270.

Briffa, M., and R.W. Elwood. 2004. Use of energy reserves in fighting hermit crabs. Proceedings of the Royal Society of London Series B, Biological Sciences 271:373–379.

Brown, A.C., and E.R. Trueman. 1996. Burrowing behaviour and cost in the sandy-beach oniscid isopod *Tylos granulatus* Krauss, 1843. Crustaceana, 69:425–437.

Burggren, W.W., and B.R. McMahon. 1988. Biology of the land crabs. Cambridge University Press, New York.

Caldwell, R.L., and H. Dingle. 1975. Ecology and evolution of agonistic behavior in stomatopods. Naturwissenschaften 62:214–222.

Christy, J. 2007. Predation and the reproductive behavior of fiddler crabs (genus *Uca*). Pages 211–231 *in* J.E. Duffy and M. Thiel, editors. Evolutionary ecology of social and sexual systems: crustaceans as model organisms. Oxford University Press, New York.

Christy, J.H. 1982. Burrow structure and use in the sand fiddler crab, *Uca pugilator* (Bosc). Animal Behaviour 30:687–694.

Christy, J.H., J.K. Baum, and P.R.Y. Backwell. 2003. Attractiveness of sand hoods built by courting male fiddler crabs, *Uca musica*: test of a sensory trap hypothesis. Animal Behaviour 66:89–94.

Clutton-Brock, T. 2009. Cooperation between non-kin in animal societies. Nature 462:51–57.

Clutton-Brock, T. 2016. Mammal societies. Wiley-Blackwell, New York.

Crane, J. 1975. Fiddler crabs of the world. Princeton University Press, Princeton, New Jersey.

Dalosto, M.M., A.V. Palaoro, J.R. Costa, and S. Santos. 2013. Aggressiveness and life underground: the case of burrowing crayfish. Behaviour 150:3–22.

Darwin, C. (1845) 2001. The voyage of the Beagle. Modern Library, New York.

Dawkins, R. 1999. The extended phenotype: the long reach of the gene. Oxford University Press, New York.

de Wilde, P.A.W.J. 1973. On the ecology of *Coenobita clypeatus* in Curaçao with references to reproduction, water economy and osmoregulation in terrestrial habitats. Studies on the Fauna of Curaçao and other Caribbean Islands 44:1–138.

Dingle, H., and R.L. Caldwell. 1978. Ecology and morphology of feeding and agonistic behavior in mudflat stomatopods (Squillidae). Biological Bulletin 155:134–149.

Dorgan, K.M. 2015. The biomechanics of burrowing and boring. Journal of Experimental Biology 218:176–183.

Duffy, J.E. 2007. Ecology and evolution of eusociality in sponge-dwelling shrimp. Pages 387–409 *in* J.E. Duffy and M. Thiel, editors. Evolutionary ecology of social and sexual systems: crustaceans as model organisms. Oxford University Press, New York.

Ebensperger, L.A., and F. Bozinovic. 2000. Communal burrowing in the hystricognath rodent, *Octodon degus*: a benefit of sociality? Behavioral Ecology and Sociobiology 47:365–369.

Ebensperger, L.A., and H. Cofre. 2001. On the evolution of group-living in the New World cursorial hystricognath rodents. Behavioral Ecology 12:227–236.

Evans, S.M., J. Prince, J. Foster-Smith, E. Drew, and R.D. Phillips. 2010. Optimal and anti-predator foraging in the sand bubbler crab *Scopimera inflata* (Decapoda: Ocypodidae). Journal of Crustacean Biology 30:194–199.

Faulkes, C.G., and N.C. Bennett. 2013. Plasticity and constraints on social evolution in African mole-rats: ultimate and proximate factors. Philosophical Transactions of the Royal Society of London Series B, Biological Sciences 368:20120347.

Faulkes, C.G., N.C. Bennett, M.W. Bruford, H.P. O'Brien, G.H. Aguilar, and J.U. Jarvis. 1997. Ecological constraints drive social evolution in the African mole-rats. Proceedings of the Royal Society of London Series B, Biological Sciences. 264:1619–1627.

Faulkes, Z. 2013. Morphological adaptations of crustaceans for digging and burrowing. Pages 276–295 *in* L. Watling and M. Thiel, editors. The natural history of the Crustacea, volume 1: Functional morphology and diversity. Oxford University Press, New York.

Ghazoul, J. 2001. Effect of soil hardness on aggression in the solitary wasp *Mellinus arvensis*. Ecological Entomology 26:457–466.

Gould, J.L., and C.G. Gould. 2007. Animal architects: building and the evolution of intelligence. Basic Books, New York.

Greenaway, P. 2003. Terrestrial adaptations in the Anomura (Crustacea: Decapoda). Memoirs of Museum Victoria 60:13–26.

Grow, L. 1981. Burrowing behaviour in the crayfish, *Cambarus diogenes* Girard. Animal Behaviour 29:351–356.

Guerra-Bobo, M., and T.E. Brough. 2011. Neighbour density, body size and anti-predator hiding time in the New Zealand mud-crab *Austrohelice crassa*. Journal of the Marine Biological Association of the United Kingdom 91:691–694.

Hansell, M. 2005. Animal architecture. Oxford University Press, New York.

Hazlett, B.A. 1981. The behavioral ecology of hermit crabs. Annual Review of Ecology and Systematics 12:1–22.

Herreid, C.F., and R.J. Full. 1986. Energetics of hermit crabs during locomotion: the cost of carrying a shell. Journal of Experimental Biology 120:297–308.

Highsmith, R.C. 1983. Sex reversal and fighting behavior: coevolved phenomena in a tanaid crustacean. Ecology 64:719–726.

Horch, K. 1975. The acoustic behavior of the ghost crab *Ocypode cordimana* Latreille, 1818 (Decapoda, Brachyura). Crustaceana 29:193–205.

Hughes, D.A. 1966. Behavioural and ecological investigations of the crab *Ocypode ceratophthalmus* (Crustacea: Ocypodidae). Journal of Zoology 150:129–143.

James-Pirri, M.J., and J.S. Cobb. 1999. Behavioral interactions of postlarval and fifth instar lobsters (*Homarus americanus*) in a simulated cobble environment. Marine and Freshwater Behaviour and Physiology 32:207–222.

Kim, T.W., J.H. Christy, S. Dennenmoser, and J.C. Choe. 2009. The strength of a female mate preference increases with predation risk. Proceedings of the Royal Society of London Series B, Biological Sciences 276:775–780.

Koga, T., and S. Ikeda. 2010. Perceived predation risk and mate defense jointly alter the outcome of territorial fights. Behavioral Ecology and Sociobiology 64:827–833.

Koga, T., P.R.Y. Backwell, J.H. Christy, M. Murai, and E. Kasuya. 2001. Male-biased predation of a fiddler crab. Animal Behaviour 62:201–207.

Kolluru, G.R., Z.S. Green, L.K. Vredevoe, M.R. Kuzma, S.N. Ramadan, and M.R. Zosky. 2011. Parasite infection and sand coarseness increase sand crab (*Emerita analoga*) burrowing time. Behavioural Processes 88:184–191.

Krakauer, D.C., K.M. Page, and D.H. Erwin. 2009. Diversity, dilemmas, and monopolies of niche construction. The American Naturalist 173:26–40.

Laidre, M.E. 2007. Vulnerability and reliable signaling in conflicts between hermit crabs. Behavioral Ecology 18:736–741.

Laidre, M.E. 2009. How often do animals lie about their intentions? An experimental test. The American Naturalist 173:337–346.

Laidre, M.E. 2010. How rugged individualists enable one another to find food and shelter: field experiments with tropical hermit crabs. Proceedings of the Royal Society of London Series B, Biological Sciences 277:1361–1369.

Laidre, M.E. 2011. Ecological relations between hermit crabs and their shell-supplying gastropods: constrained consumers. Journal of Experimental Marine Biology and Ecology 397:65–70.

Laidre, M.E. 2012a. Homes for hermits: temporal, spatial and structural dynamics as transportable homes are incorporated into a population. Journal of Zoology 288:33–40.

Laidre, M.E. 2012b. Niche construction drives social dependence in hermit crabs. Current Biology 22:R861–R863.

Laidre, M.E. 2013a. Eavesdropping foragers use level of collective commotion as public information to target high quality patches. Oikos 122:1505–1511.
Laidre, M.E. 2013b. Foraging across ecosystems: diet diversity and social foraging spanning aquatic and terrestrial ecosystems by an invertebrate. Marine Ecology 34:80–89.
Laidre, M.E. 2014. The social lives of hermits. Natural History 122:24–29.
Laidre, M.E. 2017. Ruler of the atoll: the world's largest land invertebrate. Frontiers in Ecology and the Environment 15:527–528.
Laidre, M.E. 2018. Coconut crabs. Current Biology 28: R58–R60.
Laidre, M.E., and R.W. Elwood. 2008. Motivation matters: cheliped extension displays in the hermit crab, *Pagurus bernhardus*, are honest signals of hunger. Animal Behaviour 75:2041–2047.
Laidre, M.E., and A.L. Greggor. 2015. Swarms of swift scavengers: ecological role of marine intertidal hermit crabs in California. Marine Biology 162:969–977.
Laidre, M.E., and R.A. Johnstone. 2013. Animal signals: a primer. Current Biology 23:R829–R833.
Laidre, M.E., and R. Trinh. 2014. Unlike terrestrial hermit crabs, marine hermit crabs do not prefer shells previously used by conspecifics. Crustaceana 87:856–865.
Laidre, M. E., and G.J. Vermeij. 2012. A biodiverse housing market in hermit crabs: proposal for a new biodiversity index. Cuadernos de Investigación UNED 4:175–179.
Laidre, M.E., E. Patten, and L. Pruitt. 2012. Costs of a more spacious home after remodelling by hermit crabs. Journal of the Royal Society Interface 9:3574–3577.
Laland, K.N., J. Odling-Smee, and M.W. Feldman. 2000. Niche construction, biological evolution, and cultural change. Behavioral and Brain Sciences 23:131–175.
Lavalli, K.L., and D.E. Barshaw. 1986. Burrows protect postlarval lobsters (*Homarus americanus*) from predation by the non-burrowing cunner (*Tautogolabrus adspersus*), but not from the burrowing mud crab (*Neopanope texani*). Marine Ecology Progress Series 32:13–16.
Linsenmair, K.E. 2007. Sociobiology of terrestrial isopods. Pages 339–364 *in* J.E. Duffy and M. Thiel, editors. Evolutionary ecology of social and sexual systems: crustaceans as model organisms. Oxford University Press, New York.
Lucrezi, S., and T.A. Schlacher. 2014. The ecology of ghost crabs. Oceanography and Marine Biology: An Annual Review 52:201–256.
Luna, F., and C.D. Antinuchi. 2006. Cost of foraging in the subterranean rodent *Ctenomys talarum*: effect of soil hardness. Canadian Journal of Zoology 84:661–667.
Macintosh, D.J. 1979. Predation of fiddler crabs (*Uca* spp.) in estuarine mangroves. Biotropica Special Publications 10:101–110.
Mathews, L.M. 2002. Territorial cooperation and social monogamy: Factors affecting intersexual behaviours in pair-living snapping shrimp. Animal Behaviour 63:767–777.
Maynard Smith, J. 1982. Evolution and the theory of games. Cambridge University Press, New York.
McLaughlin, P.A. 2015. Crustaceans in mobile homes. Pages 145–162 *in* M. Thiel and L. Watling, editors. The natural history of the Crustacea, volume. 2: Lifestyles and feeding biology. Oxford University Press, New York.
Mejaes, B.A., A.G.B. Poore, and M. Thiel. 2015. Crustaceans inhabiting domiciles excavated from macrophytes and stone. Pages 118–144 *in* M. Thiel and L. Watling, editors. The natural history of the Crustacea, volume 2: Lifestyles and feeding biology. Oxford University Press, New York.
Moyano, M.P.S., M. Lorusso, J. Nunez, P. Ribeiro, M.A. Gavio, and T. Luppi. 2016. Male size-dependent dominance for burrow holding in the semiterrestrial crab *Neohelice granulata*: multiple tactics used by intermediate-sized males. Behavioral Ecology and Sociobiology 70:1497–1505.
Muramatsu, D. 2009. To build or not to build—or to destroy burrow hoods in a population of *Uca lactea*. Journal of Crustacean Biology 29:290–292.
Nordhaus, I., K. Diele, and M. Wolff. 2009. Activity patterns, feeding and burrowing behavior of the crab *Ucides cordatus* (Ucididae) in a high intertidal mangrove forest in North Brazil. Journal of Experimental Marine Biology and Ecology 374:104–112.
Odling-Smee, F.J., K.N. Laland, and M.W. Feldman. 2003. Niche construction: the neglected process in evolution. Princeton University Press, Princeton, New Jersey.

Odling-Smee, J., D.H. Erwin, E.P. Palkovacs, M.W. Feldman, and K.N. Laland. 2013. Niche construction theory: a practical guide for ecologists. The Quarterly Review of Biology 88:3–28.
Oka, S., T. Tomita, and K. Miyamoto. 2016. A mighty claw: pinching force of the coconut crab, the largest terrestrial crustacean. PLoS ONE 11:e0166108.
Paul, D.H. 1971. Swimming behavior of the sand crab, *Emerita analoga* (Crustacea, Anomura). Zeitschrift für vergleichende Physiologie 75:259–285.
Perez, D.M., J.H. Christy, and P.R.Y. Backwell. 2016. Choosing a mate in a high predation environment: female preference in the fiddler crab *Uca terpsichores*. Ecology and Evolution 6:7443–7450.
Rebach, S., and D.W. Dunham. 1983. Studies in adaptation: the behaviour of higher Crustacea. Wiley and Sons, New York.
Richardson, A., and P. Araujo. 2015. Lifestyles of terrestrial crustaceans. Pages 299–336 *in* M. Thiel and L. Watling, editors. The natural history of the Crustacea, volume. 2: Lifestyles and feeding biology. Oxford University Press, New York.
Rubenstein, D.I., and R.W. Wrangham (editors). 1986. Ecological aspects of social evolution. Princeton University Press, Princeton, New Jersey.
Salmon, M., and S.P. Atsaides. 1968. Visual and acoustic signalling during courtship by fiddler crabs (genus *Uca*). American Zoologist 8:623–639.
Shillaker, R.O., and P.G. Moore. 1978. Tube building by the amphipods *Lembos websteri* Bate and *Corophium bonnellii* Milne Edwards. Journal of Experimental Marine Biology and Ecology 33:169–185.
Sigmund, K. 1993. Games of life: explorations in ecology, evolution, and behaviour. Oxford University Press, New York.
Stamhuis, E.J., T. Reede-Decker, Y. van Etten, J.J. de Wiljes, and J.J. Videler. 1996. Behaviour and time allocation of the burrowing shrimp *Callianassa subterranea* (Decapoda, Thalassinidea). Journal of Experimental Marine Biology and Ecology 204:225–239.
Strassmann, J.E., and D.C. Queller. 2014. Privatization and property in biology. Animal Behaviour 92:305–311.
Suter, R.B., G.E. Stratton, and P.R. Miller. 2011. Mechanics and energetics of excavation by burrowing wolf spiders, *Geolycosa* spp. Journal of Insect Science 11:22.
Thiel, M. 1999. Extended parental care in marine amphipods: II. Maternal protection of juveniles from predation. Journal of Experimental Marine Biology and Ecology 234:235–253.
Thiel, M. 2001. Parental care behavior in the wood-boring isopod *Sphaeroma terebrans*. Pages 267–276 *in* B. Kensley and R.C. Brusca, editors. Isopod systematics and evolution. Crustacean Issues 13. Balkema, Rotterdam, The Netherlands.
Thiel, M. 2007. Social behavior of parent-offspring groups in crustaceans. Pages 294–318 *in* J.E. Duffy and M. Thiel, editors. Evolutionary ecology of social and sexual systems: crustaceans as model organisms. Oxford University Press, New York.
Vermeij, G.J. 2010. The evolutionary world: how adaptation explains everything from seashells to civilization. Thomas Dunne, New York.
von Frisch, K., and O. von Frisch. 1974. Animal architecture. Harcourt, New York.
Wada, K. 1993. Territorial behavior, and sizes of home range and territory, in relation to sex and body size in *Ilyoplax pusilla* (Crustacea: Brachyura: Ocypodidae). Marine Biology 115:47–52.
Warner, G. F. 1977. The biology of crabs. Van Nostrand-Reinhold, New York.
Weber, J.N., and H.E. Hoekstra. 2009. The evolution of burrowing behaviour in deer mice (genus *Peromyscus*). Animal Behaviour 77:603–609.
Weber, J.N., B.K. Peterson, and H.E. Hoekstra. 2013. Discrete genetic modules are responsible for complex burrow evolution in *Peromyscus* mice. Nature 493:402–U145.
White, C.R. 2001. The energetics of burrow excavation by the inland robust scorpion, *Urodacus yaschenkoi* (Birula, 1903). Australian Journal of Zoology 49:663–674.
Wong, B.B.M., C. Bibeau, K.A. Bishop, and G.G. Rosenthal. 2005. Response to perceived predation threat in fiddler crabs: trust thy neighbor as thyself? Behavioral Ecology and Sociobiology 58:345–350.
Zucker, N. 1981. The role of hood-building in defining territories and limiting combat in fiddler crabs. Animal Behaviour 29:387–395.

12

PREDATOR-INDUCED DEFENSES IN CRUSTACEA

Linda C. Weiss and Ralph Tollrian

Abstract

The capacity of an organism with a given genotype to respond to changing environmental conditions by the expression of an alternative phenotype is a fascinating biological phenomenon. Plasticity enables organisms to cope with environmental challenges by altering their morphology, behavior, physiology, and life history. Especially, predation is a major factor driving plasticity in response to seasonal fluctuations of predator populations. Therefore, many taxa have evolved strategies to adapt to this environmental challenge, including morphological defenses, life history shifts, and behavioral adaptations. The evolution of inducible defenses is dependent on 4 factors: a selective agent, a reliable cue, associated costs, and the resulting benefit. Ecologically, predator-induced defenses are of general importance because they reduce predation rates and hence dampen the dynamics of predator-prey systems to stabilize food webs. We analyze the defensive strategies in many crustacean taxa and describe how they can act in concert to reduce predation risk. Additionally, prey species may perform predation risk assessment and reduce defense expression when conspecifics are dense. With increasing numbers of conspecifics, the individual predation risk is reduced due to prey dilution, predator confusion, and increased handling times. Consequently, the need to develop a strong defense is reduced and costs for the full defenses expression can be saved.

In many cases, predation risk is detected through predator-specific chemical cues that are perceived by prey and indicate a potential hazard. This sensitivity to predator-derived chemicals plays an important role in ecological and evolutionary processes. Often the chemical nature of such cues is unknown but critical for the complete understanding of food web interactions. This chapter summarizes the current knowledge in this field and describes the diversity of known substances capable of inducing defenses in prey species. It suggests that knowing the biologically active compounds will help us further understand the signaling pathways responsible for plasticity, and it describes the neuronal and neurohormonal pathways allowing cost-benefit optimized adaptations. We conclude that future investigations will benefit from the current era of next-generation

Life Histories. Edited by Gary A. Wellborn and Martin Thiel.

sequencing and that using environmental omics - applications has the potential to unravel distinct molecular mechanisms to advance understanding of how organisms are able to cope with environmental challenges and to more deeply elucidate the ecology and evolution of phenotypic plasticity.

INTRODUCTION

Crustaceans show a wide array of defenses against predators, some of which are permanently expressed and others inducible (i.e., expressed only in response to an emerging threat). The best example of a permanent defense is the carapace: a shield that extends from the head region enveloping the body that can carry spines or be cryptically colored. Likewise, spines of zoea larvae are considered permanent antipredator defenses. Inducible defense strategies from crustaceans have predominantly been described in planktonic prey species such as the freshwater cladoceran *Daphnia*. These organisms live in the open water where physical refuges from predators are limited and seasonal variation of biotic environmental factors is high. Many inducible defense strategies have been reported from freshwater organisms. In this chapter, we focus on the environmentally controlled adaptions of phenotypic traits, often using the example of the freshwater crustacean *Daphnia*, which has a thoroughly studied ecology. We explain context-dependent forms of inducible defenses in the light of evolutionary adaptations. Also, we provide insights into the molecular mechanisms underlying defense expression, beginning with the different types of cues, their perception, neuronal integration, and the ultimate endocrine signals that control alternative phenotype expression. Finally, we suggest future research directions in the field of environmentally controlled phenotypic plasticity.

PHENOTYPIC PLASTICITY

The ability of an organism with a given genotype to respond to changing environmental conditions by the development of an alternative phenotype is a fascinating biological phenomenon (Bradshaw 1965). This mechanism enables organisms to cope with environmental challenges by altering their morphology, behavior, physiology, and life history (Whitman and Agrawal 2009, Morris and Rogers 2014). These alterations may or may not be reversible, with behavior being usually the most flexible response, while morphology and life history responses require longer timescales, but all enhance the overall fitness of the respective organism. Factors that may induce changes in phenotypic traits range from seasonal fluctuations of abiotic factors (e.g., light, temperature, oxygen availability, salinity, pH) to biotic factors (e.g., presence or absence of predators or conspecifics and heterospecifics; Fusco and Minelli 2010). Cost-benefit optimized adaptations affect the ecological success of individuals, populations, and species. Within an interspecific context, phenotypic plasticity is especially interesting in predator-prey interactions (Jeschke et al. 2008); prey organisms can adapt to an increased risk of predation and increase fitness by developing adaptive defensive strategies (Tollrian 1995, Tollrian and Harvell 1999).

Inducible Defenses

Inducible defenses are an intriguing form of phenotypic plasticity that usually decreases the likelihood of predator encounters or reduces the effects of predator attacks. Defenses may occur in the form of behavioral or morphological adaptations or shifts of life history parameters, and many taxa show a variety of different antipredatory adaptations. Inducible defenses have evolved to be nonpermanent (i.e., inducible) because of the variability of the selective agent (e.g., predator population size changes due to seasonal fluctuations). Fluctuations in defense expression likely hamper a coevolutionary response from the predator. Furthermore, a reliable cue is needed to indicate the presence of a potential danger, which is then countered by the development of an effective inducible defense.

At the same time, these defensive strategies are associated with costs, including plasticity costs, allocation costs, and environmental costs. Plasticity costs result from the organism's ability to be plastic. They are decrements for providing and maintaining the genetic and physiological facilities to detect and respond to cues. Allocation costs derive from increased energy and materials required for the development of the defenses. These costs may result, for example, from higher energy expenditures associated with escape movements or from material allocations for the development of morphological defenses. Environmental costs may be associated with the defenses and their interaction with the varying environments. These costs, also referred to as external costs, can result, for example, from changes in swimming speed due to aberrant hydrodynamics of an alternative body shape. All these costs, except plasticity costs, can be avoided when the defense is superfluous.

Inducible defenses are of ecological importance as they reduce predation rates and hence dampen the dynamics of predator-prey systems, which stabilizes food webs (Verschoor et al. 2004). Moreover, it is well conceivable that the timing of inducible defenses is crucial for the defenses to be effective. If there is a substantial time lag between the detection of the predator and the expression of the induced response, the inducibility of the defenses, as compared to having permanent ones, may not be advantageous. Behavioral defenses can be expressed rather suddenly just with the occurrence of the predatory threat. Classical behavioral adaptations are the flight response, increased alertness, or hiding behavior. In comparison, morphological defenses are slow in their expression; they need to be developed before serving as beneficial protection. Life history shifts represent a different defensive strategy in which the organism alters its reproductive development. In consequence, parameters such as body size and age at first reproduction, brood size, and the size of individual offspring are shifted. Inducible chemical defenses are predominantly known from plants that on herbivore feeding accumulate toxins influencing growth behavior and reproduction of herbivores. For instance, the dinoflagellate algae *Alexandrium minutum* accumulates toxins in response to so-called copepodamines that are released by predatory copepods. The more toxic cells are less susceptible to predation, which results in an increase of relative abundance of the defended type (Selander et al. 2015). However, some crustaceans even make use of defended plants: for example, the amphipod *Pseudamphithoides incurvaria* constructs a housing from the chemically defended seaweed *Dictyota bartayresii*. This seaweed produces a diterpene alcohol that deters feeding by fishes (Hay et al. 1990). Similarly, the decorator crab *Libinia dubia* selectively decorates its carapace with the chemically defended seaweed *Dictyota menstrualis* (Stachowicz and Hay 1999).

There is a vast diversity of these defensive strategies, which are specific to the different types of predators present, and the strategies can be formed individually but often act in concert. As such, behavioral adaptations may bridge the time lag between the slower-developed defenses in morphology and life history.

Behavioral Antipredator Adaptations

Diel vertical migration patterns can be understood as the classical antipredator defense of freshwater and marine zooplankton (Lampert 1989, Haupt et al. 2009). Organisms evade predators by migrating to a refuge zone that provides shelter from visually foraging predators (Ringelberg 2010); active migration to the surface waters is only performed during the night when there is less risk of visual predation. Moreover, predation risk can induce a size-structured depth separation of prey, as observed in *Daphnia longispina*. Small individuals, being less vulnerable to visual predation, continue to feed in surface waters during the day and obtain a competitive release from larger, downmigrating conspecifics (Hansson and Hylander 2009). Diel vertical migration comes at a cost of delayed maturation due to the decreased temperature in the deeper parts of the lake (Dawidowicz and Loose 1992). Still, this defensive strategy is favored by selection, and it has evolved several times within crustaceans, including different copepod taxa, such as *Calanopia americana* (Cohen et al. 2005), *Acartia hudsonica* (Bollens and Frost 1991), *Pseudocalanus newmani* (Frost and Bollens 1992), and a variety of *Daphnia* species (Haupt et al. 2009, Fig. 12.1).

Another type of behavioral evasion strategy involves seeking a physical refuge. This behavior is widespread and arguably the primary predator avoidance mechanism of any benthic crustacean. Some of the various refuge strategies include sheltering by means of being withdrawn or wedged into a crevice, hiding by masking vegetation, or burying into substratum (Faulkes 2013). A variety

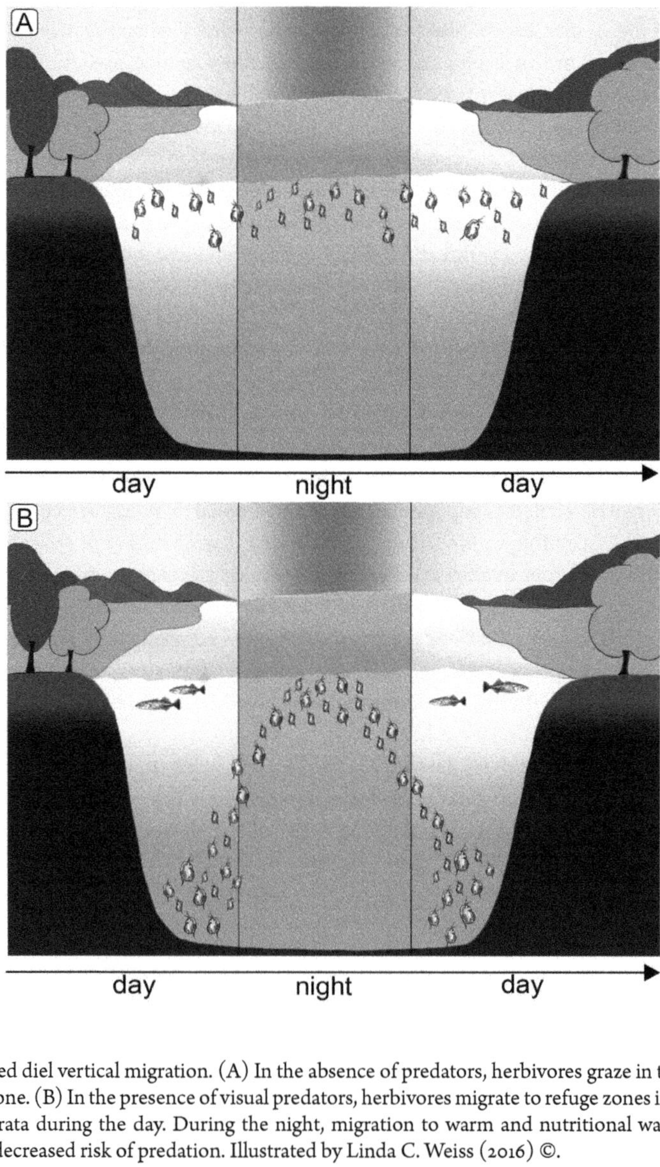

Fig. 12.1.
Predator-induced diel vertical migration. (A) In the absence of predators, herbivores graze in the nutrient-rich phototrophic zone. (B) In the presence of visual predators, herbivores migrate to refuge zones in deeper, cooler, darker water strata during the day. During the night, migration to warm and nutritional water strata allows feeding with a decreased risk of predation. Illustrated by Linda C. Weiss (2016) ©.

of sand crabs, several species of penaeid shrimp, and some slipper lobsters (Lavalli et al. 2007) are known to rapidly bury themselves in soft substrata (Lavalli and Spanier 2015). Other behavioral changes include periodicity adaptations where prey restrict their foraging behavior to safer periods, reducing the likelihood of predator encounter, but at the cost of reduced feeding. For example, the American lobster *Homarus americanus* reduces its overall activity pattern and remains sheltered in the presence of cod or sea raven. This behavior may translate into reduced lobster growth and reproduction (Wilkinson et al. 2015). Reproductive success may be lowered because mating events consist of activities such as mate search, mate assessment, courtship, mate guarding, and fertilization (Lima 1998). The time invested in mating behavior depends on the associated predation risk such that males often behave more cautiously and reduce the intensity of conspicuous courtship behaviors. To maximize fitness, organisms (such as the rock shrimp *Rhynchocinetes typus*) must balance reproduction with the likelihood of being killed by a predator. Whereas youngsters jeopardize an entire life span when taking the risk of reproduction during predator presence, older conspecifics with the expectation of reduced reproductive success can accept greater risks when facing predation (Ory et al. 2015). Similar observations have been made in amphipods where the trade-off between predator avoidance and mate acquisition was examined in *Gammarus* sp. (Dunn et al. 2008). Here the frequency of precopula pair formation was reduced in the presence of predator cues. Nevertheless, those animals that did form pairs showed a decrease in partner selection (Dunn et al. 2008, Mathis and Hoback 1997). In conclusion, reproduction is still warranted, yet at a reduced frequency, during predator avoidance behavior but at the cost of choosing the best mating partner.

Prey can reduce swimming speed as another behavioral antipredator strategy. This results in reduced chances of predator encounter or decreased detectability by tactile predators (Pijanowska and Kowalczewski 1997). Another adaptive behavioral strategy has been observed in marine planktonic crustaceans. Copepods, amphipods, and mysidaceans reduce feeding activity in the presence of fish, resulting in the reduction of gut fullness. This behavior has been proposed to be adaptive because it reduces visibility to predators (Cieri and Stearns 1999, Hamrén and Hansson 1999, Wisenden et al. 1999).

Occasionally, behavioral defenses can be combined with morphological structures. Mud crab zoeae flare their spines and flex their abdomens over their carapace following attacks (Morgan 1987). The abdomen may possess a pair of abdominal spines that become erect when the abdomen is flexed, further complicating the feeding process (Morgan 1987).

Morphological Defenses

Induction of diverse structural defenses in response to the presence of predators occurs in planktonic freshwater cladocerans and barnacles but has less frequently been observed in mobile benthic crustaceans. It is quite possible that there are more inducible morphological defenses to be found also in marine zooplankton, if studied. In general, induced morphological defense structures are any kind of antipredatory morphological traits developed in the presence of a specific predator. Such structures may appear in the form of body extensions such as spines, enlarged body appendages, or strengthened shells and carapaces, which all become effective during attack and capture by predators. Often these structures hamper prey handling by the predator, which increases escape chances (Weiss et al. 2012b). All these morphological defense structures reduce predation in a cost-benefit optimized manner, are specific to the respective predator, and counter the mode of predation. For example, the barnacle *Chthamalus anisopoma* exhibits a bent-shell morphology in response to the predatory gastropod *Acanthina angelica* (Lively 1986). This morphological adaptation, in which the rim of the aperture is oriented perpendicular to the base, is more resistant to predation. An alternative strategy was observed in the barnacle *Chthamalus fissus*, which has the capacity to develop a defensive morphology in response to the predatory snail *Mexanthina lugubris lugubris* (Jarrett 2009). This predator forages by ramming its shell spine into the barnacle's operculum. Juvenile *C. fissus* exposed to the predator develops narrower opercular openings, sometimes together with changes in shell morphology. This plasticity in operculum morphology may allow juvenile barnacles that are still too small to be preyed on by *M. lugubris lugubris* the opportunity to develop a defensive morphology before they become suitably sized prey (Jarrett 2009).

The freshwater crustacean *Daphnia* has been studied thoroughly for its ability to develop morphological defenses in response to predators. In this genus, an array of different types of inducible morphological defenses can be observed that act synergistically to provide effective protection (Fig. 12.2). An example of a morphological defense is spine formation against planktivorous fish such as the three-spined stickleback *Gasterosteus aculeatus* (Fig. 12.2A) in *D. lumholtzi* (Tollrian 1994). Fish are limited in their gape width; hence, spine development increases prey size and hampers easy ingestion by fish (Fig. 12.2B,C).

In response to the presence of the predatory phantom midge larvae *Chaoborus, Daphnia pulex* develops little spines in the neck region, termed *neckteeth* (Fig. 12.2D–G). These tiny structures are assumed to interfere with the feeding apparatus of *Chaoborus*, reducing the predation success due to increased escape chances (Fig. 12.2G). Similarly, *D. cucullata* develops helmets to defend against *Chaoborus* predation (Fig. 12.2H,I). Not only does this helmet offer protection against *Chaoborus* predation, it has been shown that helmeted *D. cucullata* are also protected against other common freshwater predators such as *Leptodora kindtii* and *Cyclops* species. The helmets act in different size classes and via different mechanisms against these predators, rendering these helmets a multitool against a diversity of predators (Laforsch and Tollrian 2004).

Another type of handling disabilities imposed by prey morphological defenses is the development of large crests in *D. longicephala* in response to the actively feeding heteropteran *Notonecta* (Grant and Bayly 1981; Fig. 12.2J–L). These crests are suspected to interfere with the tight grasp of the notonectid, again increasing escape chances for the prey organism.

Daphnia barbata has been shown to develop distinct morphological defenses based on the type of predator rather than a general response (as described for *D. cucullata*). The defenses are based on the same structures, but formed in a different way. Against *Notonecta* predation, *D. barbata* extends the helmet and spine length in a straight manner (Herzog and Laforsch 2013; Fig. 12.2J,N–O). In the case of *Triops cancriformis* predation, this species develops a bent morphology, where the helmet and the spine are bent backward in comparison to noninduced morphotypes (Fig 12.2M, P). A different defensive strategy against *Triops* predation, development of a "crown of thorns," was recently reported in *D. atkinsoni* (Petrusek et al. 2009).

A recent study by Herzog et al. (2016) describes a completely novel inducible defense observed also in *D. barbata*, giving a first report of a free-living Bilateria with the ability to flexibly respond to predation risk by abandoning bilateral symmetry. The defense is accomplished by having a twisted carapace with the helmet and the tail spine deviating from the body axis into opposing directions, resulting in a complete abolishment of bilateral symmetry. The twisted morphotype is suspected to considerably interfere with the feeding apparatus of the predator, contributing to the effectiveness of the array of defensive traits in *D. barbata* (Herzog et al. 2016).

In addition to these obvious defense structures, hidden morphological plasticity was also revealed in *D. pulex* (Laforsch et al. 2004), *D. cucullata* (Laforsch et al. 2004), and *D. magna* (Rabus et al. 2013, Riessen 2012), which show a strengthening of their armor, providing physical protection against mechanical challenges. In the case of *D. magna* such hidden defenses, together with an increased bulkiness, reduce crushability by the predator's mouthparts (Rabus and Laforsch 2011, Rabus et al. 2013).

In general, morphological defenses are specific to the predator. This means that not only their shape bears an advantage over undefended conspecifics, but also that the ontogenetic timeline of

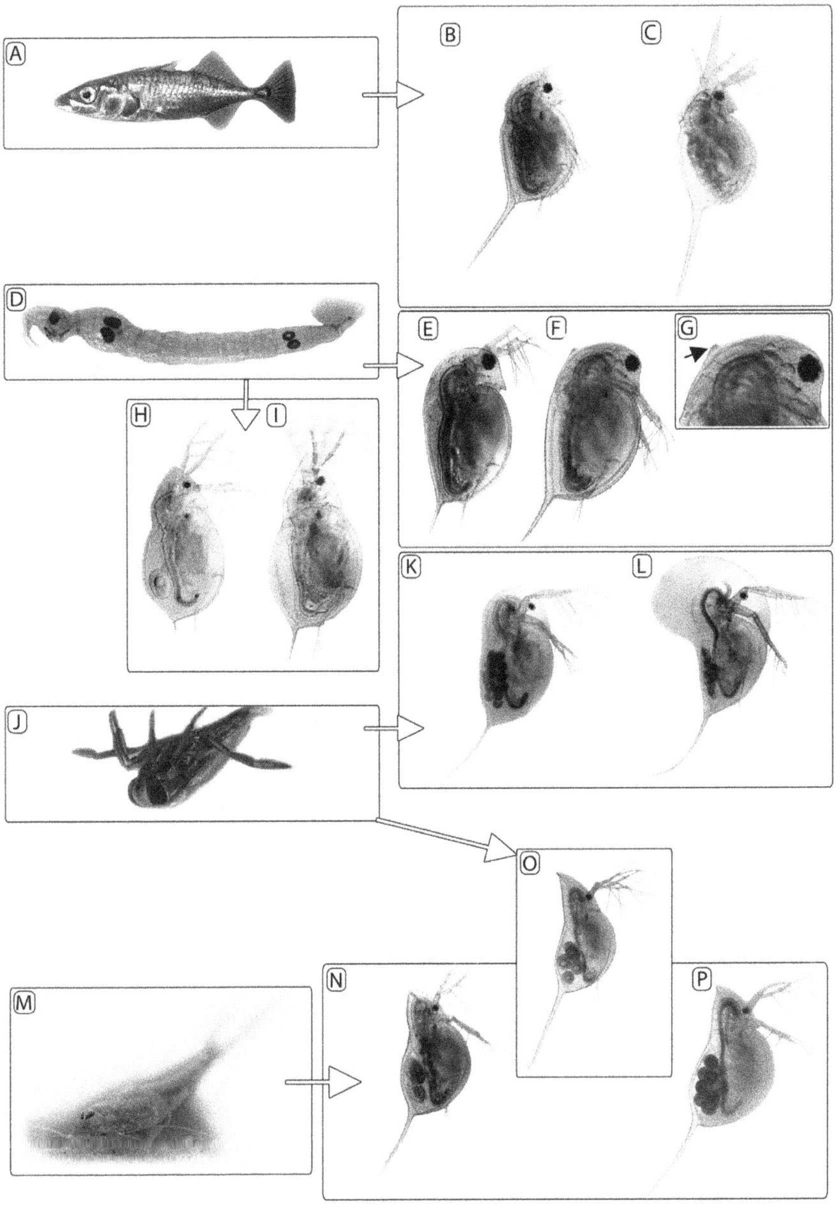

Fig. 12.2.
Predator-induced defenses in different *Daphnia* species. (A) Predator *Gasterosteus aculeatus* (three-spined stickleback) that induces helmets in *Daphnia lumholtzi*. (B) Undefended *D. lumholtzi*. (C) Defended *D. lumholtzi* with remarkably elongated head and tail spines. (D) The invertebrate predator *Chaoborus obscuripes* commonly described to induce defenses in *Daphnia pulex*. (E) Undefended *D. pulex* compared to (F) defended *D. pulex* carrying neckteeth in the dorsal head region. (G) Insert shows magnification of neckteeth displayed in (F). Likewise, *Chaoborus* induces helmet development in *Daphnia cucullata*. (H) Undefended *D. cucullata*. (I) Defended *D. cucullata* with helmet and elongated tail spine. (J) The backswimmer *Notonecta glauca* induces morphological defenses in *Daphnia longicephala*. (K) The undefended *D. longicephala* morphotype is small and inconspicuous in comparison to the defended morphotype. (L) Defended *D. longicephala* grow large crests as well as elongated tail spines. (M) The ancient predator *Triops cancriformis* that induces defenses in *Daphnia barbata*. (N) *D. barbata* (here: undefended form) develops defense modalities adapted to the predation regime. (O) *Notonecta*-defended *D. barbata* develop larger and straight helmets in comparison to (N) and to (P) the *Triops*-defended morphotype, which has larger and backward-bending helmets and tail spines. Illustrated by Linda C. Weiss (2016) ©. See color version of this figure in the centerfold.

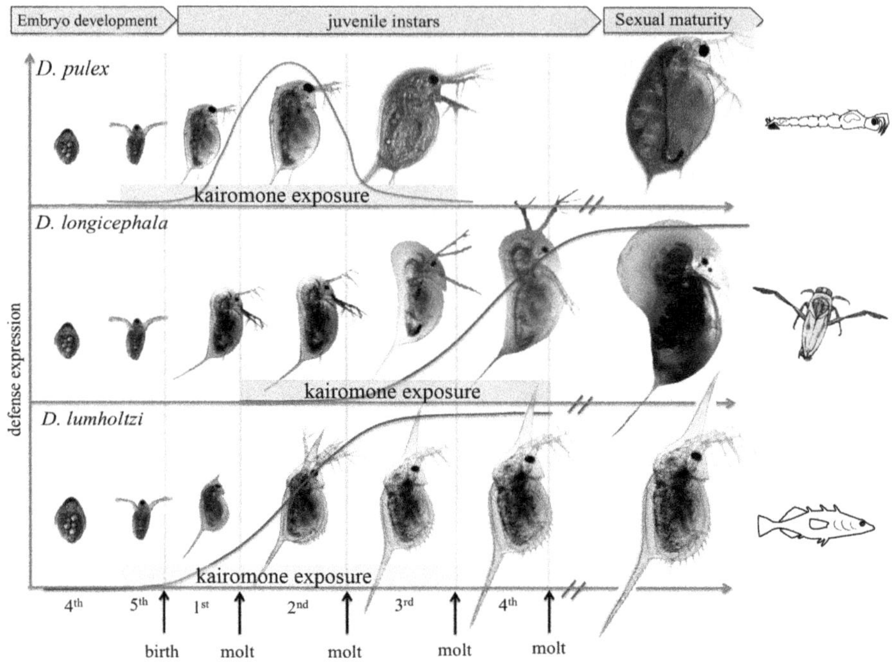

Fig. 12.3.
Development of inducible morphological defenses in *Daphnia pulex, D. longicephala,* and *D. lumholtzi.* All defenses are developed with a time lag of one instar. In *D. pulex* defensive neckteeth are developed de novo in early juvenile instars (first and second). *Daphnia longicephala* develops defensive crests in the fourth juvenile instar. Crests are continuous traits, if the predator cue persists, crests are developed in a continuous manner throughout juvenile and adult stages. In *D. lumholtzi* defensive helmets are developed continuously from the third juvenile instar if the predator cue persists. Just as crests, defensive helmets are continuous traits. Illustrated by Linda C. Weiss (2016) ©.

expression is specifically adapted to the predator's feeding habits. For example, in the case of the gape-limited ambush predator *Chaoborus,* the small juvenile stages of prey need to be defended. In contrast, if the predator is one that relies on visual perception when foraging, it is adaptive if the prey is defended only when it reaches a size that is easily detectable. In this way, *D. lumholtzi* and *D. longicephala* save costs by not having defense development in the early instars, and expressing defenses only when they reach a threatened size (Fig. 12.3).

In addition to changes of the outer morphology, predators can induce changes in pigmentation patterns. For instance, some clones of *D. pulex* become more transparent in the presence of fish predation making them less conspicuous to this visual predator (Tollrian and Heibl 2004). Such induced camouflage strategies have also been observed in shore crabs (*Carcinus maenas*), which are able to adjust their camouflage through adjustment of brightness over a period of hours; this coloration could influence detection probability by, for example, avian predators (Stevens et al. 2014).

Life History

The life histories of many zooplankton species can be plastic and highly influenced by environmental factors, such as invertebrate and vertebrate predators, primarily planktivorous fish. It appears that predation is one of the strongest forces shaping life histories in the wild (Lynch 1980). Predators are usually size selective, and fish select larger and more conspicuous specimens, such as very visible females that are carrying eggs (Brooks and Dodson 1965, Zaret and Kerfoot 1975). Small predatory invertebrates typically prey selectively on small, young individuals (Pastorok 1981, Pijanowska 1992). The freshwater cladoceran *D. pulex* is capable of shifting its resource allocation between somatic growth and reproduction, thereby adjusting its body size and reproductive effort to increase fitness in the presence of *Chaoborus* or fish (Walsh et al. 2012). *Chaoborus* larvae only prey on juvenile daphniids due to the gape limitation of their mouthparts and, consequently, *Daphnia* invest in somatic growth and seek a "size refuge" in growing large. At the same time maturation is delayed and clutch size is reduced, as resource allocation is shifted from reproduction to somatic growth (Tollrian and Harvell 1999). In contrast, fish predation induces investments into reproduction leading to earlier maturation at a smaller body size (Stibor and Lüning 1994). Shifts in body size are often transferred onto the next generation because offspring body size correlates with maternal size. Another adaptive life history defense against fish predation is the production of resting eggs (Ślusarczyk 1995) that can pass through the digestive tract of fishes, allowing for a temporal escape.

It has long been discussed whether life history adaptations are adaptive responses or constraints associated with the costs necessary to develop inducible defenses (e.g., Tollrian 1995, Riessen 2012). However, selection should generally favor the evolution of inducible defenses toward a high degree of protection at low costs, suggesting that life history shifts are actually antipredator strategies themselves rather than constraints resulting from the development of morphological traits (Tollrian 1995).

Factors Determining the Evolution of Phenotypic Plasticity

Daphnia are tremendously plastic and many types of responses are found within this 1 genus, including behavioral, morphological, and life history defenses. Thus, it can be hypothesized that each of these responses acts as one individual adaptive mechanism. All of these plastic traits have coevolved as independent adaptations in response to predation. For example, life history adaptations and morphological defenses are uncoupled and can be developed independently (Boersma et al. 1998), so that life history responses are adaptive responses themselves rather than trade-offs resulting from the cost of developing morphological defenses such as neckteeth (Tollrian 1995). Over the years, 4 main factors have emerged as the key factors explaining the evolution of inducible defenses. First, there has to be a reliable cue indicating a change in the environment; second, the environment has to be heterogeneous (e.g., with seasonal emergence of predators); third, the defense has to be advantageous; and fourth, the defense also has to bear costs that are saved when defenses are superfluous (Harvell 1990, Tollrian and Harvell 1999).

CONSTRAINTS AND COSTS OF PHENOTYPIC PLASTICITY

Adaptation to environmental conditions is a key to fitness. Phenotypic plasticity has evolved as a strategy enabling organisms to respond to variable environmental conditions. However, plastic traits are subject to constraints. For example, neckteeth as morphological defense in *D. pulex* reduce predator efficiency but do not reduce the predator's capture rate to zero. Likewise, developmental constraints include the necessity of predator-specific receptors binding to predator-specific cues. These receptors must be functional and capable to code the signal into an adaptive response. Early functionality of such a detection system allows for early defense expression (Weiss et al. 2016), but predator cue sensitivity also imposes a constraint on defense expression.

Apart from such constraints, costs associated with defense may be paid for in terms of fitness. For example, induced defended morphs are more vulnerable to infections by virulent parasites than undefended morphs (Yin et al. 2011), which can be explained by downregulation of immune-response genes (Rozenberg et al. 2015). Moreover, plastic organisms need to constantly carry the genetic machinery allowing them to detect environmental change and transform into a better-performing phenotype (Murren et al. 2015). Therefore, there should be costs associated with plasticity itself, so-called plasticity costs (Tollrian and Harvel 1999).

Here we elaborate a conceptual framework that describes the effect of plasticity on organism fitness (Fig. 12.4). In the conceptual model, fitness of a nonplastic genotype is high when predators are absent, but fitness is highly reduced with an emerging predation risk. Once the predation risk diminishes (e.g., due to seasonal fluctuations), the fitness level of such a genotype is reestablished. In the case of a defended, but not plastic, prey species, the overall fitness level is lower in comparison to an undefended organism because of the costs of the defense, but it is less reduced in the case of predator emergence in comparison to the undefended genotype. Furthermore, a plastically defended genotype might create a "moving target," rendering coevolutionary adaptations on the side of the predator less likely. A plastic genotype has a slightly reduced fitness level in comparison to a nonplastic genotype, with the reduced fitness attributable to the previously mentioned costs of plasticity. With predator emergence, adaptive phenotypes develop defenses with a reaction delay. In this reaction delay, fitness levels are as low as in the nonplastic genotype. Once the defense is expressed, fitness levels are elevated and equal to fitness levels of a permanently defended genotype. The diminishing of predation risk again increases fitness levels, but still the phenotype costs for the defense have to be paid until the defenses are finally reverted.

The previously described reaction delays can be interpreted as trade-offs of plasticity. Organisms need to acquire information that indicates environmental change, which requires sensory mechanisms. Thus, the reaction delays for induction and reversion of the defense impose major constraints for plastic defenses. Evolution might lead to a shortening of these reaction delays, for example, via epigenetic, transgenerational induction and via early sensitivity and fast development (Agrawal et al. 1999). This concept explains the evolutionary scenario of plasticity and its associated costs.

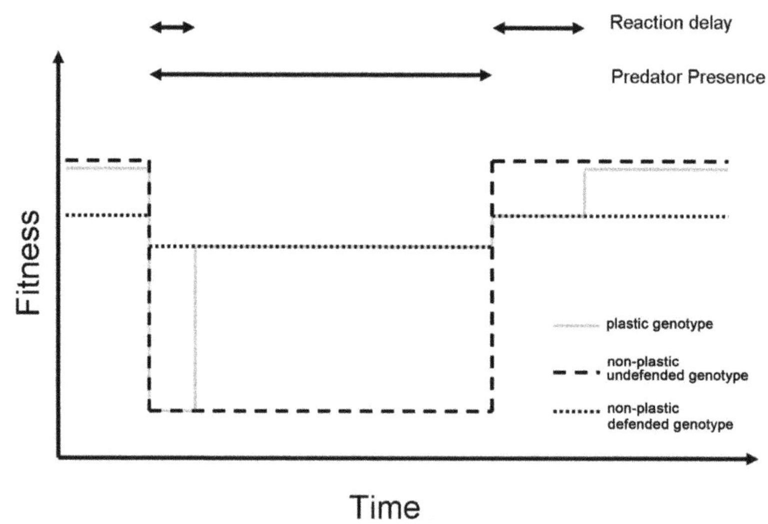

Fig. 12.4.
Conceptual model of constraints of plasticity in the context of fitness and reaction delays. Reaction delays for the induction and reversion of inducible defenses are major constraints reducing the fitness of plastic organisms. In the first reaction delay, costs arise because defenses are not yet formed, in the second reaction delay costs of defenses have to be paid until the defenses are reversed. Illustrated by Ralph Tollrian (2016) ©.

EXACT PREDATION RISK ASSESSMENT

As mentioned earlier, most defenses incur trade-offs between costs and benefits, so that the degree of defense expression increases with predator densities. However, predation risk to the individual does not solely depend on predator density but also on the density of conspecifics. Indeed, it was recently reported that the degree of defense development depended not only on predator density but was actually also adjusted to the density of the prey individuals (Tollrian et al. 2015). In *Daphnia*, the response to prey density is triggered by a chemical cue released by conspecifics. With increasing numbers of conspecifics, the individual predation risk is reduced due to prey dilution, predator confusion, and increased handling times. Handling times are defined as the time spent searching, pursuing, subduing, and consuming each prey item; dilution and confusion extend prey search and pursuit. This is most relevant in systems with Holling–type II predators (i.e., predators whose intake is saturated with increasing prey density). Therefore, the need to develop a defense to such a strong degree becomes unessential and costs for the full development of the defense can be saved. In fact, it appears that prey species use multiple sources of information of different natures in determining their actual predation risk (Tollrian et al. 2015).

NATURE OF ENVIRONMENTAL CUES

Planktonic crustaceans use a variety of environmental cues, including mechanical, visual, and olfactory stimuli (Hazlett 2011). Marine copepods react to mechanical or visual stimuli, and some marine plankton species use bioluminescence for communication (Kiørboe and Visser 1999, Haddock et al. 2010). Glowing bioluminescence is considered an attractant signal, whereas sudden flashes are considered repellent. Thus, bioluminescence should be viewed as a trait evolved to communicate either for defense against predators or for offense to attract or distract prey (Escribano and Riquelme-Bugueño 2015).

Communication between organisms of the same and different species is common in aquatic ecosystems (Dicke and Grostal 2001). Species can acquire information on food sources, potential mating partners, and predators via chemical perception. For organisms with poor vision, or in turbid environments, chemical cues are advantageous as they can be transmitted over long temporal and spatial scales. These infochemicals explicitly affect internal physiological processes such as development, growth, and reproduction in many aquatic organisms. As such, aquatic ecosystems are embedded in a network of chemical cues, which significantly complicate our current knowledge of trophic interactions (Wyatt 2011). Kairomones are defined as chemical cues that are beneficial to the receiver but not to the sender in the context of interspecific information transfer. As infochemicals, they evoke a behavioral or physiological reaction in the receiver that is adaptively favorable for the receiver but not for the sender. Sensitivity to predator-derived kairomones plays an important role in ecological and evolutionary processes that enable the prey to survive predation pressure.

It is assumed that evolution of prey sensitivity to these cues is favored if their production cannot be eluded by the predators (e.g., if they originate as metabolites from physiological processes). Otherwise, coevolution acting on the predator would potentially have stopped their production. Cues described to date with disparate chemistry include proteins that signal the development of inducible defenses in ciliates (Kusch and Heckmann 1992), aliphatic sulfates released by *Daphnia* inducing defensive coenobia in algae (Yasumoto et al. 2005), and herbivore-induced release of simple hydrocarbon chains serving as chemical defenses in brown algae (Schnitzler et al. 2000). Recently, copepodamines have been identified that induce toxin production in algal prey (Selander et al. 2015). All such chemical cues may contain specific or unspecific information about the potential predation risk. Conspecific alarm cues, which are prominent in fish, are released from damaged cells (e.g., club cells) or tissues on injury (Mathis et al. 1995, Chivers et al. 1996) and have been reported to induce defenses in *Daphnia*. Waterborne cues from macerated conspecifics alter *Daphnia* morphology but not to the same extent as kairomones, which have a distinctly stronger effect (Laforsch et al. 2006). Most likely, broadly defined alarm cues in *Daphnia* provide an unspecific signal that does not allow differentiation between fish or invertebrate predators. Kairomones or alarm cue labeling by conspecific-consuming predators may form a blend of chemicals that unequivocally indicates active predation (Stabell et al. 2003, Schoeppner and Relaya 2005, Ferrari et al. 2010). Overall, the chemical perception of such cues initiates a series of biological reactions that ultimately result in the expression of a defended phenotype.

PHYSIOLOGY UNDERLYING THE ADAPTATION TO PREDATOR-SPECIFIC CUES

The arthropod solution when adapting morphologically to environmental challenges is to molt into one of the several adaptive phenotypes. This is controlled by the temporal and spatial regulation of growth and molting through a counterbalanced action of juvenile hormones and ecdysteroids. The nervous system perceives, integrates, and transforms the kairomone signal into developmental changes and initiates endocrine signals when life history parameters are shifted (Barry 2002, Miyakawa et al. 2010, Weiss et al. 2012a).

Whereas juvenile hormones control growth and suppress reproduction, ecdysone derivatives induce molting and embryogenesis (Martin-Creuzburg et al. 2007, Nakatsuji et al. 2009, Sumiya et al. 2014). Thus, in crustaceans a proper balance between methyl farnesoate (a juvenile hormone) and 20-hydroxyecdysone (the active form of ecdysone) is important for the normal progression of oogenesis. An imbalance toward ecdysone would result in life history shifts where somatic growth is traded with reproduction. Consequently, animals molt earlier at a smaller size. Conversely, an overabundance of juvenile hormones increases somatic growth and postpones the molting event.

From an evolutionary perspective, it appears that juvenile hormone signaling has emerged as a major, highly conserved endocrine pathway in arthropod development that also regulates the response to changing environmental conditions (Dennis et al. 2014). These include caste determination in social insects, photoperiodic adaptation of reproductive modes in aphids (Le Trionnaire et al. 2008), seasonal pattern formation in butterflies (Truman et al. 1974), as well as the development of color in caterpillars (Nijhout 2013), horns in scarab beetles (Emlen et al. 2006, Gotoh et al. 2011), and morphological defenses in *Daphnia* (Oda et al. 2011, Miyakawa et al. 2013, Dennis et al. 2014). All of the above are mediated by juvenile hormones often in combination with dopamine signaling pathways (Fig. 12.5; Nijhout 2013, Weiss et al. 2015).

Predator Perception

In general, chemical cues bind to chemoreceptors that are located on some kind of chemoreceptive organ. For example, the receptors for the detection of predator cues released by the backswimmer *Notonecta glauca* were recently shown to be located on the first antennae of *D. longicephala*. Chemoreceptors located here presumably bind to the different predator cues, which changes membrane conductance in the dendrites to cause neuron excitation in the brain, integration, and subsequently excitation or inhibition of neurosecretory cells to control hormone release, resulting in development of adaptive phenotypes (Derby and Weissburg 2014, Schmidt and DeForest 2011).

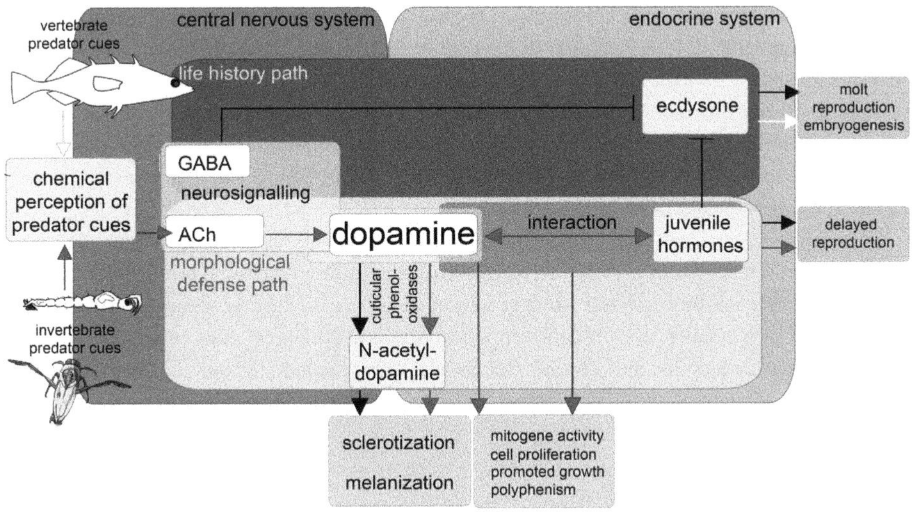

Fig. 12.5.
Conceptual pathway of elements controlling the development of inducible defenses in *Daphnia*. The network consists of successive components: chemical perception of predator cues, changes in neuronal signaling in the central nervous system, and neurohormonal changes. Black arrows indicate physiological pathways running without environmental predator cues. Gray arrows mark the individual components of the pathway transforming predator cues (e.g., *Chaoborus/Notonecta* into morphological defenses). White arrows mark the individual components of the pathway transforming predator cues (e.g., fish into life history shifts). The arrows do not indicate any interaction between the signaling components but rather refer to their involvement in the respective cascade. Based on Weiss et al. (2015).

Neuronal Control of Predator-Induced Phenotypic Plasticity

Whereas hormones serve as the switch that control alternative developmental pathways of phenotypic plasticity (Nijhout 2013), the central nervous system is responsible for perception and coding of environmental cues. In the case of predator-specific chemical cues, the kairomones are perceived, neuronally integrated, and transformed into hormonal changes that result in the development of a defended phenotype (Weiss et al. 2015). Chemoreceptors specific to predator cues, neurotransmitters for chemical information transfer, and hormones that ultimately have the capability to regulate gene expression lead to changes in the phenotype. In *Daphnia*, dopamine transmits the signal from the nervous onto the endocrine system (Weiss et al. 2015; Fig. 12.5). Together with juvenile hormones, dopamine stimulates growth and delays reproduction. Furthermore, dopamine itself can be converted via cuticular phenoloxidases into N-acteyl-dopamine, which is a component involved in the cross-linking of orthoquinones responsible for cuticle sclerotization, as well as melanization processes after molting. Dopamine itself is centrally controlled by neurotransmitters where acetylcholine acts as the stimulatory agent in the regulation of morphological defenses and inhibitory signaling by gamma-aminobutyric acid (GABA) is involved in the development of life history shifts (Weiss et al. 2012a). In general, acetylcholine in the brain alters neuronal excitability, influences synaptic transmission, induces synaptic plasticity, and coordinates firing of groups of neurons. As a result, it changes the state of neuronal networks throughout the brain and modifies their response to internal and external inputs, which is the classical role of a neuromodulator.

The neurotransmitter GABA modulates life history responses against fish predation, including a reduced body length and clutch size but increased delay until maturity (Weiss et al. 2012a). This indicates the presence of an underlying GABAergic neuronal control of the neurophysiology of fish kairomone transmission. The observed responses could potentially be explained by a relieved inhibition. In general, GABA is known to have inhibitory functions. In the absence of vertebrate predators, *Daphnia* life history shifts could be inhibited by GABAergic signals by inhibiting the release of ecdysone (Käuser et al. 1988, Beckerman et al. 2013). This results in a "general purpose" life history. On the perception of fish cues, this GABAergic inhibition appears to be relieved and specific life history parameters change. This would elicit the adjustment of life history parameters in a fast and time-efficient manner (Fig. 12.5).

Taken together, ecdysone, and dopamine in combination with juvenile hormone, counterbalance growth and reproduction. A fish-induced imbalance toward ecdysone, for example, via relief of the GABAergic inhibition of ecdysone secretion, results in life history shifts, in which animals mature earlier at a smaller size. In contrast, *Chaoborus* and *Notonecta* cues, which are mediated by acetylcholine, dopamine, and juvenile hormones, induce somatic growth. This appears to be controlled via differential cholinergic transmission, which could stimulate additional dopamine release.

CONCLUSIONS AND FUTURE DIRECTIONS

The picture of the evolution of phenotypic plasticity has become more and more complete. Phenotypically plastic characters have been described within their environmental context. In *Daphnia*, the enormous phenotypic flexibility is speculated to be regulated by the "ecoresponsive" *Daphnia* genome (Colbourne et al. 2011). This contains different sets of paralogous genes activating evolutionarily conserved and highly derived physiological pathways utilizing different sets of recently diverged genes (Colbourne et al. 2011). These pathways must be carefully regulated within and between different life stages in order to create the defended phenotype that matches the present environmental condition.

Bradshaw (1965) was one of the first authors to recognize the importance of genetic variation and differential gene expression in plasticity. In the era of next-generation sequencing, these are now being frequently addressed. Often, affected genes are those that regulate the metabolic pathways that contribute to the different phenotypes. This allows us to start deciphering the molecular mechanisms underlying phenotypic plasticity traits, which is fundamental for understanding the conditions that might favor or disfavor the evolution of plasticity on an interorganismal scale. For example, one plausible molecular model of plasticity suggests that duplicates from gene families may exhibit environment-specific expression levels (Colbourne et al. 2011). Many current investigations analyze the timely correlated expression levels of gene clusters and their control. However, differential gene expression does not seem to be solely responsible for plasticity. In fact, epigenetic traces may play a pivotal role especially in transgenerational adaptations. More thorough studies on DNA methylation as well as investigation of histone modifications in, for example, environmentally controlled sex determination and phenotypic plasticity, will not only contribute to our understanding of epigenetics per se but also reveal how genes are regulated environmentally. In turn, it is then important to determine tissue-specific differential gene expression and whether these genes are translated into functional proteins. *Daphnia* is an excellent candidate for studying environmental influences on the diversity of context dependent-developmental programs including life history traits, because clonal lines are genetically identical. Taken together, the description of genomic, epigenomic, transcriptomic, and proteomic responses to environmental stimuli will reveal not only the pathways of plasticity but also their evolution.

REFERENCES

Agrawal, A.A., C. Laforsch, and R. Tollrian. 1999. Transgenerational induction of defenses in animals and plants. Nature 401:60–63.

Barry, M.J. 2002. Progress toward understanding the neurophysiological basis of predator-induced morphology in *Daphnia pulex*. Physiological and Biochemical Zoology 75:179–186.

Beckerman, A.P., J. de Roij, S.R. Dennis, and T.J. Little. 2013. A shared mechanism of defense against predators and parasites: chitin regulation and its implications for life-history theory. Ecology and Evolution 3:5119–5126.

Boersma, M., P. Spaak, and L. de Meester. 1998. Predator-mediated plasticity in morphology, life history, and behavior of *Daphnia*: the uncoupling of responses. The American Naturalist 152:237–248.

Bollens, S.M., and B.W. Frost. 1991. Diel vertical migration in zooplankton: rapid individual response to predators. Journal of Plankton Research 13:1359–1365.

Bradshaw, A.D. 1965. Evolutionary significance of phenotypic plasticity in plants. Advances in Genetics 13:115–155.

Brooks, J.L., and S.I. Dodson. 1965. Predation, body size, and composition of plankton. Science 150:28–35.

Chivers, D.P., B.D. Wisenden, and R.J.F. Smith. 1996. Damselfly larvae learn to recognize predators from chemical cues in the predator's diet. Animal Behavior 52:315–320.

Cieri, M.D., and D.E. Stearns. 1999. Reduction of grazing activity of two estuarine copepods in response to the exudate of a visual predator. Marine Ecology Progress Series 177:157–163.

Cohen, J.H., and R.B.J. Forward. 2005. Photobehavior as an inducible defense in the marine copepod *Calanopia americana*. Limnology and Oceanography 50:1269–1277.

Colbourne, J.K., M.E. Pfrender, D. Gilbert, et al. (66 more authors). 2011. The Ecoresponsive Genome of *Daphnia pulex*. Science 331:555–561.

Dawidowicz, P., and C.J. Loose. 1992. Metabolic costs during predator-induced diel vertical migration of *Daphnia*. Limnology and Oceanography 37:1589–1595.

Dennis, S.R., G.A. LeBlanc, and A.P. Beckerman. 2014. Endocrine regulation of predator-induced phenotypic plasticity. Oecologia 176:625–635.

Derby, C.D., and M.J. Weissburg. 2014. The chemical senses and chemosensory ecology of crustaceans. Pages 263–292 *in* C. Derby, and M. Thiel, editors. The natural history of the Crustacea, volume 3: Nervous systems and control of behavior. Oxford University Press, New York.

Dicke, M., and P. Grostal. 2001. Chemical detection of natural enemies by arthropods: an ecological perspective. Annual Review of Ecology and Systematics 32:1–23.

Dunn, A.M., J.T., Dick, and M.J. Hatcher. 2008. The less amorous Gammarus: predation risk affects mating decisions in *Gammarus duebeni* (Amphipoda). Animal Behaviour 76:1289–1295.

Emlen, D.J., Q. Szafran, L.S. Corley, and I. Dworkin. 2006. Insulin signaling and limb-patterning: candidate pathways for the origin and evolutionary diversification of beetle horns. Heredity 97:179–191.

Escribano, R., and R. Riquelme-Bugueño. 2015. Planktonic crustaceans: lifestyles in the water column. Pages 262–278 *in* M. Thiel, and L. Watling, editors. The natural history of the Crustacea, volume 2: Lifestyles and feeding biology. Oxford University Press, New York.

Faulkes, Z. 2013. Morphological adaptations for digging and burrowing. Pages 276–295 *in* L. Watling, and M. Thiel, editors. The natural history of the Crustacea, volume 1: Functional morphology and diversity. Oxford University Press, New York.

Ferrari, M.C.O., B.D. Wisenden, and D.P. Chivers. 2010. Chemical ecology of predator-prey interactions in aquatic ecosystems: a review and prospectus. Canadian Journal of Zoology 88:698–724.

Frost, B.W., and S.M. Bollens. 1992. Variability of diel vertical migration in the marine planktonic copepod *Pseudocalanus newmani* in relation to its predators. Canadian Journal of Fisheries and Aquatic Sciences 49:1137–1141.

Fusco, G., and A. Minelli A. 2010. Phenotypic plasticity in development and evolution: facts and concepts. Philosophical Transactions of the Royal Society B: Biological Sciences 365:547–556.

Gotoh, H., R. Cornette, S. Koshikawa, Y. Okada, L.C. Lavine, D.J. Emlen, and T. Miura. 2011. Juvenile hormone regulates extreme mandible growth in male stag beetles. PLoS ONE 6:e21139.
Grant, J.W.G., and I.A.E. Bayly. 1981. Predator induction of crests in morphs of the *Daphnia carinata* king complex. Limnology and Oceanography 26:201–218.
Haddock, S.H., M.A. Moline, and J.F. Case. 2010. Bioluminescence in the sea. Marine Science 2:443–493.
Hamrén, U., and S. Hansson. 1999. A mysid shrimp (*Mysis mixta*) is able to detect the odour of its predator (*Clupea harengus*). Ophelia 51:187–191.
Hansson, L.-A., and S. Hylander. 2009. Size-structured risk assessments govern *Daphnia* migration. Proceedings of the Royal Society B: Biological Sciences 276:331–336.
Harvell, C.D. 1990. The ecology and evolution of inducible defenses. The Quarterly Review of Biology 65:323–340.
Haupt, F., M. Stockenreiter, M. Baumgartner, M. Boersma, and H. Stibor. 2009. *Daphnia* diel vertical migration: implications beyond zooplankton. Journal of Plankton Research 31:515–524.
Hay, M.E., J.E. Duffy, and W. Fenical. 1990. Host-plant specialization decreases predation on a marine amphipod: an herbivore in plant's clothing. Ecology 71:733–743.
Hazlett, B.A. 2011. Chemical cues and reducing the risk of predation. Pages 355–370 *in* T. Breithaupt, and M. Thiel, editors. Chemical communication in crustaceans. Springer, New York.
Herzog, Q., and C. Laforsch. 2013. Modality matters for the expression of inducible defenses: introducing a concept of predator modality. BMC Biology 11:113.
Herzog, Q., M. Rabus, B.W. Ribeiro, and C. Laforsch. 2016. Inducible defenses with a "twist": *Daphnia barbata* abandons bilateral symmetry in response to an ancient predator. PLoS ONE 11:e0148556.
Jarrett, J.N. 2009. Predator-induced defense in the barnacle *Chthamalus fissus*. Journal of Crustacean Biology 29:329–333.
Jeschke, J.M., C. Laforsch, and R. Tollrian. 2008. Animal prey defenses. Pages 189–194 *in* S.J. Jørgensen, and B.D. Fath, editors. General ecology. Elsevier, Oxford, England.
Käuser, G., H.M. Brandtner, H.-J. Bidmon, and J. Koolman. 1988. Ecdysone synthesis and release by the brain-ring gland complex of blowfly larvae. Journal of Insect Physiology 34:563–569.
Kiørboe, T, and A. Visser. 1999. Predator and prey perception in copepods due to hydromechanical signals. Marine Ecology Progress Series 179:81–95.
Kusch, J., and K. Heckmann. 1992. Isolation of the Lembadion-factor, a morphogenetically active signal, that induces Euplotes cells to change from their ovoid form into a larger lateral winged morph. Developmental Genetics 13:241–246.
Laforsch, C., and R. Tollrian. 2004. Embryological aspects of inducible morphological defenses in *Daphnia*. Journal of Morphology 262:701–707.
Laforsch, C., W. Ngwa, W. Grill, and R. Tollrian. 2004. An acoustic microscopy technique reveals hidden morphological defenses in *Daphnia*. Proceedings of the National Academy of Sciences of the USA 101:15911–15914.
Laforsch, C., L. Beccara, and R. Tollrian. 2006. Inducible defenses: the relevance of chemical alarm cues in *Daphnia*. Limnology and Oceanography 51:1466–1472.
Lampert, W. 1989. The adaptive significance of diel vertical migration of zooplankton. Functional Ecology 3:21–27.
Lavalli, K.L., and E. Spanier. 2015. Predator adaptations of decapods. Pages 190–228 *in* M. Thiel, and L. Watling, editors. The natural history of the Crustacea, volume 2: Lifestyles and feeding biology. Oxford University Press, New York.
Lavalli, K.L., E. Spanier, and F. Grasso. 2007. Behavior and sensory biology of slipper lobsters. Pages 133–181 *in* K.L. Lavalli, and E. Spanier, editors. The biology and fisheries of slipper lobsters. CRC Press, Boca Raton, Florida.
Le Trionnaire, G., J. Hardie, S. Jaubert-Possamai, J.C. Simon, and D. Tagu. 2008. Shifting from clonal to sexual reproduction in aphids: physiological and developmental aspects. Biology of the Cell 100:441–451.
Lima, S.L. 1998. Stress and decision-making under the risk of predation: recent developments from behavioral, reproductive, and ecological perspectives. Advances in the Study of Behaviour 27:215–290.

Lively, C.M. 1986. Predator-induced shell dimorphism in the acorn barnacle *Chthamalus anisopoma*. Evolution 40:232–242.

Lynch, M. 1980. Predation, enrichment, and the evolution of cladoceran life histories: a theoretical approach. Pages 367–376 *in* W.C. Kerfoot, editor. The evolution and ecology of zooplankton communities. American Society of Limnology and Oceanography Special Symposium No. 3.

Martin-Creuzburg, D., S.A. Westerlund, and K.H. Hoffmann. 2007. Ecdysteroid levels in *Daphnia magna* during a molt cycle: determination by radioimmunoassay (RIA) and liquid chromatography—mass spectrometry (LC–MS). General and Comparative Endocrinology 151:66–71.

Mathis, A., D.P. Chivers, and R.J.F. Smith. 1995. Chemical alarm signals: predator deterrents or predator attractants? The American Naturalist 145:994–1005.

Mathis, A., and W.W. Hoback. 1997. The influence of chemical stimuli from predators on precopulatory pairing by the amphipod, *Gammarus pseudolimnaeus*. Ethology 103:33–40.

Miyakawa, H., M. Imai, N. Sugimoto, Y. Ishikawa, A. Ishikawa, H. Ishigaki, Y. Okada, S. Miyazaki, S. Koshikawa, R. Cornette, and T. Miura. 2010. Gene up-regulation in response to predator kairomones in the water flea, *Daphnia pulex*. BMC Developmental Biology 10:45.

Miyakawa, H., H. Gotoh, N. Sugimoto, and T. Miura. 2013. Effect of juvenoids on predator-induced polyphenism in the water flea, *Daphnia pulex*. Journal of Experimental Zoology Part A: Ecological Genetics and Physiology 319:440–450.

Morgan, S.G. 1987. Morphological and behavioral antipredatory adaptations of decapod zoeae. Oecologia 73:393–400.

Morris, M, and S.M. Rogers. 2014. Integrating phenotypic plasticity within an ecological genomics framework: recent insights from the genomics, evolution, ecology, and fitness of plasticity. Pages 73–105 *in* C.R. Landry, and N. Aubin-Horth, editors. Ecological genomics: ecology and the evolution of genes and genomes. Advances in Experimental Medicine and Biology 781.

Murren, C.J., J.R. Auld, H. Callahan, C.K. Ghalambor, C.A. Handelsman, M.A. Heskel, J.G. Kingsolver, H.J. Maclean, J. Masel, H. Maughan, D.W. Pfennig, R.A. Relyea, S. Seiter, E. Snell-Rood, U.K. Steiner, and C.D. Schlichting. 2015. Constraints on the evolution of phenotypic plasticity: limits and costs of phenotype and plasticity. Heredity 115:293–301.

Nakatsuji, T., C.Y. Lee, and R.D. Watson. 2009. Crustacean molt-inhibiting hormone: structure, function, and cellular mode of action. Comparative Biochemistry and Physiology Part A: Molecular and Integrative Physiology 152:139–148.

Nijhout, H.F. 2013. Arthropod developmental endocrinology. Pages 123–148 *in* A. Minelli, G. Boxshall, and G. Fusco, editors. Arthropod biology and evolution. Springer-Verlag Berlin, Heidelberg, Germany.

Oda, S., Y. Kato, H. Watanabe, N. Tatarazako, and T. Iguchi. 2011. Morphological changes in *Daphnia galeata* induced by a crustacean terpenoid hormone and its analog. Environmental Toxicology and Chemistry 30:232–238.

Ory, N.C., T.C. van Son, and M. Thiel. 2015. Mating rock shrimp hedge their bets: old males take greater risk, but only after careful assessment of the investment scenario. Behavioral Ecology and Sociobiology 69:1975–1984.

Pastorok, R.A. 1981. Prey vulnerability and size selection by *Chaoborus* larvae. Ecology 62:1311–1324.

Petrusek, A., R. Tollrian, K. Schwenk, A. Haas, and C. Laforsch. 2009. A "crown of thorns" is an inducible defense that protects *Daphnia* against an ancient predator. Proceedings of the National Academy of Sciences of the USA 106:2248–2252.

Pijanowska, J. 1992. Anti-predator defense in three *Daphnia* species. Internationale Revue der gesamten Hydrobiologie und Hydrographie 77:153–163.

Pijanowska, J., and A. Kowalczewski. 1997. Predators can induce swarming behavior and locomotory responses in *Daphnia*. Freshwater Biology 37:649–656.

Rabus, M., and C. Laforsch. 2011. Growing large and bulky in the presence of the enemy: *Daphnia magna* gradually switches the mode of inducible morphological defences. Functional Ecology 25:1137–1143.

Rabus, M., T. Söllradl, H. Clausen-Schaumann, and C. Laforsch. 2013. Uncovering ultrastructural defences in *Daphnia magna*—an interdisciplinary approach to assess the predator-induced fortification of the carapace. PLoS ONE 8:e67856.

Riessen, H.P. 2012. Costs of predator-induced morphological defences in *Daphnia*. Freshwater Biology 57:1422–1433.
Ringelberg, J. 2010. Diel vertical migration of zooplankton in lakes and oceans: causal explanations and adaptive significances. Springer, London, England.
Rozenberg, A., M. Parida, F. Leese, L.C. Weiss, R. Tollrian, and J.R. Manak. 2015. Transcriptional profiling of predator-induced phenotypic plasticity in *Daphnia pulex*. Frontiers in Zoology 12:18.
Schmidt, M., and M. DeForest, Jr. 2011 Neuronal processing of chemical information in crustaceans. Pages 123–147 *in* M. Thiel, and T. Breithaupts, editors. Chemical communication in crustaceans. Springer, New York.
Schnitzler I., G. Pohnert, M. Hay, and W. Boland. 2000. Chemical defense of brown algae (*Dictyopteris* spp.) against the herbivorous amphipod *Ampithoe longimana*. Oecologia 126:515–521.
Schoeppner, N.M., and R.A. Relaya. 2005. Damage, digestion, and defense: the roles of alarm cues and kairomones for inducing prey defenses. Ecology Letters 8:505–512.
Selander, E, J. Kubanek, M. Hamberg, M.X. Andersson, G. Cervin, and H. Pavia. 2015. Predator lipids induce paralytic shellfish toxins in bloom-forming algae. Proceedings of the National Academy of Sciences of the USA 112:6395–6400.
Stabell, O.B., F. Ogbebo, and R. Primicerio. 2003. Inducible defenses in *Daphnia* depend on latent alarm signals from conspecific prey activated in predators. Chemical Senses 28:141–153.
Stachowicz, J.J., and M.E. Hay. 1999. Reducing predation through chemically mediated camouflage: indirect effects of plant defenses on herbivores. Ecology 80:495–509.
Stevens, M., A.E. Lown, and L.E. Wood. 2014. Color change and camouflage in juvenile shore crabs *Carcinus maenas*. Frontiers in Ecology and Evolution 2:14.
Stibor, H., and J. Lüning. 1994. Predator-induced phenotypic variation in the pattern of growth and reproduction in *Daphnia hyalina* (Crustacea: Cladocera). Functional Ecology 8:97–101.
Sumiya, E., Y. Ogino, H. Miyakawa, C. Hiruta, K. Toyota, S. Miyagawa, and T. Iguchi. 2014. Roles of ecdysteroids for progression of reproductive cycle in the fresh water crustacean *Daphnia magna*. Frontiers in Zoology 11:60.
Ślusarczyk, M. 1995. Predator-induced diapause in *Daphnia*. Ecology 76:1008–1013.
Tollrian, R. 1994. Fish-kairomone induced morphological changes in *Daphnia lumholtzi* (Sars). Archiv für Hydrobiologie 130:69–75.
Tollrian, R. 1995. Predator-induced morphological defenses: costs, life history shifts, and maternal effects in *Daphnia pulex*. Ecology 76:1691–1705.
Tollrian, R., and D.C. Harvell. 1999. The evolution of inducible defenses: current ideas. Pages 307–321 *in* R. Tollrian, and D.C. Harvell, editors. The ecology and evolution of inducible defenses. Princeton University Press, Princeton, New Jersey.
Tollrian, R., and C. Heibl. 2004. Phenotypic plasticity in pigmentation in *Daphnia* induced by UV radiation and fish kairomones. Functional Ecology 18:497–502.
Tollrian, R., S. Duggen, L.C. Weiss, C. Laforsch, and M. Kopp. 2015. Density-dependent adjustment of inducible defenses. Scientific Reports 5:12736.
Truman, J.W., L.M. Riddiford, and L. Safranek. 1974. Temporal patterns of response to ecdysone and juvenile hormone in the epidermis of the tobacco hornworm, *Manduca sexta*. Developmental Biology 39:247–262.
Verschoor, A.M., M. Vos, and I. van der Stap. 2004. Inducible defences prevent strong population fluctuations in bi-and tritrophic food chains. Ecology Letters 7:1143–1148.
Walsh, M.R., J.P. DeLong, T.C. Hanley, and D.M. Post. 2012. A cascade of evolutionary change alters consumer-resource dynamics and ecosystem function. Proceedings of the Royal Society B: Biological Sciences 279:3184–3192.
Weiss, L.C., S. Kruppert, C. Laforsch, and R. Tollrian. 2012a. *Chaoborus* and *Gasterosteus* anti-predator responses in *Daphnia pulex* are mediated by independent cholinergic and GABAergic neuronal signals. PloS ONE 7:e36879.
Weiss, L.C., C. Laforsch, and R. Tollrian. 2012b. The taste of predation and the defenses of prey. Pages 111–126 *in* C. Bronmark, and L.-A. Hansson, editors. Chemical ecology in aquatic systems. Oxford University Press, New York.

Weiss, L.C., F. Leese, C. Laforsch, and R. Tollrian. 2015. Dopamine is a key regulator in the signalling pathway underlying predator-induced defenses in *Daphnia*. Proceedings of the Royal Society B: Biological Sciences 282:20151440.

Weiss, L.C., E. Heilgenberg, L. Deussen, S.M. Becker, S. Kruppert, and R. Tollrian. 2016. Onset of kairomone sensitivity and the development of inducible morphological defenses in *Daphnia pulex*. Hydrobiologia 779:135–145.

Whitman, D.W., and A. Agrawal. 2009. What is phenotypic plasticity and why is it important. Phenotypic Plasticity of Insects 10:1–63.

Wilkinson, E.B., J.H. Grabowski, G.D. Sherwood, and P.O. Yund. 2015. Influence of predator identity on the strength of predator avoidance responses in lobsters. Journal of Experimental Marine Biology and Ecology 465:107–112.

Wisenden, B.D., A. Cline, and T.C. Sparkes. 1999. Survival benefit to antipredator behavior in the amphipod *Gammarus minus* (Crustacea: Amphipoda) in response to injury-released chemical cues from conspecifics and heterospecifics. Ethology 105:407–414.

Wyatt, T.D. 2011. Pheromones and behavior. Pages 23–38 *in* M. Thiel, and T. Breithaupt, editors. Chemical communication in crustaceans. Springer, New York.

Yasumoto, K, A. Nishigami, M. Yasumoto, F. Kasai, Y. Okada, T. Kusumi, and T. Ooi. 2005. Aliphatic sulfates released from *Daphnia* induce morphological defense of phytoplankton: isolation and synthesis of kairomones. Tetrahedron Letters 46:4765–4767.

Yin, M., C. Laforsch, J.N. Lohr, and J. Wolinska. 2011. Predator-induced defense makes Daphnia more vulnerable to parasites. Evolution 65:1482–1488.

Zaret, T.M., and W.C. Kerfoot. 1975. Fish predation on *Bosmina longirostris*: body-size selection versus visibility selection. Ecology 56:232–237.

13

LIFE HISTORY ADAPTATION IN PREY

Gary A. Wellborn

Abstract

Predation is a powerful agent of life history evolution in prey species, as demonstrated in diverse examples in crustaceans. Ubiquitous size- and age-selective predation mediates trade-offs among reproductive effort, survival, and growth, which cause evolution of constitutive and phenotypically plastic shifts in age and size at maturity. In accord with predictions of life history theory, comparative studies demonstrate that contrasting forms of selective predation generate divergent evolutionary changes in age- and size-specific allocation of reproductive effort within populations and species. Predation risk also influences egg and offspring size, and some crustaceans exhibit phenotypic plasticity in offspring size in response to chemical cues of predators. Because age-selective predation impacts the relative benefits of earlier versus later reproductive investment, predation may also shape senescence and life span of crustaceans. Additionally, individual differences in risk-taking behavior, sometimes termed "personalities," have been examined in several crustaceans, and these may arise through among-individual variation in reproductive value. Finally, in some crustacean groups limb autotomy is a common, but costly, antipredator defense, and life history perspectives on autotomy suggest individuals may balance costs and benefits during predator encounters. Much of our understanding of predation's role in life history evolution of prey derives from studies of crustaceans, and these organisms continue to be promising avenues to elucidate mechanisms of life history evolution.

INTRODUCTION

Predation is a principal driver of life history evolution in crustaceans and may influence multiple features of a species' life history, including age and size at maturity, allocation of reproductive effort across ontogeny, size and number of eggs in broods, and life span. Because nearly all species are at

Life Histories. Edited by Gary A. Wellborn and Martin Thiel.

risk of predation during at least some period in their life, predation is likely to influence the life history phenotype of essentially all crustaceans. These effects of predation on life histories are not only universal but are also diverse in their expression across crustacean taxa because of the enormous taxonomic diversity of crustaceans, as well as diversity of their ecological settings and lifestyles (Schram 1986).

Predators are agents of natural selection on prey life histories when the intensity of predation is nonuniform across prey age or size (Law 1979, Taylor and Gabriel 1993). For example, predators that selectively consume larger, older individuals within a prey population generally produce natural selection in the prey population for allocation of physiological resources toward greater reproductive investment at younger age and smaller body size (Law 1979, Roff 2001). Evolutionary responses to selection on life history traits require the presence of adequate genetic variance in traits under selection, and life history traits often harbor significant evolutionary potential to respond to selection (Baer and Lynch 2003, Fisk et al. 2007, Stoks et al. 2016). The combination of apparent ubiquity of nonuniform predation across size and age of prey species, coupled with genetic variance in life history traits, suggests that (1) populations often respond readily to predation-mediated selection on life histories and (2) these responses are to some degree pervasive across species.

Prey populations may adapt to predation risk by evolving constitutive life history traits, or genetic and ecological circumstances may favor phenotypically plastic life history responses in which development of the life history phenotype is dependent on environmental cues that correlate with predation risk (Tollrian 1995, Lass and Spaak 2003; see also Chapter 12 in this volume). Plastic developmental responses are beneficial when predation intensity is spatially or temporally variable over sufficient scales to be favored over constitutive development of the life history phenotype. Habitats characterized by stable predation intensity on a prey population do not benefit from plasticity, and therefore constitutive life history phenotypes are expected in these environments (Forsman 2015).

Body size substantially affects ecological and evolutionary dynamics in many species, and is a central link between levels of predation risk and the trade-offs among life history traits that shape life history evolution. Fecundity, for example, is often closely correlated with body size in many plant and animal groups and almost comprehensively so within crustacean species (Sainte-Marie 1991, Corey and Reid 1991). Likewise, survival is another strongly size-dependent characteristic, which may increase or decrease with body size, depending on the foraging biology of primary predators in an environment (Zaret 1980, Wellborn 1994, Sparkes 1996, Zhao et al. 2006, Ory et al. 2012). Also, ecological interactions with resources are commonly dependent on size (Wirtz 2012). Because size is correlated with many critical ecological (White et al. 2007) and life history (Atkinson and Hirst 2007) processes, it is often considered the most ecologically significant feature of an organism's phenotype (Werner and Gilliam 1984, Brown et al. 2007).

Body size often shapes the nature and consequences of predator-prey interactions in a community. The form of selective predation experienced by a prey population is a primary determinant of the prey's evolutionary response to predation. Two contrasting forms of size-selective predation are most relevant. Positive size selection imposes greatest mortality on larger individuals within a population, and negative size selection causes greater mortality for smaller prey individuals. Positive size-selective predation is especially common when predators search visually and are larger than their prey (Zaret 1980), whereas negative size-selective predation generally involves handling constraints on prey capture and consumption (Pastorok 1981). Prey morphological defenses, such as spines, can also reduce susceptibility to predation by imposing handling constraints for predators (Barnhisel 1991).

This chapter explores the scope of predation's influences across life history and phenotype in crustaceans. A primary focus of the chapter is consideration of the allocation of reproductive effort across ontogeny, perhaps the most fundamental life history trait. Reproductive allocation is shaped by trade-offs among survival, fecundity, and growth, and it influences maturation size, number and size of broods, and other traits. The influence of predation on senescence and life span is examined, as is predation's role in the evolution of behavioral variation within a population. Limb autotomy is a principal antipredator defense in many crustaceans, and the chapter concludes with a life history perspective on autotomy.

REPRODUCTIVE EFFORT

Reproductive effort refers to an individual's proportional allocation of total available resources into a brood during a bout of reproduction. Because any resources devoted to reproduction are not available for other fitness-related functions, investment in reproduction generates fitness trade-offs that shape evolution of life histories (Williams 1966, van Noordwijk and Jong 1986, Roff 2002). Investment in current reproduction causes a decline in future reproductive prospects through multiple mechanisms, including reduced investment in somatic maintenance and defense, which may reduce survival and slow growth, in turn lowering fecundity (Reznick et al. 2000, Dmitriew 2011). Sarma et al. (2002), who examined these "costs of reproduction" in several cladoceran species using laboratory life table experiments, found that individuals investing more heavily in current reproduction often suffer reduced subsequent survival and reproduction, especially under resource stress. Costs of reproduction produce a trade-off between level of resource investment in current reproductive effort versus expected future reproduction, and many recent studies have explored underlying physiological pathways that mediate resource allocation trade-offs in the context of life histories (Harshman and Zera 2006, Cox et al. 2010, Cox et al. 2014). Predation markedly influences evolution of age- and size-specific allocation of reproductive effort across the life span of prey species, and these allocation trade-offs shape both maturation size and age and allocation of reproductive effort across successive broods.

Maturation Size and Age

Maturation size and age are aspects of the more general life history problem of optimal allocation of reproductive effort over an organism's life span (Stearns 1992). Fitness is often highly sensitive to changes in maturation age because earlier reproduction allows an individual's offspring to themselves begin reproducing earlier (Roff 1992). We expect strong selection favoring early reproduction in growing populations, where this compounding effect is greatest, including populations that increase seasonally in numbers and have multiple and overlapping generations, such as many species of cladocerans, copepods, and amphipods. The benefits of earlier maturation must, however, be weighed against its accompanying costs. Initiation of reproduction requires diversion of physiological and other resources into production and care of offspring, rather than into other fitness-related processes such as growth, somatic maintenance, and defense. This diversion of resources causes a decline in prospects for future reproductive success due to reductions in subsequent survival, growth, and fecundity. Additionally, earlier reproduction may produce less fit offspring because of decreased provisioning of eggs, for example, or reduced quality of parental care (Sato and Suzuki 2010). In general, we expect life histories favored by natural selection to be those that balance a life history phenotype's benefits against its costs such that fitness is maximized (Stearns 1992).

Two observations suggest that size and age at maturity in crustaceans are likely to vary appreciably among populations and species that experience differing regimes of size-selective predation. First, rates of predation mortality appear to be very commonly, and perhaps universally, size-dependent, as discussed above. Second, fecundity increases with body size in crustaceans (Sainte-Marie 1991, Corey and Reid 1991), thereby providing a potential fitness benefit for delaying reproduction, and, instead, investing current resources into growth to a larger size. All else equal, theories of optimal life histories indicate earlier reproduction is likely to be favored when predation mortality increases with body size, and later reproduction is favored when mortality decreases with body size (Taylor and Gabriel 1992, Taylor and Gabriel 1993).

Although it is difficult to test life history theories by precisely predicting the optimal size and age of maturity for any single population because of the significant challenges of empirically quantifying all relevant costs and benefits across all possible sizes and ages, comparative and experimental studies make clear that predation often significantly shapes size and age at maturity in crustaceans and other taxa. For example, Fisk et al. (2007) studied evolutionary changes in life histories of *Daphnia melanica* in historically fishless mountain lakes subjected to anthropogenic introductions of trout, which selectively consume larger *D. melanica*. Consistent with expectations from theory, a laboratory culture experiment demonstrated that *Daphnia* collected from lakes with introduced trout had evolved earlier age and smaller size at maturity compared to *Daphnia* collected from historically fishless lakes (Fig. 13.1). We expect the direction of life history evolution to be reversed when prey species incur selective predation on smaller, rather than larger, individuals within the population, and this is the case. For example, in a laboratory selection experiment, Spitze (1991) examined evolutionary dynamics in a mixture of clones of *Daphnia pulex* in both the presence and absence of a common negatively size-selective predator in freshwater lakes, the larval dipteran *Chaoborus* (*C. americanus*, in this study). *Daphnia* populations used in the study naturally coexist with *Chaoborus*, and when experimentally released from negatively size-selective predation in predator-free cultures, *D. pulex* rapidly evolved earlier investment in reproduction, including earlier age at reproduction and increased fecundity. We can surmise, therefore, in the natural habitat

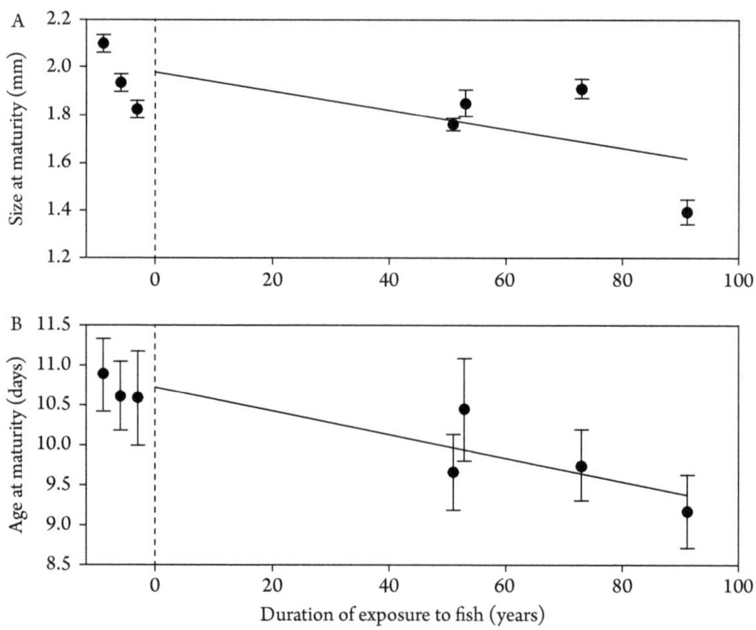

Fig. 13.1.
Evolutionary change in size and age at maturity in laboratory-reared *Daphnia melanica* obtained from lakes with introduced trout (right of dashed line), which selectively consume larger *Daphnia* and thus select for smaller maturation size and age. Magnitude of evolutionary change increases with time since trout introductions, and rates of change are as high as 4,000 darwins (change by factor of *e* per million years). *Daphnia* from lakes without introduced trout (left of dashed line) are shown for comparison. From Fisk et al. (2007), under Creative Commons license CC-BY.

selective predation by *Chaoborus* on smaller individuals maintains a delayed investment in reproduction and later age at maturation in *D. pulex*. Comparative studies of natural prey populations and species that face differing forms of size-selective predation provide additional support for the substantial influence of predation on evolution of maturation size and age (Wellborn and Broughton 2008). Some populations of the freshwater isopod *Lirceus fontinalis* coexist with a salamander that selectively consumes smaller isopods, and *L. fontinalis* in these populations delay maturity to a larger size, compared to populations that do not coexist with the salamander (Sparkes 1996).

Numerous studies have examined developmentally plastic life history responses of crustacean prey when exposed to chemical cues of predators, termed *kairomones*, and these studies demonstrate phenotypically plastic responses that largely mimic predictions for evolved constitutive responses (Stibor and Lüning 1994, Riessen 1999, Spaak et al. 2000, Declerck and Weber 2003, Pauwels et al. 2010; see also Chapter 12 in this volume). For example, *Daphnia galeata* in Lake Constance, Germany, coexist with fish, but risk of predation varies substantially across large temporal shifts in abundance of the juvenile fish that feed on *D. galeata*; these juvenile fish selectively consume larger individuals within the population (Sakwinska 2002). A laboratory experiment tracked developmental plasticity of life history traits in hours-old *D. galeata* exposed to water from an aquarium holding juveniles of the predatory fish *Leuciscus leuciscus*. Compared to controls that were not exposed to fish chemical cues, exposed *D. galeata* matured at a significantly earlier age and smaller body size, and these differences were largely attributable to maturation at an earlier instar. Furthermore, *D. galeata* exposed to fish cues went on to produce smaller offspring in larger clutches relative to female size (Sakwinska 2002). These phenotypically plastic responses mimic expectations for constitutive genetic change in life histories (Taylor and Gabriel 1992), and this developmental plasticity is clearly an adaptive response to temporally fluctuating predation intensity.

Mechanisms Mediating Phenotypically Plastic Change in Maturation Size and Age

When life history traits of prey are the result of phenotypically plastic responses to chemical or visual cues of predators, the proximate mechanisms causing change in life history are often not readily apparent (Ball and Baker 1996, Noonburg and Nisbet 2005). For example, prey may respond to the presence of predators by altering behavior in ways that reduce their predation risk such as a reduction in foraging rate, and mature at a smaller size as a consequence of a reduced growth rate (Abrams and Rowe 1996). In this case, the proximate cause of smaller size at reproduction is reduced growth due to lower feeding rates. Alternatively, reduced maturation size may be a direct physiological effect of the predator cue triggering more rapid development in the prey. The distinction between these alternatives is important because age at maturity may be greater when growth in size is slowed in response to a predator and there is no accompanying increase in size-specific development rate, a condition with potentially large fitness costs. Fitness costs of the response to predators may be reduced if prey respond by increasing development rate to reach maturity at a smaller size and relatively earlier age.

Some experimental studies have sought to tease apart these "behavioral" and "physiological" mechanisms of life history response to predator cues, including a laboratory study of responses of the cladoceran *Daphnia magna* to manipulation of both food level and presence or absence of fish cues (Beckerman et al. 2007). This study demonstrated that although *Daphnia* behaviorally reduced energy intake in the presence of fish chemicals, *Daphnia* also simultaneously increased their developmental rate in the presence of the fish kairomone, and matured at an earlier age for a given growth rate. Other studies also lend support to the development rate mechanism (Stibor and Machacek 1998, Sakwinska 2002). One key physiological process in maturation is production of yolk to provision eggs, and chemical cues of predators may initiate adaptive shifts in yolk production. In *D. magna*, exposure to fish chemical cues caused early initiation of yolk production as well as earlier maturity (Stibor 2002). Conversely, exposure to chemical cues of the predatory invertebrate *Chaoborus* resulted in later maturity, as well as reduced rates of yolk protein synthesis, although initiation of yolk synthesis was not affected by *Chaoborus* chemical cues (Stibor 2002). Collectively, these studies suggest life histories can exhibit plastic physiological and developmental responses to environmental cues of predators, and these may be attuned to specific predator taxa (Beckerman et al. 2010).

Allocation of Reproductive Effort

Most crustaceans are iteroparous (i.e., produce 2 or more broods over their lifetime) and continue to grow in body size and increase brood size between reproductive events (see Chapter 4 in this volume). This condition generates a complex interplay among benefits and trade-offs associated with growth, survival, and reproduction. Resolution of the evolutionary outcome of this interplay is often called the "general life history problem," and the optimal allocation of reproductive effort across life span depends on all the factors involved (León 1976, Kozlowski and Wiegert 1986). The concept of reproductive value, which is calculated from life tables, is helpful in considering the general life history problem (Williams 1966). The reproductive value for an individual of a specific age equals its expected contribution to future population size and thus is the sum of the 2 components of reproductive value: the individual's current reproduction and its expected future reproduction. Expected future reproduction is commonly termed *residual reproductive value*. In the optimal life history, reproductive effort in each size or age class is that which realizes current and residual reproductive values in each size or age class such that lifetime fitness is maximized (Stearns 1992). A simple prediction arising from this construction is that a decrease in residual reproductive value, perhaps due to a size-selective predator, increases the relative value of current reproduction.

Positive Size-Selective Predation and Allocation of Reproductive Effort

Many prey populations experience positive size-selective predation in which larger individuals are most at risk of being consumed (Peer et al. 1986, Wellborn 1994, Zhao et al. 2006, Hart and Bychek 2011). Predators may disproportionately consume larger individuals within a population because these prey provide greater energetic reward (Werner and Hall 1974) or because they are more likely to be detected (O'Brian et al. 1976). When predation mortality disproportionately affects larger individuals in a population, the optimal life history shifts toward greater allocation of reproductive effort early in life, even at the cost of reduced future reproduction, growth, and survival (Fig. 13.2). In this circumstance, the probability of living to a larger size is low, and the fitness-maximizing life history shifts resources into early reproduction (Law 1979, Taylor and Gabriel 1993), including earlier maturation and greater reproductive effort, and consequently invests fewer resources into additional growth (Rinke et al. 2008).

Predatory fish and some predatory invertebrates and birds often selectively consume larger individuals in crustacean populations. For example, in a North American stream, trout prey throughout the year on the benthic amphipod *Gammarus pseudolimnaeus*, as evidenced by the occurrence of these amphipods in stomach contents of more than half of adult trout examined (Newman and Waters 1984). Mean size of consumed *Gammarus* was about 63% greater than that of the stream population, indicating strong size-selective predation. In some freshwater habitats, the backswimmer *Notonecta* is an important predator that disproportionately preys on larger zooplankton individuals. In a laboratory experiment that measured preference among differently sized cladoceran species, *Notonecta hoffmani* consistently preferred the larger species (Scott and Murdoch 1983). For example, larger *Ceriodaphnia reticulata* were consumed at about 3 times the rate of smaller individuals. In the Bay of Fundy, the tube-dwelling amphipod *Corophium volutator* is

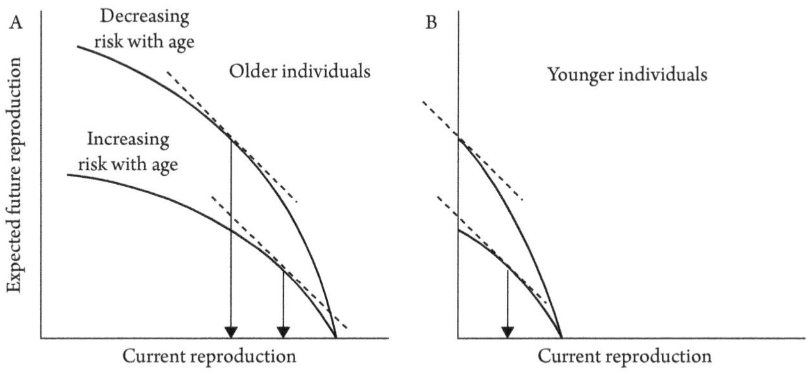

Fig. 13.2.
Comparison of optimal reproductive effort for populations with increasing versus decreasing predation risk as age increases. Curves represent trade-offs between investment in current versus future reproduction. Dashed lines have slope of −1 and depict equal-fitness solutions under the assumption that fitness equals the sum of current and future reproduction. Arrows show optimal life histories. (A) In mature adults, the optimal life history for a population subject to high adult mortality is to invest heavily in current reproduction because there is a comparatively low probability of surviving to produce a later clutch, but in a population with low adult predation risk with high assurance of survival, the optimal life history invests relatively less in current reproduction to achieve greater future reproduction. (B) In younger individuals, populations with high adult predation risk initiate reproduction at a younger age than do populations with low adult predation risk. Redrawn from Wellborn (1994).

subject to a few weeks of intense predation by migrating shorebirds, principally sandpipers, which cause a sharp decline in *C. volutator* density (Peer et al. 1986). The birds consume mostly large individuals, with amphipods recovered from digestive tracks of sandpipers being 48% larger than the mudflat population as a whole.

Positive size-selective predation, because it disproportionately increases mortality on larger individuals, typically causes natural selection for greater reproductive effort early in life. Earlier maturation is one manifestation of this shift in reproductive effort, but increased proportional allocation of resources, including greater brood size, should also be commonly favored by selection (Gadgil and Bossert 1970, Law 1979, Michod 1979, Charlesworth 1980, Kozlowski and Wiegert 1987, Taylor and Gabriel 1993). For example, when reared in the presence of fish chemical cues, *Daphnia hyalina* and *Diaphanosoma brachyurum* produced about twice as many eggs as individuals reared without the chemical cue (Dawidowicz et al. 2010). Similarly, *Daphnia hyalina* reared in the chemical presence of fish or *Notonecta* had significantly greater reproductive effort than individuals in predator-free controls (Stibor and Lüning 1994), and Engelmayer (1995) obtained similar experimental results for *D. pulex*.

Negative Size-Selective Predation and Allocation of Reproductive Effort

Predation that falls primarily on smaller individuals in a population, known as negative size selective predation, generally selects for a shift in reproductive effort toward larger size and later age (Taylor and Gabriel 1992). In freshwater systems, small or juvenile fish commonly impose negative size selective predation on crustacean prey (Gélinas et al. 2007), as do some invertebrates, including midge larvae in the genus *Chaoborus* (Pastorok 1981, Campbell 1991). For example, in a detailed laboratory study, the efficiency of prey capture attempts by the larval midges *C. americanus* and *C. trivittatus* was about 90% for the smallest cladoceran individuals but declined steadily to near 0%

for large individuals, and similar results were observed for *Chaoborus* predation on copepods (Swift and Fedorenko 1975). Predatory fish may also exhibit negative size selection. A laboratory size-preference study of the marine fish *Scartella cristata* found these predators consumed the smaller of two shrimp individuals (*Lysmata wurdemanni*) in about 80% of trials (Baeza 2006).

Because predation risk declines with increased body size under negative size selective predation, prey may sometimes escape predation risk by growing into a "size refuge" from predation in which prey grow beyond the capture capabilities of predators (Swift and Fedorenko 1975, Wellborn 1994). In ecological circumstances for which a size refuge is possible, selection can favor rapid growth and delayed reproductive effort until a mostly invulnerable size is reached (Taylor and Gabriel 1993, Gårdmark and Dieckmann 2006, Urban 2007). There is some empirical support for this prediction. Predation by *Chaoborus* on *Daphnia pulex* declines substantially with *Daphnia* body size, and when predation is experimentally removed, growth rate of *D. pulex* declined (i.e., slower growing clones became more prevalent), implying elevated growth rate in *Chaoborus*-adapted *Daphnia* was an adaptation to negative size selective predation by *Chaoborus* (Spitze 1991). On the other hand, selection does not always favor rapid growth into a size refuge. When mortality or other costs are high, the alternatives of early maturation and reproduction (Gårdmark and Dieckmann 2006) or a slow, cryptic lifestyle to avoid detection by predators (Urban 2007) are also possible.

Corophium *Amphipods and Shorebird Predation*

Comparative population studies contribute substantially to our understanding of predation's effect on life history evolution and provide instructive insights into the interplay of predator selectivity, prey growth, and allocation of reproductive effort. In the Bay of Fundy, Canada, some populations of the intertidal amphipod, *C. volutator*, routinely experience episodic, intense, and positively size-selective predation by migrating shorebirds for a few weeks in late summer (Peer et al. 1986), whereas other nearby populations are not visited by birds due to features of local geography (Hilton et al. 2002). In all populations, *C. volutator* have 2 generations per year. An overwintering generation initiates reproduction in spring, and their offspring, which constitute the summer generation, grow, mature, and reproduce during summer. These summer offspring form the next overwintering cohort.

Although their seasonal life cycles are similar, bird-visited and bird-free populations differ substantially in salient features of their life histories (Hilton et al. 2002). Bird-visited sites have at least two major life history adaptations in response to episodic predation. First, the overwintering cohort at bird sites invests heavily in offspring production in the spring, producing about twice as many eggs in the first clutch than do their counterparts in no-bird sites. Egg size does not differ between sites, implying that these larger broods for *Corophium* at bird-visited sites reflect a much greater reproductive effort. Because the migrating birds do not arrive until late summer, this increased reproductive effort of amphipods during spring, well before birds arrive, is presumably a constitutive adaptation of these populations. In the bird-visited populations, the overwintering cohort's large investment in reproductive effort into the first clutch allows their offspring (the new summer cohort) time to grow, mature, and reproduce before the highly predictable arrival of the predatory birds. Reproduction in the summer cohort just before birds arrive is advantageous because their young offspring are small and consequently less susceptible to predation by the positively size-selective birds (Peer et al. 1986). Thus, *Corophium* at bird-visited sites concentrate reproductive effort in the spring clutch because offspring from later clutches are unlikely to survive predation by birds.

The second adaptation in *Corophium* at bird-visited sites is that females of the summer cohort mature appreciably earlier than those at bird-free sites, and this early maturation is likely to facilitate reproduction early in the season, before birds arrive. A consequence of this early maturation in bird-visited populations, however, is that these females have much a smaller body size at maturation, about 60% smaller than females at bird-free sites, and these smaller females produce smaller clutches than females at bird-free sites (Hilton et al. 2002). In contrast to *Corophium* life histories at bird-visited sites, overwintering females at bird-free sites spread out reproductive effort across multiple clutches and survive to produce larger second and third clutches of larger offspring (Hilton et al. 2002).

Alternative Life Histories Among Hyalella *Amphipod Species*

Interspecific comparative studies provide a fruitful avenue for gaining insight into predation's role in shaping prey life histories (Moore et al. 2016). Freshwater amphipods in the genus *Hyalella* are diverse in North America. They include a few described species along with multiple putative species known through genetic and ecological studies but that are not yet formally described (Witt et al. 2006, Wellborn and Broughton 2008, Major et al. 2013). A particularly salient feature of this diversity is that species having broad geographic distributions fall into one of two phenotypic types, or "ecomorphs," that differ in body size and life history traits (Fig. 13.3), and also differ in ecological distribution (Wellborn and Broughton 2008). Small-ecomorph species occupy lakes and other bodies of water containing *Lepomis* sunfish, such as bluegill (*L. macrochirus*), which commonly

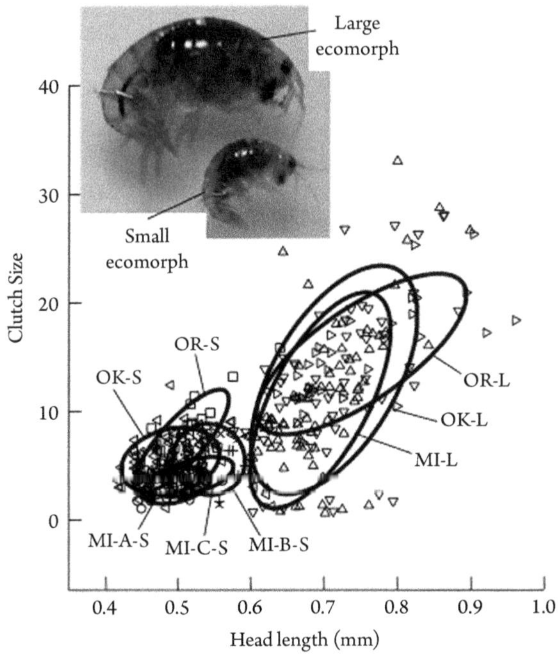

Fig. 13.3.
Regional ecomorphs of *Hyalella* amphipods in North America. Large-ecomorph populations in Oregon (OR-L), Oklahoma (OK-L), and Michigan (MI-L) mature at a larger body size and attain substantially larger adult body size than do small-ecomorph populations in Oregon (OR-S), Oklahoma (OK-S), or Michigan, where three species (MI-A-S, MI-B-S, MI-C-S) cooccur in many lakes. Note that small-ecomorph species initiate reproduction at about 4.3 mm head length, and large-ecomorph species initiate reproduction at about 6.0 mm head length; adult body sizes of large and small ecomorphs are effectively nonoverlapping. Inset photo shows females at size of maturity in large and small ecomorphs from Oklahoma. From Wellborn and Broughton (2008), with permission from John Wiley and Sons.

consume *Hyalella* (Wellborn 1995, Wellborn et al. 2005). Large-ecomorph *Hyalella* species occupy a variety of habitats, including fishless marshes, lakes lacking *Lepomis* fishes, and, in some lakes, very shallow areas that provide a refuge from predation by fish (Wellborn 1994, Wellborn et al. 2005). Phylogenetic analysis indicates multiple origins of *Hyalella* ecomorphs in North America, but their precise evolutionary history is unresolved (Wellborn and Broughton 2008).

Lepomis sunfish are positively size-selective predators on *Hyalella* amphipods, with larger individuals considerably more likely to be consumed than smaller individuals (Wellborn 1994, Wellborn and Cothran 2007). This strong size selective predation by *Lepomis* has likely driven the evolution of small body size in the small-ecomorph species, which are among the smallest in this species-rich genus (Wellborn and Broughton 2008). One species, *Hyalella spinicauda,* which coexists with *Lepomis,* is particularly small in body size, and is significantly smaller than 2 other common small-ecomorph species, *H. wellborni* and *Hyalella* sp., that cooccur with *H. spinicauda* in lakes within in the Great Lakes region of North America (Wellborn and Cothran 2004). In contrast to *H. wellborni* and *Hyalella* sp., the very small *H. spinicauda* is largely invulnerable to fish predation (Wellborn and Cothran 2007), suggesting that this species may essentially achieve a "size refuge" by maturing at a very small size and growing slowly after maturity. Large-ecomorph species do not occur in the same habitat with *Lepomis,* presumably due to intense predation, but they commonly face predation from predatory invertebrates including larval dragonflies and hemipteran water bugs, which are both negatively size selective on *Hyalella,* and can reach high densities in the absence of *Lepomis* (Wellborn 1994). The contrasting patterns of size-selective predation across habitat types experienced by *Hyalella* cause correspondingly contrasting size-specific patterns of overall mortality in their natural habitats (Wellborn 1994). In lakes containing *Lepomis* sunfish, small-ecomorph *Hyalella* species experience increasing risk of mortality as they grow. Conversely, mortality of large-ecomorph *Hyalella* is greatest for small juveniles and then declines as they grow larger.

Observed life histories of large-bodied and small-bodied species in *Hyalella* conform closely to predictions from life history theory. Small-ecomorph species face increased mortality as they grow, and these species initiate reproduction at a small body size (Wellborn and Broughton 2008). Large-ecomorph species, which experience declining mortality risk as they grow, invest in growth while they are small and delay reproduction until they reach a large body size, at which they achieve a size refuge from predation (Wellborn 1994). As a result, adult body mass of large-ecomorph species is about threefold greater than in small-ecomorph species (Wellborn and Broughton 2008). Moreover, evidence suggests that small-ecomorph species have greater per-brood reproductive effort than large-ecomorph species (Wellborn 1994), as anticipated by theories of reproductive allocation (Law 1979). Because small-ecomorph species have lower expected future reproduction due to lower survival, selection in this case favors high investment in current reproductive effort, even at a cost to future reproduction.

EGG SIZE AND BROOD SIZE

Because optimal egg size for a reproducing female depends in part on the relationship between an offspring's size and its probability of surviving to maturity (Smith and Fretwell 1974), predation risk can cause selection on offspring size. More generally, offspring size is a critical life history trait that is likely to be under intense selection in animals due to the inevitable trade-off between the size and number of progeny (Guisande et al. 1996, Fox and Czesak 2000). For example, a selection experiment with *Daphnia pulicaria* populations from ecologically disparate habitat types reveals high heritability for offspring size, suggesting that strong stabilizing selection maintains egg size within a narrow range in each population (Baer and Lynch 2003).

Studies of crustaceans and other taxa suggest a potentially prominent role of predation risk in shaping evolution of offspring size. In some habitats, *D. pulex* face predation by the larval dipteran *Chaoborus* that selectively consumes smaller *Daphnia* individuals, whereas in other habitats the backswimmer *Notonecta* is an abundant predator that selectively consumes larger individuals. When experimentally exposed to chemical cues of each predator, *D. pulex* in the *Chaoborus*-exposure treatment produced larger (and fewer) offspring than predator-free controls, whereas those in the *Notonecta*-exposed treatment produced smaller (and more) offspring (Lüning 1992). These results are consistent with a hypothesis of adaptive plasticity in which larger, but fewer offspring are produced when predation risk primarily affects smaller individuals, while smaller and more numerous offspring are produced when larger individuals are at greater risk. Ituarte et al. (2014) examined plastic phenotypic responses of larval and juvenile stages of the freshwater shrimp *Palaemonetes argentinus* when exposed only as embryos to kairomones of a predatory fish. Larvae of predator-exposed embryos were larger and had longer rostra, characteristics Ituarte et al. (2014) suggest deter predation by small fish. Similar characteristics appear to reduce predation risk in some decapod zoeae (Morgan 1987). Comparable patterns were discovered in 2 recent studies of fish. A split-brood experiment in a cichlid found that females exposed to a predator produced heavier eggs and larger offspring (Segers and Taborsky 2012), and visual predator cues induced larger offspring in poeciliid *Gambusia holbrooki* (Mukherjee et al. 2014).

Freshwater pelagic ecosystems often have pronounced seasonal dynamics, and species' life history traits, including offspring size, can vary seasonally. In Lake Constance, near the European Alps, the native pelagic community includes a large predatory cladoceran, *Bythotrephes longimanus*, which consumes herbivorous cladocerans in the genera *Daphnia* and *Bosmina*. Whitefish prey on *Bythotrephes*, with adult fish selectively consuming larger *Bythotrephes*. In contrast, juvenile whitefish consume smaller individuals, when they grow large enough to consume them at all, due to the limited mouth gape size of juvenile fish. In early summer, when adult fish are the primary predators, *Bythotrephes* initiate reproduction at a small body size, and produce large clutches of small eggs (Straile and Hälbich 2000), life history features that are presumably adaptive under positive size-selective predation. In midsummer, however, a rapid life history shift occurs in which females mature at a larger body size, and produce smaller clutches of larger eggs. Straile and Hälbich (2000) attribute this shift to the increasing predatory impact of abundant juvenile whitefish growing sufficiently large to consume smaller *Bythotrephes*. That is, the life history shift coincides with a transition from positive to negative size-selective predation on *Bythotrephes*. This pattern is at least consistent with adaptive plasticity in *Bythotrephes*' life history across the changing seasonal dynamics of the Lake Constance ecosystem.

LIFE HISTORY EVOLUTION IN DYNAMIC FOOD WEBS

Predator-prey interactions occur in the context of ecological communities, as do the evolutionary dynamics these interactions generate, but much of life history theory implicitly ignores indirect interactions propagated though food web linkages. When these linkages are strong, feedback through the food web can affect a prey species' evolutionary response to predation (Abrams and Rowe 1996, Cressler et al. 2010). For example, Day et al. (2002) modeled the dynamics of a species in which small and large individuals both share the same resource and predator, but the predator disproportionately consumes small individuals such that larger individuals can reach a relative size refuge. Prey in their model incur the common trade-off between allocation to growth versus reproduction. The model found that outcome of life history evolution in prey depends on the strength of linkages across the food web. For example, when predator density is largely independent of prey density in the model, increased predation on small prey causes prey to evolve rapid growth into the size refuge. When predator density is entirely dependent on prey density, however, there is no evolutionary change in prey because prey consumption reduces the predator's food source and therefore the predator's density and its impact on the prey. Although this result is the extreme case, it points to the potential importance of feedback via the food web. A second finding of the model is that resource levels in the environment can alter the evolutionary outcome for prey. Greater productivity of the prey's resource increasingly favors the life history strategy of delaying reproduction in order to grow rapidly into a size refuge, rather than the alternative strategy of foregoing rapid growth and investing in reproduction at a small size.

Although I am not aware of direct tests of these predictions, some research in crustaceans is instructive. Multiple populations of *Daphnia ambigua* are adapted to exploit a transient period of high-quality food in the spring, whereas other populations are adapted to a temporally changing, often low-quality, food resource (Walsh et al. 2014). These distinct population types are associated with differences in seasonal patterns of predation by the planktivorous fish *Alosa pseudoharengus*, which are anadromous in some lakes, but landlocked in others (Post et al. 2008). *Daphnia* in lakes with anadromous *Alosa* have large populations in spring, but decline precipitously in early summer when juvenile fish feed heavily on zooplankton. A laboratory experiment found that *Daphnia* populations from lakes with a dependable high-quality resource develop more rapidly, mature at a larger size, and have greater fecundity than populations from lakes that often experience a low-quality food resource (Walsh et al. 2014). This result supports the conclusion that *Daphnia* that develop and reproduce in high-quality resource environments have evolved life history adaptations that favor growth to a larger maturation size, while those occupying lower quality resource environments have intrinsically lower growth rates and mature at a smaller body size.

SENESCENCE AND LIFE SPAN

Senescence, or aging, refers to the decline in physiological function of an organism with age. Predation mortality is an important driver of evolution of senescence rates and life span in crustaceans and other metazoans because it affects the relative fitness benefits of investment in earlier versus later reproduction (Williams 1957, Dudycha 2001, Ricklefs 2010a). The evolutionary logic is straightforward. In a population subject to high rates of predation, alleles that generate high reproductive effort early in life are favored, even at the cost of reduced investment in physiological maintenance and longevity, because the probability of reproduction later in life is low. This effect is strengthened by a decline in the intensity of selection with age. That is, the strength of natural selection acting on alleles promoting late-life fitness is lower than that on alleles promoting early fitness because fewer older individuals survive to be exposed to selection (Hamilton 1966). These mechanisms, which constitute what is generally termed the *evolutionary theory of aging* (Charlesworth 1993, Kirkwood 2002), are a logical extension of life history theories addressing maturation age and age-specific allocation of reproductive effort. Not surprisingly then, among-species variation in rates of senescence tends to be correlated with maturation age and age-specific reproductive effort. Specifically, high rates of senescence are likely in taxa that mature earlier and have higher reproductive effort (Williams 1957, Stearns 1992). For example, Boonekamp et al. (2014) conducted a lifelong brood manipulation experiment to examine the prediction that an increase in reproductive effort results in reduced longevity because the increased investment in reproduction necessarily entails reduced investment in somatic maintenance and other longevity mechanisms. Birds in a brood-enlarged treatment suffered 3 times greater actuarial senescence (increase in mortality rate with age) than birds in the brood-reduced treatment, a result supporting a strong negative association between high investment in reproduction, especially early in life, and life span. The evolutionary theory of aging focuses on ultimate causes of senescence, but the proximate physiological causes may include many processes, including oxidative stress (Selman et al. 2012), insulin signaling (Hughes and Reynolds 2005), and telomere shortening (Monaghan and Haussmann 2006).

Senescence is difficult to study in wild populations, but available evidence supports predictions of the evolutionary theory of aging and suggests predation is a potentially key cause of variation among taxa in rates of senescence. In a particularly extensive study, evaluation of senescence across 124 taxonomic families of terrestrial vertebrates found later senescence is strongly correlated with later onset of maturity, even when effects of body size were statistically removed (Ricklefs 2010a). Furthermore, the study indicated that greater extrinsic mortality (mortality unrelated to aging, such as predation) is significantly associated with earlier age of senescence. Other studies are also consistent with the evolutionary theory of aging, including experimental studies. For example, a laboratory selection experiment in *Drosophila* manipulated adult mortality rate while holding larval and adult densities constant (Stearns et al. 2000). Experimental lines subject to higher adult mortality, imposed by removal of adults, evolved earlier peak fecundity and higher intrinsic mortality (senescence), leading to shorter life span, as predicted.

Although crustaceans likely offer rich opportunities to build on our understanding of the evolution of senescence, few studies have been conducted. These few, however, point to much promise for future research. Dudycha (2001) tested the association between rates of mortality and rates of senescence in the closely related sister species, *Daphnia pulex* and *D. pulicaria*. These species are regionally sympatric but sort between local habitats, with *D. pulex* occurring in ponds and *D. pulicaria* common in deep lakes, and the study examined six lake and five pond populations. Per-capita mortality rates, measured within natural environments, differed substantially between the habitat types, with *D. pulex* in ponds experiencing approximately sixfold greater daily mortality than *D. pulicaria* in lakes. Rates of senescence were determined for all populations by assessing survivorship and fecundity in standardized predator-free laboratory cultures. Consistent with predictions of the evolutionary theory of aging, senescence occurred much earlier in *D. pulex* (Fig. 13.4). A separate study in this system largely confirmed these differences between species in rates of senescence, and additionally found that *D. pulex* from ponds had higher juvenile growth and greater early-life increase in fecundity than *D. pulicaria* (Dudycha and Tessier 1999), suggesting the greater early investment in reproduction by *D. pulex* is obtained at the cost of reduced longevity.

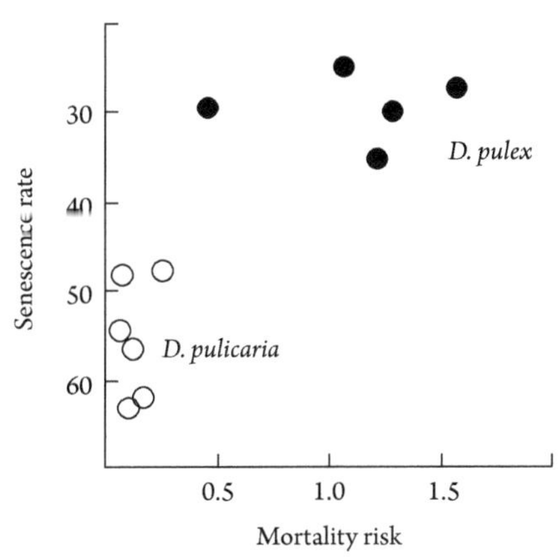

Fig. 13.4.
Comparison of senescence rate between *Daphnia* sister species under different mortality regimes (Dudycha 2001). *Daphnia pulicaria* occurs in lakes, and *D. pulex* occupies pond habitats where they experience high predation mortality. Senescence rate was quantified in predator-free laboratory cultures and quantified as the age (days) at which 75% of lifetime fitness is attained. *Daphnia pulex* senesce more rapidly than *D. pulicaria*, and this result is consistent with the prediction that higher mortality risk selects for earlier and greater reproductive investment and, consequently, earlier senescence. Redrawn from Dudycha (2001).

Given the prevalence of phenotypically plastic, predator-induced life history shifts in prey species within crustaceans, it is reasonable to expect concomitant changes in timing or rate of senescence, and some recent studies support this prediction. Dawidowicz et al. (2010) conducted a laboratory life table experiment to test effects of exposure to fish chemical cues in clonal lineages of *Daphnia hyalina* and *Diaphanosoma brachyurum*, with lineages derived from seven lakes of diverse trophic status. Results were similar in both species. Lineages exposed to fish cues initiated reproduction at an earlier age, and had higher reproductive effort early in life, with *Daphnia*, for example, producing a mean of 6.9 eggs by age 10 days with exposure, versus 3.8 eggs in controls. Consistent with the prediction that greater early investment in reproduction is obtained at the expense of longevity, exposure to chemical cues also lead to an approximately 20% reduction in life span in both species, and the effect was similar across the seven clonal linages. A later study investigated the temporal dynamics of senescence in these species, including age at onset and rate of decline in viability (Pietrzak et al. 2015), parameters that, in general, differ broadly among taxa, and this variation may reflect adaptive differences (Ricklefs 2010b). Pietrzak et al. (2015) quantified the dynamics of senescence in *Daphnia* and *Diaphanosoma* exposed to chemical cues of predatory fish, and found that reduced life span in *Daphnia* was consistent with early onset of senescence and steady mortality, but in *Diaphanosoma* reduced life span resulted from later onset, but accelerated aging. Reasons for these species differences in patterns of senescence are unclear, but this and other studies in crustaceans (see Chapter 7 in this volume) signal much opportunity for examining the mechanisms by which extrinsic mortality drives species variation in longevity.

Evolution of Individual Differences in Risk-Taking Behavior

Models of life history evolution typically ignore among-individual variation in how organisms experience trade-offs, but consideration of this variation leads to interesting consequences, including an adaptive explanation for animal personalities. Individuals within a population may differ in age, size, physiological state, or local environmental conditions, for example, and these differences may affect optimal behavior patterns. Fitness-enhancing activities, such as foraging, often entail a fitness cost such as elevated predation risk, and optimal behavior under these competing demands often depends on individual phenotype. Clark (1994) explored a model in which behavioral options involve both risks and rewards and concluded the optimal behavior for each decision should depend on an individual's residual reproductive value. An individual with comparatively high fat reserves, for example, has greater expected future fecundity than an undernourished individual, and consequently should be relatively refrained in risk-taking behavior. Other models examine ecological circumstances generating positive feedback in which high-condition individuals, for example, benefit from more risky behavior, as is the case when high-condition individuals can better evade predators (Luttbeg and Sih 2010). Among-individual variation in residual reproductive value can produce a polymorphic population in which some individuals consistently exhibit more risky behavior than others (Clark 1994, Wolf et al. 2007, Luttbeg and Sih 2010), and many studies document such individual differences, or "personalities" (Carere and Maestripieri 2013). Moreover, individual differences in behavioral type, such as boldness, may manifest across different functional contexts, such as risk-taking under predation threat and aggressiveness in territorial interactions (Sih et al. 2004a). Correlated personalities across multiple functional contexts, known as "behavioral syndromes" (Sih et al. 2004a), are reported from a broad range of vertebrates and some invertebrates (Sih et al. 2004b, Mather and Logue 2013).

Several studies examine personalities and behavioral syndromes in crustaceans (Gherardi et al. 2012). For example, the fiddler crab *Uca mjoebergi* retreats to burrows when threatened by an aerial predator, but some individuals reemerge quickly while others remain in burrows considerably longer, and this boldness or shyness is consistent over repeated trials with a simulated predator

(Reaney and Backwell 2007). Although these behavioral types do not differ in body size, bold individuals seeking a new burrow engaged in more fights with territorial males, and were more successful in displacing the territory holder than were shy males. Additionally, bold males spent more time actively courting females and garnered more matings. Mowles et al. (2012) experimentally investigated the nature of behavioral syndromes in the hermit crab *Pagurus bernhardus* by examining 3 behavioral contexts: the rate of recovery from "startle response," which is roughly an index of boldness; exploratory behavior when encountering a novel shell; and aggressiveness in contests to procure a shell from another individual. Each of these behaviors was assessed in both the absence and presence of perceived predation risk achieved through a chemical cue. Considered separately, individual exploratory and aggression behaviors were not consistent between perceived safe and risky environments, but boldness was correlated between predation treatments. Considered across contexts, individuals exhibited consistent levels of both boldness and exploratory behavior in the experiment, and did so in both risky and safe treatments, indicating a behavioral syndrome across these behavioral types. The experiment also pointed to correlated behaviors across boldness-aggression contexts, but only in the safe environment. These and the few other studies of behavioral types and syndromes in crustaceans (Gherardi et al. 2012) suggest rich possibilities for taxonomically broader investigations. One unanswered question is the causal processes driving expression of behavioral types and syndromes, including what role predation may play in these behavioral polymorphisms (Ory et al. 2015, Kight et al. 2013).

AUTOTOMY

Autotomy, the voluntary shedding of a body part, is an effective antipredator defense adaptation, but also entails potentially significant costs (Maginnis 2006). Both benefits and accompanying costs of autotomy are often dependent on the environment (Kuo and Irschick 2016), suggesting the natural history of autotomy is best evaluated in a life history context. Autotomy of limb appendages is common among crustaceans and has been documented in numerous taxa, especially within decapods (Juanes and Smith 1995, Fleming et al. 2007). In contrast to forced removal of an appendage by a predator, autotomy is characterized by the controlled self-amputation of a body part at a defined breakage plane, and it occurs for the purpose of defense (Fleming et al. 2007, Hopkins and Das 2015). The wound caused by autotomy at the breakage plane can heal quickly and with minimal trauma (Hopkins 1993). In the blue crab (*Callinectes sapidus*), for example, autotomy of a cheliped precipitates immediate closure of the wound by a membrane. A rapid but mild physiological response, including elevated heart rate and elevated hemolymph flow to the affected area, follows, and these physiological responses subside within 3 hours (McGaw 2006, and see Hopkins and Das 2015). Such adaptions that limit trauma may be essential for effective autotomy because prey often must immediately invoke escape behavior to avoid the predator. In crustaceans, autotomy is largely restricted to walking legs and to chelipeds (anterior appendages carrying a large claw), which are the most commonly autotomized limbs (Shirley and Shirley 1988, Fleming et al. 2007), perhaps because chelipeds are often held outstretched and used to fend off threats. Some crabs utilize "attack autotomy," in which the crab defensively attacks and grasps a predator with its powerful chelipeds before one or both chelipeds are autotomized, leaving the chelipeds attached to the predator as the crab escapes (Robinson et al. 1970). In crustaceans, autotomy is often followed by regeneration of the limb beginning with the subsequent molt, and multiple molts may be required to fully restore the limb to its preautotomized size (Hopkins and Das 2015). This ability to regenerate a lost limb may enhance the likelihood autotomy will evolve in a taxon.

The ecology of autotomy in crustaceans is perhaps best understood in porcelain crabs. These anomuran crabs autotomize chelipeds when attacked by a predator, such as larger brachyuran crabs. In a laboratory experiment, the porcelain crabs *Petrolisthes cinctipes* and *P. manimaculis* were observed in interactions with potential predators, primarily purple shore crabs (*Hemigrapsus nudus*), with which they coexist along the coast of California (Wasson et al. 2002). When a predator attacked a porcelain crab, a cheliped was autotomized in about one-third of the interactions overall, and in two-thirds of interactions when the predator grasped the prey exclusively by a prey's cheliped, a condition that allows the prey to freely retreat from the predator after autotomy (Fig. 13.5). Autotomy, in the experiment, was very effective. Only 1 of 58 prey individuals failed to escape the predator after autotomy, and in most instances the predator did not pursue the porcelain crab but instead consumed the autotomized cheliped (Wasson et al. 2002). Autotomy in porcelain crabs is sometimes also effective against predation by rockfish, which engulf their prey whole. In rockfish stomach contents, about two-thirds of porcelain crabs were represented by their entire body, but one-third were represented only by a cheliped, suggesting autotomy allowed successful avoidance of ingestion in some cases (Knope and Larson 2014). The propensity to autotomize chelipeds depends on characteristics of the predator encountered. Whereas autotomy of chelipeds is common in interactions with a larger predator, in encounters with smaller predators, porcelain crabs instead use their formidable chelipeds to fight off predation attempts (Wasson and Lyon 2005).

Limb autotomy in crustaceans entails a variety of potential fitness costs, many of which can be substantial, and their magnitudes generally depend on environmental factors (Maginnis 2006, Fleming et al. 2007). Although detrimental impacts of the wound created at the moment of autotomy are mitigated by rapidly acting physiological adaptations (McGaw 2006, Hopkins and Das 2015), loss of the autotomized limb has manifold consequences. One immediate cost is that loss of a limb to autotomy prevents that limb from providing an antipredator function in a subsequent attack, at least until the limb is regenerated, and this cost may be especially great when the autotomized limb is a cheliped, which are used to fend off attacks (Figiel and Miller 1995, Wasson and Lyon 2005). There are also costs associated with the mechanical function of the limb in locomotion and feeding (Maginnis 2006), but some taxa can at least partially compensate for a lost limb. For example, blue crabs with one cheliped autotomized were able to consume soft-shell clams at the same rate as that of intact crabs by using walking legs to manipulate and position clams for crushing by their remaining cheliped (Smith and Hines 1991). Nonetheless, crabs that lost both chelipeds, while able to consume clams by using only walking legs, did so at only one-third the rate of intact crabs.

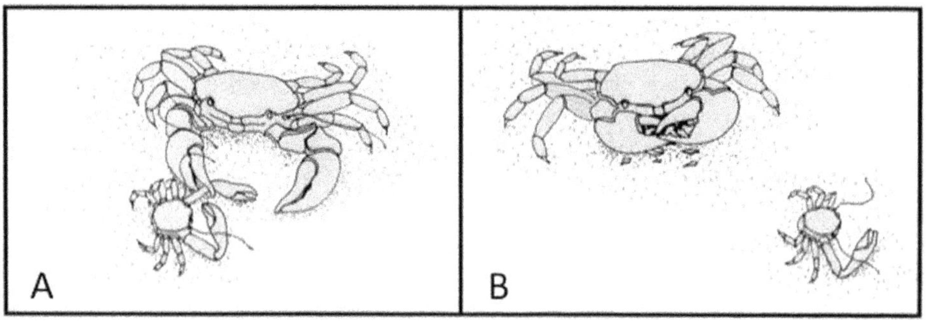

Fig. 13.5.
Limb autotomy in porcelain crabs is an effective defense against predation by larger crabs. (A) Larger predatory crab grasps cheliped of a porcelain crab and triggers autotomy. (B) Larger crab pauses to consume autotomized cheliped, allowing porcelain crab to escape. From Wasson et al. (2002), with permission from Oxford University Press.

Crustaceans with autotomized limbs commonly experience reduced body growth and smaller body size, which may affect multiple fitness components (Juanes and Smith 1995). A field study of tagged spiny lobsters (*Panulirus argus*) investigated the impact of naturally occurring autotomy, and found that juveniles with missing limbs grew at only about 60% of the rate of intact individuals (Davis 1981). Similarly, a laboratory experiment showed cheliped autotomy in the porcelain crab *Petrolisthes laevigatus* caused a substantial reduction in growth rate (Barría and González 2008; Fig. 13.6). Limb autotomy affects growth primarily by reducing the percentage increase in body size at the time of molting, or "molt increment" (Juanes and Smith 1995). Evidence suggests diversion of physiological resources to limb regeneration contributes substantially to the decline in molt increment. A study of the effect of cheliped regeneration following experimentally induced autotomy in juvenile American lobsters (*Homarus americanus*) found the body size increase at molting of individuals regenerating lost chelipeds was about 35% less than intact, control individuals (Cheng and Chang 1993). Molt increments of individuals that were autotomized, but not regenerating (because autotomy occurred after a critical point in the molt cycle), did not differ from intact individuals, suggesting reduced body growth in lobsters was associated with the process of regeneration, and not autotomy itself.

Smaller body size of individuals that experience autotomy can also cause reduced reproduction and survival. Because female fecundity generally increases with body size within populations in Crustacea (Sainte-Marie 1991), smaller adult body size due to autotomy may decrease female reproductive success (Juanes and Smith 1995, Fleming et al. 2007), and in males, greater body (Aquiloni and Gherardi 2008) and chela size (Mariappan et al. 2000) are associated with mating success (see also Sekkelsten 1988, Abello et al. 1994). In an experimental study, crayfish (*Procambarus clarkii*) with induced autotomy of chelae survived at a rate only 75% of that for intact individuals, but this reduced survival was not found in low-density treatments, as may be expected if antagonistic interactions decline with density (Figiel and Miller 1995).

A life history perspective on autotomy would integrate benefits and costs of autotomy across relevant ecological circumstances to generate testable predictions. Kuo and Irschick (2016) used this approach to understand among-population variation in propensity for tail autotomy in a lizard. They reasoned that individuals within populations that experience greater predation risk,

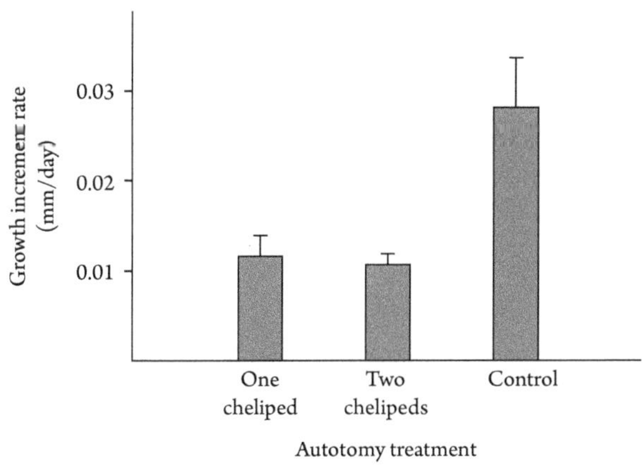

Fig. 13.6.
Effect of cheliped autotomy on subsequent growth in the porcelain crab *Petrolisthes laevigatus*. Experimental autotomy of one or both chelipeds caused substantial and statistically significant reduction in growth compared to intact control individuals. Increment rate is the increase in body size at molting divided by the intermolt period. Redrawn from data in Barría and González (2008).

for example, should more readily autotomize their tail when compared to individuals from low-risk populations. This expectation emerges from the substantial costs of autotomy. If there are sometimes instances in which an individual autotomizes a limb, but would have survived without autotomy, these individuals acquire the autotomy's costs, but without its benefits. We expect natural selection to adjust propensity to autotomize a body part to be relatively higher, for example, in populations with high predator density, with predators that are more dangerous, or when costs incurred by losing the body part are relatively low, as may be the case when food resources are abundant, making subsequent regeneration of the limb less costly. When evaluated among their 5 study populations, Kuo and Irschick (2016) found strong empirical support for predictions of a model that integrated multiple environmental factors, with propensity to autotomize increasing with both predator density and food abundance, and decreasing with intensity of male-male aggression, which causes unbeneficial autotomy. As Kuo and Irschick (2016) point out, their general approach can be readily extended to crustacean systems. Such studies hold much promise for a more comprehensive understanding of the evolutionary ecology of autotomy.

CONCLUSIONS AND FUTURE DIRECTIONS

Predation shapes crustacean life history phenotypes in manifold ways, including age and size at maturation, body size of offspring and adults, brood size, and life span. Studies in crustaceans supply much of our empirical understanding of ecological, genetic, and evolutionary mechanisms by which predators affect life histories of their prey. Future research will likely integrate the molecular physiology and genetic architecture of trade-offs with the ecological interactions that generate life history evolution in natural populations. Such integrative studies can illuminate causes and constraints in evolution and bring a deeper understanding of the diversity of phenotypes observed in nature. We will also gain greater understanding of crustacean life histories through empirical studies that integrate the predator-prey interaction with interactions across the broader food web and ecosystem, as theoretical models make clear that characteristics of the food web and ecosystem may substantially affect the course of phenotypic evolution in a prey population. Finally, it will be important to consider prey life history evolution in the context of the competing demands faced by prey individuals, such as mating, foraging, and other activities that may impose selection on life history traits in ways that moderate or reinforce the direction and magnitude of selection imposed by predators. Crustacean research is contributing to our growing knowledge in each of these nascent approaches. Undoubtedly, crustacean systems will continue to provide fertile opportunities for uncovering mechanisms of life history evolution across levels of biological organization from ecology to physiology to genomics.

REFERENCES

Abello, P., C.G. Warman, D.G. Reid, and E. Naylor. 1994. Chela loss in the shore crab *Carcinus maenas* (Crustacea: Brachyura) and its effect on mating success. Marine Biology 121:247–252.

Abrams, P.A., and L. Rowe. 1996. The effects of predation on the age and size of maturity of prey. Evolution 50:1052–1061.

Aquiloni, L., and F. Gherardi. 2008. Mutual mate choice in crayfish: large body size is selected by both sexes, virginity by males only. Journal of Zoology 274:171–179.

Atkinson, D., and A.G. Hirst. 2007. Life histories and body size. Pages 33–54 *in* A. Hildrew, D. Raffaelli, and R. Edmonds-Brown, editors. Body size: the structure and function of aquatic ecosystems. Cambridge University Press, Cambridge, England.

Baer, C.F., and M. Lynch. 2003. Correlated evolution of life-history with size at maturity in *Daphnia pulicaria*: patterns within and between populations. Genetics Research, Cambridge 81:123–132.

Baeza, J.A. 2006. Testing three models on the adaptive significance of protandric simultaneous hermaphroditism in marine shrimp. Evolution 60:1840–1850.

Ball, S., and R.L. Baker. 1996. Predator-induced life history changes: antipredator behavior costs or facultative life history shifts? Ecology 77:1116–1124.

Barnhisel, D.R. 1991. The caudal appendage of the cladoceran *Bythotrephes cederstroemi* as a defense against young fish. Journal of Plankton Research 13:529–537.

Barría, E.M., and M.I. González. 2008. Effect of autotomy and regeneration of the chelipeds on growth and development in *Petrolisthes laevigatus* (Guérin, 1835) (Decapoda, Anomura, Porcellanidae). Crustaceana 81:641–652.

Beckerman, A.P., G.M. Rodgers, and S.R. Dennis. 2010. The reaction norm of size at maturity under multiple predator risk. Journal of Animal Ecology 79:1069–1076.

Beckerman, A.P., K. Wieski, and D.J. Baird. 2007. Behavioral versus physiological mediation of life history under predation risk. Oecologia 152:335–343.

Boonekamp, J.J., M. Salmons, S. Bouwhuis, C. Dijkstra, and S. Verhulst. 2014. Reproductive effort accelerates actuarial senescence in wild birds: an experimental study. Ecology Letters 17:599–605.

Brown, J.H., A.P. Allen, and J.F. Gillooly. 2007. The metabolic theory of ecology and the role of body size in marine and freshwater ecosystems. Pages 1–15 *in* A. Hildrew, D. Raffaelli, and R. Edmonds-Brown, editors. Body size: the structure and function of aquatic ecosystems. Cambridge University Press, Cambridge, England.

Campbell, C.E. 1991. Prey selectivities of threespine sticklebacks (*Gasterosteus aculeatus*) and phantom midge larvae (*Chaoborus* spp.) in Newfoundland lakes. Freshwater Biology 25:155–167.

Carere, C., and D. Maestripieri. 2013. Animal personalities. University of Chicago Press, Chicago, Illinois.

Charlesworth, B. 1980. Evolution in age-structured populations. Cambridge University Press, Cambridge, England.

Charlesworth, B. 1993. Evolutionary mechanisms of senescence. Genetica 91:11–19.

Cheng, J-H., and E.S. Chang. 1993. Determinants of postmolt size in the American lobster (*Homarus americanus*). I. D_1^3 is the critical stage. Canadian Journal of Fisheries and Aquatic Sciences 50:2106–2111.

Clark, C.W. 1994. Antipredator behavior and the asset-protection principle. Behavioral Ecology 5:159–170.

Corey, S., and D.M. Reid. 1991. Comparative fecundity of decapod crustaceans I. The fecundity of thirty-three species of nine families of caridean shrimp. Crustaceana 60:270–294.

Cox, R.M., M.B. Lovern, and R. Calsbeek. 2014. Experimentally decoupling reproductive investment from energy storage to test the functional basis of a life-history trade-off. Journal of Animal Ecology 83:888–898.

Cox, R.M., E.U. Parker, D.M. Chaney, A.L. Liebl, L.B. Martin, and R. Calsbeek. 2010. Experimental evidence for physiological costs underlying the trade-off between reproduction and survival. Functional Ecology 24:1262–1269.

Cressler, C.E., A.A. King, and E.E. Werner. 2010. Interactions between behavioral and life-history trade-offs in the evolution of predator-defense plasticity. The American Naturalist 176:276–288.

Davis, G.E. 1981. Effects of injuries on spiny lobster, *Panulirus argus*, and implications for fisheries management. Fishery Bulletin 78:979–984.

Dawidowicz, P., P. Predki, and B. Pietrzak. 2010. Shortened lifespan: another cost of fish-predator avoidance in cladocerans? Hydrobiologia 643:27–32.

Day, T., P.A. Abrams, and J.M. Chase. 2002. The role of size-specific predation in the evolution and diversification of life histories. Evolution 56:877–887.

Declerck, S., and A. Weber. 2003. Genetic differentiation in life history between *Daphnia galeata* populations: an adaptation to local predation regimes? Journal of Plankton Research 25:93–102.

Dmitriew, C.M. 2011. The evolution of growth trajectories: what limits growth rate? Biological Reviews of the Cambridge Philosophical Society 86:97–116.

Dudycha, J.L. 2001. The senescence of *Daphnia* from risky and safe habitats. Ecology Letters 4:102–105.

Dudycha, J.L., and A.J. Tessier. 1999. Natural genetic variation of life span, reproduction, lifespan, and juvenile growth in *Daphnia*. Evolution 53:1744–1756.

Engelmayer, A. 1995. Effects of predator-released chemicals on some life history parameters of *Daphnia pulex*. Hydrobiologia 307:203–206.

Figiel, C.R., and G.L. Miller. 1995. The frequency of chela autotomy and its influence on the growth and survival of the crayfish *Procambarus clarkii* (Girard, 1852) (Decapoda, Cambaridae). Crustaceana 68:472–483.

Fisk, D.L., L.C. Latta, IV, R.A. Knapp, and M.E. Pfrender. 2007. Rapid evolution in response to introduced predators I: rates and patterns of morphological and life-history trait divergence. BMC Evolutionary Biology 7:22.

Fleming, P.A., D. Muller, and P.W. Bateman. 2007. Leave it all behind: a taxonomic perspective of autotomy in invertebrates. Biological Reviews 82:481–510.

Forsman, A. 2015. Rethinking phenotypic plasticity and its consequences for individuals, populations and species. Heredity 115:276–284.

Fox, C.W., and M.E. Czesak. 2000. Evolutionary ecology of progeny size in arthropods. Annual Review of Entomology 45:341–369.

Gadgil, M., and W.H. Bossert. 1970. Life historical consequences of natural selection. The American Naturalist 104:1–24.

Gårdmark, A., and U. Dieckmann. 2006. Disparate maturation adaptations to size-dependent mortality. Proceedings of the Royal Society B: Biological Sciences 273:2185–2192.

Gélinas, M., B. Bernadette-Alloul, and M. Slusarczyk. 2007. Alternative antipredator responses of two coexisting *Daphnia* species to negative size selection by YOY perch. Journal of Plankton Research 29:775–789.

Gherardi, F., L. Aquiloni, and E. Tricarico. 2012. Behavioral plasticity, behavioral syndromes, and animal personality in crustacean decapods: an imperfect map is better than no map. Current Zoology 58:567–579.

Guisande, C., J. Sánchez, I. Maneiro, and A. Miranda. 1996. Trade-off between offspring number and offspring size in the marine copepod *Euterpina acutifrons* at different food concentrations. Marine Ecology Progress Series 143:37–44.

Hamilton, W.D. 1966. The moulding of senescence by natural selection. Journal of Theoretical Biology 12:12–45.

Harshman, L.G., and A.J. Zera. 2006. The costs of reproduction: the devil is in the details. Trends in Ecology and Evolution 22:80–86.

Hart, R.C., and E.A. Bychek. 2011. Body size in freshwater planktonic crustaceans: an overview of extrinsic determinants and modifying influences of biotic interactions. Hydrobiologia 668:61–108.

Hilton, C., S.J. Walde, and M.L. Leonard. 2002. Intense episodic predation by shorebirds may influence life history strategy of an intertidal amphipod. Oikos 99:368–376.

Hopkins, P.M. 1993. Regeneration of walking legs in the fiddler crab *Uca pugilator*. American Zoologist 33:348–356.

Hopkins, P.M., and S. Das. 2015. Regeneration in crustaceans. Pages 168–198 *in* E.S. Chang, and M. Thiel, editors. The natural history of the Crustacea, volume 4: Physiology. Oxford University Press, New York.

Hughes, K.A., and R.M. Reynolds. 2005. Evolutionary and mechanistic theories of aging. Annual Review of Entomology 50:421–425.

Ituarte, R.B., M.G. Vázquez, M.A. González-Sagrario, and E.D. Spivak. 2014. Carryover effects of predation risk on postembryonic life-history stages in a freshwater shrimp. Zoology 117:139–145.

Juanes, F., and L.D. Smith. 1995. The ecological consequences of limb damage and loss in decapod crustaceans: a review and prospectus. Journal of Experimental Marine Biology and Ecology 193:197–223.

Kight, C.R., M. David, and S.R.X. Dall. 2013. The evolution of animal personality variation. eLS. John Wiley & Sons, Ltd, Chichester, England.

Kirkwood, T.B.L. 2002. Evolution of ageing. Mechanisms of Ageing and Development 123:737–745.

Knope, M.L., and R.J. Larson. 2014. Autotomy in porcelain crabs in an effective escape mechanism from rockfish predation. Marine Ecology 35:471–477.

Kozlowski, J., and R.G. Wiegert. 1986. Optimal allocation of energy to growth and reproduction. Theoretical Population Biology 29:16–37.

Kozlowski, J., and R.G. Wiegert. 1987. Optimal age and size at maturity in annuals and perennials with determinant growth. Evolutionary Ecology 1:231–244.
Kuo, C.-Y., and D.J. Irschick. 2016. Ecology drives natural variation in an extreme antipredator trait: a cost-benefit analysis integrating modeling and field data. Functional Ecology 30:953–963.
Lass, S., and P. Spaak. 2003. Chemically induced anti-predator defences in plankton: a review. Hydrobiologia 491:221–239.
Law, R. 1979. Optimal life histories under age-specific predation. The American Naturalist 114:399–417.
León, J.A. 1976. Life histories and adaptive strategies. Journal of Theoretical Biology 60:301–335.
Luttbeg, B., and A. Sih. 2010. Risk, resources and state-dependent adaptive behavioral syndromes. Philosophical Transactions of the Royal Society B: Biological Sciences 365:3977–3990.
Lüning, J. 1992. Phenotypic plasticity of *Daphnia pulex* in the presence of invertebrate predators: morphological and life history responses. Oecologia 92:383–390.
Maginnis, T.A. 2006. The costs of autotomy and regeneration in animals: a review and framework for future research. Behavioral Ecology 17: 857–872.
Major, K., D.J. Soucek, R. Giordano, M.J. Wetzel, and F. Soto-Adams. 2013. The common ecotoxicology laboratory strain of *Hyalella azteca* is genetically distinct from most wild strains sampled in eastern North America. Environmental Toxicology and Chemistry 32:2637–2647.
Mariappan, P., C. Balasundaram, and B. Schmitz. 2000. Decapod crustacean chelipeds: an overview. Journal of Biosciences 25:301–313.
Mather, J.A., and D.M. Logue. 2013. The bold and the spineless: invertebrate personalities. Pages 13–35 *in* C. Carere, and D. Maestripieri, editors. Animal personalities. University of Chicago Press, Chicago, Illinois.
McGaw, I.J. 2006. Cardiovascular and respiratory responses associated with limb autotomy in the blue crab, *Callinectes sapidus*. Marine and Freshwater Behaviour and Physiology 39:131–141.
Michod, R.E. 1979. Evolution of life histories in response to age-specific mortality factors. The American Naturalist 113:531–550.
Monaghan, P., and M.F. Haussmann. 2006. Do telomere dynamics link life- style and lifespan? Trends in Ecology and Evolution 21:47–53.
Moore, M.P., R. Riesch, and R.A. Martin. 2016. The predictability and magnitude of life-history divergence to ecological agents of selection: a meta-analysis in livebearing fishes. Ecology Letters 19:435–442.
Morgan, S.G. 1987. Morphological and behavioral antipredatory adaptations of decapod zoeae. Oecologia 73:393–400.
Mowles, S.L., P.A. Cotton, and M. Briffa. 2012. Consistent crustaceans: the identification of stable behavioural syndromes in hermit crabs. Behavioral Ecology and Sociobiology 66:1087–1094.
Mukherjee, S., M.H. Heithaus, J.C. Trexler, J. Ray-Mukhrjee, and J. Vaudo. 2014. Perceived risk of predation affects reproductive life-history traits in *Gambusia holbrooki*, but not in *Heterandria formosa*. PLoS ONE 9:e88832.
Newman, R.M., and T.F. Waters. 1984. Size-selective predation on *Gammarus pseudolimnaeus* by trout and sculpins. Ecology 65:1535–1545.
Noonburg, E.G., and R.M. Nisbet. 2005. Behavioural and physiological responses to food availability and predation risk. Evolutionary Ecology Research 7:89–104.
O'Brian, W.J., N.A. Slade, and G.L. Vinyard. 1976. Apparent size and the determinant of prey selection by bluegill sunfish (*Lepomis macrochirus*). Ecology 57:1304–1310.
Ory, N.C., D. Dudgeon, C.P. Dumont, L. Miranda, and M. Thiel. 2012. Effects of predation and habitat structure on the abundance and population structure of the rock shrimp *Rhynchocinetes typus* (Caridea) on temperate rocky reefs. Marine Biology 159:2075–2089.
Ory, N.C., T.C. van Son, and M. Thiel. 2015. Mating rock shrimp hedge their bets: old males take greater risk, but only after careful assessment of the investment scenario. Behavioral Ecology and Sociobiology 69:1975–1984.
Pastorok, R.A. 1981. Prey vulnerability and size selection by *Chaoborus* larvae. Ecology 62:1311–1324.
Pauwels, K., R. Stoks, and L. De Meester. 2010. Enhanced anti-predator defence in the presence of food stress in the water flea *Daphnia magna*. Functional Ecology 24:322–329.

Peer, D.L., L.E. Linkletter, and P.W. Hicklin. 1986. Life history and reproductive biology of *Corophium volutator* (Crustacea: Amphipoda) and the influence of shorebird predation on population structure in the Chignecto Bay, Bay of Fundy, Canada. Netherlands Journal of Sea Research 20:359–373.
Pietrzak, B., P. Dawidowicz, P. Predki, and M.J. Danco. 2015. How perceived predation risk shapes patterns of aging in water fleas. Experimental Gerontology 69:1–8.
Post, D.M., E.P. Palkovacs, E.G. Schielke, and S.I. Dodson. 2008. Intraspecific variation in a predator affects community structure and cascading trophic interactions. Ecology 89:2019–2032.
Reaney, L.T, and P.R.Y. Backwell. 2007. Risk-taking behavior predicts aggression and mating success in a fiddler crab. Behavioral Ecology 18:521–525.
Reznick, D., L. Nunney, and A. Tessier. 2000. Big houses, big cars, superfleas, and the costs of reproduction. Trends in Ecology and Evolution 15:421–425.
Ricklefs, R.E. 2010a. Life-history connections to rates of aging in terrestrial vertebrates. Proceedings of the National Academy of Science of the USA 107:10314–10319.
Ricklefs, R.E. 2010b. Insights from comparative analysis of aging in birds and mammals. Aging Cell 9:273–284.
Riessen, H.P. 1999. Predator-induced life history shifts in *Daphnia*: a synthesis of studies using meta-analysis. Canadian Journal of Fisheries and Aquatic Sciences 56:2487–2494.
Rinke, K., S. Hülsmann, and W.M. Mooij. 2008. Energetic costs, underlying resource allocation patterns, and adaptive value of predator-induced life-history shifts. Oikos 117:273–285.
Robinson, M.H., L.G. Abele, and B. Robinson. 1970. Attack autotomy: a defense against predators. Science 167:300–301.
Roff, D.A. 1992. The evolution of life histories. Chapman & Hall, New York.
Roff, D.A. 2001. Age and size at maturity. Pages 99–112 *in* C.W. Fox, D.A. Roff, and D.J. Fairbairn, editors. Evolutionary ecology concepts and case studies. Oxford University Press, New York.
Roff, D.A. 2002. Life history evolution. Sinauer Associates, Boston, Massachusetts.
Sainte-Marie, B. 1991. A review of the reproductive bionomics of aquatic gammaridean amphipods: variation of life history traits with latitude, depth, salinity, and superfamily. Hydrobiologia 223:187–227.
Sakwinska, O. 2002. Response to fish kairomone in *Daphnia galeata* life history traits relies on shift to earlier instar at maturation. Oecologia 131:409–417.
Sarma, S.S.S., S. Nandini, and R.D. Gulati. 2002. Cost of reproduction in selected species of zooplankton (rotifers and cladocerans). Hydrobiologia 481:89–99.
Sato, T., and N. Suzuki. 2010. Female size as a determinant of larval size, weight, and survival period in the coconut crab, *Birgus latro*. Journal of Crustacean Biology 30:624–628.
Schram, F.R. 1986. Crustacea. Oxford University Press, Oxford, England.
Scott, M.A., and W.M. Murdoch. 1983. Selective predation by the backswimmer, *Notonecta*. Limnology and Oceanography 28:352–366.
Segers, F.H.I.D., and B. Taborsky. 2012. Juvenile exposure to predator cues induces a larger egg size in fish. Proceedings of the Royal Society B: Biological Sciences 279:1241–1248.
Sekkelsten, G.I. 1988. Effect of handicap on mating success in male shore crabs *Carcinus maenas*. Oikos 51:131–134.
Selman, C., J.D. Blount, D.H. Nussey, and J.R. Speakman. 2012. Oxidative damage, aging, and life-history evolution: where now? Trends in Ecology and Evolution 27:570–577.
Shirley, S.M., and T.C. Shirley. 1988. Appendage injury in Dungeness crabs, *Cancer magister*, in southeastern Alaska. Fishery Bulletin 86:156–160.
Sih, A., A.M. Bell, and J.C. Johnson. 2004a. Behavioral syndromes: an ecological and evolutionary overview. Trends in Ecology and Evolution 19:372–378.
Sih, A., A.M. Bell, J.C. Johnson, and R.E. Ziemba. 2004b. Behavioral syndromes: an integrative overview. Quarterly Review of Biology 79:241–277.
Smith, C.C., and S.D. Fretwell. 1974. The optimal balance between size and number of offspring. The American Naturalist 108:499–506.
Smith, L.D., and A.H. Hines. 1991. The effect of cheliped loss on blue crab *Callinectes sapidus* Rathbun foraging rate on soft-shell calms *Mya arenaria* L. Journal of Experimental Marine Biology and Ecology 151:245–256.

Spaak, P., J. Vanoverbeke, and M. Boersma. 2000. Predator-induced life-history changes and the coexistence of five taxa in the *Daphnia* species complex. Oikos 89:164–174.
Sparkes, T.C. 1996. Effects of predation risk on population variation in adult size in a stream-dwelling isopod. Oecologia 106:85–92.
Spitze, K. 1991. *Chaoborus* predation and life history evolution in *Daphnia pulex*: temporal pattern of population diversity, fitness, and mean life history. Evolution 45: 82–92.
Stearns, S.C. 1992. The evolution of life histories. Oxford University Press, Oxford, England.
Stearns, S.C., M. Ackermann, M. Doebeli, and M. Kaiser. 2000. Experimental evolution of aging, growth, and reproduction in fruitflies. Proceedings of the National Academy of Science of the USA 97:3309–3313.
Stibor, H. 2002. The role of yolk protein dynamics and predator kairomones for the life history of *Daphnia magna*. Ecology 83:362–369.
Stibor, H., and J. Lüning. 1994. Predator-induced phenotypic variation in the pattern of growth and reproduction in *Daphnia hyalina* (Crustacea: Cladocera). Functional Ecology 8:97–101.
Stibor, H., and J. Machacek. 1998. The influence of fish-exuded chemical signals on the carbon budget of *Daphnia*. Limnology and Oceanography 43:997–1000.
Stoks, R., L. Govaert, K. Pauwels, B. Jansen, and L. De Meester. 2016. Resurrecting complexity: the interplay of plasticity and rapid evolution in the multiple trait response to strong changes in predation pressure in the water flea *Daphnia magna*. Ecology Letters 19:180–190.
Straile, D., and A. Hälbich. 2000. Life history and multiple antipredator defenses of an invertebrate pelagic predator, *Bythotrephes longimanus*. Ecology 81:150–163.
Swift, M.C., and A.Y. Fedorenko. 1975. Some aspects of prey capture by *Chaoborus* larvae. Limnology and Oceanography 20:418–425.
Tollrian, R. 1995. Predator-induced morphological defenses: costs, life history shits, and maternal effects in *Daphnia pulex*. Ecology 75:1691–2017.
Taylor, B.E., and W. Gabriel. 1992. To grow or not to grow: optimal resource allocation for *Daphnia*. The American Naturalist 139:248–266.
Taylor, B.E., and W. Gabriel. 1993. Optimal adult growth of *Daphnia* in a seasonal environment. Functional Ecology 7:513–521.
Urban, M. 2007. The growth-predation risk trade-off under a growing gape-limited predation threat. Ecology 88:2587–2597.
van Noordwijk, A.J., and G. de Jong. 1986. Acquisition and allocation of resources: their influence on variation in life history tactics. The American Naturalist 128:137–142.
Walsh, M.R., K.J. La Pierre, and D.M. Post. 2014. Phytoplankton composition modifies predator-driven life history evolution in *Daphnia*. Evolutionary Ecology 28:397–411.
Wasson, K., and B.E. Lyon. 2005. Flight or fight: flexible antipredator strategies in porcelain crabs. Behavioral Ecology 16:1037–1041.
Wasson, K., B.E. Lyon, and M. Knope. 2002. Hair-trigger autotomy in porcelain crabs is a highly effective escape strategy. Behavioral Ecology 13:481–486.
Wellborn, G.A. 1994. Size-biased predation and prey life histories: a comparative study of freshwater amphipod populations. Ecology 75:2104–2117.
Wellborn, G.A. 1995. Predator community composition and patterns of variation in life history and morphology among *Hyalella* (Amphipoda) populations in southeast Michigan. American Midland Naturalist 133:322–332.
Wellborn, G.A., and R.E. Broughton. 2008. Diversification on an ecologically constrained adaptive landscape. Molecular Ecology 17:2927–2936.
Wellborn, G.A., and R.D. Cothran. 2004. Phenotypic similarity and differentiation among sympatric cryptic species in a freshwater amphipod species complex. Freshwater Biology 49:1–13.
Wellborn, G.A., and R.D. Cothran. 2007. Niche diversity in crustacean cryptic species: complementarity in spatial distribution and predation risk. Oecologia 154:175–183.
Wellborn, G.A., R. Cothran, and S. Bartholf. 2005. Life history and allozyme diversification in regional ecomorphs of the *Hyalella azteca* (Crustacea: Amphipoda) species complex. Biological Journal of the Linnean Society 84:161–175.

Werner, E.E., and J.F. Gilliam. 1984. The ontogenetic niche and species interactions in size-structured populations. Annual Reviews of Ecology and Systematics 15:393–425.
Werner, E.E., and D.J. Hall. 1974. Optimal foraging and the size selection of prey by bluegill sunfish (Lepomis macrochirus). Ecology 55:1042–1052.
White, E.P., S.K.M. Ernest, A.J. Kerkhoff, and B.J. Enquist. 2007. Relationships between body size and abundance in ecology. Trends in Ecology and Evolution 22:323–330.
Williams, G.C. 1957. Pleiotropy, natural selection, and the evolution of senescence. Evolution 11:398–411.
Williams, G.C. 1966. Natural selection, the costs of reproduction, and a refinement of Lack's principle. The American Naturalist 100:687–690.
Wirtz, K.W. 2012. Who is eating whom? Morphology and feeding type determine the size relation between planktonic predators and their ideal prey. Marine Ecology Progress Series 445:1–12.
Witt, J.D.S., D.L. Threloff, and P.D.N. Hebert. 2006. DNA barcoding reveals extraordinary cryptic diversity in an amphipod genus: implications for desert spring conservation. Molecular Ecology 15:3073–3082.
Wolf, M., G.S. van Doorn, O. Leimar, and F.J. Weissing. 2007. Life-history trade-offs favour the evolution of animal personalities. Nature 447:581–584.
Zaret, T.M. 1980. Predation and freshwater communities. Yale University Press, New Haven, Connecticut.
Zhao, X., M.G. Fox, and D.C. Lasenby. 2006. Effect of prey density, prey mobility and habitat structure on size selection and consumption of amphipods by a benthic feeding fish. Archiv für Hydrobiologie 165:269–288.

14

CANNIBALISM IN CRUSTACEANS

Bronwyn Bleakley

Abstract

Cannibalism is widespread in animal taxa, but perhaps nowhere more so than in crustaceans. It receives considerable research attention because it structures populations, influences the management of economically important species, and affects species of conservation concern on both the endangered and invasive ends of the spectrum. Crustaceans are particularly likely to engage in cannibalism because they molt. Molting is an energetically expensive process, and cannibalism may provide critical nutrition, even in typically herbivorous species. In addition, molting is a dangerous process that puts individuals at risk, and molting creates distinct size classes, with smaller individuals experiencing greater risk of depredation. The occurrence of cannibalism in crustaceans is influenced by many environmental factors, including habitat type and complexity and the availability of refugia, the availability of alternative prey, and the size structure of populations. In addition, the occurrence of cannibalism may be influenced by genetics and kin selection. While all these factors have been found to influence cannibalism in a range of crustacean species, there is significant variation within and among species in the likelihood of cannibalism and which factors are most influential in any given group. Despite much research on the proximate causes of cannibalism in crustaceans, many avenues of research remain, including the genetics of cannibalism and the degree to which kin selection might influence the evolution of crustacean cannibalism.

INTRODUCTION

Cannibalism, the killing and ingestion of a conspecific, has long been a focus of evolutionary, ecological, and applied studies of behavior (Fig. 14.1; Fox 1975, Polis 1981, Elgar and Crespi 1992). Cannibalism attracts such attention because it is taxonomically widespread, greatly affects population dynamics, including the success of invasive species, and influences the management of economically important

Life Histories. Edited by Gary A. Wellborn and Martin Thiel.

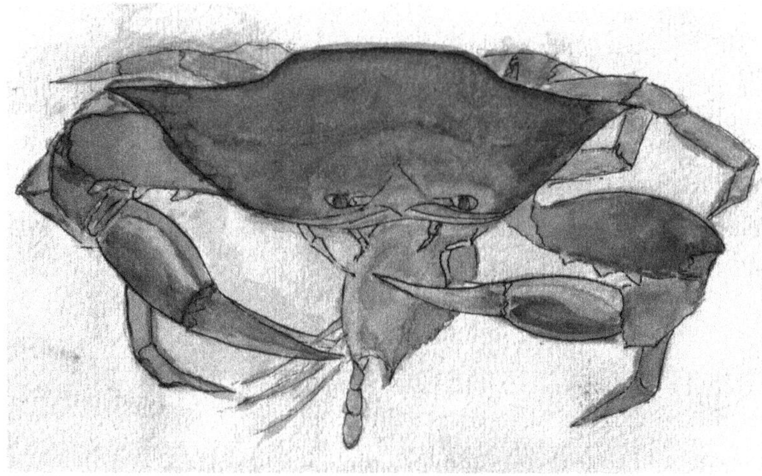

Fig. 14.1.
Intraspecific predation. An adult blue crab, *Callinectes sapidus*, kills and consumes a juvenile blue crab. Based on Smithsonian Environmental Research Center 2014. Illustrated by Bronwyn H. Bleakley ©. See color version of this figure in the centerfold.

species and species of conservation concern. Cannibalism may vary dramatically among closely related species and influences both intra- and interspecies competition and predation (e.g., Dick et al. 1999). It affects many commercially important species, especially among crustaceans (Lovrich and Sainte-Marie 1997, Fernández 1999, Zmora et al. 2005, Beal 2012, Daly and Swingle 2013), and yet it can be difficult to reduce in commercial and managed populations (Jormalainen and Shuster 1997, Borisov et al. 2007). Microhabitat distribution and community composition and organization are influenced by cannibalism (Dick et al. 1995, Dick and Platvoet 2000, Kuroda et al. 2005, Amaral et al. 2009, Casariego et al. 2011). Conversely, the frequency of cannibalism may depend on population structure and habitat complexity because the particular social partners (potential predators or potential prey) an individual interacts with are a property of the population structure (Jormalainen and Shuster 1997, McGrath et al. 2007, MacNeil et al. 2008, Bleakley et al. 2013).

Whether cannibalism regularly occurs in a species ultimately reflects the specific balance between costs and benefits of engaging in cannibalism. Three general approaches have been used to describe the expression and evolution of cannibalism. First, evolutionary studies of cannibalism typically focus on describing the costs incurred or benefits obtained by cannibalistic individuals. Direct fitness consequences for cannibals within populations accrue from nutritional benefits, reduced competition, injury or death sustained during the cannibalistic interaction, and pathogen transmission (Meffe and Crump 1987, Dong and DeAngelis 1998, Pfennig et al. 1998, Wagner et al. 1999, Williams and Hernandez 2006, van Huis et al. 2008). Second, explorations of the specific proximate cues that trigger cannibalism in individuals (e.g., stress or intra- vs. interfamilial interactions; Richardson et al. 2010) are used to predict when cannibalism will occur and to reduce economic and conservation losses in managed populations (Daly et al. 2012a, Daly and Swingle 2013, Rosewarne et al. 2013). Third, ecological studies of cannibalism focus on the interplay between cannibalism and population dynamics, particularly the role of cannibalism in shaping intraspecific competition (reviewed in Dick et al. 1999, Claessen et al. 2004, MacNeil et al. 2004, Platvoet et al. 2009).

Most crustaceans, even those that are typically herbivorous, exhibit predatory behavior (reviewed in MacNeil et al. 1997, Kneib et al. 1999). The propensity to engage in at least some predation, including cannibalism, is likely driven by supplementary dietary needs for nitrogen, vitamins, and other nutrients that are required for growth and reproduction (Wolcott and O'Connor 1992, Kennish 1996, Linton and Greenaway 2007). Conspecifics are a high-quality food source that provides the correct nutrients in the right ratios (reviewed in Parsons et al. 2013). Prey consumption may provide the greatest energetic payoff, and cannibalism is often the most successful strategy when there is a low abundance of other prey (Linton and Greenaway 2007, Griffen et al. 2015). Crustaceans exhibit a number of traits that are associated with the expression of cannibalism: they grow through distinct size classes and are susceptible to attack while molting, often occur at high densities, and may suffer nutritional stress as a result of competition (Fox 1975, Polis 1981). Cannibalism may be an important mechanism for population regulation, especially when refugia are limited or cannibalism is a density-dependent occurrence (reviewed in Kneib et al. 1999). As such, cannibalism is likely widespread, if not ubiquitous, across crustacean taxa, including branchiopods (e.g., *Triops* spp.; Waterkeyn et al. 2011), copepods (Uye and Liang 1998, Ohman and Hirche 2001, Camus and Zeng 2009), and ostracods (Rossi et al. 2011), but is best described within the Malacostraca, particularly decapods, stomatopods, and peracarids. This chapter will describe in detail the proximate factors that lead to cannibalism throughout the Crustacea.

NUTRITION AND FOOD AVAILABILITY

Cannibalism often results from stress or changes in nutritional needs across life stages (Fox 1975, Polis 1981). Among feeding strategies, carnivory provides the greatest energetic payoff (reviewed in Griffen and Riley 2015). For example, cannibalism is prevalent in adult red land crabs (*Gecarcinus lateralis*) that are fed a nitrogen-poor, plant-based diet. Supplemental feeding of red crabs with soybeans, which are nitrogen fixers, greatly decreases cannibalism, suggesting cannibalism is facultatively expressed during nutritionally stressful times such as growth and pre-ecdysis (Wolcott and Wolcott 1984, Wolcott and O'Connor 1992). Additionally, calcium may be a limiting nutrient in some aquatic systems, and cannibalism may allow individuals to recover vital nutrients, especially after molting (reviewed in Stevens et al. 2014). However, calcium supplementation does not decrease cannibalism in juvenile red king crabs, *Paralithodes camtschaticus* (Daly et al. 2012a).

Hunger increases rates of intraspecific predation in both the burrowing crab *Neohelice granulata* and in *Cyrtograpsus angulatus* (Luppi et al. 2001). Hunger level of the predator influences cannibalism in many species, including *Hemigrapsus penicillatus* (Okamoto and Kurihara 1989) and porcelain crabs, *Porcellana platycheles* (Amaral et al. 2009). Likewise, hungry or starved snow crabs (*Chionoecetes opilio*) readily feed on both exuviae and prey (Dutil et al. 1997, Lovrich and Sainte-Marie 1997). In many cases, the presence of alternative prey greatly reduces cannibalism. Asian shore crabs (*H. sanguineus*) show weak cannibalistic tendencies and readily eat other food items when they are available (Griffen et al. 2015). Similarly, the presence of alternative prey reduces cannibalism in *N. granulata*, the Dungeness crab *Cancer magister*, the isopod *Saduria entomon*, and the gammarids (Dick et al. 1993, Sparrevik and Leonardsson 1998, Fernández 1999, Luppi et al. 2002). Effective aquaculture of mud crabs *Scylla serrata* requires dietary supplementation with alternative prey (Suprayudi et al. 2002).

GENETICS

Relatively little is known about the genetics of cannibalism (Stevens 1989, Stevens 1994, Cheng and Jefferson 2008, Ellen et al. 2010). In particular, very little is known about the genetics of cannibalism in invertebrates beyond a set of classic experiments performed in experimental lines of flour beetles (*Tribolium castaneum*). These beetles were selected for high and low levels of cannibalism and then maintained stably without selection for 60 or more generations. Both adult and larval flour beetles will consume eggs and adults will consume pupae, but there is no adult-adult cannibalism in *T. castaneum*. In this system, egg cannibalism was influenced by a set of genes distinct from the genes that influence pupal cannibalism. Adult and larval cannibalism are genetically correlated, suggesting that aggression and predation are underlain by the same mechanisms, although the behavior changes ontogenetically. Furthermore, differences between the high and low cannibal lines appear associated with two quantitative trait loci of large effect (reviewed in Stevens 1989).

Genetic effects on egg cannibalism among potentially inbred laboratory populations are difficult to extrapolate to cannibalism resulting from mobile individuals (adult or juvenile) attacking other mobile individuals in large outcrossed populations, which comprise the vast majority of crustacean cannibalism (e.g., Fig. 14.1). However, there is often significant variation in cannibalistic behavior both among and within species of crustaceans, with some individuals frequently killing and consuming conspecifics and others rarely or never engaging in cannibalism (Lovrich and Sainte-Marie 1997, Griffen and Mosblack 2011, Lee et al. 2013). For example, Asian shore crabs *H. sanguineus* vary dramatically in their individual dietary choices (Fig. 14.2A; Griffen and Mosblack 2011). Individual differences in dietary choice result from genetic, environmental, and ontogenetic differences among individuals. Socorro isopods *Thermosphaeroma thermophilum* vary substantially in their likelihood of attacking a conspecific based on an individual's sex and body size, as well as the body size of the potential victim (Fig. 14.2B; Bleakley et al. 2013). Body size is a heritable trait in Socorro isopods, thus providing a genetic mechanism for consistent cannibalistic behavior within an individual (Shuster et al. 2005, Bleakley et al. 2013). Consistent individual differences in intraspecific predatory behavior across environments and family effects on the latency to attack and the incidence of cannibalism have been used to estimate heritability of cannibalism in other arthropods, particularly spiders (Hvam et al. 2005, Johnson et al. 2010). The presence of such differences between individuals and among families in crustacean species suggests that the propensity to behave as a cannibal is at least partially heritable in crustaceans (sensu Boake 1994).

One difficulty in measuring genetic effects on cannibalism is that the expression of cannibalism within any given interaction is binary, but an individual will not be cannibalistic in every interaction with conspecifics over the course of its lifetime. As such, cannibalism is best described as a quantitative trait. Quantitative genetic models of threshold traits describe cannibalism with normally distributed additive genetic variation but an associated underlying "liability," or "switchpoint," leading to binary patterns of expression (threshold models reviewed in Roff 1996, 1997, and Schausberger and Croft 2001). Alternatively, the relevant genetic variation may reside in the phenotypic value of the threshold with the switchpoints for different individuals being normally distributed (Hazel et al. 1990, 2004). Under both models, individuals essentially have a genetic propensity to engage in cannibalism that is switched on by the environment, with some individuals requiring less environmental input to engage in cannibalism than others. Environmental influences may be derived from information about potential victims. For example, an individual may only attack a potential victim if that victim is very small, while another individual will attack much larger social partners. Male Socorro isopods alter their aggression, which often leads to cannibalism, in response to characteristics of their social partners, while female Socorro isopods do not (Fig. 14.2B; Bleakley et al. 2013). When behavior is influenced by characteristics of social partners, it is termed an *interacting phenotype*. Quantitative genetic models for interacting phenotypes exist but have yet to specifically model cannibalism, although such models could provide valuable predictions for empirical studies of cannibalism in crustaceans (Moore et al. 1997, McGlothlin et al. 2010, Bleakley et al. 2013).

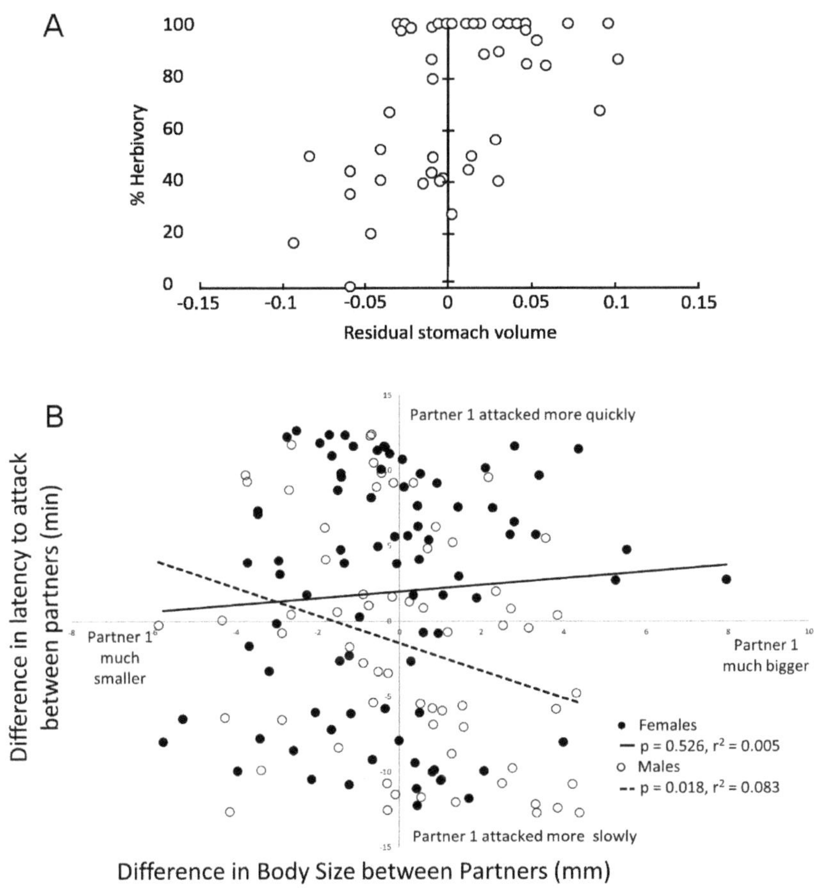

Fig. 14.2.
Individuals vary in their dietary choices and likelihood of engaging in cannibalism. (A) Individual differences in dietary choice for Asian shore crabs allowed to forage in field enclosures. Gut contents are shown as percent herbivory. Residual stomach volume corrects for the size of the crab. Individual differences in dietary choice result from genetic, environmental, and ontogenetic differences among individuals. From Griffen and Mosblack (2011), with permission from John Wiley and Sons. (B) Variation in Socorro isopod latency to attack based on the relative body size of an individual compared to its social partner. These differences in feeding preferences and aggressive behavior may result from differences in the availability of particular foods, physiological state, and other environmental factors, as well as genetic differences among individuals. From Bleakley et al. (2013), with permission from John Wiley and Sons.

SEX AND REPRODUCTIVE STATUS

Males and females may use different proximate cues to initiate cannibalism, have different nutritional needs, and often differ dramatically in body size. Because cannibalism is strongly associated with size and many species of crustaceans show larger male size dimorphism, cannibalism is thought to be more common in male crustaceans than females (Fox 1975, Polis 1981). Male snow crabs engage in size-associated cannibalism more frequently than female snow crabs (Kolts et al. 2013), as do the "skeleton shrimp" amphipods *Caprella penantis*, and *C. grandimana*, (Martínez-Laiz and Guerra-García 2015). *Gammarus tigrinus, Chionoecetes opilio,* and *T. thermophilum* females experience greater degrees of intraspecific predation by large males (Jormalainen and Shuster 1997,

MacNeil et al. 2008, Kolts et al. 2013). *Gammarus duebeni celticus* and *G. pulex* also show greater cannibalism by males, particularly on newly molted individuals. Males of all these species are more likely to attack other males but will also attack females. Likewise, females will attack juveniles but at a lower rate than males (Dick 1995, Jormalainen and Shuster 1997, Kolts et al. 2013). In contrast, only female *Heterocypris incongruens* ostracods were found to engage in cannibalism (Rossi et al. 2011), and *N. granulata* females were more likely to engage in cannibalism because of sex-specific characteristics of their chelipeds (Luppi et al. 2001). In many cases, however, no differences are observed between males and females in rates of cannibalism. Red claw crabs *Sesarma bidens* and their congener *S. dehaani* do not differ in rates of cannibalism according to sex. Kneib et al. (1999) measured the survival of juvenile red claw crabs paired with adult male or female conspecifics and found no differences in the survival of juveniles paired with adult males or adult females. No sex biases in predator or prey rates for males and females were found for blue swimmer crabs *Portunus pelagicus* (Marshall et al. 2005) or among juvenile *Romaleon setosum* that do not differ in size (Cerda and Wolff 1993).

Males often engage in male-male competition to obtain exclusive access to females and such contests can end in injury and ultimately cannibalism (Jormalainen 1998, Jormalainen and Shuster 1999). Many species of crustaceans mate directly after females molt, with males engaging in precopulatory mate-guarding behavior as the females approach ecdysis. Males sometimes eat females in precopula, particularly if males have been denied foraging opportunities (Ward 1985, Dick et al. 1993, Dick 1995, Dick et al 1995, Jormalainen and Shuster 1999). Male Socorro isopods may struggle to hold large females and intersexual conflict over mating can lead to male predation on females (Jormalainen 1998, Jormalainen and Shuster 1999). Similar intersexual conflict and cannibalism has been observed in multiple species of gammarid amphipods (Ward 1985, Dick 1995). Conversely, male casualties are more frequent when males are similarly sized or smaller than the females they attempt to guard in coconut crabs *Birgus latro* (Helfman 1979, cited in Sato and Yoseda 2010). Cannibalism of females on males occurs early in the breeding season for parasitic fish-tongue isopods *Ichthyoxenus fushanensis* with a reversal later in the breeding season (Tsai and Dai 2003). Sexual cannibalism in this species appears to be associated with limited resources in small hosts and is therefore plastic in response to the availability of food (Tsai and Dai 2003).

Eating a potential mate, particularly when males eat females in populations with male-biased operational sex ratios, can have significant negative fitness consequences for males (Jormalainen and Shuster 1999). In shore crabs (*Carcinus maenas*) male feeding behavior is inhibited by 20-hydroxyecdysone (crustecdysone), which females excrete (Hayden et al. 2007, reviewed in Hardege and Terschak 2010). Female shore crab feeding responses are not inhibited by exposure to the hormone, and intermolt females will eat soft-shelled females even when exposed to crustecdysone. Excretion of this pheromone may have evolved as a signal to deter precopula cannibalism in this species (Hayden et al. 2007). Reproductive condition can also affect the likelihood of cannibalism, although there is no clear pattern across taxa for when females should be most cannibalistic. For example, *Romaleon setosum* females feed very little while carrying eggs, potentially to avoid egg cannibalism (Cerda and Wolff 1993). Conversely, cannibalism is highest in *Gammarus* amphipods during reproductive periods, possibly because juveniles are more readily available (Dick et al. 1995).

SIZE ASYMMETRIES, GROWTH, AND MOLTING

Wherever animals exhibit asymmetries in risk, cannibalism can occur (Fox 1975, Smith 1979, Polis 1981). Size asymmetries are common, with larger individuals expected to attack smaller individuals more often, such as in blue swimmer crabs and Dungeness crabs (Fernández 1999, Marshall et al. 2005). However, cannibalism may be especially prevalent in crustaceans because molting creates asymmetries in risk that are independent of body size (Fox 1975, Polis 1981, Dick 1995). Individuals that are molting are both at risk of cannibalism and incapable of engaging in cannibalism (reviewed in Marshall et al. 2005). Furthermore, molting animals may change their behavior to mitigate the risks of cannibalism. For example, spiny lobsters (*Panulirus argus*) molt in the dark to avoid aggressive conspecifics (Lipcius and Herrnkind 1982). Reproductive condition can also impose risks on females. Female *R. polyodon* carrying eggs are less mobile and more vulnerable to depredation themselves and may therefore limit their movements and interactions with other crabs (Cerda and Wolff 1993).

Both the risk of being depredated by a conspecific and the potential to engage in cannibalistic behavior are frequently size-specific and stage-dependent. Ecological models of cannibalism focus on size-dependent cannibalism and many models use age-structured populations that correlate to size classes (Cushing 1992, Crowley and Hopper 1994, Fagan and Odell 1996, Claessen et al. 2004, Huss et al. 2010). In crabs, the incidence of cannibalism rises as juveniles become larger, especially as larger individuals become able to prey on smaller individuals (e.g., Luppi et al. 2001). For example, adult southern king crabs *Lithodes santolla* are well documented to engage in cannibalism but juveniles do not (Stevens and Swiney 2005, Vinuesa et al. 2013). Whereas individuals of all sizes may prey on all available (i.e., smaller) size classes in some species, as in Socorro isopods and *Gammarus tigrinus* (Fig. 14.3A; Dick et al. 1993, Jormalainen and Shuster 1997, Lovrich and Sainte-Marie 1997, MacNeil et al. 2008), others, such as red king crabs, direct predation toward particular size classes and ignore size classes that are too small (Fig. 14.3B). Nutritional requirements and feeding preferences often exhibit ontogenetic shifts (e.g., Lovrich and Sainte-Marie 1997).

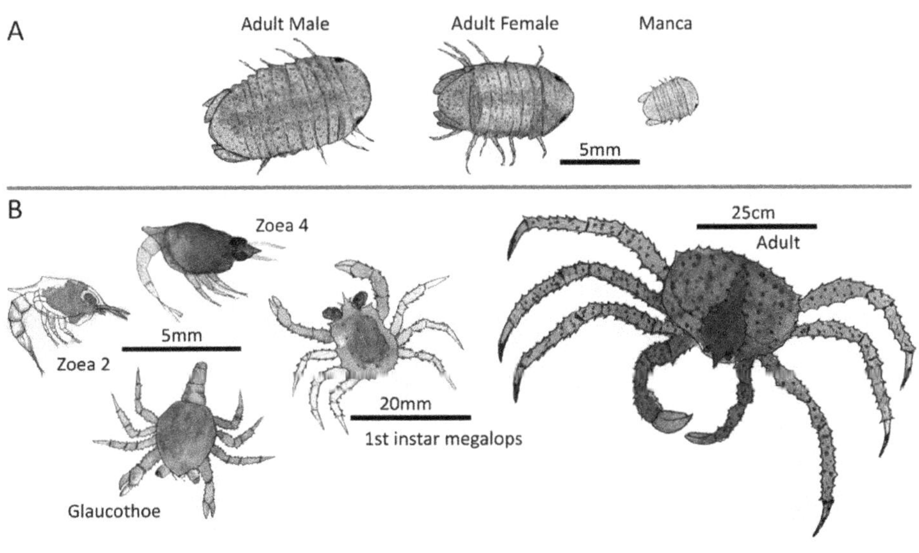

Fig. 14.3.
Size asymmetry and profitability of prey influence the incidence of cannibalism. (A) Cannibalism in Socorro isopods (*Thermosphaeroma thermophilum*), which display size-structured cannibalism. Any smaller individual is at risk, although manca (direct developing juveniles) and small females are at greatest risk of depredation. (B) Effects of size asymmetry and profitability on cannibalism in red king crabs (*Paralithodes camtschaticus*). Smaller individuals within a developmental cohort and across close developmental classes are susceptible to cannibalism; however, adults likely do not eat larvae because extreme size differences reduce the profitability of cannibalism among those classes. Solid lines indicate most common direction of attack. Dotted line indicates direction of attack when the size difference between the aggressor and victim is small enough to make the attack profitable. Illustrated by Bronwyn H. Bleakley ©. See color version of this figure in the centerfold.

Megalopae and first instar juvenile mangrove crabs *Ucides cordatus* differ somewhat in their propensity to engage in cannibalism. Megalopae eat other megalopae, likely because of differences in the timing of molting. However, first instar juveniles do not appear to cannibalize megalopae but will cannibalize other first instar juveniles under some environmental conditions (Ventura et al. 2011). Large adult male burrowing crabs and *Cyrtograpsus angulatus* prey on large juveniles but not small juveniles, potentially because they do not provide enough energy to be profitable. Large juveniles of each species prey on small juveniles (Méndez Casariego et al. 2009).

Differences in cannibalistic behavior directed toward different smaller size classes may depend on the relative difference in body size between predator and prey, which determines the profitability of eating particular prey and the risk of injury from fighting with and subduing prey (Fox 1975, Lovrich and Sainte-Marie 1997). Early instar snow crabs were consumed by smaller adult males but not by larger adult males, probably reflecting the benefits of consumption to the smaller males (Lovrich and Sainte-Marie 1997, Sainte-Marie and Lafrance 2002). When provided with a range of conspecific prey sizes, large snow crabs consume mostly individuals in intermediate size classes. In predation trials, large snow crabs were provided with intraspecific prey from a range of size classes. Intermediately sized prey suffered the greatest mortality. In addition, the number of nonfatal injuries to prey crabs peaked at nearly twice the size of the most frequently consumed crabs, suggesting that larger potential prey are more difficult and dangerous to pursue. Cannibalism therefore peaks at an intermediate size class in these crabs (Fig. 14.4; Dutil et al. 1997). Male isopods modulate their latency to attack based on the relative body size of their potential victims. Although male Socorro isopods will attack other large males, they attack relatively smaller males much faster than larger males (Fig. 14.2B; Bleakley et al. 2013). Relative body size influences cannibalism in a few other arthropods as well (e.g., Hack et al. 1997). However, little is known about the relative importance of an individual's relative body size compared to potential prey versus an individual's absolute body size (Bleakley et al. 2013).

Vulnerability to cannibalism and the risk of mortality is typically size-dependent (e.g., Ruiz et al. 1993, Jormalainen and Shuster 1997). As a result, adults often depredate juveniles. Cannibalism by adults is most frequent among young-of-the-year in Dungeness crabs (Fernandez et al. 1993).

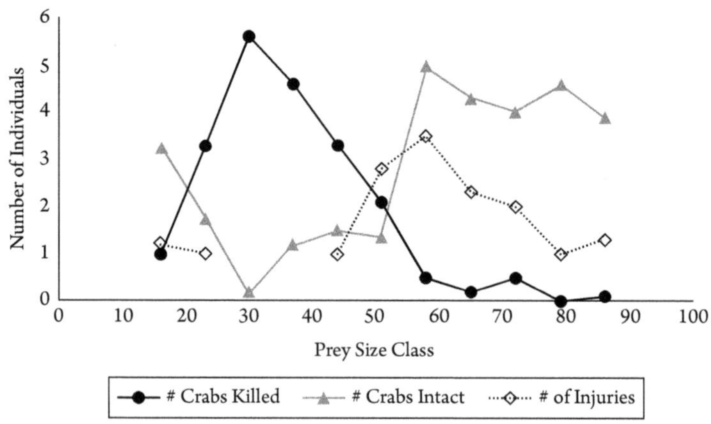

Fig. 14.4.
Smaller individuals are often at greater risk for cannibalism. The black line and filled circles denote the mean number of snow crabs killed. The gray line with filled triangles shows the mean number of crabs left alive in any trial based on size. Small and larger crabs survived more often than crabs of medium size. The dotted line with open diamonds shows the mean number of injuries in surviving prey of various sizes. From Dutil et al. (1997), with permission from Elsevier.

Three percent of food items found in the gastrointestinal tract of adult stomatopods (*Gonodactylus oerstedii*) were juvenile stomatopods (Reaka 1987). Juvenile red claw crabs and their congener (*S. bidens*, and *S. dehaani*) exhibit 95%–100% survival in the absence of adults, whereas only 75% of juveniles survive in the presence of conspecific adults (Kneib et al. 1999). Mortality in snow crab juveniles is highest when large adult males are paired with the smallest size classes and declines as the juveniles increase in size (Dutil et al. 1997). *Romaleon setosum* adults eat tens of thousands of smaller juveniles whenever they are readily available, such that cannibalism removes at least 10% of the juveniles produced in the population, but rarely eat other adults (Cerda and Wolf 1993). Adult blue crab cannibalism can be associated with a staggering 75%–97%-mortality in juveniles (Hines et al. 1987). Unsurprisingly, cannibalism of juveniles by adults is also widespread in amphipods, including *Gammarus* spp. (reviewed in MacNeil et al. 1997).

Earlier, and therefore smaller, larval stages are also vulnerable to cannibalism by larger, later stage larvae. Australian giant crabs *Pseudocarcinus gigas*, mangrove crabs *Ucides cordatus*, and red shrimp *Penaeus marginatus* all exhibit significant larval cannibalism (Gopalakrishnan 1976, Gardner and Maguire 1998, Ventura et al. 2008). Red king crab *Paralithodes camtschaticus* juveniles impose significant mortality on glaucothoe (postlarvae), irrespective of juvenile size (Stevens and Swiney 2005). Recognition and management of juvenile-larval cannibalism is especially important in species that are cultured. Southern king crabs are an important, and in some locations, endangered fisheries resource. Cultures of this crab are critically important to maintaining fisheries by producing juveniles that can be released to recruit in natural populations. However, in controlled experiments in culture, every stage of larvae and juvenile attacked earlier stages. Attacks were exclusively unidirectional, with the larger or older stage attacking the smaller or younger stage, even when there was a single large individual and many small individuals present in the trial (Sotelano et al. 2012). The production of red king crabs in areas that have also experienced fishery collapse is greatly increased by size-grading the larvae and juveniles, such that larger individuals do not have access to smaller individuals (Daly et al. 2012a). In contrast, blue king crabs *Paralithodes platypus* show comparatively less larval cannibalism in cultures and may be easier to stock than their congener (Daly and Swingle 2013).

Cannibalism can be the principle source of mortality in juveniles in size-structured populations of crustaceans. Juveniles suffer the greatest mortality in crabs, isopods, and gammarids (Botsford and Wickham 1978, Hines et al. 1987, Jormalainen and Shuster 1997, Lovrich and Sainte-Marie 1997). Recruitment for benthic species with pelagic larvae is structured by both pre- and post-settlement processes. Settlement may be initiated by conspecific cues indicating the location of appropriate resources. However, the specific locations larvae choose to settle often minimizes the risk of predation, including intraspecific predation. Thus, the risk of cannibalism influences settlement patterns in many species (Caddy 1986, Lovrich and Sainte-Marie 1997, Eggleston et al. 1998). For example, adult-juvenile cannibalism inhibits recruitment in stomatopods (Reaka 1987) and in Dungeness crab (Fernandez et al. 1993, Fernández 1999). Patterns of settlement may result directly from juveniles seeking out particular habitats to avoid cannibalism. Red king crab glaucothoe and juveniles actively seek complex habitat structure, and although the structure does not reduce cannibalism in the postlarval stage, it appears to allow greater survival when the larvae metamorphose (Stevens and Swiney 2005).

Size asymmetry is not the only risk asymmetry that can lead to cannibalism, however. Injury can leave even large individuals vulnerable. Hermit crabs *Clibanarius digueti* increase movement in response to conspecific alarm cues, using odor cues to locate and consume a dead or injured conspecific (Tran 2014). Although individuals may avoid injury, all crustaceans must molt, leaving the animal vulnerable to attack until its new exoskeleton can harden. As such, molt stage greatly affects the likelihood that an individual will be attacked (Polis 1981). Mud crabs respond to molting and injury in conspecifics with increased activity and investigative behavior, suggesting that these conditions signal vulnerability to cannibalism. These crabs increased tactile and feeding behavior when exposed to the scent of a molting conspecific and increased tactile behavior when exposed to the scent of an injured conspecific, although both responses depended on the sex and size of the crab (Wall et al. 2009). Intermolt *G. d. celticus* and *G. pulex* killed and consumed molting congeners in all combinations of adult males, females, and juveniles. However, smaller individuals did not attack larger molting individuals and males distinguished between molting females and molting males, attacking the latter more often (Dick 1995).

The effect of a potential prey item's molt stage varies. Porcelain crabs show no juvenile cannibalism associated with molt stage. Edible crabs *Cancer pagurus* cannibalize each other as juveniles only when a prey individual is vulnerable because of molting or injury but engage in intense size-mediated cannibalism across size classes (Amaral et al. 2009). *Saduria entomon* isopods exhibit markedly reduced survival when molting individuals are housed with intermolt individuals regardless of the availability of refugia, suggesting molting individuals are particularly susceptible to cannibalism (Sparrevik 1999). Similarly, blue swimmer crabs (*Portunus pelagicus*) experience increasing mortality as they approach molt. Mortality peaks for these crabs during and shortly after ecdysis, even when refuges are abundant (Fig. 14.5; Marshall et al. 2005). Molt-associated cannibalism, like size-dependent cannibalism, can hamper conservation and economic management efforts. For example, the endangered white-clawed crayfish *Austropotamobius pallipes* requires access to refugia in aquaculture settings to prevent molt-associated cannibalism between adults (Rosewarne et al. 2013).

Molt stage changes behavior, altering both an individual's risk of depredation and opportunity to engage in cannibalism (Lipcius and Herrnkind 1982). Increased vulnerability may lead to increased defensive behavior on the part of a potential victim. Northern krill *Meganyctiphanes norvegica* molt in deep water to reduce the risk of cannibalism (Villefranche-sur-Mer 1999). Blue crabs *Callinectes sapidus* show distinct habitat preferences based on molt stage, with intermolt adult males and premolt reproductive females co-occurring in open water basins, while juvenile females avoided those areas (Hines et al. 1987). Smaller, pre-reproductive females could be at risk

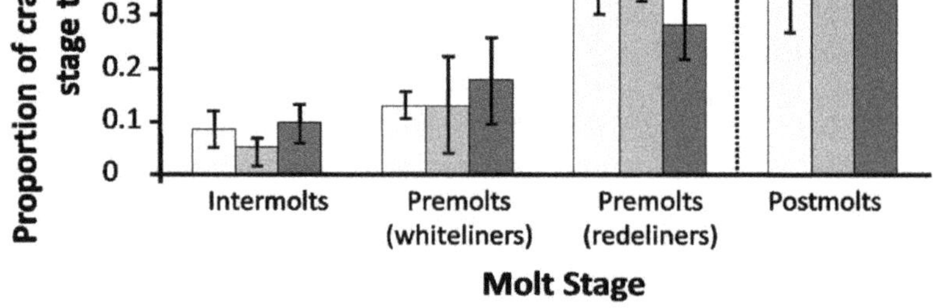

Fig. 14.5.
Molt stage and antipredator behavior influence the risk of cannibalism. Juvenile blue-swimmer crabs, *Portunus pelagicus*, are at the greatest risk of mortality during and directly after ecdysis, regardless of refuge utilization. Refuge quantity influences the mortality rate differently depending on molt stage. Mortality ± Standard Error. Modified from Marshall et al. (2005), with permission from Elsevier.

of cannibalism by overly aggressive males attempting to engage in mate-guarding and mistakenly eating the female when the female resists guarding, as has been found in premolt female Socorro isopods (Jormalainen and Shuster 1999). Predator avoidance behavior improves in red king crab juveniles from molt stage 2 to molt stage 4, reducing the risk of cannibalism (Stoner et al. 2013). Likewise, use of refuges improves survival in molting blue swimmer crabs (Fig. 14.5; Marshall et al. 2005).

KIN RECOGNITION AND KIN SELECTION

Many species live in populations where they encounter and interact with relatives. Under some conditions, they may eat relatives. For example, *Heterocypris incongruens* ostracods eat their clones, some harpacticoid copepods (*Harpacticus* sp.) will eat their nauplii in highly competitive dense populations (Dahms and Qian 2006, Rossi et al. 2011), and offspring consume the mother in the semelparous isopod *Schizidium tiberianum* (Warburg and Cohen 1991). However, eating relatives and thereby decreasing indirect fitness, which accrues through the reproduction of relatives passing on other copies of an individual's genes, is a significant cost of cannibalism in many species (Pfennig 1997). While *H. incongruens* benefit substantially enough from reduced competition in high-density swarms to outweigh any lost indirect fitness (Rossi et al. 2011), kin selection typically favors kin recognition, which prevents individuals from eating relatives (Pfennig 1997). Female *G. pulex* amphipods carrying late-stage embryos significantly decrease consumption of juveniles as they near the release of the embryos from the brood pouch, suggesting that eating their own young may impose a net fitness cost (Lewis et al. 2010). Crustaceans that provide extended parental care, such as red swamp crayfish *Procambarus clarkii,* become more aggressive toward offspring if maternal contact with offspring is prevented for more than 24 hours (Hazlett 1983, Figler et al. 1997). The avoidance of filial cannibalism in these cases is dependent on indirect mechanisms of kin recognition, timing of emergence and familiarity, rather than direct kin recognition (Elwood 1994). The ability to directly recognize kin is predicted to be selected for in animals that live in large or highly structured groups (Penn and Frommen 2010), and in crustaceans, among those groups where members routinely leave and then return to the family group (reviewed in Thiel 2007). For example, the sponge-dwelling shrimp *Synalpheus regalis* and several congeners exhibit monogynous eusocial colonies whose defense requires kin recognition (Duffy et al. 2002). However, this type of social organization is quite rare in crustaceans.

The degree to which kin recognition and kin selection might structure cannibalism in crustaceans is poorly understood; there are relatively few well documented examples of crustacean species recognizing kin. While rare, kin recognition is found in a range of crustaceans, including isopods, amphipods, copepods, and decapods, but is best understood in amphipods (Thiel 2007). Because some species of amphipods occur in very large aggregations, they are at greater risk of eating relatives than species that disperse. Concomitantly, at least limited kin recognition has, indeed, been found in some of these species, suggesting that kin selection favors individuals that can recognize and avoid eating relatives (reviewed in Beermann et al. 2015). Rock-pool amphipods *Apherusa jurinei* perform specialized maternal care that includes brood flushing, which may inadvertently eject embryos from the brood pouch. Females may eat loose embryos or insert them into the brood pouch and resume parental care. In such cases, females are less likely to eat their own young than unrelated young (Patterson et al. 2008). Kin recognition has also been identified in copepods, such as the Mediterranean rock-pool copepod *Tigriopus fulvus*. Female *T. fulvus* recognize their offspring and preferentially feed on unrelated nauplii (Gallucci and Ólafsson 2007). In contrast, the harpacticoid copepod *Harpacticus* sp. shows no maternal recognition of related nauplii, modulating its feeding based on the density of the nauplii and nutritional stress (Dahms and Qian 2006). This last case may be characteristic of many crustaceans, but the opportunity to investigate kin recognition and cannibalism in amphipods remains a potentially productive avenue of research.

HABITAT-MEDIATED CANNIBALISM

Population regulation is the maintenance of the population at a stable density between lower and upper limits of population viability. Density-dependent cannibalism must be more common or otherwise have greater impacts at higher densities and have fewer effects at low densities to be maintained through time. This is because frequent cannibalism in low-density populations could reduce or eliminate entire size classes from the population (reviewed in Lovrich and Sainte-Marie 1997). These density-dependent effects can be the result of high densities of prey (e.g., juveniles), high densities of predators (e.g., adults), or both (Kneib et al. 1999). For example, cannibalism within developmental stages of mangrove crabs *Ucides cordatus* increases at higher densities (Ventura et al. 2011). Poor survivorship was found in both megalopae and first instar juveniles at high densities. Juveniles reared at a moderate density of 200 individuals per square meter survived equally well as those reared individually. However, juveniles reared at 500 individuals per square meter exhibited significantly lower survivorship rates than those in the lower-density conditions, likely due to increased cannibalism. Likewise, cannibalism increases with stocking density in southern king crab (Sotelano et al. 2012, Sotelano et al. 2016). At high densities, when populations overshoot their carrying capacity, fairy shrimp cannibalize nauplii to increase reproductive output of drought resistant cysts (Dumont and Ali 2004). In contrast high densities of predators may, nonintuitively, yield less mortality of smaller prey when predators compete and interfere with each other reducing individual consumption of smaller conspecifics (Moksnes et al. 1997, Méndez Casariego et al. 2009).

Despite the widespread positive association between cannibalism and density, not all species will engage in density-dependent cannibalism. Mortality of early or late benthic instar snow crabs does not vary with density (Lovrich and Sainte-Marie 1997). Likewise, although co-occurring grapsid crabs *Sesarma bidens* and *S. dehaani* consume juveniles of both their own and the other species, neither does so in a density-dependent manner (Kneib et al. 1999). Similarly, mangrove crabs *U. cordatas*, which display a low level of aggressiveness compared to closely related species, engage in very little cannibalism even at high densities and under stressful rearing conditions (Ventura et al. 2008). In contrast, many species engage in cannibalism most often or exclusively at high densities. For example, mud crabs, blue crabs, and Turkish narrow-claw crayfish *Astacus leptodactylus* in aquaculture hatcheries all engage in density-dependent cannibalism (Quinitio et al. 2001, Zmora et al. 2005, Farhadi et al. 2014). *Romaleon setosum* engage in significantly greater cannibalism in the winter and summer when recruitment pulses increase the density of juveniles (Cerda and Wolff 1993). Rarely, crustaceans may take advantage of the dilution effect to avoid filial cannibalism. For example, several species of mysid shrimp engage in reproductive synchrony in large aggregations. Females that synchronize the release of reproductive cysts limit cannibalism by other adults because any female reproducing at that time is at higher risk of eating her own young in the swarm. However, this benefit appears to be maximized in small, rather than very large, groups (Johnston and Ritz 2001).

Environmental Heterogeneity and Availability of Refuges

Environmental heterogeneity in both distribution of predators and the availability of refugia will change effective density and allows individuals to segregate. Spatial distribution and availability of refuges, particularly for juveniles and small adults, can be crucial in mediating the risk of cannibalism (Fernandez et al. 1993, Dittel et al. 1995, Kneib et al. 1999). Juvenile blue crab survive best when they are at low density and have access to refuges (Fig. 14.6; Dittel et al. 1995). Red king crabs exhibit the greatest survival in complex habitat and low densities, regardless of intraspecific or interspecific predation (Pirtle et al. 2012, Daly et al. 2012b, Long et al. 2015). Increased habitat complexity reduces both cannibalism and intraguild predation in complexes of gammarid species (MacNeil et al. 2008). Socorro isopods exhibit much lower levels of cannibalism in a natural pool and in managed populations that are provided both rock and algal structure compared to populations provided with no or minimal habitat structure (Jormalainen and Shuster 1997). *Hemigrapsus penicillatus* crabs show strong preferences for habitats with the right aperture holes to best evade cannibalism (Kurihara et al. 1989). Broadly, the availability of refugia and the complexity of the microhabitat decrease cannibalism for many decapod species (Heck and Wilson 1987, Navarrete and Castilla 1990, Heck and Hambrook 1991, Dumbauld et al. 1993, Luppi et al. 2001, Zmora et al. 2005, Ventura et al. 2008, Amaral et al. 2009).

Photoperiod, light intensity, background color, and water pH may also affect the incidence of cannibalism, in large part because these factors structure which individuals are able to interact, in much the same way as the availability of refugia. Dark background color, which simulates the effects of refugia, improves larval survival in mud crabs (Rabbani and Zeng 2005). Molting rate and survival were greatest in the lowest light condition for *Palaemonetes argentinus* (Díaz et al. 2003). However, neither photoperiod nor light intensity influenced cannibalism in Australian giant crab *Pseudocarcinus gigas* (Gardner and Maguire 1998). In addition, sufficient light is required for proper molting and will therefore trade off with the benefits of reduced cannibalism in low-light conditions. European lobsters (*Homarus gammarus*) reared in low pH water harden more slowly after molt because low seawater alkalinity greatly reduces physiological uptake of calcium carbonate from the water. As a result, molting individuals experienced greater risk of predation from conspecifics. Calcium is not typically a limiting nutrient in marine environments (Middlemiss et al. 2016); however, as ocean acidification increases with changes in global temperature and carbon dioxide levels, cannibalism could increase in marine species that become calcium limited.

Habitat type interacts with density to shape the occurrence of cannibalism. Juvenile blue crabs experience high mortality from cannibalism at low densities (showing inverse density-dependence) in sandy habitat but low mortality from cannibalism at low densities in grass habitat.

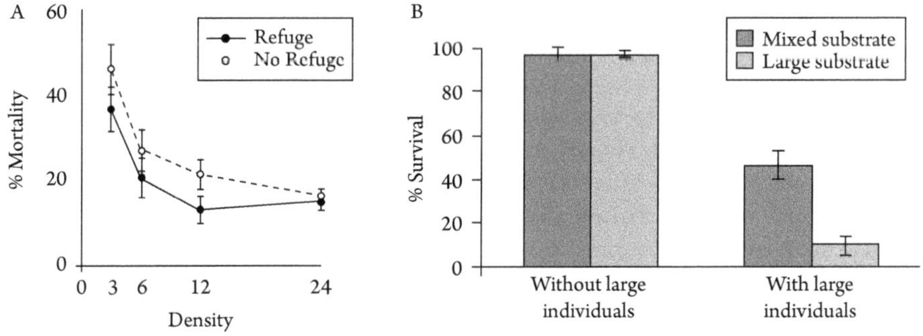

Fig. 14.6.
Population density and the availability of refuges influence the incidence of cannibalism. (A) Mean percent mortality ± standard error is shown for different densities of juvenile blue crabs housed in laboratory tanks with and without access to refuges. From Dittel et al. (1995), with permission from Bulletin of Marine Science. (B) Small *Gammarus pulex* preferentially use small refuges in the presence of large *G. pulex* and survive better in small refuges because large cannibals are excluded. From McGrath et al. (2007).

Per-predator consumption decreases when more predators are present in grass habitat, eliminating prey-density dependence of cannibalism entirely (Moksnes et al. 1997). Shore crabs engage in similar interference competition between cannibals, where higher densities of juveniles decreased per capita predation rates. Despite the decrease in individual cannibalism rates, prey mortality increases with increased predator density, and in field experiments as many as 71% of larvae may be lost due to density-dependent cannibalism in this species (Moksnes 2004). Juvenile blue crabs aggregate in spatially complex habitats, yielding five times as many predators as found in mud habitats. However, the availability of refugia buffered larval exposure to cannibalistic juveniles, mitigating density-dependent mortality (Moksnes and Heck 2006). Similarly, although juvenile density did not influence rates of cannibalism in *Sesarma bidens* or *S. dehaani*, the density of conspecific adults combined with availability of refugia and competition greatly affected the probability of survival in both species (Kneib et al. 1999).

Cannibalism can stabilize population fluctuations, limit competition, and allow the coexistence of competitive species (Hatcher et al. 2014). Complex population dynamics that result from a combination of many of the previously mentioned environmental factors for cannibalism are found in invasive/native species complexes of gammarid amphipods. *G. d. celticus* is native to and widespread in western Europe. *Gammarus tigrinus* is native to North America and invasive in western Europe. *Dikerogammarus villosus* is also invasive in the Netherlands but arrived well after *G. tigrinus* (Dick et al. 1999, Platvoet et al. 2009). *Dikerogammarus villosus* is a much stronger predator, with adults preying on all size classes of *G. tigrinus* regardless of molt stage while *G. tigrinus* is only able to prey on juvenile intermolt *D. villosus*. Both species are cannibalistic. The presence of complex habitat greatly increases the survival of *G. tigrinus*, in part because it reduces cannibalism by *G. tigrinus* adults. Cannibalism in both species may reduce the number of adults that ultimately survive, reducing intraguild predation and allowing the coexistence of the two species (Platvoet et al. 2009). Conversely, the invasive *G. pulex* largely replaced the native *G. d. celticus* in Ireland because cannibalism reduced the competitiveness of the native species. The frequency of males eating females was quite low in both species and did not differ between the two. However, the frequency of male-male cannibalism was much higher in the native *G. d. celticus*, greatly increasing the total predation pressure on the species (Dick et al. 1993, Dick et al. 1999).

COMPARISON WITH OTHER TAXA

The behavioral ecology of cannibalism, including the effects of cannibalism on fitness, the inclusion of cannibalism in behavioral syndromes (correlated sets of traits), sexual conflict over mating, the role of kin selection in structuring cannibalism, and the degree to which community structure influences and is influenced by cannibalism, have been better explored for many noncrustacean arthropods (Wagner et al. 1999, Fromhage and Schneider 2005, Johnson and Sih 2005, Rudolf 2007). For example, cannibalism is heritable in both ladybird beetles *Harmonia axyridis* and fishing spiders *Dolomedes triton*. Strong genetic correlations, or behavioral syndromes, for aggressive behaviors can lead to maladaptive precopulatory cannibalism in fishing spiders (Johnson and Sih, 2005) but cannibalism alone confers fitness benefits in ladybird beetles (Wagner et al. 1999). Studies of kin selection in other arthropods, including extensive work in social insects, are well developed (reviewed in Queller and Strassman 1998). The presence of kin recognition in some cannibalistic species of crustaceans, such as rock-pool amphipods, suggest that kin selection could be found within diverse crustaceans and bears further investigation. Last, the genetic influences on cannibalism remain poorly understood in general and wholly unexplored in crustaceans, offering many opportunities for research.

FUTURE DIRECTIONS

Studies of cannibalism in other taxa, particularly other arthropods, suggest many avenues of investigation to better understand cannibalism in crustaceans, including those highlighted above. The influence of many environmental factors on the incidence of cannibalism, including nutritional stress, the availability of refugia, and the size structure of populations, have been well explored for some crustaceans, particularly decapods that are of economic interest (Table 14.1). However, much work remains for other groups of crustaceans, including species that are not reared through aquaculture and more diverse nondecapods and nongammarids. In addition, few direct comparisons of closely related species that differ in their incidence of cannibalism have been undertaken for crustaceans. Such comparisons could elucidate genetic causes of cannibalism, as well as provide more information about the environmental correlates of cannibalism. Studies that investigate the role of cannibalism in structuring species interactions could be broadened from those described for *Gammarus* amphipods (e.g., Dick et al. 1993).

Table 14.1. Summary of influences on the likelihood of cannibalism.

NUTRITION AND FOOD AVAILABILITY	NOTES	SPECIES	REFERENCE
Access to calcium		*Holthuisana transversa*	Middlemiss et al. 2016
	No influence	*Paralithodes camtschaticus*	Daly et al. 2012a
Food limitation		*Ichthyoxenus fushanensis*	Tsai and Dai 2003
Hunger		*Chionoecetes opilio*	Dutil et al. 1997, Lovrich and Sainte-Marie 1997
		Cyrtograpsus angulatus	Luppi et al. 2001
		Hemigrapsus penicillatus	Okamoto and Kurihara 1989
		Neohelice granulata	Luppi et al. 2001
		Porcellana platycheles	Amaral et al. 2009
Lack of alternate prey		*Cancer magister*	Fernández 1999
		Gammarus spp.	Dick et al. 1993
		Hemigrapsus sanguineus	Griffen et al. 2015
		Neohelice granulata	Luppi et al. 2002
		Saduria entomon	Sparrevik and Leonardsson 1998
		Scylla serrata	Suprayudi et al. 2002
Nutritional stress	Nitrogen limitation	*Gecarcinus lateralis*	Wolcott and Wolcott 1984, Wolcott and O'Connor 1992
		Harpacticus spp.	Dahms and Qian 2006
	No effect	*Ucides cordatas*	Ventura et al. 2008

(continued)

Table 14.1. **(Continued)**

GENETICS			
Genetics	Probable	*Hemigrapsus sanguineus*	
Genetics	Probable	*Thermosphaeroma thermophilum*	Shuster et al. 2005, Bleakley et al. 2013
SEX AND REPRODUCTIVE STATUS			
Intersexual conflict		*Ichthyoxenus fushanensis*	Tsai and Dai 2003
	Female biased	*Birgus latro*	Helfman 1979
	Male biased	*Gammarus spp.*	Ward 1985, Dick et al. 1993, Dick 1995, Dick et al. 1995
	Male biased	*Thermosphaeroma thermophilum*	Jormalainen 1998, Jormalainen and Shuster 1999
Reproductive condition		*Romaleon setosum*	Cerda and Wolff 1993
		Gammarus pulex	Lewis et al. 2010
		Gammarus spp.	Dick 1995
		Romaleon setosum	Cerda and Wolff 1993
		Streptocephalus proboscideus	Dumont and Ali 2004
Sex	Male biased	*Caprella penantis*	Martínez-Laiz and Guerra-García 2015
	Pheromonal inhibition of males	*Carcinus maenas*	Hayden et al. 2007
	Male biased	*Chionoecetes opilio*	Kolts et al. 2013
		Gammarus duebeni celticus	Dick 1995
		Gammarus duebeni celticus	Dick et al. 1993, Dick et al. 1999
	Male biased	*Gammarus duebeni celticus*	Dick 1995
		Gammarus pulex	Dick 1995
	Male biased	*Gammarus pulex*	Dick 1995
	Male biased	*Gammarus tigrinus*	MacNeil et al. 2008
	Female biased	*Heterocypris incongruens*	Rossi et al. 2011
	Female biased	*Neohelice granulata*	Luppi et al. 2002
	No difference	*Portunus pelagicus*	Marshall et al. 2005
	No difference	*Romaleon setosum*	Cerda and Wolff 1993
		Scylla serrata	Wall et al. 2009

(*continued*)

Table 14.1. **(Continued)**

	No difference	*Sesarma bidens*	Kneib et al. 1999
	No difference	*Sesarma dehaani*	Kneib et al. 1999
	Male biased	*Thermosphaeroma thermophilum*	Jormalainen and Shuster 1997
SIZE ASYMMETRY, GROWTH, AND MOLTING			
Antipredator behavior		*Callinectes sapidus*	Hines et al. 1987
		Meganyctiphanes norvegica	Villefranche-sur-Mer 1999
		Paralithodes camtschaticus	Stoner et al. 2013
Conspecific injury		*Cancer pagurus*	Amaral et al. 2009
	Chemical cues	*Scylla serrata*	Wall et al. 2009
Developmental stage		*Callinectes sapidus*	Moksnes and Heck 2006
		Lithodes santolla	Stevens and Swiney 2005, Vinuesa et al. 2013
		Romaleon setosum	Cerda and Wolff 1993
		Ucides cordatus	Ventura et al. 2011
		Ucides cordatus	Ventura et al. 2011
Molt stage		*Austropotamobius pallipes*	Rosewarne et al. 2013
		Callinectes sapidus	Hines et al. 1987
		Cancer pagurus	Amaral et al. 2009
		Gammarus duebeni celticus	Dick 1995
		Gammarus pulex	Dick 1995
		Meganyctiphanes norvegica	Villefranche-sur-Mer 1999
		Panulirus argus	Lipcius and Herrnkind 1982
		Paralithodes camtschaticus	Stoner et al. 2013
	No effect	*Porcellana platycheles*	Amaral et al. 2009
		Saduria entomon	Sparrevik 1999
		Scylla serrata	Wall et al. 2009
		Thermosphaeroma thermophilum	Jormalainen and Shuster 1999
		Ucides cordatus	Ventura et al. 2011
Profitability		*Chionoecetes opilio*	Dutil et al. 1997, Lovrich and Sainte-Marie 1997, Sainte-Marie and Lafrance 2002
Size asymmetry	Adults eat larvae	*Callinectes sapidus*	Hines et al. 1987

(continued)

Table 14.1. **(*Continued*)**

		Cancer magister	Fernandez et al. 1993, Fernández 1999
	Adults eat larvae	*Cancer magister*	Botsford and Wickham 1978, Fernandez et al. 1993, Fernández 1999
		Cancer pagurus	Amaral et al. 2009
		Chionoecetes opilio	Lovrich and Sainte-Marie 1997
		Cyrtograpsus angulatus	Méndez Casariego et al. 2009
		Gammarus duebeni celticus	Dick 1995
		Gammarus pulex	Dick 1995
		Gammarus spp.	Dick et al. 1993
	Adults eat larvae	*Gammarus spp.*	MacNeil et al. 1997
		Gammarus tigrinus	MacNeil et al. 2008
	Adults eat larvae	*Gonodactylus oerstedii*	Reaka 1987
	Larger larvae prey on smaller larvae	*Lithodes santolla*	Sotelano et al. 2012
		Neohelice granulata	Méndez Casariego et al. 2009
	Larger larvae prey on smaller larvae	*Paralithodes camtschaticus*	Stevens and Swiney 2005, Daly et al. 2012a
	Limited larval interstage larval cannibalism	*Paralithodes platypus*	Daly and Swingle 2013
	Larger larvae prey on smaller larvae	*Penaeus marginatus*	Gopalakrishnan 1976
		Portunus pelagicus	Marshall et al. 2005
	Larger larvae prey on smaller larvae	*Pseudocarcinus gigas*	Gardner and Maguire 1998
	Adults eat larvae	*Romaleon setosum*	Cerda and Wolff 1993
	Adults eat larvae	*Sesarma bidens*	Kneib et al. 1999
	Adults eat larvae	*Sesarma dehaani*	Kneib et al. 1999

(*continued*)

Table 14.1. **(Continued)**

	Adults eat larvae	Stomatopods	Reaka 1987
		Ucides cordatus	Ventura et al. 2011
	Larger larvae prey on smaller larvae	*Ucides cordatus*	Ventura et al. 2008
KIN RECOGNITION AND KIN SELECTION			
Kin recognition/ familiarity	Reduction	*Apherusa jurinei*	Patterson et al. 2008
	Reduction	*Procambarus clarkii*	Hazlett 1983, Figler et al. 1997
	Reduction	*Tigriopus fulysus*	Gallucci and Ólafsson 2007
Kin selection	Reduction	*Anisomysis mixta australis*	Johnston and Ritz 2001
	Reduction	*Gammarus pulex*	Lewis et al. 2010
	Reduction	*Paramesopodopsis rufa*	Johnston and Ritz 2001
	Reduction	*Tenagomysis tasmaniae*	Johnston and Ritz 2001
HABITAT			
Availability of juvenile conspecific prey		*Gammarus spp.*	Dick 1995
Competition	Intraspecies	*Cyrtograpsus angulatus*	Méndez Casariego et al. 2009
	Interspecies	*Gammarus tigrinus*	Platvoet et al. 2009
	Intraspecies	*Neohelice granulata*	Méndez Casariego et al. 2009
	Intraspecies	*Sesarma bidens*	Kneib et al. 1999
	Intraspecies	*Sesarma dehaani*	Kneib et al. 1999
Density		*Paramesopodopsis rufa*	Johnston and Ritz 2001
		Anisomysis mixta australis	Johnston and Ritz 2001
		Astacus leptodactylus	Farhadi et al. 2014
		Callinectes sapidus	Zmora et al. 2005
	No effect	*Chionoecetes opilio*	Lovrich and Sainte-Marie 1997
		Harpacticus spp.	Dahms and Qian 2006
		Lithodes santolla	Sotelano et al. 2012, Sotelano et al. 2016

(continued)

Table 14.1. **(Continued)**

		Paralithodes camtschaticus	Long et al. 2015
		Romaleon setosum	Cerda and Wolff 1993
		Scylla serrata	Quinitio et al. 2001
	No effect	*Sesarma bidens*	Kneib et al. 1999
	No effect	*Sesarma dehaani*	Kneib et al. 1999
		Streptocephalus proboscideus	Dumont and Ali 2004
		Tenagomysis tasmaniae	Johnston and Ritz 2001
	No effect	*Ucides cordatas*	Ventura et al. 2008
		Ucides cordatus	Ventura et al. 2011
Habitat complexity/ refuge availability		*Austropotamobius pallipes*	Rosewarne et al. 2013
		Callinectes sapidus	Zmora et al. 2005
		Callinectes sapidus	Moksnes and Heck 2006
		Cancer magister	Dumbauld et al. 1993
		Chasmagnathus granulata	Luppi et al. 2001
		Cyrtograpsus angulatus	Luppi et al. 2001
		Dyspanopeus sayi	Heck and Hambrook 1991
		Gammarus spp.	MacNeil et al. 2008
		Gammarus tigrinus	Platvoet et al. 2009
		Hemigrapsus penicillatus	Kurihara et al. 1989
		Holthuisana transversa	Middlemiss et al. 2016
		Palaemonetes argentinus	Díaz et al. 2003
		Paralithodes camtschaticus	Stevens and Swiney 2005
		Paralithodes camtschaticus	Pirtle et al. 2012, Daly et al. 2012b
		Pseudocarcinus gigas	Gardner and Maguire 1998
	No effect	*Saduria entomon*	Sparrevik 1999
		Scylla serrata	Rabbani and Zeng 2005
		Thermosphaeroma thermophilum	Jormalainen and Shuster 1997
Habitat type		*Callinectes sapidus*	Moksnes et al. 1997
		Carcinus maenas	Moksnes 2004
Social environment		*Thermosphaeroma thermophilum*	Bleakley et al. 2013

CONCLUSIONS

Crustaceans are vitally important components of many ecosystems and are model taxa for understanding invasive species, and many are of vital economic importance. Molting generates highly size-structured populations in crustaceans. Frequent encounters with larger individuals in such size-structured populations coupled with the risks inherent in shedding their protective exoskeletons make crustaceans particularly susceptible to cannibalism. However, there is significant variation in the expression of cannibalism across species. Understanding the biology and ecology of crustaceans therefore requires detailed investigations of the causes and consequences of cannibalism within any given species.

REFERENCES

Amaral, V., J. Paula, S. Hawkins, and S. Jenkins. 2009. Cannibalistic interactions in two co-occurring decapod species: effects of density, food, alternative prey and habitat. Journal of Experimental Marine Biology and Ecology 368:88–93.

Beal, B.F. 2012. Ocean-based nurseries for cultured lobster (*Homarus americanus* Milne Edwards) postlarvae: initial field experiments off the coast of eastern Maine to examine effects of habitat and container type on growth and survival. Journal of Shellfish Research 31:167–176.

Beermann, J., J.T. Dick, and M. Thiel. 2015. Social recognition in amphipods: an overview. Pages 85–100 *in* L. Aquiloni, and E. Tricarico, editors. Social recognition in invertebrates: the knowns and the unknowns. Springer International Publishing, Switzerland.

Bleakley, B., S. Welter, K. McCauley-Cole, S. Shuster, and A. Moore. 2013. Cannibalism as an interacting phenotype: precannibalistic aggression is influenced by social partners in the endangered Socorro isopod (*Thermosphaeroma thermophilum*). Journal of Evolutionary Biology 26:832–842.

Boake, C.R.B. 1994. Applications of quantitative genetics to behavioral evolution. Pages 305–325 *in* C.R.B. Boake, editor. Quantitative genetic studies of behavioral evolution. University of Chicago Press, Chicago.

Borisov, R., A. Epelbaum, N. Kryakhova, A. Tertitskaya, and N. Kovatcheva. 2007. Cannibalistic behavior in red king crabs reared under artificial conditions. Russian Journal of Marine Biology 33:227–231.

Botsford, L.W., and D.E. Wickham. 1978. Behavior of age-specific, density-dependent models and the northern California Dungeness crab (*Cancer magister*) fishery. Journal of the Fisheries Board of Canada 35:833–843.

Caddy, J. 1986. Modelling stock-recruitment processes in Crustacea: some practical and theoretical perspectives. Canadian Journal of Fisheries and Aquatic Sciences 43:2330–2344.

Camus, T., and C. Zeng. 2009. The effects of stocking density on egg production and hatching success, cannibalism rate, sex ratio and population growth of the tropical calanoid copepod *Acartia sinjiensis*. Aquaculture 287:145–151.

Casariego, A.M., J. Alberti, T. Luppi, P. Daleo, and O. Iribarne. 2011. Habitat shifts and spatial distribution of the intertidal crab *Neohelice* (*Chasmagnathus*) *granulata* Dana. Journal of Sea Research 66:87–94.

Cerda, G., and M. Wolff. 1993. Feeding ecology of the crab *Cancer polyodon* in La Herradura Bay, northern Chile. II. Food spectrum and prey consumption. Marine Ecology Progress Series 100:119–125.

Cheng, H.W., and L. Jefferson. 2008. Different behavioral and physiological responses in two genetic lines of laying hens after transportation. Poultry Science 87:885–892.

Claessen, D., A.M. de Roos, and L. Persson. 2004. Population dynamic theory of size–dependent cannibalism. Proceedings of the Royal Society B: Biological Sciences 271:333–340.

Crowley, P.H., and K.R. Hopper. 1994. How to behave around cannibals: a density-dependent dynamic game. The American Naturalist 143:117–154.

Cushing, J.M. 1992. A size-structured model for cannibalism. Theoretical Population Biology 42:347–361.

Dahms, H.-U., and P.-Y. Qian. 2006. Kin recognition during intraspecific predation of *Harpacticus* sp. (Copepoda, Harpacticoida). Zoological studies 45:395–403.

Daly, B., and J.S. Swingle. 2013. High-density nursery culture of recently-settled blue king crabs (*Paralithodes platypus*): comparisons to red king crabs (*Paralithodes camtschaticus*). Aquaculture 416:196–200.

Daly, B., J.S. Swingle, and G.L. Eckert. 2012a. Increasing hatchery production of juvenile red king crabs (*Paralithodes camtschaticus*) through size grading. Aquaculture 364:206–211.

Daly, B., G.L. Eckert, T.D. White, and B. Sainte-Marie. 2012b. Predation of hatchery-cultured juvenile red king crabs (*Paralithodes camtschaticus*) in the wild. Canadian Journal of Fisheries and Aquatic Sciences 70:358–366.

Díaz, A.C., L.G. Sousa, E.I. Cuartas, and A.M. Petriella. 2003. Growth, molt and survival of *Palaemonetes argentinus* (Decapoda, Caridea) under different light-dark conditions. Iheringia, Série Zoologia 93:249–254.

Dick, J.T. 1995. The cannibalistic behaviour of two *Gammarus* species (Crustacea: Amphipoda). Journal of Zoology 236:697–706.

Dick, J.T., and D. Platvoet. 2000. Invading predatory crustacean *Dikerogammarus villosus* eliminates both native and exotic species. Proceedings of the Royal Society B: Biological Sciences 267:977–983.

Dick, J.T., R.W. Elwood, and W.I. Montgomery. 1995. The behavioural basis of a species replacement: differential aggression and predation between the introduced *Gammarus pulex* and the native *G. duebeni celticus* (Amphipoda). Behavioral Ecology and Sociobiology 37:393–398.

Dick, J.T., W.I. Montgomery, and R.W. Elwood. 1993. Replacement of the indigenous amphipod *Gammarus duebeni celticus* by the introduced *G. pulex*: differential cannibalism and mutual predation. Journal of Animal Ecology 62: 79–88.

Dick, J.T., W.I. Montgomery, and R.W. Elwood. 1999. Intraguild predation may explain an amphipod replacement: evidence from laboratory populations. Journal of Zoology 249:463–468.

Dittel, A. I., A.H. Hines, G.M. Ruiz, and K.K. Ruffin. 1995. Effects of shallow water refuge on behavior and density-dependent mortality of juvenile blue crabs in Chesapeake Bay. Bulletin of Marine Science 57:902–916.

Dong, Q., and D.L. DeAngelis. 1998. Consequences of cannibalism and competition for food in a smallmouth bass population: an individual-based modeling study. Transactions of the American Fisheries Society 27:174–191.

Duffy, E.J., C.L. Morrison, and K.S. Macdonald. 2002. Colony defense and behavioral differentiation in the eusocial shrimp *Synalpheus regalis*. Behavioral Ecology and Sociobiology 51:488–495.

Dumbauld, B.R., D.A. Armstrong, and T.L. McDonald. 1993. Use of oyster shell to enhance intertidal habitat and mitigate loss of Dungeness crab (*Cancer magister*) caused by dredging. Canadian Journal of Fisheries and Aquatic Sciences 50:381–390.

Dumont, H.J., and A.J. Ali. 2004. Stage-specific cannibalism and spontaneous cyst hatching in the freshwater fairy shrimp *Streptocephalus proboscideus* Frauenfeld. Hydrobiologia 524:103–113.

Dutil, J., J. Munro, and M. Peloquin. 1997. Laboratory study of the influence of prey size on vulnerability to cannibalism in snow crab (*Chionoecetes opilio* O. Fabricius, 1780). Journal of Experimental Marine Biology and Ecology 212:81–94.

Eggleston, D.B., L.L. Etherington, and W.E. Elis. 1998. Organism response to habitat patchiness: species and habitat-dependent recruitment of decapod crustaceans. Journal of Experimental Marine Biology and Ecology 223:111–132.

Elgar, M.A., and B. Crespi. 1992. Cannibalism, ecology and evolution among diverse taxa. Oxford University Press, Oxford, England.

Ellen, E.D., V. Ducrocq, B.J. Ducro, R.F. Veerkamp, and P. Bijma. 2010. Genetic parameters for social effects on survival in cannibalistic layers: combining survival analysis and a linear animal model. Genetics Selection Evolution 42:27.

Elwood, R.W. 1994. Temporal-based kinship recognition: a switch in time saves mine. Behavioural Processes 33:15–24.

Fagan, W.F., and G.M. Odell. 1996. Size-dependent cannibalism in praying mantids: using biomass flux to model size-structured populations. The American Naturalist 147:230–268.

Farhadi, A., C. Gardner, and P. Kochanian. 2014. Reducing cannibalism of narrow clawed crayfish *Astacus leptodactylus* Eschscholtz 1823 through management of photoperiod and stocking density. Asian Fisheries Science 27:286–196.

Fernández, M. 1999. Cannibalism in Dungeness crab *Cancer magister*: effects of predator-prey size ratio, density, and habitat type. Marine Ecology Progress Series 182:221–230.

Fernandez, M., D. Armstrong, and O. Iribarne. 1993. First cohort of young-of-the-year Dungeness crab, *Cancer magister*, reduces abundance of subsequent cohorts in intertidal shell habitat. Canadian Journal of Fisheries and Aquatic Sciences 50:2100–2105.

Figler, M.H., G.S. Blank, and H.V. Peeke. 1997. Maternal aggression and post-hatch care in red swamp crayfish, *Procambarus clarkii* (Girard): the influences of presence of offspring, fostering, and maternal molting. Marine and Freshwater Behaviour and Physiology 30:173–194.

Fox, L.R. 1975. Cannibalism in natural populations. Annual Review of Ecology and Systematics 6:87–106.

Fromhage, L., and J.M. Schneider. 2005. Safer sex with feeding females: sexual conflict in a cannibalistic spider. Behavioral Ecology 16:377–382.

Gallucci, F., and E. Ólafsson. 2007. Cannibalistic behaviour of rock-pool copepods: an experimental approach for space, food and kinship. Journal of Experimental Marine Biology and Ecology 342:325–331.

Gardner, C., and G.B. Maguire. 1998. Effect of photoperiod and light intensity on survival, development and cannibalism of larvae of the Australian giant crab *Pseudocarcinus gigas* (Lamarck). Aquaculture 165:51–63.

Gopalakrishnan, K. 1976. Larval rearing of red shrimp *Penaeus marginatus* (Crustacea). Aquaculture 9:145–154.

Griffen, B.D., and H. Mosblack. 2011. Predicting diet and consumption rate differences between and within species using gut ecomorphology. Journal of Animal Ecology 80:854–863.

Griffen, B.D., and M.E. Riley. 2015. Potential impacts of invasive crabs on one life history strategy of native rock crabs in the Gulf of Maine. Biological Invasions 17:2533–2544.

Griffen, B.D., M. Vogel, L. Goulding, and R. Hartman. 2015. Energetic effects of diet choice by invasive Asian shore crabs: implications for persistence when prey are scarce. Marine Ecology Progress Series 522:181–192.

Hack, M.A., D.J. Thompson, and D.M. Fernandes. 1997. Fighting in males of the autumn spider, *Metellina segmentata*: effects of relative body size, prior residency and female value on contest outcome and duration. Ethology 103:488–498.

Hardege, J. D., and J.A. Terschak. 2010. Identification of crustacean sex pheromones. Pages 373–392 *in* T. Breithaupt and M. Thiel, editors. Chemical communication in crustaceans. Springer, New York.

Hatcher, M.J., J.T.A. Dick, and A.M. Dunn. 2014. Parasites that change predator or prey behaviour can have keystone effects on community composition. Biology Letters 10:20130879.

Hayden, D., A. Jennings, C. Müller, D. Pascoe, R. Bublitz, H. Webb, T. Breithaupt, L. Watkins, and J. Hardege. 2007. Sex-specific mediation of foraging in the shore crab, *Carcinus maenas*. Hormones and Behavior 52:162–168.

Hazel, W.N., R. Smock, and M.D. Johnson. 1990. A polygenic model for the maintenance and evolution of conditional strategies. Proceedings of the Royal Society B: Biological Sciences 242:181–187.

Hazel, W.N., R. Smock, and C.M. Lively. 2004. The ecological genetics of conditional strategies. The American Naturalist 163:888–900.

Hazlett, B.A. 1983. Parental behavior in decapod Crustacea. Pages 171–193 *in* S. Rebach, and D.W. Dunham, editors. Studies in Adaptation: The Behaviour of Higher Crustacea. Wiley, New York.

Heck, K.L., and J.A. Hambrook. 1991. Intraspecific interactions and risk of predation for *Dyspanopeus sayi* (Decapoda: Xanthidae) living on polychaete (*Filograna implexa*, Serpulidae) colonies. Marine Ecology 12:243–250.

Heck, K.L., and K.A. Wilson. 1987. Predation rates on decapod crustaceans in latitudinally separated seagrass communities: a study of spatial and temporal variation using tethering techniques. Journal of Experimental Marine Biology and Ecology 107:87–100.

Helfman, G.S. 1979. Coconut crabs and cannibalism. Natural History 88:76–83.

Hines, A.H., R.N. Lipcius, and A.M. Haddon. 1987. Population dynamics and habitat partitioning by size, sex, and molt stage of blue crabs *Callinectes sapidus* in a subestuary of central Chesapeake Bay. Marine Ecology Progress Series 36:55–64.

Huss, M., T. van Kooten, and L. Persson. 2010. Intra-cohort cannibalism and size bimodality: a balance between hatching synchrony and resource feedbacks. Oikos 119:2000–2011.

Hvam, A., D. Mayntz, and R.K. Nielsen. 2005. Factors affecting cannibalism among newly hatched wolf spiders (Lycosidae, *Pardosa amentata*). Journal of Arachnology 33:377–383.

Johnson, J.C., and A. Sih. 2005. Precopulatory sexual cannibalism in fishing spiders (*Dolomedes triton*): a role for behavioral syndromes. Behavioral Ecology and Sociobiology 58: 390–396.

Johnson, J.C., K. Kitchen, and M.C. Andrade. 2010. Family affects sibling cannibalism in the black widow spider, *Latrodectus hesperus*. Ethology 116:770–777.

Johnston, N., and D. Ritz. 2001. Synchronous development and release of broods by the swarming mysids *Anisomysis mixta australis, Paramesopodopsis rufa* and *Tenagomysis tasmaniae* (Mysidacea: Crustacea). Marine Ecology Progress Series 223:225–233.

Jormalainen, V. 1998. Precopulatory mate guarding in crustaceans: male competitive strategy and intersexual conflict. Quarterly Review of Biology 73:275–304.

Jormalainen, V., and S.M. Shuster. 1997. Microhabitat segregation and cannibalism in an endangered freshwater isopod, *Thermosphaeroma thermophilum*. Oecologia 111:271–279.

Jormalainen, V., and S.M. Shuster. 1999. Female reproductive cycle and sexual conflict over precopulatory mate-guarding in *Thermosphaeroma* (Crustacea, Isopoda). Ethology 105:233–246.

Kennish, R. 1996. Diet composition influences the fitness of the herbivorous crab *Grapsus albolineatus*. Oecologia 105:22–29.

Kneib, R., S. Lee, and J. Kneib. 1999. Adult–juvenile interactions in the crabs *Sesarma* (*Perisesarma*) *bidens* and *S.* (*Holometopus*) *dehaani* (Decapoda: Grapsidae) from intertidal mangrove habitats in Hong Kong. Journal of Experimental Marine Biology and Ecology 234:255–273.

Kolts, J.M., J.R. Loworn, C.A. North, J.M. Grebmeier, and L.W. Cooper. 2013. Effects of body size, gender, and prey availability on diets of snow crabs in the northern Bering Sea. Marine Ecology Progress Series 483:209–220.

Kurihara, Y., K. Okamoto, and S. Takeda. 1989. Preference of the grapsid crab *Hemigrapsus penicillatus* (De Haan) for an appropriate aperture. Marine and Freshwater Behaviour and Physiology 14:169–179.

Kuroda, M., K. Wada, and M. Kamada. 2005. Factors influencing coexistence of two brachyuran crabs, *Helice tridens* and *Parasesarma plicatum*, in an estuarine salt marsh, Japan. Journal of Crustacean Biology 25:146–153.

Lee, K.W., H.U. Dahms, H.G. Park, and J.H. Kang. 2013. Population growth and productivity of the cyclopoid copepods *Paracyclopina nana, Apocyclops royi* and the harpacticoid copepod *Tigriopus japonicus* in mono and polyculture conditions: a laboratory study. Aquaculture Research 44:836–840.

Lewis, S.E., J.T. Dick, E.K. Lagerstrom, and H.C. Clarke. 2010. Avoidance of filial cannibalism in the amphipod *Gammarus pulex*. Ethology 116:138–146.

Linton, S.M., and P. Greenaway. 2007. A review of feeding and nutrition of herbivorous land crabs: adaptations to low quality plant diets. Journal of Comparative Physiology B 177:269–286.

Lipcius, R., and W. Herrnkind. 1982. Molt cycle alterations in behavior, feeding and diel rhythms of a decapod crustacean, the spiny lobster *Panulirus argus*. Marine biology 68:241–252.

Long, W.C., S.B. van Sant, and J.A. Haaga. 2015. Habitat, predation, growth, and coexistence: could interactions between juvenile red and blue king crabs limit blue king crab productivity? Journal of Experimental Marine Biology and Ecology 464:58–67.

Lovrich, G.A., and B. Sainte-Marie. 1997. Cannibalism in the snow crab, *Chionoecetes opilio* (O. Fabricius) (Brachyura: Majidae), and its potential importance to recruitment. Journal of Experimental Marine Biology and Ecology 211:225–245.

Luppi, T.A., E.D. Spivak, and K. Anger. 2001. Experimental studies on predation and cannibalism of the settlers of *Chasmagnathus granulata* and *Cyrtograpsus angulatus* (Brachyura: Grapsidae). Journal of Experimental Marine Biology and Ecology 265:29–48.

Luppi, T., E. Spivak, K. Anger, and J. Valero. 2002. Patterns and processes of *Chasmagnathus granulata* and *Cyrtograpsus angulatus* (Brachyura: Grapsidae) recruitment in Mar Chiquita coastal lagoon, Argentina. Estuarine, Coastal and Shelf Science 55:287–297.

MacNeil, C., J.T. Dick, and R.W. Elwood. 1997. The trophic ecology of freshwater *Gammarus* spp. (Crustacea: Amphipoda): problems and perspectives concerning the functional feeding group concept. Biological Reviews of the Cambridge Philosophical Society 72:349–364.

MacNeil, C., D. Platvoet, and J.T. Dick. 2008. Potential roles for differential body size and microhabitat complexity in mediating biotic interactions within invasive freshwater amphipod assemblages. Fundamental and Applied Limnology/Archiv für Hydrobiologie 172:175–182.

MacNeil, C., J. Prenter, M. Briffa, N.J. Fielding, J.T. Dick, G.E. Riddell, M.J. Hatcher, and A.M. Dunn. 2004. The replacement of a native freshwater amphipod by an invader: roles for environmental degradation and intraguild predation. Canadian Journal of Fisheries and Aquatic Sciences 61:1627–1635.

Marshall, S., K. Warburton, B. Paterson, and D. Mann. 2005. Cannibalism in juvenile blue-swimmer crabs *Portunus pelagicus* (Linnaeus, 1766): effects of body size, moult stage and refuge availability. Applied Animal Behaviour Science 90:65–82.

Martínez-Laiz, G., and J. Guerra-García. 2015. Dietary analysis of caprellids *Caprella penantis* and *Caprella grandimana* (Crustacea: Amphipoda) in southern Spain. Marine Biology 162:2057–2066.

McGlothlin, J.W., A.J. Moore, J.B. Wolf, and E.D. Brodie, III. 2010. Interacting phenotypes and the evolutionary process III: social evolution. Evolution 64:2558–2574.

McGrath, K.E., E.T. Peeters, J.A. Beijer, and M. Scheffer. 2007. Habitat-mediated cannibalism and microhabitat restriction in the stream invertebrate *Gammarus pulex*. Hydrobiologia 589:155–164.

Meffe, G.K., and M.L. Crump. 1987. Possible growth and reproductive benefits of cannibalism in the mosquito fish. The American Naturalist 129:203–212.

Méndez Casariego, A., J. Alberti, T. Luppi, and O. Iribarne. 2009. Stage-dependent interactions between intertidal crabs: from facilitation to predation. Journal of the Marine Biological Association of the United Kingdom 89:781–788.

Middlemiss, K.L., M.A. Urbina, and R.W. Wilson. 2016. Effects of seawater alkalinity on calcium and acid–base regulation in juvenile European lobster (*Homarus gammarus*) during a moult cycle. Comparative Biochemistry and Physiology Part A: Molecular & Integrative Physiology 193:22–28.

Moksnes, P.-O. 2004. Self-regulating mechanisms in cannibalistic populations of juvenile shore crabs *Carcinus maenas*. Ecology 85:1343–1354.

Moksnes, P.-O., and K.L. Heck. 2006. Relative importance of habitat selection and predation for the distribution of blue crab megalopae and young juveniles. Marine Ecology Progress Series 308:165–181.

Moksnes, P.-O., R. Lipcius, L. Pihl, and J. van Montfrans. 1997. Cannibal-prey dynamics in young juveniles and postlarvae of the blue crab. Journal of Experimental Marine Biology and Ecology 215:157–187.

Moore, A.J., E.D. Brodie, and J.B. Wolf. 1997. Interacting phenotypes and the evolutionary process. I. Direct and indirect genetic effects of social interactions. Evolution 51:1352–1362.

Navarrete, S.A., and J.C. Castilla. 1990. Resource partitioning between intertidal predatory crabs: interference and refuge utilization. Journal of Experimental Marine Biology and Ecology 143:101–129.

Ohman, M., and H.-J. Hirche. 2001. Density-dependent mortality in an oceanic copepod population. Nature 412:638–641.

Okamoto, K., and Y. Kurihara. 1989. Feeding habit and food selection of the grapsid crab *Hemigrapsus penicillatus*. Japanese Journal of Ecology 39:195–202.

Parsons, W., W. Zhong, and V.H. Rudolf. 2013. Mating status and kin recognition influence the strength of cannibalism. Animal Behaviour 85:365–369.

Patterson, L., J.T. Dick, and R.W. Elwood. 2008. Embryo retrieval and kin recognition in an amphipod (Crustacea). Animal Behaviour 76:717–722.

Penn, D.J., and J.G. Frommen. 2010. Kin recognition: an overview of conceptual issues, mechanisms and evolutionary theory. Pages 55–85 *in* P.M. Kappeler, editor. Animal behaviour: evolution and mechanisms. Springer, Berlin, Germany.

Pfennig, D.W. 1997. Kinship and cannibalism. Bioscience 47:667–675.

Pfennig, D.W., S.G. Ho, and E.A. Hoffman. 1998. Pathogen transmission as a selective force against cannibalism. Animal Behaviour 55:1255–1261.

Pirtle, J.L., G.L. Eckert, and A.W. Stoner. 2012. Habitat structure influences survival and predator-prey interactions of early juvenile red king crab, *Paralithodes camtschaticus*. Marine Ecology Progress Series 465:169–184.

Platvoet, D., J.T. Dick, C. MacNeil, M.C. van Riel, and G. van der Velde. 2009. Invader-invader interactions in relation to environmental heterogeneity leads to zonation of two invasive amphipods, *Dikerogammarus villosus* (Sowinsky) and *Gammarus tigrinus* Sexton: amphipod pilot species project (AMPIS) report 6. Biological Invasions 11:2085–2093.

Polis, G.A. 1981. The evolution and dynamics of intraspecific predation. Annual Review of Ecology and Systematics 12:225–251.

Quinitio, E., F.D. Parado-Estepa, O. Millamena, E. Rodriguez, and E. Borlongan. 2001. Seed production of mud crab *Scylla serrata* juveniles. Asian Fisheries Science 14:161–174.

Rabbani, A.G., and C. Zeng. 2005. Effects of tank colour on larval survival and development of mud crab *Scylla serrata* (Forskål). Aquaculture Research 36:1112–1119.
Reaka, M.L. 1987. Adult-juvenile interactions in benthic reef crustaceans. Bulletin of Marine Science 41:108–134.
Richardson, M.L., R.F. Mitchell, P.F. Reagel, and L.M. Hanks. 2010. Causes and consequences of cannibalism in noncarnivorous insects. Annual Review of Entomology 55:39–53.
Roff, D.A. 1996. The evolution of threshold traits in animals. Quarterly Review of Biology 71:3–34.
Roff, D.A. 1997. Evolutionary quantitative genetics. Chapman and Hall, London, England.
Rosewarne, P., R. Mortimer, and A. Dunn. 2013. Size-dependent impacts of the endangered white-clawed crayfish (*Austropotamobius pallipes*) (Lereboullet) on the littoral community. Knowledge and Management of Aquatic Ecosystems 409:06.
Rossi, V., G. Benassi, F. Belletti, and P. Menozzi. 2011. Colonization, population dynamics, predatory behaviour and cannibalism in *Heterocypris incongruens* (Crustacea: Ostracoda). Journal of Limnology 70:102–108.
Rudolf, V.H.W. 2007. Consequences of stage-structures predators: cannibalism, behavioral effects, and trophic cascades. Ecology 88:2991–3003.
Ruiz, G.M., A.H. Hines, and M.H. Posey. 1993. Shallow water as a refuge habitat for fish and crustaceans in non-vegetated estuaries: an example from Chesapeake Bay. Marine Ecology Progress Series 99:55–68.
Sainte-Marie, B., and M. Lafrance. 2002. Growth and survival of recently settled snow crab *Chionoecetes opilio* in relation to intra-and intercohort competition and cannibalism: a laboratory study. Marine Ecology Progress Series 244:191–203.
Sato, T., and K. Yoseda. 2010. Influence of size-and sex-biased harvesting on reproduction of the coconut crab *Birgus latro*. Marine Ecology Progress Series 402:171–178.
Schausberger, P., and B.A. Croft. 2001. Kin recognition and larval cannibalism by adult females in specialist predaceous mites. Animal Behaviour 61:459–464.
Shuster, S.M., M.P. Miller, B.K. Lang, N. Zorich, L. Huynh, and P. Keim. 2005. The effects of controlled propagation on an endangered species: genetic differentiation and divergence in body size among native and captive populations of the Socorro Isopod (Crustacea: Flabellifera). Conservation Genetics 6:355–368.
Smith, J.M. 1979. Game theory and the evolution of behaviour. Proceedings of the Royal Society B: Biological Sciences 205:475–488.
Smithsonian Environmental Research Center. 2014. Blue crab cannibalism. https://www.youtube.com/watch?v=2D-HQqQLnZA (accessed on November 1, 2016).
Sotelano, M.P., G.A. Lovrich, M.C. Romero, and F. Tapella. 2012. Cannibalism during intermolt period in early stages of the Southern king crab *Lithodes santolla* (Molina 1872): effect of stage and predator–prey proportions. Journal of Experimental Marine Biology and Ecology 411:52–58.
Sotelano, M.P., G.A. Lovrich, and F. Tapella. 2016. Cannibalism among *Lithodes santolla* (Molina 1782) juveniles: effect of stocking density, stage and molt condition. Aquaculture International 24:1025–1037.
Sparrevik, E. 1999. Sediment texture and cannibalism affect survival during moult in *Saduria entomon* (Isopoda). Marine Biology 133:437–441.
Sparrevik, E., and K. Leonardsson. 1998. Recruitment in the predacious isopod *Saduria entomon* (L.): alternative prey reduces cannibalism. Journal of Experimental Marine Biology and Ecology 221:117–130.
Stevens, B.G., and K.M. Swiney. 2005. Post-settlement effects of habitat type and predator size on cannibalism of glaucothoe and juveniles of red king crab *Paralithodes camtschaticus*. Journal of Experimental Marine Biology and Ecology 321:1–11.
Stevens, B.G., A. Dunham, J. Kittaka, N. Kovatcheva, S. Persselin, and G. van der Meeren. 2014. Aquaculture and stock enhancement of king crabs. Pages 403–448 *in* B.G. Stevens, editor. King crabs of the world: biology and fisheries management. CRC Press, Boca Raton, Florida.
Stevens, L. 1989. The genetics and evolution of cannibalism in flour beetles (genus *Tribolium*). Evolution 43:169–179.

Stevens, L. 1994. Genetic analysis of cannibalism behavior in *Tribolium* flour beetles. Pages 206–224 *in* C.R.B. Boake, editor. Quantitative genetic studies of behavior. University of Chicago Press, Chicago, Illinois.

Stoner, A.W., L.A. Copeman, and M.L. Ottmar. 2013. Molting, growth, and energetics of newly-settled blue king crab: effects of temperature and comparisons with red king crab. Journal of Experimental Marine Biology and Ecology 442:10–21.

Queller, D.C., and J.E. Strassmann. 1998. Kin selection and social insects social insects provide the most surprising predictions and satisfying tests of kin selection. Bioscience 48:165–175.

Suprayudi, M.A., T. Takeuchi, K. Hamasaki, and J. Hirokawa. 2002. Effect of *Artemia* feeding schedule and density on the survival and development of larval mud crab *Scylla serrata*. Fisheries Science 68:1295–1303.

Thiel, M. 2007. Social behavior of parent-offspring groups in crustaceans. Pages 294–338 *in* J.E. Duffy, and M. Thiel, editors. Evolutionary ecology of social and sexual systems: crustaceans as model organisms. Oxford University Press, New York.

Tran, M.V. 2014. The scent of cannibalism: The olfactory basis of cannibalism in hermit crabs. Journal of Experimental Marine Biology and Ecology 457:8–14

Tsai, M.L., and C.F. Dai. 2003. Cannibalism within mating pairs of the parasitic isopod, *Ichthyoxenus fushanensis*. Journal of Crustacean Biology 23:662–668.

Uye, S.-i., and D. Liang. 1998. Copepods attain high abundance, biomass and production in the absence of large predators but suffer cannibalistic loss. Journal of Marine Systems 15:495–501.

van Huis, A., G. Woldewahid, K. Toleubayev, and W. van der Werf. 2008. Relationships between food quality and fitness in the desert locust, *Schistocerca gregaria*, and its distribution over habitats on the Red Sea coastal plain of Sudan. Entomologia Experimentalis et Applicata 127:144–156.

Ventura, R., U.A. Da Silva, G. Perbiche-Neves, A. Ostrensky, W.A. Boeger, and M.R. Pie. 2008. Larval cannibalism rates in the mangrove crab *Ucides cordatus* (Decapoda: Ocypodidae) under laboratory conditions. Aquaculture Research 39:263–267.

Ventura, R., U.A. da Silva, A. Ostrensky, and K. Cottens. 2011. Intraspecific interactions in the mangrove crab *Ucides cordatus* (Decapoda: Ocypodidae) during the metamorphosis and post-metamorphosis periods under laboratory conditions. Zoologia (Curitiba) 28:1–7.

Villefranche-sur-Mer, F. 1999. Vertical migration behaviour in the northern krill *Meganyctiphanes norvegica* is influenced by moult and reproductive processes. Marine Ecology Progress Series 190:253–262.

Vinuesa, J.H., M.A. Varisco, and P. Balzi. 2013. Feeding strategy of early juvenile stages of the southern king crab *Lithodes santolla* in the San Jorge Gulf, Argentina. Revista de Biología Marina y Oceanografía 48:353–363.

Wagner, J.D., M.D. Glover, J.B. Moseley, and A.J. Moore. 1999. Heritability and fitness consequences of cannibalism in *Harmonia axyridis*. Evolutionary Ecology Research 1:375–388.

Wall, D., B. Paterson, and R. Mohan. 2009. Behaviour of juvenile mud crabs *Scylla serrata* in aquaculture: response to odours of moulting or injured crabs. Applied Animal Behaviour Science 121:63–73.

Warburg, M., and N. Cohen. 1991. Reproductive pattern, allocation, and potential in a semelparous isopod from the Mediterranean region of Israel. Journal of Crustacean Biology 11:368–374.

Ward, P.I. 1985. The breeding behaviour of *Gammarus duebeni*. Hydrobiologia 121:45–50.

Waterkeyn, A., P. Grillas, M. Anton-Pardo, B. Vanschoenwinkel, and L. Brendonck. 2011. Can large branchiopods shape microcrustacean communities in Mediterranean temporary wetlands? Marine and Freshwater Research 62:46–53.

Williams, T., and O. Hernandez. 2006. Costs of cannibalism in the presence of an iridovirus pathogen of *Spodoptera frugiperda*. Ecological Entomology 31:106–113.

Wolcott, D.L., and N.J. O'Connor. 1992. Herbivory in crabs: adaptations and ecological considerations. American Zoologist 32:370–381.

Wolcott, D.L., and T.G. Wolcott. 1984. Food quality and cannibalism in the red land crab, *Gecarcinus lateralis*. Physiological Zoology 57:318–324.

Zmora, O., A. Findiesen, J. Stubblefield, V. Frenkel, and Y. Zohar. 2005. Large-scale juvenile production of the blue crab *Callinectes sapidus*. Aquaculture 244:129–139.

15

THE LIFE CYCLE OF SYMBIOTIC CRUSTACEANS: A PRIMER

J. Antonio Baeza, Emiliano H. Ocampo, and Tomás A. Luppi

Abstract

In the subphylum Crustacea, species from most major clades have independently evolved symbiotic relationships with a wide variety of invertebrate and vertebrate hosts. Herein, we review the life cycle disparity in symbiotic crustaceans. Relatively simple life cycles with direct or abbreviated development can be found among symbiotic decapods, mysids, and amphipods. Compared to their closest free-living relatives, no major life cycle modifications were detected in these clades as well as in most symbiotic cirripeds. In contrast, symbiotic isopods, copepods, and tantulocarids exhibit complex life cycles with major differences compared to their closest free-living relatives. Key modifications in these clades include the presence of larval stages well endowed for dispersal and host infestation, and the use of up to 2 different host species with dissimilar ecologies throughout their ontogeny. Phylogenetic inertia and restrictions imposed by the body plan of some clades appear to be most relevant in determining life cycle modifications (or the lack thereof) from the "typical" ground pattern. Furthermore, the life cycle ground pattern is likely either constraining or favoring the adoption of a symbiotic lifestyle in some crustacean clades (e.g., in the Thecostraca).

INTRODUCTION

A symbiotic lifestyle is one of the most important environmental adaptations in crustaceans (Baeza 2015). Symbiosis here is defined as dissimilar organisms living together sensu de Bary (1879). These associations most often involve small organisms (hereafter symbionts or symbiotic guests) and their large partners that serve as hosts (Baeza 2015).

In crustaceans, some degree of dependence between pairs or among small assemblages of species has evolved multiple times independently in tropical, subtropical, and temperate habitats (Baeza 2015). Indeed, symbiotic crustaceans have a long evolutionary history; parasitization of

Life Histories. Edited by Gary A. Wellborn and Martin Thiel.

ostracods by pentastomids, crinoids by copepods, and decapods by bopyrid isopods is evident in the fossil record and extends back to at least the Jurassic (de Baets and Littlewood 2015, Nagler et al. 2016). Many species of shrimps, mysids, amphipods, isopods, crabs, cirripeds, copepods, pentastomids, branchiurans, and tantulocarids, among others, engage in symbiotic associations with other invertebrates (e.g., sponges, corals, sea anemones, oysters, sea urchins, ascidians) or vertebrates, or both (Bruce 1976, Castro 1976, 2015, Baeza 2015).

Crustaceans are among the most diverse marine invertebrates (Brusca and Brusca 2003), and studies conducted during the last decades in symbiotic representatives of this species-rich clade have revealed most impressive morphologies (Schmitt et al. 1973), colorations (Limbaugh et al. 1961), nourishment tactics (Ďuriš et al. 2011), reproductive strategies (Shuster 2007), social interactions (Duffy et al. 2000, Hernández et al. 2012), and modes of interspecific communication (Vannini 1985, Becker et al. 2005). In this chapter, we provide an overview of the life cycle of representatives from all major crustacean clades that have adopted a symbiotic lifestyle, and where possible, we compare the life cycle of symbiotic species to that of their closest free-living relatives. This comparison aims to reveal life cycle adaptations to the symbiotic mode of life.

SYMBIOTIC CRUSTACEANS AND THEIR LIFE CYCLES

The systematic arrangement available at the World Register of Marine Species (http://www.marinespecies.org/) consulted during March 2018 is used herein as the taxonomic hierarchy to explore the life cycle of symbiotic crustaceans.

Superclass Multicrustacea, Class Malacostraca, Subclass Eumalacostraca

In the subclass Eumalacostraca within the diverse superclass Multicrustacea and class Malacostraca, the symbiotic lifestyle is pervasive in the superorder Eucarida, order Decapoda, and superorder Peracarida.

Order Decapoda

In the Decapoda, four (*Globospongicola, Paraspongicola, Spongicola, Spongicoloides,* and *Spongiocaris*) of six genera of stenopodid shrimps (infraorder Stenopodidea) in the families Spongicolidae and Eunicicolidae have adopted a symbiotic lifestyle. All representatives belonging to the 5 genera listed live as heterosexual pairs entrapped in the atrium of deep-water hexactinellid glass-sponges (Saito and Takeda 2003; Fig. 15.1). Also in the family Stenopodidae, most species are free-living but a few are facultative symbionts of sea anemones or sponges (e.g., *Stenopus hispidus*).

In the family Spongicolidae, shrimps are generally characterized by a reduced armor of the body and appendages and by a somewhat depressed body form (Holthuis 1993). Within the family, the life cycle of *Spongiocaris japonica* that lives in the atrium of *Euplectella oweni* is the best understood (Saito and Koya 2001, Saito et al. 2001). Ovigerous females of *S. japonica* (Fig. 15.1A) molt after hatching their offspring to spawn new eggs, likely fertilized by the male sharing the same host individual (Saito and Konishi 1999, Saito 2002). Development is direct and thus, juvenile shrimp hatch directly from eggs without intermediate planktonic stages (Saito and Konishi 1999; Fig. 15.1B). Recently hatched shrimps are mobile and leave their host individuals through the mesh of the sponge's osculum or lateral oscula (Saito and Konishi 1999). Likely, small, recently hatched juveniles disperse from the parental host individual and roam around in search of empty host individuals (and future sexual partners). These small juveniles can be found on the atrium of sponges either in small groups or solitarily. Larger juveniles and adults most often form heterosexual pairs (Saito et al. 2001; Fig. 15.1C); *S. japonica* appears to be socially monogamous (Saito 2002).

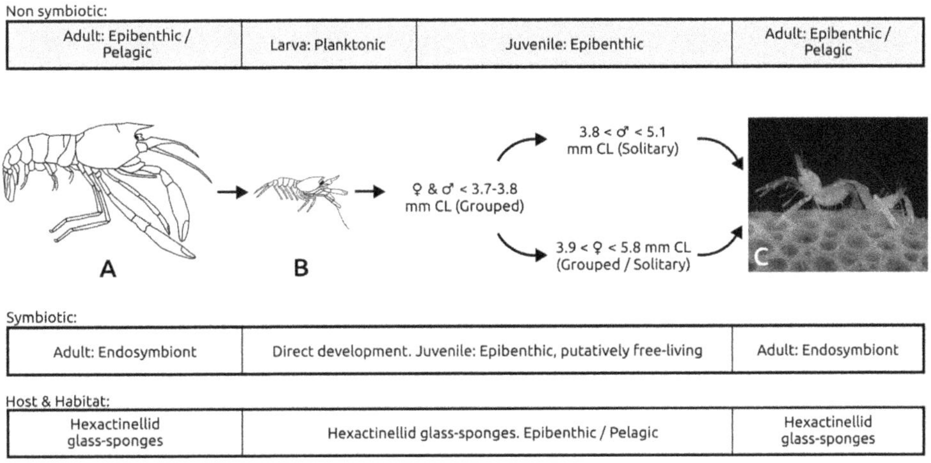

Fig. 15.1.
Stenopodidae: life cycle of the symbiotic shrimp *Spongiocaris japonica*. (A) Ovigerous females. Modified from Kubo (1942). (B) First juvenile. From Saito and Konishi (1999), with permission from The Crustacean Society. *S. japonica* have direct development, without free-living larval stages. Small juveniles disperse from the parental host and roam in search of empty hosts. These small juveniles can be found on the atrium of sponges either in groups or solitarily. (C) Paired adult shrimps of *Spongiocaris tuerkayi*, a close related species of *S. japonica*, in the surface of a glass-sponge host. Modified from Komai et al. (2016), with permission from Magnolia Press. The top and bottom rows show the life cycle of symbiotic as compared to nonsymbiotic species in the same clade as well as the host species used by the different ontogenetic phases in symbiotic species.

In the infraorder Caridea, representatives from several families engage in obligatory ectosymbiotic or endosymbiotic associations with a variety of hosts (de Grave 1999, Fransen 2006). *Thor amboinensis* (family Thoridae) is one of a few species with a relatively well-documented life cycle; it has a circumtropical distribution and lives symbiotically with several sessile macroinvertebrates (Baeza and Piantoni 2010). As in all carideans, parental females brood embryos (for an unknown time period) before they hatch as larvae. There are eight larval stages before the first decapodite in this species (Bartilotti et al. 2016). Nothing is known about the trophic ecology and bathymetric distribution of the different larval stages. In Hawaii, this species completes its larval development and undergoes metamorphosis without induction from its host species. After metamorphosis, juveniles grow and invariably mature as males first to then change sex to females later in life. Thus, this species is a protandric hermaphrodite. Limited information suggests that *T. amboinensis* features pure-search promiscuity as a mating system (Baeza and Piantoni 2010).

In the infraorder Brachyura (true crabs), the symbiotic lifestyle is also widespread (Ng et al. 2008). In the superfamily Trapezoidea, most genera (e.g., *Trapezia*) are obligate ectosymbionts of corals (Castro 1976, Castro et al. 2004). Representatives belonging to various other families develop facultative or obligatory associations with sessile or vagile macroinvertebrates in tropical and subtropical regions.

Information on the early life history of many symbiotic brachyurans is missing. *Mithraculus sculptus* and *M. forceps* (family Mithracidae) are two of the few species in which the life cycle is relatively well documented (Fig. 15.2). *M. sculptus* and *M. forceps* are mostly free-living species but engage in facultative symbioses with the coralline alga *Neogoniolithon strictum* in the Gulf of Mexico and the coral *Oculina arbuscula* in the western Atlantic, respectively (Stachowicz and Hay 1996, 1999).

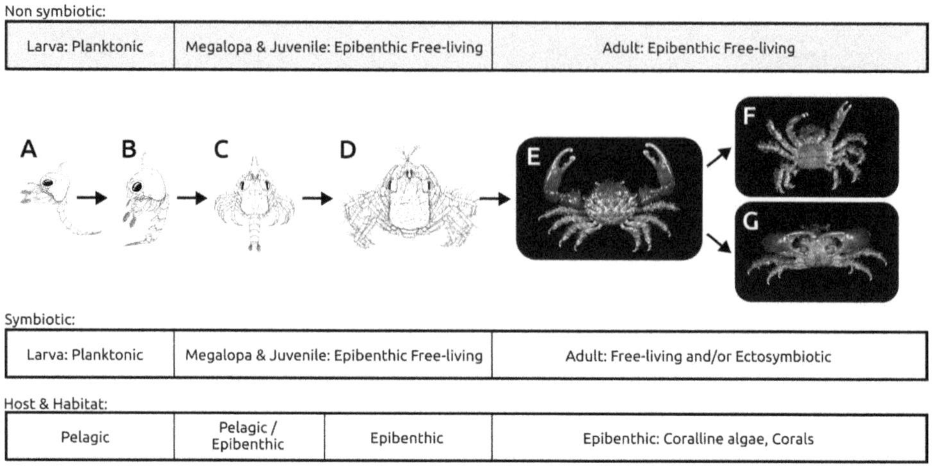

Fig. 15.2.
Decapoda Brachyura: life cycle of *Mithraculus sculptus*. (A) Zoea I larva. (B) Zoea II larva. (C) Megalopa. (D) First crab instar. (E) Dorsal view of adult crab. (F) Ventral view of adult female. (G) Ventral view of adult male. The top and bottom rows show the life cycle of symbiotic as compared to nonsymbiotic species in the same clade as well as the host species used by the different ontogenetic phases in symbiotic species. (A–D) Modified from Rhyne et al. (2006), with permission from Cambridge University Press. (E–G) Photographs by J. Poupin and L. Corbari © (from Guadeloupe, specimen deposited in the French National Museum of Natural History, MNHN IU-2013-6619).

Mithraculus sculptus and *M. forceps* feature 2 free-swimming zoea larval stages followed by a single megalopa stage (Rhyne et al. 2006; Fig. 15.2A–C). The megalopa molts into a fully benthic first crab instar with a gross morphology similar to that of older juvenile and adult crabs (Fig. 15.2D). After maturity, males attain a larger average and final body size than females. Also, males exhibit much larger chelipeds than females, while females have abdomens much wider than males (Fig. 15.2E–G; Baeza et al. 2012). Ovigerous females of *M. sculptus* produce 60 to 1,223 embryos that are brooded underneath the abdomen (Cobo and Okamori 2008).

Similarly, crabs in the superfamily Cryptochiroidea are obligate symbionts of scleractinians that live in self-constructed "galls" or "pits" in these coral colonies (Terrana et al. 2016). With a few exceptions, the life history of most members in this family is unknown (Potts 1915, Kotb and Hartnoll 2002, Terrana et al. 2016). Among gall-crabs, *Hapalocarcinus marsupialis* is a cosmopolitan tropical species that forms galls in various pocilloporid corals (Fig. 15.3A).

In *H. marsupialis*, larval development is not fully known but two zoea larvae stages have been observed in the laboratory (Gore et al. 1983; Fig. 15.3B). The symbiotic confamiliar *Troglocarcinus corallicola* is believed to have six to seven larval stages (Scotto and Gore 1981). Whether *H. marsupialis* megalopae sense chemical cues originating from coral hosts to settle is unknown. After larval metamorphosis, the gross morphology of the male and female juvenile crabs is similar (Terrana et al. 2016). Nonetheless, sexual dimorphism in terms of body size is considerable in adults of this and related species. Males do not appear to grow, remain tiny (about 1 mm carapace width; Fig. 15.3C–D), do not form galls, and apparently roam (continuously) among host individuals in search of receptive females (Castro 1976, Terrana et al. 2016). In contrast, females attain large body sizes (Fig. 15.3E–F) and induce the development of the famous galls in corals (Fig. 15.3G–J). It is not known how long the gall-forming process takes but it occurs while a solitary female living within the gall grows, matures, spawns eggs, and eventually becomes isolated from the outside world. Females of *H. marsupialis* experience considerable changes in morphology with increasing body size and age.

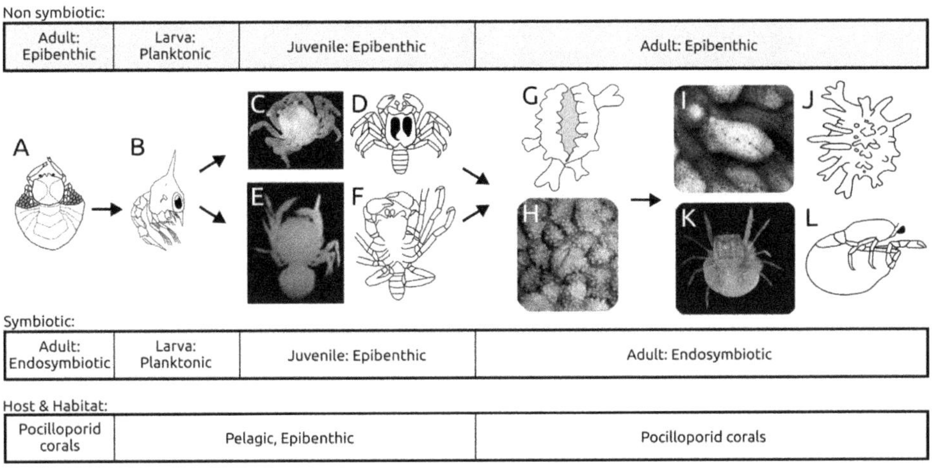

Fig. 15.3.
Decapoda Brachyura: life cycle of the gall-crab *Hapalocarcinus marsupialis*. (A) Ovigerous female. Modified from Gerald McCormak, Cook Islands Biodiversity Database, Version 2007.2. http://cookislands.bishopmuseum.org/species.asp?id=7372, accessed August 30, 2016. (B) Zoea I. Modified from Gore et al. (1983). (C) Tiny male ≈ 1 mm carapace width (CW). Photograph by Moorea Biocode, under Creative Commons license (BY-NC-SA). (D) Schematic drawing of dorsal view of male. Modified from Potts (1915). (E) Pubescent females 2–2.5 mm CW. From Terrana et al. (2016), with permission from Springer. (F) Schematic drawing of ventral view of pubescent female. Modified from Potts (1915). (G) Schematic drawing of open gall. Modified from Potts (1915). (H) Open galls in *Seriatopora* sp. Photograph by Matthieu Sontag, under Creative Commons license (BY-SA). Development of galls is induced by the gall crab. Tiny males and pubescent female mate inside open galls. Schematic drawing of open gall. From Potts (1915). (I) Close galls in *Seriatopora* sp. Photograph by Christian von Mach ©. (J) Schematic drawing of closed gall. Modified from Potts (1915). (K) Reproductively functional females 2.5–5.3 mm CW. These females are always inside close galls alone. Photograph by rosspolynesiaexpedition2014, under Creative Commons license (BY-NC). (L) Schematic drawing of lateral and dorsal view of mature female. Modified from Potts (1915). The top and bottom rows show the life cycle of symbiotic as compared to nonsymbiotic species in the same clade as well as the host species used by the different ontogenetic phases in symbiotic species.

The female abdomen exhibits positive allometric growth during puberty and greatly expands at the onset of maturity morphing into a hypertrophied "brooding pouch" (Kotb and Hartnoll 2002). Small females already have swollen spermathecae filled with spermatozoids, suggesting that mating with males occurs before functional maturity. Limited information suggests that females have the ability to store sperm to fertilize consecutive egg batches after the gall has fully closed (Terrana et al. 2016; Fig. 15.3K,L). After the eggs hatch, the first zoea larvae likely leave the parental "prison" through small pores in the gall (Kotb and Hartnoll 2002).

Lastly, the superfamily Pinnotheroidea (pea-crabs) is recognized for its symbiotic lifestyle with a wide variety of hosts (Schmitt et al. 1973). Pea-crabs can exhibit relatively complex postlarval life histories characterized by the alternation of free-living and symbiotic stages with remarkable shifts in morphology during ontogeny. *Calyptraeotheres garthi* (family Pinnotheridae), a parasitic castrator that inhabits the brooding chamber of slipper limpets, is the only species with all larvae (Ocampo et al. 2011) and postlarval stages (Ocampo et al. 2016) properly described (Fig. 15.4). Larval development comprises 5 zoea larvae and 1 megalopa (Fig. 15.4B–G). Nothing is known about the trophic ecology and bathymetric distribution of the different larval stages in this species. Limited information suggests that the invasive stage in this and various other related species is the first crab instar (Ocampo et al. 2011, 2014, 2016). However, in *Holotheres halingi* it is the megalopa that appears to be the invasive stage (Hamel et al. 1999).

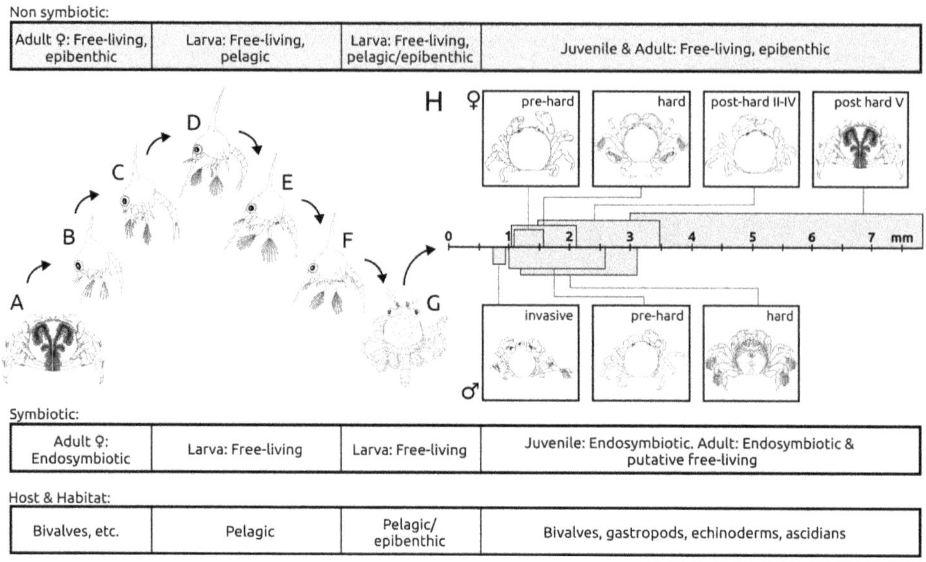

Fig. 15.4.
Decapoda Brachyura: life cycle of the pea-crab *Calyptraeotheres garthi*. (A) Ovigerous female. Modified from Ocampo et al. (2016), with permission from John Wiley and Sons. (B–G) larval series from Zoea I to Zoea V and megalopa. Modified from Ocampo et al. (2011), with permission from Springer. (H) Juvenile and adult development. The extent of each bar on the scale shows the carapace width range (mm) of each stage. Modified from Ocampo et al. (2016), with permission from John Wiley and Sons. The top and bottom rows show the life cycle of symbiotic as compared to nonsymbiotic species in the same clade as well as the host species used by the different ontogenetic phases in symbiotic species.

The invasive first crab instar of *C. garthi* exhibits various morphological features that seem to represent adaptations to host localization and colonization: a compressed body shape, hard carapace, and well-developed swimming setae on the second and third pair of ambulatory legs. The invasive crab later molts into a prehard form that loses all swimming traits (Pearce 1966, Jones 1977). The external appearance of prehard crabs (i.e., a soft exoskeleton, rounded carapace, and slender claws and ambulatory legs) suggests that it is at this stage when *C. garthi* individuals adopt an endosymbiotic lifestyle for the first time in their lives. After this prehard stage, the crabs molt into a hard-stage form that exhibits putative adaptations to live, at least temporarily, outside of host individuals. In this hard stage, various features similar to those found in invasive crabs are "reexpressed" (e.g., long natatory setae on the second and third pair of pereopods). Additionally, hard-stage crabs exhibit a well-calcified exoskeleton, stout claws, and a pigmented tegument (Fig. 15.4H).

In males of *C. garthi*, no form other than the hard stage has ever been found (Ocampo et al. 2016), similar to what was reported for other pea-crab species (Jones 1977). Thus, the hard form appears to be the terminal stage in males of this species, and the presence of well-developed gonopods at this stage suggests that the males are sexually mature (Ocampo et al. 2016; Fig. 15.4H). In contrast, *C. garthi* females undergo a much more complex metamorphosis-like process, passing through a series of posthard soft-shelled forms (II to V) before finally attaining a hard form (stage V) and becoming sexually mature. Females in posthard stages lose all "free-life" body features observed in hard-stage forms (Fig. 15.4H).

The life cycle above appears to occur in a few other species in the same subfamily Pinnotherinae (Møller Christensen and McDermott 1958, Pearce 1966). Pea crabs in the subfamily Pinnothereliinae (members of subfamilies Pinnixulalinae and Pinnixinae, sensu Palacios-Theil et al. 2016) do not exhibit such a complex life cycle. A final species to highlight in the subfamily Pinnotherinae is

Tunicotheres moseri that dwells in the atrial chamber of Caribbean ascidians. Females of *T. moseri* exhibit abbreviated development and retain embryos, zoea larvae, megalopae, and the first crab instars within their brood pouches (Hernández et al. 2012). Offspring remain protected within the female abdominal brood chamber and abandon the female and host during the first juvenile instar (Ambrosio and Baeza 2016).

Superorder Peracarida

In the superorder Peracarida, the "standard (= free-living)" life cycle involves direct development and the brooding of embryos by parental females in a ventral chamber or "marsupium" from which juveniles hatch with a gross morphology similar to that of adults. Representatives from the orders Cumacea, Mysida, Tanaidacea, and others, are almost invariably free-living. One notable exception in the order Mysida is *Heteromysis harpax*, which forms family groups in the interior of gastropod shells occupied by hermit crabs (Vannini et al. 1994). On the other hand, many species belonging to the orders Amphipoda and Isopoda are symbiotic.

In the order Amphipoda, members from the superfamily Caprelloidea, family Cyamidae (whale-lice) cling to the external surface of whales and dolphins (Gruner 1975), and many species in the suborder Hyperiidea are obligate symbionts of cnidarians, salps, and other gelatinous zooplankton (Laval 1980). All the amphipods above exhibit direct development. In a few species, juveniles remain in the parental host for long time periods (Thiel 2000).

In the order Isopoda, members from the suborder Cymothoida almost invariably exhibit a parasitic (or carnivorous and bloodsucking) lifestyle at some point in time during their lives. In the suborder Cymothoida (superfamily Cymothooidea), examples include representatives from the families Aegidae, Cymothoidae, Gnathiidae, and Tridentellidae. In the superfamily Bopyroidea, examples include representatives from the families Bopyridae, Dajidae, and Entoniscidae. In the superfamily Cryptoniscoidea, examples include species from the families Cryptoniscidae and Hemioniscidae.

In the superfamily Cymothooidea, representatives of the family Cymothoidae (fish-lice) are marine and freshwater parasites of fishes that cling to the skin, fins, gills, or mouth (i.e., the tongue-eating louse *Cymothoa exigua*) of their host individuals (Bunkley-Williams and Williams 1998). Some species even bore into muscle (Smit et al. 2014). The life cycle of *Anilocra pomacentri*, an ectoparasite of the fish *Chromis nitida*, is well known (Adlard and Lester 1995). In *A. pomacentri*, embryos develop through 4 ontogenetic stages within the marsupium of the adult female (Fig. 15.5A–C). Offspring are released from the marsupium of the parental female as a modified "manca" larval stage (also called a "pullus II" stage) that swims efficiently thanks to heavily setose pleopods (Fig. 15.5D). The released swimming mancae are infective immediately to suitable hosts and exhibit behaviors that likely help this infective stage to locate the latter (Lester 2005). Early infection of the host appears to be followed by simultaneous growth of both host and parasitic isopod. Juvenile mancae develop into males that later in life change sex to females and remain permanently attached to host individuals (Fig. 15.5E). In *A. pomacentri*, individuals are often seen solitarily when attached to hosts, suggesting that copula is short without long-term "monogamous" association between males and females. However, in *Nerocila californica*, males and females inhabit host individuals in pairs and remain in close contact (Brusca 1978 in Adlard and Lester 1995). Adult female isopods consume host blood, and feeding is limited only to those periods that correlate with the onset of vitellogenesis (Adlard and Lester 1995).

In the same superfamily Cymothooidea, representatives from the family Gnathiidae are parasites early but not late during their ontogeny. In gnathids, adult specimens find refuge in cavities in mud banks, coral colonies, dead barnacles, or sponges and do not appear to feed (Kensley and Schotte 1989). In contrast, larval stages temporarily attach to fishes and consume their blood (Lester 2005).

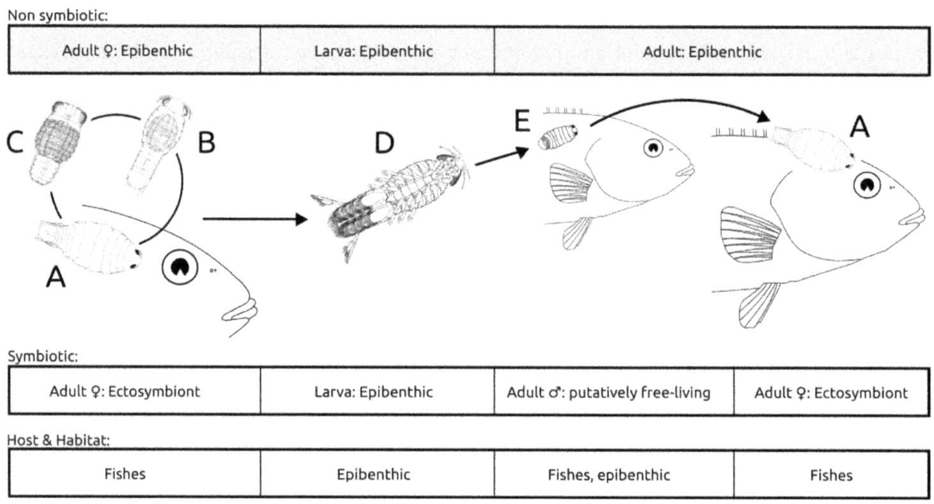

Fig. 15.5.
Isopoda: life cycle of *Anilocra pomacentri.* (A) Reproductive female attached to host. (B, C) Prehatching stages I and II, respectively. (D) Free-living manca stage. (E) Mobile male. The top and bottom rows show the life cycle of symbiotic as compared to nonsymbiotic species in the same clade as well as the host species used by the different ontogenetic phases in symbiotic species. Modified from Adlard and Lester (1995), with permission from CSIRO Publishing.

The life cycles of *Gnathia trimaculata, G. africana,* and *G. maxillaris* have been described in detail. In these species, embryos develop within the marsupium of free-living benthic females (Smit et al. 2003, Ota et al. 2012, Hispano et al. 2014). The three species exhibit six larval stages, consisting of three unfed ("zuphea") and three fed ("praniza") stages that alternate. Recently hatched unfed zuphea stages seek and attach to fish hosts and feed on their blood (Fig. 15.6). Once attached to their host, the body of the zuphea expands due to the immense volume of consumed blood; at this point in time, the zuphea turns into a praniza. After the praniza molts, it becomes a zuphea that once again searches for another host fish on which to feed. The cycle above is repeated twice more up to the point in which the third praniza established a free-living lifestyle and finds refuge in the benthic environment to begin maturing (Lester 2005). In *G. maxillaris,* three to six days after its last blood meal, the sex of the third and final praniza stage can be determined by the presence of either testes or ovaries. In *G. trimaculata,* female larvae molt at 19 to 29 days into adult females (Fig. 15.6A) while male larvae molt into adult males five to 17 days postfeeding (Fig. 15.6B; Ota et al. 2012). In *G. africana,* fertilization of eggs by the male takes place within 24 hours of completion of the female molt. Interestingly, in *G. maxillaris,* fertilization of a future female praniza III by an adult male has been observed in the laboratory (Hispano et al. 2014). Females die within a few days of the birth of zuphea I, whereas males live 1–2 months after they have molted into adults.

Overall, the life cycle of symbiotic isopods belonging to the superfamily Cymothooidea has been modified considerably in contrast to that reported for symbiotic amphipods. Further life cycle modification and body shape elaboration occurs in the superfamily Bopyroidea in which species exhibit an "aberrant" adult morphology. All known life cycles in the superfamily Bopyroidea include two crustacean hosts (Williams and Boyko 2012).

Representatives of the species-rich family Bopyridae inhabit the visceral cavity, branchial chamber, or abdominal surfaces of several benthic decapods and pelagic mysidaceans and euphausiaceans (Kensley and Schotte 1989). A well-known life cycle is that of *Orthione griffenis,* symbiotic with the mud-shrimp *Upogebia pugettensis* in the northeastern Pacific (Fig. 15.7). In this

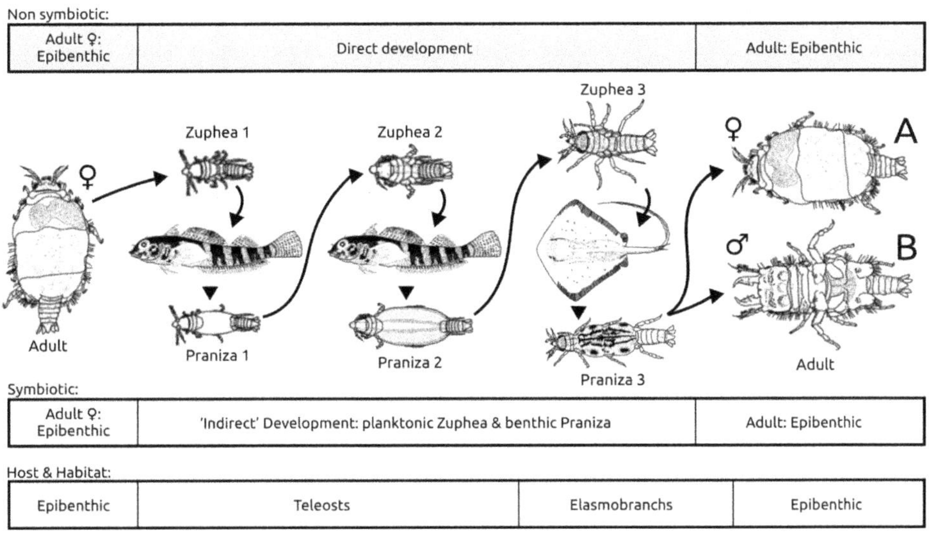

Fig. 15.6.
Isopoda: life cycle of *Gnathia trimaculata*. The first zuphea seeks and attaches to a teleost fish host individual and turns into the first praniza. After the praniza molts, it becomes a second zuphea that once again searches for another host individual on which to feed. The third zuphea seeks an elasmobranch host. This stage later molts into the third praniza that establishes a free-living lifestyle and finds refuge in the benthic environment. The top and bottom rows show the life cycle of symbiotic as compared to nonsymbiotic species in the same clade as well as the host species used by the different ontogenetic phases in symbiotic species. Modified from Ota et al. (2012), with permission from Springer.

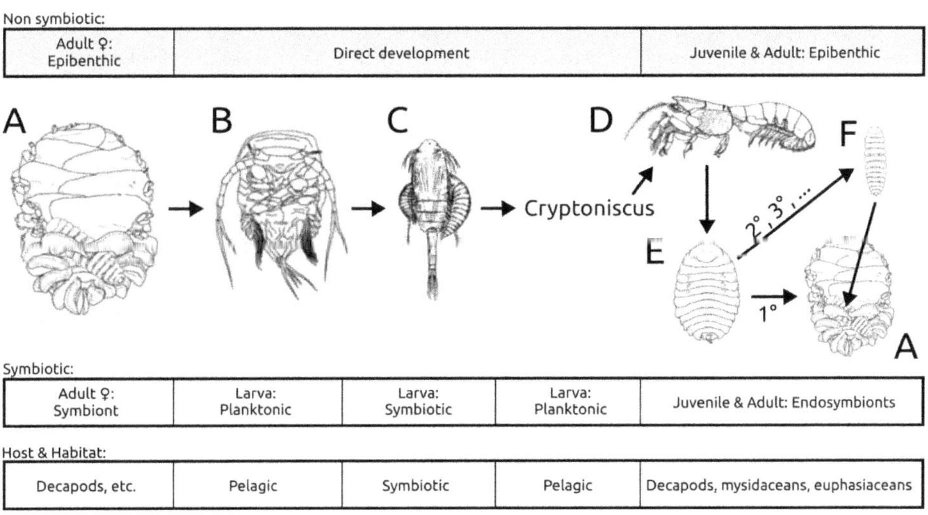

Fig. 15.7.
Isopoda: life cycle of *Orthione griffenis*. The female (A) releases epicaridium larvae (B) that parasitize calanoid copepod intermediate hosts. The epicaridium larva metamorphoses into a microniscus larva (C) and then into a cryptoniscus larva that settles onto a definitive mud shrimp host (D). The first juvenile isopod (bopyridium) to parasitize a host becomes female (E); subsequent isopods become male(s) (F) and live on the female (A). The top and bottom rows show the life cycle of symbiotic as compared to nonsymbiotic species in the same clade as well as the host species used by the different ontogenetic phases in symbiotic species. Modified from Williams and Boyko (2012).

species, eggs develop in the marsupium of parental females (Fig. 15.7A) and hatch as a nonfeeding "epicaridium" larva (Fig. 15.7B). This larva swims proficiently, seeking a copepod host (Fig. 15.7C). Once a copepod is found, the epicaridium attaches to it, perforates its exoskeleton, and starts feeding on this intermediate host with the help of styliform suctorial mouthparts and clawed pereopods. After a few days, the epicaridium larva metamorphoses, turning into a "microniscus" larva that still remains attached to the copepod host. The microniscus stage continues growing to then metamorphose into a "cryptoniscus" larva that leaves the copepod and swims until it finds its terminal second host (*U. pugettensis*). The cryptoniscus metamorphoses into the first juvenile stage or bopyridium after infecting *U. pugettensis* (Fig. 15.7D). The first bopyridium that parasitizes an individual of *U. pugettensis* becomes a female (Fig. 15.7E) while subsequent conspecific parasitic isopods turn, instead, into functional males that attach to the now larger female (Fig. 15.7F; Williams and An 2009). Thus, sex in *O. griffenis* seems to be environmentally determined, a condition also proposed for other bopyrids (Reinhard 1949). Sex determination nevertheless appears to be genetically controlled in most bopyroideans (Williams and Boyko 2012). In *O. griffenis*, sexual dimorphism is considerable; females are "giants" while males can be considered "dwarfs." Adult females, but not necessarily males, also have remarkable body modifications from the "classical" free-living body plan in this group. In females, body pigmentation is lacking, the body axis is slightly distorted (resulting in partial loss of bilateral symmetry), and the body outline is oblong or oval with greatly enlarged oostegites that generously enclose the brood pouch (Fig. 15.7A). On the other hand, the gross morphology of males resembles that of free-living isopods (Markham 2004; Fig. 15.7F).

In the superfamily Cryptoniscoidea, representatives from the families Cryptoniscidae and Hemioniscidae, among a few others, are also endosymbiotic, inhabiting the visceral cavity, branchial chamber, or abdominal surfaces of benthic decapods (including sessile barnacles), pelagic mysidaceans, and euphausiaceans (Kensley and Schotte 1989). As in the superfamily Bopyroidea, species belonging to this clade also use crustaceans as both intermediate and definitive hosts. The body plan of adult females is also the most modified within endosymbiotic isopods. Females are "featureless, bloated" (sensu Crisp 1968) individuals resembling a sac with no evident pereopods, oostegites, and external segmentation. In contrast, males in this group appear to be neotenic as they retain larval morphology (Williams and Boyko 2012). The life cycle of the *Hemioniscus balani*, a parasitic castrator of barnacles in northern Europe is relatively well studied. In *H. balani*, females inhabit the mantle cavity of several barnacles and resemble a brood sac that impairs host reproduction. Male reproductive behavior and timing of female insemination is completely unknown. Once the embryos fully develop into larvae, the female body ruptures to liberate epicaridium larvae that seek and attach to calanoid copepods. The next stage, the cryptoniscus, settles on hard rock substrata. The male then enters the mantle cavity of one of its barnacle hosts and attaches itself to the ovary. There, it matures and changes sex to female later in life (Crisp 1968, Arnott 2001).

In general, isopods belonging to the superfamily Bopyroidea and Cryptoniscoidea are recognized because of their complex life cycles. The life cycles of most species in the speciose Isopoda that have adopted a symbiotic lifestyle have not been studied. Symbiosis is absent or is much less common in other isopod suborders such as the Asellota, Calabozoidea, Limnoriidea, Oniscidea, Microcerberidea, Sphaeromatidea, and Valvifera.

Superclass Multicrustacea: Subclass Thecostraca

In the superclass Multicrustacea, subclass Thecostraca, symbiosis is obligatory in 3 major taxa: the infraclasses Ascothoracida and Facetotecta, and the superorder Rhizocephala in the infraclass Cirripedia (Høeg et al. 2005).

In the ground pattern of the Thecostraca, larval development encompasses 6 instars of pelagic nauplii and a terminal cypridiform larva adapted for attachment (Grygier 1987, Høeg 1995, Høeg et al. 2004). Little is known about the life cycle in the infraclasses Ascothoracida and Facetotecta. The enigmatic Facetotecta appears to be endosymbiotic, having an invasive stage similar to that of rhizocephalans (Høeg and Kolbasov 2002). However, organisms used as hosts by the Facetotecta are yet to be discovered (Glenner et al. 2008). In the infraclass Ascothoracida, members establish ecto- and endoparasitic associations with cnidarians and echinoderms (Grygier 1987, Grygier and Høeg 2005). The life cycle of this taxon is poorly known. In several species, reproductive females brood eggs within their mantle cavity and retain hatched nauplii larvae that are later released as "a-cyprid" larvae (sensu Høeg et al. 2004, Høeg et al. 2009). In the highly specialized order Dendrogastrida (e.g., including the genus *Dendrogaster*), parasitic in various species of Asteroidea, there are two consecutive a-cyprid larval stages. In several dendrogastrids, the first a-cyprid larva is retained within the mantle cavity and only the second a-cyprid larva abandons the parental brooding chamber, spends some time in the plankton, and later locates a new host or mate (Kolbasov et al. 2008). In the ascothoracid *Gorgonolaureus muzikae* (Order Laurida) that is parasitic in gorgonian corals in Hawaii, at least four discrete developmental stages are recognized: protander, late protander, young female, and mature female. All these stages are found within galls that they construct in the host coral. During development, the body of *G. muzikae* becames bigger, the two gall valves fuse, and the mouth appendages gain in complexity. Females attain sexual maturity during the last stage when they start incubating embryos within a brood chamber (Grygier 1981).

In the infraclass Cirripedia, superorder Rhizocephala, members are highly modified during adulthood and use a wide variety of crustaceans, mostly decapods, as hosts. The superorder Rhizocephala is divided in two orders, Kentrogonida and Akentrogonida. The life cycle of *Loxothylacus panopaei*, in the order Kentrogonida, symbiotic with the brachyuran crab *Rhithropanopeus harrisii* in the northwestern Atlantic, is well known (Glenner 2001). Figure 15.8 depicts the life cycle of a generalized rhizocephalan. In *L. panopaei*, male and female nauplius larvae hatch from a pouch-like structure (= "externa") that protrudes from the host's body (Fig. 15.8A). After two days in the plankton, the free-swimming nauplii (Fig. 15.8B,C) develop into cyprids. The male cyprid (Fig. 15.8D) metamorphoses into a dwarf male (a trichogon) after settling at or near the mantle opening of a recently emerged "externa" on a different host individual (Fig. 15.8D,F). The dwarf male then travels through the mantle cavity and inserts itself into one of two receptacles available on the virgin externa. Here the trichogon sheds its cuticle, become permanently embedded in the female tissue, and starts spermatogenesis (Glenner et al. 2010). In extreme cases, these minute males become no more than male tissue packages as their body complexity is drastically reduced (Høeg et al. 2005). Notably, only when a dwarf male has become established, do female externa mature, start oogenesis, and produce larvae. Female cyprids (Fig. 15.8C) often settle on places with thin, unsclerotized cuticle (e.g., gill lamella in the branchial chamber) of uninfested host crabs (Høeg 1995). Once settled, a kentrogon stage develops underneath the carapace of the cyprid (Fig. 15.8I). This kentrogon penetrates the integument of the host gills with the aid of a cuticle-reinforced stylet and injects a "vermigon" into the open, blood-filled space of the gill lamella. The vermigon develops a rootlet branching system after ten to 12 days postinjection into the host. Lastly, after another month, this female parasite protrudes a virgin externa at the ventral part of the host abdomen (Fig. 15.8L).

Several differences exist between the life cycles of species in the orders Kentrogonida and Akentrogonida. In the Akentrogonida, considerable variation in life cycle exists but this is not necessarily the case in the Kentrogonida. In the Kentrogonida but not Akentrogonida, the settled cyprid invariably passes through a metamorphic molt resulting in a new and highly specialized instar: the kentrogon in females, and the homologous, but morphologically very different, trichogon in males (Høeg 1995). In all Akentrogonida but only in a few Kentrogonida, offspring hatch as fully developed cyprids rather than nauplii larvae (Høeg 1995). Also, in the Kentrogonida, many male cyprids can settle on a single externa, but only two will eventually succeed in reproducing, one in each receptacle (Høeg 1987), and the externa can reproduce normally with only 1 male. In

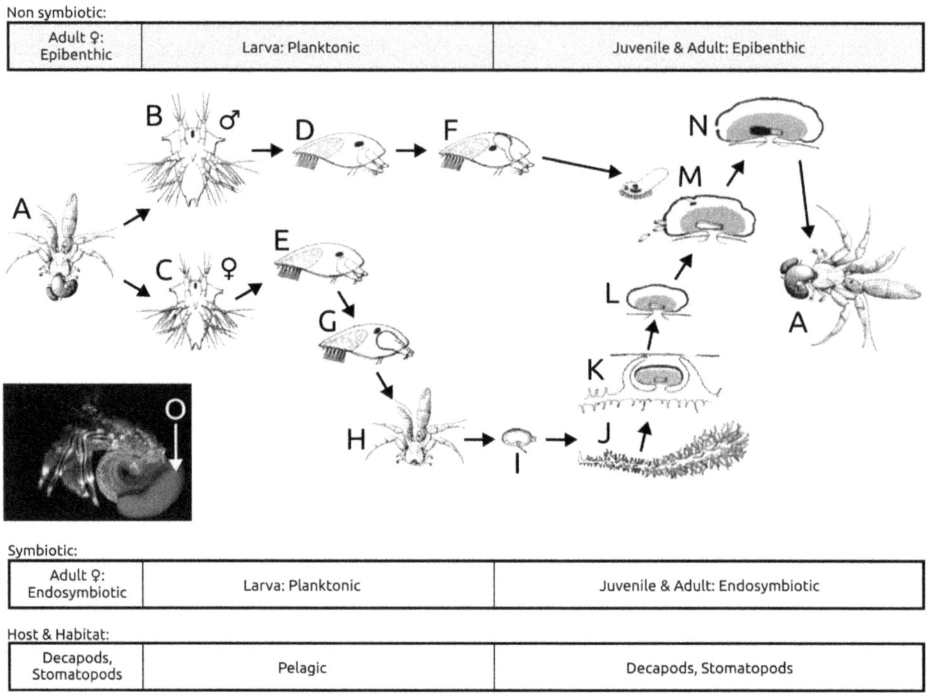

Fig. 15.8.
Cirripedia: life cycle of a generalized rhizocephalan. (A) Sexually mature externa (arrowhead) on host abdomen. (B, C) Male and female nauplius larva, respectively. (D, E) Male and female cyprid larva, respectively. (F, G) Settling cyprid on host (H). (I) Kentrogon invades the host and inoculates embryonic cells that develop into the early (J) and then late (K) interna. This interna finally emerges as a virgin externa (L). (M, N) Paired receptacles in the externa need to receive one or two trichogons to reach sexual maturity. (O) Mature externa of *Peltogaster postica* (arrowhead) and its hermit crab host, *Pagurus nigrivittatus*. The top and bottom rows show the life cycle of symbiotic as compared to nonsymbiotic species in the same clade as well as the host species used by the different ontogenetic phases in symbiotic species. (A–N) Modified from Høeg and Lützen (1995), with permission from Taylor and Francis. (O) Modified from Yoshida et al. (2014), with permission from The Crustacean Society.

contrast, in the Akentrogonida, the number of implanted males may range from 1 to more than 10 (Høeg 1985, 1991, Høeg and Lützen 1995). Høeg (1995) provides additional details on the life cycle of the Rhizocephala.

In the Thecostraca, infraclass Cirripedia, members from the superorder Acrothoracica are miniature gonochoric cirripeds that excavate burrows into calcareous substrata, including other thoracican barnacles, exoskeletons of dead or live corals, bryozoans, crinoids, and gastropod shells used by hermit crabs (Williams and McDermott 2004). The life cycle of the coral-dweller *Berndtia purpurea* is relatively well known (Utinomi 1961). In *B. purpurea*, embryo development takes place in the female mantle cavity and larvae hatch as nauplii that further develop and molt an undetermined number of times before becoming a cypris. Nauplii in the Acrothoracica can be either lecithotrophic or planktotrophic (Kolbasov et al. 2008, but see Kolbasov and Høeg 2007). In *B. purpurea*, the metamorphosed female cypris appears to settle on soft tissue of its coral host and forms a cavity before starting to excavate by its own power. In acrothoracicans, males are generally attached to the female body, but the dwarf males of *B. purpurea* most often attach to the anterior end of the wall in the female's burrow. In each female burrow, solitary, paired, or small groups of up to 6 males can be found. Adult males are characterized by their relatively long coiled penis and the absence of a mantle cavity.

Lastly, many species in the superorder Thoracica feature 6 naupliar larval stages prior to the cypris. Most taxa have planktotrophic nauplii, but lecithotrophy is prevalent or obligatory in several families. After metamorphosis of the cypris, most species adopt a sessile free-living suspension-feeding lifestyle. However, symbiosis, including parasitism, has evolved independently several times in this taxon (e.g., family Pyrgomatidae in the superfamily Balanoidea, superfamily Coronuloidea, family Poecilasmatidae, genus *Octolasmis*, among others; see Baeza 2015). In symbiotic Thoracica, the characteristic cirriped pattern of development of 6 naupliar stages followed by a cyprid larva is observed.

Superclass Multicrustacea: Subclass Copepoda

In the superclass Multicrustacea, subclass Copepoda, infraclass Neocopepoda, superorder Podoplea, symbiosis is obligatory in two orders, Monstrilloida, Siphonostomatoida, and one suborder, oecilostomatoida in the order Cyclopoida. In the Monstrilloida, the naupliar, preadult, and adult phases are planktonic but during their postnaupliar and juvenile phases, they are endosymbiotic with gastropods, polychaetes, or, rarely, equinoderms (Suárez-Morales 2011). Members of the Siphonostomatoida parasitize invertebrates and fishes (Barel and Kramers 1977). Members of the Poecilostomatoida are usually ectosymbiotic and find refuge on the buccal cavity of mollusks and equinoderms or gills of bony fishes. However, 3 genera in the Poecilostomatoida—*Clavisodalis, Echinirus,* and *Echinosocius*—have adopted an endosymbiotic lifestyle living in the esophagus of sea urchins (Dojiri and Cressey 1987). In the superorder Gymnoplea, the order Calanoida constitutes mostly free-living species. However, several species from this order have adopted a symbiotic lifestyle.

In parasitic copepods, the life cycle can be quite complex, because they involve one or two hosts coupled with either subtle or considerable changes in body morphology. Importantly, compared to the life cycle of their free-living counterparts, some parasitic copepods exhibit abbreviated life cycles with 6 naupliar dispersive stages and five copepodid stages preceding the adult phase. In parasitic copepods, the nauplius larva is often lecithotrophic and the first copepodid is the ontogenetic stage that most frequently establishes the symbiotic relationship. Also, the life cycle of symbiotic copepods most often involves a single host species (Boxshall 2005).

Nonetheless, there are remarkable deviations from the "basic" symbiotic life cycle above. For instance, the life cycle of *Lernaeocera branchialis* (family Pennellidae) involves two nauplius larvae and two rather than a single host species. Adult females live on whiting (*Merlangius merlangus*; Fig. 15.9A,B) and produce eggs that take more than 13 days to hatch as nauplius I larvae (Fig. 15.9C). An additional nauplius II stage (Fig. 15.9D) is required before larvae turn into nonfeeding copepodids (Fig. 15.9E). These infective copepodids attach to the gills of flounder *Platichthys flesus* and there pass through four additional copepodid stages before turning into sexually active adults (Fig. 15.9F–I). The entire life cycle lasts a minimum of 25 days (Whitfield et al. 1988, Brooker et al. 2007; Fig. 15.9).

Still in the parasitic Copepoda, species belonging to the family Monstrillidae represent a second remarkable example of life cycle complexity characterized by the existence of a free-swimming pelagic adult phase that does not feed. In addition, most larval and all but the last juvenile (copepodid) stages feature an endosymbiotic lifestyle (Suárez-Morales 2011). In the Monstrillidae, eggs carried by the parental females hatch into lecithotrophic nauplii that seek, attach, and tunnel into the tissue of a mollusk or polychaete. Once in the hemolymph of the host, the infective nauplius metamorphoses into a second naupliar larva. This second endoparasitic stage resembles a simple sac with two rootlike processes used for nourishment. Further development of the parasite takes place within the host individual. Lastly, when the parasite reaches the copepodid stage, it abandons the host. After a single molt, this copepodid morphs into a free-living adult that lacks all cephalic appendages other than the antennules (Huys et al. 2007, Suárez-Morales 2011).

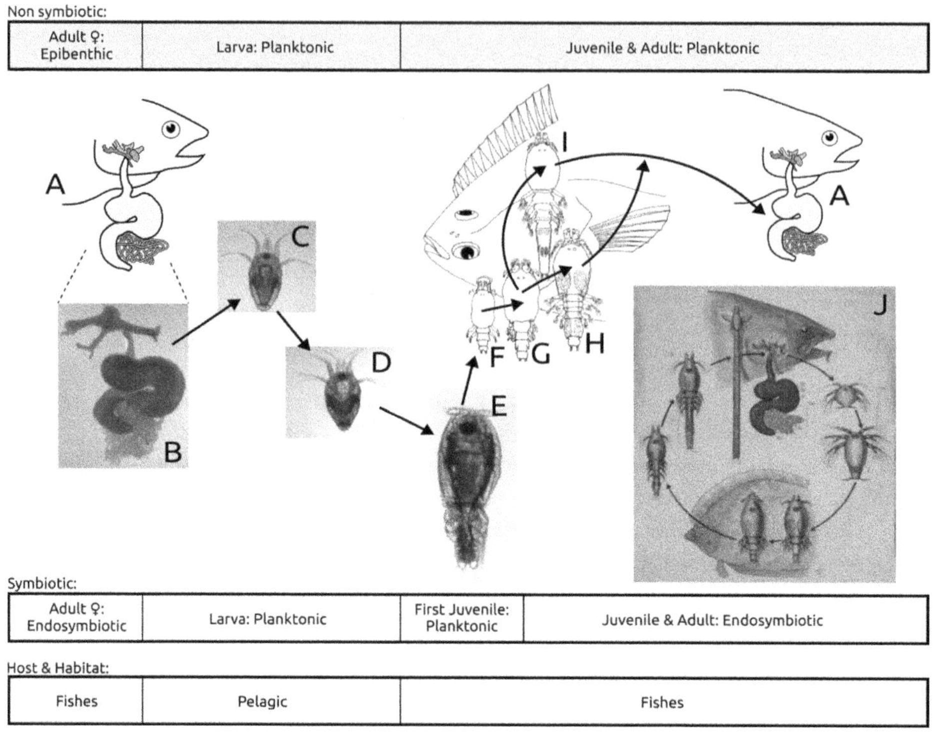

Fig. 15.9.
Copepoda: life cycle of *Lernaeocera branchialis.* Some stages have been omitted in the figure for simplicity. (A) Egg-laying female in gill chamber of a demersal fish. (B) Gravid adult female. (C, D) Free-swimming nauplius I and II larvae, respectively. (E) Free-swimming infective copepodid. (F–I) Stages on gills of flounder *Platichthys flesus.* (F) Chalimus I. (G) Chalimus III. (H) Adult immature male. (I) Adult immature female. (J) The same life cycle depicted in an old German poster. The top and bottom rows show the life cycle of symbiotic as compared to nonsymbiotic species in the same clade as well as the host species used by the different ontogenetic phases in symbiotic species. (A) Modified from (J). (B–E) Modified from Brooker et al. (2007), with permission from Elsevier. (F–I) Modified from Sproston (1942), with permission from Cambridge University Press. (J) Courtesy of Institute of Biology/Comparative Zoology, Humboldt University of Berlin ©.

Superclass Multicrustacea: Subclass Tantulocarida

Tantulocarids are minute parasites of deep-sea benthic crustaceans with a highly modified adult form (Boxshall and Lincoln 1987). The life cycle in this taxon includes a common asexual (parthenogenetic) and a rare sexual phase. A tantulus larva with a well-defined trunk and cephalothorax is produced by asexual females (Fig. 15.10A,B). This larva spends time in the benthic environment searching for a suitable host (a copepod, peracarid, or ostracod). When a host individual is found, the tantulus attaches to it with the aid of an oral stylet that punctures the host's cuticle (Fig. 15.10C,E). The larva develops into a new asexual (parthenogenetic) female after shedding its trunk (Boxshall and Lincoln 1987; Fig. 15.10G). The new trunk of this female expands to harbor the growing asexual larvae until they hatch (Fig. 15.10H). During the sexual phase, the tantulus larva attaches to the host but does not shed its trunk. Instead, the larval trunk develops a saclike expansion that harbors developing sexual adult males or females (Fig. 15.10D,F). The walls of the saclike expansion rupture once adults attain sexual maturity. The remaining portion of the sexual cycle is not well known. In the Tantulocarida, sexual dimorphism is considerable; males but not females feature well-developed swimming appendages (Fig. 15.10I,J) and paired clusters of chemosensory

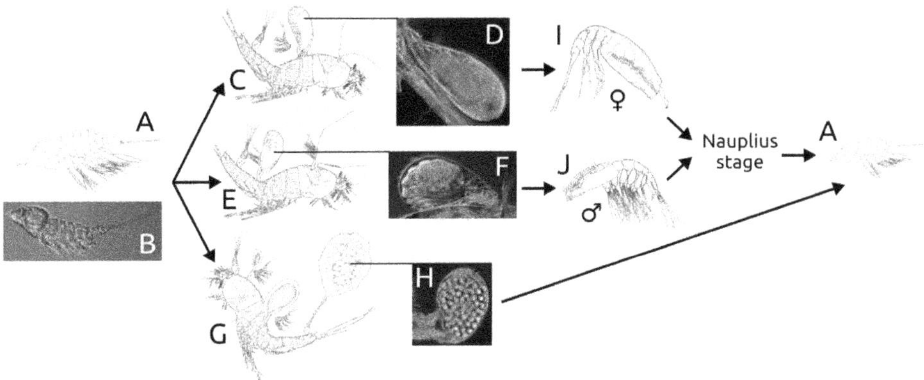

Symbiotic:

Larva: Free-living	Sexual juvenile & Parthenogenetic adult: Ectosymbionts	Sexual adult: Free-living	Larva: Free-living	Larva: Free-living

Host & Habitat:

Epibenthic	Copepods, peracarids, ostracods	Sexual adult: Epibenthic	Larva: Pelagic	Larva: Epibenthic

Fig. 15.10.
Tantulocarida: life cycle of *Arcticotantulus kristenseni.* (A, B) Tantulus larva. (C) Development of sexual females on hosts. (D) Sexual female attached to host with developing abdomen. (E) Development of sexual males. (F) Male in late stage of development attached to host. (G) Parthenogenetic female on host. (H) Parthenogenetic females with embryos inside. (I, J) Free-living adult female and male, respectively. The bottom rows summarize the life cycle. (A, B, D, F, H) Modified from Knudsen et al. (2009), with permission from Zootaxa. (C, E, G, I, J) From Huys et al. (2014), with permission from Johns Hopkins University Press.

aesthetascs. Males seem to actively search for receptive females and inseminate them using a well-developed penis-like intromittent abdominal organ (Huys et al. 1992, 1993).

Superclass Oligostraca: Class Ichthyostraca: Pentastomida and Branchiura

Pentastomids (tongue-worms) are gonochoric parasites with a wormlike body shape that require two different hosts to complete their life cycle. As adults, pentastomids dwell in the respiratory tracts of reptiles, birds, and mammals, including humans, while larvae live in the internal organs of vertebrates or arthropods (Tappe and Büttner 2009). Two genera, *Linguatula* and *Armillifer*, are involved in the majority of human infections by pentastomids. The life cycle of *A. agkistrodontis* was successfully observed under laboratory conditions (Chen et al. 2010; Fig. 15.11). This pentastomid is native to China and occasionally infects humans, causing them to develop clinical symptoms including long-term high fever, abdominal pain, diarrhea, mild anemia, hepatosplenomegaly, eosinophilia in both the bone marrow and blood, and multiple polyps in the colon. Adult *A. agkistrodontis,* and other species such as *A. armillatus,* inhabit the lungs of snakes that are used as final hosts (Chen et al. 2010, Tappe et al. 2011). During their lifetime, females of these species release hundreds of thousands of eggs into the intestinal and respiratory systems of their hosts (Fig. 15.11A,B). These eggs are later excreted and reach water bodies, infecting intermediate hosts (i.e., rats and mice; Fig. 15.11E) when swallowed. Once in the intestine of their intermediate host, eggs hatch and larvae invade the mucous layer of the small intestine and the venules before reaching the liver, spleen, and mesenteric system (Fig. 15.11C, D). There, the larvae develop into the infective nymphal stage after approximately four months. Adults develop in the final hosts (snakes), which become infected by eating infected viscera of intermediate hosts (Fig. 15.11F–J). The overall development of *A. agkistrodontis* larvae into adults in final host snakes takes approximately 10 months. Humans

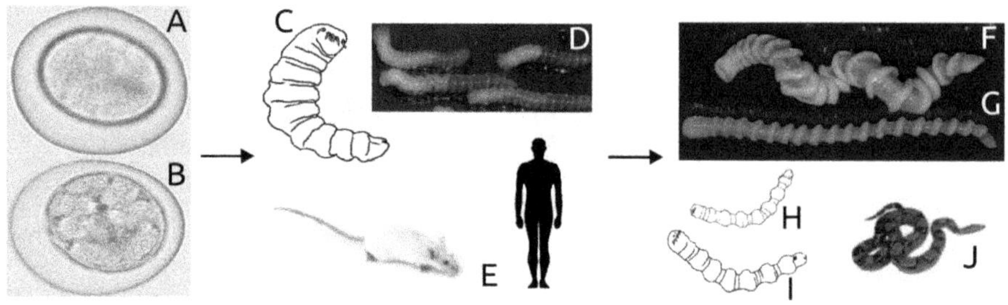

Symbiotic:

Embryos: Free- living/ Endosymbiotic	Larva: Endosymbiotic	Adult: Endosymbiotic

Host & Habitat:

Vertebrates	Vertebrates	Arthropods, vertebrates

Fig. 15.11.
Pentastomida: life cycle of *Armillifer agkistrodontis* and *A. armillatus*. (A, B) Immature and mature egg, respectively. (C) Larva of *A. agkistrodontis*. (D) Larva of *A. armillatus*. (E) Intermediate hosts (e.g., rodent and human) of *A. agkistrodontis*. (F) Gravid *A. armillatus* female (G) Preadult *A. armillatus* female. (H) Adult male of *A. agkistrodontis*. (I) Adult female of *A. agkistrodontis*. (J) Final host (snake). The bottom rows summarize the life cycle. (A–C, E, H–J) from Chen et al. (2010). (D, F, G) from Tappe et al. (2011).

are infected by consuming snake gallbladder and snake blood, as well as by drinking contaminated water with infectious eggs. In humans, eggs develop into infectious larvae after 4 months but cannot develop into adults (Chen et al. 2010).

Much less morphologically aberrant, the Branchiura (fish lice), class Ichthyostraca, live on the external surface of marine and freshwater fish and amphibians. Examples are *Argulus coregoni* infecting the rainbow trout *Oncorhynchus mykiss* (Bandilla et al. 2005) and *Argulus ambystoma* infecting the salamander *Ambystoma dumerilii* (Poly 2003). The only genus in the subclass Branchiura for which the life cycle is well studied is *Argulus* (Fig. 15.12). In general, females deposit their eggs on submerged surfaces while temporarily leaving "their" host individuals. These eggs hatch into a first swimming copepodid-form larval stage with setose cephalic appendages and two pairs of swimming legs (Fig. 15.12A). Thus, the first larval stage is well endowed for dispersal. The second larva is the first parasitic ontogenetic stage and resembles the adult in terms of gross morphology. In this second larval stage, the setae on the antenna are replaced by strong claws (Møller et al. 2007). After host attachment, larvae feed on the host fish's blood and mucus. The successive larval stages appear to leave and find new hosts at intervals. During larval development, most morphological changes are gradual. However, between the fourth and fifth larval stage, the maxillule morphs considerably from a long limb with a powerful distal claw into a short but powerful circular sucker (Shimura 1981, Boxshall 2005). In general, early larval stages are generalists as they attach to and grow on any available host. However, later during ontogeny, they switch to more suitable hosts (salmonids; Mikheev et al. 2015). Various molts succeed each other until sexual maturity is attained. In this gonochoric species, copulation most often occurs on the external surface of host individuals but occasionally happens off host individuals (Pasternak et al. 2000). Little is known about the mating system of *Argulus*. However, in *A. coregoni*, males (Fig. 15.12C) invest energy in mate searching while females (Fig. 15.12B) are rather stationary and invest in body size and hence increased fecundity (Bandilla et al. 2008). Gravid females leave hosts to deposit eggs in several different, consecutive clutches, returning to another host individual in between oviposition events (Fig. 15.12B,D; Mikheev et al. 2015). Pheromone production by reproductive females appears to aid males in finding sexual partners, and also causes an aggregated population distribution (Mikheev et al. 2015).

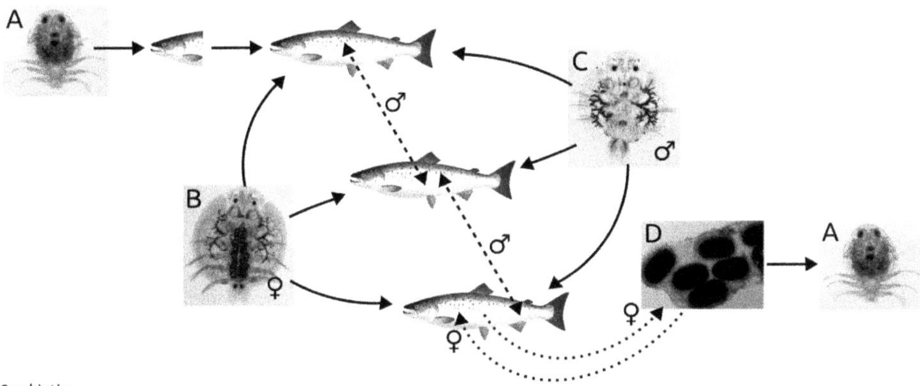

Symbiotic:

Larva: 1st stage Free-living; later stages ectosymbiotic	Adult: Ectosymbiotic & putative free-living	Embryos: Free-living	Larva: 1st stage Free-living

Host & Habitat:

Pelagic: Fishes, Amphibians	Fishes, Amphibians	Benthic	Pelagic

Fig. 15.12.
Branchiura: life cycle in fish-infecting *Argulus* spp. (A) Metanauplius, first stage larva of *Argulus bengalensis*. Modified from Banerjee et al. (2015), with permission from Springer. (B, C) Adult female and male of *A. japonicus*, respectively. Males may switch among hosts while searching for a mate (dashed lines). Females of *A. japonicus* may switch between host individuals solely after laying a clutch of eggs (D) in a suitable substrate (dotted lines). (B–D) Modified from Avenant-Oldewage and Everts (2010), with permission from Elsevier. The bottom rows summarize the life cycle.

Superclass Oligostraca: Class Ostracoda

In the Ostracoda, symbiotic associations are most often reported in the family Entocytheridae (subclass Podocopa, order Podocopida; Hart and Hart 1974). This family comprises approximately 200 freshwater species ectosymbiotic with crayfishes, amphipods, isopods, and crabs (Mesquita-Joanes et al. 2012). In the same subclass Podocopa, a few other genera in various families (e.g., Paradoxostomatidae, Pontocyprididae, and Cyprididae) also live symbiotically with echinoderms (Maddocks 1987), amphipods (Hart 1971), sponges (Martens and Harrison 1993), and even toads (Seidel 1995). Lastly, a few species in the subclass Myodocopa, family Cypridinidae, have been reported to live anchored to the gills of fishes (Bennett et al. 1997).

Both free-living and symbiotic ostracods exhibit direct development and determinate growth with a fixed number of immature juvenile instars and a terminal molt to adulthood (Smith 2014). The number of juvenile instars varies among the different ostracod clades. Most free-living ostracods have eight juvenile instars before they attain maturity (Smith 2014); in symbiotic ostracods, the number of instars is typically reduced (Kretzler 1984). In several symbiotic entocytherids, the first instar molts inside the egg (Smith and Kamiya 2005); thus, this group typically exhibits seven instead of eight juvenile instars (Aguilar-Alberola and Mesquita-Joanes 2011). *In Entocythere(=Ankylocythere) heterodonta*, adulthood is attained more quickly, after passing only though four juvenile instars (Rioja 1940). The life cycle of symbiotic ostracods is gradual; instars maintain the overall bivalved shape and progressively add appendages during development (Fig. 15.13). For instance, the first three juvenile instars (A3–A1) in the entocytherid *Uncinocythere occidentalis* have reduced pediform appendages (= legs; Smith and Kamiya 2005) and the first pair of antennae is used to attach to their urchin host's tegument (Kretzler 1984). During adulthood, *U. occidentalis* have well-developed legs ending in clasping organs that

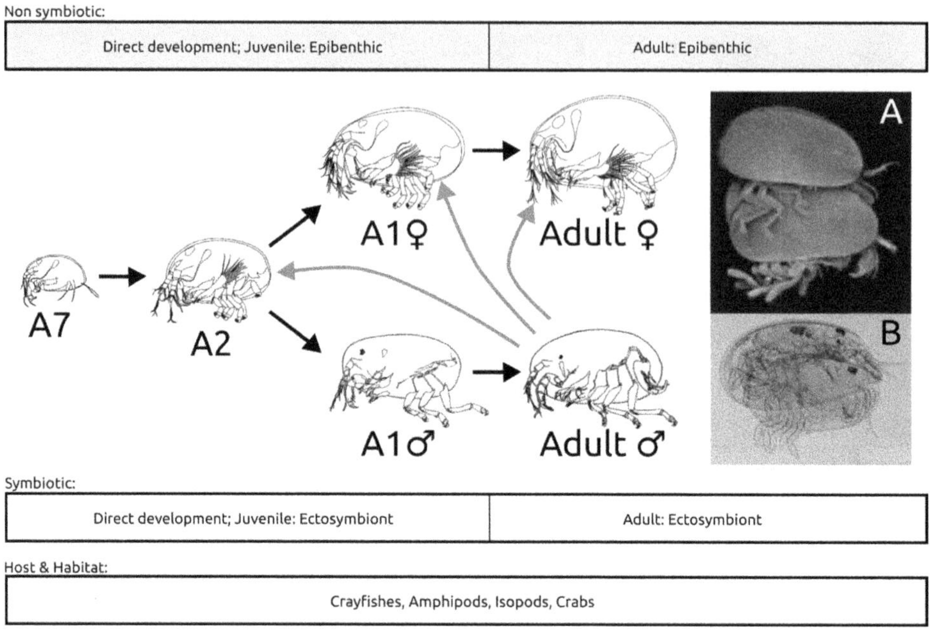

Fig. 15.13.
Ostracoda: life cycle of *Uncinocythere occidentalis*. Juveniles are named from the first to the last stage (A7–A1; A6–A3 are not shown for simplicity). Gray arrows indicate adult males inseminating female instars. (A) Micrograph of adult male (top) and A1 female (bottom). (B) Micrograph of adult male (top) and A2 female (bottom). The top and bottom rows compare the life cycle of symbiotic to nonsymbiotic species in the same clade. A7, A2, A1 female, and Adult female modified from Smith and Kamiya (2005), with permission from Springer. A1 male and Adult male modified from Hart et al. (1985). (A, B) modified from Mestre et al. (2013).

enable individuals to cling to the host epidermis (Smith and Kamiya 2005). In most symbiotic ostracods, the female and male sexual organs develop towards the end of the juvenile phase.

Asexual reproduction and parthenogenesis have been observed in several free-living ostracods (Smith 2014). However, all entocytherids and most representatives of other families of symbiotic ostracods are gonochoric (Hart and Hart 1974, Cohen and Morin 1990). Sexual dimorphism with respect to body size is notable in ostracods, including symbiotic species; females attain much larger body sizes than males (Aguilar-Alberola and Mesquita-Joanes 2011; Fig. 15.13A,B). Symbiotic species in the subclass Podocopa laid eggs over setose parts of their crustacean host's exoskeleton (Hart and Hart 1974). It is not known if ostracods parasitizing fishes (e.g., *Sheina orri*) brood eggs, although free-living representatives from the same family (Cypridinidae) do brood embryos (Cohen and Morin 1990). Females of the entocytherid *Hartiella dudichi* deposit between 1 and 16 eggs in packets on the gill of their amphipod hosts. It is not known, however, how many packets females deposit during their lifetime (Roelofs 1968). In entocytherids, copulation appears to occur between adult males and females in the last instars (i.e., A1; Cohen and Morin 1990). Nevertheless, in *U. occidentalis,* A2 juvenile females can form copulatory or guarding pairs with mature males (Aguilar-Alberola and Mesquita-Joanes 2011; Fig. 15.13B). We lack information about the mating strategies in symbiotic ostracods.

LIFE CYCLE ADAPTATIONS OF SYMBIOTIC CRUSTACEANS

This review demonstrates considerable life cycle disparity in symbiotic crustaceans. At one extreme, symbiotic amphipods, mysids, and various isopods have a relatively simple direct or abbreviated development life cycle. In all these species, benthic juveniles recruit directly into the host residence from parental brood chambers. Interestingly, the same simple (i.e., direct or abbreviated development) life cycle occurs in a few caridean eusocial shrimps (e.g., *Synalpheus regalis*; Duffy et al. 2000) and brachyuran crabs (e.g., *T. moseri*). At the other extreme, very complex life cycles occur in parasitic copepods and many isopods.

In between these extremes, many symbiotic crustaceans exhibit indirect development, spending their larval life in the pelagos. These crustaceans actively search for and settle in or on host individuals once they have reached the last larval or first postlarval stage. After settlement, they establish permanent or semipermanent relationships with their host individuals, growing, maturing, reproducing, and dying in or on their host partners. This life cycle is also exhibited by many symbiotic decapods. However, several other symbiotic crustaceans with indirect development deviate from this pattern (e.g., in the Branchiura; the fish lice *Argulus* spp.; Fig. 15.12).

One of the main goals of this review is to identify life cycle adaptations in crustaceans engaging in symbiotic relationships. Admittedly, inferences regarding adaptations using our broad comparative approach are limited considering the small number of studies describing the entire life cycle of symbiotic crustaceans. Furthermore, internal relationships within the species-rich Crustacea are not completely stable. Still, the information provided herein offers us some insight about putative life cycle adaptations in this group. We treat these insights as hypotheses that need to be further explored rather than conclusions.

In the Decapoda, no major life cycle modifications occur in species that have adopted an ectosymbiotic lifestyle. Brachyuran and anomuran crabs, caridean shrimps, and mysids exhibit a developmental pattern most similar to that reported for their closest free-living relatives. Certainly, this review may have missed subtle but otherwise relevant modifications with adaptive value due to the scarcity of published information. As in ectosymbionts, no major life cycle differences from that of free-living species seem to have occurred in endosymbiotic decapod crustaceans. For instance, in the Cryptochiridae, the few species in which the life cycle has been examined exhibit indirect development as well as the "typical" larval stages also found in free-living species (Scotto and Gore 1981).

Also in some other symbiotic decapods, a few major deviations from the ground pattern have been reported. For instance, in sponge-dwelling stenopodid shrimps (i.e., *Spongiocaris japonica*) and in a few pea-crabs (Pinnotheridae; i.e., *T. moseri*), development is either direct or abbreviated (Saito and Konishi 1999, Hernández et al. 2012). This direct or abbreviated life cycle can be interpreted, at first glance, as an adaptation to the symbiotic lifestyle, perhaps to avoid dispersal given refuge (host) scarcity or a clumped distribution (e.g., as suggested for symbiotic chthamalophilids rhizocephalans with abbreviated larval development; Bocquet-Védrine 1972). However, this conclusion needs to be considered carefully. For example, in the case of the symbiotic Stenopodidea, hosts occur in deep waters, and thus, direct development in this clade might not necessarily represent an adaptation to low host availability but might alternatively be driven by characteristics of the deep sea.

In the case of the pea-crab *T. moseri*, abbreviated development might be interpreted in a more robust manner as an adaptation to low host availability imposed by its symbiotic lifestyle. *T. moseri* is a shallow-water tropical species (and not a deep-water species like symbiotic stenopodid shrimps) that use ascidians as hosts. However, the different ascidians used by this pea-crab might not necessarily be scarce or highly aggregated. For instance, the ascidians *Phallusia nigra* in the southern Caribbean and *Styela plicata* in the Gulf of Mexico appear to quite abundant in nature (but see Ambrosio and Baeza 2016). Furthermore, many other species in the Pinnotheridae with an endosymbiotic lifestyle are known to have indirect development (Ocampo et al. 2011). Certainly, additional studies in these 2 groups are warranted because they might shed light on adaptations early during the ontogeny in symbiotic crustaceans.

In line with the absence of major life cycle modifications in the Decapoda (other than the few exceptions above), in the parasitic Rhizocephala within the Thecostraca, the life cycle (e.g., larval development) resembles that observed in free-living relatives (Høeg 1995). Glenner and Høeg (1994) noticed early that only after metamorphosis of the cypris larva, events in the ontogeny of the Rhizocephala differ drastically from that of free-living relatives. Only the process of host invasion and the adult parasite represent truly unique features in the Rhizocephala, something also noticed by Høeg (1995).

Importantly, within the Cirripedia, a notable condition is that the cyprid stages, all of them structurally very similar in the different groups comprising this clade, settle on a remarkable variety of substrata (including dissimilar host species), and then metamorphose into highly dissimilar entities that include, depending on the clade, rhizocephalan kentrogon invasive stages, juvenile suspension feeders, or dwarf males (Høeg 1995). As noticed by Høeg (1995) and even earlier by Crisp (1984), this extreme flexibility of the cyprids within a relatively rigid morphological framework contributes to the success and diversification of the Cirripedia. Considering the above, the ground pattern life cycle in the Cirripedia might be interpreted as a preadaptation favoring the evolution of symbiotic interrelationships in this group. Indeed, the Thecostraca is a clade in which parasitism has evolved several times independently (i.e., Rhizocephala and Facetotecta) and has reached a climax of morphological but not necessarily life cycle specialization (Høeg 1995).

In contrast to the little deviation in the life cycle ground pattern in the Decapoda and Thecostraca, symbiotic Peracarida exhibit remarkable life cycle deviations and major differences compared to their free-living relatives. A clear example of life cycle modification in the Peracarida (superfamily Cymothooidea) is that of *Anilocra pomacentri*, in which the parental female releases offspring from its brood pouch as a modified "manca" (= "pullus II") larval stage that swims efficiently thanks to heavily setose pleopods (Adlard and Lester 1995). Similarly, the parasitic family Gnathiidae that repeatedly engages in parasitism early but not late during ontogeny release "zuphea" larvae and also exhibit "praniza" stages instead of the "typical" manca stage characteristic of free-living species in this group. Indeed, the modifications in this remarkably different life cycle of the family Gnathiidae provide benefits to localizing and infecting host individuals.

In the same Peracarida, significant life cycle modification (as well as body shape elaboration) also occurs in parasitic isopods belonging to the superfamilies Bopyroidea and Cryptoniscoidea. In contrast to the superfamily Cymothooidea that is characterized by infesting only a single host species early during their ontogeny, all known life cycles in the Bopyroidea and Cryptoniscoidea involve two crustacean hosts (Williams and Boyko 2012). In general, isopods belonging to the superfamily Bopyroidea and Cryptoniscoidea are recognized because of their complex life cycles, a trait also shared by some parasitic copepods (e.g., *Lernaeocera branchialis* [family Pennellidae]) and tantulocarids that use 2 hosts to complete their life cycle (Brooker et al. 2007). Figure 15.14 summarizes the life cycle and the extent of modification suffered by different crustacean clades.

CONCLUSIONS AND FUTURE DIRECTIONS

The existence of complex life cycles in parasites, including the use of several hosts during different ontogenetic stages, is a puzzling phenomenon and an area of active research in evolutionary biology (Poulin and Cribb 2002). The probability of completing a life cycle is thought to decrease with increasing number of hosts and time (Poulin and Cribb 2002). Furthermore, parasites that use a single host during their lifetime do not expose themselves to the problems of interacting with different environments (host species) with dissimilar physiologies and immune systems (Morand et al. 1995). The crux of the issue then is whether the benefits of adopting complex life cycles are expected to overcome these (significant) costs. Unfortunately, the information available about the life cycle of parasitic crustaceans is still too limited to start exploring such important evolutionary questions in this group. Studies comparing the costs and benefits of complex versus simple

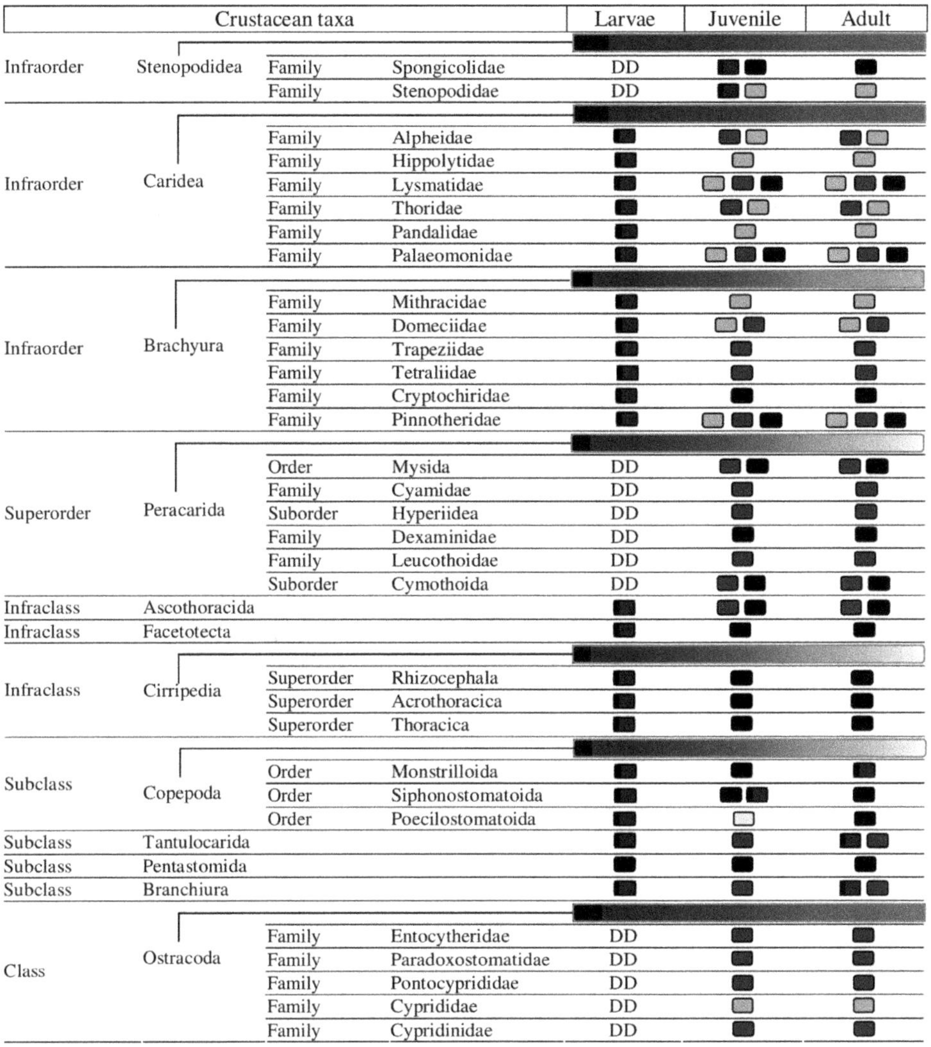

Fig. 15.14.
Overview of life cycles in symbiotic crustaceans. Boxes indicate free-living (very light gray squares), facultative ectosymbiont (light gray squares), obligate ectosymbiont (dark gray squares), or endosymbiont (black squares) condition of each life phase (i.e., larva, juvenile, adult) in major symbiotic crustacean taxa. The degree of specialization (in terms of such features as morphological adaptations; variations in number of stages; and exploitation of one, two, or more hosts during different phases) in symbiotic crustaceans commonly increase from facultative to obligate relationships, being particularly extreme, even bizarre, in some endosymbiotic groups. Horizontal bars constitute a qualitative measurement of that degree of symbiotic specialization in major crustacean taxa compared to closely related (phylogenetically) free-living groups from the same taxon. Exclusively symbiotic taxa (e.g., Ascothoracida, Tantulocarida) do not show the horizontal bar because of the lack of free-living species to compare with. A great degree of variation in shading of the horizontal bars indicates that this taxon pertains slightly as highly specialized taxa/species. Bars that are poorly variable in color indicate taxa/species that engage in symbiotic relationships in the taxon exhibit little modifications with respect to free-living relatives. DD = direct development (absence of larvae).

life cycles within monophyletic clades of parasitic crustaceans are warranted, because these will help reveal the conditions favoring complex life cycles not only in crustaceans but also in other invertebrates.

In summary, the literature review demonstrates cases of remarkable convergence with respect to life cycles in symbiotic crustaceans (e.g., between parasitic isopods, copepods, and tantulocarids). In other words, phylogenetic inertia and restrictions imposed by the body plan of some clades appear to be most relevant in determining life cycle modifications (or the lack thereof) from the "typical" ground pattern. Furthermore, there is indication that the life cycle ground pattern is likely either constraining or favoring the adoption of symbiosis within the Crustacea (e.g., in the Thecostraca).

Symbiotic crustaceans exhibit not only remarkable diversity in morphology but also in terms of life cycles. We believe that the path forward to identifying detailed life cycle adaptations in this group includes (1) the development of robust and comprehensive phylogenies of several clades, combined with (2) detailed studies describing the life cycle of the same species in those clades. This integrative approach can be most useful for proposing and testing specific hypotheses about the evolution of life history traits in this species-rich and morphologically disparate clade of marine, freshwater, and terrestrial invertebrates. Such studies will reveal not only adaptations in terms of life cycles but also other interesting modifications, likely serving to solve other relevant ecological and behavioral issues in species that have adopted a symbiotic lifestyle. For instance, the evolution of parasitic isopods is believed to follow the path from free-living forms to ectoparasites (e.g., *Anilocra* and allies), to species inhabiting the branchial and buccal cavity (e.g., *Ceratothoa imbricata*), culminating in endoparasitic cymothoids such as the remarkable *Artystone trysibia* (Adlard and Lester 1995). The combination of molecular phylogenetic with natural history studies suggested above would (1) permit us to test whether or not the evolutionary progression of symbiotic lifestyles suggested for isopods is correct and (2) help us explore other outstanding yet unresolved issues in marine invertebrates that have adopted a symbiotic lifestyle.

REFERENCES

Adlard, R.D., and R.J.G. Lester. 1995. The life cycle and biology of *Anilocra pomacentri* (Isopoda: Cymothoidae) an ectoparasitic isopod of the coral reef fish, *Chromis nitida* (Perciformes: Pomacentridae). Australian Journal of Zoology 43:271–281.

Aguilar-Alberola, J.A., and F. Mesquita-Joanes. 2011. Population dynamics and tolerance to desiccation in a crustacean ostracod adapted to life in small ephemeral water bodies. Limnologica 41:348–355.

Ambrosio, L.J., and J.A. Baeza. 2016. Territoriality and conflict avoidance explain asociality (Solitariness) of the endosymbiotic pea crab *Tunicotheres moseri*. PLoS ONE 11:e0148285.

Arnott, S.A. 2001. Infection of intertidal barnacles by the parasitic isopod *Hemioniscus balani* in north-east England. Journal of the Marine Biological Association of the UK 81:171–172.

Avenant-Oldewage, A., and L. Everts. 2010. *Argulus japonicus*: sperm transfer by means of a spermatophore on *Carassius auratus* (L). Experimental Parasitology 126:232–238.

Baeza, J.A. 2015. Crustaceans as symbionts: an overview of their diversity, host use and life styles. Pages 163–189 *in* M. Thiel, and L. Watling, editors. The Natural History of the Crustacea, volume 2: Lifestyles and feeding biology. Oxford University Press, New York.

Baeza, J.A., J.R. Anderson, A.J. Spadaro, and D.C. Behringer. 2012. Sexual dimorphism, allometry, and size at first maturity of the Caribbean King Crab, *Mithrax spinosissimus*, in the Florida Keys. Journal of Shellfish Research 31:909–916.

Baeza, J.A., and C. Piantoni. 2010. Sexual system, sex ratio and group living in the shrimp *Thor amboinensis* (De Man): relevance to resource-monopolization and sex-allocation theories. Biological Bulletin 219:151–165.

Bandilla, M., T. Hakalahti, P.J. Hudson, and E.T. Valtonen. 2005. Aggregation of *Argulus coregoni* (Crustacea: Branchiura) on rainbow trout (*Oncorhynchus mykiss*): a consequence of host susceptibility or exposure? Parasitology 130:169–76.

Bandilla, M., T. Hakalahti, and E.T. Valtonen. 2008. Patterns of host switching in the fish ectoparasite *Argulus coregoni*. Behavioral Ecology and Sociobiology 62:975–982.

Banerjee, A., S. Manna, and S. Kumar Saha. 2015. A histological evaluation of development and axis formation in freshwater fish ectoparasite *Argulus bengalensis* Ramakrishna, 1951 (Crustacea: Branchiura). Parasitology Research 114:2199–2212.

Barel, C.D.N., and P.G.N. Kramers. 1977. A survey of the echinoderm associates of the north-east Atlantic area. Zoologische Verhandelingen 156:1–159.

Bartilotti, C., J. Salabert, and A. Dos Santos. 2016. Complete larval development of *Thor amboinensis* (De Man, 1888) (Decapoda: Thoridae) described from laboratory-reared material and identified by DNA barcoding. Zootaxa 4066:399–420.

Becker, J.H.A., L.M. Curtis, and A.S. Grutter. 2005. Cleaner shrimp use a rocking dance to advertise cleaning service to clients. Current Biology 15:760–764.

Bennett, M.B., S.M. Heupel, S.M. Bennett, and A.R. Parker. 1997. *Sheina orri* (Myodocopa: Cypridinidae), anostracod parasiticon the gills of the epaulette shark *Hemiscyllium ocellatum* (Elasmobranchii: Hemiscyllidae). International Journal of Parasitology 27:275–281.

Bocquet-Védrine, J. 1972. Conditions ecologiques necessaires a l'instauration du parasitisme des Cirripedes Opercules par les Rhizocephales. Comptes Readus Hebdomadaires des Seances de l'Academie des Sciences 275:67–69.

Boxshall, G.A. 2005. Branchiura (fish lice). Pages 145–147 *in* K. Rohde, editor. Marine parasitology. CABI Publishing, Wallingford, England and CSIRO Publishing, Collingwood, Australia.

Boxshall, G.A., and R.J. Lincoln. 1987. The life cycle of the Tantulocarida (Crustacea). Philosophical Transactions of the Royal Society B: Biological Sciences 315:267–303.

Brooker, A.J., A.P. Shinn, and J.E. Bron. 2007. A review of the biology of the parasitic copepod *Lernaeocera branchialis* (L., 1767) (Copepoda: Pennellidae). Advances in Parasitology 65:297–341.

Bruce, A.J. 1976. Coral reef Caridea and "commensalism." Micronesica 12:83–98.

Brusca, R.C. 1978. Studies on the cymothoid fish symbionts of the eastern Pacific (Isopoda, Cymothoidae) I. Biology of *Nerocila californica*. Crustaceana 34:141–154.

Brusca, R.C., and G.J. Brusca. 2003. Invertebrates. Sinauer Associates, Sunderland, Massachusetts.

Bunkley-Williams, L., and E. Williams. 1998. Isopods associated with fishes: a synopsis and corrections. Journal of Parasitology 84:893–896.

Castro, P. 1976. Brachyuran crabs symbiotic with scleractinian corals: a review of their biology. Micronesica 12:99–110.

Castro, P. 2015. Symbiotic Brachyura. Pages 543–581 *in* P. Castro, P. Davie, D. Guinot, F.R. Schram, and J.C. von Vaupel Klein, editors. Treatise on Zoology—Anatomy, Taxonomy, Biology. The Crustacea. Vol 9C-I. Decapoda: Brachyura. Brill, Leiden, The Netherlands.

Castro, P., P.K.L. Ng, and S.T. Ahyong. 2004. Phylogeny and systematics of the Trapeziidae Miers, 1886 (Crustacea: Brachyura), with the description of a new family. Zootaxa 643:1–70.

Chen, S-H., Q. Liu, Y-N. Zhang, J-X. Chen, H. Li, Y. Chen, P. Steinmann, and X-N. Zhou. 2010. Multi-host model-based identification of *Armillifer agkistrodontis* (Pentastomida), a new zoonotic parasite from China. PLoS Neglected Tropical Diseases 4:e647.

Cobo V. J., and C. M. Okamori. 2008. Fecundity of the spider crab *Mithraculus forceps* (Decapoda, Mithracidae) from the northeastern coast of the state of São Paulo, Brazil, Iheringia Série Zoologia 98:84–87.

Cohen, A.C., and J.G. Morin. 1990. Patterns of reproduction in ostracodes; a review. Journal of Crustacean Biology 10:184–211.

Crisp, D.J. 1968. Distribution of the parasitic isopod *Hemioniscus balani* with special reference to the East Coast of North America. Journal of the Fisheries Research Board of Canada 25:1161–1167.

Crisp, D.J. 1984. Overview of research on marine larvae, 1940–1980. Pages 103–126 *in* J.D. Costlow, and R.C. Tipper, editors. Marine biodeterioration: an interdisciplinary dtudy. Proceedings of the Symposium on Marine Biodeterioration. Uniformed Services University of Health Sciences, E. & F.N Spon, London, England.

de Baets, K., and D.T.J. Littlewood. 2015. The importance of fossils in understanding the evolution of parasites and their vectors. Advances in Parasitology 90:1–51.

de Bary, A. 1879. Die Erscheinung der Symbiose: Vortrag gehalten auf der Versammlung Deutscher Naturforscher und Aerzte zu Cassel. Verlag von Karl J. Trübner, Strasburg, Germany.

de Grave, S. 1999. Pontoniinae (Crustacea: Decapoda: Palaemonidae) associated with bivalve molluscs from Hansa Bay, Papua New Guinea. Bulletin of the Royal Institute of Natural Science in Belgium 69:125–141.

Dojiri, M., and R.F. Cressey. 1987. Revision of the Taeniacanthidae (Copepoda: Poecilostomatoida) parasitic on fishes and sea urchins. Smithsonian Contributions to Zoology 447:1–250.

Duffy, J.E., C.L. Morrison, and R. Ríos. 2000. Multiple origins of eusociality among sponge-dwelling shrimps (*Synalpheus*). Evolution 54:503–516.

Ďuriš, Z., I. Horká, P.J. Juračka, A. Petrusek, and F. Sandford. 2011. These squatters are not innocent: the evidence of parasitism in sponge-inhabiting shrimps. PLoS ONE 6:e21987.

Fransen, C.H.J.M. 2006. On Pontoniinae (Crustacea, Decapoda, Palaemonidae) collected from ascidians. Zoosystema 28:713–746.

Glenner, H. 2001. Cypris metamorphosis, injection and earliest internal development of the Rhizocephalan *Loxothylacus panopaei* (Gissler) Crustacea: Cirripedia: Rhizocephala: Sacculinidae. Journal of Morphology 249:43–75.

Glenner, H., and J.T. Høeg. 1994. Metamorphosis in the Cirripedia Rhizocephala and the homology of the kentrogon and trichogon. Zoologica Scripta 23:161–173.

Glenner, H., J.T. Høeg, M.J. Grygier, and Y. Fujita. 2008. Induced metamorphosis in crustacean y-larvae: towards a solution to a 100-year-old riddle. BMC Biology 6:21.

Glenner, H., J.T. Høeg, J. Stenderup, and A.V. Rybakov. 2010. The monophyletic origin of a remarkable sexual system in akentrogonid rhizocephalan parasites: a molecular and larval structural study. Experimental Parasitology 125:3–12.

Gore, R.H., L.E. Scotto, and J.K. Reed. 1983. Early larval stages of the Indo-Pacific coral gall-forming crab *Hapalocarcinus marsupialis* Stimpson, 1859 (Brachyura, Hapalocarcinidae) cultured in the laboratory. Crustaceana 44:141–150.

Grygier, M.J. 1981. *Gorgonolaureus muzikae* sp. nov. (Crustacea: Ascothoracida) parasitic on a Hawaiian gorgonian, with special reference to its protandric hermaphroditism. Journal of Natural History 15:1019–1045.

Grygier, M.J. 1987. New records, external and internal anatomy, and systematic position of Hansen's y-larvae (Crustacea: Maxillopoda: Facetotecta). Sarsia 72:261–278.

Grygier, M.J., and J.T. Høeg. 2005. Ascothoracida (Ascothoracids). Pages 149–154 *in* K. Rohde, editor. Marine Parasitology. CABI Publishing, Wallingford, England and CSIRO Publishing, Collingwood, Australia.

Gruner, A.G. 1975. Caprellidea II. Fam. Cyamidae. Crustaceorum Catalogus 5:79–104.

Hamel, J.F., P.K.L. Ng, and A. Mercier. 1999. Life cycle of the pea-crab *Pinnotheres halingi* sp. nov., an obligate symbiont of the sea cucumber *Holothuria scahra* Jaeger. Ophelia 50:149–175.

Hart, C.W. 1971. New species of parasitic ostracod of the genus *Acetabulastoma* (Paradoxostomatidae, Paradoxostominae) with a discussion of the copulatory appendage homologies. Notulae Naturae 388:1–10.

Hart, C.W., Jr., L.C. Hayek, J. Clark, and W.H. Clark. 1985. The life history and ecology of the entocytherid ostracod *Uncinocythere occidentalis* (Kozloff and Whitmann) in Idaho. Smithsonian Contributions to Zoology 419:1–22.

Hart, D.G., and C.W. Hart. 1974. The ostracod family Entocytheridae. The Academy of Natural Sciences of Philadelphia monographs 18. Fulton Press, Lancaster, Pennsylvania.

Hernández, J.E., J.A. Bolaños, J.L. Palazón, G. Hernández, C. Lira, and J.A. Baeza. 2012. The enigmatic life history of the symbiotic crab *Tunicotheres moseri* (Rathbun, 1918) (Crustacea, Brachyura, Pinnotheridae): implications for its mating system and population structure. Biological Bulletin 223:278–290.

Hispano, C., P. Bultó, and A.R. Blanch. 2014. Life cycle of the fish parasite *Gnathia maxillaris* (Crustacea: Isopoda: Gnathiidae). Folia Parasitologica 61:277–284.

Høeg, J.T. 1985. Cypris settlement, kentrogon formation and host invasion in the parasitic barnacle *Lernaeodiscus porcellanae* (Müller) (Crustacea: Cirripedia: Rhizocephala). Acta Zoologica (Stockh) 66:1–45.

Høeg, J.T. 1987. Male cypris metamorphosis and a new male larval form, the trichogon, in the parasitic barnacle *Sacculina carcini* (Crustacea: Cirripedia: Rhizocephala). Philosophical Transactions of the Royal Society B: Biological Sciences 317:47–63.

Høeg, J.T. 1991. Functional and evolutionary aspects of the sexual system in the Rhizocephala (Tecostraca: Cirripedia). Pages 208–227 *in* R.T. Bauer, and J.W. Martin, editors. Crustacean sexual biology. Columbia University Press, New York.

Høeg, J.T. 1995. Sex and the single cirripede: a phylogenetic perspective. Pages 195–207 *in* F.R. Schram, and J.T. Høeg, editors. New frontiers in barnacle evolution. Crustacean issues 10. A.A. Balkema, Rotterdam, The Netherlands.

Høeg, J.T., and G.A. Kolbasov. 2002. Lattice organs in y cyprids of the Facetotecta and their significance in the phylogeny of the Crustacea Thecostraca. Acta Zoologica 83:67–79.

Høeg, J.T., and J. Lützen. 1995. Life cycle and reproduction in the Cirripedia Rhizocephala. Oceanography and Marine Biology, An Annual Review 33:427–485.

Høeg, J.T., H. Glenner, and J. Shields. 2005. Cirripedia Thoracica and Rhizocephala (barnacles). Pages 154–165 *in* K. Rohde, editor. Marine parasitology. CABI Publishing, Wallingford, England and CSIRO Publishing, Collingwood, Australia.

Høeg, J.T., N.C. Lagersson, and H. Glenner. 2004. The complete cypris larva and its significance in thecostracan phylogeny. Pages 197–215 *in* G. Scholtz, editor. Evolutionary and developmental biology of Crustacea. Crustacean issues 15. A.A. Balkema, Lisse, The Netherlands.

Høeg, J.T., M. Pérez-Losada, H. Glenner, G.A. Kolbasov, and K.A. Crandall. 2009. Evolution of morphology, ontogeny and life cycles within the Crustacea Thecostraca. Arthropod Systematic Phylogeny 67:179–217.

Holthuis, L.B. 1993. The recent genera of the caridean and stenopodidean shrimps (Crustacea, Decapoda) with an appendix on the Order Amphionidacea. National Natuurhistorisch Museum, Leiden, The Netherlands.

Huys, R., G.A. Boxshall, and R.J. Lincoln. 1993. The tantulocaridan life cycle: the circle closed? Journal of Crustacean Biology 13:432–442.

Huys, R., S. Ohtsuka, G.A. Boxshall, and T. Itô. 1992. *Itoitantulus misophricola* gen. et sp. nov.: first record of Tantulocarida (Crustacea: Maxillopoda) in the north Pacific region. Zoological Science 9:875–886.

Huys, R., J. Llewellyn-Hughes, S. Conroy-Dalton, P.D. Olson, J.N. Spinks, and D.A. Johnston. 2007. Extraordinary host switching in siphonostomatoid copepods and the demise of the Monstrilloida: integrating molecular data, ontogeny and antennulary morphology. Molecular Phylogenetics and Evolution 43:368–378.

Huys, R., J. Olesen, A. Petrunina, and J. Martin. 2014. Tantulocarida. Pages 122–127 *in* J.W. Martin, J. Olesen, and J.T. Høeg, editors. An atlas of crustacean larvae. Johns Hopkins University Press, Baltimore, Maryland.

Jones, J.B. 1977. Post-planktonic stages of *Pinnotheres novaezelandiae* Filhol, 1886 (Brachyura: Pinnotheridae). Journal of Marine and Freshwater Research 11:145–158.

Kensley, B., and M. Schotte. 1989. Guide to the marine isopod crustaceans of the Caribbean. Smithsonian Institution Press, Washington D.C..

Kolbasov, G.A., and J.T. Høeg. 2007. Cypris larvae of acrothoracican barnacles (Thecostraca: Cirripedia: Acrothoracica). Zoologischer Anzeiger 246:127–151.

Kolbasov, G.A., M.J. Grygier, J.T. Høeg, and W. Klepal. 2008. External morphology of the two cypridiform ascothoracid larva instars of *Dendrogaster*: the evolutionary significance of the two-step metamorphosis and comparison of lattice organs between larvae and adult males (Crustacea, Thecostraca, Ascothoracida). Zoologischer Anzeiger 247:159–183.

Komai, T., de Grave, S., and T. Saito. 2016. Two new species of the stenopodidean shrimp genus *Spongiocaris* Bruce & Baba, 1973 (Crustacea: Decapoda: Spongicolidae) from the Indo-West Pacific. Zootaxa 4111:421–447.

Kotb, M.M.A., and R.G. Hartnoll, 2002. Aspects of the growth and reproduction of the coral gall crab *Hapalocarcinus marsupialis*. Journal of Crustacean Biology 22:558–566.

Knudsen, S.W., M. Kirkegaard, and J. Olesen. 2009. The tantulocarid genus *Arcticotantalus* removed from Basipodellidae into Deoterthridae (Crustacea: Maxillopoda) after the description of a new species from Greenland, with first live photographs and an overview of the class. Zootaxa 2035:41–68.

Kretzler, J.E. 1984. *Echinophilus xiphidion*, new species (Ostacoda, Paradoxostomatidae) parasitic on regular echinoids of the Northeastern Pacific. Journal of Crustacean Biology 4:333–340.

Kubo, I. 1942. A new commensal shrimp, *Spongicola japonica*, n. sp. Annotationes Zoologicae Japonenses 21:90–94.

Laval, P. 1980. Hyperiid amphipods as crustacean parasitoids associated with gelatinous zooplankton. Oceanography and Marine Biology, An Annual Review 18:11–56.

Lester, R.J.G. 2005. Isopoda (isopods). Pages 138–144 *in* K. Rohde, editor. Marine Parasitology. CABI Publishing, Wallingford, England and CSIRO Publishing, Collingwood, Australia.

Limbaugh, C., H. Pederson, and F.A. Chace, Jr. 1961. Shrimps that clean fishes. Bulletin of Marine Science 11:237–257.

Maddocks, R.F. 1987. An ostracode commensal of an ophiuroid and other new species of *Pontocypria* (Podocopida: Cypridacea). Journal of Crustacean Biology 7:727–737.

Markham, J.C. 2004. New species and records of Bopyridae (Crustacea: Isopoda) infesting species of the genus *Upogebia* (Crustacea: Decapoda: Upogebiidae): the genera *Orthione Markham*, 1988, and Gyge Cornalia & Panceri, 1861. Proceedings of the Biological Society of Washington 117:186–198.

Martens, K., and K. Harrison. 1993. An interesting association between a freshwater sponge (Porifera) and an ostracod (Crustacea) in Lake Titicaca (Bolivia). Verhandlungen Internationale der Vereinigung Theoretische und Angewande Limnologie 25:923–924.

Mesquita-Joanes, F., A.J. Smith, and F.A. Viehberg. 2012. The ecology of Ostracoda across levels of biological organisation from individual to ecosystem: a review of recent developments and future potential. Pages 15–35 *in* D.J. Horne, J.A. Holmes, J. Rodriguez-Lazaro, and F.A. Viehberg, editors. Developments in quaternary sciences. Elsevier, Oxford, England.

Mestre, A., J.A. Aguilar-Alberola, D. Baldry, H. Balkis, A. Ellis, J.A. Gil-Delgado, K. Grabow, G. Klobucar, A. Kouba, I. Maguire, A. Martens, A. Mülayim, J. Rueda, B. Scharfs, M. Soes, J. Monrós, and F. Mesquita-Joanes. 2013. Invasion biology in non-free-living species: interactions between abiotic (climatic) and biotic (host availability) factors in geographical space in crayfish commensals (Ostracoda, Entocytheridae). Ecology and Evolution 3:5237–5253.

Mikheev, V.N., A.F. Pasternak, and E.T. Valtonen. 2015. Behavioural adaptations of argulid parasites (Crustacea: Branchiura) to major challenges in their life cycle. Parasites and Vectors 8:394.

Morand, S., F. Robert, and V.A. Connors. 1995. Complexity in parasite life cycles: population biology of cestodes in fish. Journal of Animal Ecology 64:256–264.

Møller, O.S., J. Olesen, and D. Waloszek. 2007. Swimming and cleaning in the free-swimming phase of *Argulus* larvae (Crustacea, Branchiura)—appendage adaptation and functional morphology. Journal of Morphology 268:1–11.

Møller Christensen, A., and J.J. McDermott 1958. Life-history and biology of the oyster crab, *Pinnotheres ostreum* Say. Biological Bulletin 114:146–179.

Nagler, C., C. Haug, U. Resch, J. Kriwet, and J.T. Haug. 2016. 150 million years old isopods on fishes: a possible case of palaeo-parasitism. Bulletin of Geosciences 91:1–12.

Ng, P.K.L., D. Guinot, and P.J.F. Davie. 2008. Systema Brachyurorum: part I. An annotated checklist of extant brachyuran crabs of the world. Raffles Bulletin of Zoology 17:1–286.

Ocampo, E.H., J.D. Nuñez, M.S. Lizarralde, and M. Cledón. 2011. Larval development of *Calyptraeotheres garthi* (Fenucci, 1975) (Brachyura, Pinnotheridae) described from laboratory reared material, with notes of larval character use on Pinnotheridae systematic. Helgoland Marine Research 65:347–359.

Ocampo, E.H., J.D. Nuñez, M. Cledón, and J.A. Baeza. 2014. Parasitic castration in slipper limpets infested by the symbiotic crab *Calyptraeotheres garthi*. Marine Biology 161:2107–2120.

Ocampo, E.H., E.D. Spivak, J.A. Baeza, and T.A. Luppi. 2016. Ontogenetic changes in the external anatomy of the parasitic castrator crab *Calyptraeotheres garthi*: implications for the timing of host colonization and sexual behavior. Biological Journal of the Linnean Society.

Ota, Y., O. Hoshino, M. Hirose, K. Tanaka, and E. Hirose. 2012. Third-stage larva shifts host fish from teleost to elasmobranch in the temporary parasitic isopod, *Gnathia trimaculata* (Crustacea; Gnathiidae). Marine Biology 159:2333–2347.

Palacios-Theil, E., J. Cuesta, and D. Felder. 2016. Molecular evidence for non-monophyly of the pinnotheroid crabs (Crustacea: Brachyura: Pinnotheroidea), warranting taxonomic reappraisal. Invertebrate Systematics 30:1–27.

Pasternak, A.F., V.N. Mikheev, and E.T. Valtonen. 2000. Life history characteristics of *Argulus foliaceus* L. (Crustacea: Branchiura) populations in Central Finland. Annals Zoologica Fennica 37:25–35.

Pearce, J.B. 1966. The biology of the mussel crab, *Fabia subquadrata* from the water of the San Juan Archipelago, Washington. Pacific Science 20:3–35.

Poly, W.J. 2003. *Argulus ambystoma*, a new species parasitic on the salamander *Ambystoma dumerilii* from Mexico (Crustacea: Branchiura: Argulidae). Ohio Journal of Science 103:52–61.

Potts, F.A. 1915. *Hapalocarcinus*, the gall forming crab, with some notes on the related genus *Cryptochirus*. Carnegie Institute of Washington, Papers from the Department of Marine Biology 8:35–69.

Poulin, R., and T.H. Cribb. 2002. Trematode life cycles: short is sweet? Trends in Parasitology 18:176–183.

Reinhard, E.G. 1949. Experiments on the determination and differentiation of sex in the bopyrid *Stegophryxus hyptius* Thompson. Biological Bulletin 96:17–31.

Rhyne, A.L., Y. Fujita, and R. Calado. 2006. Larval development and first crab of *Mithraculus sculptus* (Decapoda: Brachyura: Majoidea: Mithracinae) described from laboratory-reared material. Journal of the Marine Biological Association of the UK 86:1133–1147.

Rioja, E. 1940. Estudios carcinologicos. V. Morfologia de un ostracodo epizoario observado sobre *Cambarus (Cambarellus) montezumae Sauss.* de Mexico, *Entocythere heterodonta* n. sp. y descripcion de algunos de sus estados lavarios. Annales del Instituto de Biología, Mexico 11:593–609.

Roelofs, H.M.A. 1968. Etude du developpement de l'Ostracode marin *Sphaeromicola dudichi* Klie, 1938. Bulletin Zoologisch Museum 1:39–43.

Saito, T. 2002. Development of the external sexual characters in the deep-sea sponge-associated shrimp, *Spongicola japonica* Kubo (Decapoda: Stenopodidea: Spongicolidae). Journal of Natural History 36:819–829.

Saito, T., and K. Konishi. 1999. Direct development in the sponge-associated deep-sea shrimp *Spongicola japonica* (Decapoda: Spongicolidae). Journal of Crustacean Biology 19:46–52.

Saito, T., and Y. Koya. 2001. Gonadal maturation and embryonic development in the deep-sea sponge-associated shrimp, *Spongicola japonica* Kubo (Crustacea: Decapoda: Spongicolidae). Zoological Science 18:567–576.

Saito, T., and M. Takeda. 2003. Phylogeny of the family Spongicolidae (Crustacea: Stenopodidea): evolutionary trend from shallow-water free-living to deep-water sponge-associated habitat. Journal of the Marine Biological Association of the UK 83:119–131.

Saito, T., I. Uchida, and M. Takeda. 2001. Pair formation in *Spongicola japonica* (Crustacea: Stenopodidea: Spongicolidae), a shrimp associated with deep-sea hexactinellid sponges. Journal of the Marine Biological Association of the UK 81:789–797.

Schmitt, W.L., J.C. McCain, and E.S. Davidson. 1973. Decapoda I. Brachyura I. Family Pinnotheridae. Pages 1–160 *in* H.-E. Gruner, and L.B. Holthuis, editors. Crustaceorum Catalogus, volume 3. W. Junk B.V., The Hague, The Netherlands.

Scotto, L.E., and R.H. Gore. 1981. Studies on decapod Crustacea from the Indian River region of Florida. 23. The laboratory cultured zoeal stages of the coral gall-forming crab *Troglocarcinus corallicola* Verrill, 1908 (Brachyura, Haplocarcinidae) and its familial position. Journal of Crustacean Biology 1:486–505.

Seidel, B. 1995. Behavioural and ecological aspects of the association between *Cyclocypris globosa* (Sars, 1863) (Cypridoidea, Cyclocypridinae) and *Bombina variegata* (L., 1758) (Anura, Bombinatoridae) in temporary pools in Austria. Crustaceana 68:813–823.

Shimura, S. 1981. The larval development of *Argulus coregoni* Thorell (Crustacea: Branchiura). Journal of Natural History 15:331–348.

Shuster, S.M. 2007. The evolution of crustacean mating systems. Pages 29–47 *in* E.J. Duffy, and M. Thiel, editors. Evolutionary ecology of social and sexual systems: crustaceans as model organisms. Oxford University Press, New York.

Smit, N.J., L. Basson, and J.G. van As. 2003. Life cycle of the temporary fish parasite, *Gnathia africana* (Crustacea: Isopoda: Gnathiidae). Folia Parasitologica 50:135–142.

Smit, N.J., N.L. Bruce, and K.A. Hadfield. 2014. Global diversity of fish parasitic isopod crustaceans of the family Cymothoidae. International Journal for Parasitology: Parasites and Wildlife 3:188–197.

Smith, R.J. 2014. Ostracoda: Podocopa. Pages 165–168 *in* J.W. Martin, J. Olesen, and J.T. Høeg, editors. An atlas of crustacean larvae. Johns Hopkins University Press, Baltimore, Maryland.

Smith, R.J., and T. Kamiya. 2005. The ontogeny of the entocytherid ostracod *Uncinocythere occidentalis* (Kozloff & Whitman, 1954) Hart, 1962 (Crustacea). Hydrobiologia 538:217–229.

Sproston, N.G. 1942. The developmental stages of *Lernaeocera branchialis* (Linn.). Journal of the Marine Biological Association of the UK 25:441–466.

Stachowicz, J.J., and M.E. Hay. 1996. Facultative mutualism between an herbivorous crab and a coralline alga: advantages of eating noxious seaweeds. Oecologia 105:377–387.

Stachowicz, J.J., and M.E. Hay. 1999. Mutualism and coral persistence: the role of herbivore resistance to algal chemical defense. Ecology 80:2085–2101.

Suárez-Morales, E. 2011. Diversity of the Monstrilloida (Crustacea: Copepoda). PLoS ONE 6:e22915.

Tappe, D., and D.W. Büttner. 2009. Diagnosis of human visceral pentastomiasis. PLoS Neglected Tropical Diseases 3:e320.

Tappe, D., M. Meyer, A. Oesterlein, A. Jaye, M. Frosch, C. Schoen, and N. Pantchev. 2011. Transmission of *Armillifer armillatus* ova at Snake Farm, The Gambia, West Africa. Emerging Infectious Diseases 17:251–254.

Terrana, L., G. Caulier, G. Todinanahary, G. Lepoint, and I. Eeckhaut. 2016. Characteristics of the infestation of *Seriatopora* corals by the coral gall crab *Hapalocarcinus marsupialis* Stimpson, 1859 on the great reef of toliara, Madagascar. Symbiosis 69:113–122.

Thiel, M. 2000. Population and reproductive biology of two sibling amphipod species from ascidians and sponges. Marine Biology 137:661–674.

Utinomi, H. 1961. Studies on the Cirripedia Acrothoracica. III. Development of the female and male of *Berndtia purpurea* Utinomi. Publications of the Seto Marine Biological Laboratory 9:413–446.

Vannini, M. 1985. A shrimp that speaks crab-ese. Journal of Crustacean Biology 5:160–167.

Vannini, M., R.K. Ruwa, and G. Innocenti. 1994. Notes on the behaviour of *Heteromysis harpax*, a commensal mysid living in hermit crab shells. Ethology, Ecology and Evolution Special Issue 3:137–142.

Whitfield, P.J., M.W. Pilcher, H.J. Grant, and J. Riley. 1988. Experimental studies on the development of Lernaeocera branchialis (Copepoda: Pennellidae): population processes from egg production to maturation on the flatfish host. Hydrobiologia 167–168:579–586.

Williams, J.D., and J. An. 2009. The cryptogenic parasitic isopod *Orthione griffenis* Markham, 2004 from the eastern and western Pacific. Integrative and Comparative Biology 49:114–126.

Williams, J.D., and C.B. Boyko. 2012. The global diversity of parasitic isopods associated with crustacean hosts (Isopoda: Bopyroidea and Cryptoniscoidea). PLoS One 7:e35350.

Williams, J.D., and J.J. McDermott. 2004. Hermit crab biocoenoses: a worldwide review of the diversity and natural history of hermit crab associates. Journal of Experimental Marine Biology and Ecology 305:1–128.

Yoshida, R., M. Hirose, and E. Hirose. 2014. Hermit crab host prevalence by species of Peltogastridae (Cirripedia: Rhizocephala): hosts vary with locations on the Pacific coast in mainland Japan. Journal of Crustacean Biology 34:467–480.

16

DAPHNIA AS A MODEL FOR ECO-EVOLUTIONARY DYNAMICS

Matthew R. Walsh, Michelle Packer, Shannon Beston, Collin Funkhouser, Michael Gillis, Julian Holmes, and Jared Goos

Abstract

Much research has shown that variation in ecological processes can drive rapid evolutionary changes over periods of years to decades. Such contemporary adaptation sets the stage for evolution to have reciprocal impacts on the properties of populations, communities, and ecosystems, with ongoing interactions between ecological and evolutionary forces. The importance and generality of these eco-evolutionary dynamics are largely unknown. In this chapter, we promote the use of water fleas (*Daphnia* sp.) as a model organism in the exploration of eco-evolutionary interactions in nature. The many characteristics of *Daphnia* that make them suitable for laboratory study in conjunction with their well-known ecological importance in lakes, position *Daphnia* to contribute new and important insights into eco-evolutionary dynamics. We first review the influence of key environmental stressors in *Daphnia* evolution. We then highlight recent work documenting the pathway from life history evolution to ecology using *Daphnia* as a model. This review demonstrates that much is known about the influence of ecology on *Daphnia* life history evolution, while research exploring the genomic basis of adaptation as well as the influence of *Daphnia* life history traits on ecological processes is beginning to accumulate.

INTRODUCTION

Arthropods have long been noted in the scientific community for their diversity, complex body plans, and widespread distribution. Crustaceans are a very large subphylum of arthropods that consists of an evolutionarily diverse group of animals (Schram 2013). Many species of crustaceans have been adopted and utilized as model organisms in a variety of contexts. For example, the amphipod *Parhyale hawaiensis* is commonly used for developmental and genetic studies (Rehm

Life Histories. Edited by Gary A. Wellborn and Martin Thiel.

et al. 2009), and the stages of embryonic development are used as a basis for comparison across crustaceans and other taxa. *Tigriopus* copepods are a frequent model for ecotoxicology and neurobiological studies (Raisuddin et al. 2007). They are used to evaluate the biological consequences of metals, organic pollutants (e.g., pesticides, polychlorinated biphenyls), and endocrine-disrupting chemicals (Raisuddin et al. 2007). Additionally, *Artemia* brine shrimp serve as excellent models for toxicological assays (Nunes et al. 2006). There thus exists a diverse array of crustaceans that serve as models for applications spanning a diversity of disciplines.

Daphniidae, order Cladocera, is gaining popularity as a model taxon in ecological and evolutionary research (Cox and Hebert 2001, Adamowicz et al. 2004, Harris et al. 2012; Fig. 16.1). Species from the genus *Daphnia* make up a significant proportion of zooplankton found in freshwater and saline habitats around the world, with a species richness currently exceeding 350 (Kotov 2015). The organisms exhibit numerous characteristics that make them well suited as a model organism: small size, ease of propagation, inexpensive maintenance, and easy quantification of many traits (e.g., life history, behavior, morphology). These features make *Daphnia* particularly convenient for laboratory study (Froehlich et al. 2009, Colbourne et al. 2011). For instance, members of this genus carry developing eggs in their brood chamber, which allows accurate assessments of several life history parameters, such as age at maturation, clutch size, and interclutch interval. They exhibit cyclical parthenogenesis (i.e., they transition between asexual and sexual reproduction in response to specific changes in the environment). Sexual reproduction results in the production of diapausing resting eggs that can persist for decades or centuries and allow for the study of paleolimnology of lakes (Pollard et al. 2003). The recently published *Daphnia* genome allows examination of the genes that underlie the phenotypic traits (Colbourne et al. 2011). Their rapid parthenogenetic life cycle lends the genus to the study of epigenetics without the confounding genetic factors usually associated with sexually reproducing organisms (Moest et al. 2015). Finally, species of *Daphnia* are a key component of aquatic food webs as they fill a central position in these food chains. They are the primary grazers on phytoplankton and thus exert significant pressure on phytoplankton abundance, composition, and rates of primary production, along with correlated impacts on nutrient cycling (Carpenter et al. 1987, 1992). They are also an important food item for upper tropic levels.

Many species of *Daphnia* experience variation in common ecological conditions that are presumed to exert selection and drive evolutionary changes (see Hairston et al. 1999, Walsh and Post 2011). This includes variation in predation, competition, nutrient availability, seasonality, habitat duration, and pathogens. As a result, many studies have used *Daphnia* as a model to address fundamental questions regarding the link between ecological selective pressures and the trajectory of adaptation in nature (see Table 16.1). A key goal of this chapter is to review our current understanding of the impact of ecologically driven selection on evolution in *Daphnia* (Fig. 16.2).

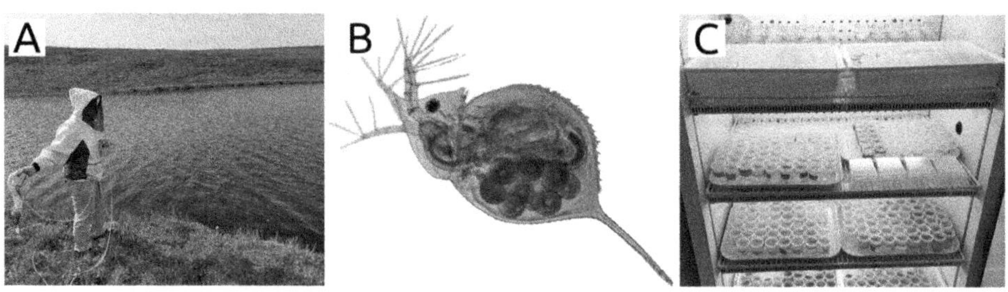

Fig. 16.1.
Daphnia as an experimental organism. (A) Sampling for *Daphnia* near Toolik Lake Long-Term Ecological Research Site (Alaska). (B) Photograph of *Daphnia lumholtzi*. (C) Experimental jars containing *Daphnia* in an environmental chamber. Photographs by Michelle Packer ©.

We assess the relationship between a broad array of ecological variables (i.e., predation, seasonality, competition) and the resulting selection on aspects of *Daphnia* behavior, life history, and physiology. For each section we evaluate the strength of the evidence for evolution, as well as the methods researchers have used to examine the evolutionary consequence of a given selective force (e.g., population comparisons, within-population variation, cline analysis). In addition, we give an overview of the current understanding of the molecular and genomic mechanisms that underlie evolutionary shifts in *Daphnia* traits.

A second objective of this chapter is to review what we currently know about the influence of *Daphnia* evolution on ecological processes (Fig. 16.2). It is now well established that evolution can occur rapidly in a natural setting, typically within a period of years to decades (Hairston et al. 1999, Hendry and Kinnison 1999, Reznick and Ghalambor 2001, Duffy et al. 2012). Such "contemporary" evolution may, in turn, alter the dynamics of ecological systems and thus sets the stage for an ongoing feedback between ecological and evolutionary forces in the wild (Post and Palkovacs 2009, Matthews et al. 2011, Schoener 2011, Reznick 2013). Previous reviews have focused on the use of *Daphnia* as a model organism for understanding ecological interactions (Lampert 2006), physiology (Altshuler et al. 2011), and genomics and epigenetics (Harris et al. 2012, Miner et al. 2012). Our goal is to build on the use of *Daphnia* as a model organism by focusing on the central role these crustaceans may play in reciprocal interactions between ecology and evolution. To do so, we ask how and why individuals of the genus *Daphnia* evolve in aquatic environments, whether their adaptation affects their environment, and then conclude with a discussion of future directions.

ECOLOGICAL DRIVERS OF EVOLUTION

This section explores the extent to which divergent ecological conditions are associated with overall shifts in trait values (i.e., evolution) and changes in responses to exposure to environmental cues (i.e., phenotypic plasticity). This review is not intended to be exhaustive but is instead meant to identify the general patterns of *Daphnia* adaptation in response to common environmental stressors.

Predation

Daphnia are well known to coexist with a variety of predators, including gape-limited invertebrate predators such as phantom midge larvae, as well as many planktivorous species of fish (Riessen 1999). Differences among phenotypes in predator-induced mortality have long been assumed to represent a key selective force on organismal traits (Charlesworth 1980). Not surprisingly, a large body of research has explored the responses of *Daphnia* to predators in aquatic environments.

Early work focused on the patterns of phenotypic plasticity that many species of *Daphnia* exhibit when exposed to predators. This work revealed contrasting responses to gape-limited invertebrate predators versus nongape-limited vertebrate predators. For instance, *Daphnia* species respond to the presence of chemical cues emitted from invertebrate *Chaoborus* by growing faster (and larger) and delaying maturation (Weider and Pijanowska 1993, Riessen 1999). Conversely, *Daphnia* raised in the presence of fish chemical cues typically mature earlier and invest more heavily into reproduction (Stibor 1992, Riessen 1999). Such divergent patterns of plasticity are typically interpreted as adaptive responses to contrasting patterns of size-specific mortality. That is, *Daphnia* invest in growth in the presence of *Chaoborus* to attain a sufficient size to exceed the gape width of the invertebrate predator. In the presence of fish, *Daphnia* invest in reproduction at the expense of growth because larger individuals are more susceptible to predatory fish.

Table 16.1. Studies evaluating relationships between ecological forces and evolution in *Daphnia*. Result denotes the evidence for evolution (i.e., trait evolution or evolved differences in phenotypic plasticity) or for the potential for evolution to occur (i.e., clonal variation).

Species	**Ecological factor**	**Approach**	**Traits**	**Result**	**Reference**
D. pulex	Invertebrate predation	Experimental evolution	Life history	Trait evolution	Spitze 1991
D. magna	Vertebrate predation	Population comparison	Life history	Trait evolution	de Meester and Boersma 1999
D. magna	Vertebrate predation	Clone comparison	Life history	Clonal variation	Weider and Pijanowska 19993
D. magna	Vertebrate predation	Population comparison	Behavior	Trait evolution Phenotypic plasticity	de Meester 1996
D. magna	Vertebrate predation	Resurrection	Behavior	Trait evolution Phenotypic plasticity	Cousyn et al. 2001
D. pulex	Invertebrate predation	Population comparison	Morphology	Trait evolution Phenotypic plasticity	Parejko and Dodson 1991
D. mendotae, D. cucullata	Vertebrate predation	Species comparison	Life history	Phenotypic plasticity	Spaak et al. 2000
D. pulex	Invertebrate predation	Clone comparison	Life history	Clonal variation	Tollrian 1995
D. magna	Vertebrate predation	Population comparison	Life history Morphology Behavior	Clonal variation Phenotypic plasticity	Boersma et al. 1998
D. pulex	Invertebrate predation	Clone comparison	Life history, Morphology	Clonal variation	Spitze 1992
D. ambigua	Vertebrate predation	Population comparison	Life history	Trait evolution	Walsh and Post 2011
D. ambigua	Vertebrate predation	Population comparison	Life history	Phenotypic plasticity	Walsh and Post 2012
D. galeata	Resource quality	Resurrection	Life history	Trait evolution	Hairston et al. 1999
D. galeata	Resource quality	Resurrection	Life history	Phenotypic plasticity	Hairston et al. 2001

D. pulicaria	Resource quality	Population comparison	Life history	Trait evolution Phenotypic plasticity	Sarnelle and Wilson 2005
D. ambigua	Resource quality	Population comparison	Life history	Phenotypic plasticity	Walsh et al. 2014
D. pulicaria	Temperature	Clone comparison	Life history	Clonal variation	Palaima and Spitze 2004
D. magna	Temperature	Population and clone comparison	Life history	Clone variation	Mitchell and Lampert 2000
D. magna	Temperature	Population and clone comparison	Physiology	Trait evolution Clone variation	Geerts et al. 2015a
D. magna	Temperature	Temporal comparison	Life history	Trait evolution	Carvalho 1987
D. pulex	Temperature	Experimental evolution	Life history	Trait evolution	Scheiner and Yampolsky 1998
D. magna	Temperature	Experimental evolution	Life history	Trait evolution Phenotypic plasticity	van Doorslaer et al. 2009
D. magna	Temperature	Experimental evolution Resurrection	Physiology	Trait evolution	Geerts et al. 2015b
D. dentifera	Pathogen	Population comparison	Disease resistance	Trait evolution	Duffy and Sivars-Becker 2007
D. magna	Pathogen	Resurrection	Disease resistance	Trait evolution	Decaestecker et al. 2005
D. magna	Pathogen	Clone comparison	Disease resistance	Clone variation	Vale and Little 2012
D. magna	Pathogen	Clone comparison Temporal comparison	Disease resistance	Clone variation	Mitchell et al. 2004b
D. dentifera	Pathogen	Temporal comparison	Disease resistance	Trait evolution	Duffy et al. 2012

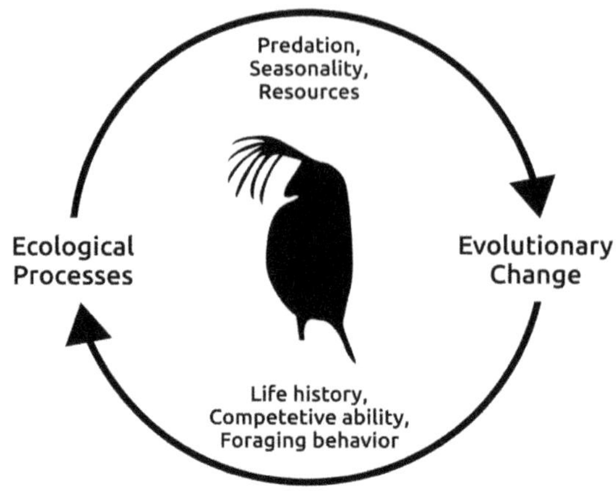

Fig. 16.2.
Daphnia as a mediator of ecoevolutionary dynamics. This chapter focuses on the ecological drivers of adaptation as well as the influence of *Daphnia* traits on the properties of populations, communities, and ecosystems.

These patterns of phenotypic plasticity set the stage for investigations into selection on predator-induced plasticity (Table 16.1). Research that has incorporated multiple clones in experiments has consistently documented genetic variation in the magnitude and direction of phenotypic responses to predators (Spitze 1991, Boersma et al. 1998). This variation is a precursor to evolution by natural selection and indicates that *Daphnia* contain sufficient variation to respond to predator-driven selection. Several studies have indeed shown that contrasting predation regimes or temporal shifts in predation intensity are associated with genetic shifts in phenotypic plasticity (Table 16.1). For instance, Boersma et al. (1998) demonstrated that clones from habitats with fishes were more responsive (in terms of the magnitude of phenotypic changes in the presence of fish chemical cues) than clones from "fishless" environments. Walsh and Post (2012) showed that variation in fish predator communities are associated with evolved differences in predator-induced plasticity. Here, predictable increases in mortality due to migration by anadromous populations of a predatory fish (the alewife, *Alosa pseudoharengus*) has driven evolutionary increases in plastic responses in *Daphnia ambigua* compared to populations that experience consistently strong (or weak) patterns of fish predation.

Predator-driven changes in plasticity are not limited to life history traits. Several studies have revealed evolutionary changes in the behavior of *Daphnia*. Fishes are visually oriented predators and *Daphnia* typically descend to deeper depths to avoid these predators during the day (i.e., negative phototaxis). Research has shown that *Daphnia* from lakes with fishes are more negatively phototactic in the laboratory when compared with clones from lakes that lack predators (de Meester 1996). Cousyn et al. (2001) conducted a "resurrection" experiment to examine selection on *Daphnia* behavior (see Fig. 16.3). These researchers hatched resting eggs from lake sediment from time periods that differed in the intensity of fish predation (based on fish stocking records) and showed that increased fish predation was correlated with increased antipredator behavior in prey.

Research has begun to consider the underlying molecular mechanisms for predator-induced plasticity. Schwarzenberger et al. (2009) evaluated patterns of plasticity in gene expression of several candidate genes in *D. magna* that were exposed to chemical cues produced by fish and invertebrate predators. This approach revealed strong upregulation of genes involved in protein biosynthesis, protein catabolism, and protein folding. Exposure to invertebrate predator cues was associated with upregulation of the genes involved with protein synthesis and catabolism but downregulation of cyclophylin. Given that *Daphnia* differ in their life history responses to fish

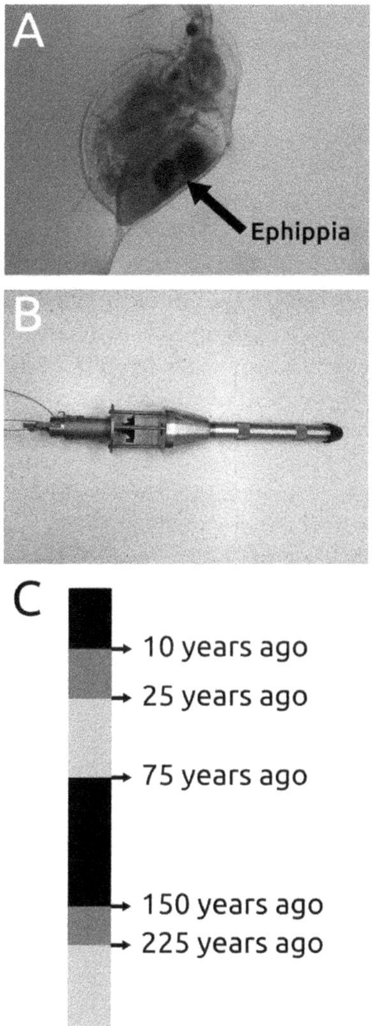

Fig. 16.3.
Resurrection ecology. Many species of *Daphnia* transition between asexual and sexual reproduction when environmental conditions deteriorate. Sexual reproduction results in the production of weather-resistant resting eggs (i.e., ephippia) that sink to the bottom of lakes and accumulate in the sediment. These eggs remain viable for decades. As a result, this naturally occurring egg bank can be used to test how *Daphnia* have evolved in response to a change in environmental conditions over time within a given lake. To perform a resurrection experiment, lake sediment is first obtained via a sediment corer, the age of the sediment layers (and eggs) are determined via isotopic analyses, and then the eggs can be hatched and reared for experimental study in the laboratory. (A) *Daphnia* with an ephippia, (B) lake sediment corer, and (C) hypothetical example of the results of isotope analyses for a sediment core. Photographs by Matthew R. Walsh ©.

versus invertebrate predator cues (Riessen 1999), these contrasting gene expression responses could signal that cyclophylin is linked to the expression of life history traits (Tollrian and Leese 2010). Similarly, Rozenberg et al. (2015) used RNA-Seq to quantify patterns of gene expression in response to predator cues. This approach revealed 230 differentially expressed genes (158 upregulated, 72 downregulated). Some of the strongly upregulated genes were connected with cuticle function, vitellogenin, chromatin reorganization, and regulation of the cell cycle, while downregulated gene classes included C-type lectins and proteins involved in lipogenesis. Such results clearly provide a connection to known phenotypic responses as they illustrate changes in the expression of genes associated with morphology (cuticle strengthening) and resource allocation when *Daphnia* are exposed to predator cues.

In addition to considering how predators may influence the degree of phenotypic plasticity in *Daphnia*, research has also tested the influence of predation on the evolution of overall trait values (Table 16.1). De Meester et al. (1999) showed that *Daphnia* from lakes with fish were smaller in size and allocated more energy toward reproduction than *Daphnia* from lakes without fish (see also Walsh and Post 2011). Spitze (1991) used a laboratory selection approach to demonstrate that predation by *Chaoborus americanus* rapidly selects for faster growth and a larger body size (but also earlier maturation and increased clutch size). Research has also begun to consider the genomic changes that underlie responses to divergent patterns of predation (Scoville and Pfrender 2010, Latta et al. 2012). For instance, in high ultraviolet (UV) alpine environments, *Daphnia* are typically pigmented and highly visible. The introduction of fishes in alpine lakes in the Sierra Nevada mountains is associated with evolutionary shifts in allocation towards pigmentation. *Daphnia* typically increase the amount of pigmentation when they are exposed to UV radiation (i.e., exhibit phenotypic plasticity), but *D. melanica* that are found with introduced fishes are no longer plastic because they consistently produce uniformly low melanin pigmentation, making them more transparent and less visible to fishes. Scoville and Pfrender (2010) showed that this evolutionary loss in plasticity is associated with changes in the upregulation of two genes in the melanin pathway. Collectively, this body of work has clearly shown that predators impose selection and drive evolutionary shifts in the life history and behavioral characteristics of *Daphnia*. For more details, see Chapter 12 and Chapter 13 in this volume.

Resource Quality

Lakes contain a diverse array of phytoplankton, ranging from high-quality green algae (*Scenedesmus*) to grazer-resistant cyanobacteria (i.e., *Anabaena*; Vijverberg 1989). This diversity results in large spatial and temporal variation in resource quantity and quality. Grazer-resistant species of phytoplankton, such as cyanobacteria, are generally considered to be low-quality food for *Daphnia* for multiple reasons. Cyanobacteria's colonial and filamentous morphology negatively affects the foraging rates of *Daphnia* (DeMott 1989) and is considered nutritionally deficient when compared to other phytoplankton species (von Elert and Wolffrom 2001, von Elert et al. 2002, Martin-Creuzberg et al. 2005, Brett et al. 2009). Furthermore, cyanobacteria can produce intracellular secondary metabolites that may be toxic to *Daphnia* (DeMott et al. 1991). An increased proportion of cyanobacteria in the diet of *Daphnia* is associated with declines in the survival, growth, and reproduction of *Daphnia* (Hairston et al. 1999, Rohrlack et al. 2001, Lurling 2003, Sarnelle et al. 2010). These negative impacts on fitness point to a relationship between resource quality and evolutionary change in *Daphnia* (Hairston et al. 1999, Sarnelle and Wilson 2005, Blom et al. 2006, Lemaire et al. 2012).

Research has indeed shown that variation in resource quality, in terms of phytoplankton composition, can act as a significant selective pressure on the traits of *Daphnia* (Table 16.1). Sarnelle and Wilson (2005) showed that *D. pulicaria* from lakes with high concentrations of cyanobacteria experienced smaller fitness declines in response to increasing amounts of cyanobacteria than *Daphnia* from less productive lakes that were characterized by consistently low concentrations of cyanobacteria. Hairston et al. (1999) resurrected eggs from the sediment of Lake Constance, Germany, before and after a period of eutrophication (i.e., high cyanobacteria) and found that increased concentrations of cyanobacteria were associated with the evolution of increased resistance to cyanobacteria in *Daphnia galeata* (see also Hairston et al. 2001). That is, clones from the posteutrophication period that were fed cyanobacteria were less sensitive and exhibited smaller declines in growth than clones from the pre-eutrophication period. Walsh et al. (2014) further showed that the negative impacts of cyanobacteria on the expression of life history traits in *Daphnia ambigua* were reduced in populations that experienced high concentrations of cyanobacteria when compared with populations of *Daphnia* that were typically absent from lakes during the summer (when cyanobacteria are dominant). This small but growing body of work has revealed evidence for local adaptation in response to variation in resource quality and that declines in food quality appear to consistently modify the shape of the reaction norms of *Daphnia*; increased densities of cyanobacteria are associated with declines in phenotypic plasticity.

Although resource quality is certainly a function of phytoplankton composition within lakes, *Daphnia* are often exposed to changes in nutrient availability that also affect the quality of their food. *Daphnia*, like most heterotrophs, are often nutritionally imbalanced with their food and thus may be limited by certain nutritional components. Ecological stoichiometry (ES) is a framework that abstracts organisms and their environments into atoms of elements, particularly carbon (C), nitrogen (N), and phosphorus (P), in order to study the role of the balance of energy and biologically important nutrients in shaping ecological interactions (Sterner and Elser 2002). A key principle of ES is the relatively homeostatic nature of heterotroph demand. Because all traits require elemental resources, organismal content is a composite trait that represents the whole individual. Organismal nutritional demand is, therefore, a function of evolutionary history and selection pressures imposed on traits (Woods et al. 2004, Kay et al. 2005, González et al. 2011). Because elements are a common currency on which all biology is based, ES provides a useful, taxon-universal framework from which links can be made between higher-order environmental processes, such as biogeochemical cycles, and those at the organismal level.

Since the inception of ES, *Daphnia* have been an important model in which stoichiometric predictions have been tested. Central to ES is the idea that nutritional imbalances result in growth penalties. Perhaps the most well-known line of stoichiometric inquiry is that of the growth rate hypothesis. Put simply, this hypothesis suggests that variation in C:N:P ratios in organisms is the result of differences in ribosomal RNA (rRNA) content to meet the demands of organismal growth (Elser et al. 1996, Weider et al. 2005a). As such, rapidly growing organisms such as *Daphnia* likely have high P demands. Indeed, multiple empirical studies have demonstrated clear links between *Daphnia* life history and P content in food (e.g., DeMott et al. 1998, Boersma 2000, Hood and Sterner 2014). Similar studies in other crustacean taxa have also demonstrated clear plastic responses in life history to elemental availability (e.g., Cothran et al. 2012, Danger et al. 2013).

The life history consequences associated with changes in elemental supply have also been shown to be genetically determined, and microevolutionary patterns within populations may be influenced by environmental element supply. Multiple studies within *Daphnia* have demonstrated significant genotype-by-environment (G × E) interactions in key life history parameters. For example, Weider et al. (2005b) examined competitive responses in two *D. pulex* genotypes (clones) that vary in their intergenic spacer (IGS) length in response to environmental P supply. IGS should be positively associated with growth rate due to a greater potential for rRNA production to support high growth. In examining competition between the two clones in two environments that differ in phosphorus availability, the researchers found that the clone with longer IGS outcompeted the shorter IGS clone under high P (low C:P) conditions, while the opposite pattern was observed under low P (high C:P) conditions. Using these same clones, Jeyasingh and Weider (2005) observed significant differences in responses to predator cues that were driven by environmental P supply. Additionally, these genotypes also differed in their C and P use efficiencies, as well as the expression of genes under different P environments (Roy Chowdhury et al. 2014). The role of nutritional environment as a selective force was further studied in *Daphnia* by Frisch et al. (2014) by combining ES with resurrection ecology. This study resurrected clones spanning approximately

1,600 years from one Minnesota lake and observed significant shifts in population genetic structure that roughly correspond to historical changes in P supply within lake sediments and human population density. Additionally, they observed significant differences in phosphorus use between modern and ancient clones, with modern clones generally being less P efficient than ancient clones, suggesting that increases in environmental P supply have selected for inefficient P use. Together, these studies illustrate the important role of nutritional quality in potentially driving evolutionary shifts within and between populations.

Similar to the connection between predator cues and trait plasticity, research has also explored the shifts in gene expression in response to contrasting resource conditions (Schwarzenberger et al. 2010, Jeyasingh et al. 2011, Dudycha et al. 2012) and exposure to various environmental contaminants (Shaw et al. 2007, Vandegehuchte et al. 2010, Jansen et al. 2013). This work includes manipulations of cyanobacteria (Schwarzenberger et al. 2010), phosphorus availability (Jeyasingh et al. 2011), and diet complexity (Dudycha et al. 2012). For instance, Schwarzenberger (2009) identified 3 candidate genes that were upregulated when *D. magna* were exposed to toxic cyanobacteria (i.e., *Microcystis*). These genes are linked to glycolysis, protein catabolism, and the tricarboxylic acid cycle. Dudycha et al. (2012) found 14 genes that were differentially expressed between diet treatments that consisted of a simple high quality laboratory diet versus a more complex diet collected from a natural lake. This included genes that are known to influence gut processes and resource allocation.

Temperature

Many *Daphnia* species are widely distributed and are likely to experience significant daily or seasonal variation in ambient temperatures. Interest in temperature as an agent of selection has risen dramatically in recent years due to observed rapid increases in temperature associated with global climate change. Latitudinal variation in temperature (Conover and Present 1990), as well as changes in temperature due to such factors as global climate change, both present the opportunity for temperature-mediated evolution. Many studies have revealed genetic variation across a suite of traits in response to manipulations of temperature in *Daphnia* (see Palaima and Spitze 2004, Mitchell and Lambert 2000). Research has also shown that seasonal changes in temperature are typically associated with shifts in genotype frequencies in lakes (Hebert 1987, Mitchell et al. 2004a). Studies documenting evolutionary shifts in response to variation in temperature in the traits of *Daphnia* are also now beginning to accumulate (Table 16.1) utilizing a diversity of approaches. Geerts et al. (2015a) demonstrated that heat tolerance decreases with increasing latitude in populations of *Daphnia magna* across northern Europe. Carvalho (1987) compared clones of *Daphnia magna* before and after seasonal shifts in temperature and found that the transition from spring to summer was associated with selection for clones that exhibited increased survivorship and fecundity at high temperatures. These complementary approaches to the study of adaptation (i.e., latitudinal cline, seasonal shifts in temperature) provide some evidence for genetic responses to temperature.

The results of several laboratory selection experiments offer the clearest evidence for a link between temperature and *Daphnia* evolution (Table 16.1). Scheiner and Yampolsky (1998) performed a quasinatural selection experiment in the laboratory and found that manipulations of temperature primarily acted on mean trait values and not levels of temperature-induced phenotypic plasticity. Van Doorslaer et al. (2009a, 2009b) performed a series of selection experiments to isolate the selective influence of temperature. This work showed that multiple generations of rearing at a higher temperature selected for greater intrinsic rates of increase (and associated changes in other life history traits), although such responses depended upon the population dynamics of the experimental units (van Doorslaer et al. 2009a). In a particularly noteworthy study, Geerts et al. (2015b) combined selection experiments with a resurrection ecology approach to test the influence of rising

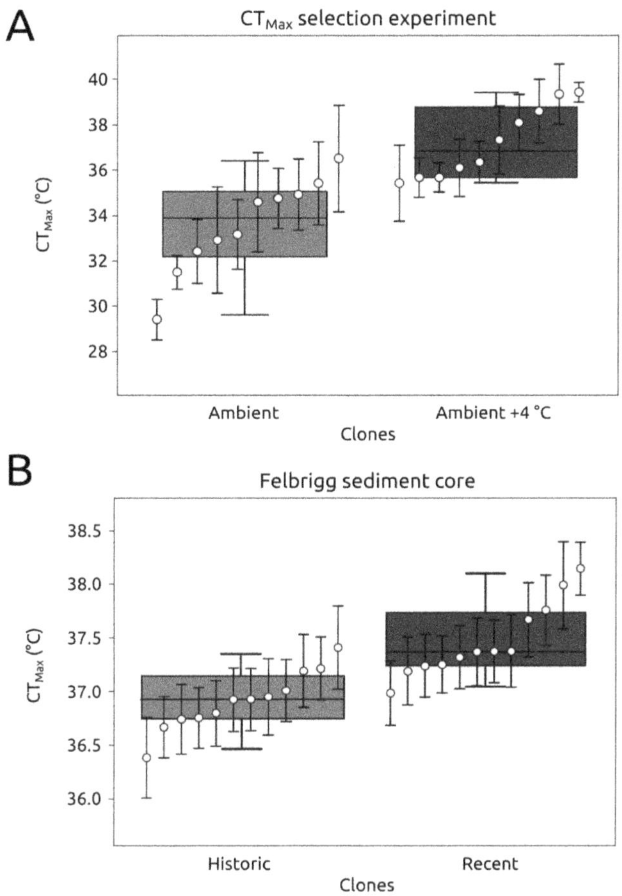

Fig. 16.4.
Average critical thermal maxima (CT_{Max}) for *Daphnia* clones hatched from sediment in a selection experiment and resurrection study performed in Geerts et al. (2015b). The selection experiment exposed clones of *Daphnia* to increased water temperatures (+4°C), while the resurrection study compared clones from two different time periods. The authors assessed variation in thermal performance by measuring CT_{Max} for all clones. (A) Average CT_{Max} values for *D. magna* clones grouped by selection temperature (subject to different selection temperatures: ambient (left box plot) or ambient + 4°C (right box plot). (B) Average CT_{Max} values of *D. magna* clones hatched from sediment cores of different dates. "Historic" clones dated to 1955–1965 (left box plot); "recent" clones dated to 1995–2005 (right box plot). In each plot clones are ordered within population according to increasing CT_{Max}. Error = ±1 standard error of mean. Modified from Geerts et al. (2015b), with permission from Nature Publishing Group.

temperatures on thermal tolerance in *Daphnia*. Results showed that experimental increases in temperature (+4°C) rapidly selected for genetic increases in thermal tolerance within two years (Fig. 16.4). The authors then compared the thermal tolerance of clones hatched from two different time periods. Clones hatched from sediment from 1995 to 2005 exhibited higher thermal tolerance than clones from the 1960s (sediment dated as between 1955 and 1965; Fig. 16.4). The authors argued that these genetic shifts in thermal tolerance are driven by recent increases in temperatures and the severity and frequency of heat waves. Such results provide evidence that *Daphnia* exhibit the capacity to respond to naturally occurring changes in temperature and may, in turn, continue to adapt to future increases in water temperature.

Parasitism and Disease

Daphnia are susceptible to a wide array of parasites, including fungi, bacteria, and protozoans. Parasites exhibit the clear potential to alter the fitness of *Daphnia* and thereby impose selection via castration, decreases in fecundity, and shifts in the timing of maturation (Ebert et al. 2004, Chadwick and Little 2005, Decaestecker et al. 2005, Clerc et al. 2015). Any evolutionary consequences of pathogens, however, will likely depend on the specific interactions between host and parasite. The reason is that the fitness consequences of parasites have been shown to depend on parasite type and transmission strategy (Ebert et al. 2000). For instance, sharp declines in fecundity were observed when *Daphnia* were infected by endoparasites, but a less severe pattern of infection was noted when hosts were infected by epibionts (Decaestecker et al. 2005). Transmission strategy (vertical or horizontal) also resulted in contrasting responses in the *Daphnia* host: vertically transmitted parasites decreased overall host density significantly more than parasites that were horizontally transmitted (Ebert et al. 2000, Vizoso and Ebert 2005).

Daphnia typically harbor genetic variation in their susceptibility to pathogens (Mitchell et al. 2004b, Vale and Little 2012). Vale and Little (2012) demonstrated host genotypes of *D. magna* varied in their susceptibility to infection by *Pasteuria ramosa*, as well as parasite load after being infected (see also Mitchell et al. 2004b). Yet the impact of the pathogen on *Daphnia* fecundity or fecundity compensation (i.e., reproducing earlier) did not differ among the genotypes (Vale and Little 2012). Conversely, other research showed that host genotype of *D. magna* did not differ in susceptibility but did vary in the degree of fecundity compensation when infected by *Glugoides intestinalis* (Chadwick and Little 2005). In contrast with other ecological selective pressures (i.e., predators, resource quality), there is little evidence to suggest that *Daphnia* parasites affect the evolution of trait plasticity. However, it is known that environmental conditions strongly influence the susceptibility of *Daphnia* to disease (Duffy et al. 2011, Engelbrecht et al. 2013). For instance, exposure to invertebrate predator cues increased the susceptibility of *Daphnia dentifera* to a yeast pathogen (*Metschnikowia bicuspidata*; Duffy et al. 2011). The authors argue that *Daphnia* responded to the predator cue by growing faster and attaining a larger body size. As a result, the observed increase in infection in the presence of the chemical cue was likely due to an increase in surface area to contract the spores (Duffy et al. 2011).

One interesting trend apparent in the literature is that parasites and pathogens can rapidly drive evolutionary shifts in *Daphnia* (Decaestecker et al. 2007, Duffy and Sivars-Becker 2007, Duffy et al. 2009, 2012). For example, Duffy et al. (2012) compared the resistance of *Daphnia dentifera* from lakes with and without recent outbreaks of pathogenic yeast (*Metschnikowia bicuspidata*). They found that *Daphnia* from lakes that recently experienced an epidemic were more resistant to infection to the yeast pathogen than *Daphnia* from lakes without recent epidemics (Duffy and Sivars-Becker 2012). Work by Duffy et al. (2012) further showed that the trajectory of parasite-driven evolution depends on the ecological context. Highly productive lakes with low predation pressure were associated with large yeast epidemics and rapid evolutionary increases in *Daphnia* resistance to infection. Conversely, low productivity and high predation leads to small disease outbreaks and the evolution of increased susceptibility in *Daphnia* (Duffy et al. 2012). In another study, Decaestecker et al. (2007) took advantage of the fact that the dormant stages of *Daphnia* and their parasites remain viable in lake sediment to track the coevolutionary dynamics of *Daphnia*-parasite interactions. This approach demonstrated that *D. magna* can rapidly evolve increased resistance to the bacterial endoparasite *Pasteuria ramosa* within a few years.

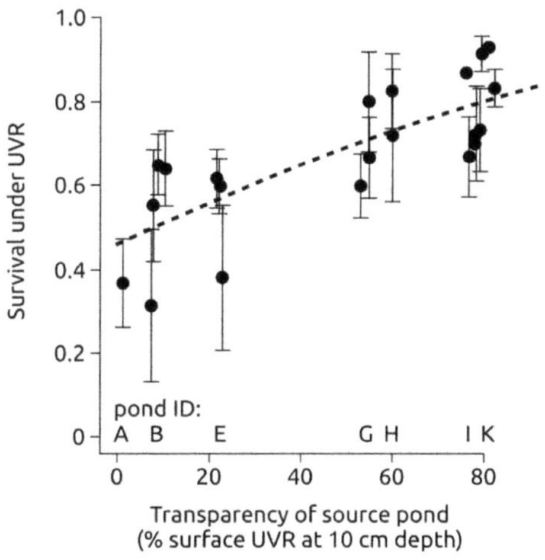

Fig. 16.5.
Influence of ultraviolet radiation on evolution in *Daphnia*. Survival (mean ± standard error) following exposure to ultraviolet radiation for *D. melanica* clones from seven ponds that vary in water transparency. Modified from Miner and Kerr (2011), with permission from The Royal Society.

Ultraviolet Radiation

UV radiation is a form of short-wavelength radiation that has the potential to damage DNA, alter gene transcription and expression, and increase mutation rates (Cadet et al. 2005, Pfeifer et al. 2005). Miner and Kerr (2011) evaluated populations of *Daphnia melanica* in Olympic National Park in Washington State for adaptation to differences in UV radiation levels. They found that populations of *Daphnia* from ponds with higher UV radiation transparency were more tolerant (i.e., exhibited higher survival) of UV radiation in the laboratory (Fig. 16.5). Genotypes of *Daphnia* from high UV radiation environments also exhibited faster rates of DNA repair than clones from habitats characterized by low UV radiation (Miner et al. 2015). There was, however, no evidence for a cost of adaptation to increased UV radiation via declines in growth in the absence of UV radiation (Miner and Kerr 2011).

Summary of the Pathway from Ecology to Evolution

The sections above clearly illustrate that several environmental stressors select for evolutionary changes in the traits of *Daphnia*. The approaches that researchers have utilized include comparisons among natural populations, selection experiments in the laboratory, latitudinal clines, temporal evaluations, and resurrection studies to show that predators, phytoplankton composition, environmental temperature, and pathogens are all drivers of evolution in *Daphnia* (Table 16.1). For some ecological selective pressures (i.e., predation and resource quality) this includes work exploring genetic shifts in overall trait values and patterns of phenotypic plasticity across a broad range of traits (e.g., life history, behavior). Evolutionary studies on responses to temperature and pathogens generally focus on specific traits that are expected to respond to that particular stressor (i.e., thermal tolerance for temperature, resistance for pathogens).

ECOLOGICAL CONSEQUENCES OF ADAPTATION

Given that there are now many examples of *Daphnia* evolution (Table 16.1) and also that *Daphnia* have well-known ecological influences on phytoplankton abundance and composition (Carpenter et al. 1992), a rapidly growing body of research has begun to dissect the contributions of *Daphnia* adaptation to the ecology of aquatic environments. The first signal indicating the likely importance of the pathway from *Daphnia* adaptation to ecology was provided by work exploring the ecological significance of genetic variation among clones of *Daphnia* (Weider et al. 2005b, 2008). As described above, Weider et al. (2005b) compared the competitive ability of clones that differ in growth rate (and rRNA structure) and found that the outcome of competition depended strongly on resource quality and light intensity; one clone dominated under high nutrients while the other was a stronger competitor under low resource conditions (see also Weider et al. 2008). Similar to the presence of genetic variation in responses to environmental cues, contrasting ecological effects among clones foreshadowed that evolved differences in *Daphnia* traits may represent a significant agent of ecological change.

Recent work has taken the key next step and shown that genetically based differences among populations of *Daphnia* have significant consequences for the properties of populations, communities, and ecosystems (Steiner et al. 2007, Pantel et al. 2011, Walsh et al. 2012, Pantel et al. 2015). Pantel et al. (2011) showed that the invasion success and ultimate zooplankton community composition differed among genetically distinct populations of *Daphnia pulex* × *pulicaria*. Walsh et al. (2012) demonstrated that predator-driven shifts in the life history traits of *Daphnia ambigua* are associated with differential impacts on phytoplankton abundance and rates of primary production (see Fig. 16.6 for conceptual figure of ecoevolutionary dynamics in this system). Walsh et al. (2012) also showed that the experimental effects of evolution on ecology measured in the laboratory parallel the trends observed naturally in lakes. Pantel et al. (2015) used a "common gardening" approach (sensu Matthews et al. 2011) to experimentally create evolutionary divergent *Daphnia* populations and then test the reciprocal influence of evolution on ecology. Here, the authors manipulated the presence and absence of predators and macrophytes for 2 months to create locally adapted populations of *Daphnia magna*. They then showed that these experimental populations have contrasting effects on zooplankton community composition and that the magnitude of this effect of evolution on ecology was similar to that of adding common ecological drivers of communities (i.e., adding predators or macrophytes).

In addition to the above examples illustrating a connection between *Daphnia* adaptation and ecological processes, evolutionary shifts related to organismal nutrition may also alter the ecology of natural systems. Herbivorous consumers and primary producers are coupled through grazing and the recycling of resources (Lehman 1980). Again, ES provides a useful framework from which to understand the relationship between adaptation and consumer-driven nutrient recycling (CNR). Because ES is based on the first principles of conservation of matter, it follows that any elemental resources not used for organismal processes must be released back into the environment. CNR by *Daphnia* is well studied via stoichiometric theory. Key to our understanding CNR was the realization that organisms do not recycle different nutrients (e.g., N and P) at the same rates (Elser and Urabe 1999). Shifts in CNR were observed due to changes in fish predation in natural systems by Elser et al. (1988), where the dominance of *Daphnia* in zooplankton communities resulted in P limitation of phytoplankton, while those dominated by copepods were N limited. This shift in CNR was later attributed to differences between *Daphnia* and calanoid copepods in their organismal stoichiometry, where *Daphnia* are more P rich than copepods and thus must recycle consumed P at lower rates (Andersen and Hessen 1991, Sterner et al. 1993).

Recently, more attention has been paid to the ecological consequences of microevolutionary shifts in nutrient use within populations. As illustrated earlier in the chapter (see Resource Quality section), genetic differences in resource use often have dramatic impacts on *Daphnia* life histories

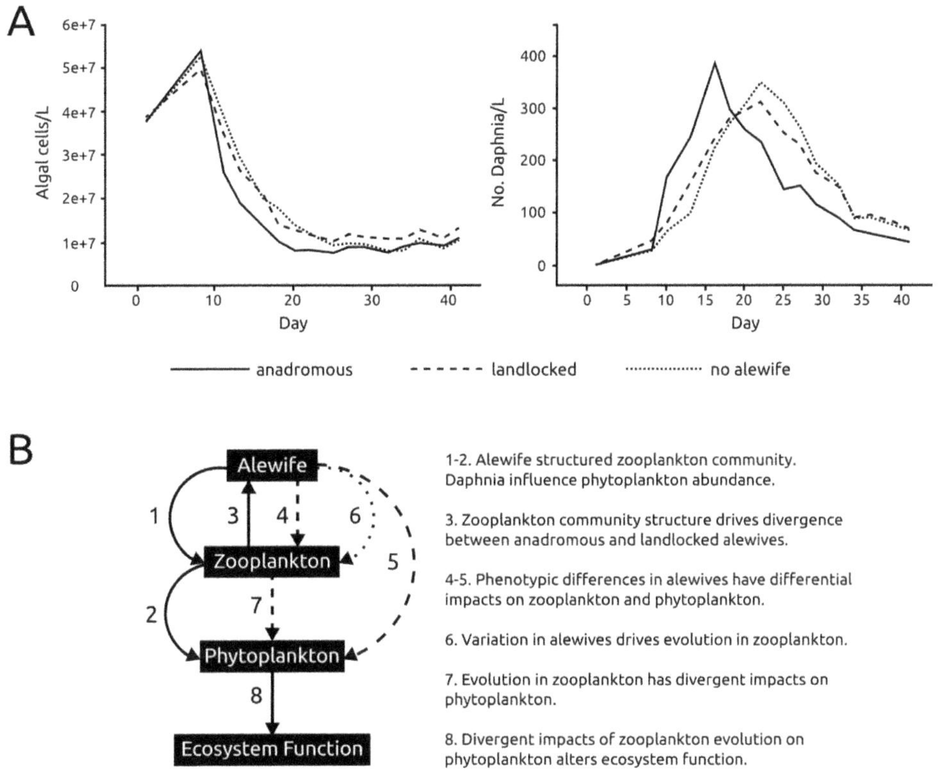

Fig. 16.6.
Eco-evolutionary feedbacks between a fish predator, the alewife (*Alosa pseudoharengus*), and *Daphnia ambigua* in lakes in Connecticut. Alewives are either permanent residents in lakes (i.e., landlocked) or they migrate between ocean and freshwater for spawning (i.e., anadromous; Post et al. 2008). This intraspecific variation in alewives is associated with life history evolution in *Daphnia* (Walsh and Post 2011, 2012). Evolution in *Daphnia*, in turn, alters consumer resource dynamics and rates of primary production (Walsh et al. 2012). (A) Differences in algal density and *Daphnia* abundances in mesocosms that contained *Daphnia* from lakes with anadromous, landlocked, and no alewives. *Daphnia* from lakes with anadromous alewives exhibited significantly faster rates of population growth, which, in turn, lead to faster declines in algal abundances. Error = ±1 standard error of mean. (B) A conceptual model illustrating the ecoevolutionary feedbacks in this system. Ecological interactions are depicted by solid arrows. Evolutionary interactions are depicted by dotted arrows. Pathways from evolution to ecology are depicted by dashed arrow. Modified from Walsh et al. (2012).

in response to nutrient availability. Less is known about how variation in resource use driven by environmental or genetic sources may result in shifts in nutrient excretion and thus CNR. In *Daphnia*, some interesting patterns have begun to emerge. In a comparison of ancient and modern genotypes resurrected from lake sediments, Roy Chowdhury (2016) found significant differences in resource use between genotype ages in response to P availability. These data also indicate that P-inefficient *Daphnia* recycle P at faster rates, resulting in greater algal growth. Similarly, Lind and Jeyasingh (2015) observed that *Daphnia* genotype and P supply interact to influence primary producer growth. Together, these results suggest that microevolutionary shifts in nutrient use can significantly alter the recycling of nutrients within natural systems, potentially having large, cascading effects on ecosystems. Additionally, these studies illustrate that the explicit examination of organismal demand and use may provide for a more thorough understanding of fundamental eco-evolutionary dynamics within natural systems.

The literature described above has clearly shown the potential for genetic variation or local adaptation to significantly alter ecological processes. Given the recent rise of studies connecting *Daphnia* phenotypes with the genes that encode these phenotypes, it is important to begin to consider the roles that genes and gene expression play in the ongoing interplay between ecological and evolutionary forces. One recent experimental study using rotifer-algae interactions connected trait evolution, gene expression, and ecological dynamics (Becks et al. 2012). Even though Becks et al. (2012) did not use *Daphnia*, their experiments on zooplankton-algae interactions are the first to incorporate a genomic perspective in the study of ecoevolutionary interactions. In predator-prey microcosms, cyclical changes in the abundance of the predator (i.e., rotifers) caused fluctuating selection on a prey (i.e., algae) defense trait (clumping). During these predator-prey cycles, genes that were first upregulated during time periods when the prey defense trait was not prevalent were then downregulated as the defense trait increased in prevalence. In addition, the genes that changed most strongly in expression were those directly associated with the defense characteristic (or its fitness cost). Such results clearly link rapid shifts in gene expression with observed changes in ecological dynamics. Interestingly, the patterns of expression for individual genes differed across multiple predator-prey cycles. The authors argue that such varying expression patterns indicate that specific phenotypes may potentially be produced by multiple transcription networks.

CONCLUSIONS

In this chapter, we have highlighted the many benefits of *Daphnia* as an experimental organism that allow them to provide insights into eco-evolutionary dynamics. Researchers have used characteristics of *Daphnia* such as an ease of culture and experimentation, a well-defined ecological role in freshwater environments, a published genome (Colbourne et al. 2011), and the ability to resurrect historic propagules from previous decades or generations (i.e., Frisch et al. 2014) to show that *Daphnia* (1) exhibit a clear capacity to evolve in response to common ecological selective pressures (Table 16.1) and (2) has served as a guide to begin to dissect the reciprocal influence of evolution on ecological processes (Fig. 16.3). Throughout this chapter, we have also highlighted recent advances in our understanding of genomics and ecological stoichiometry in the context of *Daphnia* eco-evolutionary dynamics. Researchers are thus now well positioned to leverage these complementary frameworks and subdisciplines to comprehensively evaluate the interaction between ecology and evolution from genes to ecosystems. This includes research that advances beyond phenotypes when evaluating the pathway from evolution to ecology and instead quantifies the connection between evolution at the genomic level and the properties of population, communities, and ecosystems (see Becks et al. 2012 for a non-*Daphnia* example). Research can also leverage genomics with stoichiometry to evaluate the simultaneous ecological importance of genes and elements. One challenge in the evaluation of reciprocal interactions between ecology and evolution is that our current understanding of the pathway from evolution to ecology primarily stems from laboratory or mesocosm experiments often with limited complexity. Natural systems consist of multiple trophic levels with a multitude of direct and indirect interactions within and across trophic levels. As a result, the extent to which evolutionary changes result in altered ecological properties and initiates an interplay of ecological and evolutionary processes in nature is an open question. To better understand the potential for evolution to influence ecology, research needs to more explicitly consider and incorporate the complexity typically observed in nature and perform experiments in natural systems. Ideally, such a study would explore the feedback from evolution to ecology by tracking the traits (or genes or elemental composition) and fitness of individuals. Given their small size, short life span, and immense local populations sizes, such experiments would be very challenging to implement with *Daphnia*. Nevertheless, the totality of the evidence to date collectively illustrates that *Daphnia* as a study organism offers researchers the opportunity to explore all facets of interactions between ecological and evolutionary forces. It seems likely *Daphnia* will continue to be at the forefront of research on eco-evolutionary dynamics in nature.

ACKNOWLEDGMENTS

The authors thank the University of Texas at Arlington and the National Science Foundation for providing support during the development and writing of this chapter.

REFERENCES

Adamowicz, S.J., P.D.N. Hebert, and M.C. Marinone. 2004. Species diversity and endemism in the *Daphnia* of Argentina: a genetic investigation. Zoological Journal of the Linnean Society 140:171–205.

Altshuler, I., B. Demiri, S. Xu, A. Constantin, N.D. Yan, and M.E. Cristescu. 2011. An integrated multi-disciplinary approach for studying multiple stressors in freshwater ecosystems: *Daphnia* as a model organism. Integrative and Comparative Biology 51:623–633.

Andersen, T., and D.O. Hessen. 1991. Carbon, nitrogen, and phosphorus content of freshwater zooplankton. Limnology and Oceanography 36:807–814.

Becks, L., S.P. Ellner, L.E. Jones, and N.G. Hairston. 2012. The functional genomics of an eco-evolutionary feedback loop: linking gene expression, trait evolution, and community dynamics. Ecology Letters 15:492–501.

Blom, J.F., H.I. Baumann, G.A. Codd, and F. Jüttner. 2006. Sensitivity and adaptation of aquatic organisms to oscillapeptin J and (D-Asp3,(E)-Dhb7) microcystin-RR. Archiv für Hydrobiologie 167:547–559.

Boersma, M. 2000. The nutritional quality of phosphorus limited algae for *Daphnia*. Limnology and Oceanography 45:1157–1161.

Boersma, M., P. Spaak, and L. de Meester. 1998. Predator-mediated plasticity in morphology, life history, and behavior of *Daphnia*: the uncoupling of responses. The American Naturalist 152:237–248.

Brett, M.T., M. Kainz, S.J. Taipale, and H. Seshan. 2009. Phytoplankton, not allochthonous carbon, sustains herbivorous zooplankton production. Proceedings of the National Academy of Sciences of the USA 106:21197–21201.

Cadet, J., E. Sage, and T. Douki. 2005. Ultraviolet radiation-mediated damage to cellular DNA. Mutation Research/Fundamental and Molecular Mechanisms of Mutagenesis 571:3–17.

Carpenter, S.R., J.F. Kitchell, J.R. Hodgson, P.A. Cochran, J.J. Elser, M.M. Elser, D.M. Lodge, D. Kretcher, X. He, and C.N. von Ende. 1987. Regulation of lake primary productivity by food web structure. Ecology 68:1863–1876.

Carpenter, S.R., K.L. Cottingham, and D.E. Schindler. 1992. Biotic feedbacks in lake phosphorus cycles. Trends in Ecology and Evolution 7:332–336.

Carvalho, G.R. 1987. The clonal ecology of *Daphnia magna* (Crustacea: Cladocera): II. Thermal differentiation among seasonal clones. Journal of Animal Ecology 1987:469–478.

Chadwick, W., and T.J. Little. 2005. A parasite-mediated life-history shift in *Daphnia* magna. Proceedings of the Royal Society of London B: Biological Sciences 272:505–509.

Charlesworth, B. 1980. Evolution in age structured populations. Cambridge University Press, Cambridge, England.

Clerc, M., D. Ebert, and M.D. Hall. 2015. Expression of parasite genetic variation changes over the course of infection: implications of within-host dynamics for the evolution of virulence. Proceedings of the Royal Society of London B: Biological Sciences 282:20142820.

Colbourne, J.K., M.E. Pfrender, D. Gilbert, W.K. Thomas, A. Tucker, T.H. Oakley, S. Tokishita, A. Aerts, G.J. Arnold, M.K. Basu, D.J. Bauer, C.E. Cáceres, L. Carmel, C. Casola, J.H. Choi, J.C. Detter, Q. Dong, S. Dusheyko, B.D. Eads, T. Fröhlich, K.A. Geiler-Samerotte, D. Gerlach, P. Hatcher, S. Jogdeo, J. Krijgsveld, E.V. Kriventseva, D. Kültz, C. Laforsch, E. Lindquist, J. Lopez, J.R. Manak, J. Muller,

J. Pangilinan, R.P. Patwardhan, S. Pitluck, E.J. Pritham, A. Rechtsteiner, M. Rho, I.B. Rogozin, O. Sakarya, A. Salamov, S. Schaack, H. Shapiro, Y. Shiga, C. Skalitzky, Z. Smith, A. Souvorov, W. Sung, Z. Tang, D. Tsuchiya, H. Tu, H. Vos, M. Wang, Y.I. Wolf, H. Yamagata, T. Yamada, Y. Ye, J.R. Shaw, J. Andrews, T.J. Crease, H. Tang, S.M. Lucas, H.M. Robertson, P. Bork, E.V. Koonin, E.M. Zdobnov, I.V. Grigoriev, M. Lynch, and J.L. Boore. 2011. The ecoresponsive genome of *Daphnia pulex*. Science 331:555–561.

Conover, D.O., and T. Present. 1990. Countergradient variation in growth rate: compensation for length of the growing season among Atlantic silversides from different latitudes. Oecologia 83:316–324.

Cothran, R.D., A.R. Stiff, P.D. Jeyasingh, and R.A. Relyea. 2012. Eutrophication and predation risk interact to affect sexual trait expression and mating success. Evolution 66:708–719.

Cousyn, C., L. de Meester, J.K. Colbourne, L. Brendonck, D. Verschuren, and F. Volckaert. 2001. Rapid, local adaptation of zooplankton behavior to changes in predation pressure in the absence of neutral genetic changes. Proceedings of the National Academy of Sciences of the USA 98:6256–6260.

Cox, A.J., and P.D.N. Hebert. 2001. Colonization, extinction, and phylogeographic patterning in a freshwater crustacean. Molecular Ecology 10:371–386.

de Meester, L. 1996. Evolutionary potential and local genetic differentiation in a phenotypically plastic trait of a cyclical parthenogen, *Daphnia magna*. Evolution 50:1293–1298.

de Meester, L., M. Boersma, and P. Spaak. 1999. Environmental stress and local adaptation in *Daphnia magna*. Limnology and Oceanography 4:393–402.

Danger, M., J. Arce Funck, S. Devin, J. Heberle, and V. Felten. 2013. Phosphorus content in detritus controls life-history traits of a detritivore. Functional Ecology 27:807–815.

Decaestecker, E., S. Declerck, L. de Meester, and D. Ebert. 2005. Ecological implications of parasites in natural *Daphnia* populations. Oecologia 144:382–390.

Decaestecker, E., S. Gaba, J.A. Raeymaekers, R. Stoks., L. van Kerckhoven, D. Ebert, and L. de Meester. 2007. Host-parasite 'red queen' dynamics archived in pond sediment. Nature 450:870–873.

DeMott, W.R. 1989. The role of competition in zooplankton succession. Pages 195–252 *in* U. Sommer, editor. Plankton ecology: succession in plankton communities. Springer, Berlin, Germany.

DeMott, W.R., R.D. Gulati, and K. Siewertsen. 1998. Effects of phosphorus-deficient diets on the carbon and phosphorus balance of *Daphnia magna*. Limnology and Oceanography 43:1147–1161.

DeMott, W.R., Q. Zhang, and W.W. Carmichael. 1991. Effects of toxic cyanobacteria and purified toxins on the survival and feeding of a copepod and three species of *Daphnia*. Limnology and Oceanography 36:1346–1357.

Duffy, M.A., and L. Sivars-Becker. 2007. Rapid evolution and ecological host-parasite dynamics. Ecology Letters 10:44–53.

Dudycha, J.L., C.S. Brandon, and K.C. Deitz. 2012. Population genomics of resource exploitation: insights from gene expression profiles of two *Daphnia* ecotypes fed alternate resources. Ecology and Evolution 2:329–340.

Duffy, M.A., S.R. Hall, C.E. Cáceres, and A.R. Ives. 2009. Rapid evolution, seasonality, and the termination of parasite epidemics. Ecology 90:1441–1448.

Duffy, M.A., J.M. Housley, R.M. Penczykowski, C.E. Caceres, and S.R. Hall. 2011. Unhealthy herds: indirect effects of predators enhance two drivers of disease spread. Functional Ecology 25:945–953.

Duffy, M.A., J.H. Ochs, R.M. Penczykowski, D.J. Civitello, C.A. Klausmeier, and S.R. Hall. 2012. Ecological context influences epidemic size and parasite-driven evolution. Science 335:1636–1638.

Ebert, D., M. Lipsitch, and K.L. Mangin. 2000. The effect of parasites on host population density and extinction: experimental epidemiology with *Daphnia* and six microparasites. The American Naturalist 156:459–477.

Ebert, D., H.J. Carius, T. Little, and E. Decaestecker. 2004. The evolution of virulence when parasites cause host castration and gigantism. The American Naturalist 164:S19–S32.

Elser, J.J., and J. Urabe. 1999. The stoichiometry of consumer-driven nutrient recycling: theory, observations, and consequences. Ecology 80:735–751.

Elser, J.J., D. Dobberfuhl, N.A. Mackay, and J.H. Schampel. 1996. Organism size, life history, and N: P stoichiometry: Towards a unified view of cellular and ecosystem processes. BioScience 46:674–684.

Elser, J.J., M.M. Elser, N.A. MacKay, and S.R. Carpenter. 1988. Zooplankton-mediated transitions between N- and P-limited algal growth. Limnology and Oceanography 33:1–14.

Engelbrecht, W., O. Hesse, J. Wolinska, and C. Laforsch. 2013. Two threats at once: encounters with predator cues alter host life-history and morphological responses to parasite spores. Hydrobiologia 715:93–100.

Frisch, D., P.K. Morton, P.R. Chowdhury, B.W. Culver, J.K. Colbourne, L. Weider, and P.D. Jeyasingh. 2014. A millennial-scale chronicle of evolutionary responses to cultural eutrophication in *Daphnia*. Ecology letters 17:360–368.

Froehlich, T., G.J. Arnold, R. Fritsch, T. Mayr, and C. Laforsch. 2009. LC-MS/MS-based proteome profiling in *Daphnia pulex* and *Daphnia longicephala*: the *Daphnia pulex* genome database as a key for high throughput proteomics in *Daphnia*. BioMed Central Genomics 10:171.

Geerts, A.N., L. de Meester, and R. Stoks. 2015a. Heat tolerance and its evolutionary potential along a latitudinal gradient in *Daphnia magna*. Evolutionary Ecology Research 16:517–528.

Geerts, A.N., J. Vanoverbeke, B. Vanschoenwinkel, W. van Doorslaer, H. Feuchtmayr, D. Atkinson, B. Moss, T.A. Davidson, C.D. Sayer, and L. de Meester. 2015b. Rapid evolution of thermal tolerance in the water flea *Daphnia*. Nature Climate Change 5:665–668.

González, A.L., J.M. Fariña, A.D. Kay, R. Pinto, and P.A. Marquet. 2011. Exploring patterns and mechanisms of interspecific and intraspecific variation in body elemental composition of desert consumers. Oikos 120:1247–1255.

Hairston, N.G., C.L. Holtmeier, W. Lampert, L.J. Weider, D.M. Post, J.M. Fischer, C.M. Cáceres, J.A. Fox, and U. Gaedke. 2001. Natural selection for grazer resistance to toxic cyanobacteria: evolution of phenotypic plasticity? Evolution 55:2203–2214.

Hairston, N.G., W. Lampert, C.E. Cáceres, C.L. Holtmeier, L.J. Weider, U. Gaedke, J.M. Fischer, J.A. Fox, and D.M. Post. 1999. Lake ecosystems: rapid evolution revealed by dormant eggs. Nature 401:446–446.

Harris, K.D.M., N.J. Bartlett, and V.K. Lloyd. 2012. *Daphnia* as an emerging epigenetic model organism. Genetics research international 2012:147892.

Hebert, P.D.N. 1987. Genotypic characteristics of cyclic parthenogens and their obligately asexual derivatives. Pages 175–195 *in* S.C. Sterns, editor. The evolution of sex and its consequences. Birkhäuser Verlag, Basel, Switzerland.

Hendry, A.P., and M.T. Kinnison. 1999. Perspective: the pace of modern life: measuring rates of contemporary microevolution. Evolution 1999:1637–1653.

Hood, J.M., and R.W. Sterner. 2014. Carbon and phosphorus linkages in Daphnia growth are determined by growth rate, not species or diet. Functional Ecology 28:1156–1165.

Jansen, M., L. Vergauwen, T. Vandenbrouck, D. Knapen, N. Dom, K.I. Spanier, A. Cielen, L. de Meester. 2013. Gene expression profiling of three different stressors in the water flea *Daphnia magna*. Ecotoxicology 22:900–914.

Jeyasingh, P.D., A. Ragavendran, S. Paland, J.A. Lopez, R.W. Sterner, and J.K. Colbourne. 2011. How do consumers deal with stoichiometric constraints? Lessons from functional genomics using *Daphnia pulex*. Molecular Ecology 20:2341–2352.

Jeyasingh, P.D., and L.J. Weider. 2005. Phosphorus availability mediates plasticity in life-history traits and predator–prey interactions in *Daphnia*. Ecology Letters 8:1021–1028.

Kay, A.D., I.W. Ashton, E. Gorokhova, A.J. Kerkhoff, A. Liess, and E. Litchman. 2005. Toward a stoichiometric framework for evolutionary biology. Oikos 109:6–17.

Kotov, A.A. 2015. A critical review of the current taxonomy of the genus *Daphnia* O. F. Muller, 1785 (Anomopoda, Cladocera). Zootaxa 3911:184–200.

Lampert, W. 2006. *Daphnia*: model herbivore, predator and prey. Polish Journal of Ecology 54:607–620.

Latta, L.C., L.J. Weider, J.K. Colbourne, and M.E. Pfrender. 2012. The evolution of salinity tolerance in *Daphnia*: a functional genomics approach. Ecology Letters 15:794–802.

Lehman, J.T. 1980. Release and cycling of nutrients between planktonic algae and herbivores. Limnology and Oceanography 25:620–632.

Lemaire, V., S. Brusciotti, I. van Gremberghe, W. Vyverman, J. Vanoverbeke, and L. de Meester. 2012. Genotype x genotype interactions between the toxic cyanobacterium Microcystis and its grazer, the waterflea *Daphnia*. Evolutionary Applications 5:168–182.

Lind, P.R., and P.D. Jeyasingh. 2015. Genotypic differences in phosphorus use physiology in producers (*Chlamydomonas reinhardtii*) and consumers (*Daphnia pulex*) interact to alter primary and secondary production. Evolutionary Ecology 29:551–563.

Lurling, L. 2003. *Daphnia* growth on microcystin-producing and microcystin-free Microcystis aeruginosa in different mixtures with the green alga *Scenedesmus obliquus*. Limnology and Oceanography 48:2214–2220.

Martin-Creuzberg, D., A. Wacker, and E. von Elert. 2005. Life history consequences of sterol availability in the aquatic keystone species *Daphnia*. Oecologia 144:362–372.

Matthews, B., A. Narwani, S. Hausch, E. Nonaka, H. Peter, M. Yamamichi, K.E. Sullam, K.C. Bird, M.K. Thomas, T.C. Hanley, and C.B. Turner. 2011. Toward an integration of evolutionary biology and ecosystem science. Ecology Letters 14:690–701.

Miner, B.E., and B. Kerr. 2011. Adaptation to local ultraviolet radiation conditions among neighbouring *Daphnia* populations. Proceedings of the Royal Society B: Biological Sciences 278:1306–1313.

Miner, B.E., L. de Meester, M.E. Pfrender, W. Lampert, and N.G. Hairston. 2012. Linking genes to communities and ecosystems: *Daphnia* as an ecogenomic model. Proceedings of the Royal Society B: Biological Sciences 279:1873–1882.

Miner, B. E., P.M. Kulling, K.D. Beer, and B. Kerr, B. 2015. Divergence in DNA photorepair efficiency among genotypes from contrasting UV radiation environments in nature. Molecular ecology 24:6177–6187.

Mitchell, S.E., and W. Lampert. 2000. Temperature adaptation in a geographically widespread zooplankter, *Daphnia magna*. Journal of Evolutionary Biology 13:371–382.

Mitchell, S.E., J. Halves, and W. Lampert. 2004a. Coexistence of similar genotypes of *Daphnia magna* in intermittent populations: response to thermal stress. Oikos 106:469–478.

Mitchell, S.E., A.F. Read, and T.J. Little. 2004b. The effect of a pathogen epidemic on the genetic structure and reproductive strategy of the crustacean *Daphnia magna*. Ecology Letters 7:848–858.

Moest, M., A.C. Chiaia-Hernandez, M.P. Frey, J. Hollender, and P. Spaak. 2015. A mixture of environmental organic contaminants in lake sediments affects hatching from *Daphnia* resting eggs. Environmental Toxicology and Chemistry 34:338–345.

Nunes, B.S., F.D. Carvalho, L.M. Guilhermino, and G. Van Stappen. 2006. Use of the genus *Artemia* in ecotoxicity testing. Environmental Pollution 144:453–462.

Palaima, A., and K. Spitze. 2004. Is a jack-of-all-temperatures a master of none? An experimental test with *Daphnia pulicaria* (Crustacea: Cladocera). Evolutionary Ecology Research 6:215–225.

Pantel, J.H., M.A. Leibold, and T.E. Juenger. 2011. Population differentiation in *Daphnia* alters community assembly in experimental ponds. The American Naturalist 177:314–322.

Pantel, J.H., C. Duvivier, and L. de Meester. 2015. Rapid local adaptation mediates zooplankton community assembly in experimental mesocosms. Ecology Letters 18:992–1000.

Parejko, K., and S.I. Dodson. 1991. The evolutionary ecology of an antipredator reaction norm: *Daphnia pulex and Chaoborus americanus*. Evolution 45:1665–1674.

Pfeifer, G.P., Y-H You, and A. Besaratinia. 2005. Mutations induced by ultraviolet light. Mutation Research/Fundamental and Molecular Mechanisms of Mutagenesis 571:19–31.

Pollard H.G., J.K. Colbourne, and W. Keller. 2003. Reconstruction of centuries-old *Daphnia* communities in a lake recovering from acidification and metal contamination. Ambio 32: 214–218.

Post, D.M., and E.P. Palkovacs. 2009. Eco-evolutionary feedbacks in community and ecosystem ecology: interactions between the ecological theatre and the evolutionary play. Philosophical Transactions of the Royal Society B: Biological Sciences 364:1629–1640.

Post, D.M., E.P. Palcovacs, E.G. Schielke, and S.I. Dodson. 2008 Intraspecific variation in a predator affects community structure and cascading trophic interactions. Ecology 89:2019–2032.

Raisuddin, S., K.W.H. Kwok, K.M.Y. Leung, D. Schlenk, and J.S. Lee. 2007. The copepod *Tigriopus*: a promising marine model organism for ecotoxicology and environmental genomics. Aquatic Toxicology 83:161–173.

Rehm E.J., R.L. Hannibal, R.C. Chaw, M.A. Vargas-Vila, and N.H. Patel. 2009. The crustacean *Parhyale hawaiensis*: a new model for arthropod development. Cold Spring Harbor Protocols 1:373–404.

Reznick, D.N. 2013. A critical look at reciprocity in ecology and evolution: introduction to the symposium. The American Naturalist 181:S1–S8.

Reznick, D.N., and C.K. Ghalambor. 2001. The population ecology of contemporary adaptations: what empirical studies reveal about the conditions that promote adaptive evolution. Genetica 112:183–198.

Riessen, H.P. 1999. Predator-induced life history shifts in *Daphnia*: a synthesis of studies using meta-analysis. Canadian Journal of Fisheries and Aquatic Sciences 56:2487–2494.

Rohrlack, T.K., E. Dittmann, T. Borner, and K. Christoffersen. 2001. Effects of cell-bound microcystins on survival and feeding of *Daphnia* spp. Applied Environmental Microbiology 67:3523–3529.

Roy Chowdhury, P. 2016. Differences in phosphorus use between ancient and extant *Daphnia* genotypes alters algal stoichiometry and abundance. Inland Waters 6:165–172.

Roy Chowdhury, P., J.A. Lopez, L.J. Weider, J.K. Colbourne, and P.D. Jeyasingh. 2014. Functional genomics of intraspecific variation in carbon and phosphorus kinetics in *Daphnia*. Journal of Experimental Zoology Part A: Ecological Genetics and Physiology 321:387–398.

Rozenberg, A., M. Parida, F. Leese, L.C. Weiss, R. Tollrian, and J.R. Manak. 2015. Transcriptional profiling of predator-induced phenotypic plasticity in *Daphnia pulex*. Frontiers in Zoology 12:18.

Sarnelle, O., and A.E. Wilson. 2005. Local adaptation of *Daphnia pulicaria* to toxic cyanobacteria. Limnology and Oceanography 50:1565–1570.

Sarnelle, O., S. Gustafsson, and L. Hansson. 2010. Effects of cyanobacteria on fitness components of the herbivore *Daphnia*. Journal of Plankton Research 32:471–477.

Scheiner, S.M., and L. Yampolsky. 1998. The evolution of *Daphnia pulex* in a temporally varying environment. Genetics Research 72:25–37.

Schoener, T.W. 2011. The newest synthesis: understanding the interplay of evolutionary and ecological dynamics. Science 331:426–429.

Schram, F. R. 2013. Comments on crustacean biodiversity and disparity of body plans. Functional Morphology and Diversity. The Natural History of the Crustacea, 1:1–33.

Schwarzenberger, A., C. Courts, and E. von Elert. 2009. Target gene approaches: gene expression in *Daphnia magna* exposed to predator-borne kairomones or to microcystin-producing and microcystin-free *Microcystis aeruginosa*. BMC Genomics 10:527–541.

Schwarzenberger, A., A. Zitt, P. Kroth, S. Mueller, and E. von Elert. 2010. Gene expression and activity of digestive proteases in *Daphnia*: effects of cyanobacterial protease inhibitors. BMC Physiology 10:6–20.

Scoville, A.G., and M.E. Pfrender. 2010. Phenotypic plasticity facilitates recurrent rapid adaptation to introduced predators. Proceedings of the National Academy of Sciences of the USA 107:4260–4263.

Shaw, J.R., J.K. Colbourne, J.C. Davey, S.C. Glaholt, T.H. Hampton, C.Y. Chen, C.L. Folt, and J.W. Hamilton. 2007. Gene response profiles for *Daphnia pulex* exposed to the environmental stressor cadmium reveals novel crustacean metallotheoneins. BMCGenomics 8:477–497.

Spaak, P., J. Vanoverbeke, and M. Boersma. 2000. Predator-induced life-history changes and the coexistence of five taxa in a *Daphnia* species complex. Oikos 89:164–174.

Spitze, K. 1991. *Chaoborus* predation and life-history evolution in *Daphnia pulex*: temporal pattern of population diversity, fitness, and mean life history. Evolution 45:82–92.

Spitze, K. 1992. Predator-mediated plasticity of prey life history and morphology: *Chaoborus* americanus predation on *Daphnia pulex*. The American Naturalist 1992:229–247.

Steiner, C.F., C.E. Cáceres, and S.D. Smith. 2007. Resurrecting the ghost of competition past with dormant zooplankton eggs. The American Naturalist 169:416–422.

Sterner, R.W., and J.J. Elser. 2002. Ecological stoichiometry: The biology of elements from molecules to the biosphere. Princeton University Press, Princeton, New Jersey.

Sterner, R.W., D.D. Hagemeier, W.L. Smith, and R.F. Smith. 1993. Phytoplankton nutrient limitation and food quality for *Daphnia*. Limnology and Oceanography 38:857–871.

Stibor, H. 1992. Predator induced life-history shifts in a freshwater cladoceran. Oecologia 92:162–165.

Tollrian, R. 1995. Predator-induced morphological defenses: costs, life history shifts, and maternal effects in *Daphnia pulex*. Ecology 76:1691–1705.

Tollrian, R., and F. Leese. 2010. Ecological genomics: steps towards unraveling the genetic basis of inducible defenses in *Daphnia*. BMC Biology 8:51–54.

Vale, P.F., and T.J. Little. 2012. Fecundity compensation and tolerance to a sterilizing pathogen in *Daphnia*. Journal of Evolutionary Biology 25:1888–1896.

van Doorslaer, W., R. Stoks, C. Duvivier, A. Bednarska, and L. de Meester. 2009a. Population dynamics determine genetic adaptation to temperature in *Daphnia*. Evolution 63:1867–1878.
van Doorslaer, W., J. Vanoverbeke, C. Duvivier, S. Rousseaux, M. Jansen, B. Jansen, H. Feuchtmayr, D. Atkinson, B. Moss, R. Stoks, and L. de Meester. 2009b. Local adaptation to higher temperatures reduces immigration success of genotypes from a warmer region in the water flea *Daphnia*. Global Change Biology 15:3046–3055.
Vandegehuchte, M.B., D.D. Coninck, T. Vandenbrouck, W. De Coen, and C.R. Janssen. 2010. Gene transcription profiles, global DNA methylation and potential transgenerational epigenetic effects related to Zn exposure history in *Daphnia magna*. Environmental Pollution 158:3323–3329.
Vijverberg, J. 1989. Culture techniques for studies on the growth, development and reproduction of copepods and cladocerans under laboratory and in situ conditions: a review. Freshwater Biology 21:317–373.
Vizoso, D.B., and D. Ebert. 2005. Mixed inoculations of a microsporidian parasite with horizontal and vertical infections. Oecologia 143:157–166.
von Elert, E., and T. Wolffrom. 2001. Supplementation of cyanobacterial food with polyunsaturated fatty acids does not improve growth of *Daphnia*. Limnology and Oceanography 46:1552–1558.
von Elert, E., D. Martin-Creuzburg, and J.R. Le Coz. 2002. Absence of sterols constrains carbon transfer between cyano-bacteria and a freshwater herbivore (*Daphnia galeata*). Proceedings of the Royal Society B: Biological Sciences 270:1209–1214.
Walsh, M.R., and D.M. Post. 2011. Interpopulation variation in a fish predator drives evolutionary divergence in prey in lakes. Proceedings of the Royal Society B: Biological Sciences 278:2628–2637.
Walsh, M.R., and D.M. Post. 2012. The impact of intraspecific variation in a fish predator on the evolution of phenotypic plasticity and investment in sex in *Daphnia ambigua*. Journal of Evolutionary Biology 25:80–89.
Walsh, M.R., J.P. DeLong, T.C. Hanley, and D.M. Post. 2012. A cascade of evolutionary change alters consumer-resource dynamics and ecosystem function. Proceedings of the Royal Society B: Biological Sciences 279: 3184–3192.
Walsh, M.R., K.J. La Pierre, and D.M. Post. 2014. Phytoplankton composition modifies predator-driven life history evolution in Daphnia. Evolutionary Ecology 28:397–411.
Weider, L.J., and J. Pijanowska. 1993. Plasticity of *Daphnia* life histories in response to chemical cues from predators. Oikos 1993:385–392.
Weider, L.J., J.J. Elser, T.J. Crease, M. Mateos, J.B. Cotner, and T.A. Markow. 2005a. The functional significance of ribosomal (r) DNA variation: impacts on the evolutionary ecology of organisms. Annual Review of Ecology, Evolution, and Systematics 2005:219–242.
Weider, L.J., W. Makino, K. Acharya, K.L. Glenn, M. Kyle, J. Urabe, and J.J. Elser. 2005b. Genotype × environment interactions, stoichiometric food quality effects, and clonal coexistence in *Daphnia pulex*. Oecologia 143:537–547.
Weider, L.J., P.D. Jeyasingh, and K.G. Looper. 2008. Stoichiometric differences in food quality: impacts on genetic diversity and the coexistence of aquatic herbivores in a *Daphnia* hybrid complex. Oecologia 158:47–55.
Woods, H.A., W.F. Fagan, J.J. Elser, and J.F. Harrison. 2004. Allometric and phylogenetic variation in insect phosphorus content. Functional Ecology 18:103–109.

INDEX

www.ingramcontent.com/pod-product-compliance
Ingram Content Group UK Ltd.
Pitfield, Milton Keynes, MK11 3LW, UK
UKHW051048150926
475595UK00006B/183

9 780190 620271